Expert Systems and Robotics

NATO ASI Series

Advanced Science Institutes Series

A series presenting the results of activities sponsored by the NATO Science Committee, which aims at the dissemination of advanced scientific and technological knowledge, with a view to strengthening links between scientific communities.

The Series is published by an international board of publishers in conjunction with the NATO Scientific Affairs Division

A Life Sciences B Physics	Plenum Publishing Corporation London and New York
C Mathematical and Physical Sciences D Behavioural and Social Sciences E Applied Sciences	Kluwer Academic Publishers Dordrecht, Boston and London
F Computer and Systems Sciences G Ecological Sciences H Cell Biology I Global Environmental Change	Springer-Verlag Berlin Heidelberg New York London Paris Tokyo Hong Kong Barcelona

NATO-PCO DATABASE

The electronic index to the NATO ASI Series provides full bibliographical references (with keywords and/or abstracts) to more than 30000 contributions from international scientists published in all sections of the NATO ASI Series. Access to the NATO-PCO DATABASE is possible in two ways:

– via online FILE 128 (NATO-PCO DATABASE) hosted by ESRIN, Via Galileo Galilei, I-00044 Frascati, Italy.

– via CD-ROM "NATO-PCO DATABASE" with user-friendly retrieval software in English, French and German (© WTV GmbH and DATAWARE Technologies Inc. 1989).

The CD-ROM can be ordered through any member of the Board of Publishers or through NATO-PCO, Overijse, Belgium.

Series F: Computer and Systems Sciences Vol. 71

Expert Systems and Robotics

Edited by

Timothy Jordanides

Department of Electrical Engineering
California State University
1250 Bellflower Boulevard, Long Beach, CA 90840-5005, USA

Bruce Torby

Department of Mechanical Engineering
California State University
1250 Bellflower Boulevard, Long Beach, CA 90840-5005, USA

Springer-Verlag
Berlin Heidelberg New York London Paris Tokyo
Hong Kong Barcelona Budapest
Published in cooperation with NATO Scientific Affairs Division

Proceedings of the NATO Advanced Study Institute on Expert Systems and Robotics, held in Corfu, Greece, July 15–27, 1990.

Additional material to this book can be downloaded from http://extra.springer.com

ISBN-13:978-3-642-76467-7 e-ISBN-13:978-3-642-76465-3
DOI: 10.1007/ 978-3-642-76465-3

Softcover reprint of the hardcover 1st edition 1991

45/3140-543210 – Printed on acid-free-paper

Preface

The area of control systems is a fairly mature area within electrical engineering. It emphasizes the formality and generality of mathematical modeling. Recently, a significant amount of activity has occurred in the area of intelligent machines. The functions of an intelligent control system combine the high-level decision making of the digital computer with the advanced mathematical modeling and synthesis techniques of systems theory. It may be thought of as a fusion of operations research, artificial intelligence, and control theory.

The areas of intelligent machines or robotic systems is of enormous technological and economic interest in the NATO countries as competition in productivity intensifies. The readers of this volume will find some research work already accomplished in the analytical theory of intelligent machines, work in progress and of current interest and some specific examples for further research.

The papers in this volume range from the most theoretical to some descriptions of very practical working robots. Besides the main lectures and the twenty-four short papers presented, many group meetings were held at the Conference to allow the opportunity for active participation of everyone present in this Conference. The group meetings gave us the flexibility to accomplish many things which would not be feasible in a lecture setting. The task of reducing the dynamism and enthusiastic discussions of those meetings to writing was not an easy one. Nonetheless, the proceedings of those group meetings are found in Section VI. In Section I, three papers on Vision are presented, with sensors covered in Section II. The fusion of Expert Systems and Robotics as manifested by software and hardware implementations and simulation of new designs is found in Section III. In Section IV, the theory of Intelligent Machines as well as specific examples of Robot Control are found. We have found the large number of example systems included here to be invaluable in understanding Intelligent Control. The section on Applications (section VI) includes some interesting practical industrial examples.

It was the general opinion of all who attended the Conference that this Advanced Study Institute was important and valuable to their on-going research projects. Everyone saw the need of having further institutes on the same subject. The lectures and short papers presented were of exceptional quality and the editors would like to thank all the contributers for their strong efforts. We hope that the lectures, papers and group discussions summarized here will lead to further advances in the design, theory, and control of intelligent machines.

Timothy Jordanides
Director, ASI Expert Systems and Robotics

Introductory Conference Remarks

Welcome to the 1990 Advanced Study Institute on Expert Systems and Robotics. An ASI, as most of you probably know, is an international, high level, advanced "summer school" in an area of current interest to the NATO member countries. Now, some of you may wonder why we call this a summer school as many of the participants here have important positions in industry, academia, and government. I hope that you perceive this term "school" here in the broadest possible sense. After having spent 26 years in university teaching, I still consider myself a student in the sense that Socrates called himself a perennial student when he said, "γηϱασκω αει διδασκωμενος" (I am always learning as I get older).

During the next two weeks, I hope that all of you, whether lecturer or participant, will bring to this assembly the curiosity of a student along with the expertise of your particular area. Those of you who are young scientists in the area of Expert Systems and Robotics will find a wealth of information to take home with you. My hope is, that even those of you who are at the forefront of your discipline will go away in two weeks with at least one new question or one new idea that you did not think about before. Ideas are the commodity of our business. Many successful companies started on one single new idea. I would like to welcome everybody to express and share your interests and ideas with each other during the next two weeks.

We have an outstanding scientific program. The papers to be presented range from the most theoretical to some very practical working examples of robots. Besides the main lectures and the twenty-four short papers, many group meetings scheduled typically in the afternoons will allow the opportunity for active participation of everyone present in this Conference. The group meetings will give us the flexibility to accomplish many things which would not be feasible in a lecture setting:

1. Allow the opportunity for everyone in this ASI to provide their input and their expertise which eventually will be filtered by the group leader and presented to the assembly.
2. Provide an efficient forum to deal with the interdisciplinary and complex subject of Expert Systems and Robotics.
3. Foster the creation of professional links among the members of the group, as we get to know each other professionally.
4. Allow the introduction of poster sessions. Everyone here is welcome to suggest additional topics for discussion. If sufficient interest is expressed on any issue which is not covered in the agenda, we will find time to discuss it. This is your Institute, and your interest and participation will contribute to its success.

Timothy Jordanides
Director

Concluding Conference Remarks

As this ASI has come to a successful conclusion in this wonderful setting of the island of Corfu, Greece, the editorial task of categorizing the proceedings is at hand. The main areas outlined in the announcement of the Institute were:

* Expert systems in design and implementation of intelligent robots
* Computer vision as applied to robots (machine perception)
* Theory and application of intelligent control
* Methods of dealing with uncertain systems.

The constructive interaction between the participants provided the opportunity for information dissemination in some depth in the following areas:

* Vision and Image Analysis
* Robotic Sensory Systems
* Software/Hardware Systems/Simulation which encompasses expert systems considerations
* Robot Control which includes the theory of Intelligent Machines and Applications.

One of the most pleasant tasks of the Director at the conclusion of an ASI is to acknowledge the contributions of individuals and institutions to the success of the meeting. I would like to thank my co-directors, Profs. Gianni Casale and Spyros Tzafestas, co-editor Prof. Bruce Torby, Prof. Ger Honderd, and of course the NATO Scientific Affairs Committee that provided the funding without which this meeting would not have been possible. Above all, the active and enthusiastic participation of all individuals, both lecturers and participants, was pivotal to the success of this meeting. The names of all individuals present is appended in this volume.

Timothy Jordanides
Director

Table of Contents

Gestalt-based Approach to Robot Vision

Philipp W. Besslich and Henning Bässmann

University of Bremen, FB-1, P.O.Box 330440, D-2800 Bremen 33, West Germany

Abstract: The human visual system is usually able to recognize objects *as well as* their spatial relations *without* the support of depth information like stereo vision. For this reason we can easily understand cartoons, photographs and movies. It is the aim of our current research to exploit this aspect of human perception in the context of computer and robot vision. From a monocular TV image we obtain information about the type of an object observed in the scene *and* its position relative to the camera (viewpoint). This report deals with the theory of human image understanding as far as used in this system and describes the realization of a blocks world vision system based on these principles.

Keywords: Perceptual Organization, Gestalt, Mental Rotation, Geon, Hough Transformation, Monocular Vision, Multi-Step Matching

1 Human Image Understanding

Research of human image understanding falls within the scope of Gestalt psychology [17] (perceptual organization [13]), cognitive psychology [1] and cognitive science (psychology, philosophy, linguistic and artificial intelligence [1]) as an interdisciplinary topic.

One interesting idea concerning human image understanding is the approach of Biederman [8] [9]. He defined a set of 36 basic visual components which he named geons (geometrical ions). They are generalized cones described by their edges. Variation in the shape of geons is based on NAP's (nonaccidental properties, [14] [15]). These are properties of image structures which are very likely to be preserved in spite of mapping the 3-D world onto a 2-D image plane. Two important NAP's are *collinearity* and *parallelism.* For instance, points which are collinear in a 3-D world are collinear in the respective image, too. Similarly, lines which are parallel in a 3-D world are also parallel in the respective image.

In Biederman's approach the objects of a scene are recognized as an arrangement of a few geons: Recognition-by-Components (RBC). Biederman maintains that: "If an arrangement of three geons can be recovered from the input, objects can be quickly recognized even when they are occluded, rotated in depth, novel, extensively degraded, or embedded in a scene." [9].

NATO ASI Series, Vol. F 71
Expert Systems and Robotics
Edited by T. Jordanides and B. Torby

In order to determine spatial relations of an object represented by a 2-D image, human beings "rotate" the 3-D representation of this object in their mind. This phenomena of human vision is known as "mental rotation" [10] [18]. This is to say, the human visual system *searches* for the viewpoint, but does not keep all possible projections in memory. This searching procedure is fast due to parallel processing of the neural system [10]. Of similar importance is the fact that this procedure is carried out on the geon level. Hence, the search space is rather small.

It is the aim of our current research to take advantage of the RBC approach and the principle of mental rotation in the context of robot vision. To keep the realization simple, presently our objects are parts of blocks world scenes which are grabbed from a single TV camera. In this case the "geometrical ions" are closed polygons which represent the plane surfaces of the blocks.

2 Strategy of the Vision System

The basic concept of the system is shown in Fig. 1. The first step in the system is to compute a gradient image of the scene. The next step extracts *closed* polygons. This is not a trivial task since gradient operation destroys vertices of potential polygons. We have developed an algorithm (based on the Hough transform) to reconstruct those vertices. They form the base for the search for polygons. Some polygons may be "impossible", i.e. they can never be component of a visible object (e.g. a polygon with intersecting segments). A filter procedure erases those.

The remaining polygons are simple geons. The next procedure to be applied searches for relations between polygons (e.g. common segments) to obtain polyhedra. Some of the polyhedra may represent objects of the blocks world scene. Other ones are "impossible" *a priori* and are erased immediately. An example for such a case is a polyhedron which consists of overlapping polygons.

The last step is a matching procedure aimed to find correspondences between a polyhedron and a projection of a model of the blocks world object. If such a match fails, the projection parameters have to be varied until a successful match occurs. This procedure is a simple realization of the idea of mental rotation.

Usually the complexity of a search space like this is of high order, so that an efficient handling seems to be impossible. However, if we search for correspondences between *relations* of these geons that form a polyhedron and *relations* of those geons that form the current projection of the model, then the search problem becomes easily manageable. This matching procedure is illustrated in the right part of Fig. 1. It consists of three steps:

(1) An initial matching step, which compares the number and type of visible polygons of an object in the scene with the model base (in the example of Fig. 1 it selects both "chock" and "roof"),

(2) An outline matching step, which compares coarsely the outlines of an object in the scene with the models selected by the initial step and estimates the spatial parameters of the object represented by (φ_u, θ_u).

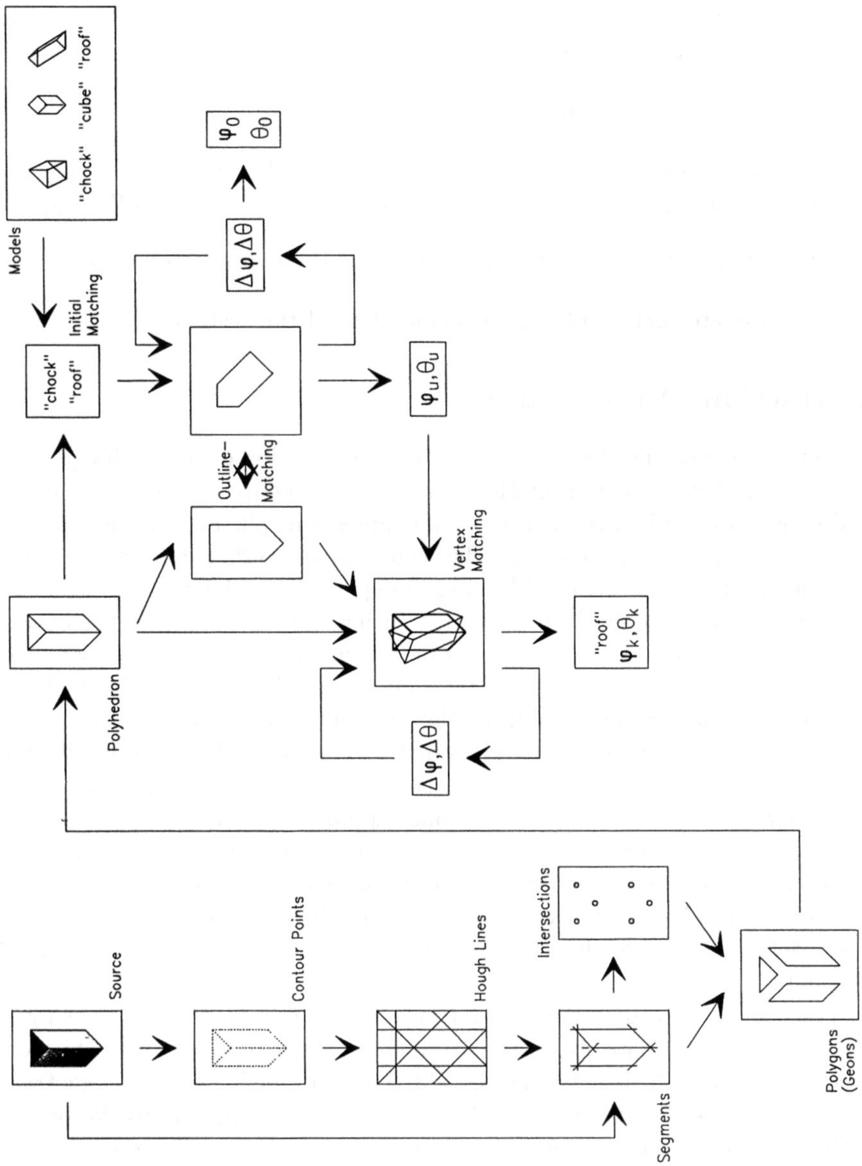

Figure 1: Basic concept of the procedure.

(3) A vertex matching step, which checks the results of the former steps and yields a better spatial matching concerning the parameters (φ_u, θ_u).

The results of the procedures are:

- the determination of the object(s) of a blocks world scene which is (are) represented by a simple 2-D image (in the example of Fig. 1 it is "roof") and
- the determination of the *spatial* parameters of these objects (φ_k, θ_k).

This report illustrates details of the implementation of the system.

3 Extraction of Polyhedra

In this section we consider details concerning the extraction of polyhedra (cf. the left part of Fig. 1). The first step of the extraction procedure is a gradient operation followed by greylevel thinning (non-maxima suppression) [2]. Fig. 2, Fig. 3 and Fig. 4 illustrate the results of these first steps. The source picture ($256 \times 256 \times 8$) of an example is shown in Fig. 2 (array: "source"[1]). Fig. 3 depicts the polar componentes (arrays: "mag" and "ang") of the gradient operation (using a 9×9 processing window). The gradient orientation of a pixel ("ang") can be determined using the "compass card" in the top right corner. The result of the thinning process is shown in Fig. 4. Note that the maximum deviation permissible in the gradient orientation of two neighboring pixels has been chosen to be relatively small. Therefore only a few gradient pixels "survive" the thinning procedure. At a first glance, this proceeding seems to be not very beneficial. It contradicts the traditional approach of aiming for curves without any fraction. Nevertheless, Gestalt psychology shows that human beings do not need closed curves to recognize an object. Since we use Gestalt principles for our purpose the extraction of polygons will turn out to be relatively simple. Moreover, by using this approach we gain the advantage of having to handle only a comparatively small value of data.

In a second step we apply the Hough transform in order to detect collinear gradient pixels. By this operation we obtain the parameters (θ, r) of straight lines which connect collinear pixels. Fig. 5 depicts the straight lines obtained by the Hough transform (lists: "inc" and "acc"). We use two different forms of Hough accumulation for *each* line. The first one increments an accumulator cell each time a fitting gradient pixel is found ("inc"). The second method uses the magnitude of the fitting gradient pixel to increase the contents of the respective accumulator cell ("acc").

The bright lines in the "inc"-picture are due to a large number of fitting gradient pixels, while the weaker lines are caused by a few fitting gradient pixels only. Hence, the incremental form is sensitive towards the *number* of collinear gradient pixels but ignores their magnitude. Contrary to this, the accumulative form is sensitive towards the *gradient magnitudes* of collinear gradient pixels. Nevertheless, a line caused by a few fitting gradient pixels only is bright in the "acc"-picture if the sum of their gradient

[1] Information is stored in various arrays and lists which are cited in inverted commas

Figure 2: Example of a source image.

Figure 3: Result of gradient operation.

thin

Figure 4: Result of non-maxima suppression.

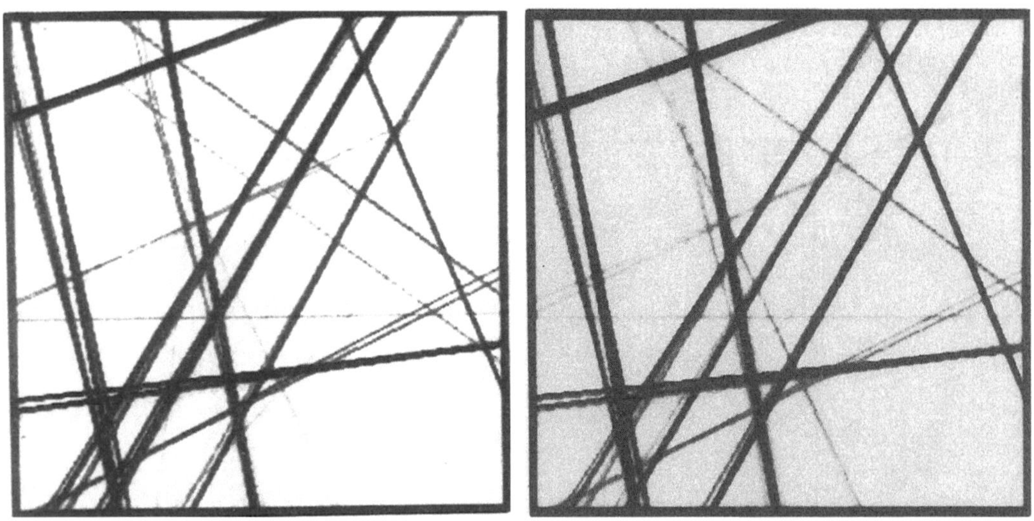

inc acc

Figure 5: Results of Hough transform.

magnitude pixels is high. On the other hand, if many gradient pixels are collinear but their magnitudes are low, then the corresponding line in the "acc"-picture will be weak.

Depending on the "inc"and "acc" values of a line it will be regarded as significant or not. Tracing along the significant lines in the *source* image we look for significant greylevel differences. This processing step produces a structured list that contains a set of straight line *segments*. The structure of this list reflects collinearity and parallelism properties of the segments. Furthermore, the segments are attributed to the mean greylevel difference between the regions which they separate. In Fig. 6 (list: "al-seg") the bright segments are those representing a high greylevel difference, the weak segments depict those of lower differences.

The next step is aimed at an enhancement of the segment list ("al-seg"). The results of this procedure are shown in Fig. 10 (list: "al-super"). Fig. 8 illustrates the enhancement process itself. Fig. 8 ("al-seg/colst") shows the typical structure of a cluster of segments. Note that Fig. 6 contains several of such clusters. Now we search for lines (θ, r) representing the respective clusters. For this purpose each segment is assigned a weight which is the product of its length and its mean greylevel difference. The procedure starts with recognizing the parameters (θ, r) of the "strongest" segment as the parameters of a clustered line. This line acts as a "condensation kernel" for the other segments in the cluster, i.e. segments which are approximately parallel and near to the line. This is an easy task since these segments are represented by the data structure mentioned above.

The first step of the "condensation process" is to pack segments, that means projecting them onto the line (Fig. 8 "prlst") followed by a merging of overlapping segments. In order to implement the projection, we need the two coordinates (y_0, x_0) and (y_1, x_1) of the termination points of a segment serving as a "condensation kernel" (cf. Fig. 7). Next we need those two lines, which are perpendicular to the "condensation kernel" and are running through the termination points. These lines are described by the parameters $((\theta + \frac{\pi}{2}), r_0')$ and $((\theta + \frac{\pi}{2}), r_1')$. Let

$$r = x \cos\theta \; + \; y \sin\theta \tag{1}$$

be the equation describing the clustered line. Then the perpendicular line running through (y_0, x_0) is given by

$$\begin{aligned} r_0' &= x_0 \cos(\theta + \frac{\pi}{2}) \; + \; y_0 \sin(\theta + \frac{\pi}{2}) \\ &= y_0 \cos\theta \; - \; x_0 \sin\theta. \end{aligned} \tag{2}$$

Similarly we find

$$r_1' = y_1 \cos\theta \; - \; x_1 \sin\theta \tag{3}$$

for the line running through (y_1, x_1).

The second step of the "condensation process" is to merge the projected segments (cf. Fig. 8 "fusion" and „al-super"). For this purpose we use the parameters r_0' and r_1' where $r_0' < r_1'$. Let r_0' be the *head* and r_1' the *tail* of the respective segment. Even in case of overlapping segments the merging is simple: Head and tail of the most left and most right cluster segment represent the new segment.

al-seg

Figure 6: Result of the extraction of segments.

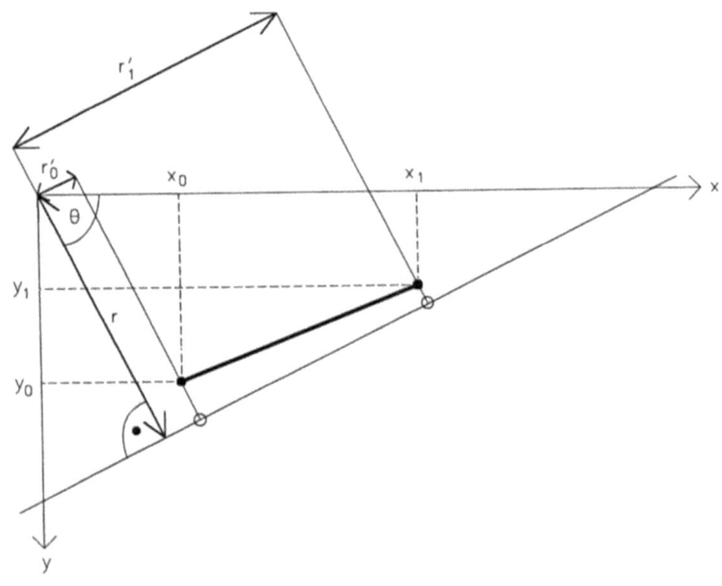

Figure 7: Projection of a segment.

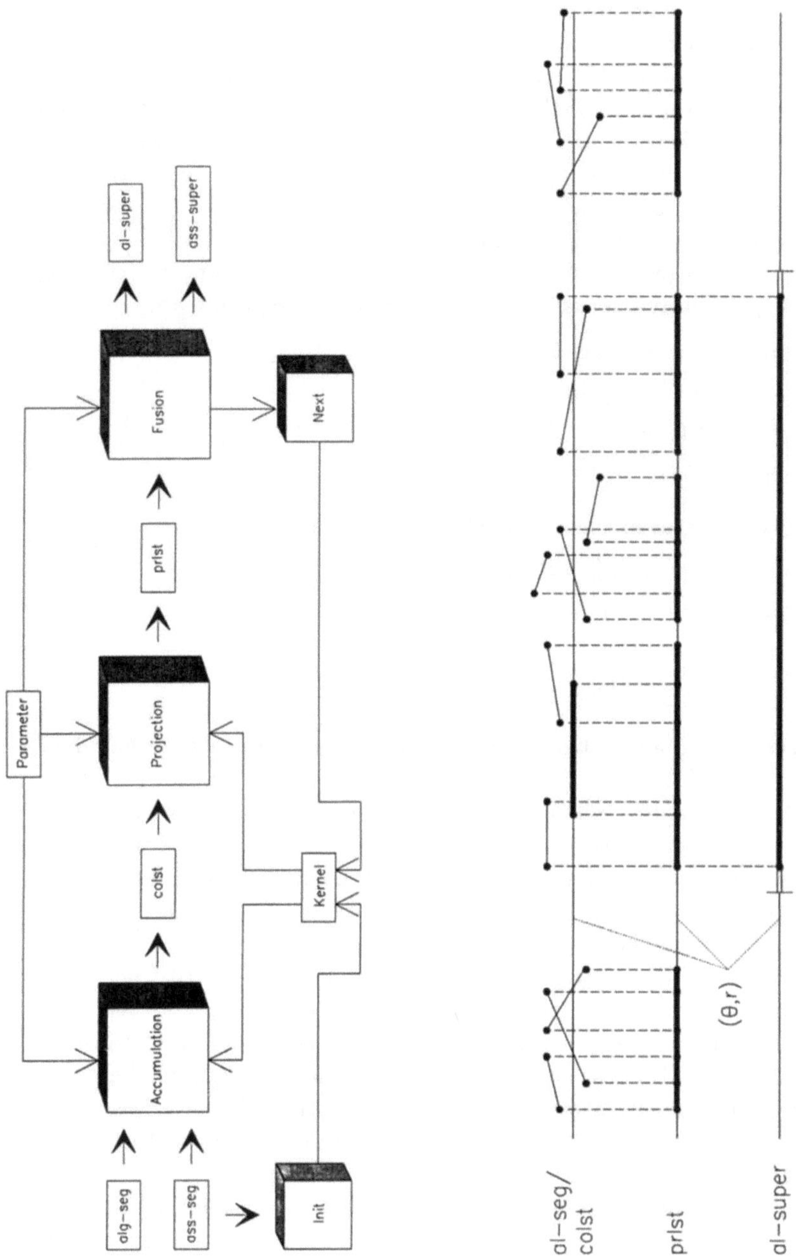

Figure 8: Enhancement of segments.

Nevertheless, the occurence of gaps rises some problems. Fig. 8 ("prlst") illustrates the situation: Which gap is allowed to be filled ?

Based on the approach of Gowda [11] we apply the following straightforward procedure: Compare the sum of the lenghts of two adjacent segments with the lenght of the gap between them. If the lenght of a gap between two adjacent segments is less then 10 percent of the sum of the lenghts of these segments, then fill the gap.

For the purpose of the search of polygon vertices its useful to extend the lenght of the "super segments" (cf. Fig. 9). Let such a segment be described by its termination points (x_0', y_0') and (x_1', y_1'). The parameters r_0' and r_1' are known (cf. Fig. 7, Eq.(2) and Eq.(3)). Then the coordinates of the projected segments are

$$y_0' = r \sin\theta + r_0' \cos\theta \tag{4}$$
$$x_0' = r \cos\theta - r_0' \sin\theta \tag{5}$$
$$y_1' = r \sin\theta + r_1' \cos\theta \tag{6}$$
$$x_1' = r \cos\theta - r_1' \sin\theta \tag{7}$$

while the coordinates of the extended segments are

$$y_0'' = r \sin\theta + (r_0' - e) \cos\theta \tag{8}$$
$$x_0'' = r \cos\theta - (r_0' - e) \sin\theta \tag{9}$$
$$y_1'' = r \sin\theta + (r_1' + e) \cos\theta \tag{10}$$
$$x_1'' = r \cos\theta - (r_1' + e) \sin\theta. \tag{11}$$

e describes the amount of extension. In our experiments we found 10 percent of the original length of the segment to be a reasonable amount for e.

The last step of the enhancement process erases the segments of which the current cluster consists from the segment list ("al-seg") so that they cannot appear as elements of another cluster. The enhancement process then continues with the next "strongest" segment (cf. Fig. 8, "next"). The whole procedure (accumulation , projection and fusion) repeats until the segment list is empty.

The result of the enhancement steps are shown in Fig. 10. The bright segments are the "strong" ones, while the weak segments are the "weak" ones.

Note that even a large number of trigonometric calculations causes no problems in computing time because they are calculated only once during the generation of the structured segment list. That means that each segment (or set of segments) in the list relates to a set of data describing the geometric properties of the segment. Two of these data are $\sin\theta$ and $\cos\theta$.

The extended segments (list: "al-super") form the base for another important step of the system. They are checked for intersections, which are the vertices of potential polygons. In order to remove spurious intersections we first run a "cleaning procedure". In this process we cluster intersections which are close neighbors. For this purpose each intersection is assigned a weight being the sum of the weights of the intersecting segments. Again we start processing the "strongest" intersection and regard it as a "condensation kernel" for close neighbors. That means that all the neighboring intersections are assigned the coordinates of the "condensation kernel". The absorbed intersections are erased from the list and the condensation process continues with the

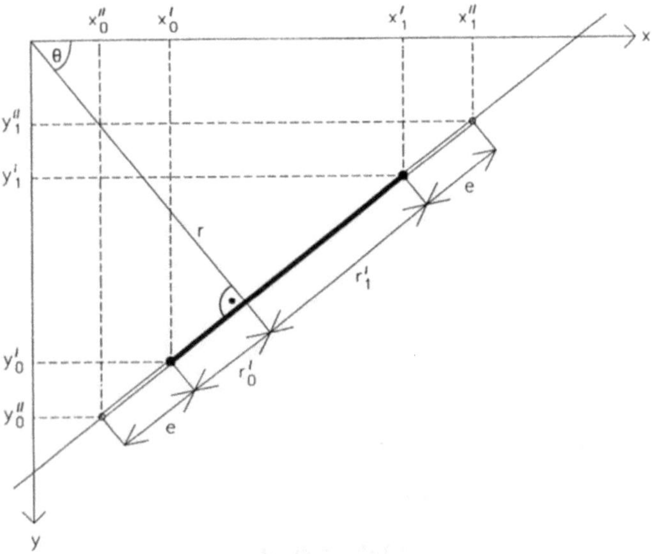

Figure 9: Extension of a segment.

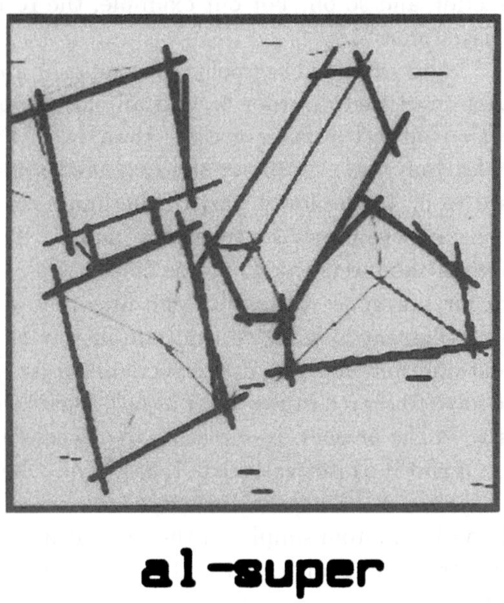

Figure 10: Result of the enhancement of segments: The extended "super segments".

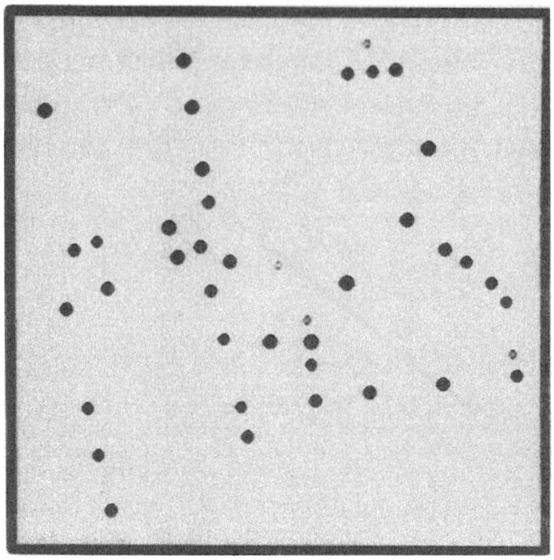

Figure 11: Result of the intersection searching procedure.

next "strongest" intersection and so on. For our example, the result of this process is illustrated by Fig. 11 (list: "cross").

Intersections ("cross") and extended segments ("al-super") are the base for a special procedure which was developed in order to find all closed polygons consisting of n vertices. n may be any natural number greater than two. Fig. 12 illustrates an example of this search for polygons. It shows the extended segments a to h having the intersecting points 0 to 5. In the lower part of the figure an array is created the row and column addresses of which are the segments indices. The cells of this array contain either the indices of the intersections or are empty. Fig. 12 shows an example for $n = 4$. The search for polygons starts with the first row of the array (segment a). The first intersecting segment is b, so we continue in row b, search again for the first intersecting segment and find c. Next, c intersects d. So far all the four segments occured. Therefore, we have to search in row d for a path back to a. If this search fails no closed polygon exists. In the present example we are successful and find a polygon with the segments a, b, c, d and the intersections $0, 1, 5, 4$. The other polygons are found in a similar way.

In order to keep the vision system simple, in the present implementation we apply the polygon search algorithm for triangles ($n = 3$) and quadrangles ($n = 4$) merely. Fig. 14 shows the results (lists: "po 3" and "po 4") of the blocks world example. Note that a weight is attributed to the polygons. The bright ones consist of segments which are "strong" themselves. On the other hand, the weak ones consist of weak segments.

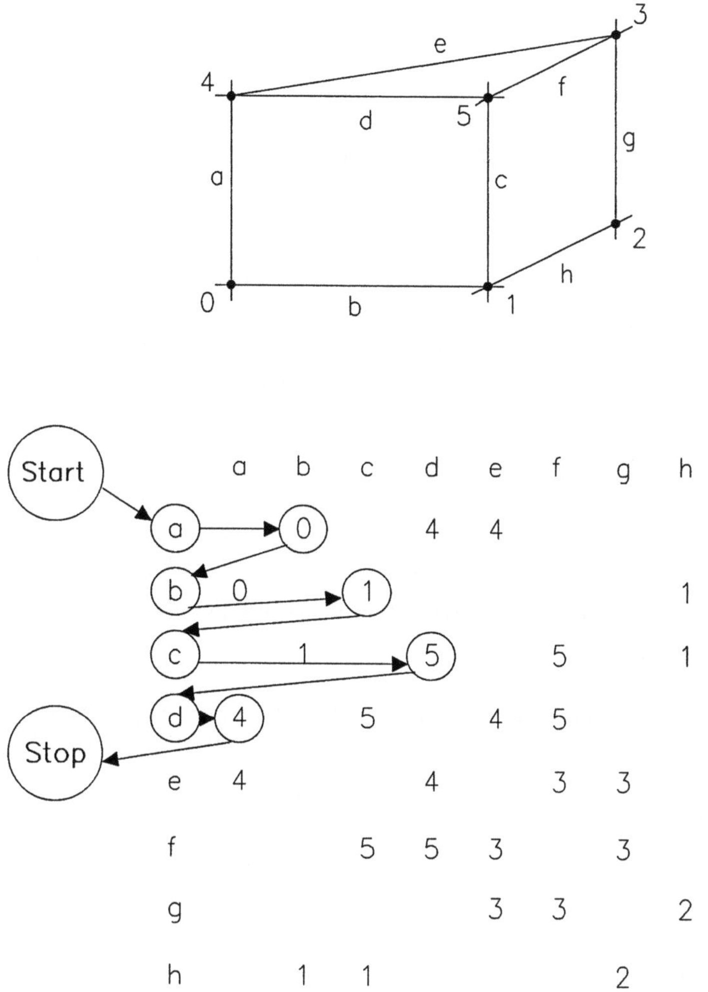

Figure 12: Example of a search for polygons.

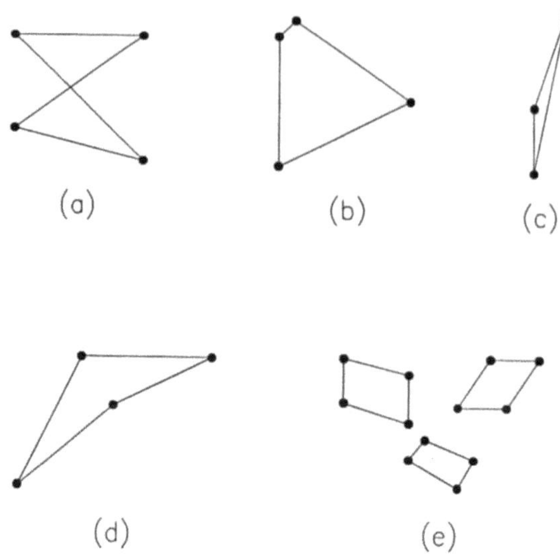

Figure 13: Examples of useless polygons.

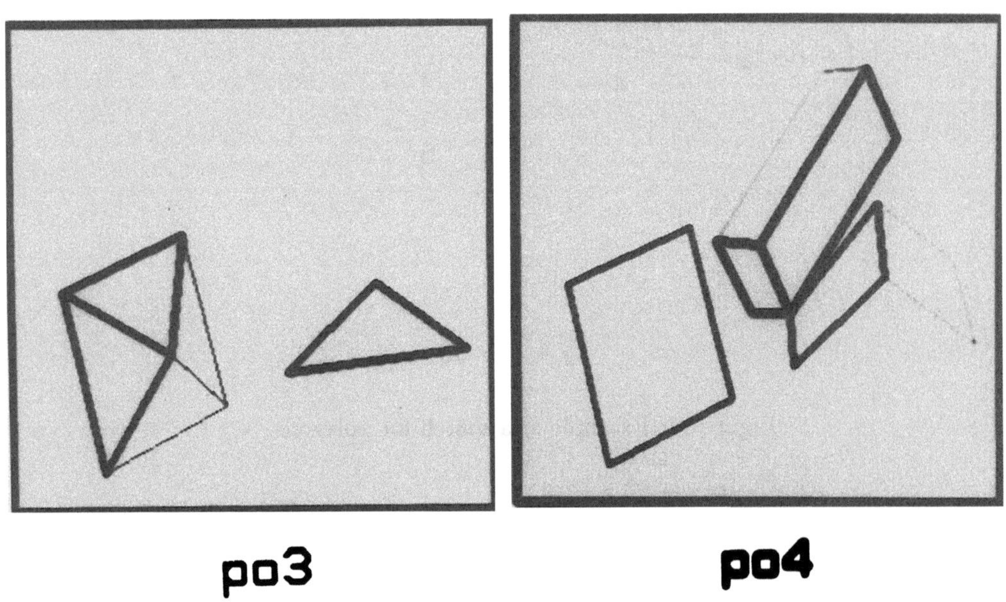

Figure 14: Results of the polygon searching procedure.

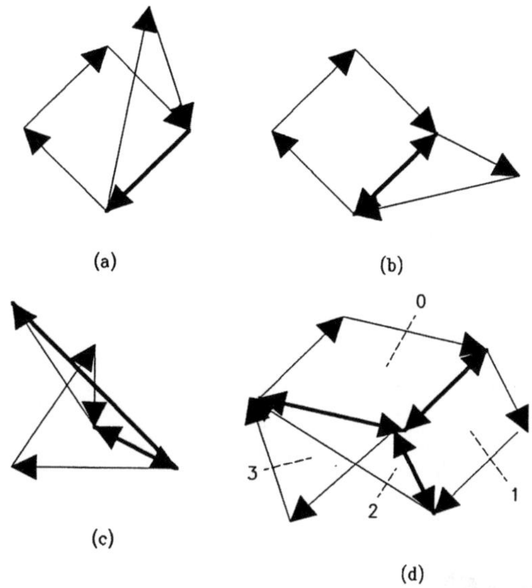

Figure 15: Four examples of constellations of triangles and squares.

Some of the polygons are not suitable for further processing. Fig. 13 shows polygons with

(a) intersecting segments,

(b) very short segments,

(c) an included area which is approximately zero,

(d) not enough parallel segments,

(e) no "legal" neighbor polygon, e.g. a polygon which at least shares two vertices with the one to be checked.

Condition (d) is based upon the knowledge that the objects of our scenes consist of polygons with at least two parallel segments (exception: $n = 3$).

Degenerated polygons according to (a) to (e) must be erased. The result of this process (lists: "po 3" and "po 4") forms the source of the next processing step (shown in Fig. 14): The extraction of polyhedra. Prior to this, two more operations must be performed: Erasing of concave polygons and giving the remaining ones a clockwise orientation. The reason for doing so is explained using the examples shown in Fig. 15. If two clockwise oriented closed polygons share a segment and the orientations of the

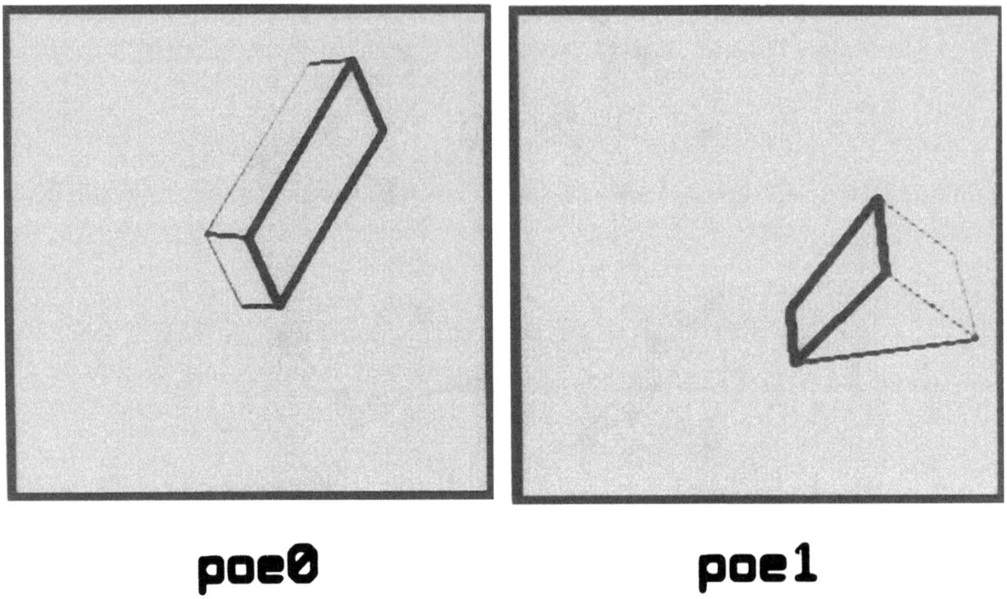

Figure 16: Result of the polyhedra searching procedure.

common segment are equal, then the two polygons intersect (a). These two polygons cannot form a polyhedron. If the orientations at the common segment are opposite, then a legal polyhedron is present (b). Unfortunately this rule does not work in the case of concave polygons (c). Since the objects in blocks world scenes consist of convex polygons only, we exclude concave polygons. For detailed information concerning the mathematics of polygons and polyhedra refer to [16].

Searching for polyhedra consists of three basic steps. In the first step a list is created which contains information concerning the neighborhood of each polygon. A valid neighboring polygon fits in the sense of Fig. 15b to the current polygon. Fig. 15d shows a simple example of a limited neighborhood. Assuming polygon 0 is the current polygon, there are three neighbors: 1, 2 and 3. Note that the neighbors 2 and 3 share the same common segment with 0. If we assume polygon 3 as the current polygon, then it has two neighbors (0 and 2) which share the same common segment. Polygon 1 is not a neighbor. Similarly we find the neighbors of the current polygons 1 and 2.

The second step extracts "raw" polyhedra from the neighborhood list. For this purpose we assign another weight to the current polygon. As such we use the number of common segments which are shared by neighbors. The current polygon 0 in the above example would fetch the weight two. We start the extraction with the polygon bearing the highest number of common segments. If there are more than one of those we first use the polygon with the highest previous weight. The neighbors of the current polygon are erased from the list and the next "strongest" polygon will become the current one.

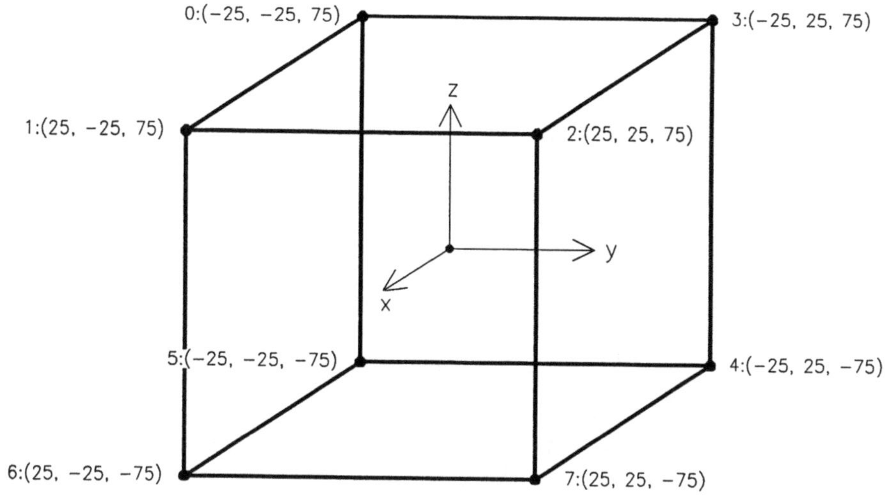

Figure 17: Model of a block.

The last step cleans the "raw" polyhedra in order to obtain the intrinsic polygon. For example, the polygons 0, 1 and 2 of Fig. 15d constitute an intrinsic polyhedron, while polygon 3 should be erased because it has fewer neighbors than polygon 2.

Fig. 16 shows the two "strongest" polyhedra obtained from the blocks world example. Due to shadows, noise or background distortion certain polyhedra may have been extracted which do not correspond to an object in the scene. They will be erased by the matching procedure (right part of Fig. 1).

4 Model Base

Fig. 17 shows the model of a block represented by its vertices 0 to 7 and centred around the origin of a right-handed Cartesian coordinate system, the coordinate values of which describe the size of a block. In the current example the block is 150 mm in height, 50 mm in width and its depth is 50 mm. In Section 2 it was stated that only a few typical combinations of polygons can occur, depending on the viewpoint from which we look at a polyhedron. For the example of Fig. 17 we have five such combinations. Fig. 18 illustrates them along with the minimum and the maximum values of the camera parameters θ and φ (they are defined in Section 5) that are allowed without leaving the specified combination. Note that for certain types of polyhedra the minimum (or maximum) of one parameter holds only if the other parameter does not take extreme values. In other words: if both parameters are minimum or maximum, the specified combination may not be valid. Hence, an algorithm which uses these parameters must take this effect into account (cf. Section 6).

	$\theta_{min}\,\theta_{max}$	$\varphi_{min}\,\varphi_{max}$
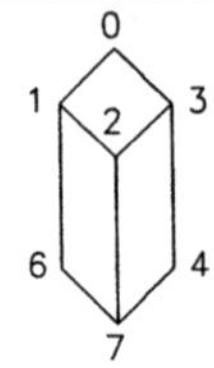	5° 85°	5° 85°
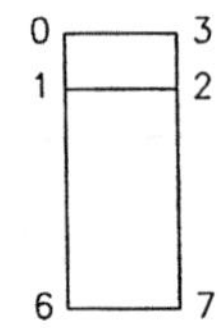	5° 85°	0° 0°
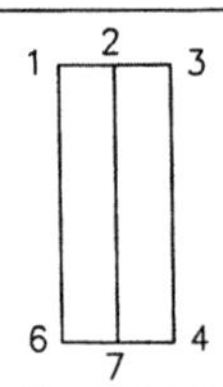	90° 90°	5° 85°
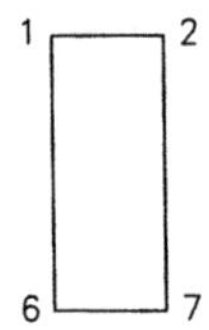	90° 90°	0° 0°
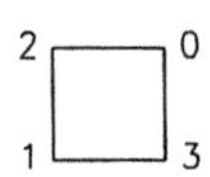	0° 0°	0° 0°

Figure 18: Possible constellations of the block shown in Fig.17.

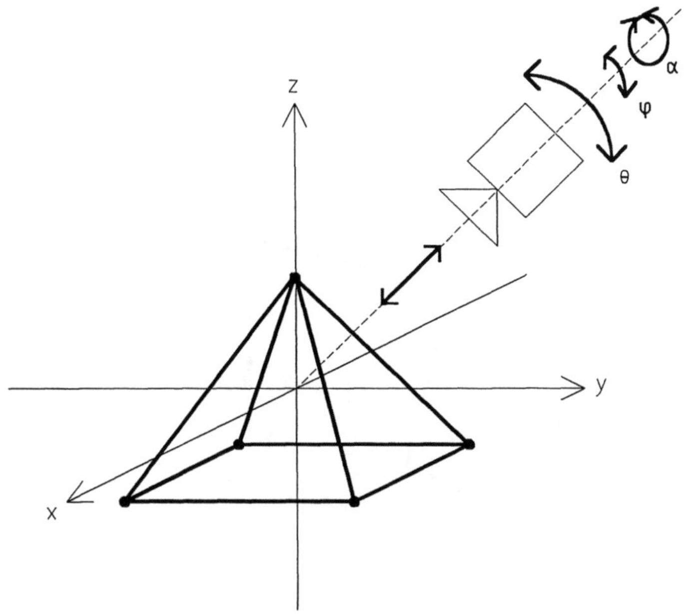

Figure 19: Projection scenario.

For each block which is a part of the blocks world the model base contains

- its vertices (Fig. 17),
- the basic viewpoint constellations (Fig. 18),
- the minimum and maximum values of the parameters of these constellations,
- the visible vertices of the block for these constellations.

5 Projection Algorithm

Fig. 19, Fig. 20 and Fig. 21 show the projection scenario. The model block (e.g. a pyramid) is represented by its vertices and centred around the origin (cf. Section 4). A camera pointing at the origin has four degrees of freedom:

(a) distance from the origin (point P in Fig. 20),

(b) rotation around the z axis (azimuth, φ in Fig. 20),

(c) rotation on the arc of the meridian (θ in Fig. 20),

(d) rotation around the optical axis of the camera (α in Fig. 21).

The vertices of the model block must now be projected onto the camera plane. For our blocks world examples, perspective projection is not necessary since the distance

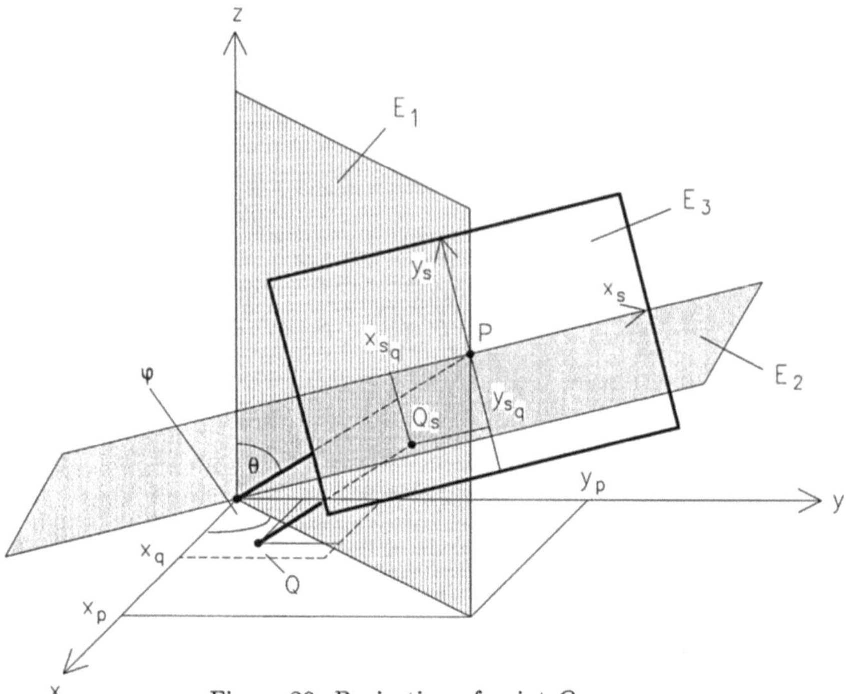

Figure 20: Projection of point Q.

between camera and object is much greater than the size of the object itself. Fig. 20 shows the projection Q_s of a point Q onto the plane E_3. Q_s is represented by the Cartesian system (x_s, y_s) with the origin P. P itself is determined in the three-dimensional world by its coordinates (x_p, y_p, z_p). The distance between the 3-D origin and P is r. In the 3-D system the coordinates of Q are (x_q, y_q, z_q). On plane E_3 the projection Q_s takes the coordinates (x_{s_q}, y_{s_q}). E_1 is the meridian plane. It is perpendicular to the (x,y) plane. Plane E_2 is perpendicular to E_1. The intersection between these two planes is a line running through both the 3-D origin and P. E_3 is the projection plane and perpendicular to E_1. The intersection between E_2 and E_3 represents the horizontal axis x_s of the projection plane and the intersection between E_1 and E_3 represents the vertical axis y_s.

For the formal treatment of the projection problem there are some cases in which the spherical coordinate system is more convenient than the Cartesian system. In such a system P is determined by

$$\begin{pmatrix} x_p \\ y_p \\ z_p \end{pmatrix} = \begin{pmatrix} r\cos\varphi\sin\theta \\ r\sin\varphi\sin\theta \\ r\cos\theta \end{pmatrix} = r\sin\theta \begin{pmatrix} \cos\varphi \\ \sin\varphi \\ 1/\tan\theta \end{pmatrix}. \tag{12}$$

In order to handle the planes E_1, E_2 and E_3 we need the plane equation

$$ax + by + cz + d = 0, \tag{13}$$

of which the corresponding vector form is

$$XX_1 + d = 0, \tag{14}$$

where

$$X = \begin{pmatrix} x \\ y \\ z \end{pmatrix} \quad \text{and} \quad X_1 = \begin{pmatrix} a \\ b \\ c \end{pmatrix}. \tag{15}$$

X_1 is perpendicular to the plane and of magnitude $\sqrt{a^2+b^2+c^2}$. Note that for a plane running through the origin we have $d = 0$. We will use this property in context with planes E_1 and E_2. On the other hand, the Hessian normal form of a plane is

$$x\cos\alpha + y\cos\beta + z\cos\gamma - r = 0 \tag{16}$$

where

$$\cos\alpha = \frac{a}{\sqrt{a^2+b^2+c^2}}, \tag{17}$$

$$\cos\beta = \frac{b}{\sqrt{a^2+b^2+c^2}}, \tag{18}$$

$$\cos\gamma = \frac{c}{\sqrt{a^2+b^2+c^2}}, \tag{19}$$

$$r = -\frac{d}{\sqrt{a^2+b^2+c^2}}. \tag{20}$$

r is the perpendicular distance between the plane and the origin. Now we can focus on the special description of the planes E_1, E_2 and E_3 in Fig. 20. E_1 is spanned by

$$Z = \begin{pmatrix} 0 \\ 0 \\ 1 \end{pmatrix} \quad \text{and} \quad P = \begin{pmatrix} \cos\varphi \\ \sin\varphi \\ 1/\tan\theta \end{pmatrix}. \tag{21}$$

Hence the normal vector of E_1 is

$$X_1 = Z \times P = \begin{pmatrix} -\sin\varphi \\ \cos\varphi \\ 0 \end{pmatrix}, \tag{22}$$

while d_1 is zero since the plane runs through the origin. Plane E_2 is spanned by

$$X_1 = \begin{pmatrix} -\sin\varphi \\ \cos\varphi \\ 0 \end{pmatrix} \quad \text{and} \quad P = \begin{pmatrix} \cos\varphi \\ \sin\varphi \\ 1/\tan\theta \end{pmatrix}, \tag{23}$$

and its normal vector is

$$X_2 = X_1 \times P = \begin{pmatrix} \cos\varphi/\tan\theta \\ \sin\varphi/\tan\theta \\ -\sin^2\varphi - \cos^2\varphi \end{pmatrix} = \begin{pmatrix} \cos\varphi \\ \sin\varphi \\ -\tan\theta \end{pmatrix}, \tag{24}$$

with $d_2 = 0$. In order to determine the projection line running through Q and Q_s we use the planes E_1' and E_2' the intersection of which is the projection line. Let E_1' be parallel to E_1 and E_2' parallel to E_2. Then the distances r_1 between E_1' and E_1 and r_2 between E_2' and E_2 can be used to compute Q_s. E_1' is described by

$$\begin{pmatrix} x_q \\ y_q \\ z_q \end{pmatrix} \begin{pmatrix} -\sin\varphi \\ \cos\varphi \\ 0 \end{pmatrix} + d = 0 \tag{25}$$

where

$$d = x_q \sin\varphi - y_q \cos\varphi. \tag{26}$$

The perpendicular distance between E_1' and E_1 can easily be calculated using the Hessian normal form

$$r_1 = -\frac{d}{\sqrt{a_2 + b_2 + c_2}} = \frac{y_q \cos\varphi - x_q \sin\varphi}{\sqrt{\sin^2\varphi + \cos^2\varphi}} = y_q \cos\varphi - x_q \sin\varphi. \tag{27}$$

Similarly, E_2' is described by

$$\begin{pmatrix} x_q \\ y_q \\ z_q \end{pmatrix} \begin{pmatrix} \cos\varphi \\ \sin\varphi \\ -\tan\theta \end{pmatrix} + d = 0, \tag{28}$$

where

$$d = -x_q \cos\varphi - y_q \sin\varphi + z_q \tan\theta. \tag{29}$$

The perpendicular distance between E_2' and E_2 is

$$\begin{aligned} r_2 &= \frac{x_q \cos\varphi + y_q \sin\varphi - z_q \tan\theta}{\sqrt{1 + \tan^2\theta}} \\ &= (x_q \cos\varphi + y_q \sin\varphi - z_q \tan\theta)\cos\theta \\ &= (x_q \cos\varphi + y_q \sin\varphi)\cos\theta - z_q \sin\theta. \end{aligned} \tag{30}$$

The parameters r_1 and r_2 allow the calculation of x_{s_q} and y_{s_q}. In the situation shown in Fig. 20 (alignment of x_{s_q} and E_2 as well as y_{s_q} and E_1) we simply have $x_{s_q} = r_1$ and $y_{s_q} = r_2$. If, however, the camera is rotated by an angle α (i.e. the projection plane E_3 is also rotated), then we have to apply a coordinate transform in order to relate x_{s_q} and y_{s_q} to r_1 and r_2 (cf. Fig. 21):

$$\begin{aligned} x_{s_q} &= r_1 \cos\alpha + r_2 \sin\alpha \\ y_{s_q} &= r_2 \cos\alpha - r_1 \sin\alpha. \end{aligned} \tag{31}$$

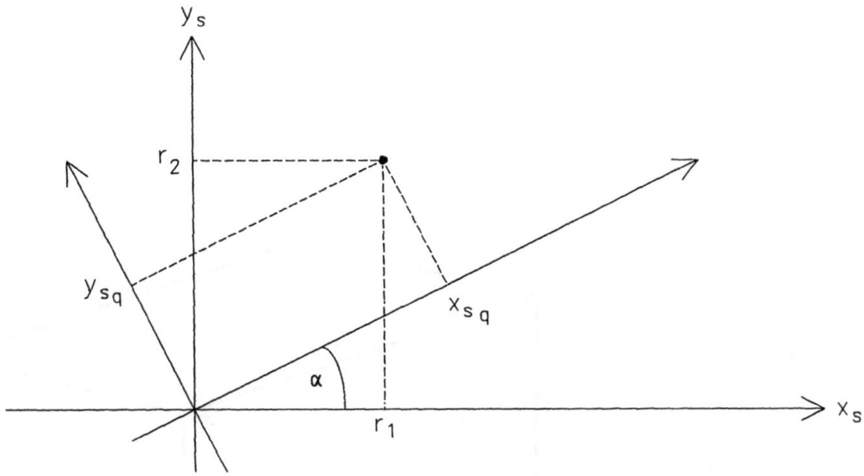

Figure 21: Rotation of the projection plane.

The last step of the projection procedure concerns the determination of hidden planes. Fig. 22 shows the back plane of a block. The normal vector of this plane is

$$N = (D - A) \times (B - A). \tag{32}$$

The angle

$$\tau = \arccos \frac{NP}{|N||P|} \tag{33}$$

between the normal vector and the projection vector provides information about the visibilty of the plane: If $\tau \leq \pi/2$ the plane is hidden. However, it is sufficient to consider the dot product NP. If

$$NP = \begin{cases} \leq \pi/2 & \text{then the plane is hidden} \\ \text{else} & \text{the plane is visible.} \end{cases} \tag{34}$$

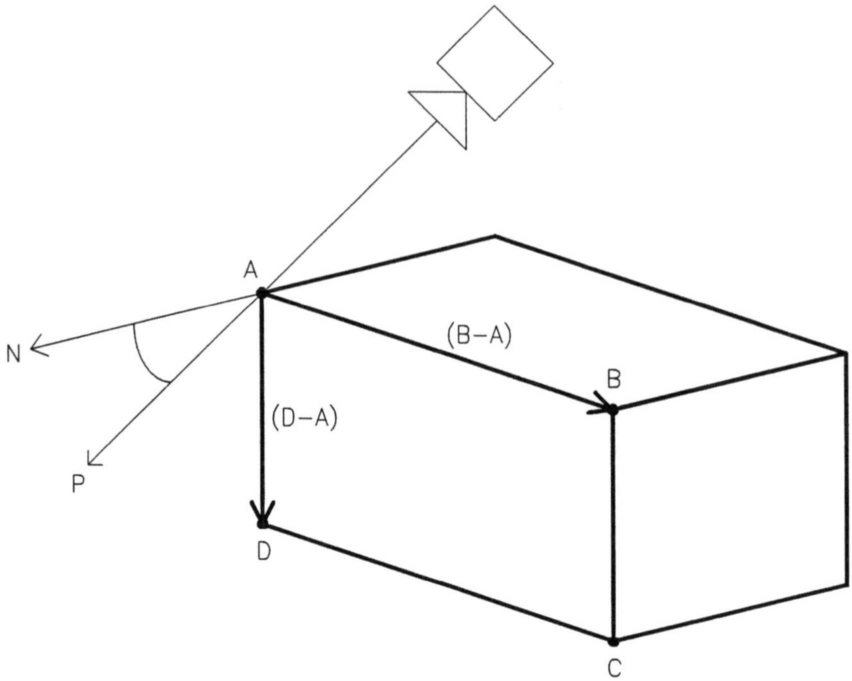

Figure 22: Determination of hidden planes.

```
(let* ((outline (find-outline scene-polyhedron)))
  (do ((theta theta-min (+ theta theta-step)))
      ((> theta theta-max))
      (do ((phi phi-min (+ phi phi-step)))
          ((> phi phi-max))
          (let* ((proj (projection theta phi current-model)))
            (when (not (out-of-border proj))
                  (let* ((outline* (find-outline* proj)))
                    (when (match? outline outline*)
                          (store-result theta phi scene-polyhedron proj))))))))
```

Figure 23: LISP procedure of the outline matching step.

6 Matching Algorithm

The matching procedure finds correspondences between the polyhedra in the scene and the projections of the models (cf. Section 2 and Fig. 1). We apply three steps of matching:

(1) An initial matching step, which compares the number and type of visible polygons of an object in the scene with the model base,

(2) An outline matching step, which compares the outlines of an object in the scene with the models extracted by the initial step,

(3) A vertex matching step, which checks the results of the former steps.

The Initial Matching Step

The initial step extracts a limited number of models from the model base. They are related to the actual polyhedron in the scene. Therefore, the following matching steps require less effort for searching. The initial step itself is very fast due to its simplicity.

The Outline Matching Step

The principle of the matching procedure used for the outline step can best be expressed by the LISP procedure which is depicted in Fig. 23.

At the beginning the *outline* of the current polyhedron in the scene is determined. Two loops implement the search for the camera parameters between their limits *theta-min* — *theta-max* and *phi-min* — *phi-max* (cf. Section 4 and Fig. 18). The increments of the searching procedure *theta-step* and *phi-step* should be relatively large, so that only a few search steps need to be performed. Now the *projection* is computed for each combination (θ, φ) and the result *proj* checked whether it describes the desired constellation or not. If it does, the *outline** of this projection is also computed. The occurence of a *match* between both these outlines is then stored for further considerations.

Fig. 24 illustrates an example of the outline matching procedure. On the left hand side of the figure the projection of a polyhedron in the scene is depicted. Its outline is emphasized. Right of it, different projections (a), (b) and (c) of two models are shown. The initial matching procedure relates both model 1 and model 2 to the scene polyhedron. The outline step rejects model 2 and determines the camera parameters (θ, φ) of the matching projection (b) of model 1.

In order to describe an outline we use two parameters: the lengths of its segments and the inner angles of its vertices. This approach is the base for a fast rotation-invariant and scale-invariant matching procedure. A successful match occurs if both

- the inner angles of the outlines of object and model and
- the *ratios* of the lengths of segments

are similar.

Fig. 25 shows an example of the initial and the outline matching step. The source image is shown in the top left corner. Next to it the current polyhedron with label 1

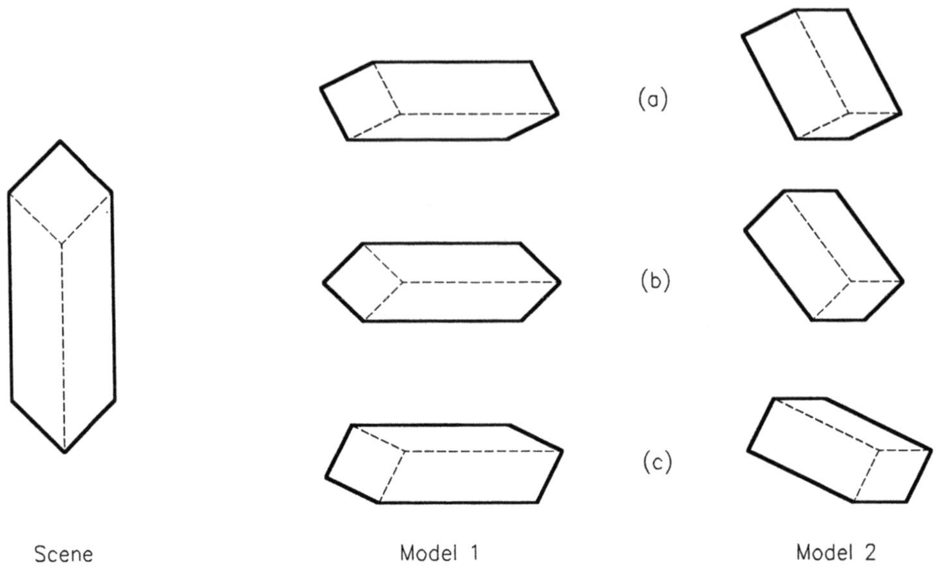

Figure 24: Principle of the outline matching step.

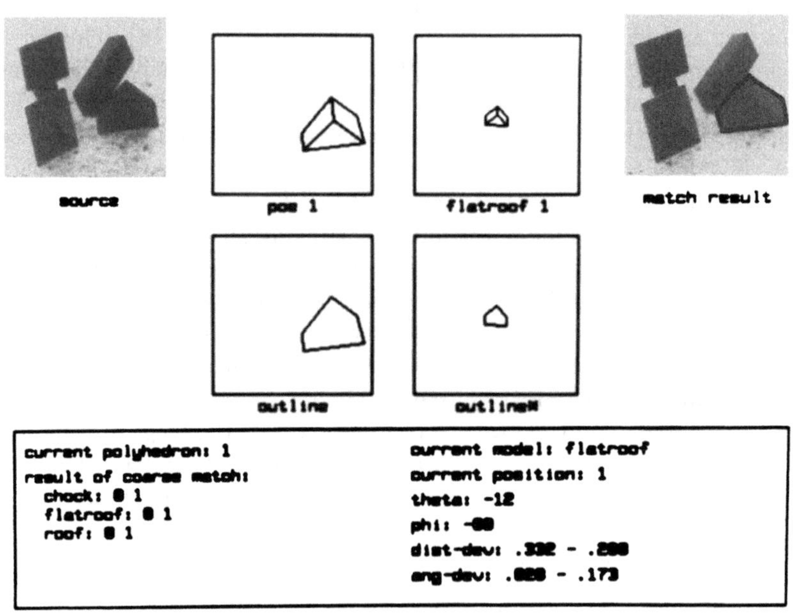

Figure 25: Example of initial and outline matching step.

("poe 1") and its outline are depicted. In the bottom row of the picture information generated by the matching procedure is notified. According to that, the initial matching step offers three possible models named "chock", "flatroof" and "roof" as well as the labels of some of the possible constellations. This approach offers the advantage to considerably simplify the search for the viewpoint. The results of the search steps are shown in the right part of Fig. 25. The model under consideration is "roof" with its constellation "1". The current camera position is determined by the values of θ and φ. As stated above, a successful match occurs if both the inner angles of the outlines of object and model as well as the *ratios* of the lengths of segments are similar. The parameters "ang-dev" and "dist-dev" are the measure of similarity. As depicted in the top right corner of Fig. 25 the current match is nearly perfect.

The Vertex Matching Step

Based on the outline features which are obtained from the previous step (i.e. the inner angles of the outlines of object and model, the *ratios* of the lengths of segments) the model outline is to be mapped into the scene domain. For this purpose we choose the longest segment of the scene outline as a reference for the respective model segment. If

- one segment of a polygon,
- the angle between this segment and an adjacent segment
- the direction of the continuation (cf. the explanations given below) and
- the length of the adjacent segment

are known, the adjacent segment is determined. In Fig. 26 (x_0, y_0) and (x_1, y_1) represent the termination points of the segment which is already known. The points (x_1, y_1) and (x_2, y_2) represent the continuing segment, but (x_2, y_2) is yet unknown. We use α and l_2 to compute this point. Furthermore the direction of continuation is required. It may be determined using the cross product

$$v = (x'_1 - x'_2)(y'_1 - y'_0) - (x'_1 - x'_0)(y'_1 - y'_2), \tag{35}$$

where (x'_0, y'_0), (x'_1, y'_1) and (x'_2, y'_2) are the termination points of both the segments in the model domain. Based on this information we obtain

$$x_2 = \begin{cases} x_1 - l_2 \cos\alpha_2 & \text{if} \quad v \geq 0 \\ x_1 + l_2 \cos\alpha_2 & \text{else} \end{cases} \tag{36}$$

and

$$y_2 = \begin{cases} y_1 - l_2 \sin\alpha_2 & \text{if} \quad v \geq 0 \\ y_1 + l_2 \sin\alpha_2 & \text{else} \end{cases} \tag{37}$$

with

$$\alpha_2 = \arctan\frac{y_1 - y_0}{x_1 - x_0} - \alpha. \tag{38}$$

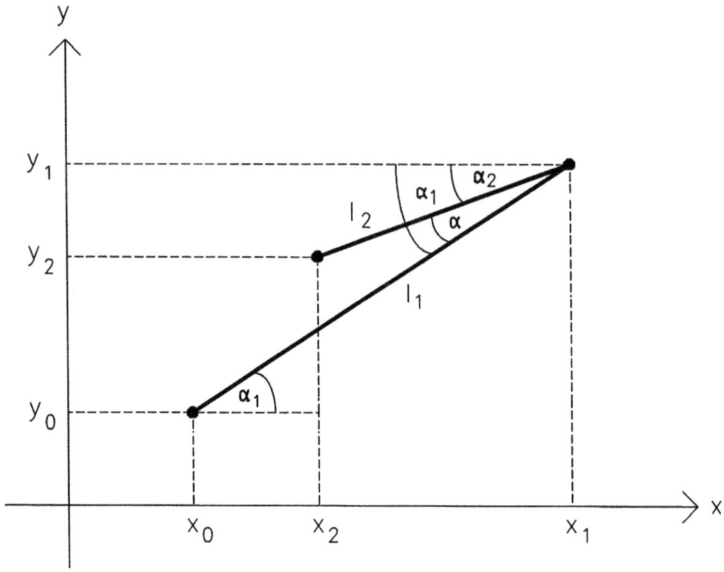

Figure 26: Computation of the coordinates (x_2, y_2) if (x_0, y_0), (x_1, y_1), l_2 and α are known.

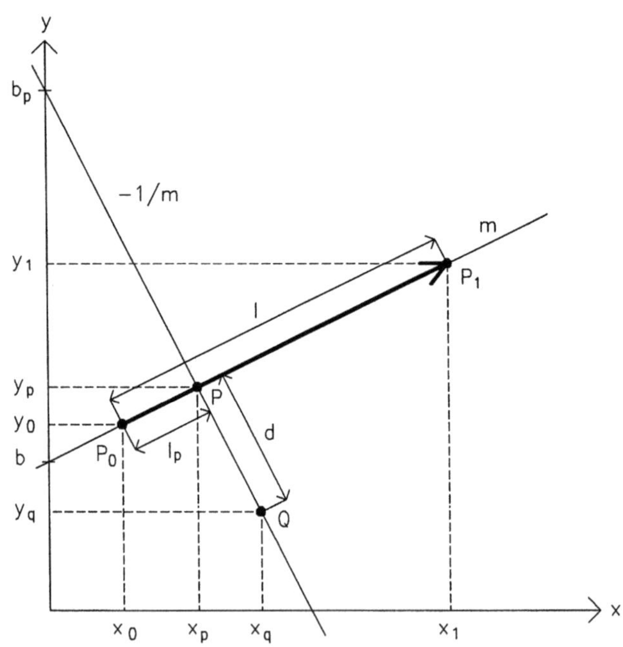

Figure 27: Computation of d and l_p.

Thus, mapping the outline is a simple task. We need a little more effort for the determination of the non-outline vertices. Fig. 27 shows one segment of an outline (P_0, P_1) and a point Q which may be a non-outline vertex. In order to obtain rotation-invariance and scale-invariance, the coordinates of Q must be expressed by l_p and d. Therefore point P has to be computed first. We obtain its coordinates as

$$x_p = \frac{y_q - y_0 + \frac{x_q}{m} + mx_0}{m + \frac{1}{m}} \tag{39}$$

and

$$y_p = y_0 + m(x_p - x_0). \tag{40}$$

The magnitudes of d and l_p are given by the Euclidian distance between P and Q and P and P_0, respectively. In order to describe the position of Q correctly, the sign of l_p must be determined. By definition we know that Q is located on the right hand side of the outline segment [3]. Thus d needs no sign. We use P_0 as reference point and define l_p as positive, if it is located on the "P_1-side". Otherwise it is negative. It is easy to show that the situation "P_1-side" is determined by the parameters x_0, x_1 and x_p. If the condition

$$((x_0 - x_1) > 0) \oplus ((x_0 - x_p) > 0) \tag{41}$$

is true, then l_p is negative ($\oplus$ is the EXOR function).

Scale-invariance is easily obtainable if we deal with the relative quantities d/l and $\pm l_p/l$ instead of d and l_p. As we are now able to determine the position of a non-outline vertex in the model domain, we have to consider the mapping of this vertex into the scene domain.

Mapping the outline segment into the scene domain is carried out by the outline matching step. Thus the segment is known and we can now compute P the coordinates of which are (cf. Fig. 28):

$$x_p = \frac{d_p}{d_0} x_0 \qquad \text{and} \qquad y_p = mx_p + b. \tag{42}$$

Fig. 29 illustrates the determination of Q in the scene domain. We obtain

$$x_q = \begin{cases} \dfrac{d_p^{\perp} + d}{d_p^{\perp}} x_p & \text{if} \qquad (m > 0) \oplus (d_0 < d_1) \\[2ex] \dfrac{d_p^{\perp} - d}{d_p^{\perp}} x_p & \text{otherwise} \end{cases} \tag{43}$$

and

$$y_q = b_p - \frac{x_q}{m}. \tag{44}$$

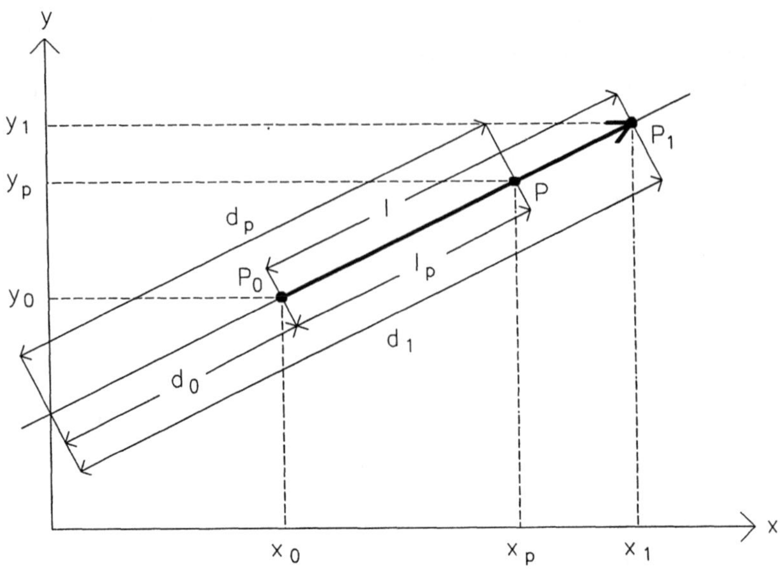

Figure 28: Computation of P in the scene domain. The relation $\pm l_p/l$ is provided by the outline matching step.

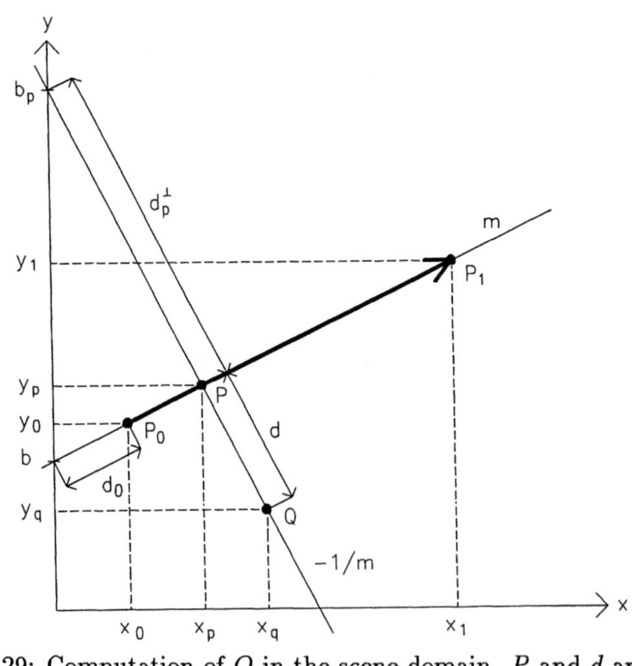

Figure 29: Computation of Q in the scene domain. P and d are known.

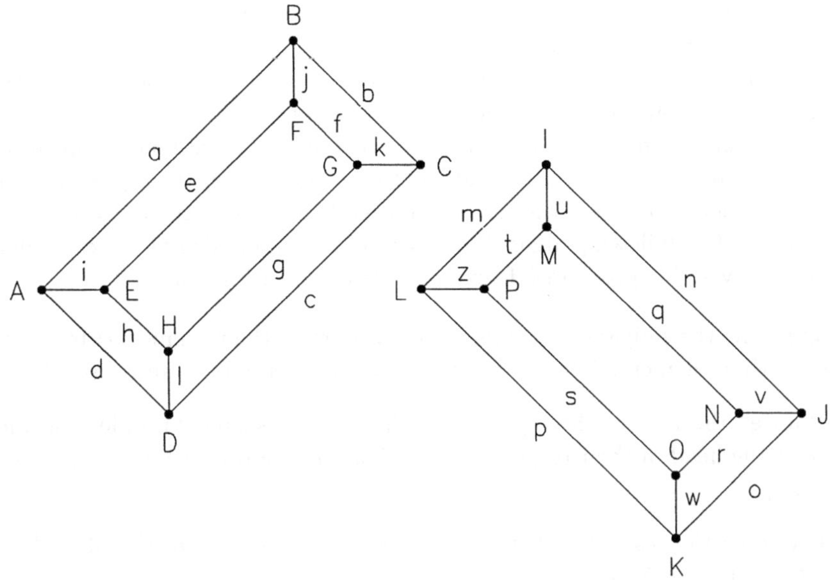

Figure 30: Example for the correspondence problem.

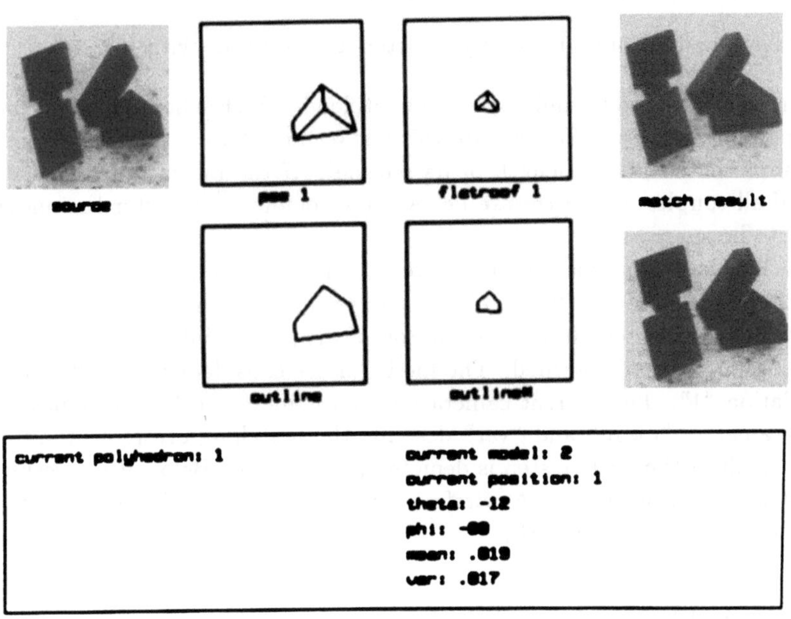

Figure 31: Result of the vertex matching step.

The condition $(m > 0) \oplus (d_0 < d_1)$ becomes evident when all the four possible combinations of the sub-conditions $(m > 0)$ and $(d_0 < d_1)$ are considered (cf. Fig. 29). Note that in the case of an outline segment which is positioned parallel to either the x-axis or the y-axis several of the computations have to be modified accordingly.

The tools for mapping a model into the scene domain are now complete and we can proceed with the actual vertex matching procedure. Preparatory we have to find the correspondences between the non-outline vertices of the current object and the current model. The following algorithm based on the knowledge of correspondences of the outlines solves this problem (cf. example depicted in Fig. 30):

(1) Determine the polygones (represented by a list of their segments) to which both the outline segments belong. Let these two segments be a and n (cf. Fig. 30).

(2) "Rotate" the lists of the segments until both the segments under consideration are at the head of their respective lists. For our example we obtain (a, j, e, i) and (n, v, q, u).

(3) Determine the respective lists of vertices. The example in Fig. 30 yields (A, B, F, E) and (I, J, N, M).

(4) From these lists of vertices erase the first two elements which represent the outline vertices of the segments under consideration. The remaining lists (in our example (F, E) and (N, M)) reflect the desired correspondences.

(5) "Rotate" the lists of segments by one step and continue with (3).

The error between model and object is quantified by the Euclidian distance between the corresponding vertices. We evaluate this distance relative to the size of the object (which is represented by the length of its outline). If the mean and variance of the distances of all couples of vertices are below a user-defined parameter the matching is considered to be successful.

Fig. 31 illustrates an example of the vertex matching step. The source image is shown in the top left corner. Next to it the current polyhedron with label 1 ("poe 1") and its outline are depicted. In the bottom row of the picture information generated by the matching procedure is notified: The model under consideration is "flatroof" with its constellation "1". The current camera position is indicated by the values of θ and φ. The parameters "mean" and "var" describe the matching error. In the top right corner the result of the outline step is depicted (cf. Fig. 25). Below it the result of the vertex matching step is shown. A wireframe which represents the model fits exactly with the object of the blocks world.

7 Realization and Conclusion

Our system is implemented on an IBM/AT. The scene is grabbed by an ordinary TV camera and digitized by a plug-in image processing board. Algorithms are implemented in C or in the LISP dialect PC-SCHEME. Since our system serves only for basic investigations, so far we dispensed with optimizing the algorithms regarding speed.

Therefore, computing time for the example shown above was approximately 15 minutes if LISP is used in the interpreting mode.

Although the system presently deals with simple objects only, it shows the principle of monocular computer vision based on the RBC approach and the idea of mental rotation. The system provides:

- Determination of the objects in a blocks world scene which is represented by a single 2-D image.
- Determination of the *spatial* parameters of these objects.

8 Acknowledgment

This work was supported in part by **Deutsche Forschungsgemeinschaft** (German Research Council) under grant No. AZ322675, Be769/7 (drittes Jahr).

References

[1] Anderson, J.R.: Cognitive Psychology and its Implications. W.H. Freeman and Company, New York, Oxford, 1985.

[2] Bässmann, H. and Besslich, Ph.W.: Konturorientierte Verfahren der digitalen Bildverarbeitung. Springer-Verlag, Berlin, Heidelberg, New York, London, Paris, Tokyo, 1989 (in German).

[3] Bässmann, H. and Besslich, Ph.W.: Monocular Computer Vision. Proc. Third International Conference on Image Processing and its Applications, Warwick, 18.-20. July, IEE Conference Publication Number 307, 107-111, 1989.

[4] H. Bässmann and Ph.W. Besslich
„Monocular Computer Vision: An Approach Exploiting the Theory of Mental Rotation"
Laser '89: 9th International Congress, Munich, 5.-7. June, 1989.

[5] H. Bässmann and Ph.W. Besslich
„On the Recognition-By-Components Approach Applied to Computer Vision"
Proc. SPIE Int. Conf. on Advances in Intelligent Robotics Systems, Session on Intelligent Robots and Computer Vision VIII: Algorithms and Techniques, Philadelphia, USA, 5.-11. Nov., 52-62, 1989.

[6] Besslich, Ph.W. and Bässmann, H.: Curve Enhancement Using Rule-Based Relaxation. Int. Cong. on Optical Science and Engineering, Hamburg, 19.-23. Sept. 1988, P.J.S. Hutzler and A.J. Oosterlinck (Eds.), Image Processing II, SPIE Proc. No. 1027, 154-160, 1989.

[7] Besslich, Ph.W. and Bässmann, H.: A Tool for Extraction of Line-Drawings in the Context of Perceptual Organization. Proceedings of the International Conference on Computer Analysis of Images and Patterns, Leipzig, GDR, 8.-10. Sept. 1989, (K. Voss, D. Chetverikov and G. Sommer, Eds.), 54-56, 1989.

[8] Biederman, I.: Human Image Understanding: Recent Research and a Theory. CVGIP 32, 29-73, 1985.

[9] Biederman, I.: Matching Image Edges to Object Memory. First Int. Conf. on Computer Vision, 384-392, 1987.

[10] Funt, B.V.: A Parallel-Process Model of Mental Rotation. Cognitive Science 7, 67-93, 1983.

[11] Gowda, K.C.: Cluster Detection in a Collection of Collinear Line Segments. Pattern Recognition 17, 221-237, 1984.

[12] Kanal, L. and Tsao Th.: Artificial Intelligence and Natural Perception. Int. Conf. on Intelligent Autonomous Systems, 60-70, 1986.

[13] Kubovy, M. and Pomerantz, J.R. (Eds.): Perceptual Organisation. Lawrence Erlbaum Associates, Hillsdale, New Jersey, 1981.

[14] Lowe, D.G.: Perceptual Organization and Visual Recognition. Kluwer Academic Publishers, Boston, 1985.

[15] Lowe, D.G.: The Viewpoint Consistency Constraint. Int. Journal of Computer Vision 1, 57-72, 1987.

[16] Lyusternik, L.A.: Convex Figures and Polyhedra. D.C. Heath and Company, Boston, 1966.

[17] Simon, H.A.: The Information Processing Explanation of Gestalt Phenomena. Computers in Human Behavior 2, 241-255, 1986.

[18] Shepard, R.N. and Metzler, J.: Mental Rotation of Three-dimensional Objects. Science 171, 701-703, 1971.

Matching: Check Points within the Framework of a Knowledge-based Visual Inspection System

Matthias F. Carlsohn, Lutz Gruber

DST - Deutsche System-Technik GmbH

Abstract

Matching within an inspection system is one of the most crucial processing steps in testing how well measures, shape or completeness of an industrial machined work-piece fits the important properties of its "golden prototype".

Especially, pure image based systems suffer from the fact that pixel-to-pixel comparisons are effected by variations in scale, illumination, positioning accuracy or tolerances of single properties of the specimen. Additionally, unprecisely defined features of the test patterns, incomplete specifications of all details to be tested, restrict automation of visual inspection still to few applications only. Knowledge-based matching using a priori information about specimen (e.g. CAD data) and inspection environment would overcome these disadvantages.

This paper presents an overview about matching algorithms in knowledge-based visual inspection. From this survey a list of assessment criteria for matching schemes is derived. Initial ideas for a generic matching concept using graph matching techniques is proposed. Discussion of the new approach is mainly devoted to control of matching by prediction graphs derived from model data.

Keywords: generic matching scheme, inspection systems, symbolic object representation, graph matching, prediction graph, perceptual grouping, compatibility, similarity measure

1 Introduction

Matching is one of the essential stages in the area of artificial intelligence. Wherever a classification or categorization of information (recognition problem) is required, it is necessary to match a sample instance to a predefined model description of the object, in

NATO ASI Series, Vol. F 71
Expert Systems and Robotics
Edited by T. Jordanides and B. Torby

computer vision, mainly constrained by their expected appearence within an image. It is a general problem in computer vision to recognize objects in a scene and compare them against a structural description of the scene.

The matching process in computer vision tries to correlate perceptions of the real world with idealized object models in order to derive decisions (e.g., how objects can be dealt with appropriately in a given situation). For that reason, matching is a kernel function in a model-based vision framework that acts as an interface between the data-driven and model-driven parts of a vision system (see Figure 1).

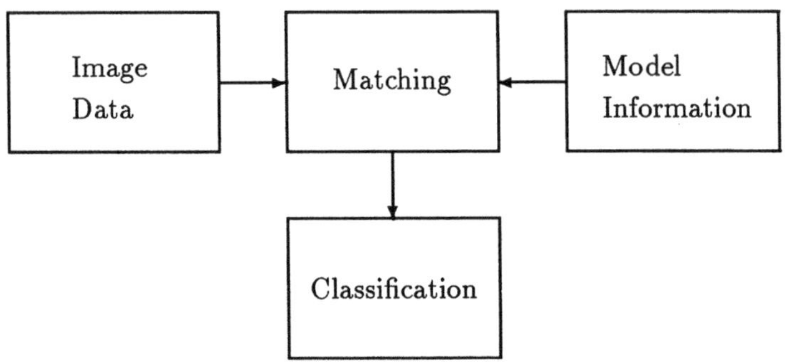

Figure 1: General framework for matching

The use of knowledge pools like CAD-data bases, originally generated for design only, needs concepts of using this information flexible for reasons of quality assurance, functionality tests and measurements. Such a concept has to consider that a priori knowledge from model data bases is represented appropriately, so that it is directly comparable to data gathered by image analysis.

The model information can cover the entire scale, from simple synthetic images on pure iconic level, up to highly symbolic models on a very abstract level of representation. Accordingly, matching algorithms cover the whole range from simple correlation techniques to complex methods for comparison of relational structures.

One of the main goals in knowledge-based matching is to achieve compatibility of image and model information. This is a prerequisite for the comparison of image and model data and the extraction of an object interpretation (or classification).

More sophisticated demands are the matching of properties and features which deviate within tolerance ranges by using compatibility measures in order to discover similarities while dissolving ambiguities, simultaneously.

2 Matching on Iconic Level

Due to the common definition matching can be considered as a comparison of features extracted from an image with a pre-existing idealised model structure. The model for iconic matching methods is a synthesized image, often called a "golden reference".

2.1 Template Matching

Template matching is a basic filtering method for detection of particular image features. A prerequisite for this method is the a priori knowledge about the appearance of this feature in the image. In this case one can detect the feature by a template operator. The template is a synthesized image of the expected structure. A similarity measure is computed which reflects how well the image data match the template for each possible template location. The point of maximal match can be selected as the location of the feature.

A standard measure between a function f(**x**) and a template t(**x**) is the square of the *Euclidian distance* given by

$$d^2(\mathbf{y}) = \sum_{x=-M}^{M} \sum_{y=-N}^{N} [f(\mathbf{x}) - t(\mathbf{x} - \mathbf{y})]^2 \qquad (1)$$

where M and N define the size of the template. If the image at point **y** is an exact match, then $d(\mathbf{y}) = 0$, and otherwise $d(\mathbf{y}) > 0$. After expansion and elimination of constant terms the remainder of (1) will be the *cross correlation* between f and t.

$$R_{ft}(\mathbf{y}) = \sum_{x=-M}^{M} \sum_{y=-N}^{N} f(\mathbf{x})t(\mathbf{x} - \mathbf{y}) \qquad (2)$$

This is maximized when the portion of the image that is covered by the template is identical to t.

These methods of linear correlation matching techniques are adequately used only in cases where the average intensity varies slowly compared to the template size. *Normalized correlation* metrics [1] are less dependent on local properties of the reference as well as the input image than the unnormalized correlation, but they are more sensitive to the signal-to-noise ratio of the images.

An extension of basic template matching techniques is proposed by Frei and Chen [4]. This matching approach compares the image against an orthogonal set of templates.

3 Knowledge-Based Matching

The previously mentioned image based matching approaches have several shortcomings. A comparison between an image and a "golden" image is strongly affected by variations in scale, illumination and positioning. These pixel-wise techniques are also very sensitive to the signal-to-noise ratio of the analysed image.

The disadvantages of pure image-based matching techniques can be avoided by use of a priori knowledge about scenes under inspection. In the area of industrial applications such a model is a complete description of the inspected object. This model can be derived from CAD-data made with design purpose of the work-piece, and offers full three-dimensional information about the work-piece.

An object model can represent properties at several levels of abstraction (see Figure 2). That implies that the matching operations have to be performed at different levels also. Such an approach is proposed in [5, 9].

The knowledge provided by an object model can be used to check against the data extracted from an image of the work-piece, i.e., to derive predictions how an object appears in the image. If image and model data are identical inside a specified tolerance range, then the prediction is verified. Such a verification implies an assignment of an interpretation (label) to the object.

Another important item in knowledge-based matching is the determination of an "optimized" control strategy. Such a strategy is essential to reduce the required amount of search while comparing image and model data. Every control strategy is strongly related to the underlying representation structure, but most of the well known search reduction techniques can be used at least.

4 Matching Against a Parameterized Model

One important class of matching schemes compares data to a parameterized model that represents geometrical or statistical properties. This approach uses a pre-existing representation to interpret a set of input data. If a set of parametrical representations is assumed, the goal is to find a *best fit* for the input data. A characteristical example is shown in Figure 3.

Parameter optimization techniques can relate geometrical structures to lower-level representations by using of merit function which measures how well the relations match. One possible application of this method is the comparison of a straight line within an image against the equation $Ax + By + C = 0$ of a straight line. The models are described by a vector of parameters $\mathbf{a} = (a_1, .., a_n)$. The merit function M must rate each set of those parameters in terms of a real number. For example M could be a function of both the parameters *a* and the image *f(x)*. The optimization problem is to find such a function that

$$M(\mathbf{a}, f(\mathbf{a}))$$

becomes a maximum.

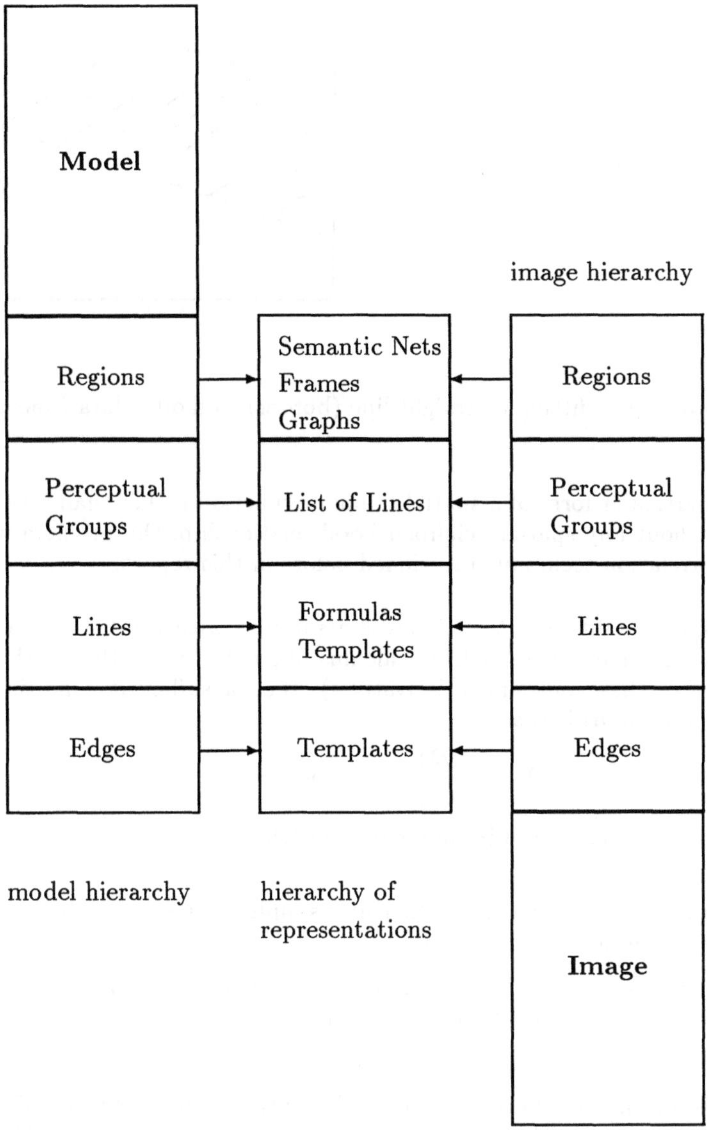

Figure 2: Object properties and image features linked by appropriate representations

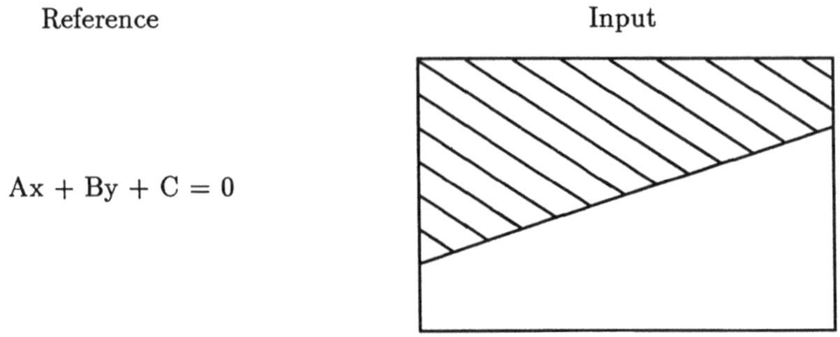

Figure 3: Matching or fitting a straight line (border) to model data (linear equation)

In case **a** is a particular form of a spatial template function rather than a vector of scalar parameters, without any spatial neighbourhood relationship, the problem would appear as an iconic correlation technique mentioned before in this report.

Formally, rank of the different techniques is associated with the behaviour of the merit function M. A fundamental result from calculus of similarities is that if M is sufficiently well behaved (i.e. has continuous derivatives), then a sufficient condition for a local maximum (or minimum) is that

$$M_{a_j} = \frac{\partial M}{\partial a_j} = 0 \quad \text{for } j = 1,..,n$$

This condition can be exploited in different ways [1].

- For some cases the equation is sufficiently simple so that a can be determined analytically (e.g. least square fitting).
- An approximate solution for a^0 can be iteratively determined by moving in the gradient direction or the direction of maximum improvement:

  $$a_j^k = a_j^{k-1} + cM_{a_j}$$

  where c is a constant. This is the most elementary of several kinds of *gradient (hill-climbing) techniques.* Here the gradient is defined with respect to M and does not mean luminance edge strength.
- If the partial derivatives are expensive to calculate, the coefficients can be perturbed (either randomly or in a structured way) and the perturbations kept if they improve M:

  (1) $\mathbf{a}' := \mathbf{a} + \Delta\mathbf{a}$

  (2) $\mathbf{a} := \mathbf{a}'$ if $M(\mathbf{a}') > M(\mathbf{a})$

5 Matching of Perceptual Groupings

Lowe's algorithm [6] uses matching of perceptual groupings to determine the viewing parameters from a single two-dimensional image. The proposed matching method compares directly the projection of a three-dimensional object model to the two-dimensional image.

This approach is based on the presumption that several aspects of an object's projection remain nearly invariant within a particular range of possible viewpoints. A process called *perceptual organization* detects groupings and structures within the image that are likely to be invariant over wide ranges of viewpoint positions. These groupings can be matched to corresponding structures of the object model using a knowledge-based matching process. The basic principles are shown in Figure 4.

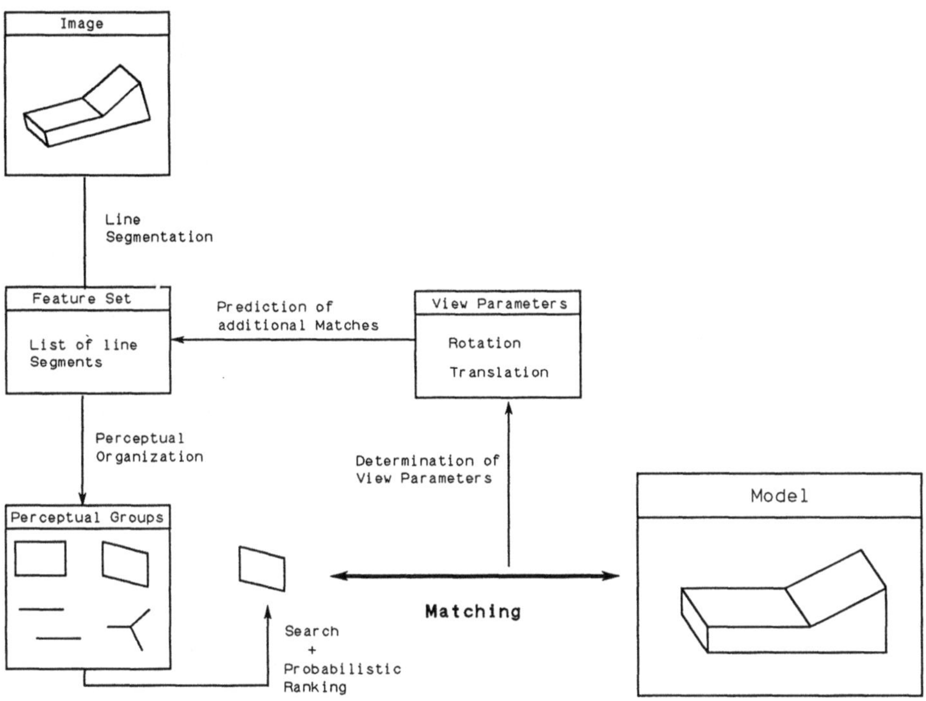

Figure 4: Principles of Lowe's knowledge based vision model

It is possible to use the principle of probabilistic reasoning to rank the potential matches in terms of their predicted reliability, thereby focusing the search on the most reliable evidence present in a particular image.

Matches of viewpoint-invariant aspects of an object are considered as an initial *trigger* for a search process and a viewpoint dependent analysis of further matches. A first estimation of the viewpoint can be achieved by the matching of invariant features. A more reliable computation is accomplished in an iterative scheme. Matches among the model and some image features can be used to constrain the three-dimensional position of the model and parts of it. These constraints in turn lead to further predictions for the location of additional image features in the image, finally leading to refined matches and to additional constraints. To define the problem more precisely:

- Given a set of known correspondences among three-dimensional points on a model surface and points in a two-dimensional image. What are the values of the unknown parameters of the projection and its model which result in the projection of the given model points transformed onto the corresponding image points.

This problem is solved by application of Newton's method iteratively (see Figure 5).

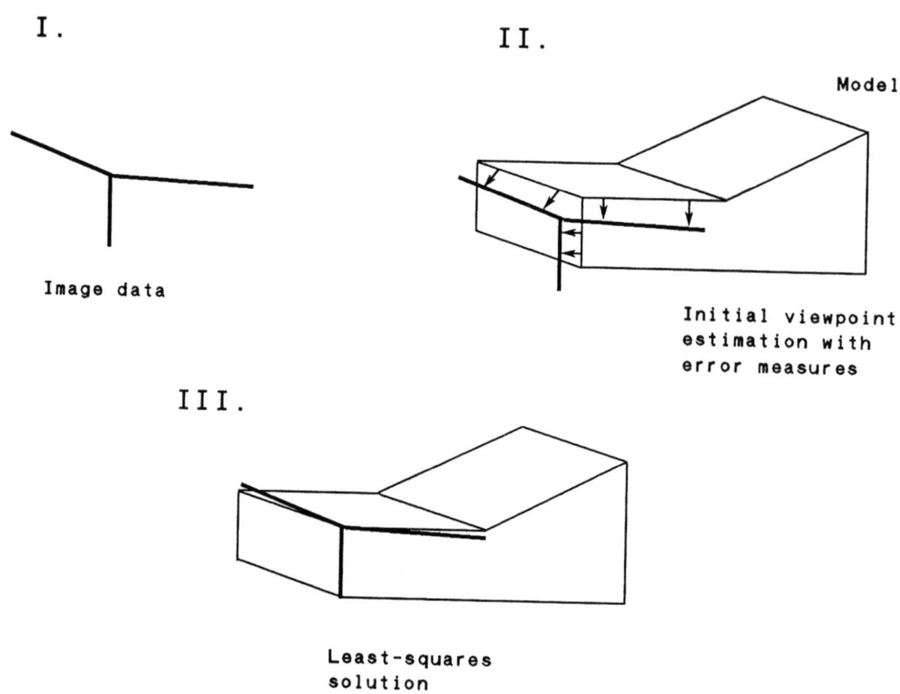

Figure 5: Application steps of Newton's method to achieve spatial correspondence among 2-D image segments and the projection of a 3-D object model

The proposed algorithm is able to apply line-to-line correspondences in addition to point-to-point correspondences. The off-set of such a correspondence is related to the distance between lines rather than in terms of the deviations in the location of endpoints.

The **object model** consists of a set of 3-D line segments associated with an elementary form of visibility specification. This measure contains a Boolean equation combining the hemispheres from which the segment is visible.

Once the initial set of hypothesized matches has been applied in order to discover the viewpoint parameters, this estimate of the viewpoint can be used to predict the locations of other model features in the image for successive propagation of the match. The predictions of these image locations can be obtained by simply projecting the model edges onto the image using the current set of viewpoint parameters.

The search effort may be reduced by using groupings with a high degree of complexity first. If only line segments are matched against each other there is a high number of potential correspondences. This number can be reduced by using more complicated perceptual groupings. On the other hand more complicated structures are less viewpoint-invariant than simpler ones. For example, each straight line which has been detected in the image is an instant of a grouping that can be matched against each straight edge of the model, but this would cause a large amount of search.

6 Matching of Relational Structures

Object models can be represented as hierarchical graph structures with the following concepts:

- **Nodes:** structural entities of the object
- **Arcs:** relations between structural entities
- **Attributes:** additional descriptions of object entities

In this case matching is the comparison of two graph structures [1, 8].

6.1 Graph-Theoretic Algorithms

There are a lot of matching algorithms dealing with relational structures that describe n-ary relations as a directed graph with labeled arcs. If the content of an image as well as the object model are represented as a graph structure, the matching process is reduced to an *isomorphism* problem.

In pure versions of the graph isomorphism problem, the nodes and arcs are unlabeled. A match is successful if there is an 1:1 correspondence among arcs and nodes of the two graph structures. The lack of expressive power in these graphs, and the requirement for 'perfect' matches, limit the usefulness of this kind of matching techniques, especially in the presence of noisy images and incomplete references.

In practice, graph nodes have attached properties and arcs are assigned by binary predicates. Therefore, matching must also consider the compatibility of node properties and arc predicates.

The following are some definitions of matching between graphs [1].

- **Graph isomorphism** - Given two graphs (V_1, E_1) and (V_2, E_2), find a 1:1 onto mapping (isomorphism) f between V_1 and V_2 such that for each $v_1, v_2 \in V_1, V_2, f(v_1) = v_2$ and that for each edge of E_1 connecting any pair of nodes $v_1, v_1' \in V_1$, there is an edge of E_2 connecting $f(v_1)$ and $f(v_1')$.

- **Sub-graph isomorphism** - Find isomorphisms between a graph (V_1, E_1) and subgraphs of another graph (V_2, E_2). This is computationally more complex than the detection of isomorphism, because one does not know in advance which subsets of V_2 have to be investigated.

- **"Double" sub-graph isomorphism** - Find all isomorphisms between sub-graphs of a graph (V_1, E_1) and sub-graphs of another graph (V_2, E_2).

Perhaps the most crucial problem with graph isomorphism algorithms is their computational complexity.

The sub-graph isomorphism problem is *NP-complete*; that means all known deterministic algorithms have running times (in the worst case) that increase exponentially with the number of nodes and arcs of the particular graph structure. The average-case complexity of known algorithms is proportional to N^3 with N the number of nodes in the graph, while an optimized algorithm of Corneil and Gotlieb [3] is proportional to N^2. Algorithms for detection of graph isomorphisms are presented in [2, 3].

Two solutions for the sub-graph isomorphism problem are presented in [10]. The first one is a simple enumerative search of the tree of possible matches between nodes. The second approach uses a parallel-iterative refinement at each stage of the search. This process removes node pairs from the graph and propagates the effects of such rejections; one rejected match can lead to more matches being rejected, like a "snow ball effect".

The problem of "double" sub-graph isomorphism may be reduced to sub-graph isomorphism by application of clique detection. A *clique* of size N is a totally connected sub-graph of size N. The detection of isomorphisms among sub-graphs of a graph A and sub-graphs of a graph B is accomplished by generating an association graph G (see section 6.1.2) from the graphs A and B first, and subsequently searching for maximum cliques in G.

Generally, matching should introduce *quantifiable similarities* between image and test pattern or object and model, respectively. A match is not merely a correspondence, but a correspondence that has been quantified according to its 'goodness'. This measure of goodness is called the *matching metric.* Such metrics are also well suited to describe the degree of correspondence among graph structures.

6.1.1 Algorithm for Sub-graph Isomorphism

In this section an algorithm for detecting sub-graph isomorphisms [10] is presented. It is a simple form of a tree search enumeration procedure. The purpose is to find all of the isomorphisms between a graph $G_1 = (V_1, E_1)$ and sub-graphs of a second graph given $G_2 = (V_2, E_2)$. The number of nodes and arcs of G_1 and G_2 are p_1, q_1 and p_2, q_2, respectively. The adjacency matrices of G_1 and G_2 are $A = [a_{ij}]$ and $B = [b_{ij}]$.

A matrix M' is defined to be a p_1 (rows) x p_2 (columns) matrix whose elements are 1's and 0's, such that each row and column contains exactly one single 1. Such a matrix $M' = [m'_{ij}]$ can be used to permute the rows and colums of B to produce a new matrix C. This matrix is defined by $C = [c_{ij}] = M'(M'B)^T$, where T denotes a transposition. If it is true that

$$\underset{\substack{(1 \leq i \leq p_{1'}) \\ (1 \leq j \leq p_1)}}{(\forall i \forall j)} \; (a_{ij} \; = \; 1) \Longrightarrow (c_{ij} \; = \; 1) \qquad (1)$$

then M' specifies an isomorphism between G_1 and G_2. In this case, for $m'_{ij} = 1$ the jth node in G_1 corresponds to the ith node in G_2.

Initially, the enumeration algorithm starts with a p_1 x p_2 matrix $M^0 = [m^0_{ij}]$ which is constructed in accordance with the rule

$$[m^0_{ij}] = \begin{cases} = 1 & \text{if the degree of the } j\text{th node of } G_2 \text{ is greater than} \\ & \text{or equal to the degree of the } i\text{th node of } G_1 \\ = 0 & \text{otherwise} \end{cases}$$

The enumeration algorithm works by generating all possible matrices M' such that for each element m'_{ij} of M', $(m'_{ij} \; = \; 1) \Longrightarrow (m^0_{ij} \; = \; 1)$. For each such matrix M', the algorithm tests for isomorphism by applying condition (1). Matrices M' are constructed systematically by changing to 0 all elements except one of the 1's in each of the rows of M^0, with respect to the definition that no column of matrix M' may contain more than one 1.

In the search tree, the terminal nodes at depth $d = p_1$ correspond to distinct matrices M'. Every terminal node at depth $d < p_1$ is represented by a matrix M. This matrix can be constructed from M^0 by changing all but one 1's to 0. The change is performed in d rows of the matrix M. The algorithm uses two binary vectors F and H to record which columns have already been investigated.

6.1.2 Association Graph Techniques

The *association graph* is an auxiliary data structure built up from two relational structures which should be matched. The advantage of an association graph is its simple and pure graph-theoretic structure which is amenable for graph-theoretic algorithms such as clique finding.

Canonically, a relational structure is a set of elements V, a set of properties P defined over the elements of V, and a set of binary relations R, defined over pairs of elements. Given two relational structures defined by (V_1, P, R) and (V_2, P, R), the association graph G is constructed as follows:

- for each $v_1 \in V_1$ and $v_2 \in V_2$, construct a node of G labeled (v_1, v_2) if v_1 and v_2 have the same properties for each $p \in P$. Thus the nodes of G denote assignments of nodes with similar properties
- two nodes (v_1, v_2) and (v_1', v_2') of G will be connected if they represent compatible assignments according to R.

Each node of the association graph G corresponds to a pair of nodes, one each from graph A and B, with compatible properties. Arcs within G hint at compatible associations of pairs of A and B. In this sense, maximal cliques indicate sets of consistent associations, i.e., sub-graph isomorphisms.

In this context, a successful match between (V_1, P, R) and (V_2, P, R) represents just a set of assignments that are mutually compatible. The 'best' match is characterized by the largest set of assignments (node correspondences) that are mutually compatible obeying the relations R, but this is the largest totally connected set of nodes (a clique) in an association graph. A clique to which no new nodes may be added without destroying the clique property is a *maximal* clique.

In this terminology, larger cliques are taken to indicate better matches since they account for more nodes. Thus the best match is determined by the largest maximal clique in the association graph.

An association graph may be searched not only for cliques as described previously, but also for *r-connected* components. The r-connected component is a set of nodes where each node is connected to at least r other ones of the same set. A clique of size n is an $n-1$-connected component.

6.2 Fuzzy Relational Matching

The method of pure association graph matching has been extended by the approach of *fuzzy relational matching* of Menhardt [7]. This approach applies the theory of fuzzy sets to the matching problem by the introduction of fuzzy relations. These kind of relations are better suited to express ambiguities of correspondences than traditional relations. Fuzzy relational matching is based on the following definitions [7]:

- The *fuzzy set* of an ordinary set A is defined by its membership function which maps the elements of A to real values of the interval [0,1].

  $$\mu : A \mapsto [0,1]$$

  where μ denotes the degree of membership.

- A *fuzzy relation* p of order t on a set A is a fuzzy subset of the cartesian product A^t

  $$p : A^t \mapsto [0,1]$$

- A *fuzzy graph* is a pair of [μ,p] where μ is a fuzzy subset and p is a binary fuzzy relation on μ with the following condition:

  $$p(x_i, x_j) \leq min(\mu(x_i), \mu(x_j)), \qquad x_i, x_j \in A$$

  This means that every edge has a lower membership degree than the associated nodes.

- A fuzzy relational structure as an extension of the graph definition is a pair $[\mu, \{p^1, .., p^n\}]$ with n fuzzy relations on μ.

The problem is to match two fuzzy relational structures $[\mu, \{p^1, .., p^n\}]$ for a scene model and $[\mu, \{p'^1, .., p'^n\}]$ for a symbolic description of the scene. The association or compatibility graph will be constructed analogical to the non-fuzzy approach previously mentioned in this paper with the slightly different goal of searching for *fuzzy cliques* now. A fuzzy clique within a fuzzy graph [μ,p] is a subset C of the original set A of μ with the following condition:

- The maximum of the weakest edge from each node of the complement of the clique to a node of the clique must be lower than the weakest edge of the clique itself.

7 Assessment Criteria for Matching Schemes

The next chapter will summarize important critria for assessing matching schemes. These criteria can be used for evaluation of existing matching schemes as well as for requirement specification of future schemes to be developed.

7.1 Compatibility between Representations

Perhaps the most important point in matching is to guarantee the comparability of object model and image data. Because of their nature, object properties are three-dimensional while image features are usually restricted to two dimensions. Therefore, it is necessary to solve the problem of achieving compatibility between different forms of representations as well on the image level as on the model level.

There are two general strategies to obtain compatible structures:

- **Projection** of object model onto the image plane.
- **Backprojection** of the extracted feature sets onto the three-dimensional object space.

7.2 Levels of Matching

Matching schemes can be classified in accordance to the level of abstraction where the matching operation is performed:

- Matching on iconic level
- Matching of elementary features
- Matching of structural units (e.g., perceptual groupings)
- Matching against a parameterized model
- Matching of relational structures
- Matching against rules

7.3 Representation Structures

Matching methods are strongly related to the underlying representation of object properties and image features. This section only offers a list of possible representations. A detailed description would exceed the scope of this paper. Appropriate representations for matching mentioned in the literature are

- Synthesized images or templates
- Parameterized models
- List of features
- REV structures
- Graph structures
- Frame structures
- Semantic nets
- Set of rules

7.4 Control Strategies for Matching Methods

Since matching methods are often involved with search processing, it is neccessary to specify adequate control strategies to reduce the required effort. The following list describes only control stategies which are mostly used in practical applications.

- An important control strategy for knowledge-based matching is the hypothesis generation and verification cycle. Initial matches are used to make further predictions about the location of other features. This stategy is used in connection with the grow-a-match paradigm where the actual state of the match is used for a stepwise extension and verification.
- Hypothesize-and-verify is a well known top-down technique where an internal modelling process makes predictions about the way objects will appear. Here, the role of perception becomes the verification instance of either predictions or hypothesis with respect to the objects which are derived from the model.
- The process of selection of predicted features which should be matched is based on search trees.

  To realize an efficient matching strategy it is necessary to reduce the search effort by elemination of non-relevant branches. This can be done by selection of the most significant features first. All the other concepts for tree searching (e.g., order of traversal) are also important for matching.
- Hierarchical decision-tree matching with ad hoc metrices is a popular way to identify input data structures as instances of reference models and thus classify the input instances accordingly [11]. Decision trees are used when it is predictable that certain features can be extracted more reliably than others and that particular relations are either easier to sense, or more important for establishing a match than others.
- Another control mechanism is related to the different levels of matching. It is not necessary that matching will be performed on one level only. It is imaginable that a matching operation is initially performed on a high level and, subsequently, additional verifications of a predicted match are performed on lower levels (top-down approach). A bottom-up approach is also possible where the results of low-level matching will be used as prerequisite for matching on higher levels.

8 A Generic Matching Concept on Symbolic Level

A matching scheme for relational structures is dependent upon the actual task description. If it is necessary to determine the orientation of an object in correspondence to its projected image, an initial match has to be performed to get the approximate location. Therefore the initial match should be related only to those features that are nearly invariant to changes in orientation. This premise is satisfied by perceptual grouping of primitives like the four sides of a polygon.

In cases where the orientation of the object is known in advance the initial match can be omitted, and matching starts immediately. The object model is a relational structure represented by a graph. The resulting object model is then called a *prediction graph*, i.e. the model provides all a priori known properties that are expected to be visible in the image. Consequently, the matching process is controlled by hypothesis derived from the model structure. Within the scene context, each model node represents one particular object property and the goal is to verify the presence of the corresponding feature in the image. The result of these hierarchical verifications is a successive generation of the *image graph*.

The next step is the comparison between prediction graph and image graph using graph matching techniques as described before in this paper. The whole matching concept is shown in Figure 6:

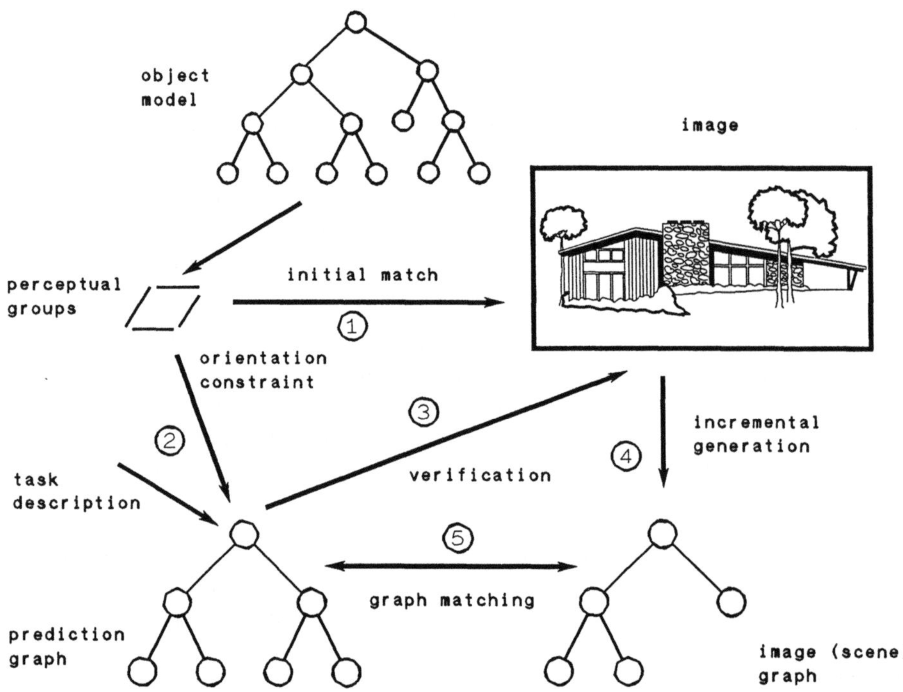

Figure 6: Generic matching scheme

STEP 1: Determination of viewing parameters

Matching of perceptual groups, i.e., those which are nearly invariant to changes in orientation (e.g., four-sided polygons). Detected groups are compared against the object model to constrain the viewing parameters.

Matching is performed by a grow-a-match strategy:

- An initial match is used to achieve a first approximation of the viewpoint.
- Confirmed correspondences lead to predictions about further feature correspondences.

STEP 2: Determination of the prediction graph

The prediction graph repesents an application dependent subset of the object model. This graph structure is determined by the viewing parameters and the task description.

The prediction graph consists of all object properties whose presence has to be tested in the image.

STEP 3: Local Matching

The term local matching refers to the comparision of object properties (nodes of the prediction graph) and the equivalent image features.

Additionally, the selection of operators for feature selection is also controlled by the prediction graph.

STEP 4: Generation of the image (scene) graph

The image graph is built from the verified correspondences on the local level (ref. results of STEP 3).

STEP 5: Global Matching

The result of STEP 4 is an image graph whose structure is compatible to the prediction graph. The global matching operation is performed by graph matching techniques (graph isomorphism, association graphs).

In order to resolve the ambiguities implied by noisy images and slight variations in location it is appropriate to use fuzzy relations and a fuzzy relational matching method as proposed e.g., by Menhardt [7] to achieve more reliable matching results. This is true especially in cases where uncertainty influences inspection either from the synthesis path or from the analysis path within the vision framework.

Acknowledgements

This work has been partially supported by the European Commission within the VIDIMUS project P2592 of the ESPRIT II programme.

References

[1] D.H.Ballard, C.M.Brown: *Computer Vision*, Prentice Hall Inc., 1982.

[2] A.T.Berztiss: *A Backtrack Procedure for Isomorphisms of Directed Graphs*, Journal of the ACM, Vol.17, No.1, 51-64, 1975.

[3] D.G.Corneil, C.C.Gotlieb: *An Efficient Algorithm for Graph Isomorphism*, Journal of the ACM, Vol.17, No.1, 51-64, 1970.

[4] W.Frei,C.C.Chen: *Fast boundary detection: a generalization and a new algorithm*, IEEE Transactions on Computers, Vol.26, No.2, 988-998, 1977.

[5] C.-E.Liedtke, M.Ender: *Wissensbasierte Bildverarbeitung*, Springer-Verlag, 1989.

[6] D.G.Lowe: *Three-Dimensional Object Recognition from Single Two-Dimensional Images*, Artificial Intelligence, Vol.31, No.3, 355-395, 1987.

[7] W.Menhardt: *Fuzzy Relational Matching*, Proceedings of the Esprit-Workshop: Advanced Matching in Vision and Artificial Intelligence, 119-133, 1990.

[8] H.Niemann, H.Bunke: *Kuenstliche Intelligenz in Bild- und Sprachanalyse*, B.G.Teubner Verlag, 1987.

[9] S.Pfleger, L.Altamirano Robles, I.Popova-Atanassova: *Focus of Attention for Blur Handling*, Proceedings of the Esprit-Workshop: Advanced Matching in Vision and Artificial Intelligence, 135-145, 1990.

[10] J.R.Ullmann: *An Algorithm for Sub-graph Isomorphism*, Journal of the Association for Computing Machinery, Vol.23, No.1, 31-42, 1976.

[11] P.H.Winston: *Learning Structural Descriptions from Examples*, in "The Psychology of Computer Vision", Ed. P.H.Winston, 1975.

the vehicle. The use of a vision system is useful when these motion sensors are in error because wheels lose contact with the ground or the vehicle moves onto a slippery floor.

In the section to follow, a description of the vision-navigation operations is presented in the most generic approach. The problem for indoor environments is then addressed, and the resulting sensing strategies that can be followed are presented. Then the mathematics to build models locally and to derive motion estimates from two successive frames are reported. Additionally, because neither sensing, nor the interpretation of sensed data are errorless, modeling of the uncertainties associated with these mathematical and physical models is also considered.

2. Generic stereo - vision - based navigation

The block diagram for vision navigation is presented in Figure 1. [4]. The first operation, called stereo vision, involves the determination of the three - dimensional coordinates in a robot - centered reference frame for a set of environment primitives P. These primitives will be used as landmarks by being tracked at subsequent robot positions. This first operation generally comprises stereo matching and point triangulation and results in the so called (here) local model. In a general AGV vision system, what follows after point triangulation is the formation of geometric representations (solid models) and the symbolic description of the primitives. The former task involves point interpolation along surface patches (not necessarily planes) and grouping of patches for the construction of 3 - D objects, while the latter involves 3 - D object recognition.

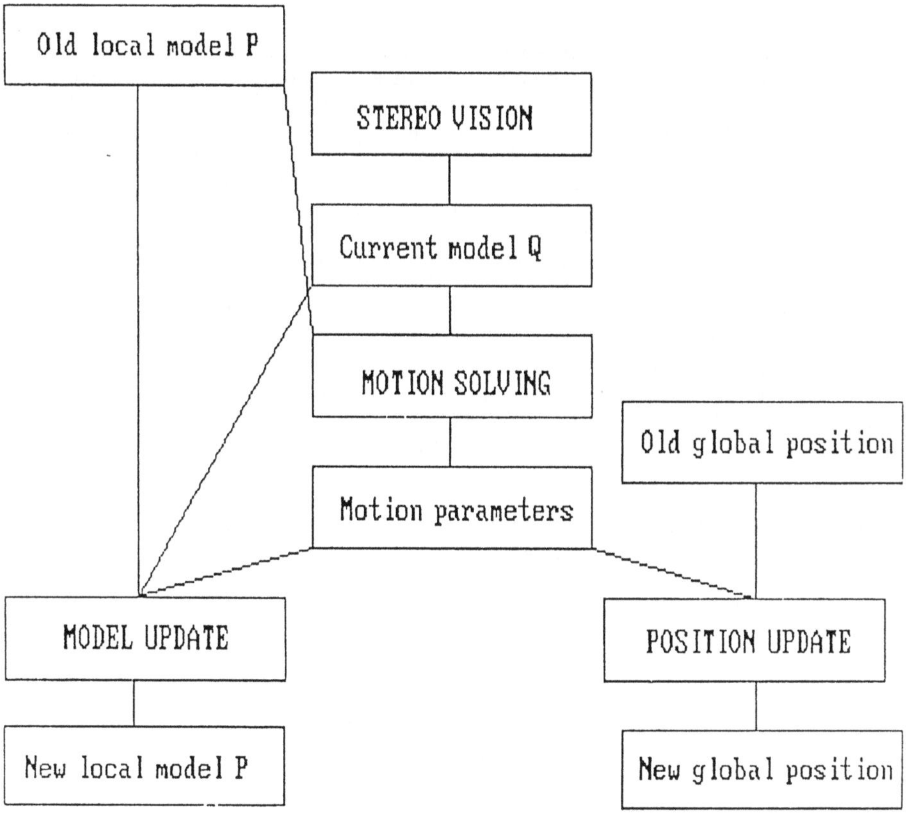

Figure 1: Stereo - vision - based navigation diagram.

STEREOSCOPIC VISION FOR AGV GUIDANCE

B. Dimitriadis and K. Papanikolaou
HITEC S.A.
18 Poseidonos St.
GR - 176 74 Kallithea
Athens GREECE
Basile_Dimitriadis_Hitec_Ltd@eurokom_ie

Abstract

This paper first presents the complete scenario for Autonomously Guided Vehicle (AGV) guidance based on stereoscopic vision. The case of indoor environment is examined in details, and the required operations are further discussed. The concept of logical sensors is followed to address the motion parameter estimation problem as a perception task that can be achieved directly by using stereo - camera data. The mathematics for the operations involved are presented focusing on uncertainty issues associated with the physical and mathematical models used.

Keywords: robot navigation / stereoscopic vision / logical sensors / error modeling / least squares / motion estimation / AGVs.

1. Introduction

Intelligent machines, able to autonomously perform useful tasks in a complex, dynamic, previously unknown environment, require a combination of advanced sensing and control with sophisticated mechanism design. Furthermore, understanding of the environment by the machine is imperative, and in order to increase the quality of perception, multiple sensors with complementary characteristics are used in combination. Among other sensing technologies, vision is becoming mandatory to autonomous robots, since it provides a rich environment description useful for truly intelligent interactions.

The general scenario for an Autonomously Guided Vehicle (AGV) is to realize the task of going from point A to point B in an unknown environment. By unknown here, we mean that only a generic model of the expected surrounding space is provided to the robot at the outset of its operations, that the environment may dynamically change, and that unexpected objects may be present possibly interfering with its motion. Given the initial world model of the environment, the robot begins its movement while simultaneously sensing is used to first acquire environment geometrical data in order to fill in the generated model with local spatial details, and to keep track of the robot's position and orientation by gathering incremental motion estimates.

Stereo - vision can be used to achieve both tasks. Stereo matching followed by triangulation of conjugate image points, performed on images that are acquired by different cameras at the same point in time, provides the local three - dimensional geometry of the surrounding space. Certain geometric features in this model, called landmarks, are also visible from additional viewpoints in space and can therefore be identified in the respective local models that are generated as the robot motion progresses. The local coordinates of these landmarks at different time frames are combined to provide the motion information of the mobile robot from frame - to - frame. Streoscopic vision - based navigation is exactly this dual operation of using the stereo vision system to update the three - dimensional local model of the environment while at the same time keeping track of the robot position and orientation.

AGVs motion parameters are usually provided by shaft encoders mounted on the wheels of

NATO ASI Series, Vol. F 71
Expert Systems and Robotics
Edited by T. Jordanides and B. Torby

For the operation of motion-parameter estimation, the local model may just comprise a set of points , characteristic of the primitives present in the scene. So, after the first stereo pair has been digitized and processed, the old local model formed by the stereo - vision operation consists of points P. Then the robot moves to a new position and a second stereo pair of images is digitized. The stereo - vision operation results to a new set of coordinates for points Q, now forming the current local model. By relating the old and current model coordinates for points common in the two models - visible from both viewpoints - the parameters of the motion, namely displacement and rotation, are derived. In essence, these parameters give the transformation (translation and rotation), between the coordinate systems of the old and the current model. Last are the model and position / orientation updates. Model update involves transformation of the old model coordinates into the current coordinate frame and combining them with the new measurements obtained. This is done because new three - dimensional measurements are acquired for the primitives present in both models, and, consequently, their positional accuracy is expected to increase. The motion estimates can be combined with measurements from other sensors (e.i. shaft encoders) to update the robot's global location. This block diagram describes a complete cycle from one robot position to the next and is repeated each time a new stereo pair of digital images is acquired.

Such a stereo - vision - based navigation procedure is the very generic that can be defined involving several abstraction levels in environment-data representation as well as a number of important machine vision operations, such as 3 - D environment model construction, scene analysis and object recognition. The stereo - vision system is used to produce both a geometrical representation of the environment as well as a symbolic description of the viewed scene. As mentioned above, for the derivation of motion parameters, common primitives in the old and current models have to be identified. A matching operation of three - dimensional features is therefore required. It involves the formation of symbolic environment descriptions for the two models, so that landmarks can be defined at the symbolic level. It is then easier to track them from frame-to-frame in the local models that are subsequently generated. Once two symbolic descriptions are matched, the inverse transformation is performed to regenerate their geometric descriptions and give space coordinates for the respective landmark.

Such an approach for landmark tracking can, however, be complex and considerably slow in the whole vision - navigation procedure requiring large time intervals between the digitization of successive stereo pairs. These intervals relate to the processing time required by the operations of the block diagram.

3 The approach for indoor scenes

We have designed and present here the stereo vision and navigation operations for an AGV restricted to an indoor environment. This implies a simplification of the motion the machine is allowed to undertake, which in turn results in a set of algorithms for stereo vision, feature tracking and motion parameter estimation that use sensoric and physical world data without construction of complex geometric or symbolic environment representations.

The robot is constrained to travel along a plane surface, which of course is the floor in the indoor environment. Its motion can therefore be described by two translations along the x - and y- axes and by a rotation about the z - axis. This implies that at least two points common in both old and current models are required to solve for the three motion parameters. Furthermore, for the description of the vehicle's motion, a two - dimensional representation of the environment is sufficient. The intersections of the objects present in the scene with the horizontal plane at the level of the cameras define this representation. The most common features that will be used in the navigation procedure are the edges of doors, walls, windows and pieces of furniture that are modeled as elongated vertical lines. The 2 - D environment model is obviously a realistic one for this scene configuration.

These motion requirements and resulting necessity for environment representations allow the use of fast operations in stereo vision. In fact, all promising primitives for navigation are displayed in the images as long vertical edges. Furthermore, forgetting about their length, all important information associated with vertical edges is provided by any point along them. In short, examination of the images at the robot's horizon plane, which projects to the central epipolar lines in the two image planes, suffices for the determination of local models, and allows the use of fast epipolar line correlation techniques. The Laplacians of the intensity profile of the central epipolar lines in the two image planes are correlated to establish correspondences among edge points. These points are then tracked along the same epipolar line in subsequent vehicle positions. Additionally, triangulation of corresponding points provides the required three - dimensional information (in reality it is 2 - D information along a certain horizontal plane).

This approach has also been inspired by our research activities in logical sensors. A logical sensor is a perception unit characterized by an algorithm and the type of output it provides from the sensor signal it receives [3]. A set of logical sensors is used to carry out a perception task. This is a demand to the sensor data acquisition and processing systems. The estimation of vehicle's position / orientation is the perception task we want to achieve using physical data from a stereo - camera arrangement. The operations for 2 - D determination and tracking of points directly performed on image data and the estimation of motion parameters from 2 - D points are the layers of logical sensors required to achieve this task.

Furthermore, a logical sensor is equipped with tests to evaluate its performance [6]. For the vision-navigation operations, acceptance of results for the different logical sensor layers can be based on the modeled uncertainties for the respective operation. In particular, error modeling for the point triangulation is presented and the accuracy measures are used in the estimation of the motion parameters. Additionally, the accuracies of the estimated parameters are computed to allow the acceptance of observations or to demand the rejection of existing outliers.

4. Stereo - vision and related uncertainty

The stereo vision system consists of two laterally displaced cameras, located at equal elevations, with their optical axes horizontal. As illustrated in Figure 2 on an overhead view, the cameras have a baseline distance equal to B and are focused at infinity using as image distances focal length f. Furthermore, the robot - centered system has its origin in the middle of the baseline and its x - axis coinciding with it. To derive the (x,y) coordinates for a point P, measurements along the rows of the left and right digital images (x_l and x_r) are carried out. The coordinates for point P are given by:

$$\begin{bmatrix} Px \\ Py \end{bmatrix} = \begin{bmatrix} X(x_l, x_r) \\ Y(x_l, x_r) \end{bmatrix} = \begin{bmatrix} \dfrac{B(x_l + x_r)}{2(x_l - x_r)} \\ \dfrac{Bf}{x_l - x_r} \end{bmatrix} \tag{1}$$

Figure 3 gives an illustration of errors associated with the stereo - vision system. The ticks on the image planes are discrete pixel boundaries. Lines radiating from corresponding pixels form 2 - D points in space. Because of uncertainty in pixel localization introduced by image quantisation, all points lying in the shaded region project on the same pixel. In other words, the estimated location of point P from image measurements falls within this diamond shape region. Uncertainty in triangulation can be modeled with multivariate normal distributions. In that case, the equi - probability density contours are given by ellipses which match adequately this diamond shape.

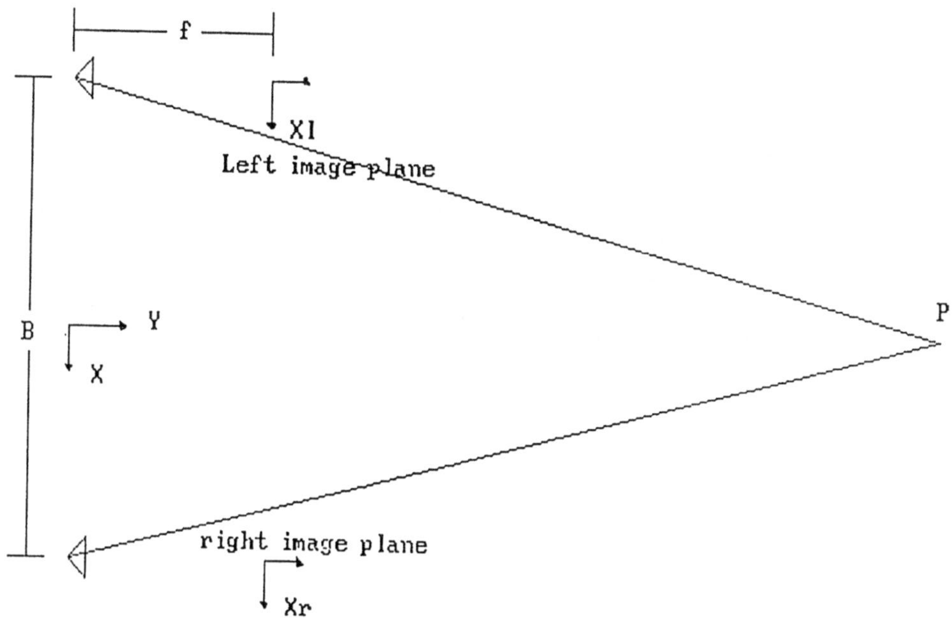

Figure 2: Stereo - vision system

We assume 1 - D normal distribution for measured image coordinates and we derive 2 - D Gaussian distributions for the computed 2 - D point coordinates. Since in (1) functions are linear with respect to x_l and x_r, they can be linearized about the mean of x_l and x_r with a Taylor series expansion. This is a commonly used strategy, especially in photogrammetric applications, yielding acceptable approximations to coordinate errors. If image measurements are described with a mean μ_{xl}, μ_{xr} and variance σ_{xl}, σ_{xr} then point p is obtained to be:

$$\begin{aligned} Px &= X(\mu_{xl}, \mu_{xr}) + \frac{\partial X}{\partial x_l}(x_l - \mu_{xl}) + \frac{\partial X}{\partial x_r}(x_r - \mu_{xr}) \\ Py &= Y(\mu_{xl}, \mu_{xr}) + \frac{\partial Y}{\partial x_l}(x_l - \mu_{xl}) + \frac{\partial Y}{\partial x_r}(x_r - \mu_{xr}) \end{aligned} \tag{2}$$

and the covariance matrix for this point is:

$$V_p = J \begin{bmatrix} \sigma^2_{xl} & 0 \\ 0 & \sigma^2_{xr} \end{bmatrix} J^T \quad (3)$$

where the J matrix is the Jacobian:

$$J = \begin{bmatrix} \dfrac{\partial X}{\partial x_l} & \dfrac{\partial X}{\partial x_r} \\ \dfrac{\partial Y}{\partial x_l} & \dfrac{\partial Y}{\partial x_r} \end{bmatrix}$$

In (3) the off - diagonal elements are taken to be zero since measurements in the left and right images are statistically independent.

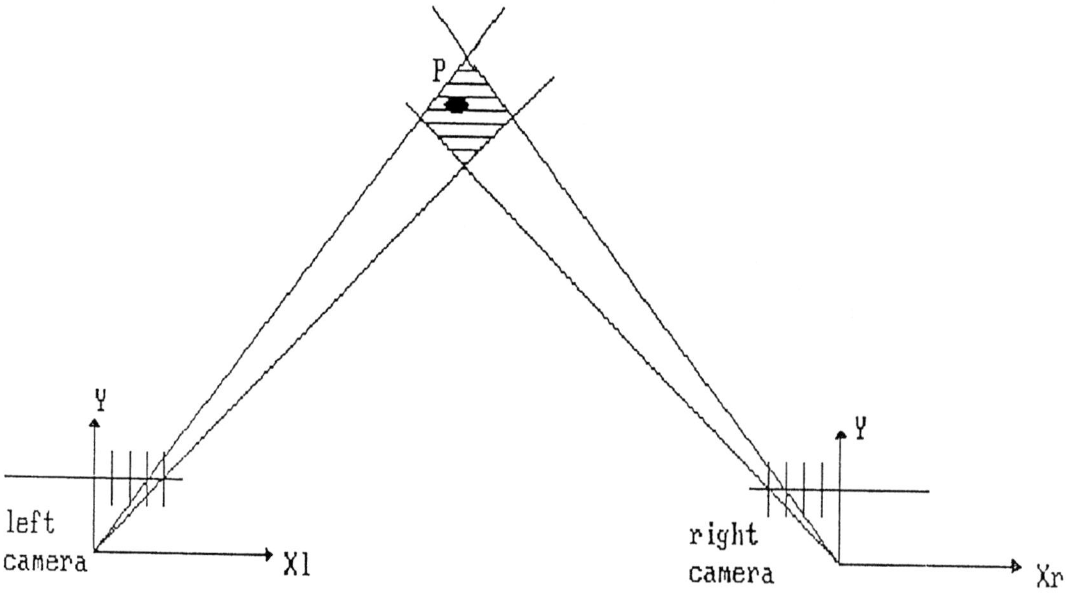

Figure 3: Errors for the stereo - vision system.

5. Estimation of motion parameters

The motion parameters to be estimated are two translations T_x and T_y along the x- and the y-axes respectively and one rotation k about the z - axis. To solve for these parameters, at least two points Pi in the old model and the corresponding two points Qi in the current model are needed. The general motion equations using translation vector T and rotation matrix R relate old and current coordinate frames as follows:

$$Q = R\,P + T \tag{4}$$

which for a pair of points (Pi, Qi) in our application become:

$$\begin{aligned} \cos k\, P_x - \sin k\, P_y + T_x &= Q_x \\ \sin k\, P_x + \cos k\, P_y + T_y &= Q_y \end{aligned} \tag{5}$$

Usually, more than two pairs of points will be used in order to improve the reliability of the solution. The least - squares approach is then used to derive the best estimates for the motion parameters. The least square principle endeavors to see that the set of adjusted variables is as close as possible to the sample values of the observations taking their stochastic properties into consideration [5]. Least squares do not require a priori knowledge of the distribution associated with the observations. However, when observations are normally distributed, the estimates from least squares will be identical to those from the method of maximum likelihood.

The vector of the independent parameters that have to be estimated is:

$$x = [T_x, T_y, k]^T$$

while the vector of observations is:

$$l = [P_x, P_y, Q_x, Q_y]$$

Since equations (4) are not linear with respect to the parameters and observations, we can linearize by applying Taylor's series. The linear form of (4) expressed with matrices is:

$$\begin{bmatrix} \cos k^o & -\sin k^o & -1 & 0 \\ \sin k^o & \cos k^o & 0 & -1 \end{bmatrix} \begin{bmatrix} \delta P_x \\ \delta P_y \\ \delta Q_x \\ \delta Q_y \end{bmatrix} + \begin{bmatrix} 1 & 0 & -(\sin k^o P_x + \cos k^o P_y) \\ 0 & 1 & (\cos k^o P_x - \sin k^o P_y) \end{bmatrix} \begin{bmatrix} \delta T_x \\ \delta T_y \\ \delta \kappa \end{bmatrix} = \begin{bmatrix} Q_x - \cos k^o P_x + \sin k^o P_y - T_x^o \\ = \\ Q_y - \sin k^o P_x - \cos k^o P_y - \quad T_y^o \end{bmatrix} \tag{6}$$

where T_x^o, T_y^o, k^o are initial estimates for the motion parameters. Approximate values for these can be obtained from internal sensors of the robot or from the previous motion estimation in the last cycle of the block diagram. We treat observation variables as constant coefficients of the unknown parameters in (5) and obtain again the least squares formalisms for them in a linear solution. These estimates are then used as initial values in (6). Using least squares formalism (6) can be rewritten:

$$Cv + A\delta x = W \tag{7}$$

The solution for δx is given as follows:

$$P_a = (C\,V_1\,C^T)^{-1}$$

$$P = \sigma_o^2\,P_a$$

where σ_o^2 is the a priori reference variance which can be set to any convenient value. The unit is often preferred. V_l is the covariance matrix of the observations, consisting of the covariance of points Pi in the old model and the covariance of points Qi in the current model.

$$V_1 = \begin{bmatrix} V_P & 0 \\ 0 & V_Q \end{bmatrix} \tag{8}$$

The current model covariance V_Q is given by (3), while the covariance of the old model is given by the a - posteriori covariance of the observations as given below and has been computed in the previous position of the vehicle. Again observations in frames **P** and **Q** are uncorrelated.

$$N = A^T P A$$

$$t = A^T P W$$

and

$$\delta x = N^{-1}\,t \tag{9}$$

The covariance of the estimated parameters is:

$$V_{\hat{x}} = \sigma_o^2\,N^{-1} \tag{10}$$

and the a - posteriori reference variance is given by:

$$\hat{\sigma}_o^2 = \frac{W^T P W - x^T N x}{r} \tag{11}$$

r is the redundancy or degree of freedom and for this solution is equal to the total number of P and Q points used in the solution minus three.

The update of the coordinates for the points **P** and **Q** is carried out using this same least squares solution. We have also been investigating the update with the use of extended Kalman filtering [1-2], but this is not presented here. The update is not affecting at this stage the geometrical accuracy of the model generated by the vision system as discussed in the second section of this presentation, but is only related to the points used in the estimation of motion parameters and is internal to the logical sensor. The updated estimates are given by the a - posteriori observation vector:

$$\hat{l} = l + v$$

where $$\upsilon = P^{-1} C^T P_a (W - A (x + \delta x)) \tag{12}$$

The a - posteriori covariance of the observation vector is given as:

$$V_l = \sigma_o^2 P^{-1} - \sigma_o^2 P^{-1} C^T P_a [I - A N_a^{-1} A^T P_a] C P^{-1} \tag{13}$$

where

$$N_a = 1 / \sigma_o N.$$

6. Concluding remarks

This paper presents the utilization of stereoscopic-vision sensing in the navigation of autonomously guided vehicles. The whole procedure is addressed in a generic fashion, but the situation in an indoor environment is discussed in details.

The task of providing vehicle position and orientation information from stereo - vision sensoric data is tackled as a perception task that is carried out by several processing layers according to the concepts of logical sensors. All the sensor layers still operate very close to the physical world and not to symbolic abstract representations. Furthermore, the major processing steps are associated with uncertainty models so as to cope with errors present in the system and to give measures of performance for the logical sensor layers involved in the procedure.

Performance tests are necessary especially in the estimation of motion parameters procedure. First, this procedure is carried out iteratively. Therefore it should be defined when the iterations are terminated. This decision can be based on the covariance of motion parameters. Secondly, outliers should be rejected from the sets of observations and not included in the motion parameter estimation. A procedure to eliminate pairs of points, from P_i and Q_i data set, has to be formulated based on accuracy of observations.

A final point that needs attention is related to the initialization and frequency of the navigation procedure. The positional accuracy of the robot at a point in time is a function of errors in the estimation of the motion parameters and the total distance traveled by the robot. The latter means that positional errors will accumulate as the operation progresses and therefore the operation should stop and rerun after the global position is initialized. The accuracy in the estimation of the motion parameters is a function of the accuracy in the determination of landmark points. The errors associated with the stereo - vision system are presented in the fourth section of this presentation and are smaller for points closer to the robot. By defining a maximum acceptable error in the estimation of motion parameters, the distances from the robot to points P_i and Q_i that are close enough to yield the required motion parameters can be determined. Then the maximum allowable distance from frame - to - frame can be computed, and combined with the velocity of the vehicle, can be used to estimate the required time intervals.

Acknowledgment

This work is partially being supported by ESPRIT II project P-2043 MARIE.

References

1. Ayache, N. and O. Faugeras.: Maintaining Representations of the Environment of a Mobile Robot. IEEE Transactions on Robotics and Automation, Vol. 5, No. 6, pp. 804-819 (1989).

2. Kriegman, D.J., E. Triendl, and T.O. Binford.: Stereo Vision and Navigation in Buildings for Mobile Robots. IEEE Transactions on Robotics and Automation, Vol. 5, No. 6, pp. 792 - 803 (1989).

3. Krose, B. et al.: Sensors: requirements, models and simulation. Deliverable for the CEC in the context of ESPRIT II project P-2043 MARIE (1990).

4. Mathies. L. and S.A. Shafer.: Error Modeling in Stereo Navigation. IEEE Transactions on Robotics and Automation, Vol. 3, No. 3, pp. 234 - 248 (1987).

5. Mikhail, E.M.: Observations and Least Squares. Harper and Row Publishers, N.Y., N.Y 1976.

6. Weller, G.A., F.C.A. Groen and L.O. Hertzberger.: A Sensor Processing Model Incorporating Error Detection and Recovery. NATO ARW on Traditional and Non - traditional Sensors, Maratea, Italy, Aug. 1989.

Hierarchical Methods for Robot Vision

Walter G. Kropatsch

Dept. for Pattern Recognition and Image Processing,
Technical University of Vienna, Treitlstr. 3, A-1040 WIEN, Austria

Abstract

There is a large variety of tasks that require robots to sense, to see, or to hear their environment. Even if such a capability is not absolutely necessary, it enables the robot to react on events from the 'outside'-world. From relatively simple tasks such as inspection up to navigation in an unknown surrounding, vision faces two major problems: the huge amount of data and the need for quick and reliable answers.

Massively parallel data and processing structures are required. Among those, image pyramids have shown great robustness and high performance. Originally intended as an ordered collection of images at multiple resolutions, many computations using an image of $n \times n$ picture elements need on the order of $O(\log n)$ parallel steps. Besides numerical computations, symbolic processing can also be done to extend the application area to mid- and high-level vision tasks. They give us hope to achieve acceptable performance in vision tasks that have to be processed with millions of bytes in an extremely short time.

Curves represent shapes at an intermediate level. Hierarchical symbolic curve representations aim at a stepwise reduction of the data to 'significant parts', like corners or curvature extrema, while preserving the major property of a curve: its connectivity. The proposed representation and symbolic computation scheme allows fast object recognition and may find applications such as in multi-scale path representation and robot guidance.

Keywords: Robot Vision / Image Pyramids / Hierarchical Curve Representation

1 Analysis of Digital Images

The general goal of image analysis (also known under the term Image Understanding in Artificial Intelligence) is to find semantic interpretations of images, in particular to localize and name objects contained in a scene and to assess their mutual relationships. The central problem is to recognize known objects reliably, independent of variations in position, orientation, and size, even when those objects are partially occluded. Although some information is lost by the image formation process, biological vision systems demonstrate that this task can be accomplished reliably and efficiently even under difficult viewing conditions.

NATO ASI Series, Vol. F 71
Expert Systems and Robotics
Edited by T. Jordanides and B. Torby

To **see** the world, the human eye is equipped with approximately 130.000.000 receptors and 12.000.000.000 neurons in the brain. 'Technical' characteristics of the biological vision system are: 1-2 milliseconds are needed to forward data from one neuron to the next; often 1 second is sufficient for a human to recognize a complex scene (like a busy street) and to react appropriately. With these time constraints it can be concluded that only several 100 massively parallel processing steps (neuron-to-neuron connections) must suffice for the task [153].

Uhr [152,154] introduced the notion 'Recognition Cone' to describe the characteristic of a recognition system similar to biological systems. The only conclusion that can be drawn from the efficiency and the known constraints of natural recognition systems is that such systems must massively parallel-process the data. Uhr's recognition cone captures the necessary structure of the system: it receives at the bottom a huge amount of visual stimuli, processes them in parallel while reducing at the same time the data amount step-by-step until final recognition takes place by consulting only a few of the remaining data.

Most of the current approaches in image analysis can be described as a sequential multi-stage process (Fig. 1). For detailed information on image processing and analy-

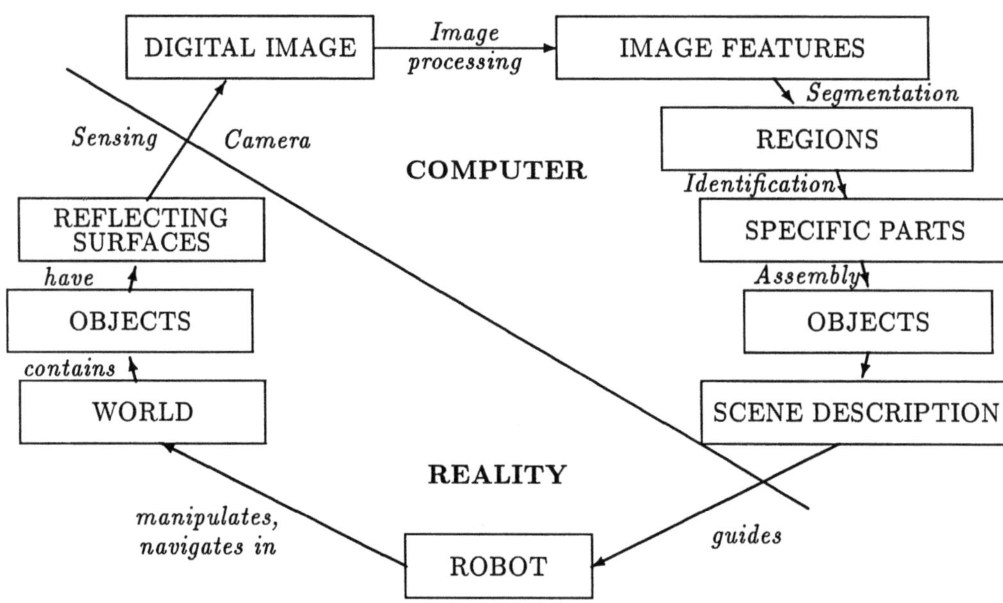

Figure 1: Processing stages in digital image analysis

sis we refer to standard books like [109], [135], [101], [142], or [72]. This multi-stage process can be seen as a step-wise translation that is guided by the interpretation goal, and constrained by its knowledge about the external world and about the available vocabularies at different processing levels. To correctly interpret a given image, it is important to know the classes of OBJECTS that can be expected in the WORLD, and what the properties of their REFLECTING SURFACES are. Only part of the WORLD can be seen within the limited frame of a particular image. Let us call this subset of OBJECTS the SCENE. In the image formation stage, the raw image data are collected by a sensing camera, which outputs an array of intensity values. Images taken at successive times or from different locations can provide also information about motion or 3D structure. This DIGITAL IMAGE is processed to produce a set of data primitives, often called FEATURES. Similar features are then assembled to more complex structures (e.g., REGIONS). A survey on image segmentation techniques can be found in [55]. SPECIFIC PARTS of the image can be derived from those regions by, e.g., geometrical properties. They are useful components for OBJECT models and SCENE DESCRIPTION. In the final phase, these complex features are matched against existing object models in a data base (e.g., OBJECTS in the WORLD vs. OBJECTS in the SCENE). The knowledge of what a ROBOT sees and where objects or obstacles are located in the scene relative to his own position enables the robot to navigate in this (visible) environment, to manipulate the objects in the WORLD, and to control his own actions.

Some of the pictorial entities, their information content, and the operations that can be performed at different processing levels are summarized in Table 1.

Table 1: Pictorial entities at different levels of processing

Entity	information content	examples for operations
Picture	imaging conditions, geometry	sampling, rectification
Pixel	gray value / color vector	enhancement, classification
Neighborhood	spatial locality	shrink, expand
(Step) edge	magnitude, orientation	edge detection and linking
Region	homogeneity, connectivity	segmentation
Boundary	shape	connecting curve segments
Image Part	specific image properties	property measurement
Object Part	specific object properties	property matching
Object	functionality	relational matching
Situation	specific configuration of objects	interpretation
Scene	visible situations of the world	description

There exist many different methods to analyze digital images, many sceneries of applications, many different vocabularies in which the final description should be expressed, but a very simple data structure at the input side. Hence the need to introduce abstract classes of operation and different levels of representation. Even the world shows many different aspects. To deal with this huge variety of data and to focus the processing to the final goal, mechanisms are necessary to agglomerate and compact the data, probably repeated in several stages, until the amount of data is small enough to make final decisions.

2 Problem Domains in Robot Vision

The problems that arise in industrial applications, where a robot system is equipped with a digital camera, can be classified in the following domains:

- Inspection
- Recognition
- Assembly
- Motion
- Navigation

These task domains can be decomposed further into smaller problems some of which are common to several tasks.

In an inspection task, a material or a surface of a product is checked for its quality. Examples include printed circuits [13], wood boards [28,77], or audio-cassette-heads [119]. These tasks require an extremely high data throughput and often a very high resolution in order to detect small defects. The models in this domain are mostly two dimensional such that template matching or classification techniques can be used.

A recognition task such as the identification of parts on a conveyour belt and the determination of their pose often starts by detecting features in the image. These features create hypotheses about the (3D-) object candidates in the image. In [26] the feature-object relation is implemented efficiently by tesselating "feature spheres". Having identified a few 2D image features with corresponding 3D object features, the next step estimates/determines the geometrical positions of the object resulting in the observed image (e.g. in [65,162]; Leung and Huang [94] use stereo in conjunction with a Hough-transform). In the final verification, all the remaining object features must be checked in order to successfully recognize the object and its pose. Characteristics of *recognition* are the static reference model for the objects and the 2D to 3D matching. Hättich and Wandres [63] recently presented a method for learning the required object model.

In an assembly task, a robot has to put together several parts often at very high precision (e.g. [49]). In addition to recognition it also involves the control of the process.

Moving objects are to be detected in a time-series of images. In a first step, features such as points or lines are detected in a few successive images. Then attributes of these features are computed and used to find the corresponding features by matching (normally only small changes are assumed). Finally, the motion and the 3D structure is calculated with the 2D coordinates of the corresponding features. Price [123] gives an overview of feature-based motion analysis methods. Weng et al. [158] derive motion parameters from line correspondencies. Peleg and Rom [118] adopt a multi-resolution approach to deal with large and small motion speeds. Small movements need high resolution but large motion often requires only low resolution and low data rate. Non-rigid motion is a new research topic proposed in [69].

A robot that navigates through its environment needs to build up/ update a 'world-model'. While in most motion approaches the camera position is fixed, navigation is a dynamic vision task. Dickmanns [33] shows applications for autonomous satellite docking, for vehicle guidance, for aircraft and space craft landing, and for ground transportation control. As in the above tasks feature extraction and model matching play an important role. But in addition, a navigating robot generates a path-plan, it estimates its trajectory [2], and predicts the view of its environment at the next step. This prediction can be compared with the observed image at a variety of different levels. Differences between prediction and observation allow the robot to update the 'world-model'. Fennema [37] describes the environment in terms of a **hierarchical collection of spatial entities ("locales")** representing objects, buildings, etc.

3 Image Pyramids

An important parameter of a digital image is its **resolution**. The spatial resolution is described in terms of the smallest dimension of the object that can just be discriminated. In an image with high resolution many details can be observed, at low resolution only large objects are recognized. A *resolution cell* is the smallest and most elementary areal constituent in a digital image. The area of a resolution cell is usually a square or a rectangle. Other shapes are possible and will be discussed in section 5.2. The position of a resolution cell within the 2D plane is determined by the coordinates (x, y) of its center. How many cells are needed to cover a given image area depends on the **size** of the cells. Table 2 compares the qualitative consequences of different cell sizes.

Table 2: Image qualities of different resolutions

	small cells	large cells
resolution	high	low
data amount	huge	smaller
computing times	(very) long	(relatively) short
details	rich and many	very few if at all
overview	bad	good
precision	high	low

Tanimoto [147] defines a **pyramid** as **a collection of images of a single scene at different resolutions**. The images of this collection can be ordered according to their cell sizes and numbered as **levels** of the resulting ordered set of images.

Two terms describe the structure of a (regular) pyramid: the **reduction factor** and the **reduction window**. The reduction factor determines the rate by which the number of cells decreases from level to level. The reduction window associates to every (higher level) cell a set of cells in the level directly below. In general, the reduction window covers the area of its associated cell. In square or rectangular im-

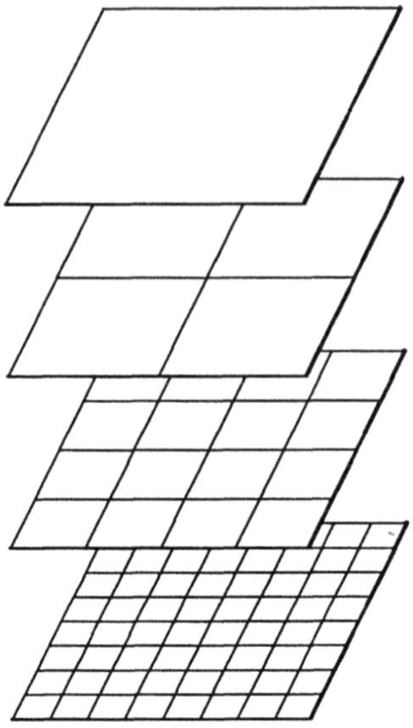

Figure 2: Classical $2 \times 2/4$ pyramid

ages, reduction windows are mostly rectangularly shaped and can be described by $(number\_of\_columns) \times (number\_of\_rows)$.

In the *classical pyramid* (Fig. 2), every 2×2 block of cells is merged recursively into one cell of the lower resolution. We formally describe this structure by $2 \times 2/4$ which specifies the 2×2 reduction window and the reduction factor of 4. This type of pyramid has been extensively studied (e.g. [149], [129]).

A quadtree (e.g., [136]) is a similar tree structure. The cells and connections of the quadtree form a substructure of the $2 \times 2/4$ classical pyramid. Pyramid and quadtree mainly differ in the fact that a quadtree has no communication links between cells of the same level.

Tanimoto's formal definitions refer to this type of pyramid [148]. He defines a cell (Tanimoto uses the term *pixel*) in a pyramid as a triple (x, y, v) which is defined in a **hierarchical domain** of $L+1$ levels:

$$\{(x, y, v) | 0 \leq x \leq 2^v, 0 \leq y \leq 2^v, 0 \leq v \leq L\} \tag{1}$$

Then a pyramid is any function whose domain is a hierarchical domain. This function assigns to every cell in the simplest case a value, but also structures of higher complexity can be stored. This definition of a pyramid will be generalized in section 5.1.

3.1 $\mathcal{O}(\log n)$ computations

We illustrate here that pyramids are efficient computational structures. The paper of Burt et al. [21] is a historical reference that elaborated the potential of pyramids for calculating image region properties efficiently.

Rosenfeld [132] distinguishes between two main flow streams in the pyramid: bottom-up summarizing and top-down delineation. Computations include intensity (gray-level pyramids), where modality detection and fitting are typical examples, and contour recognition, where blobs and ribbons need different treatment.

3.1.1 Calculating sum and mean

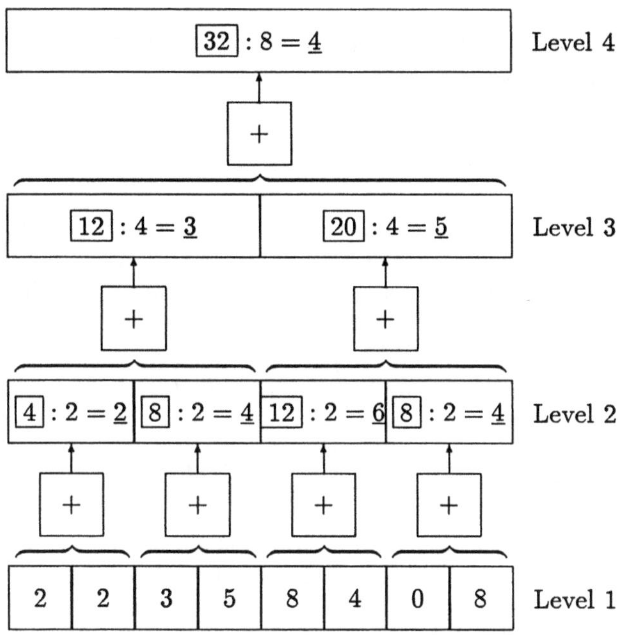

Figure 3: Computing sum $\boxed{s}$ and mean $\underline{\mu}$ in a pyramid

Let us illustrate the principles of operation by means of a simple example using a 1D (linear) pyramid. Our goal is to calculate the sum, the mean, and the variance of the 8 pixels in the base of our four level pyramid (Fig. 3). The process is very similar to **counting** the number of pixels of a binary image, as described in [141] for the GAM-pyramid. To compute the **sum**, we proceed bottom-up:

1. Four parallel additions $\boxed{+}$ produce the sums $\square$ at level 2;
2. two further additions $\boxed{+}$ are needed to fill level 3; and

3. the final result at level 4 is the sum of the two values $\boxed{12}$ and $\boxed{20}$ at level 3.

We observe that a) seven additions are needed and the sum of our 8 ($= 2^3$) pixels is computed in 3 steps; b) the results in intermediate levels represent sums of the receptive fields of all cells.

The **mean** $\underline{\mu}$ is defined as the sum divided by the number of elements. We can calculate the size of the receptive field, store it in a separate entry of every cell, and build the quotient as illustrated in Fig. 3. An alternative to the above procedure, where large values have to be stored, computes the mean $\underline{\mu}_i^{(j+1)}$ at position i of level $j+1$ using the arithmetic mean of the means $\underline{\mu}_{2i}^{(j)}, \underline{\mu}_{2i+1}^{(j)}$ in the level j below:

$$\underline{\mu}_i^{(j+1)} = \frac{\underline{\mu}_{2i}^{(j)} + \underline{\mu}_{2i+1}^{(j)}}{2}. \tag{2}$$

Using the mean as a reduction function, the levels above the base appear as a reduced version of the original image.

3.1.2 Calculating the variance

The **variance** σ^2 of the pixels measures the deviations from the mean. If all pixels have the same value, the variance is zero, the larger the difference to the mean the greater is σ^2. The variance of n values $x_1, x_2, \dots x_n$ with mean $\underline{\mu}$ is defined by ([80]):

$$\sigma^2 = \frac{1}{n-1}\left(\sum_{j=1}^{n} x_j^2 - n\underline{\mu}^2\right) \tag{3}$$

$$= \frac{n}{n-1}\left(\boxed{\frac{1}{n}\sum_{j=1}^{n} x_j^2} - \underline{\mu}^2\right) \tag{4}$$

The first term $\square$ in equ. 4 can be calculated in the same way as the mean (equ. 2) after squaring the values in the base. The result in every cell combines the precomputed values independently of the others (Fig. 4).

In the same way as sum, mean, and variance are computed in the pyramid, also **statistical moments** and **model fitting** can use the bottom-up summarization scheme that fills the cells in $\mathcal{O}(\log n)$ steps.

3.2 Numerical data reduction

In this section we concentrate on pyramids where the contents of the cells are gray values or color vectors. In other words, every level of the pyramid is an image. To fill the cells of the pyramid in a bottom-up manner, different reduction functions can be used.

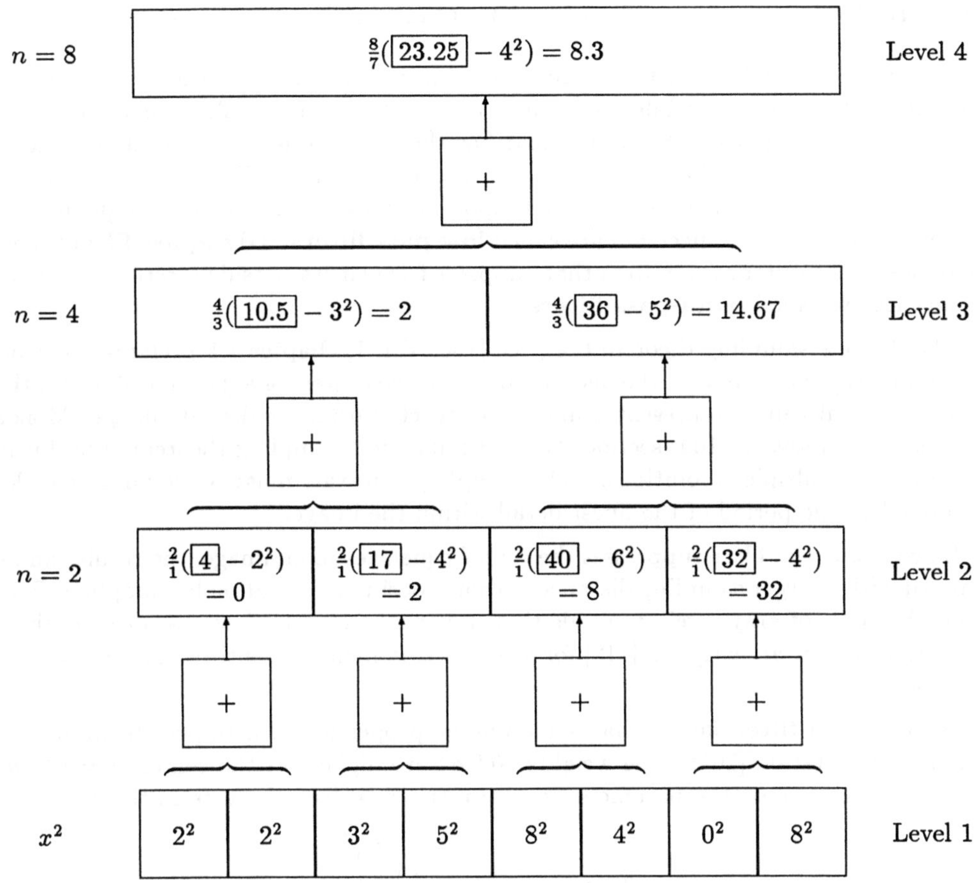

Figure 4: Computing the variance σ^2 (equ. 4) in a pyramid

3.2.1 Reduction by linear filtering, the Gaussian pyramid

Linear filters are used in digital-image processing to remove certain unwanted effects (e.g., noise) from images while enhancing others. One way to filter an image is to convolve it with a spatial-filter kernel that has the characteristics of the filter. Filters are categorized by their characteristic in the frequency domain (Fourier space). Filters that suppress high frequencies like bright spots or lines and let pass through the low frequencies of the image spectrum are called **low-pass filters. High-pass filters** have the opposite characteristic. Filters that suppress frequencies outside a certain range of frequencies are called **band-pass filters.**

The discrete sampling theorem (i.e., see [135, Vol. 1, chapter 4.1]) relates the sampling (*Nyquist*) distance (e.g., the distance between the centers of adjacent cells) with the frequencies that can be represented and reconstructed from the discrete image. Moik's physical interpretation [109, section 2.5.1] of Shannon's sampling theorem is useful in context with multiple resolutions: 'The sampling intervals must be equal or smaller than one-half the period of the finest detail within the image.'

Since a low-pass filter suppresses the high frequencies in an image, the result can be sampled with a larger sampling distance without loosing information by sampling. This forms the basis of gray level pyramids that are constructed by low-pass or smoothing filters (see [16] for an early and[112] for a recent treatment of multi-resolution low-pass transforms).

A **Gaussian filter kernel** has some unique properties with respect to multiresolution representation [161]. The weights $G(t, \sigma)$ are highest at the center ($t = 0$) and decrease depending on the parameter σ with increasing distance $|t|$ from the center:

$$G(t, \sigma) = \frac{1}{\sigma\sqrt{2\pi}} e^{-\frac{1}{2}\left(\frac{t}{\sigma}\right)^2} \tag{5}$$

Properties of the Gaussian smoothing filter $G(\sigma)$ (in this simplified notation t has been suppressed) for building **Gaussian pyramids** are investigated in [91]:

1. Repeated convolution of an image I: $G(\sigma) * (G(\sigma) * I) = G(\sigma\sqrt{2}) * I$ [30]. $G(\sigma\sqrt{2})$ is called the **equivalent weighting function** [18].

2. The mean $\mu(G(\sigma) * I) \approx \mu(I)$ remains the same except for deviations caused by the boundary. This means that **the gray value** of large homogeneous regions **does not change** as long as they are large compared to the size of the reduction window.

3. The standard deviation is reduced in the filtered image.
Since normally distributed noise is measured by the standard deviation, the amount of **noise is reduced** in the pyramid if $G(\sigma)$ is used as the reduction function.

4. Since repetitive application of $G(\sigma) * I$ is equivalent to convolution with a larger σ, the different levels of a pyramid can be built in two ways: either iteratively level-by-level from the bottom to the top or directly from the base using the equivalent

weighting function. It is shown in [91] that the iterative computation needs much less time than the direct computation.

Wells [157] makes efficient use of the separability of the Gaussian. A 2-dimensional filter with kernel $F_{n\times m}$ is called **separable** if it can be split in two 1-dimensional filters

$$F_{n\times m} = F_{n\times 1} \cdot F_{1\times m} \,. \tag{6}$$

Then the result can be obtained by convolving first the rows of I with $F_{1\times m}$ and, then, the columns of the first result with $F_{n\times 1}$.

3.2.2 The maximum pyramid

Besides linear filters, other (non-linear) reduction functions have been used.

Blanford and Tanimoto [12] made extensive experiments on their pyramid machine. They describe the detection of bright spots in a $2 \times 2/4$ pyramid structure with two different bottom-up reduction functions: *maximum* and *average*. The bright spot is found in a top-down search pass that links the apex of the pyramid to one pixel in the base.

The non-overlapping structure of their pyramid produces results that depend on the lateral position of the object region. But the width of a region, which is defined as the side length of the largest pyramidal cell completely contained in the region, will vary by at most a factor of 2 when subjected to lateral translation within the pyramid. The average, when applied to a region of width $2^j, j > 0$, at level k produces a region of width 2^{j-1} at level $k-1$. The use of the local maximum as a reduction function produces a region of width at least 2^{j-1} at level $k-1$. Averaging the first j levels and using the maximum above can detect bright spots of width 2^j or greater, $j > 0$, in lieu of regions of less width.

3.2.3 Morphological reduction

The first two operations of mathematical morphology [142], [73] are the operations of **dilation** and **erosion**. Let D denote the domain of all cell centers. The dilation of a set of cells $A \subseteq D$ with a set $B \subseteq D$ is defined by

$$A \oplus B = \{x | \text{for some } a \in A \text{ and } b \in B \,,\ x = a + b\} \tag{7}$$

The erosion of a set of cells A by B is defined by

$$A \ominus B = \{x | \text{for every } b \in B \,,\ x + b \in A\} \tag{8}$$

Classical shrink and expand operations in the 4-neighborhood take simple algebraic forms using $B = \{(0,0), (-1,0), (1,0), (0,-1), (0,1)\}$: $I^{-1} = I \ominus B$ for shrinking and $I^1 = I \oplus B$ for expanding a binary image I. Multiple **shrink** and **expand** operations are defined recursively: $I^{-k-1} = I^{-k} \ominus B$ and $I^{k+1} = I^k \oplus B$.

Rosenfeld [133] investigated these shrinking and expansion operations using a cellular pyramid. His conclusions are:

1. A cellular pyramid can compute $I^{\pm(2^k-1)}$, resampled at intervals of 2^k, in $\mathcal{O}(k)$.
2. In one dimension, a cellular triangle can compute $I^{\pm(2^k-1)}$, without resampling, in k steps each involving $\mathcal{O}(k)$ computation.
3. In two dimensions, a cellular pyramid would need $\mathcal{O}(2^k)$ computation to compute $I^{\pm(2^k-1)}$ without resampling! Hence in this case, the pyramid has no speed advantage over a conventional cellular array.
4. For a gray-scale image, expansion is generalized to local MAX (compare Blanford's and Tanimoto's maximum pyramid [12]) and shrunk to local MIN. In this case there is no speed advantage of the pyramid even in one dimension.

The purpose of the following approach is that it does not attempt to calculate a high-resolution morphological operation, but rather uses morphological filtering instead of linear filtering to reduced the image's resolution.

The morphological filters **opening** $(A \circ B)$ and **closing** $(A \bullet B)$ are combinations of dilation and erosion:

$$A \circ B = (A \ominus B) \oplus B \tag{9}$$
$$A \bullet B = (A \oplus B) \ominus B \tag{10}$$

Based on these definitions, Haralick et al [54,56] introduce a morphological reduction function. The reduction is performed in two steps: first, the input image is morphologically filtered, second, it is morphologically resampled to the reduced image size. A binary and a grayscale sampling theorem are formulated. It defines conditions on the filtering kernels that must be satisfied to preserve the relevant information after sampling. It further specifies to what precision an appropriately morphologically filtered image can be reconstructed after sampling.

3.3 Reducing redundancy between levels

In a Gaussian pyramid, large objects are represented in many levels. There are ways to remove such redundant information from the higher resolutions in order to achieve a compact representation.

Burt [18] describes a method for compressing, storing and transmitting images in a computationally efficient way. Let $G(k)$ denote a $5 \times 5/4$ Gaussian pyramid, where k denotes the different levels and $G(0)$ is the base image. The bottom-up building process is based on the reduction function $REDUCE$:

$$G(k) := REDUCE(G(k-1))\,,\; k := 1, 2, \ldots \tag{11}$$

which has two tasks:

- Filtering with a 5×5 kernel which depends on one parameter.
- Sampling (take every second pixel in every second row).

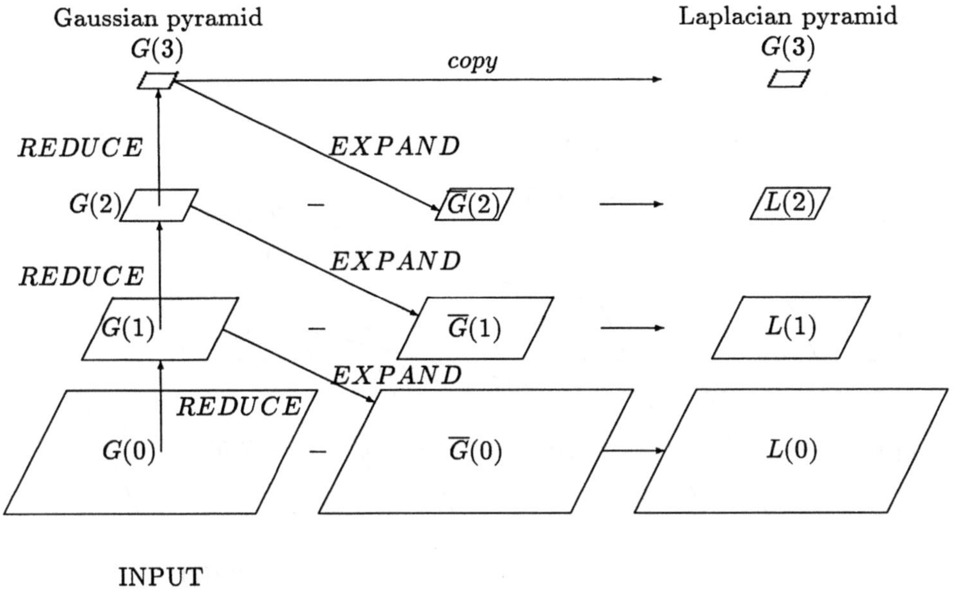

Figure 5: Building the RE Laplacian pyramid $L(0), L(1), L(2), G(3)$.

After the Gaussian pyramid is built in this way, the following steps generate the **Laplacian pyramid**:

1. Define the interpolation function $EXPAND$ using the Gaussian filter which is the reverse function of $REDUCE$.

2. The 'reduce - expand' RE Laplacian pyramid is built by:

$$L(l) := G(l) - EXPAND(G(l+1)) \text{ for } l := 0, 1, \ldots, t-1 \tag{12}$$

   t is the top level. See Fig. 5.

3. Reconstruction of $G(k)$ is exact (Fig. 6):

$$G(k) := L(k) + EXPAND(G(k+1)) \text{ for } k := t-1, t-2, \ldots, 0. \tag{13}$$

4. Hence storing $G(t), L(t-1), L(t-2), \ldots, L(0)$ is sufficient for exact reconstruction of the original image $G(0)$.

5. Quantification of $L(l)$ shows only small and visually not recognizable errors, but large reduction rates: < 1 Bit/Pixel.

An alternate way of constructing Laplacian pyramids is also defined in terms of Burt's operations [20] : the 'filter - subtract - decimate' FSD Laplacian by

$$L(l) := G(l) - F(G(l)) \text{ for } l := 0, 1, \ldots \tag{14}$$

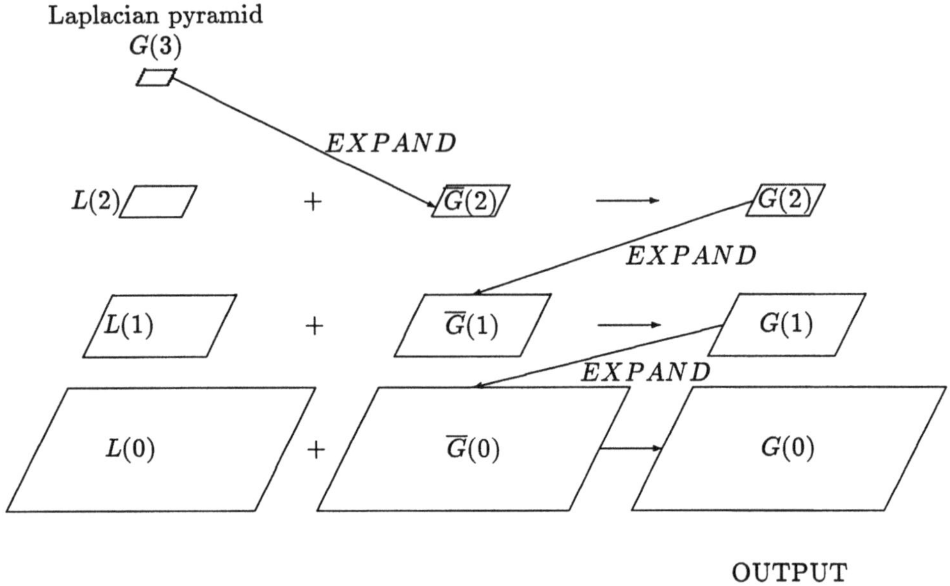

Figure 6: Reconstruction from the RE Laplacian pyramid.

The '**wavelet**'-representation [99] decomposes images $G(k)$ with orthogonal filters into images $G(k+1), L^1(k+1), L^2(k+1), L^3(k+1)$ of lower resolution which also allow exact reconstruction. Note that this representation saves the storage space for the base level of the Laplacian pyramid.

3.4 Pyramid linking and segmentation

The concept presented in this section provides an efficient method for segmenting an image through the use of a gray-level pyramid. The pyramid structure being used must be **overlapping**, i.e., the size of the reduction window must be larger than the reduction factor. The classical $4 \times 4/4$ is an example.

First, a gray-level pyramid is built bottom-up as described in the previous sections. Although mostly Gaussian pyramids are used, there is no principle restriction on the type of reduction function. In a second (**linking**) step, every cell below the apex of the pyramid chooses one of its 'fathers' to which it is most similar in gray value. This choice is possible, because in an overlapping pyramid each cell is covered by more than one cell of the next higher level. In the third step, all the cells above the base level **recompute** their gray value based on the values of their sons. This process is iterated until a stable state is reached. The final links constitute tree structures in which all the cells that link to a common father are similar in gray value.

There are different methods to cut the tree structure to get a segmented image. Typical features extracted by pyramid linking are compact homogeneous regions [67]. Grosky and Jain [50] give an overview on **pyramid linking** methods and segmentation through pyramid structures. They focus on the 4 to 1 reduction of type $n \times n/4$ pyramids.

Gross ([51], [52]) describes several variations of the pyramid linking scheme. To arrive at a segmented image two stages are distinguished: **detection** and **delineation**. In Gross' scheme, a cell of the pyramid contains three values: the gray value, the variance σ^2, and an interest measure. The later determines the presence of a blob (= compact region) in a cell of the pyramid and initiates the top-down delineation process. The instance of a blob is detected if the interest measure of the cell is greater than the interest measure of all its fathers. In the delineation stage, the outline of the detected blob is determined. Earlier versions of the boundary localization are [9]. Four top-down methods are tested in Gross' work: the local method, the global method, the updated method, and the iterated global method. Their time complexity ranges between $\mathcal{O}(\log n)$ and $\mathcal{O}(\log^2 n)$.

3.5 Further applications

3.5.1 Change detection

In [5], the Laplacian pyramid is used to indicate a significant change in a time-series of images. Let $I(t)$ denote the image taken at time t, let l_1 denote the level at which the change shall occur. The following procedure is able to initiate an **alarm** when an unusual situation occurs in the field of view:

1. $D(t) := I(t) - I(t-1)$;
2. build Laplacian pyramid $L(l,t), l := 1, 2, \ldots, l_1$ with $L(0,t) := D(t)$;
3. square level l_1: $L(l_1,t)^2$;
4. build Gaussian pyramid $G(k,t), k := 1, 2, \ldots, k_{top}$ with $G(0,t) := L(l_1,t)^2$;
5. threshold $G(k,t), k := 1, 2, \ldots, k_{top}$: alarm.

A positive value of a sample of the integrated measure $G(k_{top},t)$ indicates that the feature selected by the filter is present in the image, but its position within the image is encoded at low resolution.

3.5.2 The pyramid as a model for real-time computer vision

Burt's motivations for image recognition in pyramid machines are: '...as pattern information becomes more complex, position information becomes less precise: there is a gradual shift from encoding "where" to encoding "what".' [20]

These ideas have been further developed and are summarized in Burt's survey paper [17] at ICPR 1988. Systems for real-time computer vision must locate and analyze just

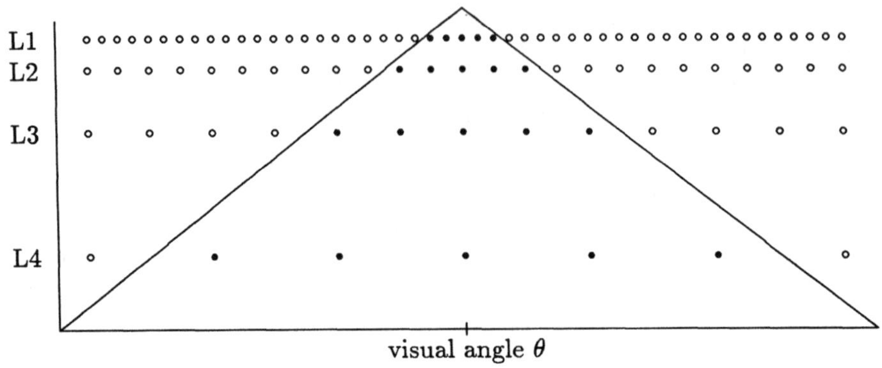

Figure 7: Formation of an electronic fovea

that information essential to the task at hand, while ignoring the vast flow of irrelevant detail. Attention mechanisms support efficient, responsive analysis by focusing the system's sensing and computing resources on selected areas of a scene. Three elements of attention in computer vision are: foveation, tracking and high level interpretation.

In human vision, acuity falls roughly in inverse proportion to the visual angle, θ, from the fovea. The non-uniform distribution of acuity in the human eye provides both a wide field of view and high acuity.

A rudimentary fovea is formed within the Laplacian pyramid (see section 3.3). After constructing the pyramid from an original image, pyramid data is discarded (set to zero) outside a central zone at each level, Figure 7. Since the region covered by each resolution level of the pyramid is double that of the next higher level, this implements a discrete approximation to a $1/\theta$ acuity distribution. In Figure 7 the width of the central zone is just 5 samples at each level, although 16 or 32 are typical in computer vision applications. This approximates what a human eye is able to resolve.

In human vision there are two distinct classes of eye movements, **saccades**, through which the eye brings critical regions of a scene into foveal vision, and **pursuit movements**, through which it follows moving objects in order to stabilize their images on the retina. **Tracking** is a key element in the control of selective, attention-based analysis in computer vision, as it is in human vision. Tracking may be said to isolate selected regions of a scene in **time** just as foveation isolates regions in **space**. It illustrates two important aspects of dynamic vision. First, relatively crude computations within a feedback loop can achieve very precise estimates of background motion. Second, when background motion is nulled through tracking, subsequent detection of object motion is greatly simplified (see also [118]).

Burt's high-level reasoning procedures are implemented as a fast, hierarchical structured search. In his object representation called **pattern tree**, the object pattern is decomposed into distinctive pattern components, or at various resolutions within a pyramid structure. The representation is constructed by first forming a full pyramid for the original pattern. Component patterns are then selected as small arrays of samples taken from this pyramid. Typically, large pattern components are represented at low resolution, while small components are respresented at high resolution. The links between components define their relative positions within the overall pattern. Search is then formulated as a sequence of simple pattern matching steps.

3.6 Bibliographical Notes

Good survey's about image pyramids and its applications can be found in [149], [1], [112]. Reference [129] is a collection of papers dealing with pyramids.

Comparisons of pyramidal architectures with different other architectures like arrays and hypercubes are given in [146], [10], [23], [24], [35], [46], [145], [153], [42], [95], [140].

References [31] and [32] describe the use of the **dynamic pyramid** to estimate the local motion in a series of images.

4 Pyramid Machines

4.1 Operations in a pyramid machine

Operations in a pyramid are mostly **local**. The new value of a cell may be computed using either external parameters (constants), or values from the cell's contents, or values from the neighboring cells in the pyramidal structure as input. These are the neighbors within the same pyramid level, the cells of the next lower level which are within the reduction window of the cell (called **sons**), and the cells of the next higher level (called **fathers** or **parents**). Every cell except the top (apex) has at least one parent cell, and may have more than one in overlapping pyramids, where a cell may be part of several reduction windows.

An important class of operations is responsible for the bottom-up information flow within the pyramid: the **reduction function**. It computes the new value of a cell exclusively from the contents of its sons. Given an image in the base of the pyramid, application of a reduction function (e.g., average) to all first-level cells fills this level. Once the cell of the first level has received a value, the same process can be repeated to fill the second level and so on to the top cell.

Further operations depend on the architecture of the pyramid. Two architecturally different types of pyramid machines are distinguished: (1) a mesh-connected multiprocessor architecture, a realization of a special, generalized cellular automaton [128], and (2) a pipeline architecture.

In a mesh-connected multi-processor pyramid machine there exists one processing element per cell. Tanimoto's machine [147] is a *bit pyramid* where every cell only contains a binary value, the basic operations are the binary operators AND, OR, and NEGATE with some additional instructions. Other pyramid machines have been proposed and/or built [100], [108], [141]. They use similar operations.

The RCA group [155] has built a pipelined preprocessor hardware to generate Gaussian and Laplacian pyramids using 2 programmable Multibus (Intel) boards with a memory update in 1/30 sec. This pyramid machine is a pipeline with five computational units:

- a filter with kernel size 5×5;
- a decimator to reduce the resolution by a factor of 4;
- an expander, to insert zeros;
- an ALU for sum and difference of images; and
- several image frame stores.

A comparison with a bit serial mesh-connected multi-processor architecture shows that the mesh machine would need 5000 chips to do the same pyramid operations as 100 chips used for the pipelined approach [20].

A further improvement of this pipeline architecture is achieved through the new concept of a *segmented pipeline*, where a buffer is introduced between elements of different data load [22].

4.2 Pyramidal and non-pyramidal architectures

Structures to which the pyramid is often compared are the cellular array, the hypercube or n-cube [42], and the prism [130]. They differ in the connections between the cells in the structure. The connections in a **cellular array** connect a cell with its three (triangular array), four (square grid with 4-connectivity), six (hexagonal array), or eight neighbors (square grid with 8-connectivity). An **n-cube** (also called hypercube) has 2^n cells and every cell is connected to n other cells. A **prism** is a stack of cellular arrays, all of the same size. It has the same number of levels as the $(n \times n/4)$ pyramid resulting in a total of $n \cdot 4^n$ cells. Besides the usual neighbor connections every cell is connected to three cells in the level above.

Fountain [42] evaluates the differences between the pyramid, the n-cube, and the (linear) array: The pyramid and the n-cube are rather evenly matched in cost, total processor performance and low-level processing times. In higher-level symbolic operations, the additional layers of the pyramid structure offer a significant improvement over the n-cube. The linear array is obviously in a lower category than the other two in cost and several aspects of performance.

Peleg and Federbush [117] describe pyramids with variable reduction factors ('contraction'). Such structures may enhance the flexibility in using appropriate cell sizes for representation and recognition.

Recently, Weems et al [156] describe an interesting pyramid-like architecture developed at the University of Massachusetts. The three levels, $512 \times 512, 64 \times 64, 4 \times 4$, differ also in the type of processors at the three levels. The highest resolution is a 1-bit serial array of processing elements (PE), the mid level is built by 32 bit PEs, and the 16 PEs at the top are LISP-processors. This architecture is the result of the experience that only very simple operations are performed at low levels while the complexity increases at higher levels.

5 Generalizing the Pyramid

5.1 The contents of a pyramidal cell

Hartley [57] allows a cell to contain more than just one (gray) value: *'... each cell in the pyramid contains a more complicated* **model** *of the region which it represents.'* Examples of hierarchical (parametrical) models for extracting edges, (smooth) curves and lines, corners, Glass and flow patterns, and textures, are used within the pyramid framework. The parameters of the models are best fitted in a least-square optimization by building the necessary sums in the pyramid in logarithmic time complexity. A main advantage of such pyramidal computations is pointed out: "relative locality".

Ahuja [3] differentiates between two broad categories of image representations: (1) those which describe the interiors of the cells and (2) those which specify their borders. Most of the pyramidal approaches belong to category (1).

A concept that falls in category (2) has been introduced for curve representation in pyramids [81]. This approach will be described in more detail in section 6. In this concept, a pyramidal cell is considered a window through which the underlying image is observed. Objects that are completely within the field of view of a cell can be recognized and described by this cell. Objects (e.g. curves) that are only partly covered must cross the boundary of the **observation window**. Hence the entire object cannot be recognized at this resolution level (of the pyramid). Information necessary to connect the observed part of the object with the parts in the adjacent cells must be passed up to the next lower resolution level. There, the cells cover a larger area and can join some parts of the level below. This process is repeated up to successively lower resolutions until the whole object is within the observation window of a cell.

More formally, this concept defines the cell that **represents** pictorial entities (e.g. primitive image parts, objects, configuration of objects) within a pyramid:

1. The region corresponding to that cell in the image space (that is a subarea of the base level in discrete representations) covers the represented pictorial entity completely.
2. No smaller cell in the pyramid fulfills property (1).

This definition assigns to every pictorial entity a unique cell in the pyramid in which it should be represented. Unfortunately, in some pyramid structures a small rigid motion (shift, rotation) of the object may cause a completely different representation (the representation cell may be many levels below or above). We therefore formulate the following goal for a pyramidal representation: pictorial entities in a pyramid should be ordered according to a specified property like size, length, etc. This would constrain the search for objects with specific properties to a bounded range of levels in the pyramid, where the entity must be represented if it appears in the image.

5.2 Different Pyramid Structures

Besides the most frequently used 'classical' $2 \times 2/4$ and $4 \times 4/4$ pyramid structures, other types of structures have also been used recently. This section starts with some new regular structures, then investigates into other possible pyramid structures and compares some of their properties, among which there is their robustness with respect to perturbations in their structure.

5.2.1 The $2 \times 2/2$ pyramid

Crowley [29], [30] computed a $2 \times 2/2$ Laplacian pyramid (he used the term DOLP[1]-transform) for detecting the peaks and ridges in the 'terrain' of gray values. In [82] the major properties of the $2 \times 2/2$ pyramid structure are presented:

- The shape of the receptive field is an octagone, the side lengths of which are expressed in the number of border cells.
- Giving equal weights to all sons, the equivalent weighting in the base is formed by Gaussian-like parabolas.
- If all cell centers of the pyramid are projected down to the base plane, the projected points are part of a grid that is a refinement of the base level of the pyramid.

  The same refinement grid solves the representation problem in [48]: the one pixel wide medial axis (skeleton) of an arbitrary shape is uniquely defined on this 'derived' grid even if the corresponding region has a diameter of an even number of pixels.
- Every son has two parents, every parent four sons.
- Using wrap around, the pyramid levels can be conveniently stored in rectangular arrays of sizes $1 \times 1, 2 \times 1, 2 \times 2, 4 \times 2, 4 \times 4, 8 \times 4, 8 \times 8, ..$
- The area of a cell is twice the area of a cell in the level below.
- Adjacent reduction windows have one cell in common (overlap).
- The grid axes of the higher levels are rotated by 45^0 with respect to the level below.

[1]Difference Of Low-Pass

5.2.2 Properties of regular structures

In pyramids the contents of a large number of high-resolution cells is propagated up to a successively smaller number of lower-resolution cells. At the first level, the contents of the base level's reduction windows is summarized, the second summarizes the local summarizations of the first level, and so on. To interpret the contents of a high-level cell it is important to know the domain in the base from which information is summarized in that cell. This region in the infinitely fine base plane is called the **receptive field** of the cell ([70,89]).

Non-overlapping structures form tree structures such that the receptive field is just the union of all leaf nodes. The refinement of the starting cell is a partition of that region. Receptive fields of overlapping structures tend to grow, the refinement of a cell is larger than the cell. The importance of overlap in pyramids is pointed out in [40]. It is also a key property for curve pyramids (see section 6).

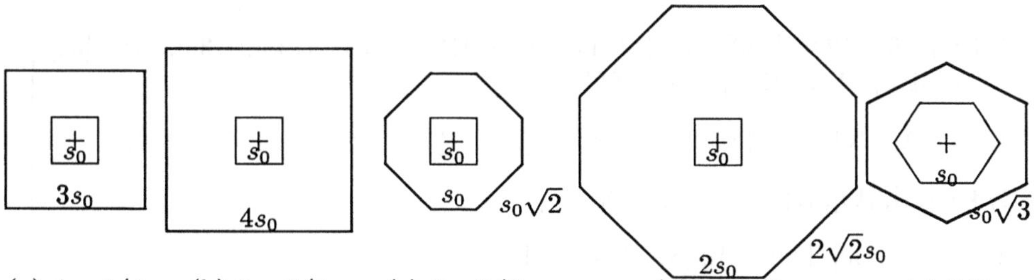

(a) $4 \times 4/4$ (b) $5 \times 5/4$ (c) $2 \times 2/2$ $(d) 3 \times 3/2$ (e) $7/4$

Figure 8: Receptive fields of 5 pyramid structures

Fig. 8 shows the sizes and shapes of the receptive fields of five overlapping pyramid structures. Table 3 summarizes the structural properties of the eight most commonly

Table 3: Properties of regular pyramid structures

Pyramid	receptive field		structural features		
	size	shape	overlap	sons	parents (weighted)
$(2m)^2/4$ (even)	$(2m-1)^2$	square	$2m(2m-1)$	$4m^2$	m^2
$(2m+1)^2/4$	$4m^2$	square	$4m^2+2m+1$	$(2m+1)^2$	$\frac{(m+1)^2}{4} + \frac{m(m+1)}{2} + \frac{m^2}{4}$
$2 \times 2/4$	1	square	0	4	1
$4 \times 4/4$	9	square	8	16	4
$5 \times 5/4$	16	square	15	25	$\frac{9}{4} + \frac{6}{2} + \frac{4}{4}$
$2 \times 2/2$	7	octagone	1	4	2
$3 \times 3/2$	28	octagone	4	9	$\frac{5}{2} + \frac{4}{2}$
7/4 hexagon	3	hexagone	1	7	$\frac{1}{2} + \frac{2}{2}$

used pyramids. It enumerates the characteristics of a cell's receptive field, its size in units of the cells area and its shape, the number of sons that two neighboring parents share (overlap), the number of sons of every cell except those in the base, and the number of parents and their frequency weight if there are more than one (e.g. $\frac{f}{n}$ means that every n-th cell has f parents).

Hartmann's hexagonal-pyramid structure [60] shows some interesting similarities with the $n \times n/2$ pyramids:

- Cell centers of different levels are in identical positions like in the $3 \times 3/2$ pyramid.
- It has the same refinement rule as the $2 \times 2/2$ pyramid to generate the higher resolution from the lower resolution: insert a new cell in between every pair of brothers.

We finish with a list of papers describing applications of the respective structure:

$2 \times 2/4$: [146], [124], [149], [143], [144], [139], [137], [154], [138], [136], [31], [32], [96], [114], [141], [153], [148], [99] .

$4 \times 4/4$: [21], [16], [66], [57], [53], [59], [58], [40], [50], [51], [79], [157], [134] .

$5 \times 5/4$: [18], [19], [1], [5], [155], [20], [17] .

$2 \times 2/2$: [29], [30], [82], [81], [87], [84], [88] .

$3 \times 3/2$: [83], [86], [85], [64], [92], [91], [113] .

7/4 **hexagonal structure** : [60], [34], [61], [62] .

5.3 Abstraction

Abstraction is a powerful data-fusion process. It enables individuals as well as processes to differentiate between **important** and less important information, methods, representations, etc. Such generalizations allow one to treat all the elements of a general class in the same way. When successively applied, they imply a hierarchical structure with different levels

- of concepts for representing knowledge about the world, e.g., the conceptual hierarchy in [7],
- of representation (see below),
- of processing stages (see below), e.g., hierarchies of invariance in cognition [8], and
- in the complexity of processing images.

Abstraction is a scientific method for the isolation of facts, for generalization, or for idealization. It has the characteristic to neglect details in favour of important features. Abstract objects form generic classes which have common properties but differ in detail. Since there is no unique way for abstraction, it may depend on functional goals (see [98,41]). Advantages of abstraction are: data reduction; reduced processing time by divide and conquer, entire generic classes are handled at once, distinguishing between class members only when necessary. As a consequence, one tries to build abstraction hierarchies that have the above advantages and that also allow the necessary distinction between individual members (e.g. [159]).

A generalized image pyramid is an example of an abstraction hierarchy in image analysis.

5.4 Representation levels

In [90] the fields of visualization and image analysis, although different in their goals, are related through common intermediate levels of representation. Visualization generates an image from a computer-stored description, digital image analysis is supposed to produce descriptions of a digital image. Descriptions at different levels of abstraction form the basis for both fields. The following levels are identified (compare Fig. 1):

1. 2D digital image with pixels;
2. image segments such as region, edge, or texton [74];
3. image segments with specific properties such as generalized cylinders;
4. fragments, parts of objects, 'GEON' [11];
5. objects, models;
6. functional areas [104];
7. natural language [150].

These seven representation levels are ordered according to several criteria: abstraction, length of description, expressive power, size of receptive field. Furthermore, numbers are more likely to appear at lower levels while symbolic descriptions dominate at higher levels.

5.5 Complexity and processing levels in image analysis

Tsotsos' model considers different levels of complexity [151]. Analysis of the complexity level is necessary to ensure that the basic space and performance constraints observed in human vision are satisfied. The **'maximum power / minimum cost principle'** ranks the many architectures that satisfy the complexity level and allows the choice of the best one. A **receptive field** is defined as the area of the visual scene in which a change in the visual stimulus causes a change in the output of the processor to which it is connected. Hexagonal images with hexagonal pixels are assumed, and lead to a **hexagonal pyramid structure**. In summary, the characteristics of a visual processing architecture are:

- spatial parallelism;
- hierarchical organization;
- localization of receptive fields;
- visual stimuli lead to logically separable maps;
- abstraction of the input token arrays.

Recently, these theoretical characteristics can be observed in practical implementations. For example, the inspection system described in [28] shows three processing levels: (1) segmentation, (2) noise cleaning and region labelling, and (3) defect classification.

5.6 Perceptual organization

The paper of Fischler and Bolles [41] addresses the problem of partitioning (perceptual organization). Most of the techniques can be characterized by one of the four paradigms (e.g., for partitioning a curve):

1. Local discontinuity detection;
2. best global description;
3. confirming evidence;
4. recursive simplification (producing a hierarchy of data sets).

The following two general principles must be satisfied by an effective technique:

1. Stability and
2. complete, concise, and complexity limited explanation.

The concept is illustrated by the selection of five points to describe a line drawing (Fig. 9). Sometimes, the complex configurations that may appear in an image have to be described by a limited vocabulary (e.g., only five points). Depending on the task that a human was given, the five points have been placed differently. The three tasks were:

1. Reconstruction (placement in Fig. 9);
2. Recognition;
3. Segmentation.

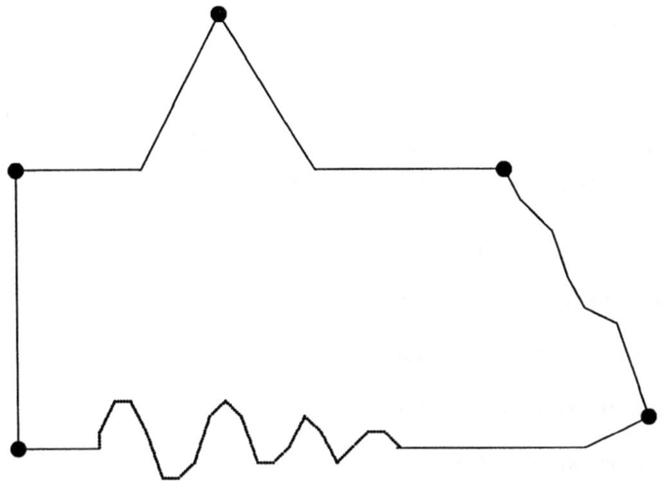

Figure 9: Describe shape by five points!

Two key points summarize the basic ideas:

Key point 1: The partitioning problem does not have a unique definition, but it is parameterized: purpose, data representation, trade off between different error types (false alarms versus misses), etc.

Key point 2: Psychologically acceptable partitions imply an explanation. This explanation must satisfy criteria for accuracy, complexity, and believability. These criteria are formulated by a set of principles which can guide the construction of an effective partitioning algorithm.

In his paper at the 10th ICPR, Aloimonos [4] extends these ideas to include active vision. He argues that visual tasks can be performed without solving the 'structure-from-motion' problem ('purposive and qualitative vision'). His system called 'Medusa' solves problems such as:

- Is there anything moving independently of me? If yes, where is it?
- Is the moving object getting closer to me?
- Track the object!
- Will the moving object eventually hit Medusa?
- Tell the motors of Medusa how to move!

5.7 Control mechanisms and evidence accumulation

Matsuyama characterizes the state-of-the-Art in the field [102]: "The expertise stored in the systems is what we, image processing researchers, have acquired and accumulated through the development of image processing techniques." Many problems are encountered in designing image analysis processes:

1. Assessment of image quality.
2. Selection of appropriate operators.
3. Determination of optimal parameters.
4. Combination of primitive operators.
5. Trial-and-Error experiments.
6. Evaluation of analysis result.

Matsuyama differentiates between IUS (image understanding systems) and ESIP (expert systems for image processing) by means of the knowledge sources mainly used in the two different kinds of systems.

The ESIP-architecture has the following major objective: formulate and describe visual information (e.g., image features, properties, and relations). ESIP should combine **qualitative symbolic reasoning** and **quantitative signal processing**. The reasoning consists of :

1. Analysis plan generation
2. Operator selection and parameter adjustment

As an extension to LLVE's single resolution image features (LLVE is the low level part of SIGMA [71,103]), Matsuyama proposes as a second strategy to represent the system's knowledge in a **hierarchical hypergraph**. Using a resolution parameter, it can be compressed for storage purposes into a single layered hypergraph which is expanded into the multilayered hypergraph during a search.

Another paper [75] surveys the planning and reasoning research being carried out in the Robot Vision Lab at Purdue. In particular, it describes a new planning system called SPAR, which uses a constraint posting approach for simultaneously fulfilling the operational, geometric and uncertainty reduction goals, and the PSEIKI system for evidential reasoning in a tangled hierarchy. The authors also mention briefly their other related research in high precision assembly under forcel torque control and robotic manipulation with structural stereopsis for depth perception.

Kitamura [78] selects the attitudes of industrial parts in the general framework of an expert system.

5.8 The robustness of pyramids

The robustness of pyramidal algorithms with respect to perturbations of the **structure** of the pyramid is studied by Meer et al [106]. The perturbations are of three different natures:

- Gaussian noise added to the generating kernels's weights;
- random local offsets of child-parent links;
- stochastic pyramid structures.

The first two perturbations apply to regular ($2 \times 2/4$) pyramid structures. Burt [16] showed that generating kernels should fulfil the following spatial constraints:

- normalisation (the sum of the weights is one);
- symmetry;
- unimodality;
- equal contribution to the next level.

Since the adding of noise to the kernels destroys Burt's constraints, several methods are given to reinstitute certain constraints on the noisy kernels. From those reinstitutions, normalisation appears to be the most important constraint and thus may help to improve the quality of low resolution pyramid levels.

Perturbations of child-parent links include offsets introduced at the children's level and offsets introduced at the parent's level. When the positions of the links were perturbed at the parent's level, normalisation was not preserved and the effects were severe. Reinstitution of normalisation improved the results in that case. The effects of perturbing the children's level were milder, because normalisation was preserved.

From these extensive experiments a general principle for constructing robust pyramid representations is concluded: the parent should control the computation so that he may normalize his weights.

Stochastic pyramids ([105]) are built by a stochastic-recursive-decimation process on the pixels of the base. Even on such severely perturbed pyramid structures, two selected algorithms show only a mild degradation. In the bimodality-detection algorithm, the pyramid fuses locally computed statistics. There is a reason why a stochastic implementation should lead to greatly different results than a regular one. In the second algorithm, object delineations, the low resolution representations show great discrepancies to the original object, but the top-down delineation process successfully recovers from incorrect root definitions.

Multiresolution algorithms appear to be very robust and only weakly dependent on the structure on which they are implemented.

6 Curves in Pyramids

Problems are encountered when thin structures such as curves and lines are to be detected in pyramids. In a gray level pyramid, curves disappear after a few reduction steps because the reduction averages the intensity over compact regions (receptive fields). Therefore an edge detector is applied to the intensity pyramid yielding an edge pyramid (e.g., [143], [58]). The problem in an edge pyramid is to combine the edges simultaneously detected at different resolutions.

Kelly was the first [76] to use a two-level ($5 \times 5/25$) multiresolution hierarchy for edge detection. He finds edges and lines first in the reduced resolution and uses them as a plan to constrain the search in the higher resolution.

Hong et al. [68] use **good continuation** as criterium to integrate local features. Hartley [59] smoothly joins polynomial curve models. The connectivity of curves was the primary goal in the concept of the curve pyramid which is presented in the sequel.

6.1 Discrete curve representations

When a curve is to be represented by discrete elements there are many variations from representations closely related with the underlying raster up to representations that capture the major properties of the curve.

Chain codes have been introduced by Freeman [43], [44] and are widely used as shape representations. They specify relative movements within a connected sequence of pixels. In another paper [160], chain codes are used to generate lines and circles. Other types of chain codes are also possible. We have introduced RULI chains that have the advantage that they can be generalized into a lower resolution by applying syntactical substitutions.

A step further away from the raster are polygonal representations. These are sequences of straight-line segments such that the next segment starts at the end point of its predecessor. So-called **vectorizers** perform (e.g. [115], [116]) the raster-to-vector conversion.

Polygons are widely used in computer graphics. Dunham and Glanz [36] present optimal approximations of planar shapes by polygons. Samet [139] stores polygons efficiently in quadtrees. Ayache and Faugeras [6] have built the system HYPER in which shapes of industrial parts are matched using polygonal approximations.

6.2 The $2 \times 2/2$ curve pyramid and the RULI-chain code

The concept of representing linear structures of digital images by **curve relations** is introduced in [81]. A cell of the underlying square grid is considered as the observation window through which a curve is observed. A curve under such an observation window connects the two points where the curve crosses the boundary of the cell. Since these two points are located on two of the four sides of the square, N, E, S, or W, the two sides are related by that curve (see example in Fig. 10).

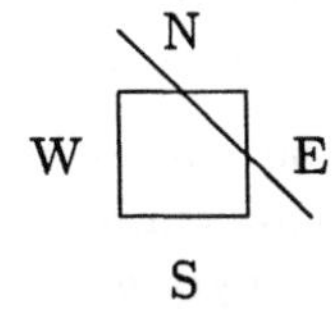

Figure 10: A NE curve relation

The set of such *curve relations* of all observed curves is stored in every cell. It approximates the observed curves.

Figure 11: Reduction in the $2 \times 2/2$ curve pyramid

The accuracy of the approximation is determined by the resolution of the cell. A lower resolution representation is derived in two steps (Fig. 11):

- subdivision of the cell contents by introducing a **diagonal** and
- merging curve relations of neighbor cells by **transitive closure**.

Iterative application of this generalization procedure yields the $2 \times 2/2$ curve pyramid.

There exists a chain code representation equivalent to the 2D concept of the curve relations: the **RULI chain code**. The four elements of this chain code $\{R, U, L, I\}$ determine the relative change in moving direction: go right (R), make a U-turn (U), go left (L), continue in the same direction (I). A RULI-chain is invariant to shifts and rotations by 90^0. The generalization of a curve represented by this RULI chain code is formulated by a 5 step formal substitution grammar. It is shown that both generalizations are identical for single curves. The grammar is further used to prove the *'length reduction property'*, i.e., that the number of codes of a non-closed curve always shrinks after generalization.

In another paper [87], three types of square grid pyramids are checked to see whether they are suited for multiresolution curve representation: the $2 \times 2/4$ pyramid, the $4 \times 4/4$ pyramid, and the $2 \times 2/2$ pyramid. Such a representation must satisfy the following goal: **the highest level up to which a curve is still represented before it disappears**

must be related to the area that is traversed by the curve. Non-overlapping pyramid structures do not satisfy this criterium. The *length reduction property* of the $2 \times 2/2$ curve pyramid is extended to **closed curves**. The number of code elements of closed chains is shown to shrink at least every second step.

Noise can have two major effects on curves in digital images: it creates a lot of new short curve segments and it cuts long curves into smaller pieces. Using the length reduction property of the $2 \times 2/2$ curve pyramid, the short, noisy, curve segments are eliminated in a top-down process that deletes all curve relations in the lower levels that are not represented in the level above [88]. Experiments with the synthetic image from [85] are reported. The only parameter of this '**structural noise filtering process**' is the level to which the pyramid is built bottom-up and which is also the starting level for the top-down deletion. In the experiments, this parameter varied from 1 to 10. The results are compared to the uncorrupted image and show that the number of deleted noisy curve segments increase by a factor of two on the average when one more level is used.

6.3 Digital straight lines

Straight lines play an important role as an intermediate shape representation. Many researchers have been interested in methods for detection of straightness from chain codes [127], [126], [120], [97] and for efficient generation of straight lines in a raster [14], [122], [15], [25] .

A necessary and sufficient condition determines whether a given chain coded curve may result from sampling a straight line [93]. An algorithm is given that checks the straightness of both a Freeman chain coded (FCC) and a RULI chain coded sequence.

Two classes of codes are defined by comparing neighboring codes: **repetition** and **single** codes. A code is called a repetition code if one of its neighbors in the sequence is the same. If both are different, it is called a single code. Repetition and single code of a chain coded straight line must be uniquely defined. Freeman [44] found furthermore that the two different codes must have neighboring orientations, e.g., FCC differ by one.

Sequences of identical codes are collected to 'runs' separated by single codes. In a reduction step, all runs are substituted by their lengths, and all single codes are eliminated. Neighboring elements of the new sequence of numbers can be compared in the same way as the original chain codes. Hence, repetition and single codes can be determined in this reduced sequence as well. It is shown that uniqueness of repetition and single codes is also required for the reduced sequence if it represents a straight line. The difference of the two must not exceed 1.

This process is repeated until the reduced sequence is so simple that the straightness can be decided immediately, e.g., if all elements are the same. It can be shown that this procedure eliminates all non-straight lines during evaluation.

One single iteration depends only on the number of the sequence's elements, and the reduced sequence is at most half as long. Therefore, the algorithm takes on the order of $n + n/2 + n/4 + n/8 + \cdots < 2n$ steps if n is the number of original chain codes. Using the repetition and single codes of all iterations the possible slopes of the line can be computed.

6.4 Corners

Polygons have been introduced as connected sequences of straight line segments. In this concept, vectorization searches for chains of maximum length that could be the result of digitizing straight lines. An alternative definition specifies a sequence of points that are connected by straight lines. This definition suggests a different vectorization strategy:

1. Find **corners** as points where the digital curve changes its direction significantly and

2. connect them by straight line segments.

Obviously this approach has problems in detecting and representing smooth curves at high resolution.

We have shown in [39] that curvature extrema can be related to corners in multiple resolutions. Consider a curve consisting of three elements, L_1, C_2, L_3. Let the circular arc C_2 connect smoothly two straight line segments L_1, L_3. Four parameters describe this sample configuration: the radius r of C_2, the angle θ between L_1 and L_3, and the lengths $l_1 > 2r, l_3 > 2r$ of L_1, L_3 respectively. If r is much larger than the side length of a resolution cell, C_2 will be recognizable in the corresponding image. If we reduce the resolution, the angle θ remains the same, as long as both L_1 and L_3 are represented. But the other three parameters, r, l_1, l_3, will decrease with respect to the reduced resolution. A smaller radius implies a larger curvature. If we continue reducing the resolution until the size of a cell approaches $(2r)^2$, C_2 will disappear, and L_1 and L_3 will form a *corner* with angle θ. This principle relates curvature extrema and corners through different resolutions.

The results of this work are described in [39,38]. The key points can be summarized as follows:

1. In a reduced resolution, curvature extrema may disappear. This causes symbolic curve representations like Richards and Hoffman's CODONs [125] to change. The modifications are local and have the effect of shrinking the length of the description. The order in which curvature extrema disappear is determined by the length of the two adjacent limbs.

2. Corners are locally detected as **non-straight** configurations of RULI-chains.

3. Corners are linked in adjacent levels of a curve pyramid. In contrast to continuous scale-space, corners need not always have a corresponding corner in the pyramid level directly below. Artifacts caused by digitization of the continuous scale space are identified as such.

4. Two sufficient conditions of the corners before digitization determine if the corners would be recognized in any possible placement and orientation. The corresponding limits depend on the size of the detector k ($k = 3$ and $k = 5$ elements have been used in the experiments):
   - Upper limits $\Theta(k)$ for the corner's angles ($\theta < \Theta(k)$ in the above example);
   - Lower limits $l_{min}(k)$ for the length of the corner's legs ($l_1 > l_{min}(k), l_3 > l_{min}(k)$ in the above example).
5. Curvature extrema become corners in a bottom-up reduction process and remain corners until they disappear.
6. Efficient top-down **matching** strategies are described and tested by simple examples (contour of a plier, Fig. 12; a contour from [41] and [121]).

In Fig. 12 three levels of a curve pyramid are built on top of a digitized image of pliers. All corners are detected within a local 3×3 observation window as non-straight curve configurations. In level 7 the four corners satisfying the necessary corner-conditions are marked (by circles). Their positions allow one to approximately identify the plier. This hypothesis is then checked in a top-town verification pass and its location and shape is refined by a coarse-to-fine strategy.

6.5 Other hierarchical approaches for digital curve assembly

Hartmann's pyramid is built on a hexagonal grid (see section 5.2). In the early version the cells contained a hierarchical contour code HCC describing a contour segment that has been detected by a Laplacian operator [60]. The cells' contents have been extended to hold also regions in the HSC (hierarchical structure code [61]). Hartmann's pyramid has been used for several years and many applications. To name one example, Drüe et al. [34] show strategies for model-based recognition similar to Burt's pattern tree.

Gauch et al.[47] use a multiresolution version of the *symmetric axis transformation* (SAT) according to Blum (see also [131]) for figure based shape descriptions. Lower resolution yields simplification. This is also true for the branching structure of the SAT. The importance of a branch is determined by its annihilation resolution (that is the level above which it does not appear any more). This definition of importance implies a hierarchical ordering of the SAT-branches. Note the analogy to our approach.

Kjell and Dyer [79] determine "long straight edge segments" at orientations of $45^0 i, i = 0, 1, 2, 3$. The hierarchical region growing procedures use a $4 \times 4/4$ pyramid :
SEGMENT → (LINK (bottom up linking), LABEL (top down label propagation)),
SMOOTH → (BUILD (bottom up weighted average), PROJECT (top down weighted average)).
They conclude: 'Features based on spatial properties of long, straight, extended edge segments are reliable features for texture description and segmentation.'

Meer, Sher, and Rosenfeld [107] present the **chain pyramid** as a solution to the curve representation problem in pyramids. Curves are represented by doubly linked lists in the cells of a $4 \times 4/4$ pyramid. A probabilistic allocation algorithm avoids the

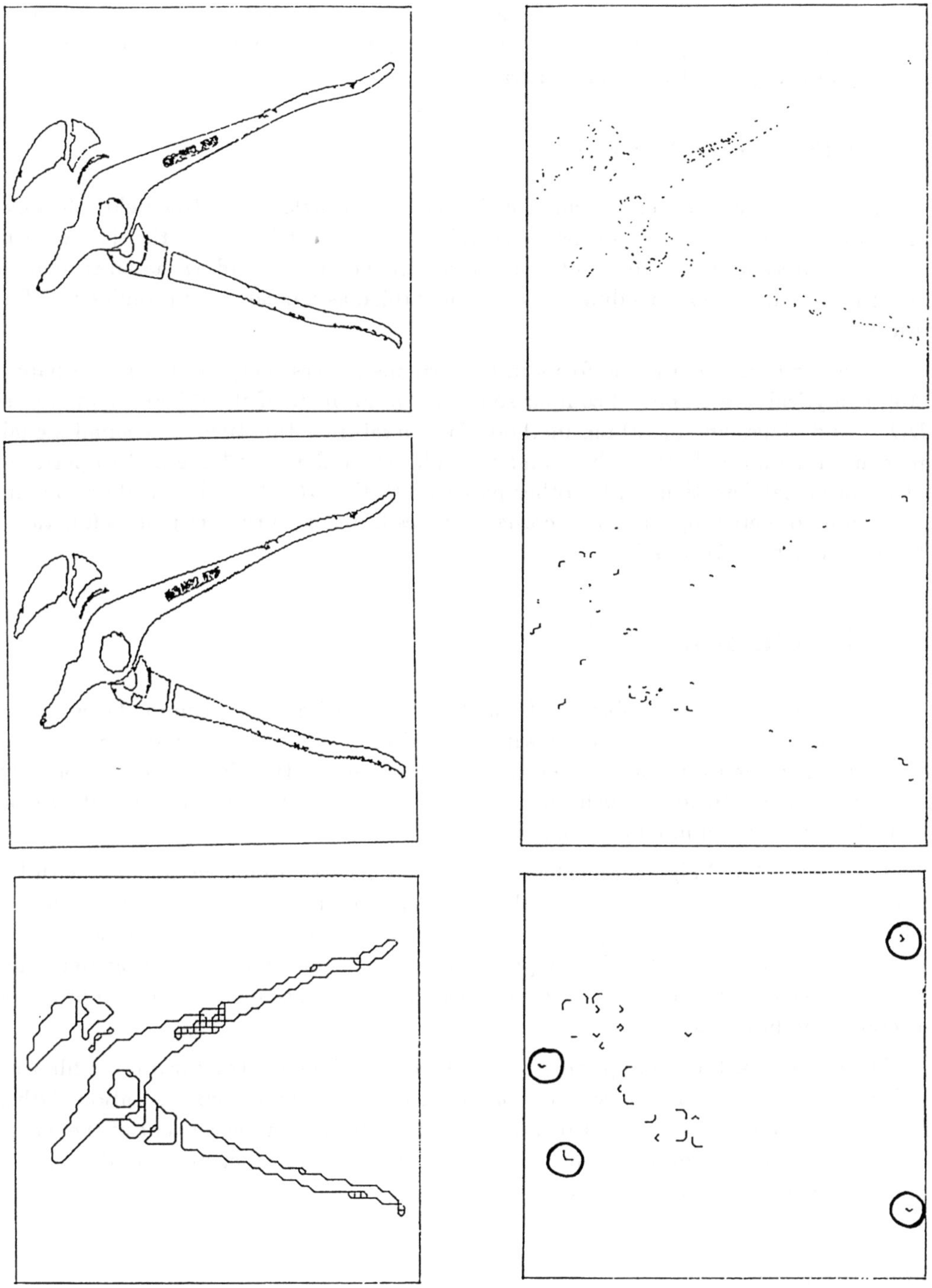

Figure 12: Levels 1,5,7 of curve pyramid of 'pliers' and detected corners

overload in the higher levels. Algorithms for preserving the local connectivity during the bottom-up building process (reduction function), for smoothing of multiscale curves, and for gap bridging between contour fragments are presented.

6.5.1 Other corner detectors

Freeman [45] detects corners in chain coded curves. Hartley and Rosenfeld [59] link curve segments in a pyramid to detect *significant* corners. Cheng and Hsu [27] detect corners of curves in digital images by measuring the degree of bending via an extended (k-step) 3×3 mask. The bending values are identified as a measure of prominence of a corner.

Curvature estimation is the focus in O'Gormans papers [111], [110]. In one paper [110], analytical and empirical comparisons have been made of the difference of slopes (DOS) and Gaussian smoothing methods for curvature estimation. For small signal angle and high noise, both analysis and examples show the DOS^+ method to perform better for signal detection. In his other paper [111], the DOS^+ method is described in more detail for detecting curve and corner features from the curvature plot as functions of two parameters, M and L.

7 Conclusion

Recipes for how specific problems of robot vision can be solved cannot be given in general. Vocabularies are required with two criteria: (1) they must describe visual situations at a highly abstract level and (2) they can be transformed into relatively primitive image detectors. Such vocabularies depend on the application and can be defined only in very limited environments.

Architectural and algorithmical constraints are derived from a comparison with biological vision systems. They prove that certain complicated vision problems can be solved even if we don't know (yet) how. General pyramidal structures satisfy several of the biological constraints. This paper shows some of the simple tasks that pyramids can perform efficiently. It studies the components of the concepts involved and proposes several generalizations.

The concept of the curve pyramid with its applications in structural noise filtering and in hierarchical corner detection is based on a pure symbolic representation. Other higher-level symbolic representations could be built on the simple concept of the 'curve relation defined in an observation window'. Applications in other problem domains of robotics seem to be straight forward.

References

[1] Adelson, E. H., Anderson, C. H., Bergen, J. R., Burt, P. J., and Ogden, J. M.: Pyramid methods in image processing. *RCA Engineer*, Vol. 29-6, Nov./Dec. 1984.

[2] Aguirre, F., Boucher, J. M., and Jacq, J. J.: Underwater navigation by video sequence analysis. In *Proc. 10th International Conference on Pattern Recognition*, pages 537–539, Atlantic City, New Jersey, USA, June 1990. IEEE Comp.Soc.Press. Vol. II.

[3] Ahuja, N., An, B., and Schachter, B.: Image representation using Voronoi tesselation. *Computer Vision, Graphics, and Image Processing*, Vol. 29:pp.286–295, 1985.

[4] Aloimonos, J.: Purposive and qualitative active vision. In *Proc. 10th International Conference on Pattern Recognition*, pages 346–360, Atlantic City, New Jersey, USA, June 1990. IEEE Comp.Soc.Press. Vol. I.

[5] Anderson, C. H., Burt, P. J., and van der Wal, G. S.: Change detection and tracking using pyramid transform techniques. *Intelligent Robots and Computer Vision*, SPIE Vol.579:pp.72–78, Sept.16-20 1985.

[6] Ayache, N. and Faugeras, O. D.: HYPER: A new approach for the recognition and positioning of two-dimensional objects. *IEEE Transactions on Pattern Analysis and Machine Intelligence*, PAMI-8(No. 1):pp.44–54, January 1986.

[7] Bajcsy, R. and Rosenthal, D. A.: Visual and conceptual focus of attention. In S. Tanimoto and A. Klinger, editors, *Structured Computer Vision*, pages 133–149. Academic Press, 1980.

[8] Ballard, D. H.: Interpolation coding: A representation for numbers in neural models. Technical Report TR–175, Dept. of CS, Univ. of Rochester, Sept. 1986.

[9] Baugher, E. S. and Rosenfeld, A.: Boundary localization in an image pyramid. *Pattern Recognition*, Vol. 19:pp.373–395, 5 1986.

[10] P. W. Besslich: Pyramidal transforms in image processing and computer vision. In S. Levialdi and V. Cantoni, editors, *Pyramidal Systems for Image Processing and Computer Vision*, volume F25 of *NATO ASI Series*, pages 215–246. Springer-Verlag Berlin, Heidelberg, 1986.

[11] Biederman, I.: Matching image edges to object memory. In *Proceedings of the First International Conference on Computer Vision*, pages 384–392, London, England, 1987.

[12] Blanford, R. P. and Tanimoto, S. L.: Bright-spot detection in pyramids. In *Proc. Eighth International Conference on Pattern Recognition*, pages 1280–1282, Paris, France, October 1986. IEEE Comp.Soc.

[13] Blanz, W. E. and Gish, S. L.: A connectionist classifier architecture applied to image segmentation. In *Proc. 10th International Conference on Pattern Recognition*, pages 272–277, Atlantic City, New Jersey, USA, June 1990. IEEE Comp.Soc.Press. Vol. II.

[14] Bresenham, J. E.: Algorithm for computer control of a digital plotter. *IBM Systems Journal*, Vol. 25(No.1):pp. 25–30, 1965.

[15] Bresenham, J. E.: Run length slice algorithm for incremental lines. In *Fundamental Algorithms for Computer Graphics*, NATO ASI Series, pages 59–104. Springer Verlag, New York/Berlin, 1985.

[16] Burt, P. J.: Fast filter transforms for image processing. *Computer Graphics and Image Processing*, Vol. 16:pp.20–51, 1981.

[17] Burt, P. J.: Attention mechanisms for vision in a dynamic world. In *Proc. 9th International Conference on Pattern Recognition*, pages 977–987, Rome, Italy, November 1988. IEEE Comp.Soc.

[18] Burt, P. J. and Adelson, E. H.: The Laplacian pyramid as a compact image code. *IEEE Transactions on Communications*, Vol. COM-31(No.4):pp.532–540, April 1983.

[19] Burt, P. J. and Adelson, E. H.: A multiresolution spline with application to image mosaics. *ACM Transactions on Graphics*, Vol. 2(No.4):pp.217–236, October 1983.

[20] Burt, P. J., Anderson, C. H., Sinniger, J. O., and van der Wal, G.: A pipelined pyramid machine. In S. Levialdi and V. Cantoni, editors, *Pyramidal Systems for Image Processing and Computer Vision*, volume F25 of *NATO ASI Series*, pages 133–152. Springer-Verlag Berlin, Heidelberg, 1986.

[21] Burt, P. J., Hong, T.-H., and Rosenfeld, A.: Segmentation and estimation of image region properties through cooperative hierarchical computation. *IEEE Transactions on Systems, Man, and Cybernetics*, Vol. SMC-11(No.12):pp.802–809, December 1981.

[22] Burt, P. J. and van der Wal, G.: An architecture for multiresolution, focal, image analysis. In *Proc. 10th International Conference on Pattern Recognition*, pages 305–311, Atlantic City, New Jersey, USA, June 1990. IEEE Comp.Soc. Vol. II.

[23] Cantoni, V.: I.P. hierarchical systems: Architectural features. In S. Levialdi and V. Cantoni, editors, *Pyramidal Systems for Image Processing and Computer Vision*, volume F25 of *NATO ASI Series*, pages 21–40. Springer-Verlag Berlin, Heidelberg, 1986.

[24] Castan, S.: Architectural comparisons. In S. Levialdi and V. Cantoni, editors, *Pyramidal Systems for Image Processing and Computer Vision*, volume F25 of *NATO ASI Series*, pages 91–108. Springer-Verlag Berlin, Heidelberg, 1986.

[25] Castle, C. M. A. and Pitteway, M. L. V.: An application of Euclid's algorithm to drawing straight lines. In *Fundamental Algorithms for Computer Graphics*, NATO ASI Series, pages 135–139. Springer Verlag, New York/Berlin, 1985.

[26] Chen, C. H. and Kak, A. C.: A robot vision system for recognizing 3-d objects in low-order polynomial time. *IEEE Transactions on Systems, Man, and Cybernetics*, Vol. SMC-19(No.6):pp.1535–1563, November/December 1989.

[27] Cheng, F.-H. and Hsu, W.-H.: Parallel algorithm for corner finding on digital curves. *Pattern Recognition Letters*, 8(No. 1):pp.47–53, July 1988.

[28] Cho, T.-H., Conners, R. W., and Araman, P. A.: A computer vision system for analysing images of rough hardwood lumber. In *Proc. 10th International Conference on Pattern Recognition*, pages 726–728, Atlantic City, New Jersey, USA, June 1990. IEEE Comp.Soc.Press. Vol. I.

[29] Crowley, J. L. and Parker, A.: A representation of shape based on peaks and ridges in the difference of low–pass transform. *IEEE Trans. Pattern Analysis and Machine Intelligence*, PAMI–6:pp.156–170, 1984.

[30] Crowley, J. L. and Stern, R. M.: Fast computation of the difference of low–pass transform. *IEEE Transactions on Pattern Analysis and Machine Intelligence*, PAMI–6:pp.212–222, 1984.

[31] Dengler, J.: Local motion estimation with the dynamic pyramid. In S. Levialdi and V. Cantoni, editors, *Pyramidal Systems for Image Processing and Computer Vision*, volume F25 of *NATO ASI Series*, pages 289–298. Springer-Verlag Berlin, Heidelberg, 1986.

[32] Dengler, J., Meinzer, H. P., and Schmidt, M.: Lokale Bewegungsanalyse mit der Dynamischen Pyramide. In G. Hartmann, editor, *Mustererkennung 1986*, Informatik Fachberichte 125, pages 276–281. Springer Verlag, 1986.

[33] Dickmanns, E. D.: Visual dynamic scene understanding exploiting high-level spatio-temporal models. In *Proc. 10th International Conference on Pattern Recognition*, pages 373–378, Atlantic City, New Jersey, USA, June 1990. IEEE Comp.Soc.Press. Vol. II.

[34] Drüe, S. and Hartmann, G.: Modellgestützte Erkennung hierarchisch codierter Objekte. In G. Hartmann, editor, *Mustererkennung 1986*, Informatik Fachberichte 125, pages 245–249. Springer Verlag, 1986.

[35] Duff, M. J. B.: Pyramids - expected performance. In S. Levialdi and V. Cantoni, editors, *Pyramidal Systems for Image Processing and Computer Vision*, volume F25 of *NATO ASI Series*, pages 59–74. Springer-Verlag Berlin, Heidelberg, 1986.

[36] Dunham, S. R. and Glanz, F. H.: Optimum uniform piecewise linear approximation of planar curves. *IEEE Transactions on Pattern Analysis and Machine Intelligence*, PAMI-8(No.1):pp.55–66, January 1986.

[37] Fennema, C. and Hanson, A. R.: Experiments in autonomous navigation. In *Proc. 10th International Conference on Pattern Recognition*, pages 24–31, Atlantic City, New Jersey, USA, June 1990. IEEE Comp.Soc.Press. Vol. I.

[38] Fermüller, C.: Hierarchisches Vergleichen von Konturen. Master's thesis, Technische Universität Graz, 1989. Diplomarbeit.

[39] Fermüller, C. and Kropatsch, W. G.: Hierarchische Kontur-Beschreibung durch Krümmung. In A. Pinz, editor, *Wissensbasierte Mustererkennung*, OCG-Schriftenreihe, Österr. Arbeitsgemeinschaft fuer Mustererkennung, pages 171–187. Oldenbourg, 1989. Band 49.

[40] Ferreti, M.: Overlapping in compact pyramids. In S. Levialdi and V. Cantoni, editors, *Pyramidal Systems for Image Processing and Computer Vision*, volume F25 of *NATO ASI Series*, pages 247–260. Springer-Verlag Berlin, Heidelberg, 1986.

[41] Fischler, M. A. and Bolles, R. C.: Perceptual organization and curve partitioning. *IEEE Transactions on Pattern Analysis and Machine Intelligence*, PAMI-8(No.1):pp.100–105, January 1986.

[42] Fountain, T.: Array architectures for iconic and symbolic image processing. *International Journal of Pattern Recognition and Artificial Intelligence*, Vol. 2(No. 3):pp. 407–424, June 1988.

[43] Freeman, H.: On the encoding of arbitrary geometric configurations. *IEEE Trans. Electron. Comput.*, EC-10:pp.260–268, 1961.

[44] Freeman, H.: Computer processing of line-drawing images. *Computing Surveys*, Vol. 6(No.1):pp.57–97, March 1974.

[45] Freeman, H.: A corner finding algorithm for chain coded curves. *IEEE Tr. on Computers*, Vol.C-23(No 3):pp.297–304, March 1977.

[46] Fritsch, G.: General purpose pyramidal architectures. In S. Levialdi and V. Cantoni, editors, *Pyramidal Systems for Image Processing and Computer Vision*, volume F25 of *NATO ASI Series*, pages 41–58. Springer-Verlag Berlin, Heidelberg, 1986.

[47] Gauch, J. M. and Pizer, S. M.: Image description via the multiresolution intensity axis of symmetry. In *Proceedings of the Second International Conference on Computer Vision*, pages 269–274, Tampa, Florida, December 1988.

[48] Gong, W. and Bertrand, G.: A fast skeletonization algorithm using derived grids. In *Proc. 9th International Conference on Pattern Recognition*, pages 776–778, Rome, Italy, November 1988. IEEE Comp.Soc.

[49] Gottschlich, S. N. and Kak, A. C.: A dynamic approach to high-precision parts mating. *IEEE Transactions on Systems, Man, and Cybernetics*, Vol. SMC-19(No.4):pp.797–810, July/August 1989.

[50] Grosky, W. I. and Jain, R.: A pyramid-based approach to segmentation applied to region matching. *IEEE Transactions on Pattern Analysis and Machine Intelligence*, PAMI-8(No.5):pp.639–650, September 1986.

[51] Gross, A. D.: *Multiresolution Object Detection and Delineation.* PhD thesis, University of Maryland, Computer Science Center, January 1986.

[52] Gross, A. D.: Multiresolution object detection and delineation. *Computer Vision, Graphics, and Image Processing*, Vol. 39(No. 1):pp.102–115, July 1987.

[53] Gross, A. D., Hartley, R. L., and Rosenfeld, A.: A fast parallel algorithm for dot linking in glass patterns. *Pattern Recognition Letters*, Vol. 3:pp.263–270, 1985.

[54] Haralick, R. M., Lin, C., Lee, J. S. J., and Zhuang, X.: Multi-resolution morphology. In *Proceedings of the First International Conference on Computer Vision*, pages 516–520, London, England, June 1987.

[55] Haralick, R. M. and Shapiro, L. G.: Image segmentation techniques. *Computer Vision, Graphics, and Image Processing*, Vol. 29(No. 1):pp.100–132, January 1985.

[56] Haralick, R. M., Zhuang, X., Lin, C., and Lee, J.: The digital morphological sampling theorem. Technical report, Department of Electrical Engineering, University of Washington, Seattle, WA 98195, February 1988.

[57] Hartley, R. L.: *Multi-Scale Models in Image Analysis* . PhD thesis, University of Maryland, Computer Science Center, 1984.

[58] Hartley, R. L.: A Gaussian-weighted multiresolution edge detector. *Computer Vision, Graphics, and Image Processing*, Vol. 30:pp.70–83, May 1985.

[59] Hartley, R. L. and Rosenfeld, A.: Hierarchical curve linking for corner detection. In S. Levialdi editor, *Integrated Technology for Parallel Image Processing*, pages 101–119. Academic Press, London, 1985.

[60] Hartmann, G.: Principles and strategies of hierarchical contour coding. In *Proceedings of the Seventh International Conference on Pattern Recognition*, pages 1087–1089, Montreal, Canada, 1984.

[61] Hartmann, G.: Unified description and recognition of continuous contours and regions. In *Proceedings of the Eighth International Conference on Pattern Recognition*, pages 1201–1203, Paris, France, October 1986.

[62] Hartmann, G.: Recognition of hierarchically encoded images by technical and biological systems. *Biological Cybernetics*, Vol. 57:pp.73–84, 1987.

[63] Hättich, W. and Wandres, H.: Automatic learning of structural models for workpiece recognition systems. In *Proc. 10th International Conference on Pattern Recognition*, pages 279–281, Atlantic City, New Jersey, USA, June 1990. IEEE Comp.Soc.Press. Vol. I.

[64] Herbst, H.: Die Auswirkungen verschiedener Reduktionsfunktionen auf Bildeigenschaften in Radar–Bildpyramiden. Master's thesis, Technische Universität Graz, 1987. Diplomarbeit.

[65] Heuser, M. and Liedtke, C.-E.: Recognition of the spatial position of industrial 3d-objects using relaxation techniques. In *Proc. 10th International Conference on Pattern Recognition*, pages 191–193, Atlantic City, New Jersey, USA, June 1990. IEEE Comp.Soc.Press. Vol. I.

[66] Hong, T.-H.: *Pyramid Methods in Image Analysis* . PhD thesis, University of Maryland, Computer Science Center, 1982.

[67] Hong, T.-H. and Rosenfeld, A.: Compact region extraction using weighted pixel linking in a pyramids. *IEEE Transactions on Pattern Analysis and Machine Intelligence*, Vol. PAMI-6(2):pp.222–229, 1984.

[68] Hong, T.-H., Shneier, M., Hartley, R. L., and Rosenfeld, A.: Using pyramids to detect good continuation. *IEEE Transactions on Systems, Man, and Cybernetics*, Vol. SMC-13(4):pp.631–635, 1983.

[69] Huang, T. S.: Modeling, analysis, and visualization of nonrigid object motion. In *Proc. 10th International Conference on Pattern Recognition*, pages 361–364, Atlantic City, New Jersey, USA, June 1990. IEEE Comp.Soc.Press. Vol. I.

[70] Hubel, D. H. and Wiesel, T. N.: Die Verarbeitung visueller Information. *Spektrum der Wissenschaft*, Wahrnehmung und visuelles System:pp. 36–47, 1986.

[71] Hwang, V. S.-S., Larry, D. S., and Matsuyama, T.: Hypothesis integration in image understanding systems. *Computer Vision, Graphics, and Image Processing*, Vol. 36:pp.321–371, November 1986.

[72] Jähne, B.: *Digitale Bildverarbeitung* . Springer, Berlin, 1989.

[73] Joo, H. and Haralick, R. M.: Application of mathematical morphology to machine vision. In H. Burkhardt, K. Höhne, and B. Neumann, editors, *Informatik Fachberichte 219: Mustererkennung 1989*, pages 1–27, Hamburg, 2.-4.10. 1989. 11.DAGM - Symposium, Springer Verlag.

[74] Julesz, B. and Bergen, J. R.: Textons, the fundamental elements in preattentive vision and perception of textures. *The Bell System Technical Journal*, Vol. 62(No. 6):pp.1619–1645, July-August 1983.

[75] Kak, A., Hutchinson, S., and A. K.M: Planning and reasoning in sensor based robotics. Presented paper at the 10. DAGM-Symposium Mustererkennung, Zürich, 27.9.-29.9., 1988.

[76] Kelly, M. D.: Edge detection in pictures by computer using planning. In B. Meltzer, and D. Michie, editors, *Machine Intelligence 6*, pages 397–409, Edinburgh, Scotland, 1971. Edinburgh University Press.

[77] Kim, C.-W. and Koivo, A. J.: Hierarchical classification of surface defects on dusty wood boards. In *Proc. 10th International Conference on Pattern Recognition*, pages 775–779, Atlantic City, New Jersey, USA, June 1990. IEEE Comp.Soc.Press. Vol. I.

[78] Kitamura, Y., Sato, H., and Tamura, H.: An expert system for industrial machine vision. In *Proc. 10th International Conference on Pattern Recognition*, pages 771–774, Atlantic City, New Jersey, USA, June 1990. IEEE Comp.Soc.Press. Vol. I.

[79] Kjell, B. P. and Dyer, C. R.: Segmentation of textured images by pyramid linking. In S. Levialdi and V. Cantoni, editors, *Pyramidal Systems for Image Processing and Computer Vision*, volume F25 of *NATO ASI Series*, pages 273–288. Springer-Verlag Berlin, Heidelberg, 1986.

[80] Kreyszig, E.: *Statistische Methoden und ihre Anwendungen.* Vandenhöck and Rupprecht, Göttingen, 1965.

[81] Kropatsch, W. G.: Hierarchical curve representation in a new pyramid scheme. Technical Report TR-1522, University of Maryland, Computer Science Center, June 1985.

[82] Kropatsch, W. G.: A pyramid that grows by powers of 2. *Pattern Recognition Letters*, Vol. 3:pp.315–322, 1985.

[83] Kropatsch, W. G.: Complementary pyramids. Presented paper at the NATO ARW in Maratea on 'Pyramidal Systems for Image Processing and Computer Vision', May 5-9, 1986.

[84] Kropatsch, W. G.: Curve representations in multiple resolutions. In *Proc. Eighth International Conference on Pattern Recognition*, pages 1283–1285. IEEE Comp.Soc., 1986.

[85] Kropatsch, W. G.: Ein Konzept mit zwei sich ergänzenden Pyramiden. In W. G. Kropatsch and P. Mandl, editors, *Mustererkennung'86*, OCG–Schriftenreihe der österreichischen Arbeitsgemeinschaft für Mustererkennung, pages 158–171. R. Oldenburg, 1986. Band 36.

[86] Kropatsch, W. G.: Grauwert und Kurvenpyramide, das ideale Paar. In G. Hartmann, editor, *Mustererkennung 1986*, Informatik Fachberichte 125, pages 79–83. Springer Verlag, 1986.

[87] Kropatsch, W. G.: Kurvenrepräsentation in Pyramiden. In W. G. Kropatsch and P. Mandl, editors, *Mustererkennung'86*, OCG-Schriftenreihe, Österr. Arbeitsgruppe fuer Mustererkennung, pages 16–51. Oldenbourg, 1986. Band 36.

[88] Kropatsch, W. G.: Elimination von "kleinen" Kurvenstücken in der $2 \times 2/2$ Kurvenpyramide. In E. Paulus, editor, *Mustererkennung 1987*, Informatik Fachberichte 149, pages 156–160. Springer Verlag, 1987.

[89] Kropatsch, W. G.: Rezeptive Felder in Bildpyramiden. In H. Bunke, O. Kübler, and P. Stucki, editors, *Mustererkennung 1988*, Informatik Fachberichte 180, pages 333–339. Springer Verlag, 1988.

[90] Kropatsch, W. G.: Wechselwirkungen zwischen Visualisierung und Bildanalyse. In W. Barth, editor, *Visualisierungstechniken und Algorithmen*, pages 99–108. Springer Verlag, 1988. Informatik Fachberichte 182.

[91] Kropatsch, W. G. and Herbst, H.: Effiziente Störungsreduktion in der Gauß-Pyramide. In F. Pichler and A. Pinz, editors, *Statistik und Mustererkennung*, OCG-Schriftenreihe, Österr. Arbeitsgemeinschaft fuer Mustererkennung, pages 22–47. Oldenbourg, 1988. Band 42.

[92] Kropatsch, W. G. and Paar, G.: Aufbau einer Pyramide auf Radarbildern. In G. Pernul and A. M. Toja, editors, *Berichte aus Informatikforschungsinstitutionen*, volume Band 37 of *OCG-Schriftenreihe*, pages 221–232, Wien–München, 1987. R. Oldenburg.

[93] Kropatsch, W. G. and Tockner, H.: Detecting the straightness of digital curves in $\mathcal{O}(n)$ steps. *Computer Vision, Graphics, and Image Processing*, Vol. 45(No. 1):pp.1–21, January 1989.

[94] Leung, M. K. and Huang, T. S.: Detecting wheels of vehicles in stereo images. In *Proc. 10th International Conference on Pattern Recognition*, pages 263–267, Atlantic City, New Jersey, USA, June 1990. IEEE Comp.Soc.Press. Vol. I.

[95] Levialdi, S.: Computer architectures for image analysis. In *Proc. 9th International Conference on Pattern Recognition*, pages 1148–1158, Rome, Italy, November 1988. IEEE Comp.Soc.

[96] Li, H., Lavin, M. A., and Le Master, R. J.: Fast Hough transform: A hierarchical approach. *Computer Vision, Graphics, and Image Processing*, Vol. 36(Nos. 2/3):pp.139–161, November/December 1986.

[97] Li, S.-X. and Loew, M. H.: Analysis and modeling of digitized straight-line segments. In *Proc. 9th International Conference on Pattern Recognition*, pages 294–296, Rome, Italy, November 1988. IEEE Comp.Soc.

[98] Lowe, D. G.: *Perceptual Organisation and Visual Recognition* . Kluwer, 1985.

[99] Mallat, S. G.: A theory for multiresolution signal decomposition: The wavelet representation. *IEEE Transactions on Pattern Analysis and Machine Intelligence*, Vol. PAMI-11(No. 7):pp. 674–693, July 1989.

[100] Maloberti, F.: Silicon implementation of multiprocessor pyramid architecture. In S. Levialdi and V. Cantoni, editors, *Pyramidal Systems for Image Processing and Computer Vision*, volume F25 of *NATO ASI Series*, pages 357–372. Springer-Verlag Berlin, Heidelberg, 1986.

[101] Marr, D.: *Vision* . Freeman, W.H., San Francisco, 1982.

[102] Matsuyama, T.: Expert systems for image processing - knowledge-based composition of image analysis processes. In *Proc. 9th International Conference on Pattern Recognition*, pages 125–133, Rome, Italy, November 1988. IEEE Comp.Soc.

[103] Matsuyama, T. and Hwang, V. S.-S.: *SIGMA, A Knowledge-Based Aerial Image Understanding System.* Plenum Press, New York, London, 1990.

[104] McKeown, D. M. and McDermott, J.: Toward expert systems for photo interpretation. In *Proc. of Trends and Applications*, pages 33–39. IEEE Comp.Soc., 1983.

[105] Meer, P.: Stochastic image pyramids. *Computer Vision, Graphics, and Image Processing*, Vol. 45(No. 3):pp.269–294, March 1989.

[106] Meer, P., Jiang, S.-N., Baugher, E. S., and Rosenfeld, A.: Robustness of image pyramids under structural perturbations. *Computer Vision, Graphics, and Image Processing*, Vol. 44(No. 3):pp.307–331, December 1988.

[107] Meer, P., Sher, C. A., and Rosenfeld, A.: The chain pyramid: Hierarchical contour processing. Technical Report TR-2072, University of Maryland, Computer Science Center, July 1988.

[108] Merigot, A., Clermont, P., Mehat, J., Devos, F., and Zavidovique, B.: A pyramidal system for image processing. In S. Levialdi and V. Cantoni, editors, *Pyramidal Systems for Image Processing and Computer Vision*, volume F25 of *NATO ASI Series*, pages 109–124. Springer-Verlag Berlin, Heidelberg, 1986.

[109] Moik, J. G.: *Digital Processing of Remotely Sensed Images* . NASA SP ;431. NASA, Washington D.C., 1980.

[110] O'Gorman, L.: An analysis of feature detectability from curvature estimation. In *Proceedings, CVPR' 88*, pages 235–240, Ann Arbor, MI, June 1988. Computer Society Conference on Computer Vision and Pattern Recognition, Computer Society Press.

[111] O'Gorman, L.: Curvilinear feature detection from curvature estimation. In *Proc. 9th International Conference on Pattern Recognition*, pages 1116–1119, Rome, Italy, November 1988. IEEE Comp.Soc.

[112] O'Gorman, L. and Sanderson, A. C.: A comparison of methods and computation for multi–resolution low– and band pass tranforms for image processing. *Computer Vision, Graphics, and Image Processing*, Vol. 37(No. 3):pp.386–401, March 1987.

[113] Paar, G.: Beibehaltung von Objekträndern in Bildpyramiden. Master's thesis, Technische Universität Graz, 1987. Diplomarbeit.

[114] Parsons, M. S.: Generating lines using quadgraph patterns. *Computer Graphics Forum*, Vol. 5:pp. 33–39, 1986.

[115] Pavlidis, T.: A hybrid vectorization algorithm. In *Proceedings of the Seventh International Conference on Pattern Recognition*, pages 490–492, Montreal, Canada, 1984.

[116] Pavlidis, T.: A vectorizer and feature extractor for document recognition. *Computer Vision, Graphics and Image Processing*, Vol.35(No.1):pp. 111–127, 1986.

[117] Peleg, S. and Federbush, O.: Custom made pyramids. In S. Levialdi and V. Cantoni, editors, *Pyramidal Systems for Image Processing and Computer Vision*, volume F25 of *NATO ASI Series*, pages 165–172. Springer-Verlag Berlin, Heidelberg, 1986.

[118] Peleg, S. and Rom, H.: Motion based segmentation. In *Proc. 10th International Conference on Pattern Recognition*, pages 109–113, Atlantic City, New Jersey, USA, June 1990. IEEE Comp.Soc.Press. Vol. I.

[119] Persoon, E., Nijholt, G., Maguire, G., and O'Brien, J.: Industrial image processing by means of an image recognition integrated system. In *Proc. 10th International Conference on Pattern Recognition*, pages 402–407, Atlantic City, New Jersey, USA, June 1990. IEEE Comp.Soc.Press. Vol. II.

[120] Pham, S.: Digital straight segments. *Computer Vision, Graphics and Image Processing*, Vol.36(No.1):pp. 10–30, 1986.

[121] Phillips, T.-Y. and Rosenfeld, A.: A method of curve partitioning using arc-chord distance. *Pattern Recognition Letters*, Vol. 5(No. 4):pp.285–288, April 1987.

[122] Pitteway, M. L. V.: Bresenham's algorithm with run-line coding shortcut. *Computer Journal*, Vol.25(No.1):pp. 114–115, 1982.

[123] Price, K.: Multi-frame feature-based motion analysis. In *Proc. 10th International Conference on Pattern Recognition*, pages 114–118, Atlantic City, New Jersey, USA, June 1990. IEEE Comp.Soc.Press. Vol. I.

[124] Ramer, U.: Extraction of line structures from photographs of curved objects. *Computer Graphics and Image Processing*, Vol.4:pp.81–103, 1975.

[125] Richards, W. and Hoffman, D. D.: CODON constraints on closed 2D shapes. *Computer Vision, Graphics and Image Proceesing*, Vol.31:pp.265–281, 1985.

[126] Ronse, C.: A simple proof of Rosenfeld's characterization of digital straight line segments. *Pattern Recognition Letters*, Vol.3:pp.323–326, September 1985.

[127] Rosenfeld, A.: Digital straight line segments. *IEEE Transactions on Computers*, Vol. C-23(No.12):pp.1264–1269, December 1974.

[128] Rosenfeld, A.: Generalized cellular automata. In M. Onoe, K. J. Preston, and A. Rosenfeld, editors, *Real-Time/Parallel Computing, Image Analysis*, pages 63–71. Plenum Press, New York, 1981.

[129] Rosenfeld, A., editor: *Multiresolution Image Processing and Analysis* . Springer, Berlin, 1984.

[130] Rosenfeld, A.: The prism machine: an alternative to the pyramid. *J. Parallel Distributed Computing*, Vol. 2:pp.404–411, 1985.

[131] Rosenfeld, A.: Axial representations of shape. *Computer Vision, Graphics and Image Processing*, Vol. 33(No.2):pp.156–173, Feb. 1986.

[132] Rosenfeld, A.: Some pyramid techniques for image segmentation. In S. Levialdi and V. Cantoni, editors, *Pyramidal Systems for Image Processing and Computer Vision*, volume F25 of *NATO ASI Series*, pages 261–271. Springer-Verlag Berlin, Heidelberg, 1986.

[133] Rosenfeld, A.: A note on shrinking and expanding operations in pyramids. *Pattern Recognition Letters*, Vol. 6(No. 4):pp.241–244, September 1987.

[134] Rosenfeld, A.: Recognizing unexpected objects: A proposed approach. *International Journal of Pattern Recognition and Artificial Intelligence*, Vol. 1(No.1):pp.71–84, April 1987.

[135] Rosenfeld, A. and Kak, A. C.: *Digital Picture Processing* , volume Vol. 1 and 2. Academic Press, New York, second edition, 1982.

[136] Samet, H.: Data structures for quadtree approximation and compression. *Communications of the ACM*, Vol. 28(No. 9):pp.973–993, September 1985.

[137] Samet, H. J.: The quadtree and related hierarchical data structures. *Computing Surveys 16*, pages pp.187–260, 1984.

[138] Samet, H. J., Shaffer, C. A., and Webber, R. E.: The segment quadtree: a linear quadtree-based representation for linear features. In *Proceedings of IEEE Comp.Soc. Conf. on Computer Vision and Pattern Recognition*, pages 385–389, San Francisco, CA, June 10-13 1985.

[139] Samet, H. J. and Webber, R. E.: Storing a collection of polygons using quadtrees. *ACM Transactions on Graphics*, Vol. 4:pp.182–222, 1984.

[140] Sanz, J. L. and Cypher, R. E.: Algorithms for massively parallel image processing architectures. In *Proc. 9th International Conference on Pattern Recognition*, pages 412–419, Rome, Italy, November 1988. IEEE Comp.Soc.

[141] Schaefer, D. H. and Ho, P.: Counting on the GAM pyramid. In S. Levialdi and V. Cantoni, editors, *Pyramidal Systems for Image Processing and Computer Vision*, volume F25 of *NATO ASI Series*, pages 125–132. Springer-Verlag Berlin, Heidelberg, 1986.

[142] Serra, J.: *Image Analysis and Mathematical Morphology* . Academic Press Inc., New York, 1982.

[143] Shneier, M.: Two hierarchical linear feature representations: Edge pyramids and edge quadtrees. *Computer Graphics and Image Processing*, Vol. 17:pp.221–224, 1981.

[144] Shneier, M.: Extracting linear features from images using pyramids. *IEEE Transactions on Systems, Man, and Cybernetics*, Vol. SMC-12:pp.569–572, 1982.

[145] Stout, Q. F.: Hypercubes and pyramids. In S. Levialdi and V. Cantoni, editors, *Pyramidal Systems for Image Processing and Computer Vision*, volume F25 of *NATO ASI Series*, pages 75–90. Springer-Verlag Berlin, Heidelberg, 1986.

[146] Tanimoto, S. L.: Programming techniques for hierarchical parallel image processors. In K. J. Preston and L. Uhr, editors, *Multicomputers and Image Processing - Algorithms and Programs*, pages 421–429. Academic Press, New York, 1972.

[147] Tanimoto, S. L.: Paradigms for pyramid machine algorithms. In S. Levialdi and V. Cantoni, editors, *Pyramidal Systems for Image Processing and Computer Vision*, volume F25 of *NATO ASI Series*, pages 173–194. Springer-Verlag Berlin, Heidelberg, 1986.

[148] Tanimoto, S. L.: From pixels to predicates in pyramid machines. In J.-C. Simon, editor, *Proceedings of the COST-13 workshop 'From the Pixels to the Features'*. Bonas, France, August 1988.

[149] Tanimoto, S. L. and Klinger, A., editors: *Structured Computer Vision: Machine Perception through Hierarchical Computation Structures.* Academic Press, New York, 1980.

[150] Truvé, S. and Richards, W.: From Waltz to Winston (via the connection table). In *Proceedings of the First International Conference on Computer Vision*, pages 393–404, London, England, 1987.

[151] Tsotsos, J. K.: A 'Complexity Level' Analysis of Vision. In *Proceedings of the First International Conference on Computer Vision*, pages 346–355, London, England, 1987.

[152] Uhr, L.: and "Recognition Cones," some test results; the imminent arrival of well-structured parallel-serial computers; positions, and positions on positions. In A. R. Hanson and E. M. Riseman, editors, *Computer Vision Systems*, pages 363–377. Academic Press, Inc., 1978.

[153] Uhr, L.: Parallel, hierarchical software/hardware pyramid architectures. In S. Levialdi and V. Cantoni, editors, *Pyramidal Systems for Image Processing and Computer Vision*, volume F25 of *NATO ASI Series*, pages 1–20. Springer-Verlag Berlin, Heidelberg, 1986.

[154] Uhr, L. and Schmitt, L.: The several steps from icon to symbol using structured cone/pyramids. In A. Rosenfeld, editor, *Multiresolution Image Processing and Analysis*, pages 86–100. Springer Verlag, Berlin, Heidelberg, New York, Tokyo, 1984.

[155] van der Wal, G. S. and Sinniger, J. O.: Real time pyramid transform architecture. In *Intelligent Robots and Computer Vision*, pages 300–305, 1985. SPIE Vol.579.

[156] Weems, C. C., Rana, D., Hanson, A. R., Riseman, E. M., Shu, D. B., and Nash, J. G.: An overview of architecture research for image understanding at the university of massachusetts. In *Proc. 10th International Conference on Pattern Recognition*, pages 379–384, Atlantic City, New Jersey, USA, June 1990. IEEE Comp.Soc.Press. Vol. II.

[157] Wells III, W.: Efficient synthesis of Gaussian filters by cascaded uniform filters. *IEEE Trans. Pattern Analysis and Machine Intelligence*, PAMI–8(Nb. 2):pp.234–239, March 1986.

[158] Weng, J., Huang, T. S., and Ahuja, N.: Estimating motion and structure from line matches: Performance obtained and beyond. In *Proc. 10th International Conference on Pattern Recognition*, pages 168–172, Atlantic City, New Jersey, USA, June 1990. IEEE Comp.Soc.Press. Vol. I.

[159] Williams, T. D.: Image understanding tools. In *Proc. 10th International Conference on Pattern Recognition*, pages 606–610, Atlantic City, New Jersey, USA, June 1990. IEEE Comp.Soc.Press. Vol. II.

[160] Wu, X. and Rokne, J. G.: Double-step incremental generation of lines and circles. *Computer Vision, Graphics and Image Processing*, Vol.37(No.3):pp. 331–344, March 1987.

[161] Yuille, A. L. and Poggio, T. A.: Scaling theorems for zero crossings. *IEEE Transactions on Pattern Analysis and Machine Intelligence*, Vol. 8(No.1):pp.15–25, January 1986.

[162] Zhang, Z. and Faugeras, O.: Building a 3d world model with a mobile robot: 3d line segment representation and integration. In *Proc. 10th International Conference on Pattern Recognition*, pages 38–42, Atlantic City, New Jersey, USA, June 1990. IEEE Comp.Soc.Press. Vol. I.

GEOMETRIC MODELING ASPECTS on VISION and ULTRASONIC RANGE SENSING for ROBOTIC APPLICATIONS

Urbano Nunes; Jorge Dias; A. Traça de Almeida
Departamento de Engenharia Electrotécnica - Universidade de Coimbra
Largo Marquês de Pombal - 3000 Coimbra - PORTUGAL

Abstract

In this paper, geometric modelling aspects related with the use of a pair of vision cameras and ultrasonic sensors mounted on a six degrees-of-freedom manipulator are described. The information for both types of sensors is used to generate the depth map of the environment. An approach is described to obtain the transformation which relates a coordinate frame associated with an ultrasonic system and the tool coordinate frame. Geometrical aspects related to the use of the ultrasonic system in tasks such as the evaluation of planar-surface equations and manipulator surface following are described, as well. In another section, a method used to calibrate the vision cameras and its optimization by recalibration is presented. The recalibration procedure is made by exploiting the capability of positioning the stereo set in different positions within the workspace.

Keywords: ultrasonic sensing, calibration, parameters estimation, vision

I. Introduction

An improvement on robot capabilities requires an increase in its processing and capabilities concerning information from its surroundings. In this way, robotics research on areas such as sensory integration and sensor fusion has been receiving increasing attention. This paper describes geometric modelling aspects related with the use of a pair of stereo cameras and ultrasonic sensors mounted on the last link of a six degrees-of-freedom manipulator (Puma 560). This work is an initial stage of a wider project, having in view the use and integration of the information from these sensors in robotic tasks, such as: sensor-based path control and environmental map construction.

Non-contact sensing is important in multiple robotic applications. Among the various sensors which belong to this class are the ultrasonic sensors, whose application in robotics has been considerable [3, 4, 7, 8, 17, 18, 20, 26] since they are inexpensive, and they give range information and some insight on surface orientation. As regards ultrasonic sensing, this paper describes approaches with the aim at enabling its use in robotic tasks such as: object detection and localization, surface description, surface inspection, sensor-based path control, etc. The paper describes approaches for the following processes:

1) Evaluation of the coordinate frame associated with the ultrasonic system mounted on the manipulator;
2) Evaluation of planar-surface normals;
3) Surface following task specification.

NATO ASI Series, Vol. F 71
Expert Systems and Robotics
Edited by T. Jordanides and B. Torby

As regards vision sensing, it is focused on the calibration of a stereo pair of cameras attached to the gripper`s manipulator. A recalibration process, made by exploiting the capability to position the stereo set in different points within the manipulator workspace, is emphasized. Vision is increasingly being used in manipulators and mobile robots. More recently, it has been pointed out that the camera calibration is one of the important problems that needs to be solved for an efficient use of vision [9, 11, 13, 16]. The camera calibration problem, and in particular the fixed camera variant, has been widely addressed. The techniques developed for the fixed problem may be divided into five major groups: 1)techniques using nonlinear optimization; 2) techniques using the computation of the perspective matrix using linear equations; 3) two-plane method; 4) techniques based on geometrical characteristics; 5) the radial-alignment-constraint method (RAC method) [23]. For a comparative survey of the different methods see [23]. In our work we exploit an adaptative calibration with the aim of achieving a dynamic adjustment of the calibration parameters by positioning the stereo set in different points. Recalibration techniques have been implemented using recursive methods and by appealing to dynamic estimation, such as Kalman filtering [9, 16]. In our case, we exploit a recursive least-squares estimation [19] for the recalibration problem. We use, as the initial condition for the recursive process, an initial calibration based on the RAC method.

In this text we follow [10, 21] notations. Thus, we represent tridimensional space points as a column Cartesian 3-vector or column homogeneous 4-vector and transformations (and coordinate systems) as column homogeneous 4x4 matrices.

I.1 Considerations about radiation and reception of sound waves

Analytical expressions for the sound near and far-field pressure produced by a rigid circular piston source are well known [14]. The expressions for the sound pressure $p(r,\theta,t)$ and for the sound intensity I at a point $\vec{p}$ in the far-field (see Figure 1a), are , respectively,

$$p(r,\theta,t) = j\frac{\rho_0 c}{2}U_0\frac{a}{r}ka\,e^{j(wt-kr)}\left[\frac{2J_1(ka\sin\theta)}{ka\sin\theta}\right] \tag{1.1}$$

and

$$I = \frac{\rho_0 c U_0^2 (ka)^2 a^2}{8r^2}\left[\frac{2J_1(ka\sin\theta)}{ka\sin\theta}\right]^2 \tag{1.2}$$

In the above expressions, ρ_0 and c denote, respectively, the density and the velocity of the sound in the transmission medium, in our case the air. The variables U_0, k and **a** denote, respectively, the particle velocity at the source, the wave number of the sound field, and the radius of the piston. All the angular dependence, as can be seen from equations (1.1) and (1.2), is defined by the bracketed terms. This dependency, which is described in terms of Bessel functions, achieves its maximum value, the unit value, for $\theta=0$. We can conclude from equations (1.1) and (1.2) the following: the sound intensity, at the points $\theta=0$, as a function of the distance **r**, follows an inverse square law; for other points, besides that factor of decrease, the intensity still decreases with the increase of the angle θ.

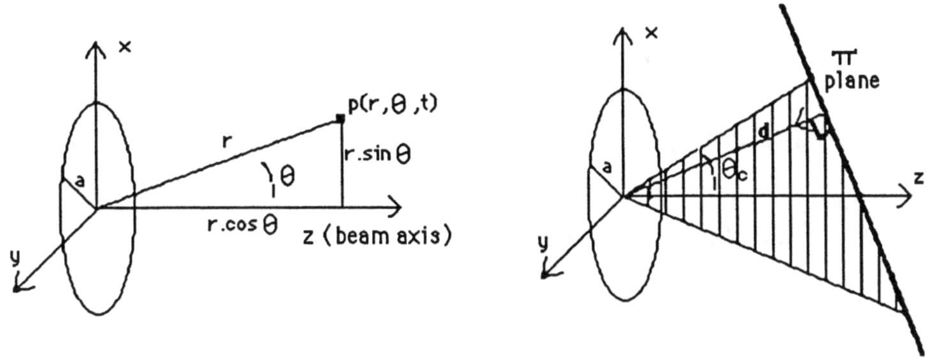

Fig.1. a) Geometry used in deriving the radiation characteristics of a circular piston [14]. b) Geometry of the distance d measured by the ultrasonic sensor, along the normal to the plane.

To analyze the echo picked up by the ultrasonic transducer it is necessary to take into account other factors. Thus, the intensity picked up still depends on the shape of the reflecting surface and on the components responsible for the absorption of the sound either due to the medium (humidity, dust content of the air, temperature, circulating currents, etc.) or due to the structure of the reflecting surface (texture, etc.). However, considering a homogeneous environment, and considering the frequency of the sound wave constant it is plausible to evaluate approximated equations relating the orientation angle (angle between the transducer`s line of sight and the normal of the reflecting surface) with the distance **r** and the echo signal picked up, that is

$$\theta=f(r,s_{echo})$$

This kind of information might be useful, as a complement to the distance information, in the application of some of the techniques described in the next sections. In these techniques, the distance d measured by the sensor is always considered the distance from the center of the transducer to the planar surface (distance taken along the surface normal), as it is illustrated in the Figure 1b.

II. Ultrasonic-sensor coordinate frame

In this section, an approach to determine the position and orientation of an ultrasonic sensor with respect to the {TOOL} frame is described. The sensor is mechanically attached to the last link of the manipulator. As the sensor measurements are unidimensional, it is enough to relate the measured distances to one coordinate axis, say the z axis of a coordinate frame, which can be assigned to the axis of the ultrasonic beam. The x and y directions can be arbitrarily assigned since form an orthonormal coordinate system with the z axis. Thus, it is only necessary to determine the z axis and the origin point of the sensor coordinate system , say {S}. These two unknowns can be solved for separately.The z axis solution can be easily obtained by using the following geometry: the axis of the ultrasonic beam placed normal to a planar surface Π, whose normal is $\vec{\mathbf{n}}_{\Pi}=\begin{bmatrix}\pi_1 & \pi_2 & \pi_3 & 0\end{bmatrix}^T$, is known with respect to a known coordinate frame, say {BASE} frame. This geometry can be achieved by analyzing the echo signal picked up by the transducer during the operation of positioning. For a given distance, the echo intensity achieves its maximum when the axis of the ultrasonic beam is normal to the surface. For this geometry, the vector $\vec{r}_3$, which represents the z axis of {S} in {TOOL} coordinates, and the normal to the plane $\vec{n}_\pi$ are in opposite directions, that is,

$$^{BASE}\vec{n}_\pi = -\,^{BASE}T_{TOOL}\begin{bmatrix}\vec{r}_3\\0\end{bmatrix} = -\begin{bmatrix}\vec{n} & \vec{s} & \vec{a} & \vec{t}\\0 & 0 & 0 & 1\end{bmatrix}\begin{bmatrix}\vec{r}_3\\0\end{bmatrix} \tag{2.1}$$

which gives,

$$\begin{cases}\pi_1 = -(n_x r_{13} + s_x r_{23} + a_x r_{33})\\ \pi_2 = -(n_y r_{13} + s_y r_{23} + a_y r_{33})\\ \pi_3 = -(n_z r_{13} + s_z r_{23} + a_z r_{33})\end{cases} \tag{2.2}$$

that is, the solution for the z axis of the sensor coordinate frame. In order to obtain the origin of {S} that is a solution for the vector $\vec{l}$, we impose another condition on the above sensor-surface geometry. That is, we assume that a point $\vec{p}$ of the surface is known with respect, say, to the {BASE} frame, and that the transducer is pointing to this point (still normal to the surface). This situation can be achieved, with a certain uncertainty, using a planar object placed over the surface (of course that object should be carefully and experimentally selected). For this geometry, the vector $\vec{p}$ in {S} coordinates, is given by

$$^{S}\vec{p} = [0 \quad 0 \quad d \quad 1]^T,$$

where d is the distance measured by the sensor. Thus, the vector $\vec{p}$ can be expressed in {BASE } coordinates as,

$$^{BASE}\vec{p} = {}^{BASE}T_{TOOL}\;{}^{TOOL}T_S\;{}^{S}\vec{p} = \begin{bmatrix}\vec{n} & \vec{s} & \vec{a} & \vec{t}\\0 & 0 & 0 & 1\end{bmatrix}\begin{bmatrix}\vec{r}_1 & \vec{r}_2 & \vec{r}_3 & \vec{l}\\0 & 0 & 0 & 1\end{bmatrix}{}^{S}\vec{p} \tag{2.3}$$

That yields the following set of equations,

$$\begin{cases}x = (n_x r_{13} + s_x r_{23} + a_x r_{33})d + n_x l_x + s_x l_y + a_x l_z + t_x\\ y = (n_y r_{13} + s_y r_{23} + a_y r_{33})d + n_y l_x + s_y l_y + a_y l_z + t_y\\ z = (n_z r_{13} + s_z r_{23} + a_z r_{33})d + n_z l_x + s_z l_y + a_z l_z + t_z\end{cases} \tag{2.4}$$

where the only unknowns are the elements of the vector $\vec{l} = [l_x \quad l_y \quad l_z]^T$ which represents the origin of {S} with respect to the {TOOL} frame. As we mentioned, there is an uncertainty associated with the normal and the point $\vec{p}$ of the plane (due both processes of determination and of calibration). A possible way to minimize that uncertainty may be by using several known points $\vec{p}_i$ of different planes. In this case it is possible to apply a least-squares estimation: 1) in the evaluation of the orientation of the frame using several planes; 2) in the evaluation of the origin of the frame using several points.

III. Evaluation of planar-surface equations by using ultrasonic sensors

The solution that follows for a planar surface consists of the evaluation of three points on the surface obtained by positioning the ultrasonic sensor at three different spatial positions (stations). For that positioning, the manipulator is used, and the following conditions are assumed: 1) the transducer is mounted on the last link of the manipulator; 2) the transformation which relates the sensor position and orientation to the {TOOL} frame is known.The z axis of the sensor frame {S}, should maintain the same direction with respect to the {BASE} coordinate frame for the three spatial positions achieved by the sensor. Under this condition, the angle θ, angle between the surface normal and the axis of the ultrasonic beam, will maintain a constant value for the different sensor positions. Following this strategy, on one hand, we obtain a simplicity for the specification of the trajectory with the guaranty that the condition yields $\theta<\theta_c$ (θ_c being the critical angle value from which there is no echo reception). On the other hand, there is uniformity on the distances measured given the constancy of the θ value. Considering that the measures are taken under the geometry described, see Figure 3.1, the points $\vec{p}_i$ (i=1,2,3) of the surface with respect to the {S} frame are expressed by

$$^{S}\vec{p}_i = \begin{bmatrix} 0 & 0 & \dfrac{d_i}{\cos\theta} & 1 \end{bmatrix}^T . \tag{3.1}$$

where d_i (i=1,2,3) represents the distance measured by the sensor at station i and θ represents the angle between the surface normal and the direction defined by the axis of the ultrasonic beam.

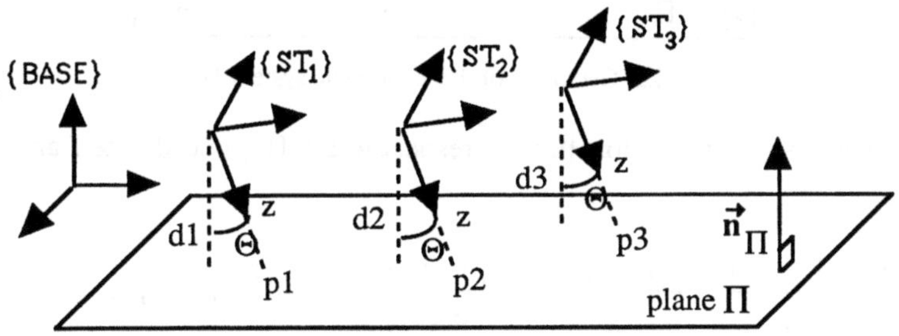

Fig.3.1 - Geometry of the sensor positioning at the three stations relatively to the plane Π. The measured distances are d1, d2 and d3.

Applying the transformation $^{BASE}\mathbf{T}_{ST_i} = \left(^{BASE}\mathbf{T}_{TOOL}\right)_i {}^{TOOL}\mathbf{T}_S$ to equation (3.1) yields,

$$^{BASE}\vec{p}_i = {}^{BASE}\mathbf{T}_{ST_i} \; {}^{ST_i}\vec{p}_i \tag{3.2}$$

this can be rewritten as,

$$^{BASE}\vec{p}_i = \begin{bmatrix} \left(n_{x_i} r_{13} + s_{x_i} r_{23} + a_{x_i} r_{33}\right)\dfrac{d_i}{\cos\theta} + n_{x_i} l_x + s_{x_i} l_y + a_{x_i} l_z + t_{x_i} \\ \left(n_{y_i} r_{13} + s_{y_i} r_{23} + a_{y_i} r_{33}\right)\dfrac{d_i}{\cos\theta} + n_{y_i} l_x + s_{y_i} l_y + a_{y_i} l_z + t_{y_i} \\ \left(n_{z_i} r_{13} + s_{z_i} r_{23} + a_{z_i} r_{33}\right)\dfrac{d_i}{\cos\theta} + n_{z_i} l_x + s_{z_i} l_y + a_{z_i} l_z + t_{z_i} \\ 1 \end{bmatrix} \tag{3.3}$$

In case there is one estimative for the angle θ, the set of equations (3.3) (with i=1,2,3) gives us the necessary data to establish the equation of the planar surface Π[22].

In some particular conditions it may be possible to estimate values for the angle θ (for a homogeneous environment, and provided that: the planar-surface dimensions and the position of the transducer relative to the surface is such that the totality of the emission cone intercepts the surface and there is no place for interferences of the sound wave due to any object). However, for the majority of the cases those conditions do not occur. In view of that, the above solution is too particular and it is necessary to exploit alternative solutions where no dependency on θ must be the goal.

A solution of this type is obtained by imposing a new restriction on the trajectory to be followed by the manipulator in order to position the ultrasonic sensor at the three stations. In the previous solution, the stations were allowed to define parallel planes whose normal was defined by the z direction of the frame {S}. In the new solution, the stations are constrained to be placed only on one of those planes. For this condition, it is possible to define a frame, say {U}, as illustrated in the Figure 3.2, such that its origin coincides with the station 1 origin, station 2 lies on the x axis, station 3 lies somewhere on the x-y plane, and the direction of the z axis of {U} coincides with the direction of the z axis of the three stations.

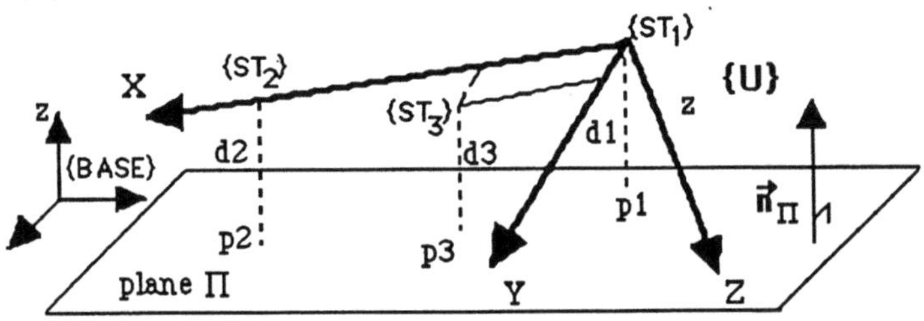

Fig.3.2 Geometry of the coordinate system {U}.

Given this geometry, the vector $\vec{\mathbf{p}}_1$ with respect to the {U} coordinate frame is known [3,4],

$$\vec{\mathbf{p}}_1 = \begin{bmatrix} \frac{d_1(d_1-d_2)}{S_{2x}} & \frac{d_1(d_1-d_3)}{S_{3y}} - \frac{S_{3x}d_1(d_1-d_2)}{S_{2x}S_{3y}} & \sqrt{d_1^2 - p_{1x}^2 - p_{1y}^2} \end{bmatrix}^T \tag{3.4}$$

where S_{2x},S_{3x} and S_{3y}, p_{1x} and p_{1y} represent, respectively, the x coordinate of the origin of the station 2, the x and y coordinates of the origin of station 3 and the x and y components of the vector $\vec{\mathbf{p}}_1$. With $\vec{\mathbf{p}}_1$ known, it is straightforward to establish the equation for the plane Π with respect to the coordinate frame {U},

$$^U\Pi = [\pi_1 \quad \pi_2 \quad \pi_3 \quad \pi_4] \qquad \text{with} \quad \begin{cases} \pi_1 = n_{\pi_x} \\ \pi_2 = n_{\pi_y} \\ \pi_3 = n_{\pi_z} \\ \pi_4 = |\vec{\mathbf{p}}_1| \end{cases} \tag{3.5}$$

whose normal is

$$\vec{\mathbf{n}}_\pi = -\frac{\vec{\mathbf{p}}_1}{|\vec{\mathbf{p}}_1|} = -\frac{\vec{\mathbf{p}}_1}{d_1} \tag{3.6}$$

By applying the transformation ${}^{BASE}T_U$, we obtain the plane equation with respect to the {BASE} frame, that is

$$ {}^{BASE}\Pi = {}^{U}\Pi . {}^{BASE}T_U^{-1} \tag{3.7} $$

The solution for the vector $\vec{p}_1$, and thus for the plane, is still more compact in the case where station 3 lies on the y axis of the frame {U} as we can conclude from equation (3.4) (equalizing S_{3x} to zero).

By using three sensors simultaneously, placed according to a geometry that defines a frame {U}, the planar surface equation is obtained by only one spatial position of the manipulator. This sensor arrangement will still allow the exploitation of control strategies for path control. This is possible by exploiting the fact that this sensor system gives us enough information to reorient the x-y plane of {U} to become parallel with the planar surface to be followed. Besides, with all sensors oriented normal to the plane, the plane equation can be estimated by reading only one sensor.

IV. Geometric considerations of the surface-following task

Let {U} be the coordinate system associated with a three sensor system as defined in the previous section. The x-y plane of {U} is parallel with the planar surface Π when all the sensor measurements are equal. Thus, equation (3.4), becomes:

$$ \vec{p}_1 = [0 \quad 0 \quad d_1]^T \tag{4.1} $$

This state may be accomplished by doing two rotations to cancel the p_{1x} and p_{1y} components. From Figure 4.1 the following equations can be written for the p_{1x} and p_{1y} components,

$$ \begin{cases} p_{1x} = d_1 \sin\phi_1 \\ p_{1y} = d_1 \sin\phi_2 \\ \text{where } d_1 = |\vec{p}_1| \end{cases} \tag{4.2} $$

where: 1) ϕ_1 is the angle between the z axis and the $\vec{p}_1$ vector after being rotated around the x axis onto the Z-Y plane; 2) ϕ_2 is the angle between the z axis and the $\vec{p}_1$ vector after being rotated around the y axis onto the X-Y plane. Another possible analysis for the p_{1x} and p_{1y} components may be done by rotating the $\vec{p}_1$ vector onto the X-Y plane (see Figure 4.1.b) as follows: 1) around the x axis to obtain the p_{1x} component; 2) around the y axis to obatin the p_{1y} component. The angle ϕ_3 in Figure 4.1.b is related with the p_{1z} component. With the equations (3.4) and (4.2) we obtain the following equations for the angles ϕ_1 and ϕ_2,

$$ \begin{cases} \sin\phi_1 = \dfrac{d_1 - d_2}{S_{2x}} \\ \sin\phi_2 = \dfrac{1}{S_{3y}}\left((d_1 - d_3) - \dfrac{S_{3x}}{S_{2x}}(d_1 - d_2) \right) \end{cases} \tag{4.3} $$

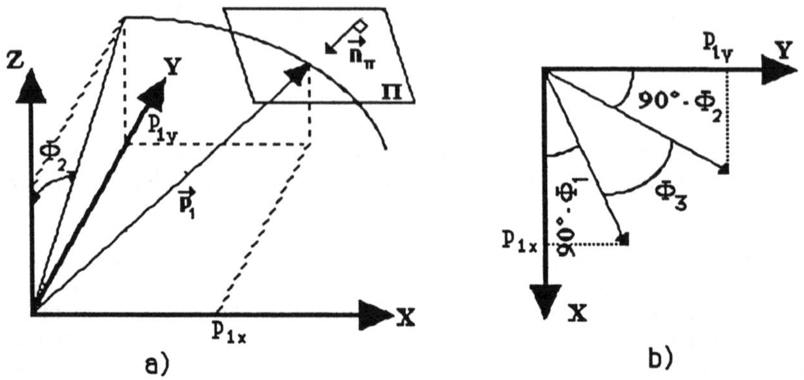

Fig. 4.1 Illustration of the angles a) ϕ_2 b) ϕ_1, ϕ_2 and ϕ_3

As Wampler [27] pointed out, and as it can be concluded from the above considerations, the surface following task may be accomplished by controlling three degrees of freedom of the task space. If we consider a velocity control, the task is accomplished by controlling the angular velocities about the x and y axis, and the linear velocity along the z axis (in our case) with respect to the {U} frame. The remaining three degrees of freedom are left free to be program specified.

If a velocity control is chosen, a damping control scheme [28], as illustrated in Figure 4.2, can be applied.

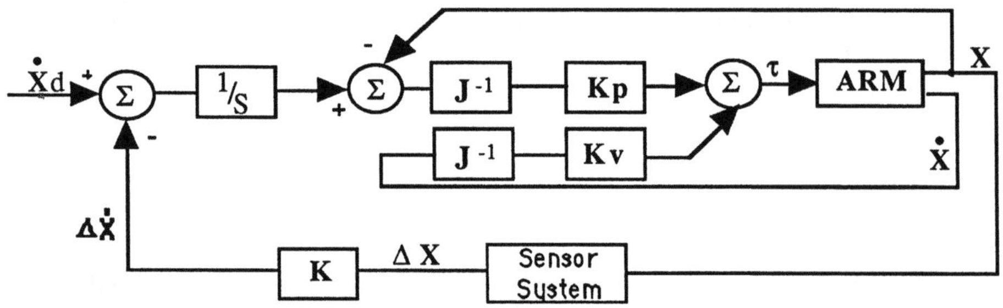

Fig.4.2 Damping control (non-contact)

In the diagram the following designations are used: **x** and $\dot{\mathbf{x}}$ are 6x1 position and velocity vectors in Cartesian coordinates (in our case in {U} coordinates, for instance); $\Delta\mathbf{x}$ is a 6x1 position error vector in Cartesian coordinates; $\dot{\mathbf{x}}_d$ is a 6x1 desired velocity vector in Cartesian coordinates; $\Delta\dot{\mathbf{x}}$ is a 6x1 correction velocity vector in Cartesian coordinates; τ is a 6x1 joint-torque vector; and **J** is the arm's Jacobian. To implement damping control, we must find out the 6x6 **K** matrix coefficients, which should be selected taking into account:

1) task constraints;
2) dynamic constraint.

As regards the item 1 for the surface following task, the selection criteria is accomplished by using the previous geometrical results. That is, the X-Y plane of {U} become parallel with the plane Π by accomplishing the following two rotations:

$$1)\ \text{a rotation about the y axis,}\ \Delta\phi_y = \phi_1 = \sin^{-1}\left(\frac{p_{1x}}{d_1}\right) \qquad (4.4)$$

$$2)\ \text{a rotation about the x axis,}\ \Delta\phi_x = -\phi_2 = -\sin^{-1}\left(\frac{p_{1y}}{d_1}\right) \qquad (4.5)$$

Finally, to position the X-Y plane at a distance d_d above the planar surface a translation Δx_z should be performed,

$$3)\ \Delta x_z = -(d_d - d_1) \qquad (4.6)$$

along the z direction of the {U} coordinate system. Thus, we may define a correction velocity vector, as

$$\Delta\dot{\mathbf{X}} = \mathbf{K}\,\Delta\mathbf{X} = \begin{bmatrix} 0 & 0 & k_{33}\Delta x_z & k_{44}\Delta\phi_x & k_{55}\phi_y & 0 \end{bmatrix}^T \qquad (4.7)$$

where

$$\Delta\mathbf{X} = \begin{bmatrix} 0 & 0 & \Delta x_z & \Delta\phi_x & \Delta\phi_y & 0 \end{bmatrix}^T \qquad (4.8)$$

is the position error vector.

To select the three K gains which must guaranty the system stability, the real-time behavior of the system must be analysed. For that purpose, we follow a similar system model and discrete analysis as that described by Wampler [27] where the dynamic behavior of the manipulator and the serial processing are modeled by a set of difference equations.

V. Camera calibration

The main problem in camera calibration is to find out, with a reasonable accuracy, the relative transformation between the 2D images and the reference of the tri-dimensional world. The set of the cameras is placed on the last link of the manipulator and the reference associated with it is known as TOOL. The transformations obtained by our calibration process are $^{TOOL}\mathbf{CALIB}_{CAM\_LEFT}$ and $^{TOOL}\mathbf{CALIB}_{CAM\_RIGHT}$ which give the relations between the references of the right and left images, respectively CAM_LEFT and CAM_RIGHT, and the 3D reference TOOL. That process also offers the transformations between the 3D references of the cameras and the reference of the last link, respectively $^{TOOL}\mathbf{T}_{CAM\_LEFT}$ and $^{TOOL}\mathbf{T}_{CAM\_RIGHT}$.

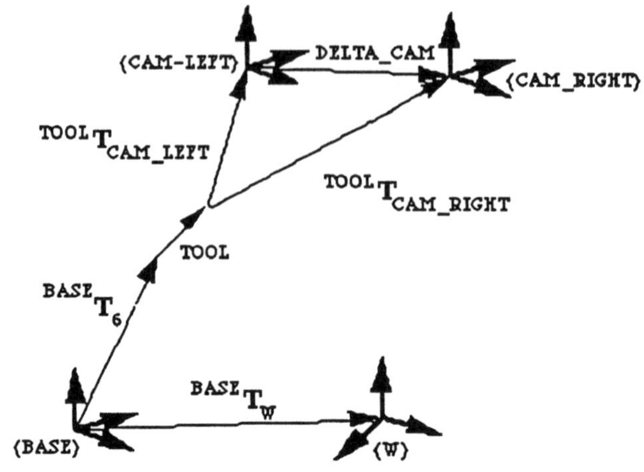

Fig. 5.1- Relation between the different references used in the system

For evaluation of ${}^{TOOL}\mathbf{CALIB}_{CAM\_LEFT}$ and ${}^{TOOL}\mathbf{CALIB}_{CAM\_RIGHT}$, we analyze the left and right images of a grid of points whose coordinates in {W} and implicitly in {TOOL} are known. By using the correspondence between the points in the 3D scene and the corresponding points in the 2D image we can write a set of equations using the model for image formation as described below. That model follows the approach presented in [23] for the image formation, and uses the same relations between the different stages of the image formation.

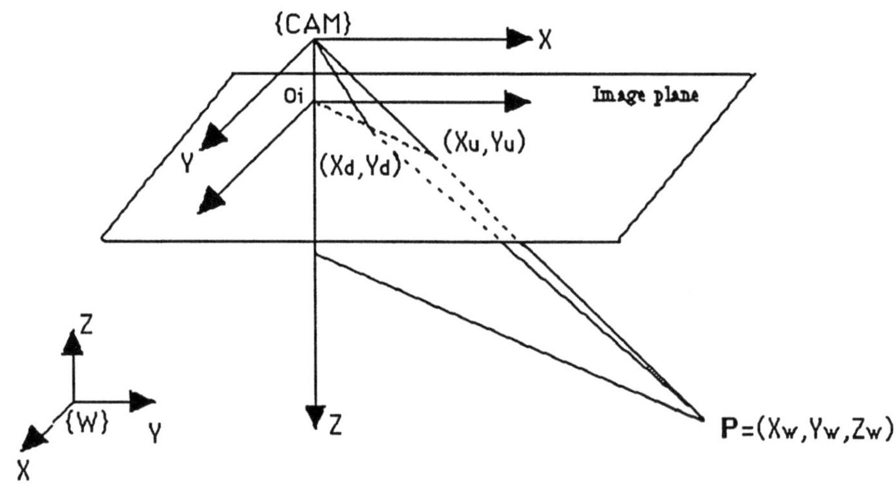

Fig.5.2 - Camera model

A point **p** in the tri-dimensional space is represented in reference {W} by (x_w, y_w, z_w) and in reference {CAM} by (x,y,z). The origin of the {CAM} reference is coincident with the center of the lens. The image plane is parallel with the X-Y plane of the {CAM} reference with an f coordinate corresponding to the focal length along the z axis.

In that image plane, a point **p** of the real world is represented by its projection, the (x_u, y_u) coordinates neglecting the lens distortion. Taking the distortion created by the lens as being radial, the projection changes from (x_u, y_u) to a point (x_d, y_d).Thus, transformation between the reference {W} and the 3D reference {CAM} is given by:

$$\begin{bmatrix} x \\ y \\ z \end{bmatrix} = \mathbf{Rot} \begin{bmatrix} x_W \\ y_W \\ z_W \end{bmatrix} + \mathbf{Trans} \tag{5.1}$$

where (x,y,z) and (x_W, y_W, z_W) are the coordinates of the point **p** expressed in {CAM} and {W} respectively. **Rot** is the (3x3) matrix representing the rotation between the references {CAM} and {W} and **Trans** the (3x1) vector representing the translation between the same references.

The relation between the points in the world and its own projections in the ideal image are given by the perspective transformation:

$$x_u = f\frac{x}{z}$$

$$y_u = f\frac{y}{z} \tag{5.2}$$

The relation between points in the ideal image to the related points in the real image is given by

$$x_u = {x_d D}/{2}$$

$$y_u = {y_d D}/{2} \tag{5.3}$$

where D is the distortion factor ([15] and [23]):

$$D = \frac{2}{1 + k_1 R_d^2} \quad \text{com} \quad R_d^2 = x_d^2 + y_d^2$$

The relation of the real-image coordinates (x_d, y_d) to computer-image coordinates (x_f, y_f) is given by:

$$\begin{cases} x_f = S_x x_d + c_x \\ y_f = S_y y_d + c_y \end{cases} \tag{5.4}$$

where

S_x, S_y - are the scale factors for the x and y axes, respectively.
(c_x, c_y) - are the computer-image coordinate for the origin in the image

The set of parameters $[S_x, S_y, C_x, C_y]$ and $[\vec{r}_1, \vec{r}_2, \vec{r}_3, T_x, T_y, T_z]$ are the intrinsic and extrinsic parameters that we need to evaluate for the LEFT and RIGHT cameras. To obtain explicit relations between extrinsic and intrinsic parameters for the cameras we can use above equations (5.1), (5.2), (5.3) and (5.4). That is,

$$\begin{cases} \frac{x_f}{S_x}z - \frac{c_x}{S_x}z - \frac{k_1\,x}{S_x^2}\left(x_f^2 + c_x^2 - 2x_fc_x\right) - \frac{k_1\,x}{S_y^2}\left(y_f^2 + c_y^2 - 2y_fc_y\right) = x \\ \frac{y_f}{S_y}z - \frac{c_y}{S_y}z - \frac{k_1\,y}{S_x^2}\left(x_f^2 + c_x^2 - 2x_fc_x\right) - \frac{k_1\,x}{S_y^2}\left(y_f^2 + c_y^2 - 2y_fc_y\right) = y \end{cases} \qquad (5.5)$$

for the relation between 2D points in the computer image and the 3D points related to the system {CAM} and,

$$\begin{cases} f\frac{x}{z} = f\frac{\vec{p}^T\,\vec{r}_1 + T_x}{\vec{p}^T\,\vec{r}_3 + T_z} \\ f\frac{y}{z} = f\frac{\vec{p}^T\,\vec{r}_2 + T_y}{\vec{p}^T\,\vec{r}_3 + T_z} \end{cases} \qquad (5.6)$$

for the relation between 3D points expressed in reference {CAM} and the same points (x_W, y_W, z_W) expressed in reference {W}. The vector (T_x, T_y, T_z) represents the origin of {CAM} expressed in reference {W} and $\vec{r}_i$ is the line i of the **Rot** matrix which describes the orientation of {CAM}. The vector $\vec{p}$ represents a point (x_W, y_W, z_W) expressed in world coordinates.

The evaluation of the intrinsic parameters $[S_x, S_y, C_x, C_y]$ and the extrinsic parameters $[\vec{r}_1, \vec{r}_2, \vec{r}_3, T_x, T_y, T_z]$ is based on the model described in [23] and follows the algorithm presented.

After manipulating the equations (5.5) and (5.6), it is straightforward to obtain an expression from which we can apply recursive methods for the estimation of the cameras' parameters using the information acquired from different positions of the stereo system. This procedure gives us a dynamic form to adjust (or filter) the cameras' parameters. The mathematical technique used to implement such an estimation is based on the covariance form of the recursive least-squares method [19].

From equation (5.5) we can obtain the matrix form:

$$\mathbf{A}\cdot\vec{m} = \vec{b} \qquad (5.7)$$

where $\vec{m}$ is the vector of unknowns. The relation between the elements of $\vec{m}$ and the intrinsic parameters is given by:

$$S_x = \frac{1}{m_1} \quad e \quad S_y = \frac{1}{m_2}$$

$$c_x = \frac{m_3}{m_1} \quad e \quad c_y = \frac{m_4}{m_2}$$

$$k_1 = \frac{m_5}{m_1^2}$$

Doing the same for equation (5.6), we can obtain two matrix equations

$$\begin{bmatrix} \vec{p}^T & 0_{(1x3)} & -\alpha\,\vec{p}^T \\ 0_{(1x3)} & \vec{p}^T & -\beta\,\vec{p}^T \end{bmatrix} \begin{bmatrix} \vec{r}_1 \\ \vec{r}_2 \\ \vec{r}_3 \end{bmatrix} = \begin{bmatrix} \alpha\,T_z - T_x \\ \beta\,T_z - T_y \end{bmatrix} \quad \Rightarrow \quad \mathbf{B}\cdot\vec{n} = \vec{c} \tag{5.8}$$

and

$$\begin{bmatrix} -1 & 0 & \alpha \\ 0 & -1 & \beta \end{bmatrix} \begin{bmatrix} T_x \\ T_y \\ T_z \end{bmatrix} = \begin{bmatrix} \vec{p}^T\,\vec{r}_1 - \alpha\,\vec{p}^T\,\vec{r}_3 \\ \vec{p}^T\,\vec{r}_2 - \beta\,\vec{p}^T\,\vec{r}_3 \end{bmatrix} \quad \Rightarrow \quad \mathbf{D}\cdot\vec{t} = \vec{e} \tag{5.9}$$

with $\vec{p} = (x_W, y_W, z_W)$, $\alpha = \frac{x}{z}$, $\beta = \frac{y}{z}$, $\vec{r}_i$ the row i of the **Rot** matrix, and T_x, T_y, T_z elements of the **Trans** vector. The vectors $\vec{n}$ and $\vec{t}$ are the unknowns representing the extrinsic parameters.

The recalibration is made in two stages: first, the recalibration of the intrinsic parameters using (5.7), and second, the recalibration of the extrinsic parameters using (5.8) and (5.9).

According to Mendel [19], if a system can be represented by a linear equation:

$$\mathbf{Z}_k = \mathbf{H}_k \cdot \mathbf{X}$$

where **H** and **Z** are matrices or vectors of the k measures read respectively in the input and output signals of the system, and **X** is the vector of the system unknown parameters, a recursive solution to obtain **X** can be expressed by:

$$\hat{\mathbf{X}}_{k+1} = \hat{\mathbf{X}}_k + \mathbf{Kw}_{k+1}\left[\vec{\mathbf{z}}_{k+1} - (\mathbf{h}_{k+1})^T \hat{\mathbf{X}}_k\right]$$

where $\vec{\mathbf{z}}_{k+1}$ and $\mathbf{h}_{k+1}$ are matrices obtained using the last (k+1) measure.

The $\mathbf{Kw}_{k+1}$ coefficient is obtained by

$$\mathbf{Kw}_{k+1} = \mathbf{P}_k\;\mathbf{h}_{k+1}\left[(\mathbf{h}_{k+1})^T\;\mathbf{P}_k\;\mathbf{h}_{k+1} + 1\right]^{-1}$$

which can be understood as a gain for the k+1 iteration and where the factor $\mathbf{P}_k$ is given by

$$\mathbf{P}_k = \left[\mathbf{I} - \mathbf{Kw}_k\;(\mathbf{h}_k)^T\right]\mathbf{P}_{k-1}$$

The application of this method to the equations (5.7), (5.8) and (5.9) is immediately. The initial conditions for $\hat{\mathbf{X}}_0$ are given by a first calibration made for one position and the initial values for $\mathbf{P}_0$ are given by:

$$\mathbf{P}_0^{-1} = \mathbf{H}_0 \cdot (\mathbf{H}_0)^T$$

where k=0 signifies that no recalibrations were made.

Adapting this method to the special conditions of (5.7), (5.8) and (5.9), the calculations have the following development:

$$\begin{array}{ccccccc} \ldots & \mathbf{P}_k & \rightarrow \mathbf{Kw}_{k+1} \rightarrow & \hat{\mathbf{X}}_{k+1} & \rightarrow \mathbf{P}_{k+1} & \ldots \\ & \Downarrow & & \Downarrow & & \\ & \text{new point } k & & \text{new} & & \\ & \text{to recalibrate} & & \text{estimation} & & \end{array}$$

VI. Experimental setup

The experimental setup is based on an integrated and evolutive system in which the addition of new sensors and actuators is possible in a modular fashion. The control system exhibits an hierarchical structure following a distributed processing model [1]. To establish the communications between the Supervisor System (a 80386 CPU personal computer) and the distributed Slave Systems, a serial network in a bus topology, the Bitbus [12], is used. This network exhibits good performance in handling short messages allowing high order rates of information transmission, (up to 2.4 Mb/s in half-duplex synchronous mode). The connections to the Puma controller, the supervisory connection, and the real-time path control communication, are assured by two software modules [5, 6]. The real-time connection is performed by a 16-bit personal computer (CPU 80286 and FPU 80287) which acts as a Local Supervisor regarding the overall control system. This unit carries out the important role of allowing the implementation and evaluation of sensor-based closed-loop control strategies.

Each of the three ultrasonic sensors used has an ultrasonic ranging module with a single transmitter and receiver transducer. The transducers have a resonant frequency of about 215KHz and the ranging modules output two signals: a digital signal whose width is directly proportional to the distance, and an amplified analog echo signal.The processing of these signals is carried out by two physically separated modules. A remote analog preprocessing module, one per each transducer, is placed on the manipulator close to the transducer. Additionally an AT-bus microcontroller-based module, with the capacity of processing information from three remote modules simultaneously, is used. The analog signal is rectified and converted to digital samples using a 12-bit ADC with serial output. At the micro-based board each serial sample is converted to parallel and stored in FIFO memories. These memories enable the parallel processing up to three analog signals (from three transducers) and allow high sampling rates. The pulse width is measured with 12-bit resolution by an interval counter also at the micro-based board.

As regards to the vision sub-system, two pairs of CCD cameras from SONY, a color pair with separated RGB signals and a b/w pair, have been used (one pair at a time). The cameras are fastened to the last link of the manipulator by a special tool designed for that purpose. The two video signal are sampled and stored in memory using an AT-bus frame grabber and the images are sampled by 512x512 pixels. The programmes for the calibration process are developed in C and run on a personal computer with a 80386 CPU. Additional image processing necessary for this process uses routines from a software library developed in our Laboratory [2].

VII. Conclusions

For building a description of the tri-dimensional world the contribution and integration of information from various sensors - complementary and redundant - is required. In this paper, the calibration of a stereo pair of cameras attached to the end-effector of a six degree-of-freedom manipulator was addressed. In particular a recalibration process based on a recursive least-squares estimation was proposed. On the other hand, an approach was described concerning integration of ultrasonic range information in the robotic system. This work represents an initial stage of a wider robotics research project having in view the integration of sensing data for tasks, such as: building local and global descriptions of the 3D world, and evaluation of sensor-based path control strategies. Besides vision and proximity sensors, we are already working on contact sensing, tactile and force sensors, which is a necessary type of information giving important physical proprieties of the world.

References

1. Almeida,A.T.; Nunes,U.C.; Dias,J.M.; Araújo, H.; Batista, J.: A Distributed System for Robotic Multi-Sensor Integration. Accepted for publication in the Int.Journal of Industrial Metrology (1990)
2. Batista, J.; Dias, J.; Araújo, H.: Biblioteca de Programas para Processamento de Imagem. In: 3º Simpósio de Electrónica das Telecomunicações, Porto, 1988 (in portuguese)
3. Brown, M.K.: Locating Object Surfaces with an Ultrasonic Range Sensor. IEEE Conf.on Robotics and Automation, pp.110-115, 1985
4. Brown, M.K.: The Extraction of Curved Surface Features with Generic Range Sensors. Int.Journ.Rob.Research, vol.5,nº1, pp.3-18, 1986
5. Capote, E.; Machado,J,; Bastos, F.; Dias,J.: Utilização do DDCMP". Int.Technical Report, DEE-University of Coimbra, 1989 (in portuguese)
6. Crisóstomo, M.: Some Aspects of Surface Following with a PUMA 560 Robot. Int.Technical Report, Man.&Eng.Systems Dep., Brunel University, 1989
7. Elfes, A.: Sonar-Based Real-World Mapping and Navigation. IEEE Journal of Robotics and Aut., Vol-RA3, nº3, pp.249-265, 1987
8. Estochen, E.L.; Neuman,C.P.; Printz, F.B.: Application of Acoustic Sensors to Robotic Seam Tracking. IEEE Trans.Ind.Electronics, vol.IE-31,nº3,pp.219-224, 1984
9. Faugeras, O. D., Toscani, G.: The Calibration Problem for Stereoscopic Vision. In: Sensor Devices and Systems for Robotics, NATO ASI Series, Vol F52, Springer-Verlag 1989
10. Fu, K. S., Gonzalez, R.C., Lee, C.S.G.: Robotics: Control, Sensing, Vision, and Intelligence. McGraw-Hill 1987, ISBN 0-07-022625-3
11. Hung, Yi-Ping.: A Simple Real-Time Method for Calibrating a Camera Mounted on a Robot for Three Dimensional Machine Vision. SPIE Vol. 1005 Optics, Illumination, and Image Sensing for Machine Vision III, 1989
12. Intel Corporation.: The Bitbus Interconnect Serial Control Bus Specification. In Distributed Control Modules Databook, pp.100-147, 1984
13. Izaguirre, A.; Pu, P.; Summers, J.: A New Development in Camera Calibration: Calibrating a Pair of Mobile Cameras. Proc. Int. Conf. on Robotics and Automation, 1985
14. Kinsler, L.E.; Frey, A.R.; Coppens,A.B.; Sanders,J.V.: Fundamental of Acoustics. John Wiley & Sons, 1982, ISBN:0-471-09410-2
15. Lenz, R.K. and Tsai, R. Y.: Calibrating a Cartesian Robot with Eye-on-Hand Configuration Independent of Eye-to-Hand Relationship. IEEE PAMI, Vol .11, No 9, Sept. 1989
16. Liang, P.; Chang, Y.L.; Hackwood, S.: Adaptive Self-Calibration of Vision-Based Robot Systems. IEEE Trans. on Sys. Man and Cyb., Vol.19, Nº4, Aug. 1989

17. Luo, R.C. and Mullen Jr.: Combined Vision/Ultrasonics for Multi-Dimensional Robotic Tracking. SPIE Vol.1003 on Sensor Fusion: Spatial Reasoning and Scene Interpretation, pp.113-121, 1988
18. Maqueira, B.; Umeagukwu,C.; Jarzynski,J.: Application of Ultrasonic Sensors to Robotic Seam Tracking. IEEE Trans.Robot.Automation, Vol.5,Nº3,pp.337-343, 1989
19. Mendel, Jerry M.: Lessons in Digital Estimation Theory. Prentice-Hall Signal Processing Series, New-Jersey, 1987, ISBN 0-13-53 0809-7 025
20. Nakajima, S. and Takahashi,Y.: An Ultrasonic Orientation Sensor with Distributed Receivers. Advanced Robotics,Vol.4,Nº2, pp.151-168, 1990
21. Paul, R. P.: Robot Manipulators: Mathematics, Programming, and Control. MIT Press, Cambridge MA, 1981
22. Penna, M.A. and Patterson, R.: Projective Geometry and its Applications to Computer Graphics. Prentice-Hall, ISBN 0-13-730649-0
23. Tsai, R.Y.: A Versatile Camera Calibration Technique for High Accuracy 3D Machine Vision Metrology Using off-the-shelf TV Cameras and Lenses. IBM RC 11413, October 1985
24. Tsai, Roger Y. and Lenz, Reimar K.: Overview of a Unified Calibration Trio for Robot Eye, Eye-to-Hand, and Hand Calibration Using 3D Machine Vision. Proc. SPIE- Sensor Fusion: Spatial Reasoning and Scene Interpretation, 7-9 Nov. 1988
25. Tsai, Roger Y., and Lenz, Reimar K.: A New Technique for Fully Autonomous and Efficient 3D Robotics Hand/Eye Calibration. IEEE J. of Robotics and Automation, Vol. -5, No. 3, June 1989
26. Tsuzuki,F.S.G.; Moscato,L.A.; Adamowski, J.C.: A Sonar Based Environmental Sensory System for Mobile Robots. IEEE Workshop SIFIR'89, pp.228-233, Zaragosa, 1989
27. Wampler, C.: Multiprocessor Control of a Telemanipulator with Spatial Proximity Sensors. Int.Journal of Robotics Research, vol.3,nº1, pp.40-50,1984
28. Whitney, D.E.: Historical Perspective and State of the Art in Robot Force Control. Int.Journal of Robotics Research, vol.6,nº1, pp.3-14, 1987

Sensor Planning for the Error Diagnosis of Robot Assembly Tasks

Jörg Raczkowsky, Kerstin Seucken

Institute for Realtime Computer Systems and Robotics, Prof. Dr.-Ing. U. Rembold, Prof. Dr.-Ing. R. Dillmann, University of Karlsruhe, P.O. Box 6980, D-7500 Karlsruhe 1, FRG

ABSTRACT: A knowledge based system for the diagnosis of assembly process errors is presented. It is part of the action monitoring system of the autonomous mobile assembly robot KAMRO. The diagnosis is supported by the multi-sensory system of the KAMRO system. The system for error diagnosis has the structure of a blackboard system. The task of the sensor system is focused to the monitoring of a product assembly. A very interesting part of the system is an approach for the automatic sensor planning for the acquisition of lacking sensor data. The first prototype of the system is being implemented with the help of the LISP based generic blackboard development shell GBB on a SUN4 computer.

Keywords: diagnosis of process errors, action monitoring, blackboard, multi-sensory system, sensor planning

INTRODUCTION

Sensor data processing for robot systems covers two essential tasks:

- Guided robot actions: This is the closed loop task to control the trajectory and compensate for small deviations within a defined model. This task requires a fast processing of sensor data and uses functions such as adaptation, standardization, etc. The knowledge about the application is only used implicitly or off line to adapt the parameters of the control.

- Monitoring of robot actions: The sensor data are not used to compensate errors directly. After recognizing an error situation in the robot working cell, the diagnosis tries to describe the deviation between the real robot cell and the world model. This will be the starting point to change the action plan with respect to the new situation in the cell.

The special point of interest in our work is the diagnosis of errors of the assembly task. As the problem is very complex, the use of an expert system is necessary. The expert system for error diagnosis has the structure of a blackboard system. The complete system is split into various sub-blackboards: domain blackboard, model blackboard, planning blackboard and control blackboard.

The Blackboard architecture [5] [10] offers several features which are very useful for robot applications. The important aspects of the concept are the high modularity based on the knowledge sources which work independently, and the explicit definition of the decision control. The control decisively influences the problem solving. This is why control knowledge is included in the knowledge of problem solving. Newell [9] organizes the knowledge into weak methods and strong methods. Nowadays the weak methods are called meta knowledge: knowing how to solve a problem.

NATO ASI Series, Vol. F 71
Expert Systems and Robotics
Edited by T. Jordanides and B. Torby

The function of the control in blackboard systems is the scheduling of the knowledge sources. Knowledge sources contribute to the problem solving on various levels of abstraction. The data driven strategy is overlayed by the other strategies.The independent knowledge sources offer the attractive possiblity of parallel processing This can be done on a multi-processor machine [4] or on a distributed computer network [12].

The Problem: Scene Analysis

We have investigated the problem of recognizing work pieces in a crowded scene. The sensors used are a camera system with a binary data picture-processing module and an ultrasonic sensor for distance measurement. The workpieces to be recognized and measured are parts of the European Assembly Benchmark [1]. The recognition is very difficult because the parts may overlap.

The examined problem is the interpretation of the measured sensor data on the feature level. The goal is a correct interpretation in order to identify the workpieces and calculate their 3D-position. The interpretation is very hard, due to the following reasons:

- The sensor data may be inaccurate or disturbed
- The sensor data may not be complete
- The sensor data may be faulty or inconsistent

A complete model of the workpieces facilitates the scene analysis. The model information will be matched with the measured data. The structure of the diagnosis board is shown in Fig 1. While processing the sensor data, the strategies used change; the lower levels use a data driven strategy, the upper levels a model-driven one.

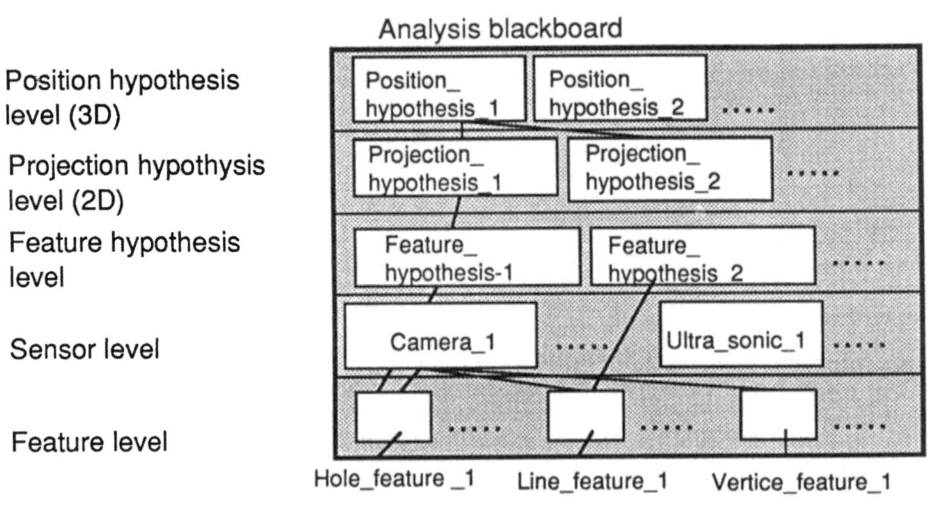

Figure 1: Analysis blackboard for the workpiece recognition

A combination of two methods is used to deal with uncertain data: the Fuzzy Set theory and the Dempster-Shafer theory[11] [13]. They are implemented on the lower levels of the diagnosis blackboard to calculate hypotheses for the identification and position of the workpieces based on the uncertain data.

The verification of the identification is model driven. For this purpose, a specially structured description of the reference workpieces, a minimal spanning tree, is used. Starting with characteristic features, a knowledge source constructs partial graphs of the workpiece. Without disturbances in the scene, the result of the graph matching will be the complete tree, but in a real scene only partial graphs will be found. The geometrical relation between the partial graphs allows the calculation of the 2D-position of the workpiece.

The missing information for determining the gripping points of the workpiece is the 3D-position which can be measured by the distance sensor, guided by the information of the vision system. The ultrasonic sensor can be used to determine the position of a workpiece surface or to detect overlapping edges. This is the starting point for sensor planning. Usually a deviation from the expected state is signaled by one (error detecting) sensor. In simple cases the diagnosis concludes successfully with the data of this single sensor. In more complex situations, the data of only one sensor is not sufficient for successful diagnosis. Additional measurements are necessary to fill this information gap . The system should automatically plan which data it needs and how it is to obtain them. The main points to consider are as follows:

- Which information is missing
- Which sensors are capable of getting the missing information
- How the sensor must be positioned so that it can obtain the searched data

Sensor Planning

It is the task of the sensor system to monitor the assembly process as well as to control knowledge acquisition for missing sensor data. Above all automated sensor planning requires recognizing the present error situation that occurs if an assembly task cannot be duly managed by the robot. An error situation arises if the touching sensor signals the work piece to not be in the expected position. The present scene then has to be analyzed; therefore, sensor planning is indispensible.

Single-stage and hierarchical planning are the classical planning methods of artificial intelligence. Single-stage planning means planning at a stretch [6],[7], being subdivided into linear and nonlinear planning. From the start, linear planning fixes the sequence, subplans put together the overall plan. This leads to a very inflexible system that is not acceptable for use in sensor planning. In the case of nonlinear planning, where only partial ordering of subplans exist, inflexibility is indeed mitigated, but problems arise with subgoal-interactions. Altogether, single-stage planning, with the great disadvantage of being forced to plan all details immediately, cannot be classified as an appropriate method for sensor planning.

Single-stage planning:	linear planning nonlinear planning
Hierarchical planning:	operator abstarction situation abstraction skeletal planning
Meta planning:	planning of strategies
Opportunistic planning:	event-driven planning

Table 1: Planning methods

Hierarchical planning proceeds on the assumption that planning goes on in several abstraction-stages, with the advantage that the whole structure is more clear. Watching hierarchical planning, we can distinguish between situation- and operator-abstraction. The basic idea of situation-abstraction is the weighting of features within the precondition of the operators. The operator-abstraction directly abstracts via the operators. Whereas situation-abstaction has the great disadvantage that forward search becomes impossible due to loose preconditions that let the searching tree grow exponential, operator-abstraction fails in understanding the abstract formulation in detail.

Furthermore, the strategy of refinement of skeletal plans, introduced by Friedland and Iwasaki [4], is to be explained here in greater detail. It is a strategy which is based on hierarchical planning and which uses the following concept: existing standard solution strategies called skeletal plans, instantiated with specific operations and data, will solve a given problem.

This planning strategy with the concept of a skeletal plan is used to model a planner for sensor application. Similar to the method used in [8], the planner is subdivided into three knowledge-acquisition stages:

1. stage: "taking a photograph of the whole scene from far away"
2. stage: "taking a photograph of some detail of this scene"
3. stage: "hypothesis and evaluation".

The planner is to be regarded as a planner for knowledge acquisition, modeling the three stages of planning as skeletal plans. Stage one requires only one single skeletal plan, whereas stage two and three require several skeletal plans. Finding out after termination of the first stage, that a complex scene in terms of overlapping work pieces is present, we know that further data for exact description of the position of the objects is necessary. Skeletal plans of stage two provide this information by taking a photograph or measuring detail-data of the objects, whereas the planner in stage three has access to the skeletal plans "hypothesis and evaluation". Refinement of skeletal plans of the second stage is reached by additionally considering the size of features.

For example: Skeletal plan A: "measure an edge".
Refinement : "measure an edge with length x".

What we miss here is a certain flexibility. This is introduced by the data structure used. Sensors just as objects, are modeled by a frame structure, where frames represent the nodes of a semantic network. Third-stage skeletal plans are strongly dependent on this data structure through which can be determined which data are missing for a complete valuation. Data within this abstraction-stage are features of work pieces like edges, curves,..., bound up with the skeletal plans of the second stage.

Run of events:
1. Touching sensor signalizes an error situation.
2. Activating planner.
3. First stage planning: taking a long-range photograph.
4. Valuation: is it a complex scene? -----no-----> 9. ready.
5. Hypothesis of position.
6. Valuation: which information is missing?
7. Second stage planning: initiate skeletal plans.
8. Valuation: data structure complete? Hypothesis of position right?
-------> Yes: 9.
-------> No: 5.
9. ready.

Friedland sets up two major steps for instantiating skeletal-planning:

1. A skeletal plan, consisting of several abstract definitions of tasks, is to be found for a certain goal. Performing tasks is the prerequisite for reaching any goal.(Choosing)
2. Instantiate the plan with the specific data of the environment.(Refinement)

Skeletal Plan "Taking a long-range photograph with a camera"

> turn camera on
> focus camera on distance x
> take a photograph of the present scene
> transmit data to computer
> turn camera off.

Figure 2: Skeleton plan: "Taking a long-range photograph with a camara"

Choosing any skeletal plan from among a great number of other plans requires that it has first be constructed. This is best done with the first level.

On the basis of an error-message, the skeleton plan "Taking a long-range photograph with a camera" is activated (see Figure 2).

Now the second step takes place by refining this skeleton plan. This also requires that a basis-instantiation for each abstract step within the plan carried out. The refining process hereby passes through a hierachy according to the saved knowledge.

Figure 3 shows all refined abstract steps of the skeleton plan constructed in Figure 2.

"turn camera on"

> press button 'on'

"focus camera on distance x"

> turn objective until adjustment of x-metres faces the arrow

"take a picture of the present scene"

> press the button 'action'
> wait for two seconds
> unlatch button

"transmit data to computer"

> transmit each point of the pixel-matrix to computer
> wait for correct transmition of each pixel

"turn camera off."

> press button 'off'

Figure 3: Refined skeletal plan: "taking a long-range picture with a camera"

Skeletal plans of the second stage are directly connected with the knowledge-saving data structure, serving mainly to complete slots of frames that objects are described through. Evaluation of the third stage beginss with the assumption that 3D-position and orientation is immediately known or calculable if all data have been determined. For this purpose, it is naturally necessary to mark all hidden features.

As we can see by this example, the method of skeletal planning is a hierachical one. Modularity permits involving new objects and new methods to the knowledge-base, since refinement and selection making knowledge is saved separately.

For the last part of sensor planning, we will see how plans can be integrated into a blackboard structure.

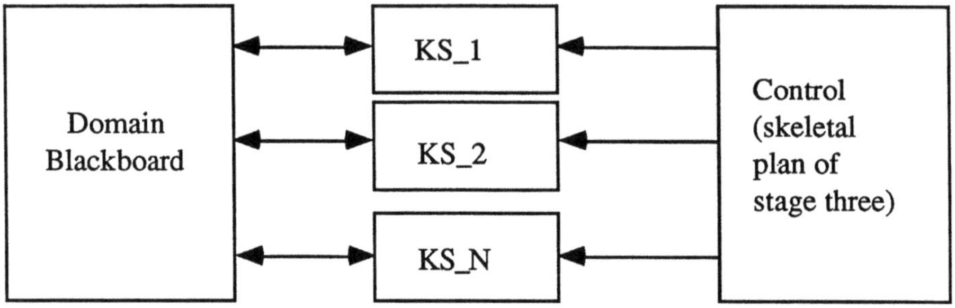

Figure 4: The implementation structure of the skeleton approach with the blackboard architecture

The control of the blackboard structure contains two skeletal plans: hypothesis and evaluation belonging to stage three. It has access to the blackboard itself, including all data within a framestructure. Slots not filled yet trigger the knowledge sources in such a way that a skeletal plan connected with an adequate feature will be activated. Different sensors can be used by those skeletal plans. If the framestructure is completely filled with data, the process of triggering knowledge sources will be stopped and the result ,that is, the3D-position and 3D-orientation, is used for proceeding the assembly task.

Implementation with GBB

The scene analysis, as an example, is implemented with the blackboard development shell GBB [2]. The shell offers two control frameworks: the simple shell and the ES-shell. The simple shell is too simple for real application. It can only be used to get some experience with GBB. The ES-shell corresponds to the BB1 control concept. It is used for the implementation.

The implemented blackboard is divided into a sensor data acquisition blackboard, a diagnosis blackboard and a blackboard to analyse the model data (see Fig. 5). Corresponding to the various levels of the diagnosis blackboard are the knowledge sources to handle the data and to generate successively higher descriptions of the scenario. Sensor models are used in the data acquisition blackboard for the correction of sensor specific errors. Additionally this model is useful for the fusion of sensor data on a low level. The model blackboard contains a specifically structured partial copy of the world model of the robot system describing both workpieces and their features. The assembly workpiece is described as a model with attributes, such as names, surfaces, number of edges, for example. The workpiece description refers to the features on the lower level: edges, holes, etc.

The strategy for the scene analysis is basically data driven. In case of some simultaneously executable knowledge sources, a best first strategy is realized. A hypothesis-and-test strategy is chosen for the construction of the 2D-position. Finally, to test all the hypotheses, a model-driven approach is used to reduce the search space. The strategies are realized using the knowledge sources and the scheduling control.

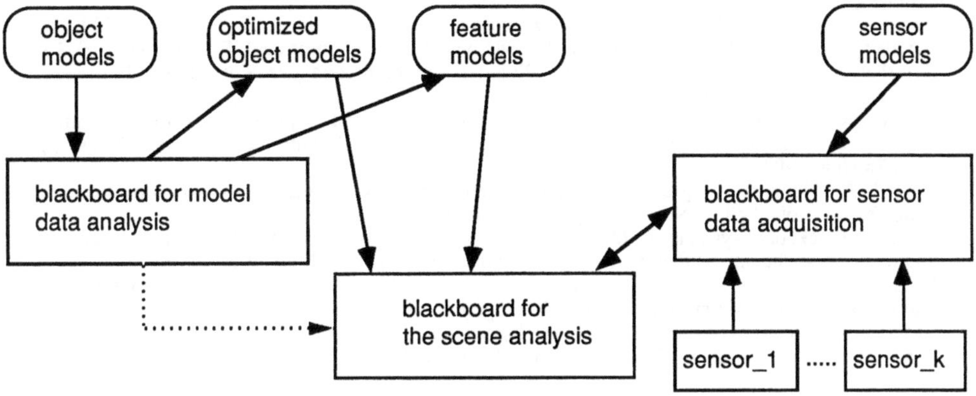

Figure 5: The implemented blackboard system

The scene analysis blackboard has a good recognition performance; however, the diagnosis time is too long for the real time requirements of robot applications. The next step in our work is the integration of other hypothesis methods for faster analysis. Furthermore, research must be done in the area of strategy and scheduling control.

Conclusion

A very important result of our research work is that we have verified that the blackboard architecture is well suited for the action monitoring task. The central data storage of the basic data and the interim results facilitate the fusion of the data of a complex sensor system, for example, a camera. In the same manner, the fusion of the data of various sensors can be accomplished. The action scheduling depends on the information placed on the blackboard. This quality is very useful for the permanent monitoring of the robot cell.

Our initial research efforts in robot sensors were concentrated on individual problem areas.The above described parts are implemented on a SUN4 computer. Our current work is to complete this system with a planning module based on the skeleton plan approach.

Acknowledgement

This research work was performed at the Institute for Realtime Computer Systems and Robotics, Prof. Dr.-Ing. U. Rembold and Prof. Dr.-Ing. R. Dillmann, Faculty for Informatics, University of Karlsruhe, 7500 Karlsruhe, Federal Republic of Germany.

References

1. Collins, K.,Palmer, A.J.,Rathmill, K.: The Development of a European Benchmark for the Comparison of Assembly Robot Programming Systems, Robot Technology and Application, S. 187-199, Springer-Verlag, 1985
2. Corkill, D.D., Gallagher, K.Q., Murray, K.E: GBB: A generic blackboard, Proceedings of the National Conference on Artificial Intelligence, pp. 1008-1014, 1986
3. Ensor, J.R., Gabbe, J.D.: Transactional blackboards, Proceedings of the International Joint Conference on Artificial Intelligence, Los Angeles, pp. 340-344, 1985
4. Friedland, P.E., Iwasaki, Y.: The concept and implementation of skeletal plans, Journal of automated reasoning 1(1985), 161-208.
5. Hayes-Roth, B.: A blackboard architecture for control, Artificial Intelligence Journal 26, pp. 251-321, 1985
6. Hertzberg, J.: Planerstellungsmethoden der Künstlichen Intelligenz, Informatik-Spektrum(1986)9: 149-161.
7. Hertzberg, J.: Planen-Einführung in die Planerstellungmethoden der Künstlichen Intelligenz, B-I-Wissenschaftsverlag, 1989.
8. Luo, Ren.C., Lin, Min-Hsiung, Scherp, Ralph.S.: The Issues and Approaches of a Robot Multi-Sensor Integration, IEEE International Conference on Robotics and Automation, Proceedings, pp.1941-1946,Raleigh, 1987
9. Newell, A.: Heuristic programming: Ill-structured problems, Progress in Operations Research, Ed.: Aronofsky, J., John Wiley, New York, pp. 360-414, 1969
10. Nii, P.H.: Blackboard systems, AI Magazine 7 (Number 3 and 4), 3: 38-53 and 4: 82-107, 1986
11. Shafer, G.: A mathematical theory of evidence, Princeton University Press, 1975
12. Velthuijsen, H., Lippolt, B.J., Vonk, C.J.: A parallel blackboard system for robot control, 10th International Joint Conference on Artificial Intelligence, pp. 1157-1159, 1985
13. Zadeh, L. A.: The Role of Fuzzy Logic in the Management of Uncertainty in Expert Systems, Approximate Reasoning in Expert Systems, Elsevier Science Publishers B. V.

COMPUTER AIDED ANALYSIS OF ROBOT ARM DYNAMICS AND ADAPTIVE CONTROL SCHEMES

A. Akmehmet[1], A. Kuzucu[2]

[1]System and Automation Department, T. Şişe ve Cam Fab. A.Ş.,Beşiktaş, Istanbul, Turkey
[2]Department of Mechanical Engineering, Istanbul Technical University, Taksim, Istanbul, Turkey

Abstract: A computer program is developed where computer oriented algorithms for robot arm dynamics are implemented together with suboptimal control algorithms for trajectory tracking. The program package allows in an interactive way the synthesis of the manipulator structure as well as its control system, and the simulation of the dynamic behaviour of the whole system under different scenarios. Two adaptive control schemes based on minimum energy terminal control are proposed and the performances of the proposed algorithms are tested with simulation studies.

Keywords: robots / modelling / computer-aided design / adaptive control / tracking / simulation

INTRODUCTION

The structure and parameters of a manipulator control system mainly depend on the structure and parameters of the manipulator. Consequently the mechanical design of a manipulator and the design of its control system are interactive processes. Since the dynamic equations of a manipulator are nonlinear and involve coupling terms, it is difficult to design a control system which is easy to tune and which leads to fast responses with high tracking performances.

Analytical derivation of the dynamic model of a manipulator becomes very cumbersome when the number of degrees of freedom increases. Researchers [2,4,6,8,10,11] proposed various computer oriented algorithms to compute the dynamic model terms recursively, based on the knowledge of joint variables. Two general algorithms, based on two basic forms of the manipulator dynamic model, are presented in this study. They compute the basic parameter matrices of any kind of open kinematic chain which consists of a series of rigid links. With those algorithms it is possible to solve both the direct and inverse problems of dynamics for manipulators.

NATO ASI Series, Vol. F 71
Expert Systems and Robotics
Edited by T. Jordanides and B. Torby

A mechanical manipulator is an open loop kinematic chain which consists of a sequence of rigid bodies (links) connected in series by joints. One end of the chain is generally fastened to a supporting base while the other end is free to move in space. The joints and links are numbered outwards starting from the base (link 0) to the end effector (link n). A body coordinate frame (x_i, y_i, z_i) is attached to each link i (i=0,...,n). Adjacent cordinate frames are related to each other by Denavit-Hartenberg parameters [3].

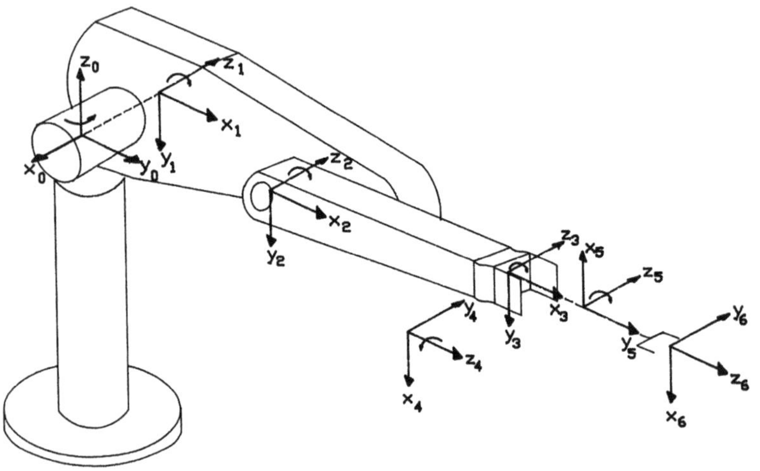

Fig.1 Link coordinate systems for a Puma type robot arm

Once the link coordinate frames have been assigned to the manipulator it is possible to obtain the Cartesian position and orientation of the manipulator end effector with given joint coordinates. This is called the direct kinematics solution. The control of manipulator also necessitates the inverse kinematics solution for which required joint coordinates are found given the desired position and orientation of the hand. The kinematic equations of a manipulator involve trigonometric functions and it is difficult to obtain an explicit solution for the inverse kinematics problem. Even if an explicit solution is available it should be used very carefully because of possible singularities.

Two control schemes based on minimum energy terminal control are also proposed in this study in accordance with the adopted manipulator dynamic model structure. The first method, similar to the well-known computed-torque technique, is based on decoupling while the second method is based on suboptimal adaptive feedback control of nonlinear systems. A simulation study has been carried out to test the performance of the proposed methods.

DYNAMIC MODEL DERIVATION

Basic Algorithm

The terms that appear in the dynamic model of a robot arm can be determined by recursive algorithms where the velocities and accelerations of each link are calculated by a forward recursion starting from the base to the end link. Then the generalized forces are obtained by a backward recursion in order to find the torques to be applied at the joints. In this study Newton-Euler formulation is used to develop two general algorithms for dynamic modelling of an n degree of freedom manipulator. Rather than calculating the velocities and accelerations, their corresponding coefficients are calculated as proposed by Stepanenko and Vukobratovic [8], making the algorithm suitable for the solution of both the direct and inverse problems of dynamics. The velocities and accelerations are also computed in each link's internal coordinate system as suggested by Luh, Walker and Paul [6], eliminating the need for a great deal of coordinate transformations.

The equation of motion for an n degree of freedom manipulator has the general structure

$$J(\mathbf{q})\ddot{\mathbf{q}}+V\dot{\mathbf{q}}+\mathbf{f}(\dot{q}_i\dot{q}_j,\mathbf{q})+\mathbf{g}(\mathbf{q})+\mathbf{h}(\mathbf{q})=\tau \quad (1)$$

where

- $\mathbf{q}$: nx1 vector of joint variables,
- $J(\mathbf{q})$: nxn symmetric, nonsingular inertia matrix,
- V : nxn diagonal viscous friction matrix,
- $\mathbf{f}(\dot{q}_i\dot{q}_j,\mathbf{q})$: nx1 vector specifying Coriolis and centrifugal effects (i,j=1, ..., n)
- $\mathbf{g}(\mathbf{q})$: nx1 vector specifying gravity effects,
- $\mathbf{h}(\mathbf{q})$: nx1 vector specifying the effects due to external forces and moments exerted on link n,
- τ : nx1 vector generalized input forces.

The recursive equations for the angular and linear velocities of the ith joint expressed in the coordinate frame of link i are given by [6]

$$R_i^0\omega = R_i^{i-1}(R_{i-1}^0\omega_{i-1} + \mathbf{z}_0\dot{q}_i(1-\epsilon)) \tag{2}$$

$$R_i^0\mathbf{v}_i = R_i^{i-1}(R_{i-1}^0\mathbf{v}_{i-1} + \mathbf{z}_0\dot{q}_i) + R_i^0\omega_i x r_i^0\mathbf{p}_i \tag{3}$$

where $\mathbf{z}_0 = (0,0,1)^T$, $R^0_i\mathbf{p}_i = (a_i,\ r_i\sin\alpha_i,\ r_i\cos\alpha_i)^T$, R^0_i is the rotation matrix between the frames i and 0, and ϵ describes the joint type ($\epsilon=0$ for a rotational joint, $\epsilon=1$ for a translational joint).

In order to have a dynamic model also suitable for the inverse dynamics solution, the angular and linear accelerations are written in the form of a vector sum suggested by Stepanenko and Vukobratovic [8]

$$R_i^0\alpha_i = \sum_{j=1}^{i}\psi_{ij}\ddot{q}_j + \theta_i \tag{4}$$

$$R_i^0\mathbf{a}_i = \sum_{j=1}^{i}\beta_{ij}\ddot{q}_j + \eta_i \tag{5}$$

$$R_i^0\mathbf{a}_{Gi} = \sum_{j=1}^{i}\lambda_{ij}\ddot{q}_j + \gamma_i \tag{6}$$

where angular and linear acceleration coefficients can be calculated by a set of recursive relations.

Applying D'Alemberts principle and using equations (4), (5) and (6) the total force and total moment vectors exerted on link i are obtained as

$$R_i^0\boldsymbol{F}_i=L_i\ddot{\boldsymbol{q}}+\boldsymbol{l}_i \tag{7}$$

$$R_i^0\boldsymbol{M}_i=N_i\ddot{\boldsymbol{q}}+\boldsymbol{n}_i \tag{8}$$

where L_i and N_i are 3xn matrices given as

$$L_i=m_i[\boldsymbol{\lambda}_{i1}\boldsymbol{\lambda}_{i2}\ldots\boldsymbol{\lambda}_{ii}\boldsymbol{0}\ldots\boldsymbol{0}] \tag{9}$$

$$N_i=I_i[\boldsymbol{\psi}_{i1}\boldsymbol{\psi}_{i2}\ldots\boldsymbol{\psi}_{ii}\boldsymbol{0}\ldots\boldsymbol{0}] \tag{10}$$

and the vectors $\mathbf{l}_i$ and $\mathbf{n}_i$ are defined as

$$\boldsymbol{l}_i=m_i\gamma_i;\boldsymbol{n}_i=I_i\theta_i+R_i^0\omega_i x(I_iR_i^0\omega_i) \tag{11}$$

Considering the forces and moments acting on link i, one can obtain the following recursive equations for the reaction forces and moments exerted on link i by link i-1

$$R_i^0\boldsymbol{f}_i=T_i\ddot{\boldsymbol{q}}+\boldsymbol{t}_i-\boldsymbol{g}_{fi}+\boldsymbol{f}_{Hi} \tag{12}$$

$$R_i^0\boldsymbol{m}_i=S_i\ddot{\boldsymbol{q}}+\boldsymbol{s}_i-\boldsymbol{g}_{mi}+\boldsymbol{m}_{Hi} \tag{13}$$

The components of the above equations are calculated starting from link n (end effector) to the base of the manipulator.

The required force or torque at joint i that the actuator should provide is equal to the projection of the reaction force or moment at the joint onto the $\mathbf{z}_{i-1}$ axis plus the friction force or moment

$$\tau_i = [R^i_{i-1}(R^0_i \mathbf{m}_i)]^T \mathbf{z}_0 + b_i \dot{q}_i, \quad \textit{if link i is rotational} \tag{14}$$

$$\tau_i = [R^i_{i-1}(R^0_i \mathbf{f}_i)]^T \mathbf{z}_0 + b_i \dot{q}_i, \quad \textit{if link i is translational} \tag{15}$$

Substituting (12) and (13) into (14) and (15) one gets the following expressions to form the inertia matrix J, and the vectors **f**, **g**, and **h**

$$J_{ij} = (R^i_{i-1} S_i)_{3j}, \quad \textit{if link i is rotational} \tag{16}$$

$$J_{ij} = (R^i_{i-1} T_i)_{3j}, \quad \textit{if link i is translational} \tag{17}$$

where the subscripts i and j on M_{ij} indicate the corresponding element of matrix M.

$$\mathbf{f}_i = (R^i_{i-1} \mathbf{s}_i)_3, \quad \textit{if link i is rotational} \tag{18}$$

$$\mathbf{f}_i = (R^i_{i-1} \mathbf{t}_i)_3, \quad \textit{if link i is translational} \tag{19}$$

$$\mathbf{g}_i = (R^i_{i-1} \mathbf{g}_{mi})_3, \quad \textit{if link i is rotational} \tag{20}$$

$$\mathbf{g}_i = (R^i_{i-1} \mathbf{g}_{fi})_3, \quad \textit{if link i is translational} \tag{21}$$

$$\boldsymbol{h}_i = (R_{i-1}^{i}\boldsymbol{m}_{Hi})_3, \quad \textit{if link i is rotational} \tag{22}$$

$$\boldsymbol{h}_i = (R_{i-1}^{i}\boldsymbol{f}_{Hi})_3, \quad \textit{if link i is translational} \tag{23}$$

where the subscript i on v_i indicates the ith component of vector **v**.

A Modified Algorithm

An alternative to the form given by (1) is to write the equation of motion as

$$J(q)\ddot{q} + V\dot{q} + C(q,\dot{q})\dot{q} + g(q) + h(q) = \tau \tag{24}$$

where C is an nxn matrix which represents, when multiplied by the vector $\dot{\mathbf{q}}$, the Coriolis and centrifugal effects. This factorization is very useful to obtain locally linear system matrices for control purposes.

In order to form C matrix, angular velocities are rewritten in the form

$$R_i^0\omega_i = \sum_{j=1}^{i} \xi_{ij}\, \dot{q}_j \tag{25}$$

The angular and linear acceleration coefficients given in Equations (4), (5) and (6) which are a function of $\dot{q}_i$ terms, can be rewritten as

$$\theta_i = \sum_{j=1}^{i-1}\sum_{k=j}^{i-1} \kappa_{ijk}\, \dot{q}_j\, \dot{q}_{k+1} \quad \textit{for} \quad i=2,\ldots,n \tag{26}$$

$$\eta_i = \sum_{j=1}^{i} \delta_{ij}\, \dot{q}_j \tag{27}$$

$$\gamma_i = \sum_{j=1}^{i} \phi_{ij}\, \dot{q}_j \tag{28}$$

The vectors t_i and s_i which appear in (12) and (13) can also be rewritten as

$$t_i=\sum_{j=1}^{n} \delta_{ij}\, \dot{q}_j \quad ; \quad s_i=\sum_{j=1}^{n} \vartheta_{ij}\, \dot{q}_j \tag{29}$$

Hence the elements of C matrix can be obtained by

$$C_{ij} = (R_{i-1}^{i}\vartheta_{ij})_3, \quad \textit{if link } i \textit{ is rotational} \tag{30}$$

$$C_{ij} = (R_{i-1}^{i}\sigma_{ij})_3, \quad \textit{if link } i \textit{ is translational} \tag{31}$$

Recursive relations for the determination of the coefficients in the basic and modified algorithms are given in [1].

CONTROL OF THE MANIPULATOR

In order to control the manipulator, two adaptive control schemes are proposed which are easy to implement due to their reduced number of tuning parameters. The system matrices used for feedback gain calculation are obtained by the algorithms that are used for the determination of dynamic model terms presented in the previous section.

Receding Horizon Computed Torque Technique

The equation of motion given by Eq.(1) can be solved for joint accelerations to give

$$\ddot{q}=J^{-1}(q)[\tau-k] \tag{32}$$

where $\mathbf{k} = V\,\dot{\mathbf{q}} + \mathbf{f}(\dot{q}_i\dot{q}_j,\mathbf{q}) + \mathbf{g}(\mathbf{q}) + \mathbf{h}(\mathbf{q})$.

Equation (32) can be written in the decoupled form

$$\ddot{q}_i = u_i \ , \quad i=1,\dots,n \tag{33}$$

where u_i is the ith element of the nx1 vector **u** defined as

$$\boldsymbol{u}=J^{-1}(q)[\tau-k] \tag{34}$$

Assuming that the desired trajectory for joint i is given by $q_{di}(t)$, $\dot{q}_{di}(t)$, and $\ddot{q}_{di}(t) = \Gamma(t)$ and defining the state variables related to joint i as

$$x_1 = q_{di} - q_i \quad ; \quad x_2 = \dot{q}_{di} - \dot{q}_i \tag{35}$$

leads to the state model

$$\frac{d}{dt}\begin{bmatrix} x_1 \\ x_2 \end{bmatrix} = \begin{bmatrix} 0 & 1 \\ 0 & 0 \end{bmatrix}\begin{bmatrix} x_1 \\ x_2 \end{bmatrix} + \begin{bmatrix} 0 \\ 1 \end{bmatrix} u_i + \begin{bmatrix} 0 \\ \gamma_i \end{bmatrix} \tag{36}$$

The minimum energy performance function

$$J=\frac{1}{2}\int_{t_0}^{t_0+T} \rho u_i^2 dt \tag{37}$$

is adjoined to the state model (36) with the terminal constraint $x(t_o+T)=0$. The solution of the minimum energy terminal control problem defined by (36) and (37) and the final state constraint leads to the control law

$$u_i = \frac{6}{T^2}(q_{di}-q_i) + \frac{4}{T}(\dot{q}_{di}-\dot{q}_i) + \gamma_i \ , \quad t \in [\, t_0 \ , t_0+T \,] \tag{38}$$

when the values of optimal feedback gains at t_o are selected as suboptimal solution [9]. Here T is called horizon time or time-to-go.

Substitution of the control law (38) into (34) gives for joint i

$$\ddot{q}_i + \frac{4}{T}\dot{q}_i + \frac{6}{T^2}q_i = \frac{6}{T^2}q_{di} + \frac{4}{T}\dot{q}_{di} + \ddot{q}_{di} \tag{39}$$

which shows that the natural frequency of joint i can be adjusted by changing only the parameter T without affecting its damping ratio which has a constant value of 0.82. Use of the control law (38) in (35) will give the generalized force vector as

$$\tau = J(q)\,[\frac{6}{T^2}(q_d - q) + \frac{4}{T^2}(\dot{q}_d - \dot{q}) + \ddot{q}_d\,] + k \tag{40}$$

Adaptive Control Scheme for Interactive Nonlinear Model

The efficiency of the manipulator control system will increase as long as the original structure of the dynamic model is preserved in the state model. Consequently a state model in the general form

$$\dot{x} = f\,(x,\ a,\ u) \tag{41}$$

where **x** is the state vector, **a** is the parameter vector, and **u** is the control vector, should be considered in order to take the coupled dynamics of the manipulator into account. Since the optimal control problem for systems modeled in the form of (41) has no available closed form solution, this nonlinear state model is approximated by a local linear model in the form

$$\dot{x} = A_k\,(x_k,a_k,U_k)\,x + B_k\,(x_k,k_k,u_k)\,u\ ;\quad t \in [\,t_k,\ t_{k+1}\,] \tag{42}$$

which is valid only in the proximity of the linearization state defined by the measured values

$\mathbf{x}_k$, $\mathbf{a}_k$, and $\mathbf{u}_k$ at the instant t_k. Then the algorithms for the determination of the suboptimal linear feedback law proposed by Kuzucu and Roch [5], are used to find the control law in the form

$$u = G_k\, x \;;\quad t \in [t_k, t_{k+1}] \tag{43}$$

where G_k is updated at each correction instant using the new values of A_k and B_k matrices.

The minimal energy terminal control policy is used again due to its attractive property of having a single adjustment parameter which is the horizon time or time-to-go given by T. The performance criterion

$$J = \int_{t_k}^{t_k+T} u^T R\, u \; dt \tag{44}$$

is adjoined to the locally linear model (42) with the terminal constraint $\mathbf{x}(t_k+1)=0$. The value of the feedback matrix at the initial time t_k is kept constant until a new correction at t_{k+1}.

System matrices of the linear model (A_k and B_k) should be computed at each correction instant in order to apply the control algorithm described above. The authors focused their attention to the possibilities of using the dynamic model coefficients computed by the algorithms given in the above section. Thus the concept of "apparent linearization" or "linearization by factorization" introduced by Pearson [7] has been adopted, considering the adaptive structure of the algorithm. Solving the equation of motion (1) for $\ddot{\mathbf{q}}$ we obtain

$$\ddot{q} = J^{-1}(q)\; u \tag{45}$$

where the control vector $\mathbf{u}$ is defined as

$$u = \tau - k \tag{46}$$

Defining the state variables as

$$x_{2i-1} = q_i \; , \quad x_{2i} = \dot{q}_i \; ; \quad i = 1,...,n \tag{47}$$

leads to a (2nx2n) A_k matrix whose elements are zero except for $A_k(i,i-1)$ i=1,3, ... ,n-1 which are equal to one and to a (2nxn) B_k matrix whose odd numbered rows are zero and even numbered rows are formed by the rows of $J^{-1}(\mathbf{q})$.

Using the modified form (24) of the equation of motion allows the introduction of C matrix into the linear model matrix A_k. Solving (24) for $\ddot{\mathbf{q}}$ one obtains

$$\ddot{q} = J^{-1}(q) \; [\; u - [\; C(q,\dot{q}) + V \;] \; \dot{q} \;] \tag{48}$$

According to (47) the even numbered rows of A_k matrix will be formed by the rows of the matrix $-J^{-1}(\mathbf{q})$ x $[C(\mathbf{q},\dot{\mathbf{q}}) + V]$ and the control vector takes the form $\mathbf{u}=\tau - [\mathbf{g}(\mathbf{q}) + \mathbf{h}(\mathbf{q})]$.

Model and Control Horizon Update Policies

The methods for the computation of the feedback gains and the open-loop components of the control vector are based on a linear model which has to approximate the nonlinear state equations of the controlled system. Linear model matrices used to generate the control vector are valid as long as the system state stays in the validity domain of the linear model. This linear model and corresponding feedback gains have to be corrected when the approximation error introduced by the linearization procedure becomes important with respect to linear terms. Therefore the system matrics, that appear in the equation of motion of the manipulator should be updated when they no more represent the actual system. There are mainly two approaches adopted for model and control parameter update policies:

i) The first one is the on-line computation of the system matrices A_k and B_k at given instants of time with a predetermined frequency as the manipulator moves along the specified trajectory. The update frequency of the system matrices can be adjusted for the desired quality of control. As a modification of that approach, the system matrices may be updated only when the system state moves away from the desired trajectories more than a tolerance value.

ii) The second approach is the off-line calculation and memorization of the system matrices for specified trajectories prior to task execution assuming that the desired trajectories are known in advance. Then during the motion of the manipulator corresponding system matrices can be picked up from the memory and kept constant as long as the system state remains in the considered region of the state space.

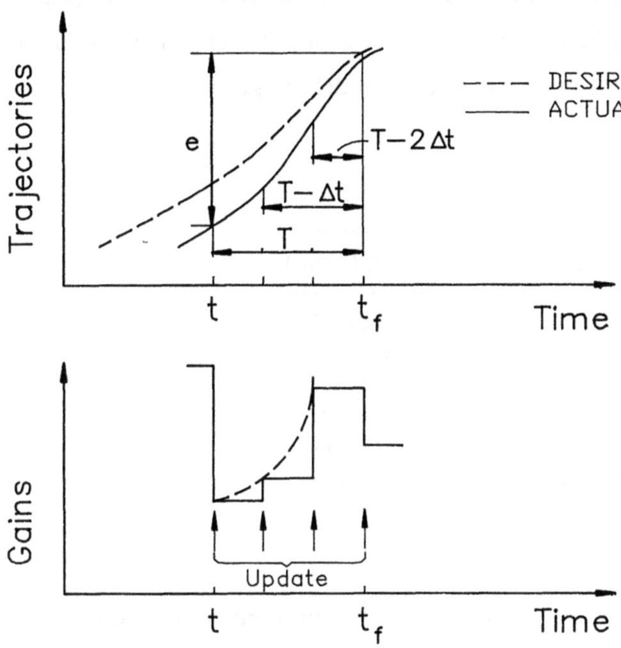

Fig.2. Change of feedback gains with variable horizon time.

In most of the adaptive control schemes found in the literature, the system matrices and the feedback gains are updated simultaneously. In this study the update frequency of the feedback gains is adjusted separately in order to emphasize the effect of changing horizon time. Depending upon the followed trajectory it may not be needed to update the system matrices as frequently as the feedback gains. Feedback gains and the control vector elements depend on the horizon time, T at which the controller should satisfy the terminal constraint. Even though the minimum energy approach leads to conservative behaviour, small horizon times will result in high feedback gains, consequently in large control magnitudes and relatively fast responses. Performances of the system can be improved by a proper selection of T and the gain update frequency for which different update policies can be adopted as shown in Fig.2.

SIMULATION RESULTS

A series of computer simulations have been carried out for a Puma type robot arm (Fig.1) with three and six degrees of freedom in order to investigate the effectiveness of the proposed control schemes. Parameters of the considered model are given in Table 1.

TABLE 1. Parameters of the manipulator model

Link	Mass (kg)	Length (m)	I_x (kgm^2)	I_y (kgm^2)	I_z (kgm^2)
1	4.0	0.20	0.0167	0.0167	0.0067
2	2.5	0.60	0.0042	0.0771	0.0771
3	1.5	0.50	0.0025	0.0325	0.0325
4	0.5	0.12	0.0009	0.0004	0.0008
5	0.5	0.12	0.0008	0.0009	0.0004
6	0.5	0.08	0.0003	0.0005	0.0003

In the first part of the simulation program off-line calculations are performed including the inverse kinematics solution. The second part of the program contains programs for simulation of the system dynamics with the selected control scheme.

The basic scenario is to have the manipulator tip to move along a straight line from point P_1 to point P_2 in 1.6 seconds (P_1 = (1,0,0), P_2 = (-0.5,0.5,0.5) expressed in meters). A parabolic velocity profile is selected for the manipulator tip and its maximum value is 1.56 m/sec.

The main parameters of the simulation are the update period of feedback gains (TGAIN), update period of system matrices (TUP) and the initial horizon time (TTG). The obtained results are summarized in Tables 2-4. The values in the last two columns of those tables evaluate the tracking quality and the control cost of the system and they are obtained by calculating the quadratic performance index

$$J = \frac{1}{2} \int_{t_0}^{t_f} [(x_d - x)^T Q (x_d - x) + u^T R u] \, dt \tag{49}$$

where $\mathbf{x}_d$ is the vector of desired states.

The first set of simulations have been carried out for the three-degrees-of-freedom model using respectively, the receding horizon computed torque technique and the adaptive feedback control. A comparison of the values given in Tables 2 and 3 shows that the adaptive feedback control gives much better results than the receding horizon computed torque technique. For both control schemes the best tracking is achieved with a horizon time of 0.04 seconds while the system matrices and feedback gains are updated each 0.01 seconds.

TABLE 2. Performances Obtained with the Receding Horizon Computed Torque Technique for the Three-Degrees-of-Freedom Model (TGAIN=0.01 sec)

TTG (sec)	TUP (sec)	max. errors on Θ_i(rad)	$\dot{\Theta}_i$ (rad/sec)	$\int e^T Q e \, dt$	$\int u^T R u \, dt$
0.04	0.01	0.085	13.92	32.04	67592
0.04	0.08	0.361	29.78	58.10	76338
0.03	0.01	0.136	14.22	17.37	15578
0.08	0.04	0.124	12.19	19.78	16047
0.16	0.01	0.241	12.79	16.39	3954
0.16	0.08	0.489	19.74	21.12	4319

A third set of simulations have been carried out for the six-degrees-of-freedom model using the same numerical values for the motion of the manipulator tip used for the three-degrees-of-freedom model. The orientation of the hand is kept fixed while following the straight line from P_1 to P_2. Performances obtained with the adaptive control scheme are presented in Table 4. It is observed that the best tracking is again achieved with the same set of control parameters as for the three-degrees-of-freedom model.

TABLE 3. Performances Obtained with the Adaptive Feedback Law for the Three-Degrees-of-Freedom Model

TTG (sec)	TUP (sec)	TGAIN (sec)	max. errors on Θ_i(rad)	$\dot{\Theta}_i$ (rad/sec)	$\int e^TQe\,dt$	$\int u^TRu\,dt$
0.04	0.01	0.01	0.00007	0.008	0.000005	14.59
0.04	0.08	0.01	0.00127	0.161	0.000848	17.03
0.04	0.01	0.04	0.08187	3.127	1.979498	5058.80
0.08	0.01	0.01	0.00027	0.011	0.000013	14.58
0.08	0.08	0.01	0.00157	0.109	0.000516	16.15
0.16	0.01	0.01	0.00210	0.044	0.000278	29.04
0.16	0.16	0.01	0.01221	0.426	0.006668	20.18
0.16	0.01	0.16	0.40008	2.899	5.094765	153.65

TABLE 4. Performances Obtained with the Adaptive Feedback Law for the Six-Degrees-of-Freedom Model

TTG (sec)	TUP (sec)	TGAIN (sec)	max. errors on Θ_i(rad)	$\dot{\Theta}_i$ (rad/sec)	$\int e^TQe\,dt$	$\int u^TRu\,dt$
0.04	0.01	0.01	0.00060	0.083	0.00039	54.5
0.04	0.08	0.01	0.01529	2.291	0.07093	69.2
0.04	0.01	0.04	0.07719	2.911	2.59373	14133.9
0.16	0.01	0.01	0.22322	5.924	7.26140	2606.6
0.16	0.16	0.01	0.42860	7.137	10.79370	2611.6
0.16	0.01	0.16	0.24714	4.559	4.95595	1685.0

Idealized torque variations for the best tracking performances are illustrated in Fig.3 for the three-degrees-of-freedom manipulator model.

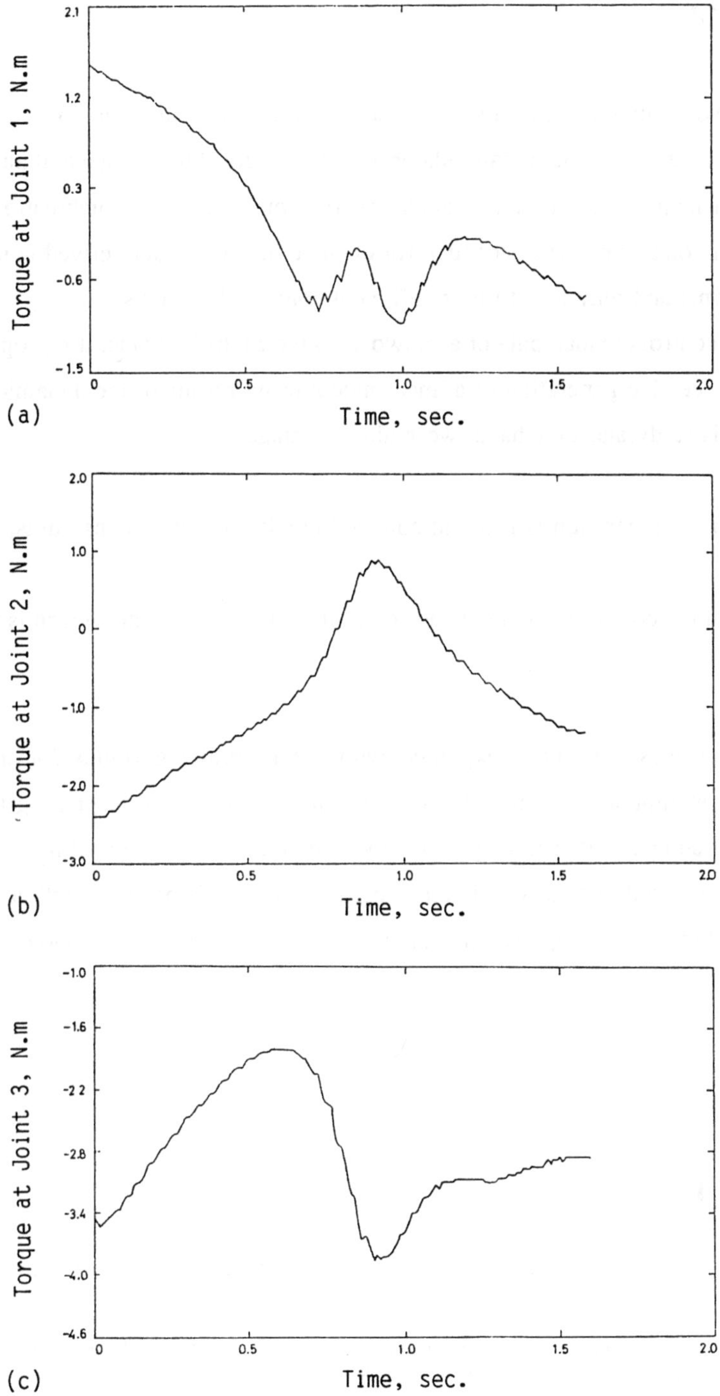

Fig.3. Idealized torques acting at joints of the three-degrees-of-freedom model obtained with the adaptive feedback law for TTG=0.04 sec., TUP=0.01 sec., TGAIN=0.01 sec.

CONCLUSION

In this study two recursive algorithms have been developed in order to form the dynamic model of an n degree of freedom manipulator arm. The algorithms are general and allow the study of a kinematic chain of any degree of freedom with any combination of joints (translational or rotational). The second form of the dynamic model derived considers also the coupling terms and may lead to more efficient control algorithms.

In order to control the manipulator arm two adaptive control schemes are proposed which are based on successive generation of a linear model using results of the dynamic modelling described in this study and they have two main advantages:

i) Feedback matrix generation is fast and compatible with real time constraints,

ii) The tuning of the control system depends on a single basic parameter which is the horizon time.

These advantages have been exploited with the general algorithms for the dynamic modelling of a manipulator in order to develop a user-friendly computer program package which can be used as a software support for the computer-aided design of industrial robots. Obviously the generality of the simulation program structure is obtained with a sacrifice on computer time but specific applications may lead to computation times suitable for real-time applications.

REFERENCES

1. Akmehmet, A.: A Study of Manipulator Dynamics and Control,MSc Thesis, Bogazici University, Istanbul, (1984)

2. Cvetkovic, V. and Vukobratovic, M.: Computer Oriented Algorithms for Modelling Active Spatial Mechanisms for Robotic Applications, IEEE Trans. Syst., Man, Cybern., SMC-12, 838-847, (1982)

3. Denavit, J, and Hartenberg, R.S.: A Kinematic Notation for Lower-Pair Mechanisms, J. Applied Mechanics, 22, 215-221, (1955)

4. Hollerbach, J. M.: A Recursive Lagrangian Formulation of Manipulator Dynamics and a Comparative Study of Dynamics Formulation Complexity, IEEE Trans., Syst., Man, Cybern., SMC-10, 730-736, (1980)

5. Kuzucu, A., and Roch, A.: Suboptimal Adaptive Feedback Control of Nonlinear Systems, Lecture Notes in Control and Information Sciences, 24, Springer-Verlag Berlin, 131-140, (1980)

6. Luh, J.Y.S., Walker, M.W., and Paul, R.P.C.: On-line Computational Scheme for Mechanical Manipulators, J. Dynamic Syst., Measurement, Contr., 102, 69-76, (1980)

7. Pearson, J.D.: Approximation Methods in Optimal Control, J. Electronics and Control, 453-469, (1962)

8. Stepanenko, Yu., and Vukobratovic, M.: Dynamics of Articulated Open-Chain Active Mechanisms, Math. Biosciences, 28, 137-170, (1976)

9. Thomas, Y., and Barraud, A.: Comande Optimale a Horizon Fuyant, Revue RAIRO, J1, 126-140, (1974)

10. Vukobratovic, M., and Potkonjak, V. : Contribution to Computer Construction of Active Chain Models via Lagrangian Form, J.Applied Mechanics, 46, 181-185, (1979)

11. Vukobratovic, M., and Potkonjak, V. : Dynamics of Manipulation Robots, Vol. I, Springer Verlag Berlin Heidelberg New York, (1982)

A Programming Environment for Sensor-Controlled Robots *

Ronny M. Blomme and Jan M. Van Campenhout

State University of Ghent, Electronics Laboratory
St.-Pietersnieuwstraat 41, B-9000 Ghent, Belgium

Abstract: This paper gives an overview of the research activities in robotics at the Electronics Laboratory of the State University of Ghent over the last few years. Two themes emerge, namely (i) the use of concurrent high-level programming languages to implement the hard real-time control functions needed for sensor driven and compliant motion; and (ii) the need for interactive graphical simulation and programming environments for sensor-equipped robots. With respect to the first theme, we describe the implementation of compliant motion on a Modula-2 based multiprocessor. Regarding the second theme, we observe that environments for programming a robot have been developed ever since robots have existed. These environments have evolved from simple *teach-by-doing* procedures towards *implicit task-level* programming and simulation systems. Although some of these complex systems are commercially available, most of them are not extendable and so researchers cannot use them for experiments and testing new ideas. Smalltalk should prove quite a suitable tool in solving this problem [15,20]. Based on two concrete developments in Smalltalk, we discuss the appropriateness of this language to implement graphical-robot-programming environments.

Keywords: Compliant motion, multiprocessor, object-oriented programming, simulation.

1 Sensor controlled motion

Robots of the first generation were used as positioning devices: the manipulator was programmed by an on-line *guiding procedure* (teach-by-doing), and no unforeseen alterations of the environment were allowed during program execution.

Off-line programming techniques and the application of different *sensors* extend the capabilities of the robot [4]. Flexibility and adaptibility have become important aspects of manipulator control systems; adaptibility requires the use of additional sensors. However, the use of sensors such as force sensors introduces two main problems:

- *task specification*: the class of possible motions is dramatically extended. Which are the relevant programming-language primitives ? How do we specify this broadened class of motions ?

*This paper is based on previous presentations published in [1,2,3]

NATO ASI Series, Vol. F 71
Expert Systems and Robotics
Edited by T. Jordanides and B. Torby

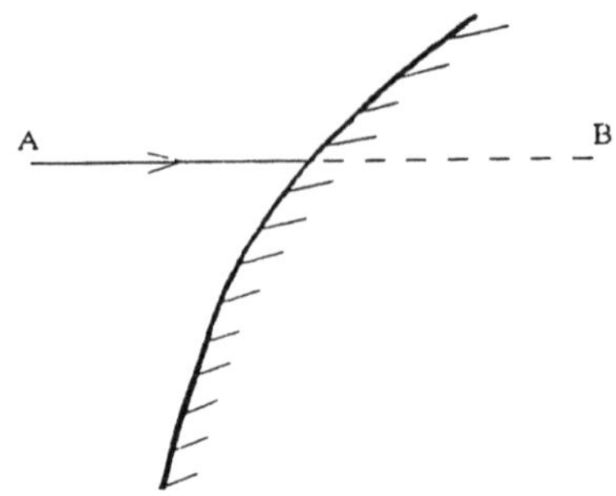

Figure 1: Move ... Until instruction

- *run-time control*: the implementation must support a very broad class of motions. What are the required control techniques ? How must the control software be structured to interpret such a large set of motion primitives, in a hard real-time way ?

Sensor-controlled motions fall into three broad classes: *look and move method*, *guarded motion* and *compliant motion*.

1.1 Look and move method

In this class, sensor and motion control are totally separated and sequentially executed. For instance, a camera determines the position and orientation of an object and, afterwards, the robot moves towards that object and can grasp it.

In the other two classes, guarded and compliant motion, both actions are intertwined.

1.2 Guarded motion

An example of guarded motion is the `Move <constraints> Until <sensor condition>` construction. While the manipulator is moving, some sensors are *monitored* without affecting the predefined movement. As soon as a condition involving sensor data is fulfilled, the motion is (gracefully) interrupted, some intermediate actions can be undertaken, and another motion primitive can be started. For instance, when the manipulator approaches an object (figure 1), the place of which is not exactly known, guarded motion can be applied. While the robot moves toward the object, a contact force sensor is monitored; as soon as a certain force threshold is exceeded during contact, the gripper has located the object and can carry on with the next task.

In fact we have not introduced a new motion primitive here. Guarded motion can be thought of as being generalized *exception handlers* [7]. The handler is a procedure which is linked with an instruction-block. When an exception occurs during the execution of this block (e.g., a force threshold is exceeded), the execution is suspended and the handler procedure is activated. The handler can withdraw the end effector, call for operator assistance, etc. On normal termination of the handler, the execution of the main program resumes with the statement following the interrupted instruction block.

In Modula-2, the language used in our implementation of the robot-control software, standard facilities are not yet provided in the "first working-draft, Modula-2 standard document" [17]. Some people have already extended Modula-2 in a non-standard way with exception-handling facilities [21]. The previous example can then be written as

```
EXCEPTION NoObject;
EXCEPTION ObjectDetected(Force);

TRY
  MoveStraight(B); (* WHILE monitoring sensors *)
  RAISE(NoObject) (* Generate an exception *)
EXCEPT
  | NoObject: WriteString('No object detected');
              HALT
  | ObjectDetected(Force): WriteString('Object detected at position');
                           WritePosition(CurrentPosition)
  | ELSE ElseHandler
END;
```

The most natural way to implement this mechanism is with parallel processes: testing the exception condition(s) or monitoring can be executed in *parallel* with the main program instructions, e.g., a motion in free space. The advantages of a parallel implementation are

- a lower response time and a higher execution speed;
- software reusability, because existing motion primitives can still be used;
- a major increase of software functionality without a similar increase of the software complexity. Indeed, the parallel operations are textually separated; no combinatorial explosion (Cartesian-product effect) takes place.

1.3 Compliant motion

When the position of the end effector is constrained and partially determined by the robot environment, one speaks of compliant motion. Typical examples are: opening a door, peg-into-hole assembly, contour tracking, etc. Mason [16] presents a more formal description of this type of motion. He states that such tasks allow only partial positional freedom. Velocity and force directions are orthogonal. "Neither pure position or pure force control is appropriate in this case, but rather a hybrid mode of control".

When a robot has to insert a peg into a hole (figure 2), it has no motion degrees of freedom in the x and y direction, nor can it rotate around these axes. These directions are force-control directions. The generated trajectory is determined by velocity constraints as well as force constraints.

It is clear that an implementation of compliant motion using pure position control is not appropriate, as unlimited forces can occur during contact of the manipulator with an object at an unknown position. At least some *passive* mechanical compliance should be present in the manipulator linkage. When only passive compliance is aimed at, no special control laws must be implemented. The most suitable solution, however, is *active* compliance implemented in the software control loop. This method offers the additional advantage of programmability. Many researchers have investigated the problem of active-compliant motion control [5,18]. Although, most commonly, force sensors are mentioned

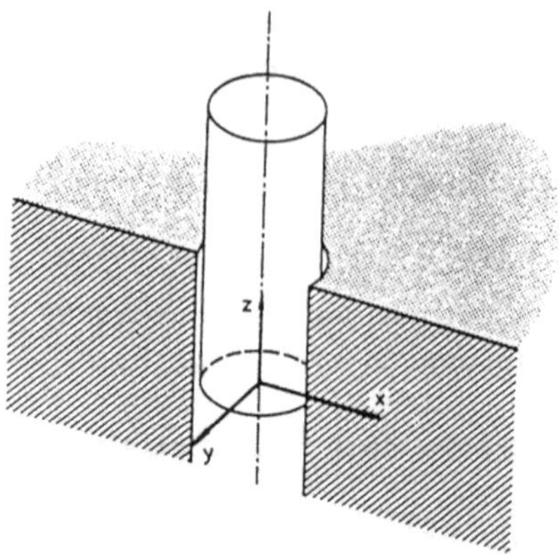

Figure 2: Peg into hole assembly (after De Schutter [12])

in the literature, these control techniques can also be applied with other kinds of sensors, e.g., proximity sensors.

With compliant motion, the trajectory is modified or even generated by the sensor readings. A combination of position and force conditions will determine the robot trajectory. An essential device in the specification of hybrid motions is the *task frame*. It is a coordinate frame in which position (or velocity) controlled directions and force controlled directions are complementary and orthogonal. In fact, sensor constraints can be defined recursively: we specify the sensor constraints in a task frame, but this frame itself can be determined by sensor data. For instance, when we want to apply a force orthogonal to a surface with a variable orientation, we define a task frame in the contact point of the force sensor and the surface. In that task frame we put our force constraints and simultaneously the task frame must track the orientation of the surface using sensor data.

In accordance to Mason's theory, the task specification should remain independent of the control strategies (control laws), e.g., servo gains are of no concern to the application programmer.

1.4 Implementation platform

The different concepts and problems mentioned so far show that advanced real-time applications, such as sensor-controlled robots, need a lot of computing power. This power is needed to perform coordinate transformations and intensive floating-point calculations, and to meet hard real-time constraints.

Since flexibility and programmability are desirable and useful features, specialized dedicated hardware is not a very suitable solution. It is also desirable that the machine controlling the real-time process supports the parallelism inherent in real-time control applications. Furthermore, it is well-known that the use of high-level programming languages is beneficial for a high programmer productivity, and more reliable and maintainable software products.

The use of a concurrent high-level language on parallel hardware thus seems appropriate to fulfill the above needs [6]. A concurrent, high-level language-based, real-time controller, designed with these considerations in mind, was proposed by Van Campenhout and Debaere [25]. We only include a brief description here.

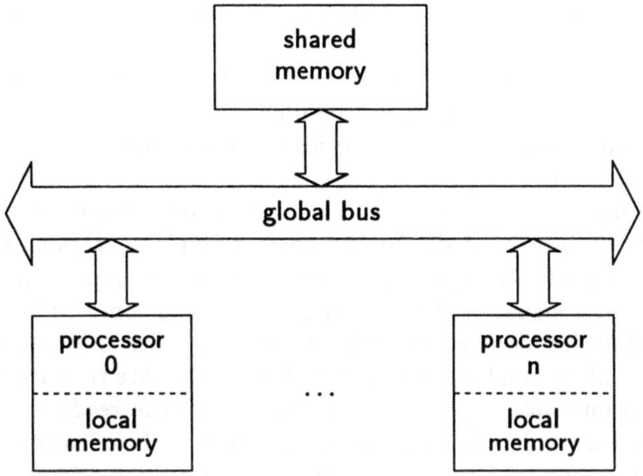

Figure 3: Shared memory architecture

1.4.1 Architecture and language

The machine is a shared-memory multiprocessor with dynamic scheduling (figure 3). Dynamic scheduling is an important aspect of real-time controllers. In our controller, processors search for available work. The decision of which process will be executed on which processor is made at *run-time*; processes can migrate from one processor to another as needed. To support this form of *dynamic load sharing*, the program code is executed from shared memory. To make this at all possible, programs are stored in a compact, intermediate representation.

Using an intermediate code with a high information density, the bus traffic is reduced and the saturation limit is moved to higher levels. Intermediate code must be interpreted. Interpretation is known to be slower than native code execution. However, the considered application heavily uses floating-point calculations. This aspect of the application renders the interpretive slowdown rather unimportant [10,11].

Modula-2 has been adopted as the high-level programming language. Being a precursor of Ada, this language and its environment offer some particular advantages, such as strong modularity, information hiding, parallelism, and access to machine-level resources. Modula-2 is simpler than Ada, and some of its implementations can be found in the public domain.

1.4.2 Real-time operating system

In many cases it is possible (though not easy) to write real-time software applications without the support of a resident, real-time operating-system kernel. Although this approach increases performance, an important advantage of a real-time operating-system kernel with built-in intertask synchronization is that it simplifies the translation from system design to code [19]. The problem is that real-time kernels are not easily ported to other hardware, and that very few support multiprocessor operation. For that reason, the real-time operating system in our implementation was entirely written in Modula-2.

This operating system supports the definition, activation and scheduling of user-defined processes.

In our control software, a major functionally independent piece of code constitutes a process [26]. This *macroscopic* form of parallelism is explicit: the parallel constructs are visible in the source program. Software for robot control can be split up in parallel processes in a rather natural way.

Processes can have to wait either for each other, or for external events, therefore different process states are needed. In the current implementation, a process can be *running*, *ready* or *waiting*. Interprocess synchronization is implemented using *signals*. Processes can wait for a signal or they can send a signal to some waiting processes.

The signal mechanism alone is not sufficient for time critical processes, i.e., processes which wait for external events, for two reasons: (i) process-state transitions involve considerable overhead and bus traffic; (ii) the precise instant for the *ready* to *running* transition depends on the rate at which other processes perform sychronization operations. Preemption is voluntary. For that reason, an efficient interrupt mechanism is implemented in our controller. A process waiting for an external event is kept on the processor, in a substate *I/O-Wait* or *running*. When an external event occurs, a transfer to the active state is generated immediately.

1.5 The implementation of motion-control primitives

In general, robot-motion control software consists of several hierarchical layers, each representing a certain level of abstraction of the underlying levels.

The highest level implements the *user (application) program*. In our implementation, the specification of a robot task takes the form of a Modula-2 process containing several procedure calls to a library of robot-motion primitives (similar to Kali [14]). Examples of such motion primitives are "move straight" and "open gripper".

The second level implements the movements by a means of *trajectory generation* in Cartesian space. The trajectory is divided into smaller path segments. Time, velocity and acceleration constraints are taken into account. The result is a sequence of points in Cartesian space with the corresponding passage times.

The third level transforms the sequence of points in Cartesian space to joint space. This is called the *inverse kinematic transformation*.

So as to obtain a sufficiently smooth sequence of setpoints for the servo algorithms, further interpolation is required in the joint space.

The lowest level in the hierarchy contains the closed loop *servo-algorithms* (PID). These algorithms transform the position error into an output proportional to the motor current. The local position feedback is derived from incremental encoders.

1.5.1 Pure position control

For pure position control, the system of processes is arranged as a pipeline. The different stages in the pipeline are coupled by means of asynchronous buffers. The synchronization of these FIFO queues is implemented using *signals*, reporting full or empty buffers. The only process that has hard real-time constraints is the servo-process (lowest level). It is implemented as an interrupt driver with a 200 Hz interrupt rate. The other processes in the pipe operate on different data, and because of the buffers, the highest level (stage) contains the most recent data. Depending on the dimension of the buffers and on the

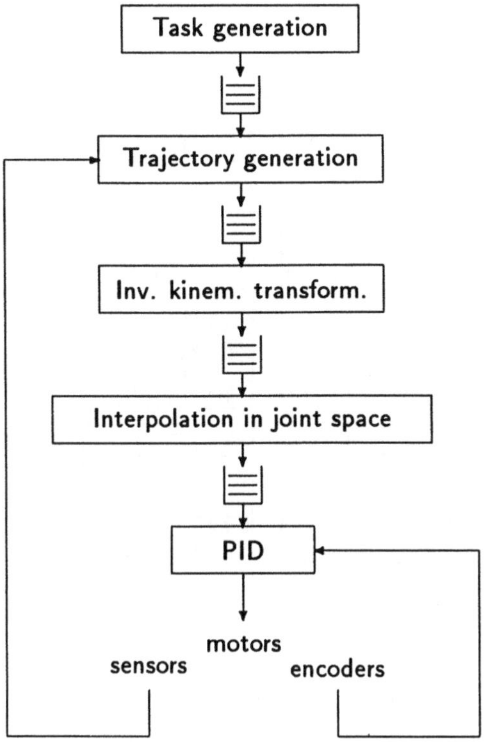

Figure 4: Pipeline parallelism

computational delays, there exists a variable time lapse between the passage through the first and the last stages in the pipeline.

1.5.2 Compliant motion

The introduction of sensor data at the trajectory-generation level in Cartesian space creates an additional feedback loop (figure 4). This loop contains a variable and large delay time which causes the correction signals to be computed on stale data. When tight and precise control must be derived from the sensor information, it is virtually impossible to keep this control loop stable.

When we want to keep the advantages of asynchronous processes, but eliminate the large delays, we have to feedback the sensor data at a lower level. The next paragraph explains how this problem can be solved.

The necessary transformations of the sensor data can be linearized, since the correction signals are usually small. In fact, we have created an inaccurate but fast control loop, and have reduced the transformations to 'simple' matrix multiplications such as (i) the transformation of errors measured in the sensor frame to corrections in the end-effector frame and (ii) the inverse Jacobian: the transformation from Cartesian-end-effector frame to joint space.

Of course, these linearized entities must be updated, since the linear approximation

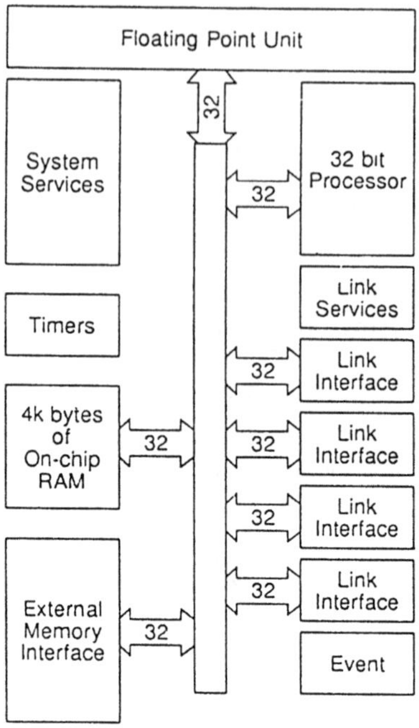

Figure 5: INMOS T800 transputer

is only acceptable in the neighborhood of the point where the linearization is made. The linearized quantities change at a slow rate relative to the frequency of use. Consequently, they can be updated in parallel and asynchronously with the control processes.

Linearization only eliminates first-order errors. Second-order terms are not eliminated and, in fact, accumulate as a result of the integrating nature of the control loop. In the sensor-controlled directions, these errors are canceled by the sensor control loop; in the other directions, the elimination of the integrated second-order errors can also be done by an asynchronous, parallel process, which uses the original slow but accurate transformations.

Thus, we have split up the global feedback loop in a linearized process, which computes rapidly varying quantities, and which has a fast but inaccurate response; and a slower but accurate process.

1.6 Transputer-based architecture

Section 1.4 describes a multiprocessor with a shared-memory architecture used as a robot-controller. A second type of parallel architecture is a distributed one. Transputers can be used as a basic component in such an architecture. The transputer links constitute the connections in a distributed network. The transputer has a RISC-architecture (figure 5) and is programmable in high-level languages. All motion-control software implemented on the shared-memory multiprocessor (Modula-2) was ported to the transputer system (Occam). In a project on mobile robots, the data acquisition for ultrasonic sensors is completely integrated in a transputer network. The same network will be used to simulate

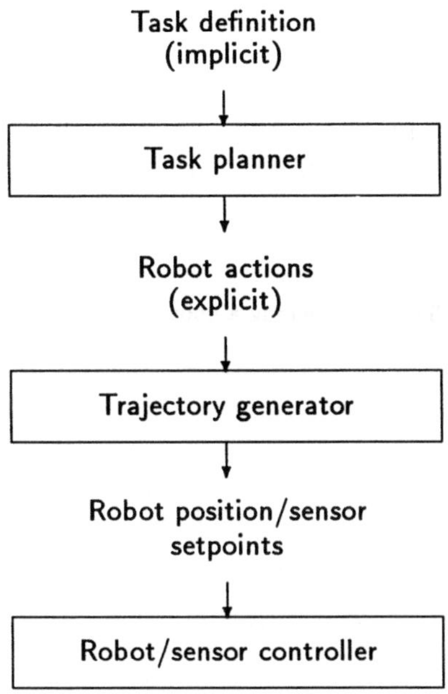

Figure 6: Data abstraction levels

a neural network for the sensor fusion problem. A neural network based on transputers will also be used to solve the low-level-motion control problem (kinematic and dynamic equations).

2 Robot programming

The first robots were programmed using a *teach-by-doing* method: the robot was shown how to complete its task (*on-line*, with a real robot). This method is highly interactive and easy to comprehend, but it is expensive and not very expressive: complex, tasks with an irregular, recurrent structure and tasks that require situation-dependent decision making are almost impossible to program.

A solution to these problems was given by the *off-line* programming techniques which use programming languages with a high abstraction level (e.g., VAL, AUTOPASS, PASRO, ...) [4]. They have a high expressive power and some have the possibility to include geometrical data extracted from CAD databases. Two classes of languages are distinguished: the *explicit* languages, which allow programming on a joint level, end-effector level, or object level; and the *implicit* languages, which need a task planner and methods of artificial intelligence to extract the robot commands (objective level).

With implicit languages it can be very difficult to predict the solution that the task planner will generate. This solution consists of explicit robot motion and sensor commands (figure 6). But even with an explicit textual description of the robot actions it is difficult to get a correct idea of what will happen during the on-line task execution. This is even more so with sensor-controlled actions. For example, opening a door or a drawer could be described as [12,22]

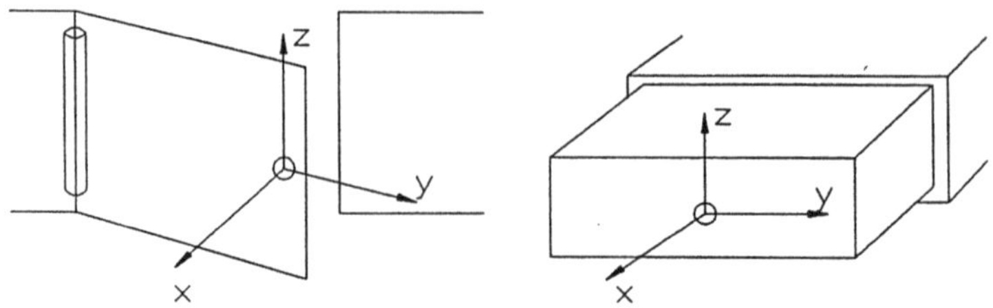

Figure 7: Opening a door or a drawer

```
grasp Handle;
AT Handle
   Vx = k;
   Fy = 0;
   Fz = 0;
   Mx = 0;
   My = 0;
   Mz = 0
DURING 3 seconds;
```

where `Handle` is a frame fixed to the handle of a door or a drawer (figure 7). The constraint that there must not be any torques in any direction will force the robot to rotate around the hinge of the door. The same constraints will generate only a translation in case of the drawer. Without any knowledge about the environment, this specification is ambiguous.

Since it is difficult to understand and to translate physical actions into a textual representation there is a need for off-line visualisation of the manipulator actions — a picture is worth a thousand words. Simulation and animation techniques are very helpful tools when programming a robot. In reference [8] the possibilities and the application of computer graphics for robot programming are discussed.

The main advantages of animation and simulation are:

- faster software development;
- fewer bugs;
- simulations can be used as learning or training facilities for the human end user, but also for the on-line, real-time control algorithms;
- experiments with a real robot are expensive and can be destructive; simulations cannot.

That simulations can be used to adapt control algorithms can be illustrated by the previous example of the force-controlled opening of a door. The door will rotate around the hinge with a certain velocity. This velocity results from our simulation and can be fed forward to the on-line control algorithms. This additional feedforward information improves the performance of the task execution [23,12].

3 The Smalltalk environment

Simulation environments that are commercially available are big complex systems. The system itself cannot be changed because the source code of these systems is not available. Because of this complexity and unavailability of the source programs, there is a need for an experimental environment that allows rapid prototyping. In this way, new ideas such as sensor models or task specification can be tested and worked out.

Smalltalk is a good candidate as a tool to solve this problem. It is based on the object-oriented paradigm and it has a graphical user interface with debugger, window manager, garbage collector, etc.

3.1 The object-oriented paradigm

In reference [24] an initial definition of object-oriented programming (OOP) is given:

> 'Object-oriented programming is a type of programming that provides a way of modularizing programs by establishing partitioned memory areas for both data and procedures that can be used as templates for spawning copies of such modules on demand'

Data encapsulation is one of the more important features of object-oriented programming. Objects are a kind of *active data* which interact through sending messages to each other or to themselves. For example, an object can be asked to visualize itself on a display. Objects can have some hidden memory (*instance variables*) and some procedures (*methods*) of which the implementation is shielded from the environment. The message-passing system activates these methods. Instance variables can be accessed only through the methods.

Objects with identical structures are grouped together: they are members (*instances*) of the same *class*. As mentioned in [24]:

> 'a class is a template for creating actual functioning objects of a given type'

Objects can be used to encode knowledge of the real world; the classes they belong to reflect the levels of abstraction of the world models. Classes can have subclasses. These contain more specific and additional information and methods. Instances of subclasses inherit all behavior of their superclass(es). *Multiple inheritance* allows objects to inherit from more than one superclass, but most implementations (including Smalltalk) do not support multiple inheritance because of the problem that two different inherited methods could have the same name.

Due to this inheritance, *polymorphism* is possible : one single method can be applied to different types of objects since a method can be applied to instances of the class itself *and* to instances of subclasses. Because of this polymorphism, different types of objects can be assigned to one variable. That is why the system can only determine at *run-time* what method should be applied (*dynamic/delayed binding*).

The capabilities of the class hierarchy and the orthogonal implementation of objects and classes constitutes the power of Smalltalk. Figure 8 shows a fragment of the class hierarchy of the basic Smalltalk system. All applications are an extension of this hierarchy.

```
Object
    Magnitude
        Character
        Date
        Time
        Number
            Float
            Fraction
            Integer
                LargeNegativeInteger
                LargePositiveInteger
                SmallInteger
```

Figure 8: Smalltalk class hierarchy

3.2 User-interface

Smalltalk-80 is a multiwindow environment and user/system interactions are menu/mouse oriented [13]. Two window types provide very important tools:

- the contents of all classes (its methods, variables, hierarchy, ...) can be inspected or edited with the *system browser*;
- the *workspace window* is an editor in which the Smalltalk interpreter can execute a piece of highlighted text.

The workspace windows also provide debugging tools and *inspectors*. These inspectors allow the examination of an object's contents and new values can be assigned to variables during debugging.

The virtual memory management and garbage collection are additional tools that make Smalltalk a very powerful interactive environment for rapid prototyping.

3.3 Results

The experimental simulation environment has been implemented in Smalltalk-80 on a Tektronix 4315 workstation by a student in electrical engineering. This environment can be used to develop a mature, robot-programming environment. It must be able to simulate robots with sensors such as force/torque sensors or cameras. Another student has implemented a simulation environment for ultrasonic sensors of mobile robots. An acoustic ray-tracing algorithm has been implemented in color Smalltalk-80 on a Tektronix 4319 workstation. Both dissertations have provided us with significant expertise in Smalltalk, its features, advantages and drawbacks on which we shall dwell in the next section.

3.4 Smalltalk: pros and cons

According to Tello [24], OOP is the programming paradigm that takes structured programming to its natural, logical conclusion. This suggests that the OOP paradigm is not an abrupt change from the traditional modular style. However, the combination of extreme modularity, reusability and inheritance results in significant advantages:

- the ability to create multiple non-interfering instances of a given function or object from the same template;
- a truly modular programming environment where redundancies in coding are kept to an absolute minimum;
- standard calling conventions for a broad range of operations that exhibit differences in behavior;
- a means of managing very large programming projects by breaking up large problems into smaller, independently functioning, highly visible parts.

The Smalltalk environment is an isolated system due to some of the following drawbacks:

- there is a lack of portability of Smalltalk source code between different Smalltalk implementations (e.g., Smalltalk-80 under UNIX, and Smalltalk-V under MSDOS);
- interfacing to other languages is not trivial;
- Smalltalk-80 runs on workstations and a graphical display is necessary. The X-window environment is not fully supported;
- getting acquainted with Smalltalk takes some time even (or especially) for programmers with experience in traditional languages. The class hierarchy is complex and the inheritance mechanism can cause subtile side-effects;
- Smalltalk-80 on UNIX workstations requires a lot of resources. The standard image is not "small" (1.5 MB);
- since Smalltalk-80 is an interpreted environment, performance could be a problem, especially for real-time applications. But this drawback is not a conceptual disadvantage and it can be relaxed using interpreter optimization, parallellism and hardware optimizations and accelerations [9].

4 Conclusion

This paper gives an overview of some research projects in robotics at the Electronics Laboratory of the State University of Ghent. Experiments have convinced us of the feasibility of asynchronous high-level language programming techniques in the design and implementation of concurrent software, in particular for the hard real-time problem of implementing sensor controlled and compliant motion.

Next, we have discussed the growing need for graphical simulation and animation systems, able to cope with robot environments and different types of sensors. Smalltalk has been found to provide a good environment for rapid prototyping in this area. The major advantages are the ease with which simulations of the robot mechanism and its environment can be set up, and the ease with which graphical representations can be constructed. The interactivity of the Smalltalk environment also allows very flexible and rapid prototyping although it may pose some efficiency and portability problems in complex environments.

5 Acknowledgments

This text presents research results of the Belgian research foundation IWONL, the Belgian National incentive-program for fundamental research in Artificial Intelligence and the Belgian programme on inter-university attraction poles initiated by the Belgian State — Prime Minister's Office — Science Policy Programming. The scientific responsibility is assumed by its authors.

References

1. Blomme, R.M., Van Campenhout, J.M.: Asynchronous Parallel Programming Techniques for Compliant Robot Motions. In: Second International Conference on Software Engineering for Real Time Systems, IEE 309, pp. 204–208.

2. Blomme, R.M.: Programming compliant robot motions in a concurrent high-level language. In: Intelligent Robots and Computer Vision VIII: Systems and Applications, SPIE conf. proc. vol. 1193, Philadelphia, pp. 229–239.

3. Blomme,R.M., Van Hauwermeiren, G.R.: Smalltalk as a Simulation Environment for Robot Manipulators. In: Modeling and Simulation, Proc. of the ESM 1990, ed. Bert Schmidt, pp. 368–373.

4. Blume, C., Jakob, W.: Programming Languages for Industrial Robots. Springer-Verlag, Berlin, 1986.

5. Brady, Hollerbach, Johnson, Lozano-Pérez and Mason (eds.): Robot Motion: Planning and Control. MIT Press, 1982.

6. Cox, I.J., Gehani, N.H.: Concurrent Programming and Robotics. In: The International Journal of Robotics Research, Vol. 8, No. 2, pp. 3–16, April 1989.

7. Cox, I.J., Gehani, N.H.: Exception Handling in Robotics. In: IEEE Computer, pp. 43–49, March 1989.

8. Dai, F., Dillmann, R.: Computergraphik für die Roboterprogrammierung. In: Informatik Forschung und Entwicklung, Springer-Verlag 1988, 3: pp. 128–138.

9. Debaere, E.H., Van Campenhout, J.M.: Interpretation and Instruction Path Coprocessing. MIT Press, 1989.

10. Debaere, E.H., Van Campenhout, J.M.: Language coprocessor to support the interpretation of Modula-2 programs. In: Microprocessors and microsystems, vol. 11 no. 6, July/August 1987.

11. Debaere, E.H., Van Campenhout, J.M.: A Shared-Memory Modula-2 Multiprocessor for Real-Time Control Applications. In: Microprocessing and Microprogramming, no. 18, North-Holland, pp. 213–220, 1986.

12. De Schutter, J.: Compliant robot motion: task formulation and control. Ph. D. Thesis K.U.Leuven, 1986.

13. Goldberg, A., Robson, D.: Smalltalk-80: the language and its implementation. Addison-Wesley, 1985.

14. Hayward, V., Hayati, S.: Kali: An Environment for the Programming and Control of Cooperative Manipulators. McGill Research Centre for Intelligent Machines, Montréal, Internal Report, March 1988.

15. LaLonde, W.R., Thomas, D.A., Johnson, K.: Smalltalk as a programming language for robotics? In: Proc. 1987 IEEE Int. Conf. on Robotics and Automation, North Carolina, 1987, pp. 1456–1462.

16. Mason, M.T.: Compliance and force control for computer controlled manipulators. In: IEEE Transactions on Systems, Man, and Cybernetics, vol. SMC-11, pp. 418–432, 1981.

17. Working Group (IST/5/13) for ISO/IEC JTCI SC22/WG13, First Working Draft Modula-2 Standard, Document D92, British Standards Institution 1989.

18. Paul, R.P.: Robot manipulators: mathematics, programming, and control. MIT Press, Cambridge, 1981.

19. Pressman, R.S.: Software Engineering: a Practitioner's Approach. McGraw-Hill Book Company, 1987.

20. Rock, S.T.: Developing robot programming languages using an existing language as a base — A viewpoint. In: Robotica, Part1, January 1989, Vol. 7, pp. 71–77.

21. Rovner, P.: Extending Modula-2 to Build Large, Integrated Systems. In: IEEE Software, November 1986, pp. 46–57.

22. Simkens, P., De Schutter, J., Van Brussel, H.: Contact force calculation in a compliant motion simulation system. In: Advances in Computing Research, Issues on Robotics and Non-linear Geometry8, JAI-press, Vol. 6, 1990.

23. Simkens, P., De Schutter, J., Van Brussel, H.: Graphical simulation of compliant motion robot tasks. In: Proceedings INCOM'89 Symposium on information control problems in manufacturing technology, Madrid, 24–27 Sept 1989, pp. 1–7.

24. Tello, E.R.: Object-oriented programming for artificial intelligence, a guide to tools and system design. Addison-Wesley, 1989.

25. Van Campenhout, J.M., Debaere, E.H.: A concurrent high-level language based real-time controller. In: Journal of the Institution of Electronic and Radio Engineers (IERE), Vol. 58, No. 2, pp. 57–62, March/April 1988.

26. Wirth, N.: Programming in Modula-2. Springer-Verlag, Berlin, 1982.

SYMBOLIC SIMULATION OF SENSORY ROBOT PROGRAMS *

Vincenzo Caglioti, and Marco Somalvico

Dipartimento di Elettronica, Politecnico di Milano, P.za L. da Vinci 32, 20133 Milano, Italy

Abstract: *The symbolic simulation of sensory robot programs is introduced, in order to allow their systematic analysis needed to cope with difficulties arising from the fact that the program execution may result in an infinite set of possible evolutions of the robot work cell. Methods are reported for the simulation of simple sensory instructions. An experimental system for symbolic simulation is described*

Keywords: *Simulation systems, symbolic simulation, sensors simulation, inference systems, manipulation programs, sensory robots*

1. Introduction

In this paper the design and implementation of a symbolic simulator of sensory robot programs devoted to the accomplishment of manipulation tasks are illustrated. The goal of a simulation system is the analysis of a program in order to determine the modifications of the work environment of the robot deriving from the execution of the program. This program analysis is relevant both to robot program development and debugging, and to the extraction of a representation of the task associated to the program, e.g., for exception handling applications [2].

Presently many simulators are available in which the motion of the robot arm is graphically represented. However, most of the simulators lack the simulation of sensor instructions. Robot programs rely on the use of sensors in order to allow the accomplishment of a given task, in spite of the uncertainty affecting the work environment. A key point in the present work is the analysis of the propagation of sensory information throughout the manipulator program.

Traditional robot simulation systems usually allow the analysis only of programs with a finite number of possible driven evolutions of the work environment. Each *state* of these evolutions can therefore be represented by the set of the numeric values of the physical variables characterizing the situation of the work environment at a given instant.

The introduction of sensor instructions in the manipulator program makes it possible to provide for an infinite set of possible evolutions driven by the execution of the program. In this case the infinite number of states does not allow representation of each state explicitly in terms of the numeric values of the descriptive physical variables. Instead, the states have to be grouped into a finite number of subsets (called nodes), within which the values of the physical variables are described symbolically.

* Work partially supported with grants of IBM-Italy

NATO ASI Series, Vol. F 71
Expert Systems and Robotics
Edited by T. Jordanides and B. Torby

Due to the symbolic nature of the representation of the values of the state variables, the analysis of the program often involves problems that are both computationally and analytically intractable. Therefore, it is necessary to introduce some restrictions, in order to develop a viable system with reasonable performances. These restrictions result in a set of "programming rules" that have to be followed by the robot programmer.

Not only rigorous analytical methods, but also short-cutting ad hoc methods (e.g., "rules of thumb") have to be employed in order to carry on the analysis of the program. This observation has led us to choose an inferential system as implementation tool, in which some inference rules activate rigorous solution methods, while other inference rules try to capture the implications of particular conditions leading to a simplification of the analysis.

The system has been designed in such a way that it can notify when, in place of an exact solution, an approximate or partial solution is provided due to analytical and computational difficulties and/or lack of simplifying conditions. In this case, the system suggests to the user to either proceed with a simplified program analysis or to redefine a simplified version of the program and start the analysis again.

In Section 2, the symbolic simulation problem is introduced and some simplifying assumptions are stated, helping to specify the scope of the simulation system. In Section 3, an inference system for symbolic simulation is described. In Section 4, the simulation of sensor instructions is illustrated. In Section 5, a practical example is reported of the simulation of a simple program containing some sensor instructions. In Section 6, some conclusions are presented.

2. The Problem of Symbolic Simulation

2.1 Some simplifying assumptions

The robot program is supposed to be written in a manipulator level language (e.g., VML, VAL, AML).

A robot program is thought of as consisting of a list of statements written in the robot programming language. The program statements can be subdivided into four classes:

- actuation statements as, e.g., "*MOVE_ARM frame*";
- sensory statements as, e.g., "*READ_IMAGE_POINT rank_order memory_variable*", which assigns to *memory_variable* the cartesian coordinates of the object vertex specified by *rank_order*, where *rank_order* = "((1 0) 3)" indicates a criterion for scanning the points of the image plane based on the order in which the points project themselves on a given direction: in our case (1 0) indicates the x-direction, while 3 indicates the third projected point;
- processing statements as, e.g., "*ADD memory_variable_1 memory_variable_2*";
- control statements as, e.g., "*IF condition THEN statements*".

The empty program is indicated by the empty list ().

The robot is regarded as an agent that produces some modifications of the *environment*. The environment is that part of the world, that is accessible to the robot by means of both its actuators and its sensors. For instance in manipulation tasks the environment is the working space of the robot workcell. Therefore the program can be viewed as an operator that modifies the environment *state*.

A *simulator* can be thought as an analyzing program (subject) that predicts the environment *evolution(s)* produced by the execution of a given program (object). An environment evolution can be given in terms of the workcell *state* as a function of time.

$$evolution = state(t)$$

The environment state is described in terms of the values of the physical variables characterizing it. In the case of manipulation tasks, physical variables describe the position of the objects which are present in the workcell together with the position of the the robot arm, their contacts and constraints. Contacts are described in terms of the contact surfaces and the contact points, while constraints are described in terms of the impeded relative motion directions between constrained objects.

The geometrical and physical features of the objects under manipulation (i.e., their *class*) is also specified within the state description, although these features are constant during a whole execution.

Since *memory variables*, i.e., variables referenced in program statements, affect the environment evolution via the program execution, it is convenient to include them in the state variables. While actuation statements modify physical variables, processing statements modify memory variables. Sensory statements modify memory variables by mapping physical ones onto memory variables.

Let us now characterize concepts about particular sets of states.

A *state item* is a type of information that can be utilized in dealing with (sets of) states. We have the following three types of *state item*:

1. *Meaning* items that specify the meaning of a list of physical variables. It consists in a Lisp-like list whose first element is an identifier. This is followed by discrete-valued parameters, and the last element is a list of physical variables. For instance: the last list in "(*contact object_1 object_2 surface_7 vertex_11* (x y z))" contains the variables specifying the coordinates of the contact point. Some legal identifiers in meaning items are: *position*, *constraint*, *class*, *photocell_level*, *control_predicate*, *image_point*, *image_edge*, *image_object*.

2. *Value* items, that specify the numeric values of physical variables referenced in meaning items. For instance:
"(*value* z 0)".

3. *Relation* items, that specify the relations between physical variables, referred to within meaning items, which are not referred to within value items. For instance:
"(*relation* $<$ $(x+y)$ 3)".

By utilizing a set of state items we can define a *state node* intended as a set of states. A *state node* is a singleton if no relation items are utilized, otherwise it is a set of states, that we assume to be an infinite set.

Now let us consider the possibly infinite computations of the robot program, which derive from two different varying causes resulting in a possibly infinite set of environment evolutions. The first varying cause is the *initial state*, which is defined by the values of the relevant physical variables at the beginning of the program execution. The second varying cause is the set of the *input interactions*, defined by the supplementary set of physical variables describing new entities, which some auxiliary tools (e.g., conveyor belt machines, parts' loaders, etc.) insert in the robot workspace.

We may consider the set of the possible states which appear at the end of a given instruction i of the program (whichever be the computations in which instruction i is executed). We assume, as our hypothesis, that the possibly infinite set of such states can be partitioned in a finite number of *state nodes*.

Two further assumption concern state nodes:

A. Let us consider a state node defined by a set of state items. The set of values of any state variable, which is referred to within one or several *relation items*, is an infinite one. These variables are called the *major* variables of the state node.

B. The variables referred to within one or several *value items* of a state node have only one possible value within the state node. These variables are called the *minor* variables of the state node.

2.2 Input interactions

The environment evolution can be written as a function of both the initial state of the workcell, and the transformations performed on the states; such transformations are produced by the program execution, and by external agents as, e.g., external machines. The robot is supposed to be the unique agent in the *interior* of the workcell. The actions produced by these external agents are thus called *input interactions*, because they bridge between the interior and the *exterior* of the workcell.

In manipulation tasks, input interactions correspond to objects entering into the workcell, e.g., by means of part feeders or conveyor belt machines. The possible results of input interactions are specified in terms both of the class (i.e., the geometrical and physical features) and of the input position of the incoming object.

In strict analogy to what happens with state variables, input variables are also described by means of *input items*. The *input items* can be divided in the same three types as the *state items*, i.e., *meaning items*, *value items*, and *relation items*. In addition, the possible results of an input interaction are organized in a finite number of sets (called *input nodes*); an *input node* is defined by a set of *input items*. Some instances of identifiers, which can appear in meaning input items, are the following ones: *class*, *relation*, *input_position*, etc.

In the present implementation of the simulation system only a finite number of possible object classes are acceptable, while an infinite number of possible input positions are allowed.

Some further assumptions now follow.

1. Input interactions can only occur in correspondence to the execution of "input statements" of the program. Therefore, no input interaction can occur independently of

the program execution. In manipulation tasks, input statements consist mainly in picking up an object from a part feeder or from a conveyor belt machine. Within this framework, the environment evolution can be written as a functional of the program, the initial state of the workcell, and the results of the input interactions.

$$evolution = state(t) = F(program,\ state(0),\ input\_interactions) \quad (1)$$

Therefore, in a program simulation system, both the initial state and the results of the input interactions must be given in order to determine in an univocal way the evolution. Non functional dependences (e.g., multi-valued "functions") would be needed in order to take into account the uncertainty of the effect of the execution of an action even when starting with a determinate state. A current approach consists in considering only "small" uncertainties around the expected values of the state variables so that the dependencies can be linearized around the expected values. In this way, a functional dependence is maintained between the expected values, while, if we arrange the state variables in *state vectors*, the variance-covariance matrices of the state vectors are updated by combining them with the Jacobian matrix [3].

In the present approach, it is supposed that the contribution to the uncertainty of the major state variables deriving from the imprecise execution of the program actions is much smaller than the contribution deriving from the *indeterminateness* of both the initial state and the result of the input interactions. Therefore, the evolution can be well approximated by a function such as that of equation (1).

2. The results of an input interaction is independent of the physical time it occurs. It possibly depends only on the results of the input interactions occurring previously during the program execution.

3. The robot environment is supposed to have no "self-evolution", i.e.,

$$self\_evolution = F(\ (),\ state(0),\ input\_interactions) = state(0)$$

These three restrictions exclude navigation tasks from the scope of this study, since the environment of a mobile robot is hardly protected against agents independent of the robot. Tasks like part transportation or part re-orientation are included. Furthermore, programs aiming at controlling changing environments are excluded as well.

Programs acting on "non-self-evolving" systems do not introduce time constraints. Therefore, the workcell evolution can be photographed at the end of discrete time intervals that correspond to the end of the execution of the program instructions.

$$evolution = (state(0), global\_state(1), \ldots,\ global\_state(n)),$$

where *global_state(i)* represents the environment state at the end of the *i*-th step of the program computation, and it is described in terms of the values of the characteristic variables (positions, contacts, constraints). It might be convenient (e.g., for error recovery purpose) to keep separate the results of the input interactions that occurred during the last execution step from the other state variables. According to this we can write:

$$evolution = ((state(0)),\ (state(1), input\_inter(1)), \ldots,\ (state(n), input\_inter(n))) \quad (2)$$

where $input\_inter(i)$ specifies the results of the input interactions that occurred during the step i of the program computation.

A state variable z is said to be a *major* variable for the step i if there is at least one state node of i for which z is a major variable. In addition, the variables describing the initial state, together with variables describing the input interactions, are called *input variables* since the initial state can be considered as an input due to the fact that it affects the evolution without being affected by it.

2.3 The symbolic simulation problem

According to the traditional approaches, simulation can be thought of as the determination of the evolution of a given system once both its initial state and the operator acting on it are given. Since we are here concerned with (possibly infinite) sets of possible initial states, we introduce a different concept of simulation: simulation consists in the determination of the *dependence* of the system evolution, both on the initial state and the input interactions, obtained as a result of the execution of a given program.

Referring to relations (1) and (2), we define the *simulation* of a given program P as the determination of a function f_P, given by

$$f_P\,(state(0), input\_interactions) = f(P, state(0), input\_interactions),$$

that correctly represents the physical relation between the initial state, the input interactions execution, and the evolution of a non-self-evolving system deriving from the execution of the program P. In practice we assume that a domain of variability of both the initial state and the results of the input interactions is given.

Given a program P, a simulation is called *numeric* if both the initial state and the input interactions may vary within a finite set. A simulation is called *symbolic* if both the initial state and the results of the input interactions may vary within infinite sets.

An example of a symbolic simulation is given by a task in which an incoming part enters into the workcell at an *indeterminate* position (and orientation) within an infinite set (e.g., within an "input" region). The evolution states subsequent to such an input interaction will possibly contain variables dependent on the above entrance position. Therefore, in general, the indeterminateness of physical (and memory) variables is propagated through the program.

Another example of input variability is given by a situation in which some incoming part is of a variable class (and, hence, it has variable geometric features). Often objects of different classes require different manipulation processes. This possibility is provided by the control statements of the robot program (e.g., "if ... then ... else ..."), which drive the execution of different program branches according to some object feature detected by means of sensors.

A successful simulation of a program should allow assessing whether there exists an allowed pair $(initial\_state, input\_interactions)$, i.e., belonging to the domain of variability, for which a meaningful condition on the state variables can be true at a given program step. Examples of questions concerning the simulation are the following ones:

1. Is a given sensory detection sufficient in order to distinguish among a certain number of possible classes of an incoming part?
2. Is the evolution state after step i dependent on the result of input interaction that occurred during step j ?
3. For what values of the initial state variables is the "then" branch of control structure selected at step i?
4. For what values of the entrance position variables of the incoming part A is there a collision between part A and part B during step i ?

Meaningful conditions can often be expressed in terms of meaning items and value items. This suggests the following definition: at a given program step i, a *virtual state node* is a set of states characterized by any collection of syntactically correct state items. A virtual state node is said to be *empty* at a given step i if for any possible pair $(initial\_state, input\_interactions)$ no state of the node can be obtained.

The problem of deciding whether a virtual state node is empty can range from very trivial cases to analytically intractable ones. Even from the mechanical point of view, troublesome problems may arise (e.g., in the determination of the motion of objects constrained in a complicated way with an object whose motion is imposed by the robot). However, the program to be simulated often is written by human programmers. Therefore, many of the unfeasible mechanical and analytical problems either can not be solved, even by the human programmer, or can be solved only by means of heuristic methods. Therefore, the restrictions that are necessarily introduced can be regarded as "programming rules" that, hopefully, do not restrict too much of the programmer's freedom.

In the next Section, the framework of the developed simulation system is further specified.

3. An Inference System for Symbolic Simulation

3.1 The output of the simulation and its construction

Let G be the program graph, representing the structure of the robot program P to be simulated. If P contains loops, then G is a directed *nonacyclic* graph. Since both the robot program P and the simulation program are supposed to terminate, every cycle present in P is supposed to be iterated a finite, bounded number of times. Therefore, the direct nonacyclic graph G can be transformed into a finite direct acyclic graph S_G by unfolding the loops contained in G. This graph that is equivalent to G from the point of view of the program representation is called the *expanded graph* of the program P.

The simulation starts with the information on the *domain* of the input variables. This domain, which is called *input domain*, is expressed by a set of information items. The identifier of all these items is "*input_domain*", while the other components are constructed by following the same syntactical rules as for *relation items*. These items are indicated as *input domain* items.

The output produced by the simulation of a program P is organized in *simulation steps* which are associated to the nodes of the expanded graph S_G. Each step corresponds to the end of the execution of a program statement. A step i is said to precede a step j if there is an arc from i to j in the expanded graph. The set of possible states at a given step i is subdivided into state nodes. Each state node S of step i is connected, through an input arc, to all the state nodes belonging to steps preceding i from which S is reachable. These state nodes are said to precede state node S. State nodes preceding a state node S will be indicated by the following notation S_1^{-1}, S_2^{-1}, etc.

The state of the simulation, at each step i, is represented both by the state of the workcell, and by the dependence of the workcell state on both the results of the input interactions and the initial state (i.e., on the input variables).

If the state node S of step i is a singleton, then the information on its dependence on the input variables is fully represented by the input arcs to S from the state nodes S_k^{-1} preceding S, and by the dependences of each of the S_k^{-1} on the input variables. If S is not a singleton (and hence it is an infinite set of state nodes), then its dependence on the input variables is described by *dependence items* in terms of the functional relations between the state variables in S and the input variables. Dependence items follow the same syntactical rules as relation items.

The set of simulation states (i.e., pairs constituted by state and dependence) are subdivided into a finite number of subsets, called *simulation nodes*, such that the simulation states of a simulation node are described by a common set of state items and a common set of dependence items. Within the set of the dependence items which concur in defining a simulation node, a set of *input domain* items is included. These items define the subset of the domain of the input variables for which the considered simulation node is reached. This subset is indicated as the *input subdomain* of the simulation node.

The additional items required to represent the dependence information relative to a simulation node possibly introduce a splitting of the starting state node. The simulation starts with a set of the possible values of the input variables, i.e., with their domain. The input variables, which describe both the initial state and the results of the input interactions, are indicated by $(x_1, \ldots, x_n)^T$. These possible values are described by the constraints:

$$I((x_1, \ldots, x_n)^T) \leq 0$$

The simulator constructs its output incrementally, by re-iterating a cycle. Each cycle starts with a starting simulation step, consisting of a finite number of simulation nodes, and constructs a new simulation step by simulating the execution of a program step. This execution is simulated separately for each starting simulation node. The eventual input interactions associated with the program step are considered. For each starting node the effects of the step execution are analyzed, and one or more arrival nodes are constructed. In addition, the functional dependence of the variables $(z_1, \ldots, z_l)^T$ of the new node on both the variables of the starting nodes and on the variables of the possible input interactions associated to the program step, is determined. This dependence can be expressed by:

$$(z_1, \ldots, z_l)^T = \mathbf{f}((y_1, \ldots, y_m)^T),$$

where $(y_1, \ldots, y_m)^T$ includes both the variables of the starting nodes and the variables of the input interactions associated with the program step, and where the value of **f** is a vector $(z_1, \ldots, z_l)^T$ (**f** is a vectorial function).

The determination of this dependence may involve the solution of difficult mechanical and physical problems, e.g., the determination of the motion of an object after a non-quasi-static inelastic collision. However, in many cases the effects of the execution of statements devoted to manipulation operations can be determined without dealing with such complex problems. In fact, often collision free motions are desired, and it is sufficient for the user that the simulation system signals the possibility of a collision, instead of analyzing the motion in case of collision. On the other hand, in some simple cases the quasi static motion of a pushing, sliding object can be predicted (see [4]).

The dependence of the variables of the new node on the input variables is determined by combining the above dependence with the dependence of $(y_1, \ldots, y_m)^T$ on the input variables $(x_1, \ldots, x_n)^T$:

$$(z_1, \ldots, z_l)^T = \mathbf{f}((y_1, \ldots, y_m)^T) = \mathbf{f}(\mathbf{g}((x_1, \ldots, x_n)^T))$$

where also **g** is a vectorial function.

Meaningful conditions on the state variables at the simulation step i can be expressed in terms of the major variables of i:

$$\mathbf{C}_i^{(z)}((z_1, \ldots, z_l)^T) \leq (0, \ldots, 0)$$

where $\mathbf{C}_i^{(z)}$ is a vectorial function. These can then be expressed in terms of the major input variables through:

$$\mathbf{C}_i^{(x)}((x_1, \ldots, x_n)^T) = \mathbf{C}_i^{(z)}(\mathbf{f}(\mathbf{g}((x_1, \ldots, x_n)^T))) \leq (0, \ldots, 0)$$

The values of the input variables, for which the above condition is true at step i, are determined by intersecting the set characterized by this last constraint with the domain of variability of the input variables.

It could be interesting to provide a representation of the range set of the state variables $(z_1, \ldots, z_l)^T$ of a node in a simulation step i in the form $D_z((z_1, \ldots, z_l)^T) \leq 0$, given the representation of its (input) subdomain $D_x((x_1, \ldots, x_n)^T) \leq 0$. To this aim it could be necessary to invert the functional dependence:

$$(z_1, \ldots, z_l)^T = \mathbf{h}((x_1, \ldots, x_n)^T)$$

All these problems are solvable in finite time in the linear case. This requires the linearity of both all the constraints expressing the "meaningful" conditions (including the "non-emptiness" conditions of the virtual nodes), and the functional relations between input variables and state variables.

In the case in which the linearity conditions fail and/or the determination of the functional relations is made difficult by the complexity of the involved mechanical problems, the system possibly provides a partial solution to the simulation problem. For any node, one of the three following cases is possible:

A. The simulator is able to recognize that the node is not empty, and it can determine the input subdomain of the node (i.e., the subset of the input domain from which the node is reachable).

B. The simulator is able to decide whether the node is empty or not, but it is not able to determine the input subdomain of the node.

C. The simulator can not even decide, whether the node is empty or not.

3.2 The structure of the simulation system

The simulation system is composed of the following modules:

A. A *user interface*, which receives as input the listing of the manipulation program to be simulated; the description of the input interactions; the domain of the input variables (which describe both the results of the input interactions and the initial states), and the geometric and physical model of the classes of the objects to be manipulated.

B. The *directing module*, which selects the instruction to be simulated, calls the instruction simulator (see the following paragraph C), and organizes the results of the instruction simulation by appropriately distributing the extracted information among the new simulation nodes.

C. The *instruction simulator*, which presides over the simulation of each program instruction. It starts with the set of the *starting* state nodes, and for each of them it determines the set of the new simulation nodes (called the *arrival* nodes), and their dependence on the input variables.

The user interface enables the introduction of the geometric models of polyhedral objects by a type of Constructive Solid Geometry.

The instruction simulator can be subdivided into three (sub)modules.

C1. A first module presides over the correct simulation of both the control statements of the program, and the processing statements.

C2. A second module tries to solve the physical and mechanical problems related to the simulation of the single actuation statements.

C3. A third module simulates the sensing instructions of the program.

The control statements of the program are simulated as follows. The branching instruction "*IF predicate THEN statements_1 ELSE statements_2*", executed at the simulation step S, is simulated by subdividing each node of the step S into two subnodes: one in which the *predicate* is true, and one in which the *predicate* is false. The cycles are simulated by unfolding them, and this reduces them to a sequence of nested branches. The sequential unfolding of the cycle is stopped as soon as a simulation step is reached in which all virtual nodes are empty.

Motion instructions are simulated correctly when the following conditions are satisfied:

- motions are collision free
- objects in contact with the moved object (i.e., the object held by the robot gripper during a motion) either move solidly with it or rest.

These conditions result in programming rules to be followed by the programmer [2]. Moreover, the second condition can also be exploited to determine the motion of objects connected to the moving object by analyzing the impeded relative motion directions.

In the presently available inplementation of the system, both conditions can be violated in some cases. Some necessary conditions for the occurrence of a collision during the motion are used in order to warn the user of the possibility of a collision. In addition, if the collision can be modeled within a quasi-static approach (namely when the inertial forces are neglectable with respect to the frictional forces) and the objects are flat, some rules of the Mason's Sensorless Manipulation [4] are utilized in order to predict the motion of a flat object pushed along the table plane.

4. Simulation of Sensing Instructions

4.1 Sensing instructions: an example of simulation

Some sensor primitives have been implemented, and they have been added to the set of the AML instructions, that can be simulated. These sensor instructions are related to the camera and to an object detector consisting of a photocell which is placed between the fingers of the robot hand.

The camera instructions activate the camera, acquire an image, and extract one or several features. In addition, the camera instructions assign to a memory variable the values of the parameters of one of the image features. The selection of the image feature among those which are present in the image is done in accordance to a criterion, which is indicated in one of the arguments of the instruction.

Observed objects are supposed to be polyhedra. Therefore, object vertices and straight line edges are the only image features which are extracted by the low level processing module attached to the vision sensor.

An example of the instruction that presides over the extraction of an image point was reported in Section 2.1. The extraction of the polar parameters ρ and θ of the straight line identified by an edge (e.g., "*READ_IMAGE_EDGE edge_rank_order memory_variable*") results in the assignment of ρ and θ to *memory_variable*. The feature selection is done in accordance to the criterion indicated in *edge_rank_order*. This argument is composed of two parts. A first part, which is analogous to *rank_order* (see Section 2.1), is used to select an image point. Once the image point is selected, the set of edges, which start at the selected image point, is considered: these edges can be distinguished from each other by means of their orientation angle. The second part of *edge_rank_order* consists in an integer which indicates the rank of the orientation angle of the edge to be selected.

An example of an instruction that allows the extraction of orientation information is the following "*READ_IMAGE_ANGLE edge_rank_order memory_variable*". This instruction assigns the orientation angle of an edge to *memory_variable*. The selection of an endpoint of the edge is performed in a similar way as in the previous case, while the selection of an edge among the edges ending at the selected point is performed otherwise. The edges can be distinguished from each other by means of their length. The edge is therefore selected based on the rank of its length in a decreasing order.

In addition, two more complex instructions have been implemented. The first one allows the localization of an object (whose geometric model is available to the system), provided that a sufficient number of edges are extracted: "*READ_OBJECT_LOCATION object*". The second instructions counts the number of the connected image contours: "*READ_CONTOUR_NUMBER memory_variable*".

Example

In this example the simulation of a camera instruction is illustrated. Let us examine the simulation of a camera instruction as, e.g., the extraction of an image point "*READ_IMAGE_POINT rank_order memory_variable*". Suppose the viewed object has three degrees of freedom: x^o, y^o, and θ. The first two variables represent the Cartesian coordinates of the origin of the reference frame attached to the object within a plane $z = cost$. The third variable represents the orientation angle of the x-axis of the reference frame attached to the object. The coordinates x^P and y^P of an object vertex P are then given by:

$$x^P = x^o + X_o^P\, cos(\theta) - Y_o^P\, sin(\theta), \quad y^P = y^o + X_o^P\, sin(\theta) + Y_o^P\, cos(\theta),$$

where X_o^P and Y_o^P are the coordinates of vertex P in the object reference frame. In the three-degrees-of-freedom case, the image-coordinates x_i and y_i of the image of point P can be written as linear functions of the point coordinates:

$$x_i = \alpha x^P \qquad y_i = \alpha y^P$$

In addition, suppose that these variables belong to the input variables. Let the domain of the input variables x^o, y^o be the rectangle $((0,0)(L,0)(L,M)(0,M))$.

As a result of the simulation, new items are created of the following form:

1. "*(image_point point P)*"
2. "*(memory_variable point* $(x_i\ y_i)$*)*"
3. "*(relation* $= (x_i)\ (\alpha(x^o + X_o^P\, cos(\theta) - Y_o^P\, sin(\theta))))$
4. "*(relation* $= (y_i)\ (\alpha(y^o + X_o^P\, sin(\theta) + Y_o^P\, cos(\theta))))$
5. "*(input_domain* $\leq\ (x^o)\ (L)$*)*"
6. "*(input_domain* $\geq\ (x^o)\ (0)$*)*"

7. "*(input_domain* $\leq$ (y^o) $(M))$"

8. "*(input_domain* $\geq$ (y^o) $(0))$"

9. "*(input_domain* $\leq$ (θ) $(360))$"

10. "*(input_domain* $\geq$ (θ) $(0))$"

The two relation items express the dependence of the two new variables x_i and y_i on the input variables. Let us now examine the virtual nodes. In our example a simulation node is represented by a particular collection of ten items of the form depicted above. Since the object vertex P, which is referred to within three of the ten items, may be any of the visible vertices of the viewed object, then there is (at least) one simulation node for each visible object vertex. Correspondingly, there is a different pair of *input domain* items, within which thee input variable θ is referred to, for each different object vertex P.

For instance, let us assume that the viewed object is a rectangular object whose vertices are A, B, C, and D. Let H be the length of the edge AB and K be the length of AD. Let the reference frame, attached to the object, have its origin in A, and its x-axis coincident with AB (then the Y-axis is coincident with AD). In addition, let *rank_order* be "(1 0) 1". This means that the left-most image vertex is selected.

We have that the left-most image vertex coincides with the image of B, C, D, or A if θ varies between, respectively, 0 and 90, 90 and 180, 180 and 270, and 270 and 360, therefore four simulation nodes are generated. The four simulation nodes share the items 2, 3, 4, 5, 6, 7, and 8. Their difference is relative to the items 1, 9, and 10. For instance in a the simulation node which corresponds to the possibility that A is the left-most visible vertex, items 1, 9, and 10 are as follows:

1. "*(image_point point A)*"

9. "*(input_domain* $\leq$ (θ) $(360))$"

10. "*(input_domain* $\geq$ (θ) $(270))$".

Notice that if the viewed object could be of a variable class, then the simulation would construct one set of nodes, similar to that one described above, for each possible object class.

Sensor instructions like that from the above example do not need a complete localization of the viewed object. Therefore they can be executed fast during the program running. Such instructions may be of practical interest to the user. In fact, the instructions can be used to extract information about the free space. The extracted information can be used in order to determine a collision free trajectory of an object which must be moved along the worktable.

4.2 The simulation of a "guarded motion"

The AML block, composed by a "*GUARDI digital_input label*" statement, and by a motion statement (indicated by "*PMOVE destination*") prescribes a guarded motion: during the motion prescribed by the "PMOVE" instruction, a binary sensor, indicated by *digital_input*, is monitored. As soon as the sensor value, which is read from the *digital input* changes (e.g., from 0 to 1), the motion is stopped, and the control is passed to the program instruction indicated by *label.* This instruction can be of practical interest, in particular when looking for an object along a line by means of a photocell or a contact sensor attached to the robot end effector.

If the sensor can be modeled as a point contact sensor attached to the robot end effector, the "GUARDI" statement can be simulated in the case in which both the object is a polyhedron, and the prescribed motion is rectilinear.

The simulation of the guarded motion instruction is illustrated by means of an example. Let O be an object lying on the worktable. Its position and orientation is specified by means of the rigid transformation which describes the reference frame attached to the object. The parameters of this transformation are indicated by the vector $\mathbf{p_O}$. The line of motion is described by means of its parameters (l_α, l_p) within which the angular parameters l_α have been distinguished from the other ones (e.g., the intercepts) l_p. For instance, a horizontal straight line can be described in terms of its orientation angle θ, and its intercept x on a given axis (which not necessarily coincides with the x-axis). Let the input variables be given by the parameters $\mathbf{x}_o$. Let us consider a starting simulation node, to which the instruction is applied. Within this simulation node, the positions both of the object O and of the robot end-effector are referred to. The input subdomain of the starting node is indicated too, and it is described by the following constraint:

$$g^{(1)}(\mathbf{x}_o) \leq 0$$

The motion line can be derived by both the initial position of the end-effector, and the motion destination which is indicated in the motion statement. Suppose the parameters of the motion line are not independent on the input variables. Then, within the starting simulation node, the relations between the input variables and both the motion line parameters, and the object position and the input variables can be written as follows:

$$(l_\alpha \;\; l_p)^T = h(\mathbf{x}_o) \tag{3}$$

$$\mathbf{p_O} = f(\mathbf{x}_o) \tag{3'}$$

Our simulation problem, in this example, can be formulated as the problem of determining the subset of the input domain for which the guarded motion stops when the point contact sensor touches a given face F of the object O.

In order to simulate the guarded motion, the motion line is described in the reference frame attached to the object. This correspond to adopting the following *relative* line parameters (i.e., relative to the object):

$$(l'_\alpha \;\; l'_p)^T = q(l_\alpha, l_p, \mathbf{p_O}) = r(\mathbf{x}_o) \tag{4}$$

A geometric method is then utilized to determine, for each face F of the object O, the constraints to be satisfied by the relative line parameters, in order that the guarded motion is stopped in contact with the face F. Notice that this information can be reconstructed off-line with respect to the program simulation as soon as the geometric model of the object is available. This method will be described in the Appendix.

Let us write the determined constraints in the following form:

$$s_F(l'_\alpha, l'_p) \leq 0 \tag{5}$$

These constraints can be rewritten in terms of the input variables, by means of equation (3):

$$g_F^{(2)}(\mathbf{x}_\circ) \leq 0$$

. These constraints concur, together with the previous one $g^{(1)}(\mathbf{x}_\circ) \leq 0$, in defining the input subdomain of the new simulation nodes

Finally, the dependence of the new state variables on the input variables has to be determined. In particular, the stopping point $\mathbf{y}$ of the guarded motion has to be expressed as a function of the input variables. The coordinates of the stopping point relative to the object (indicated by $\mathbf{y}'$) are determined by the geometric method as a function of the relative line parameters. The subsequent expression is then derived by substituting the line parameters with a function of the input variables (3):

$$\mathbf{y}' = v_F(l'_\alpha, l'_p) = v_F(h(\mathbf{x}_\circ))$$

The final expression, which relates $\mathbf{y}$ to the input variables, is derived by writing $\mathbf{y}$ as a function both of $\mathbf{y}'$ and of the object position $\mathbf{p}_\circ$, and by combining this with (3') and the last expression.

5. An Example of Simulation

An experimental system for the symbolic simulation of AML-like programs has been developed. It consists of an inferential module, written in KEE, and of a module containing Lisp functions.

In order to give a picture of the functionality of the simulation system, the simulation of a simple program, containing sensor instructions, is illustrated now. The description of both the initial state and the input interactions are first given.

In the initial state one object is present in the workcell. The initial state nodes are defined by both the (variable) class of the object, and the (variable) position and orientation.

The (initial) position of the object, specified by the variables x, y, and θ, is constrained by:

$$-150 \leq x \leq -100$$

$$-250 \leq y \leq -200$$

$$-180 \leq \theta \leq 180$$

These constraints are valid for both object classes (types) A and B. Both types of objects have an identical rectangular external shape, with minimum length 10 and maximal length 20. Due to the symmetry of both objects, only two, of the four vertices of the external contour, are distinguishable. These are called, respectively, H_A, K_A, H_B, K_B. H_A and H_B are those vertices for which the short edge is found by rotating the long edge around H_A (or H_B) by an angle of +90. The object B presents a square hole, centered on the geometric center of the object. The object center has coordinates $(10,5)$, relative to the reference frame attached to the object. The hole square has dimension 5, and the edges of the square are parallel to those of the external contour of B.

An input interaction occurs in the program as the robot grasps an object within a specified input region R_1. This input interaction has no variable results: it consists in grasping a square pin (with dimension 5) in a position, whose Cartesian coordinates are given by $(200, 200, 0)$ and whose orientation angle is 0.

We now illustrate the program to be simulated:

1. *Program: example*
2. *main: subr*
3. *read_image_point* ((1 0) 1) *pos*
4. *read_image_angle* ((1 0) 1 1) ϕ
5. *read_contour_number nc*
6. *new.point pless* $((pos(1) + 10cos(\phi) - 5sin(\phi))\ (pos(2) + 10sin(\phi) + 5cos(\phi))\ 0\ \phi)$
7. *new.point plarger* $((pos(1)+10cos(\phi)-5sin(\phi))\ (pos(2)+10sin(\phi)-5cos(\phi))\ 0\ \phi)$
8. *if* $(\phi \leq 0)$ *then pmove (pless) else pmove (plarger)*
9. *close_hand*
10. *pmove (300 0 0 0)*
11. *open_hand*
12. *if* $(nc \geq 2)$ *then begin*
13. *pmove (200 200 0 0)*
14. *close_hand*
15. *pmove (300 0 0 0)*
16. *open_hand*
17. *end*
18. *end.*

Some comments on the program are in order. During steps 3, 4, and 5, some camera instructions are executed, aimed both at assessing whether the viewed object is of class A or it is of class B (step 5), and at locating it (steps 3, and 4). In particular, steps 3 and 4 assign the coordinates of the left-most image vertex to the memory variable *pos*, and, respectively, the orientation angle of the longest edge, starting at that vertex, to the memory variable ϕ. During step 5, the number of closed contours of the viewed object is assigned to the (integer) variable nc. Subsequently, two different destinations for future motions are constructed as two different functions of the camera readings (steps 6, and 7). At step 8, a predicate is constructed dependent on the camera reading ϕ. In correspondence with the two truth values of the predicate, a motion is prescribed to one of the two above destinations. Then the robot hand is closed (step 9), a motion to a determinate destination (300 0 0 0) is executed (step 10), and the robot hand is opened (step 11).

In the second part of the program (steps 12 to 17), a sequence of statements are executed in the case nc is greater than or equal to 2. This sequence includes an input interaction in which an object is grasped, a motion to the destination (300 0 0 0), and the release of the object.

The simulation starts with two state nodes, each one corresponding to an object class. A common subset of items of the two nodes define the input domain. The input variables are x, y, and θ, which describe the position of the object. The input domain is specified by the following relations:

$$-150 \leq x \leq -100$$

$$-250 \leq y \leq -200$$

$$-180 \leq \theta \leq 180$$

The meaning items of each node constitute their distinction element: in the node corresponding to the A object, the meaning items relate these variable to the position of A, while in the node corresponding to the B object, these variables are related to the position of B.

As the step 3 is simulated, each of the two starting nodes gives rise to two "arrival" simulation nodes: one in which the leftmost point is H, and one in which the leftmost point is K. In the first case, the expression of *pos* as a function of the input variables is $pos = (x,\ y)$. In the second case, the expression of *pos* as a function of the input variables is $pos = (x - 10sin(\theta),\ y + 10cos(\theta))$.

Once the dependences of the state variables on the input ones have been determined, the input subdomains associated with the arrival simulation nodes must be determined too. The simulation determines that H is the left-most point for $-90 \leq \theta \leq 0$, while K is the left-most point for $0 \leq \theta \leq 90$. This behavior is repeated with a period of 180 degrees, due to the object symmetry.

The simulation of step 4 ends with four arrival simulation nodes: it does not introduce any further node splitting. In all of the four arrival simulation nodes, the expression of the newly created memory variable ϕ as a function of the input variables is $\phi = \theta$.

The simulation of step 5 apparently creates two arrival simulation nodes for each of the four starting simulation nodes. One node would correspond to the case in which the viewed object has two closed contours, and the other node would correspond to the case in which the viewed object has only one contour. Actually, for each of the four starting nodes only one value of nc is allowed: nodes corresponding to the A object allow $nc = 1$, while nodes corresponding to the B object allow $nc = 2$. Therefore, the possible simulation states after step 5 are partitioned into four simulation nodes.

Two new memory variables are created during steps 6 and 7. Both are destinations. They are vectors with four components: the first three components specify the destination position (in terms of the Cartesian coordinates), while the fourth specifies the orientation of the manipulator (in terms of the rotation angle around the z-axis). These variables are dependent on the input variables. Their dependence on the input variables is described in dependence items of the new nodes. The functional relations between the memory variable and the input variables are derived by combining the relations between the new memory variables (*pless* and *plarger*) and the "old" ones (*pos* and ϕ), with the relation between the old memory variables and the input variables (x, y, and θ).

In step 8, a predicate is constructed. This apparently creates two new nodes for each of the four starting nodes. But in this case, it is recognized that whatever is the starting simulation node, the predicate can only assume one truth value. In fact, in the two nodes in which H is the leftmost point, ϕ can not be greater than 0, and vice versa. Therefore the simulation of step 8 produces four simulation nodes: one for each starting node. During step 8, the new physical variable *robot_pos* is introduced: the position of the robot manipulator after the motion.

Since step 8 prescribes also a motion, then the system assesses eventual contacts between the moved object (e.g., the robot) and other objects. To do this, it expresses the robot position relative to the object position, and then it applies geometric rules in order to generate constraints in the relative position. These constraints express the contact conditions. The simulation system determines that in each node the contact conditions are always satisfied. In addition, the relative position between robot and object is expressed in terms of the input variables. The system determines that the relative position is constant (independent of the input variables).

As the *close_hand* is simulated, new items are added which express the existence of a mechanical constraint between the robot and the object that impedes any relative motion. In this way, as the robot is moved again (step 10), both *robot_pos* and *object_pos* are modified. Notice that since the input variables remain unchanged, the variable *object_pos* is no longer represented by them.

In addition, due to the invariant relative position between robot and object, the variable *robot_pos* assumes a unique possible value after step 10 (i.e., this variable is independent of the input variables, which are related to the initial position of the object).

In the second part of the program (steps 12 to 17), a new object is grasped and it is "inserted" into the hole of object B. This part of the program is executed only if $nc \geq 2$. This case corresponds with the presence of the holed object on the worktable.

6. Conclusions

The symbolic simulation problem has been introduced. Symbolic simulation allows the analysis of programs which produce an infinite set of possible evolutions of the robot work environment. The result of such an analysis can be utilized in order both to extract a task level description of a given sensory manipulation program, and to facilitate program development and debugging. A simulation system has been illustrated which aims particularly at the simulation of the sensor instructions, and at the analysis of the propagation of sensory information through the program execution. An application of symbolic simulation is being developed, consisting of on line monitoring of sensory robot programs, in order to detect both errors and faults in the program execution.

7. References

[1] J. Ben-Arie – "Probabilistic Models of Observed Features with Applications to Weighted Aspect Graphs" – Pattern Recognition Letters, vol. 11, 1990

[2] V. Caglioti, M. Somalvico – "Towards Automatic Error Recovery in Programmed Robots: Inferring the Task Intent from a Manipulator Program" – H.Haken (ed.) "Computational Systems: Natural and Artificial" Springer Series on Synergetics, Springer Verlag, 1987

[3] H. F. Durrant-Whyte – "Integration, Coordination and Control of Multi-Sensor Robot Systems" – Kluwer Academic Publishers, 1987

[4] M. Mason – "Exploration in Sensorless Manipulation" – IEEE Journal on Robotics and Automation, 1988

Appendix

A.1 Relations between relative motion line and stopping point

The constraints are determined in two phases. In the first phase the *visibility* directions of the faces of O are determined by means of a method introduced by Ben-Arie [1]. This method provides constraints on the directions of the motion line (i.e., on its angular parameters). In the second phase the constraints on the other parameters (e.g., the intercepts) of the motion line are determined.

The *visibility* directions are defined as follows:

A face F of an object O is said to be *visible* along a direction $\mathbf{k}$ if there exists a straight line segment PQ parallel to $\mathbf{k}$ such that P belongs to F, Q belongs to the exterior of the convex hull of O, and PQ has no intersection with O other than P.

We now illustrate the second phase of the method, i.e., the determination of the constraints on the non angular parameters of the motion line. For details on the first phase, see [1].

For the sake of simplicity the method will be described in two dimensions. Suppose both the object O and the motion line exist in the (x, y)-plane. The object O is a polygon, while the motion line is a horizontal. Two parameters are sufficient in order to specify the motion line: one angular parameter θ, and one intercept x on a given axis (not necessarily the x-axis)

Let O be a concave polygon. The faces of the convex hull of O, which do not belong to O, are called *window* faces. For each window face there is a set of faces that form the *concavity* of the window. Let us suppose that the face F is in the concavity of a window W (otherwise the problem is simple).

Given a direction $\mathbf{k}$, along which a face F of an object O is visible, we define the *pre-image* $p_k(F)$ of F along $\mathbf{k}$ as the set of the points of the window W from which it is possible to reach F by moving parallel to $\mathbf{k}$ without touching O. Notice that $p_k(F)$ is a segment.

Let us indicate with $k(F)$ the projection of $p_k(F)$ on F, parallel to $\mathbf{k}$. The volume internal to the trapezoid bounded by $k(F)$, $p_k(F)$, and two segments parallel to $\mathbf{k}$, is indicated by $S_k(F)$. The volume $S_k(F)$ intersects O in two vertices, not belonging to F. These points, indicated by P_k^1 and P_k^2, are called binding points.

As $\mathbf{k}$ varies within the set of the visibility directions of F, the pairs of binding points vary within a finite set of points. For each of the possible pairs of binding points, a subset of the set of the visible directions of F is defined. In our two dimensional case, this subset is an interval of values of the orientation angle of the line motion. Let us denote it with $[\theta_1, \theta_2]$, where we refer the orientation angle to the direction of the entering normal to the face F.

Let now $\mathbf{k}$ vary within the interval $[\theta_1, \theta_2]$ defined by a given pair of binding points, and consider now an oriented axis along F; let x_1 be the abscissa of the orthogonal projection of P_k^1 on F, and let $x_2 > x_1$ be the abscissa of the orthogonal projection of P_k^2 on F. Let y_1 and y_2 be the distance between F and, respectively, P_k^1 and P_k^2. We will use the intercept x with F as the second parameter of the motion line. The intercept x is bounded from below by $x_1 + y_1 \tan(\theta)$, and it is bounded from above by $x_2 + y_2 \tan(\theta)$.

$$x_1 + y_1 \tan(\theta) \leq x \leq x_2 + y_2 \tan(\theta)$$

The derived constraint concerns all of the parameters of the motion line (not only the angular ones).

Expert Systems and Multiple Criteria Decision Support

J. Clímaco [1,3], C. H. Antunes [1,3], J. P. Costa [2,3], A. G. Martins [1], A. T. Almeida [1]

[1] Department of Electrical Engineering - University of Coimbra

[2] Faculty of Economics - University of Coimbra

[3] INESC - Núcleo de Coimbra

3000 Coimbra - Portugal

Abstract: This paper aims to present the framework of a knowledge-based system which enables a re-evaluation analysis of a sample of rough solutions selected by means of an interactive multiple-criteria approach. Two case studies will be discussed, one concerning power systems expansion planning and a second one dealing with strategic telecommunication network planning. The first one will be presented in detail and the specifications of a rule-based prototype for performing the re-evaluation analysis will be discussed.

Keywords: Interactive Multiple-criteria Linear-programming / TRIMAP / Power-generation-expansion Planning / Knowledge-based Systems / Decision Support Systems / Re-evaluation Analysis / Production Systems.

1. Introduction

After the golden period of the operations research during the 60's and the 70's, where most part of the concepts and classic methods in this domain were developed, difficulties began to arise. Since the 70's the increasing complexity in the social and economical environment and the astonishing development in computer and communication sciences have created situations which require not only the modernization of the productive system but also a new type of management. In these conditions, how should the decision process be approached?

A lot have been written on this subject. In synthesis we believe that the essential ideas may be found in the principles enunciated by Ackoff [1], the key ideas of which we have summarized as follows:

1. The decision process must be interactive and participative. It is essential that managers, operations research and computer engineers join their effort into a mutually enriching collective work.

2. The plans must be continuously revised. This procedure is essential since the environment in the organizations where they are inserted changes very rapidly and this makes even more difficult to obtain reliable forecasts.

3. All the units in the organizations have to be planned in an interdependent manner, that is decisions must take into account all the interactions of the system.

NATO ASI Series, Vol. F 71
Expert Systems and Robotics
Edited by T. Jordanides and B. Torby

It was in this context that Geoffrion [6] wrote an article untitled: "Can Management Science/Operations Research Evolve Fast Enough?". As Geoffrion we believe that Decision Support Systems (DSS) will tend to come closer to the principles here formulated. The DSS are computer based systems (generally interactive) with the aim of supporting and improving the decision process especially in complex and/or ill structured tasks which require critical appreciations by the managers. They do not intend to substitute the decision maker (DM) instead they aim to be an effective support to the DM in complex and holistic tasks. These systems usually integrate modern computer science techniques including sophisticated man-machine interfaces, quantitative statistical methods, operations research methods and more recently artificial intelligent techniques.

In this paper the potential integration in a DSS of an interactive multiple-criteria decision aiding package with a knowledge base system (a re-evaluation analyzer), taking into account the following remarks, is discussed.

The results obtained by applying operations research models and methods to the study of real-world problems are always affected, among other factors, by one or more of the following sources of imprecision:

1. Approximations which are done in the mathematical modelling phase (for instance, by not considering the modular characteristics of the equipments to be installed or replaced when dealing with linear-programming models).

2. Difficulties arising in the aggregation of the decision maker's preferences when several conflicting objectives are being considered.

3. Use of heuristics to keep the computational effort within reasonable limits, due to the combinatorial explosion of most combinatoric models.

However, in most of these circumstances, especially when interactive computational tools are available, it is possible to identify some regularities by means of experimentation. This allows the identification and systematization of those factors which have stronger influence on enlarging the gap between the solutions produced by the methodologies and satisfactory real-world solutions. This kind of knowledge will lay the foundations for the implementation of knowledge-based systems aimed at improving the solutions obtained by using operations research tools.

In this paper the potentialities of this approach are discussed in the context of the use of a flexible and interactive multiple-criteria linear-programming package for studying two types of engineering problems, namely the strategic planning of telecommunication networks and the selection of power-generation-expansion planning policies.

The power-generation-expansion planning problem will be referred to in detail. The use of a rule-based system is discussed for further evaluating some selected solutions, corresponding to different expansion plans. This system enables the decision maker to define rules which support a re-evaluation analysis of the results obtained using the multicriteria linear-programming model.

In sections 2 and 3, two case studies are briefly described dealing with telecommunications network planning and power-generation-expansion planning. The TRIMAP method and the multiobjective energy-planning model are presented in section 4. The main sources of imprecision stemming from the multicriteria linear approach are discussed in section 5. The main potentialities of the rule-base approach for re-evaluating the expansion plans, and the specification of the rules implemented in a prototype version, are presented in section 6. Some conclusions are drawn in section 7.

2. Strategic Planning of Telecommunications Networks

The rapid advance of transmission and switching technologies has made increasingly attractive the introduction of a broad range of new services to be offered to the customers of telecommunication networks. This trend raises new and complex problems of telecommunication network strategic planning in the context of emerging technologies and services. The application of a computer-based multiple objective linear-programming (MOLP) package to provide decision support in a problem of telecommunications network evolution planning has been presented in [2]. The conflicting objectives considered are the net present value of the expansion of subscriber lines, to be maximized, a function representing the foreign dependence, to be minimized, and the quality of service (considering different aspects, including the availability of the various services offered to the user), to be maximized. These objective functions are to be optimized subject to technical, budgetary and market constraints.

Some sources of imprecision are the uncertainty associated with the demand of new services and the political decisions regarding the desirability of new services and technologies. The degree of foreign dependence of the technologies as well as the availability of wide band technology in the planning horizon are also critical parameters. In this context, some artificial intelligence techniques may be useful in order to test different scenarios.

3. Power-generation-expansion Planning

The planning of new units for electrical power-generation is multicriteria in nature, explicitly requiring several objective functions. Traditionally, mathematical programming models judge generation expansion planning problems by considering the total system cost as the single objective. Nowadays, however, several social measures and security considerations have also become essential to evaluate the relative merits of alternative schemes.

A model was developed considering three linear objective functions which quantify the total system cost, the environmental impact (both to be minimized), and the reliability of the supply system (to be maximized). Three categories of constraints were considered in order to satisfy "instantaneous" demand, operational capacity and budgetary restrictions.

The authors have carried out recently the development of an integrated computational tool which provides support to the decision maker in four main processes : interactive data input, automated formulation of the linear objective functions and constraints, auxiliary probabilistic computations for estimating the coefficients of the reliability objective function and dialogue mechanisms that keep the decision maker in control of the direction of search for nondominated solutions (see 4.1, and for further details see [4]).

The key role in this decision support system is played by an interactive tricriteria linear-programming algorithm named TRIMAP [3]. The decision maker intervenes very early in the process and has virtually the control of the way the search evolves through a flexible and user-friendly computational environment. This approach seems to be very useful in the domain of multicriteria power-generation planning, namely by avoiding the information overload of the decision maker.

In this paper we propose a system integrating the TRIMAP package and a knowledge-based module aimed at improving the solutions obtained by using TRIMAP. The block diagram of this system is presented in fig.1.

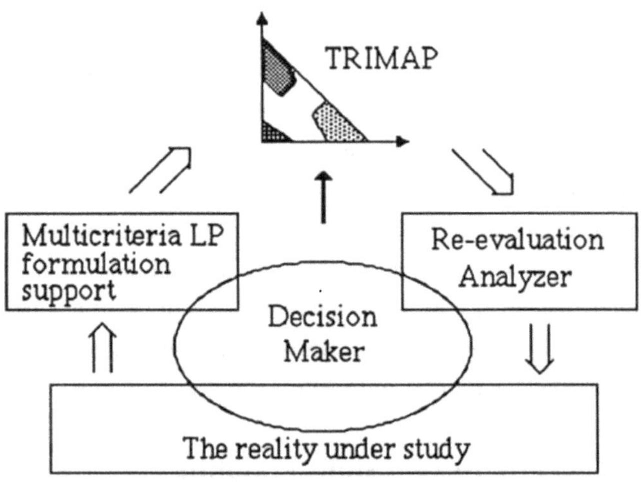

Fig. 1 - Block diagram

From the universe of nondominated solutions, a "most preferred" sample is selected by the decision maker with the help of TRIMAP, which is further evaluated by means of a re-evaluation analyzer. This module is intended to correct somehow the approximations inherent to the use of a multicriteria linear-programming model. In fact, some simplifications have been done in the mathematical modelling process in order to keep the computational effort within reasonable limits without sacrificing the adequacy of the model (see section 5). On the other hand, when planning new units of electrical power-generation the decision maker is faced with factors which are clearly qualitative in nature and highly subjective, namely, as far as the environmental impact is concerned. Any attempt to quantify this sort of factors in order to integrate them in a linear model is always controversial and leads to rough approximations.

The re-evaluation analyzer is a knowledge-based system where the domain facts and reasoning human heuristics are structured as production type rules ("condition-action" pairs). The rules can be defined or modified by the decision maker by means of a rule editor. In this way it is intended to capture his/her knowledge and expertise in evaluating the nondominated solutions, thus accommodating his/her acquired experience and enabling the emergence of new preferences and intuitions. This system is described in more depth in section 6.1.

The block which supports the formulation of the multicriteria linear-programming model (see Fig.1) is intended to facilitate the decision maker task of structuring the reality in a suited model for TRIMAP.

The decision maker competence area (see Fig.1) intersects the re-evaluation and multicriteria linear-programming blocks because the structure of these blocks allows capturing the decision maker's domain knowledge. The decision maker competence area intersects the block "reality under study", because the way in which the decision maker perceives the problem actually influences the decisions that the system is intended to support.

The method of re-evaluation analysis presented herein proves to be an interesting complementary tool to the more classical multicriteria approach for evaluating power-generation-expansion planning problems.

A similar approach has been experienced in a study dealing with the improvement of a heuristic based operations research model for rural telephone network planning [5]. The rule-based system presented therein attempts to improve the solutions generated by a heuristic model in which the equipment cost functions have been simplified in order to reduce algorithmic complexity and computational effort.

4. Multiobjective Energy Planning

This study is solely concerned with the planning of electricity generation in view of its dependence on load evolution. The problem of siting the new units and its effects on transmission network expansion planning are not considered. For further details on the modelling process see [4].

4.1. TRIMAP Main Features

Let us review briefly the main characteristics of the algorithm which constitutes the most important component of the multicriteria approach. The TRIMAP method [3] is based upon the progressive and selective "learning" of the set of nondominated solutions (those for which the improvement of an objective function value implies the worsening of at least another objective function value). The aim of the method is not the convergence to any "best compromise" solution, but to help the decision maker eliminate the subsets of the nondominated solutions which are of no interest to him. The method combines three main procedures: weighting space decomposition, introduction of constraints on the objective functions space and on the weighting space. Furthermore, the introduction of constraints on the objective function values can be translated into the weighting space. The weighting space can be interpreted as an operational means to represent the "importance" given to each objective function (when scalarizing the multicriteria problem by means of a weighted linear function of the three original objectives).

Each nondominated extreme solution can be represented on the weighting space by an indifference region (because any combination of weights within it gives rise to the same nondominated solution).

In each step of the man/machine dialogue (when a new phase of computation is being prepared) the decision maker has just to give indications about the directions to carry on the search for the nondominated solutions. At this stage, restrictions on the values of some of the objective functions may also be introduced by the decision maker. The process only ends when the decision maker considers the knowledge about the set of nondominated solutions, sufficient. There are no irrevocable decisions along the process, being always admissible to go "backwards" (for instance, by discarding additional constraints previously imposed on the objective function values). The method is not too demanding with respect to the information required from the decision maker in each interaction, the main concern being to have a simple and flexible procedure.

The TRIMAP method is intended for tricriteria problems. This fact, although being a limitation, allows for the use of graphical means which are particularly suited for the dialogue with the decision maker. This undoubtedly increases his/her capabilities of information processing by simplifying the dialogue phase.

4.2 The Multiobjective Linear Model

4.2.1 Identification of the objective functions

The three objective functions considered in this approach are: cost and environmental impact (both to be minimized), and reliability (to be maximized). Some simplifications had to be introduced because of the linear nature of the computational tools used. Three generating technologies (oil, nuclear and coal) have been considered for power-generation-expansion.

Cost : The cost objective function is composed of the fixed charges and the operational costs of both the new units and of the units already in service at the beginning of the planning period. No fixed charges are assumed for units already in service at the beginning of the planning period since they are constant in every plan.

Reliability : The linear objective function quantifying the risk of supply failure is an approximation since the probability calculations involved in reliability assessment are intrinsically nonlinear. The fundamental parameter used to model reliability of the generating units is the load-carrying-capability (LCC) of new units. The LCC is the amount of power by which the system load can increase with the addition of a new unit in order to maintain the reliability level as it was before the unit addition.

Environmental Impact : Alternative energy generating technologies are very difficult to compare from an environmental perspective. Four indices of impact evaluation (land use, large accident/sabotage, emissions/public health and effect on ecosystems) have been considered in order to construct an objective function quantifying the environmental impact of three generating technologies (oil, nuclear and coal). The first index has little to do with the energy output of the generating groups. The second has to do both with energy output and with the existence of the generating facilities themselves. The remaining ones may be related solely to energy output. In this context, two parameters were obtained that characterize each generating alternative : one penalizing the installed capacity and the other penalizing energy output.

For details see [4].

4.2.2 Identification of the constraints

Two main categories of constraints have been considered. The first one basically imposes that all the generating power available at each period of time considered must be at least equal to the load in the same interval. Both the new and the previously existent generating units must be considered. The second category includes operational constraints of the generating units: the power output of an unit cannot exceed its rated power, previously affected by some availability factor. A budgetary constraint on the maximal feasible value for the cost objective function was also imposed.

4.3 Some Results

This section presents briefly the main results of the application of TRIMAP to a case study. Further details can be found in [4]. The minimum cost value is obtained with a mix of nuclear and coal units (solution 1, in Fig. 2), whereas minimum environmental impact corresponds only to addition of oil burning units (solution 3). The maximum of reliability is a solution, where cost is completely sacrificed to reliability (solution 2). Three distinct zones can be distinguished on the weighting space: an upper sub-area where a compromise between reliability and environmental impact keeps cost at the allowed maximum, and two lower sub-areas. In the first sub-area, nuclear units appear in just a few solutions near the optimum of cost and a smooth cost transition among solutions without nuclear technology occurs. Most of the lower sub-area corresponds to additions of coal units only. In the second sub-area, near the optimum of environmental impact, a sharp transition from all-fuel plans to all-coal plans exists.

Sharp variations of cost among these three sub-areas occur. A useful feature of TRIMAP (illustrated in [4]) allows the decision maker to introduce inferior bounds on the objective function values. This enables a more exhaustive search of solution plans in restricted sub-areas of the weighting space (because of the possibility of translating the introduction of constraints on the objective function values into the weighting space).

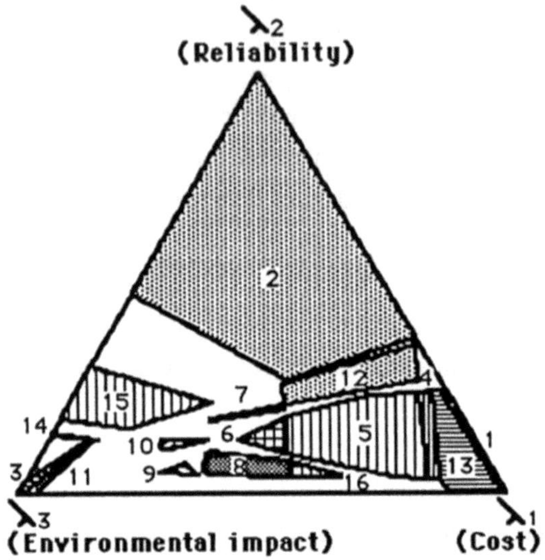

Fig. 2 - The weighting space

Fig. 2 - The weighting space

5. Some Consequences of the Multicriteria Linear Approach

An essential condition for a model to be truly useful in policy planning is that it provides a sufficiently accurate representation of the real world. The mathematical modelling of real world problems, as any process of representation of the reality, is always to some extent a simplification. This is done not only because it is impossible to capture reality in all its relevant aspects but also because models must not be too complex in order to avoid an excessive computational burden.

In our multicriteria linear model of a power-generation-expansion problem some approximations have been made. Some of them are briefly reviewed below.

The load model used is based on a peak load duration curve (LDC), which is a cumulative representation. It does not capture the time-varying nature of load and only peak values are considered. Besides, the peak LDC is approximated by a histogram where each division in the time scale represents a long period of time. It was chosen because it is well suited for use within linear-programming models and, at the same time, adequate for reliability evaluation through the loss-of-load-probability (LOLP) index. It is, however, a conservative approach that leads to oversized generation additions.

The tricriteria linear-programming model used deals with three kinds of generation technologies (oil, nuclear and coal), which are considered for the definition of the various capacity expansion plans. The consideration of generation technologies different from those used in the model, or the avoidance of one of the technologies considered in the planning process, may be of importance for evaluating alternative scenarios of evolution.

The quantifications of the environmental impacts related to different energy production technologies have been a controversial issue because of the subjectivity underlying the quantification process. In order to formulate the environmental impact objective function a table by Holdren (1977, see [7]) has been used configuring the quantitative approach discussed, that establishes a tentative ranking of several generating technologies according to different dimensions of impact evaluation. Hence, only four of these criteria have been considered - land use, large accident/sabotage, emissions/public health and effect on ecosystems - for obtaining a first ranking of the three alternative generating technologies. Further criteria, namely those which are difficult to quantify, may be considered in a re-evaluation analysis.

The linear-programming model used deals with generating group additions as continuous variables. However, the modularity of the equipments imposes a discretization of the results to be carried out . This procedure can have influence on the quality and feasibility of the solutions, enabling a more detailed analysis of the power-generation-expansion plans.

6. The Rule-Base Approach as an Improvement of the Solution Analysis Process

6.1. The General Conceptualization of the Re-evaluation Analyzer

The re-evaluation analyzer has basically the structure of those systems commonly known as "Expert Systems". It is possible to distinguish in it the main parts usually found in an expert system: the knowledge base (KB) and the inference engine (IE).

The expertise embodied in the KB of the expert system consists of facts and production rules relevant to the domain. The rules aim at describing the heuristics and/or the decision paradigms of the decision maker. As for the domain facts, they have a structure which can be either in the form of production type rules or just data describing the domain ("most preferred" nondominated solutions, statistic data, etc). Production type rules take the form of "If ... then" statements or, in other words, "condition-action" rules. The condition part of a rule is the combination of other conditions ("basic" conditions) which refer to some very specific aspect of the domain. These basic conditions are combined by means of logic "ORs" and "ANDs". The action of a rule is the combination of other "basic" actions.

A rule-editor supports the decision maker in introducing and/or performing changes in the rules contained in the KB. A debugger associated to the editor helps the user respecting the consistency of the rules. By enabling the user to change the knowledge contained in the KB, it gives the re-evaluation analyzer system (fig. 1) some adaptive characteristics. The decision maker can adapt the knowledge to his/her changing environment, thus maintaining the usefulness of the system. Last but not least, the decision maker can "customize" the KB, i.e., his/her experience and perspective of the real-world problem can be structured and integrated into the system.

At each moment of its operation, the IE selects from the KB the applicable rule(s) by evaluating the condition part of the rules. The other function of the IE is to control the execution of the action part of the rule currently selected. This may imply some needed mathematical processing to be carried out (or the procedure in charge to be invoked).

The consequence of the application of one rule can determine the next rule, thus enabling rules to be chained for obtaining suggestions of "improvement" of a selected nondominated solution. Whenever an explanation about these suggestions is requested, it will be presented to the decision maker, even though sometimes it consists only in the presentation of the applied rule(s) leading to the suggestions. The suggestions given by the system are presented to the decision maker in graphic and/or in text format, depending on their characteristics.

The framework presented in this section uses some well known artificial intelligence techniques. In the next sections we will present an example of a few rules derived from the considerations in section 5. It must be emphasized that in this paper we only present the experience with a small prototype described in 6.3.

6.2 The Main Targets of the Rules Used in the Re-evaluation Analysis Process

In this section it is intended to outline the scope of the rules in order to perform a re-evaluation analysis of some solutions selected by using TRIMAP.

As far as the load model is concerned, the rules involved aim at the evaluation of the effects that different levels of time discrimination will produce on the expansion plans. As the total duration of the higher levels of power in the LDC (see section 5) becomes smaller with a more finely defined time scale, the total risk variation allows for more conservative expansion plans, namely at the expense of additions of oil-burning groups.

Also, operational costs become lower because of the smaller amount of total energy generated. These effects are to be quantified in terms of reducing, postponing or replacing, additions of oil-burning groups by one of other available technologies, and also in terms of considering global reductions in total operational costs. Inhibiting additions of some of the generation options will allow for the analysis of the impact of some major potential political decision concerning energy planning (for example, not considering additions of oil-burning or nuclear units).

The quantification of the environmental impact related to different electric energy production technologies has been a controversial issue. There are several studies aimed at quantifying those effects according to their category. The tricriteria model used in [4] is based on this type of approach. It uses four categories of impact as listed in 4.2.1. There are several limitations to this approach that have been discussed in the literature, namely the subjectivity underlying the quantifications. Each individual has its own preferences and it may not even be possible to establish weights on the criteria satisfying different decision makers. A rule-based approach allows for the introduction of bias in the analysis corresponding to the preferences of the decision maker, and for the decomposition of the original criteria.

The linear-programming model used deals with additions as continuous variables. Discretization is unavoidable to report results according to the reality of equipment modularity. The various plans obtained at the output of the model give place to a greater number of "solutions", depending on the round-off procedures used for discretizing results. The re-evaluation analysis aims at evaluating the effects of these decisions on the various criteria, reporting them in order to emphasize the main differences among the corresponding plans.

6.3. Prototype Description

From this point onwards, we shall refer to the solution plans generated by TRIMAP as "continuous solutions", because the variables assume continuous values. The plans modified by means of the IE are oppositely referred to as "discrete solutions". The rules adopted follow.

One or more expansion plans determined by using the TRIMAP package are the starting point of the re-evaluation analysis. Firstly, the analyzer produces a set of alternative power-generation-expansion plans from the above solutions as follows. In the initial expansion plans, the generating group additions are represented as continuous variables. The rules of the KB which generate the set of alternative discrete solutions are concerned with the modularity of the equipment (in the prototype version). These rules are very simple and consist essentially of the discretization of the continuous-variable values based on some thresholds which depend on the size of the equipment modules and on the generating technologies. The system generates the feasible discrete solutions by approximating the computed continuous values, considering its most accurate lower and upper bounds as an integer number of modules. The procedure to perform this task is exemplified by the rules in section 6.4.1. The rules are made to maintain the minimum required power. The other constraints of the tricriteria model are just tested and the results presented to the DM for judgment. As explained in section 4.1. the IE determines which rule is to be applied in each circumstance.

The values of the three criteria used in TRIMAP corresponding to the expansion plans in the alternative set are then computed. By inspection of the criteria values, the DM selects a set of alternative expansion plans as restricted as possible, according to his/her preferences. Based on this restricted set, the DM can perform several re-evaluation operations. He/she may, for instance, ask the system to improve the cost value by introducing slight modifications in the selected solutions, according to pre-defined rules. Based on the information about the cost of the module equipment contained in the KB, the system generates one or more improved cost expansion plans. For the solution chosen by the DM, a module of a more expensive technology, in which concerns the investment costs, is replaced by modules of a less costly generating technology satisfying the power requirements. As for the first generated set, the minimum power requirement must be satisfied and the results of the verification of the other constraints are presented to the DM. The DM evaluates the modified solutions against the other criteria. The same procedure may be used for each of the other criteria besides investment cost.

Note that the energy produced by each unit must be recalculated for the alternative solutions generated by the re-evaluation analyzer, before the decision maker makes a final judgment.

A future version of the system which generates these expansion plans may account for other aspects besides the equipment modularity. Some examples are considering the addition of other generating technologies, decomposing some of the original criteria in order to perform a more detailed re-evaluation analysis, etc.

6.4. Prototype Rules

At each moment of its operation, the IE selects from the KB the applicable rule by evaluating the condition part of all the rules, and controls the execution of the action part of the rule currently selected.

6.4.1 Rules for generating alternative expansion plans, using modular equipment, from a solution obtained with TRIMAP

* IF the difference between the value of any variable in the continuous solution and the corresponding variable in the discrete solution is greater than 80% of the modular capacity of the generating unit considered for addition, THEN the total capacity of the unit shall be added to the value of the variable in the discrete solution (1).

* IF there are no variables in the conditions stated in rule (1) AND IF power requirements are satisfied, THEN the current solution is already a discrete solution which may be presented to the DM (2).

*IF there are no variables in the conditions stated in rule (1) AND IF power requirements are not satisfied, THEN: - several discrete solutions are generated by adding the capacity of one module to the value of each variable in the discrete solution for which there is a positive difference between the continuous and the discrete values; - the original discrete solution is discarded. (3)

6.4.2 Rules for generating alternative expansion plans from a discrete solution seeking to improve investment cost.

* Initially, one and only one generating unit module of the technology having the greatest cost by unit of power is suppressed (4).

* IF the power requirements are satisfied, THEN a valid alternative exists that shall be presented to the DM (5).

* IF the condition in rule (5) is not satisfied, several candidate plans are generated whenever possible by adding one generating module with a smaller cost per unit of power than the suppressed module. The original plan is discarded (6).

6.5 Prototype Implementation

The prototype has been implemented on a MacIntosh II microcomputer, with the aid of "Intelligent Developer" from Hyperpress Software Products, a tool enabling a fast development and implementation of small expert systems. The rules used are of "If - then" type and the knowledge is structured in different types of facts. The expert system from Intelligent Developer have also the possibility of being associated with data bases in which the facts can be aggregated.

The number of implemented rules, the underlying ideas of which are presented in section 6.4., is about sixty (a lot of them are rules of process control and knowledge structure) and the number of facts used is of the same order of magnitude.

Although the modelling of the objective functions and constraints has been simplified, we may say that this small prototype has proved the viability of building a knowledge based system with the aim of discretizing the power expansion plans computed with TRIMAP, and consequently enabling their re-evaluation regarding the objective function values. Our experience with the implementation of this prototype has also revealed some limitations of the tools used. Some of them are of quantitative nature (limitation on the number of facts and rules), and other are related to the inexistence of linking of rules with code written in procedural languages which is necessary to perform mathematical computations. It seems to us that the overtaking of these limitations is essential for the development of new and more complex and rigorous versions of the system.

7. Conclusions

A rule-based approach for evaluating power-generation-expansion plans has been presented. This system supports a re-evaluation analysis of the results obtained from a multicriteria linear-programming model. Besides contributing to correct the imprecisions stemming from the linear model, this approach enables one to capture the domain knowledge and the decision maker's expertise in evaluating the solutions.

The method of re-evaluation analysis presented is shown to be an interesting complementary tool to the multicriteria approach for evaluating alternative power-generation-expansion planning policies.

The overall system is intended to present to the decision maker various scenarios by means of an interactive and user-friendly computational environment, always keeping the user in control through a menu-driven structure of the available actions.

References

[1] Ackoff, R. L. - "Resurrecting the future of Operational Research." J. Opl Res. Soc. vol. 30, 2 pp 93 - 104, 1979.

[2] Antunes,C. H., J. Climaco and J. Craveirinha - "Planning the evolution of telecommunications networks using an interactive multiple objective linear programming package". To appear in Proceedings of AIRO 90, Sorrento, Italy, 1990.

[3] Climaco, J. and C. H. Antunes - "Implementation of a user-friendly software package - a guided tour of TRIMAP". Mathematical and Computer Modelling, vol. 12, n. 10/11, 1989.

[4] Climaco, J. , C. H. Antunes, A. Martins, J. Marta and A. Almeida - "A decision support system for power generation expansion planning with a case study". To appear in Proceedings of the 29th IEEE CDC, Honolulu, USA, 1990.

[5] Costa, J. , J. Climaco and J. Craveirinha - "A tentative approach to integrate A. I. techniques to improve a heuristic based O. R. model for rural telephone network planning". "Knowledge, Data and Computer-Assisted Decisions". Nato ASI Series, vol. F 61, pp. 221-232, Springer-Verlag, 1990.

[6] Geoffrion, A. M. - "Can MS/OR evolve fast enough?". INTERFACES 13 : 1 Feb pp 10 - 25, 1983.

[7] House, P. , J. Coleman, R. Schull, R. Matheny and J. Hock - "Comparing energy technology alternatives from an environmental perspective". DOE publication EV-0109, 1981.

Signal Representation Issues in Neural-Network Modeling of Perceptual Properties

V. C. Georgopoulos and D. Preis

Department of Electrical Engineering, Tufts University, Medford, MA 02155, USA

Abstract: This paper discusses neural-network modeling of five perceptual processes: adaptation, scaling, similarity, absolute threshold as a function of frequency and lateral inhibition. First the need for joint-domain representations is discussed. A joint-domain representation discussed in detail is the Wigner distribution. Autocorrelations of the Wigner distribution in time and frequency are taken to view interactions of the Wigner distribution with itself. These autocorrelations are then implemented using a simple neural-network. Some insight is gained in modeling the above perceptual properties with these functions.

Keywords: Neural-Networks, Perceptual Modeling, Joint-Domain Representations

Introduction

The goal of this research is the study of neural-networks as a modeling technique for various perceptual systems. In other words, we are interested in breaking the problem of perception into smaller building blocks that could be simulated by neural-network stages. This study would be useful when designing artificial intelligence or robot systems that can exhibit basic aspects of perception mechanisms with a small number of building blocks.

Perceptions can be defined as experiences caused by stimulation of the senses. The act of perception can be divided into a number of different steps. These steps include the acceptance of stimuli by the sensory receptors, the internal processing of sensory inputs through neuron pathways and finally the perception of the stimulating mechanism. There are a number of properties that perceptual systems have in common. The ones considered in this study are lateral inhibition, absolute threshold detection as a function of frequency, adaptation, scaling, and similarity. These properties are common to all five sensory systems.

All perception systems share the same major central processor. Therefore, regardless of the actual variety of inputs and outputs, a common code is used. First, the sensory receptors carry out the process of transduction - the transformation of environmental energy into electrical energy. Then the neurons produce the code. The code consists of the firing patterns

NATO ASI Series, Vol. F 71
Expert Systems and Robotics
Edited by T. Jordanides and B. Torby

of the neurons, their spatial organization and their connection to various parts of the central processor. Thus, we would like to devise a general modeling technique that could describe such mechanisms. In this way, we would be able to simulate different sensory pathways on a robotic system with similar building blocks. Artificial neural-networks lend themselves to this problem due to the fact that they are structured so that they can emulate real biological neural functions using many simple slow computational elements operating in parallel to obtain high computational throughputs.

Properties of Perceptual Systems

Adaptation is the property that allows a perceptual system to adjust itself to a given stimulus. Effectively it results in a decreased sensitivity to a stimulus when the stimulus is presented over a prolonged period of time. Examples are discussed here from various perception systems. In vision, dark and light adaptation are commonly experienced. Chromatic adaptation is a selective adaptation to a wavelength in a particular region of the visible spectrum. This occurs when looking at a large object with a particular color for a period of time and then moving one's attention to other objects. In general, a decreased sensitivity to the region of the spectrum to which the original object belongs will be noticed. In audition, an example of adaptation is in speech perception where selective adaptation occurs when a stimuli, for example /ta/ is repeated over a period of two minutes. Then a subject is less able to perceive voiced syllables such as /da/.[6] A final example of adaptation is in taste when one substance flows over the tongue and affects the perception of a similar substance. If the tongue is exposed to a salt, then the threshold for detecting a different salt is raised. There has been no evidence that shows the effect on the threshold of a different taste quality, for example sweetness.

The scaling property is how the magnitude experience is related to the physical intensity of the stimulus. In most perceptual pathways, the experience is that of compression. In the pathway for shock, the experience is that of expansion. This is done for protective purposes so that when experiencing shock we will remove ourselves from that harmful situation.

Similarity is how different stimuli are perceived to be the same. For example, a composite color consisting of a number of different wavelengths may appear to be the same with a color consisting of only one wavelength. A counter example of this property is two lines of the same length perceived to be of different lengths due to their surroundings.

Absolute threshold is the minimum stimulus energy necessary to detect the stimulus. This varies from individual to individual. It also, varies as a function of frequency or wavelength. In vision, for example, the eye in general is more sensitive (requires less stimulus energy) to the wavelength corresponding to yellow (580 nm) than it is to red (660 nm).

The most important adaptation mechanism in sensory systems is lateral inhibition. With this mechanism weak inputs to neurons are suppressed (inhibited) by strong inputs to neighboring inputs. Examples of lateral inhibition exist in all sensory pathways. In the retina lateral inhibition enhances contours and provides some gain control. It is used to detect and respond to changes (certain cells respond to moving stimuli rather than stationary). In hearing, a phenomenon called two-tone suppression is due to lateral inhibition. Here the tone-driven activity of a single neuron fiber due to a single tone can be suppressed by the presence of a second tone [12].

The properties discussed above involve complex phenomena that are functions of many variables, time, frequency, space, etc. In all perceptual systems, there is propagation of an electrical / acoustical / mechanical signal through different media. This causes frequency-dependent and space-dependent attenuations and time delays or signal dispersion. A consideration in this study is the coding of the information in the various building blocks. Due to the complex nature of the information there is a need for a joint-domain representation. The importance of spatial-time and time-frequency signal representations is illustrated by the following examples:

- In the auditory system, pitch (a function of frequency) perception is related to temporal information in the auditory-nerve-fiber discharges [13].
- In the somatosensory system, information coded as interspike intervals can be used to perceive the frequency of flutter vibration.
- In vision, there are mechanisms that shift the system from one class of receptors (rods, for night vision) to another class (cones, for day vision)as the light level changes. The size of receptive fields of the retinal output cells changes with light intensity, so that at very low intensities cells integrate light over a large spatial region and over a long period of time, trading resolution for sensitivity.As intensities increase, the retina rewires itself so as to give higher spatial resolution, but lower sensitivity [2].

There are a number of joint-domain representations that can be used. These include the Wigner distribution, the spectrogram, the Gabor transform and a number of filterbank representations. Since the Wigner distribution is the most general from which the others can be derived [8], our discussion is limited to the Wigner distribution.

Wigner Distribution

The Wigner distribution (WD) is a quadratic-signal representation introduced by Wigner in 1932 within the context of quantum mechanics. It was used by Claasen and Mecklenbräuker, in a series of papers, for time-frequency analysis[3] and is generally interpreted as signal-energy density in time and frequency.

The continuous Wigner distribution of a signal x(t) is defined by:

$$W(t,\omega) = \int_{-\infty}^{\infty} x(t+\frac{\tau}{2})x^*(t-\frac{\tau}{2})e^{j\omega\tau}d\tau$$

For a discrete signal x[n], the discrete Wigner Distribution (DWD) is used:

$$W(n,\omega) = 2 \sum_{k=-\infty}^{\infty} \exp(-2k\omega)x[n+k]x^*[n-k]$$

For numerical evaluation of the DWD, a finite number of samples must be used and therefore, the WD of a windowed signal is taken. This gives rise to the pseudo-Wigner distribution (PWD) given by:

$$W(n,m) = 2 \sum_{k=-(M-1)}^{M-1} \exp(-j2km)w[k]w^*[-k]x[n+k]x^*[n-k]$$

where 2M-1 is the length of the window used to window the signal. The algorithm used to obtain the Wigner distribution is the following [4], [11]:

- Sample the signal at a rate two times the Nyquist rate (4 times the highest frequency).
- Prewindow data - in this case we used a Gaussian window,
- Perform time-frequency distribution according to equation above.

The window used for the generation of the Wigner distribution is the t*exp(2p*k*t) window[10]. With appropriate choice of the parameters this "infinite" window can be truncated so that it has a narrow frequency response. It is essentially a double-exponential smoothing function that decreases more rapidly than a single exponential.

In this analysis, a slightly different form of the PWD was used. The signal was made analytic using the Hilbert transform. The effect of this is to make the spectrum zero for negative frequencies [7].

A record length of 50 ms of the signal s(t) was sampled at 10 KHz:

$$s(t) = \cos(2\pi 150t) + \cos(2\pi 2000t)$$

Then the signal was windowed with a Gaussian window. Figures, 1a and 1b show 10 ms of each of the individual frequency signals, 150 Hz and 200 Hz, respectively. Figure 1c shows the signal s(t) and figure 1d shows the result of windowing s(t) with a Gaussian window. This windowed signal, s'(t), is what is used for the Wigner Distribution.

Figure 2 shows the contour plot of the pseudo-Wigner distribution of s'(t). In the same figure are shown the impulse response and the frequency response of the signal s'(t). The Wigner distribution for this signal is both time limited and bandwidth limited as is the signal. As we can see, there are two Gaussians in the PWD, one at the center frequency of 150 Hz and the other at the center frequency of 2000 Hz. These Gaussian shapes are due to the characteristics of the two windows used to generate the Wigner Distribution. There is also an oscillatory term at 1075 Hz which corresponds to the frequency of half of the sum of the two tones (150+2000)/2. This characteristic of the Wigner distribution is due to quadratic content and is known as interference. The Wigner distribution is a measure of the energy density of the signal. When the signal is integrated over time and frequency, the contributions of the oscillatory terms are insignificant and the result is the total energy of the signal. Algorithms have been reported in literature to obtain the Wigner family distributions using a neural-network or an adaptive system [5], [9].

By taking the Wigner distribution of a signal we have a representation of the signal in both time and frequency. A function that gives an idea of how a signal interacts with itself is the autocorrelation function. We will examine what happens when we take the autocorrelation of the Wigner distribution first in time and in frequency separately, and then in both time and frequency at the simultaneously.

Autocorrelations of the Wigner Distribution

When performing the autocorrelation function of the PWD in time, we are essentially filtering the Wigner distribution with a time reversed version of itself, i.e., performing a matched filter function. In the upper left of figure 3, the result of this function is shown. The autocorrelation in time was performed separately for each discrete frequency. As expected the time autocorrelation function was peaked for each frequency channel at a time shift of 0 ms. The original signal s'(t) is a relatively simple one where both tones are present at the same time and over the entire 50 ms. However, if one of the tones consisted of repetitive bursts, then there would be more interesting phenomena when taking the time autocorrelation.

When performing the autocorrelation function of the PWD in frequency, we are performing a modulation function of the signal with itself. Therefore, we expect changes in the spectrum of the signal. Figure 3, in the lower right, shows the result of this function. Again a separate frequency autocorrelation was taken for each sample time. The result of this is two Gaussian terms appearing at frequency shifts of ±2150 Hz, i.e., at the sum of the two frequencies present in the original signal. There are also oscillatory terms present at the frequency shifts of ±1075 Hz. Finally, the peak of the distribution is at a 0 frequency shift for all time samples.

Taking the two-dimensional autocorrelation function (simultaneous autocorrelation in time and frequency) we find properties similar to both autocorrelation in time only and frequency only. The peak occurs at 0 time shift and 0 frequency shift. There are two Gaussians present at frequency shifts of ±2150 Hz as in the autocorrelation in frequency terms. The amount of interference terms in this 2-D autocorrelation is greatly reduced in comparison to the other autocorrelations. The 2-D autocorrelation, however, is difficult to interpret because both axis are shifts rather than actual values.

It is possible to have a neural-network to perform the autocorrelation functions in each direction (time and frequency). Since there is no additional information offered by the 2-D and since it is difficult to interpret, we will limit our discussion to the neural-network implementations of single autocorrelations. Also, to implement the 2-D autocorrelation we would require a very large number of interconnections (on the order of N^2 for each neuronal element).

Neural-Network Implementations of the Autocorrelations

Figures 4 and 5 show two implementations of a neural-network for performing the autocorrelation in time or in frequency. These implementations are shown for a 2-D signal length of only 5 points to illustrate the concept. Also only one row (or column -depending on whether the autocorrelation in time or in frequency is taken) of the array of neural processors needed is shown. The discussion will be limited to the autocorrelation in time. In the first implementation, the number of neurons is 9 (2×5-1) which is the length of the autocorrelation for a signal of length 5. The number of data points of the Wigner distribution array to which the element is connected varies depending on the neuronal element. The first neuronal element accepts information from two elements of the PWD, element 1 and element 5. These elements are multiplied together and the result is the value of the autocorrelation function for frequency i, at a time shift of -4 sample units. The second neuronal element accepts inputs from four elements of a row of the PWD. These are 1 and 4 which are multiplied together and 2 and 5 also multiplied together. The results are then summed to produce the value of the autocorrelation function for frequency i, at a time shift of -3 sample units. This process continues for all nine neuronal elements. The inputs to each neuron from a particular row of the PWD are shown at the bottom of figure 4.

In the second implementation there are 5 neurons, the same as the number of elements in a row. First, a value of an element of a row of the PWD is stored in each element . Then the various elements of the row are inputs to the neurons. At the first time instant, the only neuron that has input is the first one. This input is the value of the fifth-row element. This input is multiplied with the value stored in the neuron and the result of the neural-network at time instant 1 is the value of the autocorrelation function for frequency i, at a time shift of -4 sample units. At the second time instant, the first two neurons have inputs. The input to the first neuron is the fourth-row element and the input to the second neuron is the fifth-row element. Multiplications are performed at each neuron and the outputs are summed. The result is the value of the autocorrelation function for frequency i, at a time shift of -3 sample units. This process continues for a total of nine iterations. At the bottom of figure 5 the inputs to the neurons are shown for various times. This second implementation is the one that we will show results for in the next section.

Results of Neural-Network Implementation

Due to the nature of the Wigner distribution signal being an energy signal, when we take its autocorrelation we end up with a signal of, effectively, fourth power which has no physical significance. We would like to keep the dimensionality of the signal from increasing from second order so we take the autocorrelation of the square root of the Wigner distribution. This square-root function is performed at the neuron.

Figures 6 and 7 show the results of the autocorrelation in time and frequency of the PWD, respectively, using the neural-network type of figure 5. As we can see the general characteristics are the same as those of the autocorrelation in time and frequency by conventional methods shown in figure 3.

Discussion

We have discussed various properties of perceptual signals and how they depend simultaneously on more than one variable, such as time, frequency and space. Based on this, we discussed the need of a joint-domain signal representation. The one chosen was the Wigner distribution in time and frequency. We also showed the effects of taking various autocorrelations of the Wigner distribution signal. Neural-networks have been in literature to model perceptual systems. Here we discuss the neural-network processing of perceptual signals. Wigner distribution can be generated using neural-networks. As was discussed, the autocorrelations of the Wigner distribution can also be generated using a neural-network. It is possible to modify the weightings of the neural-networks to obtain more physiologically plausible filtering functions. These filtering functions are in general nonlinear and can be in time or in frequency or both. An example of a filtering function in frequency in audition is the use of a cochlear-filterbank models[1] showing the response of the cochlea to different frequencies. This information can be used as weightings to the neural-network.

It is important to mention that there is no physiological evidence that any of the perceptual mechanisms perform Wigner transforms. However, this representation and its autocorrelations show the signal in time and frequency and how it interacts with itself in time and in frequency. This is important to the properties mentioned in this paper: lateral inhibition, similarity, adaptation, absolute threshold as a function of frequency and scaling. The last two properties are energy-dependent nonlinear processes. Therefore, with nonlinear neural processing of the Wigner distribution, an energy-density representation, we may obtain a model of how these processes work. The other three properties are discussed in greater detail in the following subsections.

Lateral inhibition

In lateral inhibition, one stimulus inhibits another from generating neural activity. In the Wigner distribution of a signal consisting of two tones, there are oscillatory terms appearing at the frequency of half the sum of the two-tone frequencies. If the two tones are close together in frequency, due to the effects of the window used to generate the Wigner distribution and the oscillatory terms, we may have the three frequencies merging together. Furthermore, autocorrelation in frequency will result in merging of the three signals at the sum of the two frequencies and at DC.

If one of the tones consists of repetitive bursts (for example, $t_{on}=t_{off}$) , taking the Wigner distribution will result in a series of Gaussians when the tone is present and at the frequency of the tone. When taking the autocorrelation in time, due to the elongation effect of the autocorrelation, the Gaussians will merge together.

The descriptions given of the two phenomena may give some insight of how inhibition may occur either in time or frequency.

Similarity

In similarity, stimuli that are different appear to be the same. In the case of vision, a signal color may be composed of more than one wavelength and can be perceived to be the same as a color of a single wavelength different from the ones of the complex color. In the Wigner distribution and its autocorrelations in time and frequency, we found frequencies in the representations that were not part of the original signal. It may be possible to model the similarity property using the distributions discussed.

Adaptation

In adaptation, the prior history of the signal (time) is important in determining the behavior of a perceptual system to a particular stimulus. Thus, a stimulus presented over a long period of time may affect the sensitivity of the perceptual system to a newly presented stimulus. Again, this implies that the boundary of the first signal (in time and frequency) has been extended to influence signals in neighboring boundaries.

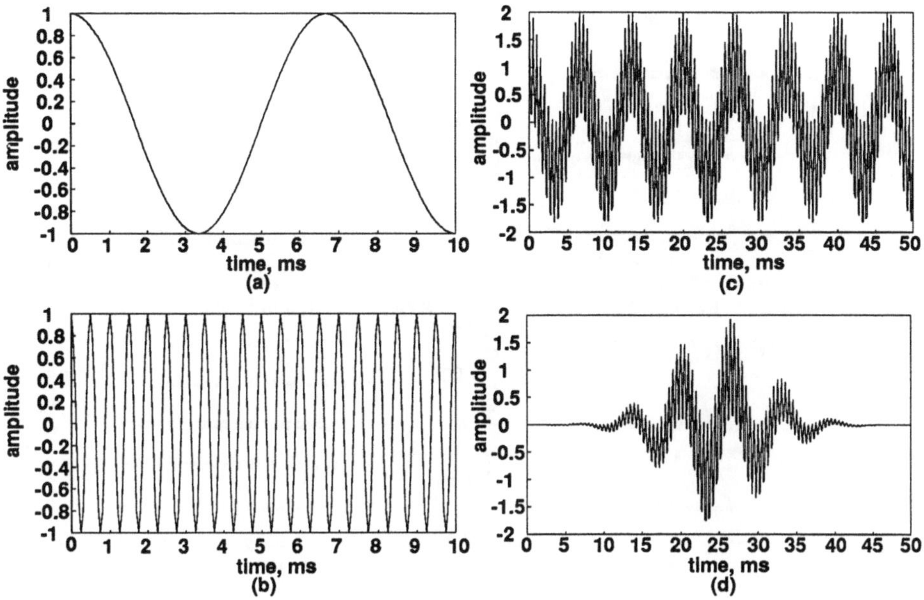

Figure 1. (a) 10 ms segment of 150 Hz signal (b) 10 ms segment of 2000 Hz signal (c) signal s(t) and (d) windowed signal s'(t).

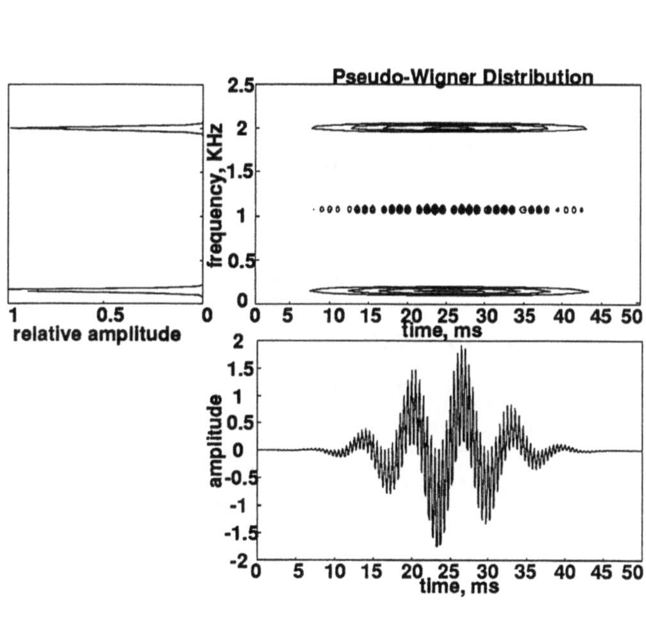

Figure 2. Contour plot of pseudo-Wigner distribution of s'(t). Also shown are the impulse response and frequency response of s'(t).

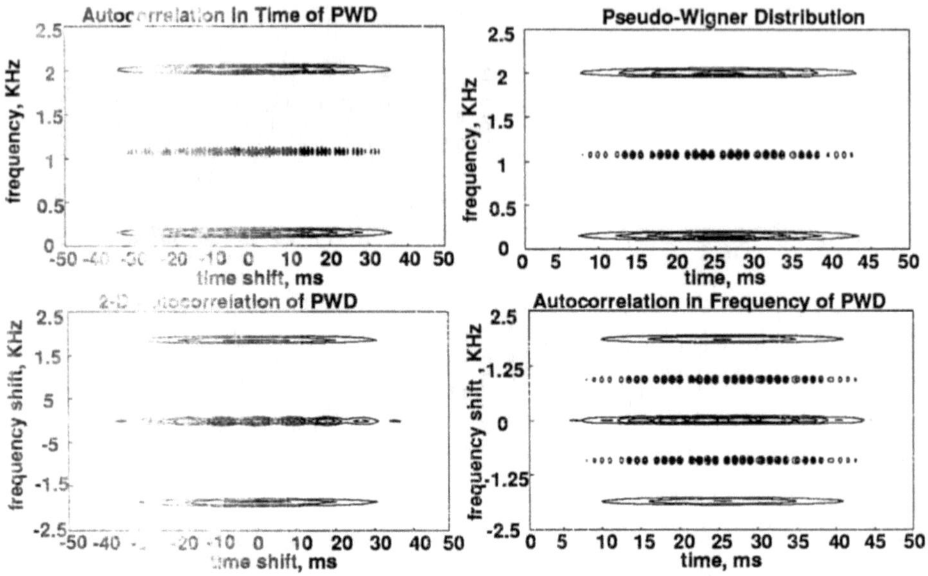

Figure 3. Contour plots of pseudo-Wigner distribution of s'(t), autocorrelation in time, autocorrelation in frequency, and 2-D autocorrelation in time and frequency of PWD.

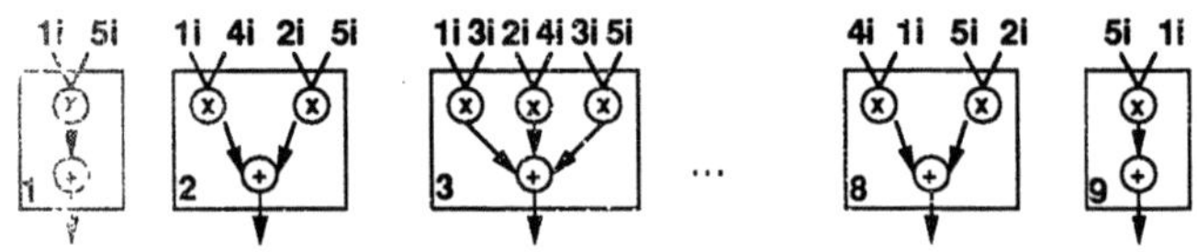

neurons

	1	2	3	4	5	6	7	8	9
pairs of inputs for each neuron	1-5	1-4	1-3	1-2	1-1				
		2-5	2-4	2-3	2-2	2-1			
			3-5	3-4	3-3	3-2	3-1		
				4-5	4-4	4-3	4-2	4-1	
					5-5	5-4	5-3	5-2	5-1

Figure 4. First neural network implementation of autocorrelation function in time or frequency of a joint-domain representation.

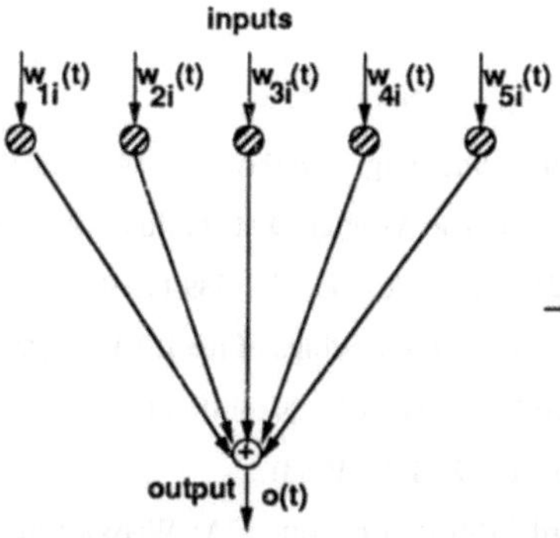

inputs to neurons wji(t)					
time	1	2	3	4	5
t=1	5				
t=2	4	5			
t=3	3	4	5		
t=4	2	3	4	5	
t=5	1	2	3	4	5
t=6		1	2	3	4
t=7			1	2	3
t=8				1	2
t=9					1

Figure 5. Second neural network implementation of autocorrelation function in time or frequency of a joint-domain representation.

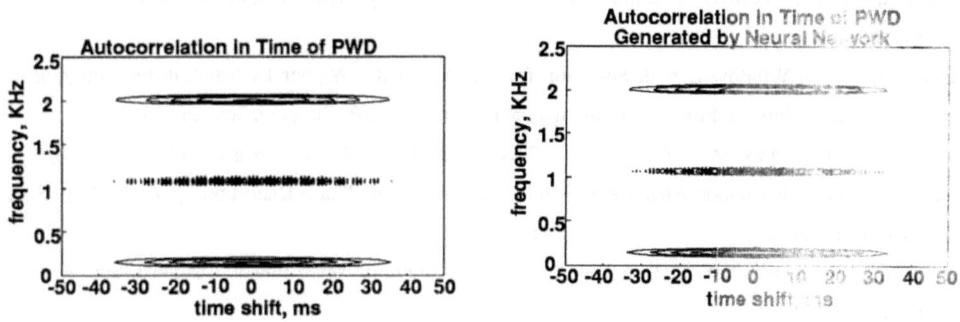

Figure 6. Result of autocorrelation function in time of the PWD of s'(t).

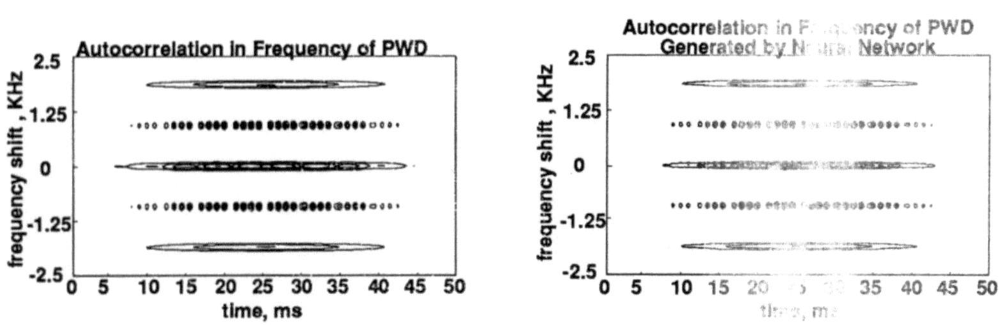

Figure 7. Result of autocorrelation function in time of the PWD of s'(t).

References

1. Allen, J.:Cochlear Modeling. IEEE ASSP Magazine. 3-29 (1985)
2. Anderson, J. A. and Rosenfeld, E. editors: Neurocomputing. Cambridge, MA: MIT Press, 1988
3. Claasen, T.A.C.M. and Mecklenbräuker, W.F.G.:The Wigner Distribution - A tool for time-frequency analysis. Parts I-III. Phillips J. Research, 35, 217-250, 276-300, 372-389 (1980)
4. Cohen, L.:Time-Frequency Distributions - A Review. Proceedings of the IEEE, 77, 941-981 (1989)
5. Daugman, J.G.:Complete Discrete 2-D Gabor Transforms by Neural-networks for Image Analysis and Compression. IEEE Transactions on ASSP, 36, 1169-1179 (1988)
6. Goldstein, E.B.:Sensation and Perception, 3rd Edition, Belmont. CA: Wadsworth Publishing Company 1989
7. Howitt, A.W.:Application of the Wigner Distribution to Speech Analysis. Massachusetts Institute of Technology, Electrical Engineering Department, S.M. Thesis, 1987
8. Jacobson, L.D. and Wechsler, H.:Joint Spatial/Spatial Frequency Representation. Signal Processing, 14, 37-68 (1988)
9. Kadambe,S. et al.:Window Length Selection for Smoothing the Wigner Distribution by Applying an Adaptive Filter Technique. Proc. of the International Conf. on ASSP, Glasgow, Scotland 1989
10. Owens, F.J. and Murphy, M.S.: A Short-Time Fourier Transform. Signal Processing, 14, 3-10 (1988)
11. Preis, D.:Wigner Distribution Analysis of Filters with Perceptible Phase Distortion. J. Audio Eng. Soc., 35, 1004-1012 (1987)
12. Sachs, M.B. and Kiang, N.Y.-S.:Two Tone Inhibition in Auditory Nerve Fibers. J. Acoust. Soc. Am. 43, 1120-1128 (1968)
13. Sejnowski, T.J.:Storing Covariance with Nonlinearly Interacting Neurons. Journal of Mathematical Biology, 4, 303-321 (1977)

Robot Control with Procedural Expert System

G.R. Meijer[1], T.L. Mai[2], E. Gaussens[2], L.O. Hertzberger[1] and F. Arlabosse[2]

[1] Computer Science Department (FWI), University of Amsterdam, Kruislaan 409, 1098 SJ Amsterdam, Netherlands

[2] Framentec S.A., Tour Fiat, 92084 Paris La Defense, France

Abstract: In this paper the use of a Procedural Expert System (PES) for the control of a robot in a partly-structured environment is presented. Performing a task in a partly-structured environment requires a certain level of autonomy of the robot. One of the key issues needed to realize this autonomy are techniques to monitor the robot activities and to handle exceptions. The control system must be reactive to new situations. Implementation architecture and realization with PES of robot control based on the Exception Handling Model is described. The experimental results show that the event-driven mechanisms needed for the control of the robot can be adequately realized with the reactive behaviour of PES.

Keywords: reactive systems; exception handling; autonomous robots; procedural expert system; knowledge-based control; planning.

1 Introduction

One of the key capabilities of an autonomous robot is to handle exceptions interrupting the pre-planned flow of robot actions [11,13,14]. Exception handling enables a robot controller to react adequately to unforeseen situations and to take measures to continue operation of the robot. In case of robot manufacturing systems, this means that production can be continued. For mobile robots, exception handling is a necessary requirement to navigate in partly-unstructured environments. It is our view that monitoring and exception-handling mechanisms have to be an integral part of the robot control process, thus interleaving planning and plan execution [12].

Exception handling encompasses several areas of research. Donald [8] and others study the problem of geometrical motion planning and adjustment. These techniques operate within a pre-planned flow of robot operations and are called local-exception handling. Global-exception handling on the other hand addresses the cases when exception handling requires a change of the pre-planned flow of operations, or when even new operations are needed [6,18]. The research work described in this paper deals with the problem of global-exception handling.

When a robot is programmed by a programmer or programming system, a sequence of robot operations is generated. This sequence is based on a model of the robot environment and on expectations how the operations of the robot will alter this model. In state of the art robot control systems, the environment model is the basis for all operations of the robot. It represents the knowledge of the programmer of the environment and how it changes under the operations of the robot. This model and the expectations are called the a-priori plan.

Executing an a-priori plan in a partly-structured or unstructured environment introduces situations in which the expectations do not hold and where exceptions cause the execution to stop. Recent research directions have therefore moved away from the a-priori plan generation

NATO ASI Series, Vol. F 71
Expert Systems and Robotics
Edited by T. Jordanides and B. Torby

and study on-line plan generation based on some external events [16]. The systems studied here are not constrained by any plan information apart from a task description and are therefore on the other end of the spectrum compared to the plan-driven robots mentioned before. In the approach taken by Brooks [5], the robot is controlled by a set of behaviours which can subsume the role of others as the environment demands so. In our view this subsumtion architecture is limited by the inability to realize goal-directed actions through the use of a-priori plan information in the behaviour of the robot. Instead, the typical case we are addressing here, is what we call autonomous systems on the shopfloor. That is, a robot which has to carry out a task in a partly-structured environment, but constrained by limits on time and resources. In these cases what we need is system autonomy within the framework of an a-priori plan.

This paper deals with a systematic way of combining plan information into a reactive system and presents an alternative approach to tackle the problem of exception handling. In the case of a partly-structured environment, only a limited amount of model information is available and as a result the a-priori plan consists only of a skeleton of task instructions. A control structure is developed in which the planning of operations is strongly driven by the events that take place in the environment of the robot and which provide the actual status of the environment. In fact, the robot controller does not have an explicit a-priori plan in terms of a sequence of robot operations. The a-priori planning activity is restricted to the definition of a set of planning functions, which can be applied in certain environment conditions. If these conditions hold true, the execution of the planning functions results in a plan of robot operations. However if an exception occurs, an alternative planning function is used, resulting in a different plan and a different flow of robot operations. The execution of the planning functions is performed by the robot controller.

In the paper the use of knowledge-based system called Procedural Expert System (PES) is described. It will be shown that the event-driven mechanisms needed for the control of the robot can be adequately realized with the reactive behaviour of PES.

The integration of an exception-handling mechanism in a planning system increases highly the complexity of the system and raises the problem of backtracking. In systems such as PES, the building of a plan and its execution are linked together and carried out in an interactive way as the system controlled by PES progresses. Plan building and execution are based on a decentralized structure. The whole planning knowledge is split into modules called Knowledge Areas (KA's). Each KA encodes a "partial plan" and the description of the agents which would start the execution of the partial plan. These agents are determined during the progress of the system and are responsible for the formation of the coming partial plans. The plan-execution level is the processing of PES while the plan building level is insured by a rule-based mechanism.

In this way, the complexity inherent in the integration of exceptions can be avoided by the use of weakly related structures and heuristics which guide the reactions of the system. A description of PES is given in [9,10]. The design of the PES based robot controller is described in [12].

An experiment was performed with the control of an assembly robot. An interface between PES and the robot controller was realized. The robot is controlled directly from the PES environment. Sensor data from the robot is coupled back to PES environment, where it can give rise to events. The capabilities of the system to cope with introduced exceptions was studied.The experiments show that exception handling at many levels of a task decomposition is possible, which makes the approach well suited for partly-structured applications. In these cases, the proper reaction to the environment conditions has a higher priority than following an a-priori plan.

The paper is constructed as follows. In the next section, the objectives of the implementation are formulated. In section three and four, an overview of the Exception Handling Model and PES are given. Section five describes the proposed system architecture. The realization of the architecture and the experimental results are presented in section 6. Finally, section 7 contains the conclusions and outlook.

2 Objectives

The application of a knowledge-based system for the supervision of an autonomous process, requires adequate interfaces between the knowledge-based system itself and the systems directly controlling the process. In many applications of knowledge-based systems, a human is involved in the process to be controlled and he interfaces directly with the knowledge-based system. This can be either to supply input data to the system, or to execute the output of the system. For instance, a diagnostic system for medical purposes interacts directly with a user to generate a diagnosis of a number of symptoms. The user supplies the actual state of the parameters selected by the system. The other way round, a knowledge-based scheduling system generates a schedule, which is supposed to be executed by an other system.

As such, the knowledge-based systems operates in an isolated environment. This is reflected in the implementation which provides a good deal of tools and mechanisms to develop user interfaces, but hardly provide adequate tools for interfacing the system to external processes. Early implementations of knowledge-based systems offered little possibilities to build interfaces with processes running outside their Prolog or Lisp shell. Recent implementations of knowledge-based systems recognize this problem and provide better possibilities for interfacing, at least by providing proper access to the interface possibilities of the implementation language. Prolog implementations like IfProlog or BIM Prolog are commercially available tools which offer good possibilities for interfacing. Also, Allegro Common Lisp offers an external language interface.

In our situation, the knowledge-based system PES was chosen for its reactive capabilities. We want to exploit these possibilities of PES in the application domain of autonomous robots. By definition, an autonomous system does not involve a human operator in the execution of a task. All control systems of the robot have to cooperate without direct human intervention. Therefore PES has to communicate in the first place directly with the other robot control functions. In particular it is the monitoring of robot operations which we want to realize by using the reactive capabilities of PES. Our objective is to realize an implementation of PES in which the monitor function is an integral part of PES through direct communication with the actuator control and sensor control functions. Before going to the proposed solution, an overview of EHM and PES is given in the next sections.

3 Exception Handling Model

3.1 Overview

In this section we describe a model which forms the frame work for the handling of exceptions of an autonomous system. This is called the Exception Handling Model (EHM). EHM describes how autonomy can be reached within the scope of an a-priori plan. Exception handling involves planning and execution of operation of a robot system which were not foreseen in advance. To cope with these situations, a planner needs a complete understanding of all the consequences of an exception. Without knowing the specific details of the application, the number of these consequences and their complexity can be very large. As a result, the planner is not always able to react adequately to exceptions in a real environment. EHM addresses this problem of general purpose planners.

The underlying assumption of EHM is that the problem of exception-handling planning for robotic systems can be reduced by combining application-driven heuristics with general-purpose task- and path-planning functions. This allows on the one hand overcoming the problems of applying general purpose planners to specific application driven domains, because the heuristic information guids the planning functions. On the other hand effective planning functions can still be freely used by the user of the system. First we introduce the basic terminology used in EHM.

In EHM the task of a robot is represented by a task decomposition structure called the task tree. The *task tree* is an ordered graph in which the root node represents the task of the robot. By decomposing the task into subtasks, additional nodes are created. At the lowest level the subtasks are no further decomposed, but directly expressed in the elementary operations of the robot. Each *elementary operation* (EO) of the robot represents a certain behaviour of the robot which is described by the control mechanism and the sensor feedback that is used. The sensor feedback is called *nominal feedback* if the value lies within the corresponding working range of the elementary operation. Any feedback outside the working range is *non-nominal feedback.* The set of working ranges are the *operating conditions* of the elementary operation.

The environment of the robot is represented by a set of *environment parameters*. These parameters describe the state of the robot, transport systems, tools, parts and obstacles. The feedback of the robots sensors can be expressed as the current value of the environment parameters. Thus the operating conditions of an elementary operation are expressed as constraints on the allowable value of the environment parameters. The robot sensors do not measure all possible environment parameters, but are limited to a selection of the parameters. These observed environment parameters of an elementary operation are called the *monitor conditions.*

Task knowledge plays an important role in EHM. In the first place EHM uses the task tree to keep track of the context information during the execution of the robot operations [11]. A *stack structure* is used to store the information concerning the tasks the robot is performing. At any moment the context of the current robot operation can be retrieved by looking into the stack structure. In the second place, the task knowledge is used to determine the *scope* of the affected a-priori plan resulting from the exception. Hence the planning horizon is limited as well.

The second important concept used by EHM is that of recovery strategies. A *recovery strategy* is a framework of planning instructions which can be executed to produce a *recovery plan* for the robot. The strategies can be regarded as blue prints for a certain behaviour of the robot, without the specifics of how this behaviour is carried out in detail. The strategies in EHM are used to encode the heuristic information needed to guide the exception-handling activity. In this way the planning complexity is reduced.

Next, we take a closer look on how the mechanisms described by EHM are used for adaptive control of autonomous robots. In such a system a continuous feedback based on the observations of the environment is necessary. This demands that given an initial task or mission, the robot postpones final decisions on which route to follow or which action to take. Only when the robot has to take action, the actual environment data has to be used to plan these actions. Even during the execution of an action, the robot has to be sensitive to unexpected events which require adaptation of the plan. This allows the robot behaviour not to be strongly constrained in a strictly predefined plan of actions. The plans become more and more precise as the robot task evolves from a given intention to a physical robotic operation. Decisions can be taken as late as possible which prevents the planner from making too many assumptions on the environment while building a plan.

In EHM this adaptive behaviour is realized by combining general-planning functions with domain-specific procedures. The combination of these two sources of information is performed by a control program. Figure 1 represents the structure of EHM.

The control program selects the proper procedures which are encoded as strategies, and uses the general-planning functions to build a recovery plan. The recovery plan allows the robot to adapt itself to the new environment conditions. As such, the control program is provided by the interpreter mechanisms of an expert system. The expert knowledge is then represented by the general-planning functions and the domain-specific procedures.

Control Program	
Domain-specific procedures	General-planning functions

Figure 1: Structure of the Exception Handling Model.

To benefit from the capabilities of EHM, we must identify the degrees of freedom the robot system has within the context of a given task or mission. For this purpose we focus on the domain of robot assembly systems, although a similar analysis can be applied to mobile robots. Assembly tasks are often underconstrained and EHM aims at exploiting the inherent degrees of freedom to allow exception handling. The degrees of freedom of a robot assembly task are:

- freedom resulting from the assembly itself
- freedom associated with the operations of the robot system.

Each of these degrees of freedom can be exploited by applying a rescheduling respectively a recovery-planning technique. Any assembly contains parts and possibly subassemblies. The order in which the assembly task can be performed is generally not uniquely defined. During the execution of the assembly task an exception can interrupt the chosen order of assembly. Although a required part of the assembly might be broken down or not available, this does not necessarily mean that no further assembly steps can be taken. The freedom in the assembly order can be used to reschedule the assembly sequence, taking into account that a certain assembly step can not be performed.

EHM describes a mechanism for the rescheduling of the assembly tasks which embeds the specific application dependent constraints by means of heuristics. These heuristics are specified in so called task rescheduling strategies. Each strategy refers to a distinct scheduling criterion. The rescheduling mechanism is controlled by a limited number of pre-defined strategies. To apply the proper strategy on the correct moment, the application needs to be modelled in terms of the strategies to follow in case a certain exception occurs. This exception mechanism is described in detail in [14,19]. In the work described in this paper however, we have concentrated mainly on exploiting the freedom of robot operations by recovery planning.

3.2 Recovery planning

For the assembly of a part, the robot has to plan a sequence of operations to perform. A typical scenario for an assembly task is 'grasp and assemble' where 'assembly' can be 'place', 'insert', or any other task. This scenario is decomposed into elementary operations which can be executed by the robot. For each elementary operation a geometrical planning has to be performed as well. The result of these planning activities is that a plan is built for the robot to perform the required assembly task. An exception during the execution of one of the operations, aborts the execution of all other planned operations. If the robot is still in an operational state, the planning capabilities of the robot can be used to generate a recovery plan.

The occurrence of the exception can have complex consequences for the state of the assembly, the parts and the resources needed for the assembly. The recovery plan has to take into account these consequences and use an appropriate scenario for the handling of the exception. Just as in the case of the rescheduling mechanisms, in EHM the scenario's are specified as strategies. Each recovery-planning strategy consists of a sequence of planning instructions, which together form the recovery plan. EHM provides a selection of the available recovery-planning strategies for each exception that might occur. This so called kernel already gives the system the capabilities to select proper recovery-planning scenario's.

To bring in the heuristic knowledge of a specific application, the kernel of recovery-planning strategies can be tailored to the needs of the application. Simply by changing the association between an exception and a strategy, or by altering the order of the strategies, the recovery-planning behaviour is adapted. Of course new strategies can be specified as well. This mechanism for recovery planning was introduced in [14,15] and is summarized in figure 2.

- update knowledge of current operating conditions
- classify the exception or exceptions in question
- perform sequence planning by using the predefined strategies
- perform motion and grasp planning to built recovery plan
- execute recovery plan
- restart normal operation

Figure 2: The recovery-planning scheme.

Three functions are added to the control system of the autonomous robot to provide the exception-handling behaviour. These are a monitor function, a diagnostic function and a function for the exception handling. For each elementary operation the monitor conditions consist of a status vector of relevant operating conditions whose value is delivered by a corresponding sensor primitive. The monitor function checks for non-nominal feedback. After detection of an exception, the diagnostic function gains additional information on the nature of the exception and updates the internal model of the robot environment.

The third function is the recovery planner which generates a recovery plan according to the selected recovery strategy. If a recovery plan can not be built, either because the planning functions do not provide a solution or the physical capabilities of the robot are inadequate, the recovery planner selects the second strategy of the strategies set. If the strategies list is exhausted, the robot cannot achieve the current task and rescheduling of tasks is needed.

4 Procedural Expert System and EHM

The planning system proposed in this application is called Procedural Expert System (PES)[1] (see reference [9] and [10]). In this section we describe its main features and its use to realize the EHM mechanism.

1 PES has been developed by Framentec within the ESPRIT project P820 - QUIC

4.1 What is a Procedural Expert System

PES is issued from M.Georgeff's work (SRI) for the NASA space shuttle. Designed to process procedural knowledge, it uses two Knowledge Bases:

1. The first contains at a given time the data describing the internal state of the system to be controlled and the knowledge the system has on the external world [1]. For a robotic application it may include sensor data as well as information on the structure of the robot or its environment.
2. The second contains the control procedures which are encapsulated into unrelated modules named *Knowledge Areas* (KA's).

Given a goal to reach and the current state of the system, PES proceeds by triggering the procedures and executing the most suitable ones. The selected action may aim to either solve the goal or to react to a new situation detected through the facts in the knowledge base.

The particularities in the PES's approach are those of a partial and hierarchical planner. *Hierarchical,* because of the hierarchy between the procedures (a general procedure calls more specific ones), and *partial* because it does not generate a whole plan of action but executes parts of a plan as soon as they have been determined (this may help in determining the following actions).

4.1.1 How does PES work

A procedure (KA) is composed of two parts:

1. A declarative part, called *invocation condition*, which describes the situation under which this KA is useful : what goal it is supposed to solve (the *Goal* clause), and what are the preconditions (the *Fact* clauses).
2. A procedural part, which can be either a rigidly-coded procedure, or a *Recursive Transitive Network* (RTN). A RTN is a graph having one starting node and several ending nodes. Arcs are subgoals to be reached (or test to be performed). Nodes define a state.
   The execution of the procedure consists then to cross down the RTN from the starting node until reaching any ending one in a depth-first, left-to-right way. To cross an arc, the action that labels it must be achieved, which can be either a test in the data base (success or fails immediately) or a (complex) subgoal to reach which may requires the *recursive* call to another KA, until reaching the KA containing the basic information, i.e., the elementary operation to be performed. So that is why the KA's can be seen as *partially* pre-defined plans of the robot.

At any given moment, several goals are active in the system and certain beliefs are held in the data base. The *interpreter* selects the applicable KA's by matching their invocation conditions with the current goals and beliefs, chooses one, and executes it. It repeats this process each time a new subgoal is activated, or a new fact is acquired.

The way PES selects one KA among several candidates can be directed by using some heuristic. However, fact-driven KA's, i.e., whose invocation condition contains only fact clauses, have higher priority than others. This allows the system to react rapidly as new beliefs are acquired in the data base.

---

[1] obviously this knowledge should be represented at the proper level of abstraction in order to be used efficiently by the planner

4.1.2 Handling Exceptions

One can see that when PES is working, it raises a stack composed of goals (or subgoals) that have been recursively called and are still unsolved. This stack forms the current system's objectives and try to fulfil them first . However, the system remain sensitive to changes in the external world and you can make it interrupt the current action to handle an unexpected event. There are two ways to do so.

1. The first one is by using fact-driven KA's. These KA's are activated as soon as the specific facts described in their invocation condition hold in the data base. Their execution, seen from the goal stack side, results in adding new goals to the stack. After these goals are solved (if you had correctly designed your KA, the exception has been handled), they are removed from the stack and the system recovers its initial objectives.
2. The other way is to use *constraints*. The idea is to associate with each goal a set of constraints. When an exception occurs, some of the constraints are violated, PES interrupts the current action and reacts by removing from the goal stack the goal associated with the violated constraints and all the subgoals that have been activated later to solve it. This is a backtrack in the whole plan that brings the system back to the last valid goal remaining in the goal stack. A new KA is selected then, either another goal-driven KA to re-try the same goal, or a fact-driven KA to perform first a recovery strategy.

The set of constraints represents here some configuration of the whole system under which the associated goal is meaningful. Its definition is strongly domain dependant. It also implies that all the exceptions are identified that would make the goal achievement fail. In this application, constraints are composed of assumptions on signals provided by logical sensors. Intuitively, they are the set of logical sensors data that should normally occur while the goal is performed.

The exception-handling mechanism implies three steps : identify/diagnose the exception, identify which step should be re-computed, and use a proper strategy to recover. **When exception classes are pre-analysed**, the PES structure and the backtracking system provide the following properties:

- the constraints represent the expected events, therefore violation of a constraint pin-points the occurrence of an exception,
- the place, i.e., the KA, where the constraint is declared corresponds to the step in the plan that should be re-processed,
- fact-driven KA's can initiate the recovery strategy either on its own, or by calling specific goal driven KA's, or even re-using another KA

4.2 Application to the EHM mechanism

4.2.1 The Task Planner

The task decomposition structure in EHM, described in section 3, is represented in PES by a set of goal-driven KA's. A goal corresponds to a task. The RTN structure of the body of a KA can encode the sequential decomposition of a task into subtasks. The mechanism of recursive calls among KA's insures the further decomposition of subtask into sub-subtasks. At the lowest level, elementary operations are encoded by KA's whose execution directly addresses the system that physically controls the arm.

At any moment, the cascade formed by the outstanding KA's which PES is executing gives the context of the current robot operation. This knowledge helps in EHM to determine what part of the plan is affected by an exception and therefore what recovery strategy to use.

One can point out that PES allows one to define more than one KA to solve a given task. By coupling the goal clause with some preconditions (fact clauses), it is possible to choose different ways or strategies to achieve a general task according to the whole state of the system. This is the way to perform task rescheduling.

4.2.2 The Exception Handler

The task level where exceptions may occur concerns the elementary operations. For each of them, and depending on the current task context, a set of sensor primitives to be monitored is defined in order to insure the execution of the EO. The conditions on these sensor primitives are expressed by constraints.

An example of a KA with this monitoring aspect is shown in figure 3. Given the task MOVE to the robot, PES selects KA1 to solve it. The execution of KA1 implies the subgoal TRANSFER. PES selects then another KA to perform TRANSFER (figure 5) and posts in a constraint stack the two constraints related to the sensor primitives: CHECK-MOTION and ROBOT-FREE. Therefore, during all the execution of TRANSFER, the state of these sensor primitives is continuously checked, and an exception is sighted as soon as a constraint-violation value is switched to TRUE. (Note that in the examples the value FALSE represents a proper operation state.)

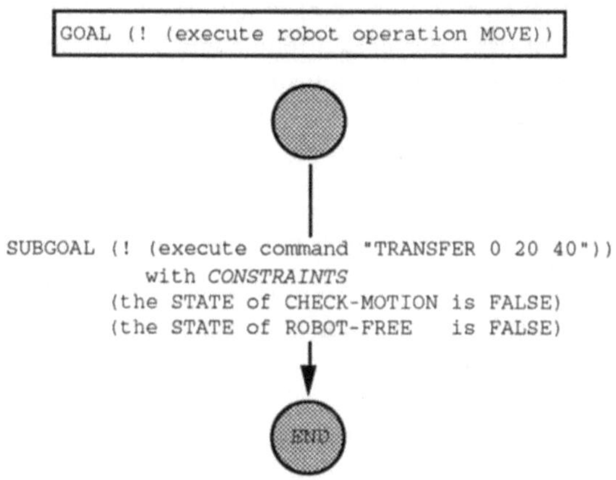

Figure 3: KA1- Monitoring aspect in the execution of MOVE

When an exception occurs, the choice of the appropriate recovery strategy to perform depends on a lot of criteria : the current EO, the exception type, the current task context, the available resources. By using fact-driven KA's to encode the recovery strategies, we can describe in the invocation condition of these KA's all the conditions under which the strategy is suitable. Therefore, the choice of the recovery strategy is left to the interpreter of PES.

5 Implementation Architecture

For robot control, two major groups of functions can be identified. The actuator control functions deal with the transformation of actuator commands into the desired motion of the actuator. Path interpolation, set-point generation, and servo control functions belong to this

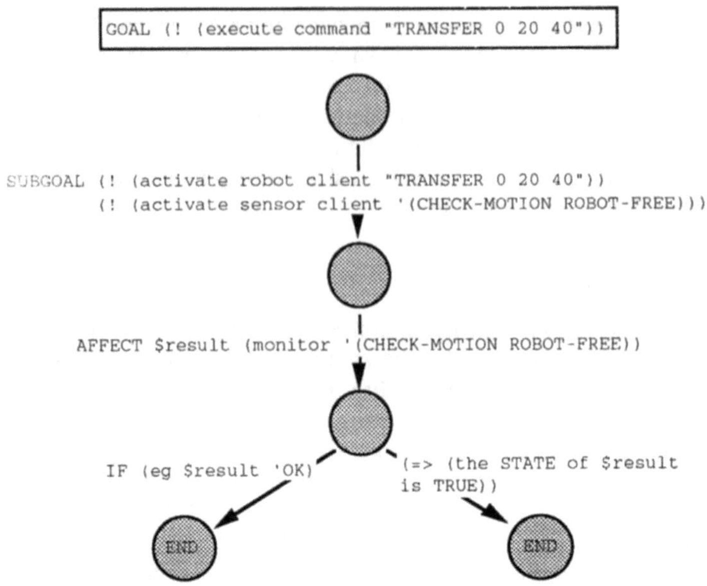

Figure 4: KA2- Execution of MOVE

category. On the other hand, there is a group of functions related to the acquisition, processing and interpretation of sensor data. For closed loop control, the sensor data is fed back to the actuator control processes to generate new actuator commands. Additionally, the sensor data is used by monitoring functions to guard the state of the autonomous system.

The implementation architecture contains three separate processes. Each process takes care of a group of functions. The processes are:

- planning and monitoring process
- actuator control process
- sensor process.

In figure 5 an overview of the implementation architecture is given, and in the next sections the individual components are described.

5.1 Planning and Monitoring

The planning and monitoring process contains the functions for planning of tasks and actions, monitoring of the execution and performing exception handling. All these functions are implemented in PES according to the implementation model of the last section. When a task instruction is received, the PES inference mechanism activates the proper KA's according to their invocation conditions. The KA's recursively invoke other KA's until a desired action of the robot is generated (figure 4). Before the action execution starts, the monitor parameters and their nominal value as expressed by constraints , are put on the constraint stack.

Next, two communication steps are taken (figure 6). In the first step, the required action is translated into a suitable interface format and sent down to the actuator control process. This communication is based on a non-blocking client-server model. Immediately after the action command has been given, a monitoring KA is invoked which establishes a second communication link with the sensor process.

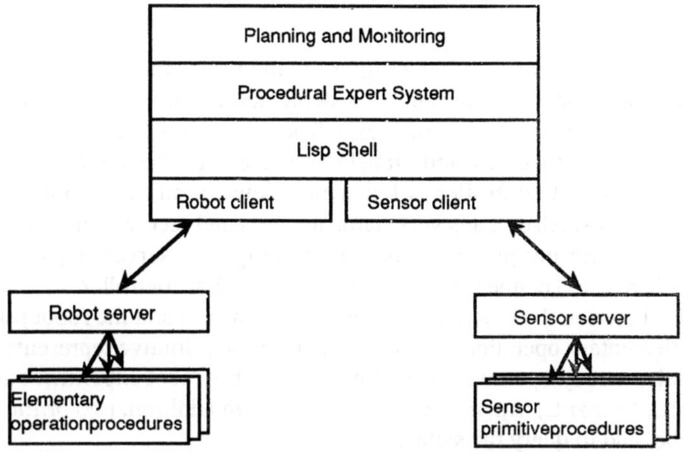

Figure 5: The implementation architecture.

The goal of this communication is to continuously update the status of the monitor parameters (stateCM, stateRFR,...) by querying the sensor process. The communication is based on a blocking client-server model. The monitor repeatedly traverses the monitor parameters and updates their value by querying the sensor process. The query process continues as long as the robot process controls the action of the robot.

When the monitor detects an exception (one of the monitor parameter values does not match its nominal value), a stop signal is directly sent to the actuator-control process while the constraint-based backtracking mechanism is released from the PES side.

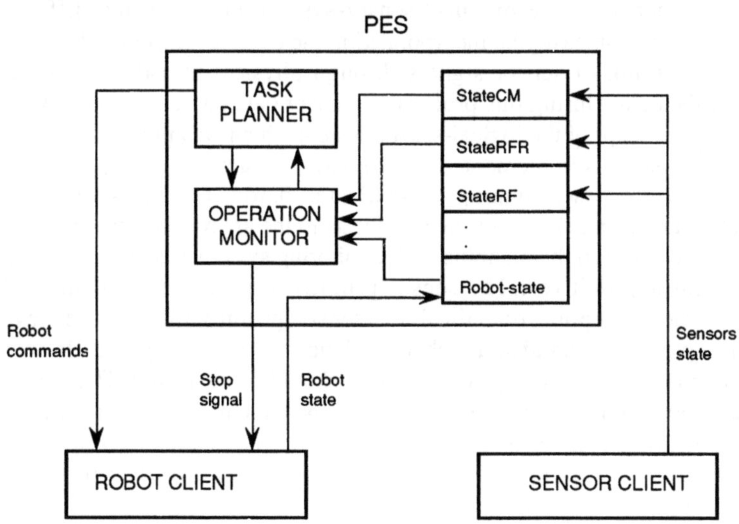

Figure 6: Communication between PES, the actuator process and the sensor process during the execution of a robot operation

5.2 Robot Server

The robot process consists of a server program (called the robot server) running on a host machine and a control process running on a dedicated robot controller. We use the robot controller on an one instruction per time basis. The robot server receives action instructions from connected client processes and translates these into the instructions for the robot controller. After the robot controller is fed with a new action, the robot server is ready to receive new instructions. Also, it answers status requests and receives interrupt messages from the planning and monitoring process. Upon an interrupt, the server stops the actions of the robot by using a hardwired connection to the PLC of the robot controller.

The interface between the planning and monitoring system and the robot server is based on the concept of elementary operations. Each communication primitive represents one elementary operation. The elementary operations used for the assembly robot experiment are: `Transfer`, `Grasp`, `Detach`, `Insert`, `Approach` and `Depart`. In addition, two primitives are used to interrupt the robot and to query the status.

On the planning and monitoring side, each communication primitive is implemented with a separate 'C' routine. These routines together form the robot client. The robot client is directly accessible from the PES environment.

5.3 Sensor System

The sensor process carries out the processing and interpretation of sensor data. The processing of sensor data in order to provide the actuator control modules with reliable information is an all but trivial matter. Because the control system modules are often only concerned with higher level sensor data abstractions, the sensor data processing is concentrated in a separate sensor system [13,20].

As a consequence of the modular sensor-system approach, the different sensor data processing strategies can easily be incorporated in the overall control model. The sensor system controls the gathering of sensor signals through physical transducers and the subsequent processing of the data. During this process the sensor data passes various levels of abstraction according to the concept of the logical-sensor system which we apply.

The sensor system is implemented in two parts; the actual server program running on the host and several processes running on dedicated hardware. The server takes care of the communication and dispatches the sensor requests to the appropriate processes.

The interface between the planning and monitoring system and the sensor system is realized by a set of sensor primitives. The sensor primitives are the equivalent of the elementary operations in the perception domain. Each sensor primitive is a higher-level, sensor-data abstraction, representing an abstract characteristic of the robot system. It does not represent how a measurement should take place, but what should be measured. The possible return value of the sensor primitive is either a Boolean value, a position or an object identifier.

The sensor primitives used are: `Check-Motion`, `Robot-Forces`, `Robot-Free`, `Plan-Position`, `Object-Available`, `Object-Orientation`, `Find-Object` and `Identify-Object`.

6 Realization and Experiment

The experimental set-up is displayed in figure 7.

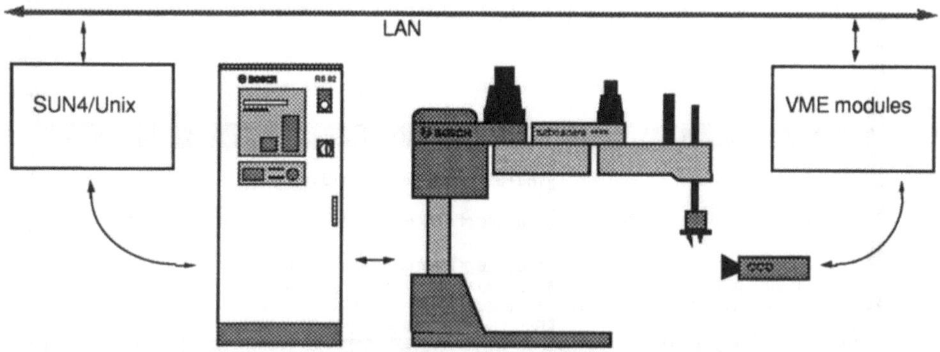

Figure 7: Experimental set-up.

The PES system used is implemented in Common Lisp employing the Allegro Common Lisp interpreter on a SUN4/Unix workstation under X-Windows. The communication links with the robot and sensors are realized with the foreign language interface. The robot controller is a Bosch Rho-8 controller and contains axis servo hardware, interpolation software and a built-in programmable logic controller. The robot controller is programmed with the robot language BAPS (Bewegungs und Ablauf Programmier Sprache). For the experiments, gripper switches and a vision sensor were used. The gripper switches are controlled by the PLC of the robot controller. To get the switches status, the sensor server queries the robot server. The vision system runs on a dedicated processor and Eltec vision hardware located in a separate VME crate. Programming and communication was initially realized with an in-house developed operating system shell. Later this was replaced by VxWorks. After some experimentation with different communication options like physical files, shared memory or pipes, we selected the named piped mechanism to implement the client server communication of the robot and the sensor system for its flexibility during program development and experimentation.

In the experiment we carried out, we traced the behaviour of the PES system during the robot operation TRANSFER and measured the reaction time of the monitor resulting from triggering the sensor system. The sensor primitives of an elementary operation were triggered at random by taking away parts, or simulating some error. The time lap between triggering of the sensor system and an interrupt of the robot motion was measured.

Figure 8 displays the user interface windows of PES directly after an exception is simulated by triggering the monitor parameter CHECK-MOTION (value is now TRUE) during the execution of the TRANSFER operation (see figures 3 and 4). In the "Constraint Stack Window" the current constraints are displayed and in the "Control Window" the decisions of PES are given. The "Execution Window" gives the status of the robot server. The user interface windows of the robot and sensor server processes are not given here.

The experiment showed that the reactive mechanism of PES is capable in coping with the events from the sensor system and to interrupt the robot actions. Both the PES implementation, and the communication method proved to be robust when multiple sensor inputs were given at one time.

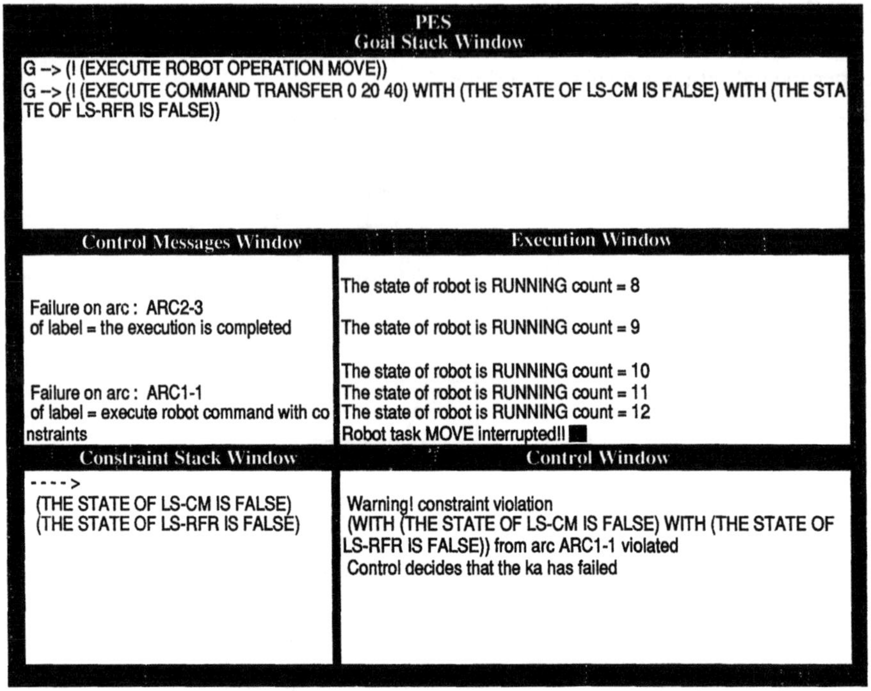

Figure 8: The user-interface during experiment with critical time values.

The experiments also showed that the current implementation prohibits adequate fast response times. The typical time lap between sensor input and robot interrupt is 5 seconds. Analysis of the different communication steps showed that the time lap is mainly due to the use of the client-server implementation with named pipes, and not to the inference steps of PES.

7 Conclusions & Further Work

We presented the realization of an exception-handling mechanisms for autonomous robots with the tool PES. Exception handling as described by EHM, provides the robot with capabilities to monitor its operations. Upon the detection of an exception, a diagnosis can be performed which leads to a classification of the exception at hand. Task rescheduling and recovery planning is performed based on heuristic knowledge of the application. This knowledge is encoded as strategies.

Due to its inherent reactive capabilities, the PES environment is well suited to realize the mechanisms of EHM. An experiment was performed with an assembly robot controlled directly from the PES environment. Sensor data from the robot is coupled back to PES environment, where it is used for monitoring. The capabilities of the system to cope with introduced exceptions was studied.

The experiments show that exception handling at many levels of a task decomposition is possible which makes the approach well suited for partly-structured applications. The communication mechanism currently used is still too slow to give acceptable response in critical situations. At this point a better performance is needed.

Further work aims to generalise the approach used here to the whole control architecture of the robot. The underlying idea of EHM is that when exceptions are pre-analysed, they can be incorporated to the whole system as simply events that may possibly occur on-line. For the task planner, they represent the factors which determine which part of the plan is going to be executed. This event-driven approach can be applied at many abstraction levels of the task decomposition. We are considering its application to the mission planning for an autonomous robot (which includes the choice of different ways to reach a general-purpose mission), and to the achievement of perception tasks by selecting the suitable set of logical sensors to be used according to factors like the required precision or the available resources.

The realisation of such a system with PES requires the use of a great number of fact-driven KA's and constraint conditions. To keep the coherency within the system, it is important to ensure that the knowledge, expressed by the preconditions of the fact-driven KA's or by the constraints, does not conflict. For this, we are considering to couple PES with an ATMS (Assumption based Truth Maintenance System). From the ATMS point of view, the set of constraints associated with a goal can be seen as the set of *assumptions* attached to a *fact*, under which the goal is meaningful (*the fact is True*). Results carried out in ATMS theory could be applied in the interpreter of PES. This remains an open research topic and further work will move in this direction.

Acknowledgements

The work on exception handling was carried out in Esprit I project 623: "Operational control for robot system integration into CIM". Started in February 1985, this project resulted in methods and tools for production planning as well as robot programming systems [7,17]. The University of Amsterdam participated in this area with the development of the Exception Handling Model (EHM) and tools to generate, and monitor the recovery procedures [21]. Currently this work is continued in Esprit II project 2202: "CIM Systems Planning Toolbox (CIM-PLATO)".

The realization of the mechanisms of EHM using PES is supported by Esprit II project 2043: "Mobile Autonomous Robot in Industrial Environment (MARIE)". In the project, numerical and symbolical techniques are studied and developed for sensing and control of an autonomous robot. The diagnosis and exception-handling system will be applied both on an inspection robot arm as on the control of a mobile platform. The use of PES in planning has been tested by Framentec, among others, in Esprit project 820 [1], and in a French space project called SATEXPERT.

References

1. Arlabosse, F et al: Esprit project 820; Toolkit specification. Project internal documentation, available at Framentec.
2. Arlabosse, F, Duong, V, Gaussens, E.: An event based planner for a real scale process control application. Proceedings IEEE 1988 workshop on AI in process engineering, August 1988.
3. Arlabosse, F, Duong, V, Lepage, P, Gaussens, E.: Blackboard and alternative architecture design for two real scale KBS industrial applications. In: Blackboard architectures and applications (eds.Jagannathen, J, Dodhiawala, R, Baum, L). Academic Press, Boston, August 1989.
4. Arlabosse, F, Jean-Bart, B, Porte, N, deRavinel, B.: An efficient problem solving architecture using ATMS. Journal of AI communications Vol I nr 4, pp 6-15, December 1988.

5. Brooks, R.A.: A robust layered control system for a mobile robot. IEEE Journal of robotics and automation, Vol RA-2, No 1, 1986.
6. Brooks, R.A.: Symbolic error analysis and robot planning. The International Journal of Robotics Research, Vol 1, No 4, pp 29-68,1982.
7. Camarinha, L, Negretto, U, Meijer, G.R.: Information Integration for Assembly Cell Programming and Monitoring in CIM. Proceedings of 21th International Symposium on Automotive Technology and Automation, Wiesbaden 6-10 November 1989.
8. Donald, B.: Error Detection and Recovery. Lecture Notes in Computer Science (eds. Goos, G., Hartmanis, J.), Vol 336, Springer Verlag 1989.
9. Georgeff, M.P, Lansky, A.L.: Procedural Knowledge. Proceedings of IEEE Special issue on Knowledge representation, 74:1383, 1986.
10. Georgeff, M.P, Lansky, A.L.: Reactive reasoning and planning. Proceedings of AAAI, p 677, 1987.
11. Gini,M. Symbolic and Qualitative reasoning for error recovery in robot programs. In: Rembold U, Hörmann K, (eds), Proceedings of the NATO International Advanced Research Workshop on Languages for sensor based Control in Robotics, Italy september 1986.
12. Meijer G.R, Hertzberger L.O, Mai T.L, Gaussens E, Arlabosse F.: Exception handling system for autonomous robots based on PES. Proceedings of Conference on Intelligent Autonomous Systems 2 (IAS-2), pp. 65. Amsterdam 11-14 december 1989, Elsevier Science Publishers.
13. Meijer G.R., Weller, G.A. Groen, Hertzberger,L.O.: Sensor based control for autonomous robots. Proceedings of IEEE international conference on control and applications, WP-3-4, Jeruzalem April 1989.
14. Meijer,G.R, Hertzberger,L.O.: Exception handling for robot manufacturing process control. Proceedings of CIM Europe conference, Madrid may 18-20 1988, IFS publications.
15. Meijer,G.R, Hertzberger,L.O.: Off-line programming of exception handling strategies. Proceedings of IFAC/SYROCO conference, Karlsruhe October 1988, VDE/VDI publications.
16. Schoppers, M.J.: Universal plans for reactive robots in unpredictable environments. Proceedings of 11th IJCAI, Milano, pp 1039-1046, 1987.
17. Spur et al.: Planning and programming of robot integrated production cells. In: Esprit '87 Achievements and Impacts, North Holland.
18. Srinivas, S.: Error recovery in robots through failure reason analyis. AFIP Proceedings of National Computer Conference, Anaheim, CA, pp 275-828, 1978.
19. Tuijnman, F, Meijer, G.R., Hertzberger, L.O.: Data modelling for a Robot Query Language. Proceedings of Conference on Intelligent Autonomous Systems 2 (IAS-2), pp. 208. Amsterdam 11-14 december 1989, Elsevier Science Publishers.
20. Weller, G.A. Groen, F.C.A, Hertzberger, L.O.: A sensor processing model incorporating error detection and recovery. Proceedings of NATO advanced research workshop on Traditional and non-traditional sensors, Maratea 28 August - 1 September 1989, NATO-ASI series.
21. Wonderen,G.M. van.: AI and high level robot control and a preliminary implementation of a High Level Interpreter. Technical report, Faculty of Mathematics & Computer Science, department of Computer Systems, July 1988.

Rapid Interference Detection

James S. Hemmerle, E. Levent Gürsöz, and Fritz B. Prinz

Engineering Design Research Center, Carnegie Mellon University, Pittsburgh, PA 15213

Abstract: An algorithm for rapid interference detection between arbitrarily shaped objects is described. The approach uses a dual representation of each object, those being a wireframe and a discretized volume (octree and raycast). Interference is detected when a portion of the wireframe of one object is contained in the volume of the other. The algorithm's speed is rooted in the speed of the point and edge inclusion tests. Tests indicate the algorithm is an order of magnitude faster than previously published work.

Keywords: Interference, Collision, Detection, Octrees, Raycast

1 Introduction

As the use of computer simulation of geometric objects grows, the need for realistic modelling and rapid and robust algorithms increases. One of the basic tests inherent in many simulations is whether two objects interfere with each other. In physical modelling of dynamic rigid bodies, the equations of motion change dramatically when objects come into contact and computational speed is critical for interactive simulations. In assembly/disassembly simulation, the operations that may be preformed hinge on whether the motion or the placement of the components cause interference of the subassemblies. Another popular application is in the area of robotic path planning, where the issue of collision detection and avoidance can play a critical role.

The method used for interference testing depends on the nature of the information desired. In an assembly application, for example, it may be important to know, not only if the pieces interfere, but also the amount of interference or the sections that interfere. In this case, the

NATO ASI Series, Vol. F 71
Expert Systems and Robotics
Edited by T. Jordanides and B. Torby

test would need to be akin to a Boolean intersection which would find the regions that are common to both parts. This is a computationally expensive operation because the details of the interference geometry are required. In some collision-avoidance algorithms, the separation distance between the objects is needed. Again, these algorithms generate more information than just whether the objects are touching or not. If one wants to know only if the objects interfere, for example in the case of testing whether a robot's path causes it to collide with an object in its envelope, these methods could be used, but other algorithms that do not calculate unneeded information would be computationally more efficient.

This paper focuses on an algorithm to rapidly detect whether two arbitrarily shaped objects interfere.

2 Representations

There are a number of tests that can be performed to tell whether two objects interfere. These tests are related to the conditions that occur when two objects overlap and which is chosen will depend on how the objects are represented. In general, two 3-dimensional objects interfere if their volumes overlap, or if a section of the boundary of one of the objects is inside the other.

There are a number candidates for representing solid objects. One popular way is through constructive-solid-geometry (CSG) trees. With this method the object is represented through a tree of Boolean operations performed on a set of primitive objects. Cameron [1] developed a three-step algorithm for detecting interference between CSG objects. He first computed bounding volumes to narrow the search space, which was then subdivided to further reduce possible interference areas. Finally an exhaustive search is preformed to settle the question. He did not report computational times for any particular case but did claim that the routine returns "within 1-2 seconds [on a Sun 3/160] in the worst cases seen".

Other methods have been developed for solid represented through their boundary shapes (B-rep modelling of solids). Recently, Gilbert and Hong [3] described a method for detecting collisions of convex polytope (linear faceted) objects. Their method relies on calculating the distance between the objects and take approximately 0.2-0.4 seconds (on an Apollo DN 4000) for two objects (one represented with 164 vertices and the other 12 vertices). The computational time appears to be proportioanl to the number of vertices in the models.

A third approach for representing solids are octrees. With this representation, the object is represented as a hierarchical tree of rectangular boxes (whose aspect ratios equal the aspect ratios of the bounding box of the object). The octree is computed through a recursive subdivision of the space into eight regions or octants. If an octant is partially in and partially out of the solid, it is further subdivided until the desired resolution is obtained (8 levels of subdivision would result in a resolution of 1/256 $= 1/2^8$ of the length of the object). Octants fully in the object are termed full or black while the octants outside the object are termed empty or white. The non-leaf nodes of the tree are called gray. (See [2] for more information on octrees.) Major et al. [5] present a solution for finding the distance between moving objects represented by octrees. They use a branch and bound method to search for the minimum distance between pairs of blocks in the two octrees. They reported times (on a Sun 3-60) averaging 40-80 seconds for octrees of levels 6-8 having 300-500 black nodes.

In a fourth representation, raycasts, the object is represented by an array of depth measurements. Two of the dimensions of the space (in our case we use the x, y position) are divided to form a grid of rectangles. In each grid location, a list, giving the range of the object in the other direction, is stored. In effect this gives the intersection points of a ray with the object.

The last two representations, octree and raycast, have a property that will be exploited in the presented method. Because of their data structures, testing whether a point is inside the object is extremely simple and computationally inexpensive. To test whether a point lies inside an object represented by an octree, one traverses the tree to find in which type of octant the point lies. To find whether a point is in the object represented by the raycast scheme, the grid box the point belongs to is calculated and then the list is searched to see if it falls in one of the solid ranges. These rapid point-inclusion tests form the basis for the interference test.

3 Method

The method presented in this paper checks for interference by testing whether the boundary of one of the objects is inside the volume of the other. Since the boundary of the objects are modelled with linear surfaces, this condition occurs if, and only if, a portion of an edge of one object is inside the other object. Thus, a dual representation of each object is used.

The volume of the object is represented by either an octree or a raycast. The boundary (or more accurately the test points) is stored as a wireframe model of the object. The test checks whether a portion of the wireframe of one object is inside the volume of the other object.

Both the wireframe and the discretized volume (octree or raycast) are defined relative to a frame attached to the object. To test for a collision between two objects, the coordinates of the entities defining the wireframe model (its vertices and edges) of each object are transformed into the frame of the other object (through a multiplication of a homogeneous matrix relating the local frames of the objects). The transformed vertices and edges are then tested for inclusion in the appropriate volumetric model. This is done in two stages. First the vertices are transformed and checked, then the edges.

Here, the implementation diverges a bit depending on whether the octree or the raycast structure is used. With the octree, the vertex is checked by traversing the tree until the octant which contains it is found. To travel from one level to the next requires three comparisons (to check which child octant to go to) and three additions (to find its coordinates relative to the child octant's center) which is extremely quick. A vertex can be transformed and tested through 6 levels in about 0.143 milliseconds (each additional level takes about 0.018 msecs). [NOTE: These and all times to follow are for a Sun 4/260.] To check an edge, the edge is tested to see if it is in more than one octant. If so, it is divided and each section is tested. The segments are further divided until they are entirely in either full or empty octants. The time needed for an edge test depends on how many times it must be divided. In the worst case, an edge segment would have to be divided into four segments when going to the next level. This occurs if the segment is in a partial (or grey) node and crosses all three coordinate planes for the octant. On the other hand, if none of the segments are in a lowest-level octant the testing time is constant for increasing octree levels.

With the raycast structure, a vertex is checked by comparing its z value to the solid ranges stored in the appropriate raycast pixel. Calculating which pixel it is in, is essentially a scaling operation and can be found as the point is being transformed (since $\mathbf{S}(\mathbf{T}p) = (\mathbf{ST})p$ where $\mathbf{S}$ is the scaling transform, $\mathbf{T}$ is the transform relating the local frames, and p is the vertex's position in its local frame). A vertex can be transformed and tested in approximately 0.070 milliseconds with this representation. The time is fairly constant for any level of subdivision since further refinement mostly changes the number of pixels in the raycast array and not the number of depth ranges which have to be searched (an additional subrange adds about 0.0015 msecs.) To check whether an edge is inside the object, the edge is discretized into points (with spacing equal to the size of the raycast pixels) and then each point is tested as above. The time for this edge test doubles when the resolution is doubled (since the number of points the edge is broken into doubles).

A number of operations are performed to increase the speed of the test (by decreasing the number of points to be transformed and tested). First a check is made to see if the bounding boxes of the objects overlap. If not, no additional tests need to be performed giving a result in about 1.4 msec. The edges are clipped at the sides of the bounding boxes to decrease the length of the line to be tested. The edge test points are generated from their transformed endpoints (as opposed to generating them once and transforming them individually, a relatively costly operation).

4 Results

As an example of results from this method, consider the case of trying to detect the collisions of the wrist and tool of a robot with the part it is trying to weld. The models for the last three joints (the wrist joints) of a Cincinnati Milicron robot, the torch, and a sample part were created within the non-manifold B-rep solid modeler NOODLES [4]. From these models, a number of octree and raycast representations with varying resolutions were created. (See Figure 1 and Table 1 for a description of the objects.)

As the robot moved the tool along the path, the method described was used to test for collision between the part and the four moving objects (three wrist links and the torch). A summary of the average times for the various situations can be found in Tables 2 - 5. At all the path points, the bounding boxes of the pairs of objects overlapped.

A number of observations may be made. First, the times for the situations when the objects collide is much quicker than non-collision situations. This is because checking can cease as soon as a collision situation is encountered. In the majority of the interference locations, a vertex of one of the objects was within the volume of the other object, so the relatively costly edge tests did not have to be performed.

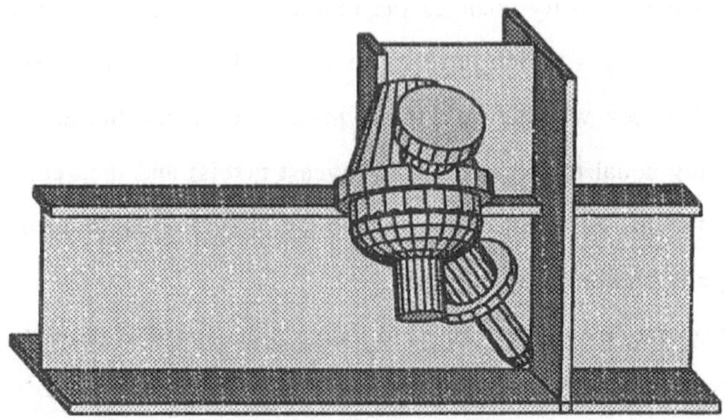

Figure 1: Objects Tested

Object	Number of Nodes	Number of Edges	Length of Edges (m.)
Part	66	192	56.16
Link4	194	282	10.12
Link5	358	611	13.80
Link6	40	60	1.33
Torch	40	72	3.04

Table 1: Model Size for Tested Objects

Generally, the tests using the raycast structure were quicker than the octree structure when both had a resolution of 1/64 of the objects. Its faster point check gives it an edge at this time. However, the raycast times significantly increased as the resolution increased (due to the increased number of test points from the edge discretization), while the octree times grew only slightly. (Unfortunately, the octree generator used could not create the corresponding finer resolution octrees. We hope to have this remedied soon.) It is expected that this trend will continue at the higher-level octrees and that at higher resolutions, the octree structure will out perform the raycast structure.

The times for non-collisions seem to correspond with the number of edges in the models. This is not too surprising since the edge test is the most costly. However, one might imagine that the times would correspond more closely with the length of the edges since this determines

Representation	Resolution	No Collision		Collision	
		Number of Pnts	Ave. Time (msec)	Number of Pnts	Ave. Time (msec)
Octree	1/32	23	35.4	19	9.2
	1/64	23	36.0	19	11.4
Raycast	1/64	22	37.3	20	13.2
	1/128	22	43.7	20	13.8
	1/256	22	56.9	20	16.9

Table 2: Timing For Test With Link4 and Part

Representation	Resolution	No Collision		Collision	
		Number of Pnts	Ave. Time (msec)	Number of Pnts	Ave. Time (msec)
Octree	1/32	19	78.9	23	11.5
	1/64	19	79.7	23	17.4
Raycast	1/64	19	73.5	23	17.4
	1/128	19	86.7	23	18.8
	1/256	19	116.3	23	20.6

Table 3: Timing For Test With Link5 and Part

Representation	Resolution	No Collision		Collision	
		Number of Pnts	Ave. Time (msec)	Number of Pnts	Ave. Time (msec)
Octree	1/32	31	15.0	11	7.6
	1/64	31	15.6	11	8.8
Raycast	1/64	31	14.9	11	8.2
	1/128	31	16.0	11	9.1
	1/256	31	19.3	11	9.9

Table 4: Timing For Test With Link6 and Part

Representation	Resolution	No Collision		Collision	
		Number of Pnts	Ave. Time (msec)	Number of Pnts	Ave. Time (msec)
Octree	1/32	39	29.1	3	19.4
	1/64	42	29.5	-	-
Raycast	1/64	42	23.2	-	-
	1/128	42	30.2	-	-
	1/256	42	44.3	-	-

Table 5: Timing for Test With Torch and Part

the number of operations. This is likely due to the clipping operations which throw out portions of the edges (or the whole edge) which are outside the bounding volumes. This gives an advantage to the objects which are compact and further separated.

The method described in this paper has a number of advantages over the other methods introduced earlier. The objects tested do not have to be convex. Nor does it depend on the object being built through a CSG tree (although the octree/raycast representations must be generated initially). Computationally, it seems to be roughly 5-10 times faster than the methods shown in [3] and [1], after taking into account machine speeds and model complexities.

5 Conclusion

A method has been developed to detect whether two arbitrarily shaped objects interfere. The approach uses a dual representation of each object, namely a wireframe and a discretized volume model. Both an octree and a raycast structure were investigated. The objects interfere if a portion of the wireframe of one of the objects is inside the volume of the other. The algorithm's speed hinges on the speed of the point inclusion test for the volume representations chosen. With the octree representation, the edge in volume tests are done directly through partitioning the edge. With the raycast representation, the edges are discretized into a number of test points. For a resolution of 1/64 of the part dimension, the raycast structure has a slight speed advantage over the octree structure, but at higher resolutions it is expected that the octree structure will be faster because of the complexity of the edge-testing methods.

References

[1] Cameron, S. *Efficient Intersection Tests for Objects Defined Constructively.* **The International Journal of Robotics Research**, vol. 8 (1989), pp. 3–25.

[2] Chen, H. and Huang, T. *A Survey of Construction and Manipulation of Octrees.* **Computer Vision, Graphics, and Image Processing**, vol. 43 (1988), pp. 409–431.

[3] Gilbert, E. G. and Hong, S. M. *A New Algorithm for Detecting the Collision of Moving Objects.* in: **Proceedings of the IEEE Conference on Robotics and Automation**. IEEE Computer Society Press, Scottsdale, AZ, 1989, pp. 8–14.

[4] Gürsöz, E. L., Choi, Y., and Prinz, F. B. *Vertex-based Representation of Non-manifold Boundaries.* in: **Geometric Modeling for Product Engineering**, edited by M. J. Wozny, J. U. Turner, and K. Preiss. Elsevier Science Publishers B. V. (North-Holland), 1990, pp. 107–130.

[5] Major, F., Malenfant, J., and Stewart, N. F. *Distance between objects represented by octre.* **Computers and Graphics**, vol. 13 (1989), pp. 497–503.

PROLOG LANGUAGE - CADSD PACKAGE INTERFACING PROCEDURES: Applications to Heuristic Identification and Control of Dynamic Systems

Hao Nguyen-Phu,
ESSTIN, Laboratoire EIA, CRAN, Parc Robert Bentz,
54500 Vandoeuvre, France

Abstract:
The development and evaluation of different ways to interface a Prolog language with interactive CADSD environments are described. The principles of a suitable knowledge-based system solver which includes both pertinent explanations and control design software are outlined. The SERMA package has been constructed using the Turbo-Prolog shell with linkage into different CADSD environments to provide computational and graphic enhancements. The covered identification domain aspects include both a new approach defined as 'Heuristic Identification' of SISO-system step responses and an expert-system advisor for the model-structure determination. The control area aspect corresponds to an intelligent front-end of in-house and commercial control design packages.

Keywords: computer aided dynamic system design / CACSD / identification / control / logic programming / prolog / knowledge based system / language interface / computer aided teaching / expert system solver / intelligent front-end

1 - INTRODUCTION

Artificial Intelligence (AI) is an area that long has attracted interest. Expert Systems (ES) or Knowledge-Based Systems (KBS) are a branch of AI that has gained increasing interest during the past few years. The purpose of this paper is to investigate the use of both expert-system techniques and Computer-Aided Dynamic System Design (CADSD) at the first local control level, especially related to the following steps, i.e., the heuristic model identification of Single-Input-Single-Output (SISO) systems [17] followed by a knowledge-based control acting as an intelligent front-end to various control software. The applications are carried out off-line and can be classified according to the 'Operator assistance' criteria. The idea of trying to combine ES with CADSD is not new. A survey of symbolic and numeric computing for control design has been outlined by James [8]. Taylor et al. [22,23] have focused on an architecture for an expert system to assist a control engineer (i) in exploiting available software to achieve an acceptable design and (ii) in coping with the complexity of computer-aided control engineering [9]. Boyle et al. [4] discussed the implementation of MAID: a KBS for multivariable control system design. When the expert system operates in an advisory role, the user works as an interface between the KBS and the CADSD programs in the earlier studies [4,5]. Other specific applications to dynamic-system identification are also reported in the literature [2,6].

NATO ASI Series, Vol. F 71
Expert Systems and Robotics
Edited by T. Jordanides and B. Torby

The overall structure of the proposed KBS is outlined in section 2. This section evaluates the environment within which our KBS helps the designer to perform a heuristic identification. Comments on the capabilities of a Prolog programming environment for different CADSD systems such as PC-MATLAB, Advanced Continuous Simulation Language (ACSL) and MATRIXx/PC are also given. Section 3 gives typical examples of these interfacing procedures under PC-DOS environment. Section 4 reviews several aspects of the realization of KBS modules such as (i) graphic identification of SISO step responses and (ii) an intelligent front-end for control software. The conclusion and a direction for future efforts are given in the last section.

2 - INTERFACING PROCEDURES

The overall structure of the proposed KBS is shown in Fig.1. The designer is in front of one computer both for the expertise and for computation and simulation work. As shown in Fig.1, the user may choose one task among the following steps: (i) IDGRA, an expert-system adviser carrying out qualitative diagnoses based upon heuristic identification of SISO step response, (ii) SERIDPA, a CAD expert-system solver performing quantitative validations for five structures or classes of models such as ARX, OE, BJ, ARMAX and PEM structures [13] and (iii) SEREG, a KBS adviser playing the role of a control engineer. There are a variety of interfacing methods which have been developed to achieve different goals. A survey of symbolic and numeric computing for control design has been presented by James [8] from minicomputers such as VAX computers, workstations to microcomputers such as IBM PC/PS or compatibles. The major coupling methods have included:

A. link between processors
B. mailboxes under VAX/VMS operating system
C. function calls
D. files
E. DOS-Call

This paper deals with different interfacing techniques for IBM PC/PS compatible machines. The choice of this type of computer was largely inspired by the fact that hardware and some software are easily accessible. Another reason is the study of feasibility. In fact, each interfacing method that can be worked out under MS-DOS operating system can be also applied under more efficient systems such as OS/2 or other Unix-like operating systems. In the case A, a link between processors implies at least two different processors or computers and a communication protocol, via a simple 9600 baud connection or via a computer network link. This case is disregarded in our study since the designer has to manipulate two keyboards and must know both the KBS system and the corresponding CADSD packages. Case B was investigated by a number of researchers. Expert systems developed at the Lund Institute of Technology [1,10,11] used VAX/VMS mailboxes to communicate with the numerical programs called

VAX/VMS sub-processes. Coupling between symbolic and numeric computing is also achieved by spawning VAX/VMS processes and using a file system (case D) to transfer data [3,7]. The data structure used for Input/Output in high-level languages is the port which is a stream-based I/O channel. The standard input and output port is the user terminal. Ports can also be connected to files or pipes. A pipe is a Unix method for stream communication between different sub-processes. In the prototype DataBase-Prolog system, the external package ProMatlab has been linked-in via the underlying Unix file system. Any function in ProMatlab can be called directly from DB-Prolog language [21]. A similar approach is to incorporate expert-system features in a design package to provide a more supportive and high-level design in a single application [20]. The function call approach can be defined as an escape mechanism which a program in one language may employ to call routines in other languages. Generally, it is a one-way approach in the sense that link via this mechanism goes from one language to the other, but not in the other direction. An extension has been proposed to support two-way communication. Backtracking mechanism was handled by deciding that all actions done by procedural programs on procedural data would not be undoable [15].

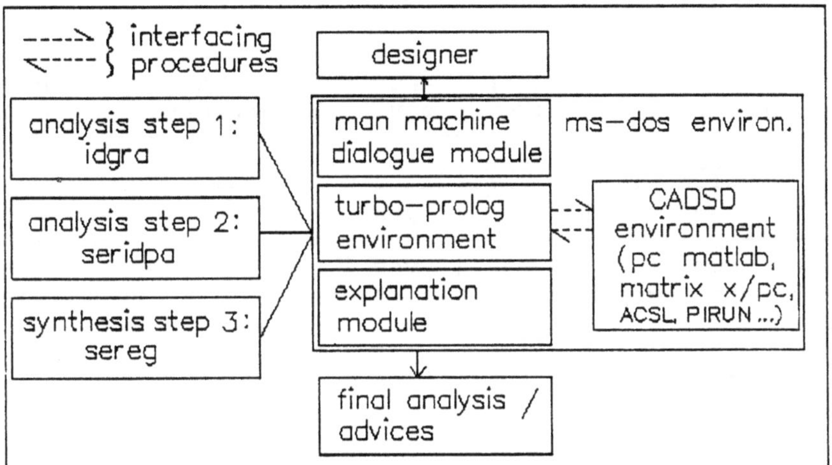

fig.1 : Architecture of the system

Under the MS-DOS operating system with the accessible main memory limited by 640 Kbytes, one can distinguish the following interfacing procedures:

(i) Type I interfacing procedure (I.P.) or linking of heterogeneous object modules by function calls;

(ii) Type II I.P. or communication by DOS-Call commands and by ASCII files;

(iii) Type III I.P. or communication by memory switching.

In fact, the function call method which is employed mainly for homogeneous object modules (i.e., conventional calls to the same type of subprograms when the entire system, including the KBS module and the numeric routines, is written in the same high-level language) can be generalized for the Turbo-Prolog language to couple the symbolic and numeric processing written in Turbo-C or Turbo-Pascal. The type II interfacing procedure is implemented by the Turbo-Prolog built-in predicate:

```
        .
        .
system('DOS string'),
        .
        .
```

provided that the main memory is sufficient for both the Prolog executable module and the corresponding CADSD environment defined by the DOS string. The type III I.P. is employed to solve the memory limitation problem: the Prolog progam or the CADSD environment resides one at a time in the main memory. The switching procedure is monitored by a DOS batch file via the presence test of a temporary file.

Ext. Prog.	Turbo-C or Turbo-Pascal	Other lang.	ACSL env.	PC-MATLAB Interpreter	MATRIXx/PC Interpreter
Type I	yes	yes/no	no	no	no
Type II	yes	yes	yes	yes	no
Type III	yes	yes	yes	yes	yes

TABLE 1. STUDY OF FEASIBILITY OF DIFFERENT I.P.

Table 1 presents the feasibility of external program calls from the Turbo Prolog environment under the MS-DOS operating system. The column 'Other language' corresponds as well as to other high-level languages (HLL) such as FORTRAN or to low-level language such as 8086-family assembler. For these languages, the type I I.P. is realizable since Turbo Prolog supports a direct link to Assembler, C, Pascal and FORTRAN programs. However, it is not demonstrated that the Turbo Prolog, version 2.0, Foreign Language Interface can be connected to Ada language. It is well known that the applications of AI through the use of KBS, written in a logic programming language, allow heuristics to be built up on the models of human experts. On the other hand, the designer can also take advantages of all the power and subtility of available CADSD packages, such as ACSL environment, matrix-based simulation languages such as PC-Matlab, or MATRIXx/PC interpreters if communication procedures are successfully implemented between heterogeneous languages.

SERMA modules (program name)	Numerical package or ext. prog.	Coupling Method	Program Status	Intended Program Use
IDGRA	PC-Matlab	type II	working	advisor for heuristic identification and tutor
IDGRA (Demo. ver.)	MATRIXx/PC	type III I.P.	under develop.	
SERIDPA	PC-Matlab + System Id. Toolbox	type II I.P.	working	advisor for SISO system Identification
SEREG	ACSL env. PC-Matlab PIRUN package...	type II and type III I.P.	prototype	intelligent front-end and tutor

TABLE 2. COUPLED CADSD PROGRAMS WITH TURBO PROLOG MODULES

In our case, the declarative programming language such as Turbo Prolog is used for step-response diagnoses (IDGRA.pro program), validations of different identified models (SERIDPA package) and as an intelligent front-end to choose appropriate controller structures and relevant compensator methods (SEREG program). Table 2 summarizes the interfacing procedures used in each case. Simulation or numeric identification steps written in a procedural programming language can be integrated in the Prolog program via standard object modules (.OBJ files) by an appropriate linker. This type of I.P. is avoided here due to the time-consuming software development on one hand and the availability of various commercial packages for this purpose on the other hand. The built-in 'system'-predicate allows the execution of DOS commands, user batch files (.BAT files) or any external executable program (.COM or .EXE files). The second solution has been chosen for SERIDPA program since PC-Matlab and the System Identification Toolbox offered a flexible, interactive environment for SISO system identifications. The following section presents some typical interfacing procedures with different CADSD packages.

3 - TYPICAL EXAMPLES OF INTERFACING PROCEDURES

3.1 INTERFACING EXTERNAL PROGRAMS WITH PROLOG

PIRUN is an in-house CADSD package constituted of different independent modules such as PIDO.bas and RUNSYS.c programs. As such, it is intended to be used as a tool by the practicing engineers to design PID-Optimal control systems with more than 10 different types of performance criteria. Besides, the designer can choose a "Minimized Settling Time defined at X %" (X= 2 or 5 % in general) criterion. The user can also add a desired criterion to obtain a set of analog PID, PI, I, P effects from a nonlinear optimization method. In the second step, the designer can experiment seven digital implementations of the previous PID values, including the most general RST controller structure with different first-order filtering effects. The RUNSYS program recommends an upper bound of the sampling rate in order to ensure the SISO-loop stability and also furnishes an optimal Runge-Kutta integration step as a function of the desired accuracy and various initial conditions of the state vector or the output and its successive derivatives. Other in-house programs or commercial packages can also be selected by the one-way link for the almost intelligent front-end cases as depicted in the figure 2.

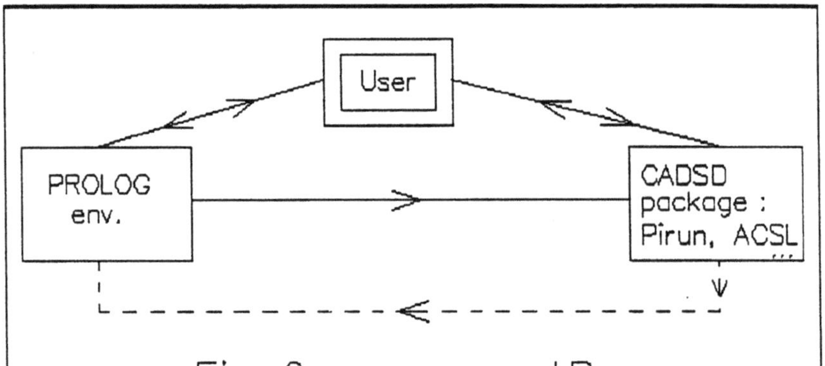

Fig 2: one-way I.P.

In fact, the user must be aware of a minimum of runtime commands or is merely guided by self-documented external programs such as PIDO or RUNSYS. The ACSL 'stop' runtime command terminates the interactive session and allows the return to the next predicate of the Prolog program if the type II interfacing procedure is employed via the built-in external-call 'system' predicate:

```
    .
    .
    system(acslcall),
    next_predicate,
    .
    .
```

where acslcall is a batch file for preparing and calling the ACSL environment.

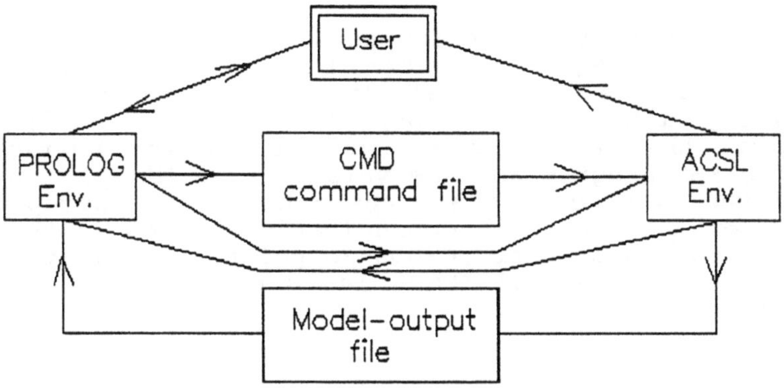

Fig.3 : I.P. from ACSL env.

If numeric data computed by the external program must be given to the Prolog program, the programmer has two possibilities: (i) the external routine can be created or modified in order to be linked-in via the type I interfacing procedure or (ii) an ASCII-format file can be used for the data exchange. A typical example of this two-way communication from the ACSL interpreter environment is given in the figure 3. The ACSL model-output file contains ASCII data for adding or modifying the Prolog dynamic database. Besides, the user does not need to know any ACSL runtime command since a .CMD command file has been already prepared before invoking the predicate 'system(acslcall)'.

3.2 INTERFACING MATRIX-BASED CONTROL SOFTWARE WITH PROLOG

There are two types of input-output sequential data files: (i) in ASCII format called .M files and (ii) a special .MAT format used only by PC-Matlab. The translation can be performed by using the 'TRANSLAT.exe' utility or other special utilities such as MAT2DAT.exe or DAT2MAT.exe written in C or FORTRAN 77 language [16]. The figure 4 shows typical I.P. from PC-Matlab environment via files. More details have been published in the previous report [18].

Another interesting matrix-based control software is the MATRIXx/PC from Integrated System Inc. The salient feature of this package is the 'Graphical Model Building and Simulation' module dealing with nonlinear control in the same environment. In counter part, the memory is not sufficient for bearing both a Turbo-Prolog KBS program and the version 5.22 MATRIXx environment because of the overlay file and the extra memory occupied by the VDI.sys driver. So, the only communication method is the type III interfacing procedure under PC-DOS as pointed out in the table 1. A demonstration version of IDGRA is currently modified in order to carry out this study.

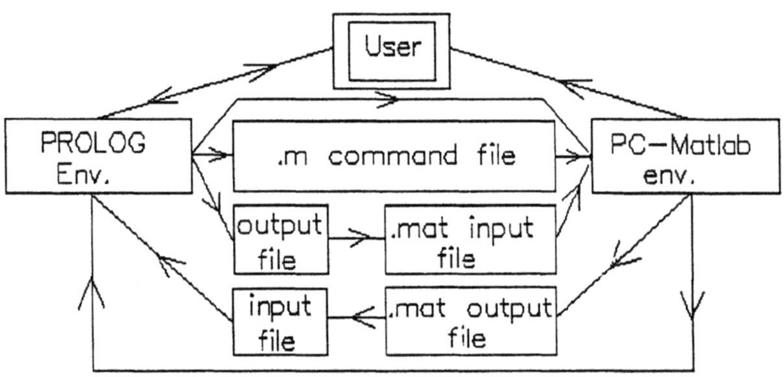

Fig.4 : I.P. from PC-Matlab

4 - REALIZATION OF KBS MODULES

4.1 IDENTIFICATION OF STEP RESPONSES

The IDGRA module has been created with the type II I.P. from PC-Matlab and via appropriate files as described in the previous sections. The main function of the set of rules is to diagnose the model of the plant according to the different aspects of step responses. A list of 27 groups of transfer functions was reported in a preceding paper [17]. It should be useful to remark once more about the flexibility of symbolic programming technique concerning the software maintenance. In fact, two other tranfer -function models have been easily appended several months later. Typical sample transactions were also presented to outline how IDGRA recognize a non-minimum phase system [12,17] or a second-order lag with a small time constant [18,19]. In the second case, IDGRA is also able to deduce, as a first approximation like a human expert, both the time constants as a function of the ordinate of the inflection point and the value of the 63% settling time.

The second identification module (SERIDPA) [17] can be used as a second step to give quantitative parameter estimations according to different numeric methods of the Matlab System Identification Toolbox (SIT) concerning ARX structures (Least Squares: LS and four-stage Instrumental Variable Method: IV4), ARMAX, Output Error (OE), Box-Jenkins (BJ) and Prediction-Error Method (PEM) structures as described in Ljung's book [14].

4.2 INTELLIGENT FRONT-END FOR CONTROL SOFTWARE

The rule knowledge base (KB) of the SEREG module is divided into two independent parts: (i) the compensator-structure KB and (ii) the control-method KB. Each KB is constituted by about 15 rules. This intelligent front-end is able to suggest the following controller structures:

- proportional controller (P),
- integral controller (I),
- proportional and integral (PI) controller,
- proportional and derivative (PD) controller,
- proportional, integral and derivative (PID) controller,
- seven different implementations of discrete PID controller,
- discrete nonlinear X / PID controller,
- discrete X / PI controller where X is the maximum (or minimum) value of the actuating signal during transients in order to reduce the time response for a positive (or negative) set-point,
- discrete RST structure controller,
- cascade lead-lag compensators ...

and the relevant output feedback design methods defined so far in the control method KB as:

- Chien-Hrones-Reswick (CHR) method,
- Ziegler-Nichols (ZN) methods, - Modified ZN methods,
- Naslin graphic method,
- Method of moments,
- Parameter-optimization method in using PIDO program,

- Root-locus method,
- Frequency-response method,
- Trial-and-error method ...

The simulation of the compensated plant is performed for the linear cases by the PIDO program or the SERMA3 KBS package acting as an intelligent tutor of conventional design techniques, such as the CHR or ZN methods. For nonlinear cases, RUNSYS program written in C language or ACSL executable modules are used to provide different discrete step responses of the system in order to investigate respectively different implementations of PID controllers or nonlinear Bang-bang effects with X / PI, X / PID compensators. The user can answer only by 'yes' or 'no' and can also request an explanation by striking the key '?'. According to the answers, the KBS guides one towards the appropriate software among five different programs or CADSD packages.

4.3 IMPLEMENTATION AND AVAILABILITY

The inference mechanism used in a Prolog language is backward chaining. Horn clauses are used here to recognize the current situation and create a goal to be achieved by the inference engine. It is well known that the process of contructing KBS environment is an intuitive approach, rather than a rigorous and well-structured design strategy. Nevertheless, it should be worthwhile to remark that the rules must be placed according to the top-down and left-right analysis of the Turbo-Prolog inference engine. Therefore, diagnostic rules should be introduced in the Prolog program according to the following recommendations:

- for the same predicate name, more complex premises must be placed first to avoid the masked solutions (as an example, the PID diagnostic rule must be placed before PI or PD diagnostic rules, which are themselves just before the P-controller selection rule); a typical session of SEREG is given in the appendix 1;

- an 'unresolved case' must be defined at the end of each diagnostic predicate in order to prevent incoherent or contradictory user's answers;

- the explanation module can be built separately from the main KBS module via the Prolog modular programming . This module can also be connected with external programs as a function of explanatory needs.

The package runs on a 16 MHz 386SX IBM-Compatible microcomputer with 1.8 Mbyte RAM memory and a 80387SX Math Co-Processor. The plots can be easily obtained on an Epson-like printer. In order to reduce the loading time from the hard disk, the CADSD environment is initially loaded into a virtual disk during the autoexec.bat starting process. Our aim is to make the SERMA package widely used in the academic and industrial communities. The interested reader should contact Dr. H. Nguyen-Phu for more details. A free demonstration diskette will be sent upon request.

5 - CONCLUSION

The possibility of using Turbo-Prolog in communication with external programs or interactive CADSD environments was explored and found to be almost adequate in a dual role, on one hand, behaving as a pertinent tutoring system, and on the other hand, and as an intelligent front-end system to in-house control design programs as well as commercial packages. The type II interfacing procedure via files is fully consistent and simple to perform, and this two-way-numeric data exchange is automatic and transparent to the user. The type I and type III interfacing procedures are generally time-consuming approaches for software development. Furthermore, the type III I.P. is more difficult to perform since the update of the Prolog dynamic database is necessary at each memory switching.

Further work is required to investigate and implement more efficient explanatory modules in taking into consideration incoherent or contradictory user's answers. Work is also needed to improve the coupling between the Prolog KBS, or other expert system shells, and commercial CADSD packages under an appropriate multitasking and/or multi-user operating system on workstations or equivalent 386-or-higher machines.

6 - ACKNOWLEDGEMENTS

This research was partly funded by the French Department of Industry under contract CLEO-47. This paper is a synthesized and extended result of the SERMA project, which spanned three years and involved researchers at several sites. I would like to thank all of the participants for their contributions, and especially the graduate students at the Nancy I University who helped to develop and test the SERMA software.

The multi-aspect support of the EE department at the California State University Long Beach (CSULB) is also gratefully acknowledged during my one-year sabbatical leave. I would like to thank Dr. Raymond T. Stefani and Dr. Barry Shahian, Professors of Electrical Engineering, for their helpful discussions concerning the use of the CADSD packages employed at the School of Engineering for teaching purpose.

7 - REFERENCES

1. Arzén K.E.: An Architecture for Expert System Based Feedback Control. Automatica 25, 813-827 (1989)

2. Betta A. and Linkens D.A.: Intelligent knowledge-based for dynamic system identification. IEE Proc. 137 D, 1-12 (1990)

3. Birdwell J.D.: Evolution of a Design methodology for LQG/LTR. IEEE Control Systems Magazine, 73-78 (1989)

4. Boyle J.M., Pang G.K.H. and Macfarlane A.G.J. (1988), The development and implementation of MAID: a knowledge based system for use in control system design. International workshop INRIA-C2A 1988, Paris.

5. Gentil S., Barraud A.Y. and Szafnicki K.: SEXI: an expert Identification package. Automatica 26, 803-809 (1990)

6. Haest M., Bastin G., Gevers M. and Wertz V.: ESPION: an expert system for system identification. Automatica 26, 85-95 (1990)

7. James J.R. (1986), Considerations concerning the construction of an expert system for control system design. Ph.D. Thesis, Rensselaer Polytechnic Institute, Troy, NY

8. James J.R.: Expert system shells for combining symbolic and numeric processing in CADCS. Proc. IFAC Symposium on CADCS 1988, Beijing.

9. James J.R., Taylor J.H. and Frederick D.K. (1985), An expert system architecture for coping with complexity in computer aided control engineering. Proc. of the IFAC 'CAD in Control and Engineering System' Conference. Lyngby, Denmark, pp. 47-52.

10. Larson J.E. and Persson P.: Knowledge representation by scripts in an expert interface. Proc. of the American Control Conference 1986, Seattle, WA

11. Larson J.E. and Persson P.: The knowledge database used in an expert system interface for IDPAC. Proc. IFAC Workshop on AI in Real-time Control 1988, Swansea, pp. 107-112

12. Laub A.J. and Little J.N.: Control System Toolbox User's Guide, The Mathwork, Inc. 1986

13. Ljung L.: System Identification Toolbox User's guide, The MathWorks, Inc. 1986

14. Ljung L.: System Identification: Theory for the User (1987), Prentice Hall, Inc.

15. Magel K.: Interfacing Prolog to Pacal. Proc. of Fall Joint Comp. Conf. 1986, pp. 229-236.

16. Moler C., Little J. and Bangert S. (1987), Pc-Matlab User's guide Version 3.2-PC, The Mathworks, Inc.

17. Nguyen-Phu H., Thanh T.H., Rinck C., Chaabane M. and Humbert C.: A knowledged-based expert system for use in process identification. Proc. SEE International Workshop "E.S. and Electrical and Power Systems" 1989, Avignon.

18. Nguyen-Phu H. et al (1989), SERMA Reference Guide, EIA-ESSTIN, Final report, French Department of Industry - Contract N°47-CLEO: ES&CAT.

19. Nguyen-Phu H. and Weiler P. (1989), SERMA1 module: IDGRA User's guide, EIA-ESSTIN.

20. Pang G.K.H., Vidyasagar M. and Heunis A.J.: Development of a new generation of interactive CACSD environments. IEEE Control System Magazine, 40-44 August (1990)

21. Tan C.T. and Maciejowski J.M.: DB-Prolog: A database programming environment for CACSD. IEEE CACSD Workshop 1989, Florida

22. Taylor J.H. and Frederick D.K.: An Expert system architecture for computer-aided control engineering. IEEE Proc. 72, 1795-1805 (1984)

23. Taylor J.H., Frederick D.K. and James J.R. (1984), An expert system scenario for computer-aided control engineering. Proc. of the American Control Conference. San Diego, CA., pp 120-128.

APPENDIX 1

A PARTIAL SESSION WITH SEREG:

```
WELCOME ! *** (SERMA Package module 3: SEREG.pro)
(C) EIA - ESSTIN by Dr. H. Nguyen-Phu et al
-----------------------------------------------------

Hello, my name is SEREG [Version 1.5 07/15/1990].
I'll propose an appropriate choice of controllers
among the following groups:

  1. digital time delay controller,
```

2. lag compensator,
3. lead feedback compensator,
4. lead compensator,
5. lead-lag compensators,
6. PID controller,
7. discrete X/PID controller,
8. discrete X/PI controller,
9. discrete RST structure controller,
10. cascade PD-PID compensators,
11. cascade I-PID compensators,
12. lead feedback compensator,
13. state variable feedback compensator,
14. PI controller,
15. PD controller,
16. Integral controller,
17. Proportional controller.

My design method knowledge base contains so far:

1. Set-point Chien Hrones Reswick (CHR) method,
2. Load-disturbance CHR method,
3. Modified ultimate oscillation Ziegler-Nichols method,
4. Ultimate oscillation Z.N. method,
5. Step-response Z.N. method,
6. Method of moments,
7. PID Parameter optimization method,
8. Root locus compensation method,
9. Frequency-response compensation methods,
10. Trial-and-error method.

You can answer only by yes [y] or no [n] and give numerical data if necessary. You can also request an explanation by striking the key: [?]

Press any key to continue ...

```
is it necessary to employ sophisticated compensators ? [n]
is it important to employ conventional and simple
                                        controllers ? [y]
...
is it necessary to give good relative damping by a
                                           D-action ? [y]
is it important to employ a cascade lead compensator ? [n]
is it necessary  to  employ  cascade  compensators ?  [n]
is it important to give  good phase margin
                      (or gain margin) by a P-action ? [y]
is it important that at least steady-state step error
                                vanishes by an I-action ? [y]
...
is it necessary that steady-state ramp error vanishes
                                  for type 1 systems ? [n]
...
is it a second order lag or a higher order system ?    [y]
```

```
I do SUGGEST a proportional, integral and derivative (PID)
controller !

                Press a key to investigate design methods ...

is it necessary to employ classic tuning rules such as
                       CHR, Naslin or Zn methods ... ? [y]
is the step response measurement available ?           [y]
is it important to choose the load-disturbance rejection
                                      as a criterion ? [y]
is the oscillation measurement available ?             [n]
is it important to choose the maximum overshoot D1(%) as
                     a qualitative performance index ? [y]
is it necessary that D1 is approximately equal to 20%? [y]


According to your answers, I PROPOSE the following design:
Set-point Chien Hrones Reswick (CHR) method.

                Press a key to call SERMA3.bat ...


        ---------------------------------------

            ANOTHER SESSION WITH SEREG:


WELCOME ! ...
  .
  .
  .
is it necessary to employ sophisticated compensators ? [y]
is it necessary to employ nonlinear compensators
                (Bang-bang effect) during transients ? [?]
```

EXPLANATION

In fact, in order to reduce the response time of the closed loop, nonlinear X/PID controllers can be used. During transients, the maximum (or the minimum) value of the actuating signal is applied to the plant.
When the switching point expressed in percent/set-point is reached, the controller is switched into a discrete time PID during the steady state.

SUGGESTION:
This effect can be investigated by the X-PID.csl program.

Press a key to continue ...

```
   .
   .
is it necessary to employ nonlinear compensators
                (Bang-bang effect) during transients ? [y]
is a short rise time important ?                       [y]
is it important to give good phase margin
                      (or gain margin) by a P-action ? [y]
   .
   .
   .
is it important to use deadbeat control design meyhod? [n]
is it necessary to implement 6 different PID control
                                       algorithms ?  [n]
is it a delayed first order lag or delayed higher order
                                               lag? [n]
is it important to use pole-assignment design method ? [n]
   .
   .
   .
I SUGGEST a discrete X/PI controller !

                Press a key to install ACSL environment ...
```

Intelligent Cooperation for Robotics

Eugénio Oliveira, Rui F.Silva and Carlos Ramos

Faculty of Engineering University of Porto,Rua dos Bragas 4099 Porto codex, PORTUGAL

Abstract : The use of Multi Agent Systems as a Distributed Artificial Intelligence paradigm for Robotics is the principal aim of our present work. In order to make different Intelligent Systems to cooperate in solving non trivial tasks, special architectures need to be designed. New functionalities are attached to each particular Agent suitable for Robotics, enabling cooperation at the Knowledge and Intelligence Levels.

Each Agent is implemented by means of a set of different specialized processes. Besides the Intelligent Systems itself, a Monitor, an Acquaintance plus Self-knowledge Modules and a Communication Module, are fundamental modules that make it possible for different Agents to cooperate by exchanging competences, information and results. Which kind of knowledge to represent, either static or dynamic, involving the overall system, how to represent it, to share it, to maintain it and to infer upon it are subjects that have been adressed and several conclusions have already been drawn. An Assembly Robotics testbed is being used for that purpose, and several Intelligent Agents developed such as: An High Level Planner, a World Descriptor and an Object Identifier.

Each one of these Intelligent Systems, incorporating its own reasoning capabilities (therefore considered as Semi-autonomous Agents), has been enhanced with several modules according to the previously referred architecture. We claim Robotics to be a good example for Intelligent Systems cooperation and we intend to apply the aforementioned principles to different classes of Robotics applications.

Keywords: Multi-Agents, Cooperation, Knowledge, Robotics, Distributed Artificial Intelligence

NATO ASI Series, Vol. F 71
Expert Systems and Robotics
Edited by T. Jordanides and B. Torby

1. Introduction

The principal aim of our research in the Distributed Artificial Intelligence (DAI) field using the Multi-Agent Systems paradigm is the coordination of different intelligent skills, specific to each one of various semi-autonomous agents, exhibiting intelligent behavior inside a global and coherent community facing a complex task.

An architecture of a decentralized set of active units which are able to exchange many different types of information may be materialized by different semi-autonomous Agents, each one with a certain degree of intelligence.

In order to be considered as semi-autonomous, an agent must have its own problem solving capabilities. Some default reasoning may also be needed to avoid the possibility of being idle whenever a requested information is missing in useful time. It is not desirable that an agent could remain in a waiting state long after a request has been made. This agent shall focus its attention on another problem and from time to time check if the answer has already arrived.

Complex problems like those adressed in robotics may need the concurrence of several skills, competences and abilities in order to reach an acceptable solution. This does not necessary imply the need of a fixed decomposition of a problem into sub-problems attaching to each computing agent one of these sub-problems, waiting for all the partial solutions and then to fuse them into the final one.

A richer form of cooperation goes beyond a rigid problem decomposition and task sharing, and it may imply "simultaneous" efforts to solve the same question by more than one Agent. Result Sharing techniques must then also be considered.

The main idea related with the use of a community of Agents that are semi-autonomous is to propose an architecture through which these Agents try to solve the problem by themselves but, at the same time, are able to put questions to each other, to accept their answers and to progress together towards a better solution.

A collection of semi-autonomous Agents shall then run in "parallel" their problem solving capabilities and cooperate, for example, through messages communication facilities.

In a cooperative "community", agents are supposed to formulate queries, to receive specific requests from others and to accept useful information (answers to their questions or voluntary information from another agent). Therefore, cooperation is seen as a much more rich concept than communication. In fact, what is involved is not only the exchange of data, but also the possibility to assign tasks to others and to redirect the path to be followed according to the incoming information. The framework of Distributed Artificial Intelligence (Huhns [3], Bond et

Gasser [1], Oliveira et al [5]) seems to be adequate to solve many aspects of the problem of cooperation between agents whenever some kind of knowledge transfer is needed.

2. Robotics Workcell

Whenever a robot arm is faced with a non-trivial task like the assembly of parts, two main factors should be taken into account that suggest DAI and a Multi-agents approach as paradigms:

* the diversity of expertise required at different levels and
* the set of constraints under which these Agents must perform, therefore cooperate, to solve a problem.

The inherently decomposability of many tasks, is another indicator in this same direction of cooperation. Each one of a given task in order to be executed in a not completely determined robotic environment, implies the planning of the work to be done by an intelligent system. Such "intelligence" needs to be supplied by a lot of significant inputs about the world, either geometrical or symbolic, which are given not only by computer Vision, but also by a priori knowledge about possible expectations (Models).

Moreover, an accurate action, performed by a physical device like a robot arm, has to be supervised at each step, by special agents monitoring its actions, and correcting, on-line, its deviations. Therefore considering that all these aspects are involved , it is obvious that a community of agents, cooperating among them, provide a possible, and elegant solution for a robotic workcell design.

3. Cooperative Community

The community of agents we are implementing, is made up of units which are not contradictory or antagonists but complementary. This does not mean that there is no competition between them. In fact, due to a certain degree of expertise overlapping, one among several candidates may be chosen to answer a question or solve a sub-problem.

As an example we can imagine that, whenever a gripper looses an object during an action, either a set of proximity Sensors plus Low Level Planning or the Vision Agent plus High level Planning Agents may try to command the regrasping of that lost object.

In our approach, we are playing with agents which are semi-autonomous. This means that if some requested information is not obtained or is missing in useful time, an agent must use their own default reasoning capabilities in order to avoid the situation of being idle. They guarantee that, with their own degree of "intelligence", they are able to arrive at a solution, possibly poor, even if they don't get any help from others.

Nevertheless, the principal aim of the community is to exchange knowledge and data among agents in order to solve tasks under more or less severe constraints. This cooperation takes place in a physical environment, using communication links between different processes or machines.

The definition of protocols, both at a knowledge level as well as at communication level are then crucial.

Agents Profile

Let us characterize Agents better. In our perspective what are they? What are they composed of? How do they cooperate?

An Agent is an active computational unit with its own capabilities, able to be executed separately, trying to solve a task the best it can. Moreover, an Agent has the "perception" of being a part of a bigger "community" of units with different skills and objectives but, nevertheless, all of them interested to cooperate to get better solutions to the problem being solved.

Each Agent has its own inference mechanism, knowledge (or just data if it is not an "intelligent" one) and communication capability.

It can sense the world and exchange data, knowledge, goals, part of plans, beliefs (with associated measures of uncertainty) and commands.

In order to do this, an Agent must know the profile of its surrounding community, which means to have a stereotyped representation of differents aspects of the other Agents.

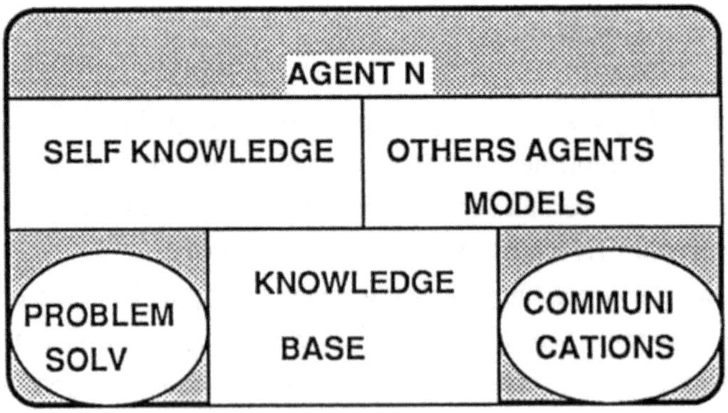

fig. 1 Principal modules of an Agent

Moreover, their own Self-knowledge, knowledge about the contents and limits of their own skills and competences, as well as the knowledge about Acquaintances, enable them to select other agents for the sake of cooperation, send queries to them and pay attention to external help.

Acquaintances

Semi-autonomous Agents in a cooperative community need to know each other at least in what concerns the possible benefits of mutual cooperation.

How shall the image of Agent A_i be presented to Agent A_k?

There have been interesting proposals for such models (Bond ; Les Gasser [1]) and we will summarize their content on the items listed below:

- a name to be recognized and an address to be located.
- its possible role. Is it a creator of the Agents or just a member of the organization ?
- skills. The Agent competence and domain knowledge.
- goals. List of goals other Agents know this one is able to achieve.
- plans. Some information about how each specific Agent can reach a conclusion.

Therefore a specific Agent A_k can be activated, selected, or required by another A_i on the basis of any of the above mentioned characteristics.

It is also our opinion that each specific Agent shall contain models of its acquaintances avoiding the existence of a separate entity with all that knowledge about the community. The latter possibility could indeed lead to a bottleneck during cooperation.

Another important aspect of the intended modelization of the "community" is the inclusion of Model(s) of human Users which almost inevitably will interact with the system. It is therefore important to build up User Models in order to improve Human Computer Interface. This matter, User Models in the Agents Community, is the subject of another on-going research taking place in our group. Good approaches to User Modelling can also be found in [6].

4. Reasoning Process

Since they are autonomous and intelligent, each Agent must be provided with its own reasoning mechanism. Different Agents may use different reasoning methods. Therefore the distributed system is composed of heterogeneous modules and their reasoning methods may depend on the domain of expertise.

In such a distributed system there may exist intelligent strategies, heuristics based, but it is also likely to find non AI systems that use simply coded algorithms. Moreover, in time critical environments such as may be the case in Robotics, the reasoning process can be performed in an **incremental** way. When an Agent is asked for a solution, it may have a default value to present, otherwise it starts a reasoning process. It then tries to achieve the fastest solution and give it back. As time goes by, given solutions improve in quality until a time limit as been reached.

Uncertainty

The heterogeneity of Agents may not only be present in the reasoning methods but also in the methodologies to deal with uncertainties. These ones can range from numerical methods to symbolic ones. Numerical methods can directly use either Bayes theorem, Certainty Factors model or be based on evidential theory (Dempster-Shaffer theory, for instance).

In a cooperative environment each Agent can get information from other Agents using different representation methods for uncertainty. Therefore there is a need to transform such representations such that the receiving Agent is able to admit them. How can it be achieved? Although inexact reasoning models may be very different in appearance, they have some characteristics in common at the conceptual level. For example, each model has 3 corresponding codes for propositions which are true, false, or unknown. We can use 3 value mappings and interval mappings to perform the desired transformation. Such a triplet is mapped from one model to another. Partly true and partly false intervals are also mapped in a similar way. This method guaranties that the transformation is correct from a qualitative point of view. Such a facility is of great use also in a robotics environment.

As an example let us suppose that the Agent modeling the world have a measure of uncertainty using intervals to deal with preconditions to grasp an object. If the location of that specific object have been extracted by the Vision Agent with a single figure to represent the uncertainty of that location, a process of matching between the two informations must be performed.

Truth Maintenance

The need of TMS or ATMS [4] arises when different (eventually contradictory) informations are generated during system information gathering or inference. As we have mentioned earlier, concerning incremental reasoning, there may arrive to each Agent different answers for the same problem at different stages of the inference process. This implies the need to have a sort of direct-dependency network with previous conclusions which may be easily updated.

We can deal with this non-monotonic aspect using either **TMS** (Truth Maintenance Systems) or **ATMS** (Assumptions based Truth Maintenance Systems).
The purpose of TMS is threefold:

i) serves as a cache to all inferences made by the Problem Solver;

ii) enables the Problem Solver to make non-monotonic inferences (Unless information in the contrary, infers A);

iii) guaranties that there will be no contradiction.

Generally speaking, conventional TMSs are more convenient for finding just one solution. In an ATMS instead of justifications associated with each datum a set of assumptions are stored representing the contexts on which the datum is valid. These assumptions are calculated by the ATMS from the justifications given by the Problem Solver. They are primitive data from which other data is derived. They are much more efficiently manipulated than the data sets they represent. There is no need for all the database to be consistent; it is easy to refer to contexts and move to different points in the search space.

In a Robotics environment the need for TMS may arrive, for example, when the Vision Agent gives erroneous information to the Planner. Whenever the latter starts reasoning with that information, planning some paths or grasping actions, the Perception (Sensors) Agent can give to it information that contradicts the one given by the Vision Agent. If the Acquaintance Model states that more credibility should be given to Sensors, this implies that actions already started should be revised.

5. Cooperation in Assembly Robotics

In this section we specify a testbed for cooperation in a Robotic environment. Some situations where cooperation is desired are shown, our Agents are described and a very simple example is also presented.

An implementation schema for cooperation

Each Agent is seen as a set of different processes running under Unix and controlled by its own Monitor. This Monitor is in charge of managing different modules : An input queue, where the requests to each specific Agent are, an Agenda where different sub-tasks to be performed are kept, and the double aspect of knowledge : Self-knowledge and knowledge about the Acquaintances.

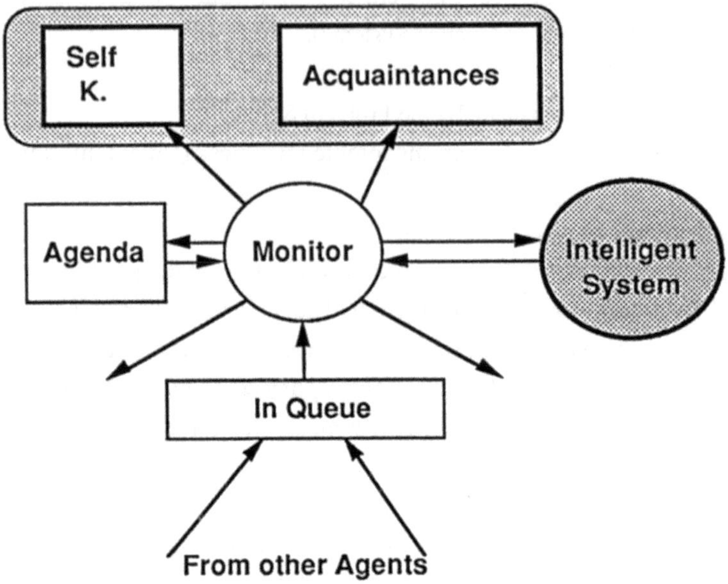

Fig. 2: Flow of information inside an Agent

Agent queries and answers are transmitted by sending messages to the receiver Agent queue according to indications got from the Acquaintance module. The Monitor takes the received message, analyzes it and depending on its quality, a goal to be executed, a fact to be inputted, a plan to be performed, places it in the Agenda for consideration by the Agent's Problem solver.

As it was told before, in order to be considered as semi-autonomous, an Agent must also have some default reasoning capabilities avoiding the need of being idle when the requested information is missing in useful time.

Message format includes, other than the content, the sender identification, the message type and possible time constraints. On the other hand, the receiver agent Monitor handles messages, and assigns priorities, according to their type, the importance of the sender and the Agenda of the receiver.

The possibility to redirect the requests is another important feature. An agent has the opportunity to inquire a set of agents. The number of elements of this set may be one, some or all other agents of the community (broadcasting).

Another important feature is the agent freedom to formulate questions to other ones and, meanwhile, go on thinking about another problem. The agent shall be able to verify, whenever desirable, if the requested answer has already arrived. Some problems can occur if the agent remains in a waiting state for the answer because the desired answer could not become available. One must pay attention to avoid a situation where a circular chaining of requests leads to a "cooperative deadlock".

To conclude, we shall notice that important cooperative features, like default reasoning, broadcasting and requests to specific agents, must also be present in a cooperative robotic testbed. Nevertheless, we shall see that default reasoning has poor credibility in this case due to the highly dynamic environment and the irreversible consequences of previous actions. This implies a real need for planning activities and correct agenda creation by the cooperative layer.

Cooperation needs in Assembly Robotics

In the Framework of our Robotics workcell it is possible to identify the need of several Agents such as action Planners (High level and Low-level), Vision, other sensors (laser, proximity sensors, etc...) and the World Representation Agent.

Below, some situations are depicted where cooperation between Agents, namely Planning, Vision and Model+World Representation (MWR) Agents is desirable.

Whenever the Planner knows, a priori, which objects are present in a specific scene, but not where they are, symbolic objects features (like name, function, color, weight, material etc...) may be sent to the MWR Agent. This Agent associates symbolic features with numeric dimensions of the modeled object and sends this informations to the Vision Agent. This definitively improves the objects search strategy, because Vision looks only for the desired objects on the image.

On the other hand if a Planner completely ignores which objects are present on the scene, Vision starts to work and, while objects are identified, their names are sent to the MWR Agent and to the Planner to enable them to act. Let us suppose that the Planner does not know yet the task to be performed (the cell is able to perform several tasks) and Vision identifies one object on the present scene. This may be enough for the Planner to start with a planning tasks. This event triggers, by itself, a new request to Vision in order to be informed about locations of specific expected objects.

There are also situations where time is a critical constraint and planning tasks impose a quicker answer on Vision. If this is the case, Vision process must speed up by applying faster processing (using for example Hough Transform for only one pixel by each 4 or 16). Clearly such an improvement leads to fast answers but some accuracy is lost. Nevertheless nothing prevents Vision to go on elaborating a better solution.

Contingency analysis or exception handling (what to do when unexpected obstacles are met or an object loose) is another field where cooperation may be of great importance.

Agents in our testbed

To test cooperation in a Robotics environment we have already developed three main cooperative Agents: Vision Agent; Planning Agent ; Model + World Descriptor (MWR) Agent.

Our Vision Agent is a simple 3D object recognizer using one image and some a priori knowledge about objects to locate. The camera position, relative to the robot is assumed to be known. Vision processing begins with Hough Transform application to detect lines (Hough [2]). It uses then a methodology known as "Preview Verify and Validate " in order to recognize objects with the help of Models.

The High level Planning Agent gets the task to be done in a simplified language. For example a specific task is presented as follows: is_goal(put_on(small_cube,orange_object)).

Positions of the objects and the complete state of the world are provided by the user or by another Agent. At this moment we have considered only high-level planning but it is our intention to enhance the system with trajectory and grasp planning. A third Agent is responsible for object models and world representation. The object models are represented as frames with slots reserved for each symbolic or numeric features, for example: name:box; material: plastic; color: yellow; weight: 100; edges: [list]; faces: [list]; size: small; position : [coordinates]. Other interesting features to be considered are possible grasp positions for each object .

A simple example of Cooperation

Below, an example of cooperation between the three agents described beforehand is represented

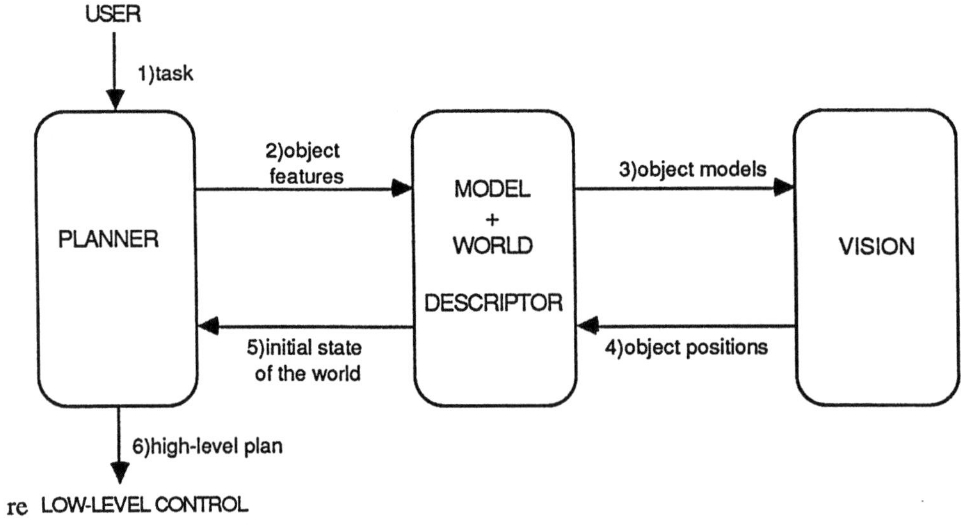

fig. 3 : Cooperation needs for three Agents

This simple example concerns a situation with three different boxes whose images have been acquired and processed and then transformed: a white box; an orange plastic box; a small red wood box.

Goals are specified by the user as follows:

is_goal(on(white,floor)). is_goal(on(plastic,white)). is_goal(on(small,plastic)).

The objects attributes like "white" "small" and "plastic" have different nature (color, size, material).

A message with such features is sent to the MWR Agent which uses them to index the object model. Geometrical information is found and given to Vision, therefore helping to locate each specific object. Finally MWR uses the object positions to obtain the initial state of the world sending to Planner the following information:

on(white,floor). on(plastic,floor). on(small,white).

Now the Planner can establish the high-level plan to execute the task:

put_on(small,floor). put_on(plastic,white). put_on(small,plastic).

Since our laboratory testbed is equipped with an academic Robotic system (APRA) controlled by a PC/AT (MSDOS) using LM programs, the previous example has been executed after transferring object positions and the derived plan to PC files. A very simple low-level planner (on the PC) has successfully executed the planned task.

Self-Knowledge

Each specific agent knows which goals it is able to perform. We can divide all the possible goals and subgoals into two main categories :

* those which are of exclusive competence of that Agent (independent goals) and,

* those which depend from other ones (dependent).

An independent goal or subgoal can be performed immediately by the Agent without any need of external information. On the other hand, if the agent needs external information a goal or subgoal is said to be dependent.

Knowledge about goals, subgoals, external information (Data and Knowledge D+Ki) and the produced information (Data and Knowledge D+Ko) is then closely interrelated in the respective module (Self-knowledge). It is up to the Monitor to manage such rich information in order to drive the incoming requests and to plan the problem solving activity of the receiving Agent.

The previous example of section 2 (with three agents) leads us to this possible sketch of knowledge (for only two Agents) :

Self-Knowledge for Vision Agent

IS --> Object Recognizer

Goal 1 --> locate objects

subgoal 1.1 - to get object models

input: D+K = geometrical models

from: MWR

subgoal 1.2 - to acquire and process the image(s)

input: D+K = urgency

from: (Planner, User)

subgoal 1.3 - to generate hypothesis and to verify them

Result: D+K = spatial positions of located objects

Goal 2 --> verify object positions

subgoal 2.1 - to get object models

input: D+K = geometrical models

from: (MWR)

subgoal 2.2 - to get the desired positions

input: D+K = positions

from: (MWR, User)

subgoal 2.3 - to acquire and process the image(s)

input: D+K = urgency

from: (Planner, User)

subgoal 2.4 - to verify

Result: D+K = y/n

Self-Knowledge for Planning Agent

IS --> High-level Planner

Goal 1 --> to generate the high-level plan

subgoal 1.1 - to know scene objects

input: D+K = object symbolic features

from: (User,Vision)

subgoal 1.2 - to know the state of the world

input: D+K = object relationships

from: (Models+World Descriptor,User)

subgoal 1.3 - to know the desired objectives

input: D+K = final state of the world

from: (User)

subgoal 1.4 - to plan

Result: D+K = high-level actions to execute the plan

Goal 2 --> to replan

subgoal 2.1 - to know the actual state of the world

input: D+K = object relationships

from: (Models+World Descriptor,User)

subgoal 2.2 - to know if the objectives are modified

input: D+K = final state of the world

from: (User)

subgoal 2.3 - to plan

Result: D+K = high-level actions

The field "from:" is related with the Acquaintances described below. The Monitor should be able to associate the field "input:" with the appropriated acquaintances in order to conclude the set of agents that must be inquired.

Acquaintances

In the Self-knowledge Module, an agent has information about its own capabilities in terms of possible goals it can achieve. But it also must know which other Agents are able to answer its needs and which data or knowledge will be required from itself.

In what concerns goals, subgoals, Data+Knowledge and results, the acquaintances are not just the union of knowledges about the other agents of the community. Only the direct interactions with each particular agent have to be represented.

For example, the high-level Planner knows that the MWR agent is able to provide the symbolic state of the world (see item 5 of figure 3), but it ignores that the last one will ask Vision about object positions in order to be able to answer the request (4 of the same figure). In such a case it is more important to the Planner to know that it is convenient to send symbolic features of objects like name or color (2 of the same figure).

As an example, what is needed for MWR Acquaintances Model is:

from Vision : "it is able to get object positions"; D+K = Object Models

from Planner : "it is able to know symbolic features of objects"

The Agenda

In our example, agents have not so many different tasks to do. On the other hand, if some other agents (laser, low-level planner, etc...) are included in the community the number of requests may dramatically increase.

Therefore, a scheduler to plan the goals may be provided to maintain and to select the "waiting" tasks. An agenda can be implemented based on goal priorities. These priorities are dynamically modified. For example, if important information is not available to perform a task, its priority could be decreased and other items of agenda selected.

The agenda acts as a priority queue where the items may modify their positions dynamically.

6. Conclusions

We are trying to implement a distributed Architecture of several Agents for Assembly Robotics using a Distributed Artificial Intelligence paradigm. For the moment we already have identified specific features which are crucial for our Agents, their Reasoning process, Acquaintances and communications mechanisms. Such Agents are now being developed in which concerns their own competence and specific needs for cooperation. Finally, in order to integrate such Agents into a "cooperative community" able to perform an assembly task, real time experiments upon complex tasks will be done on the next future.

7. Acknowledgements

We would like to acknowledge our Esprit Project 2256 partners (Krupp AE, Queen Mary WC, Framentec, NTU Athenes, ISPRA, EDRC, Univ. Amesterdam, Volmac, CERN, IRIDIA, Iberduero, Labein, Ambar). All of them are contributing for a general architecture for cooperative systems. Special thanks are still due to T.Wittig (Krupp AE), A. Mamdani (QMW College), F.Arlabosse (Framentec) G.Stassinopoulos (NTU Athenes), N.Avouris (ISPRA), in particular, are also responsible for our interest in the subject. Nick Jennings was the first who wrote Archon Technical Notes about Self-knowledge and Acquaintances Models.

8. References

1. Bond; Les Gasser (eds) : Readings in Distributed Artificial Intelligence, Morgan Kaufman
2. Hough, P. V. C. : Method and means for recognizing complex patterns, U.S. Pattent 3.069.654, 1962
3. Hunhs,M. (ed) : Distributed Artificial Intelligence, Morgan Kaufmann 1987
4. Kleer, Johan de : An assumption-based TMS, Artificial Intelligence N.28 1986
5. Oliveira E.; Ramos C; Camacho R. : AI Systems cooperation for Robotics, First Workshop on Robotics and CIM, Lisbon, 1989
6. Walhster, Kobsa: Dialogue-Based User Models, Procedings IEEE vol 74 n. 7 July 1986

The Oxford Robot World Model

Bill Triggs and Stephen Cameron
Programming Research Group, Oxford University,
11 Keble Rd, Oxford OX1 3QD, U.K.
{bill,stephen}@robots.oxford.ac.uk

ABSTRACT

After reviewing the advantages of supplying a robot with a geometric model of its surroundings, we discuss the design of a modular object-oriented database to support such a model, which is being implemented as part of the Oxford Autonomous Guided Vehicle project.

1. Model Based Robotics

The Oxford Autonomous Guided Vehicle [1, 2] is a mobile robot truck or forklift designed to operate with a degree of autonomy in a structured and fairly well known factory-like environment. It has a number of advanced sensor systems for navigation, obstacle avoidance and pallet acquisition, and a corresponding set of task and path planning strategies, but at the heart of the software is a model-based representation of the vehicle and its environment, supported by a purpose-built geometric database.

There are several good reasons for providing a robot with such a world model:

(i) It provides an effective, easy to use, high level interface to the robot: task level behaviour such as 'pick up the pallet at x and carry it to y' is natural to a robot which 'understands' its surroundings.

(ii) It encourages a correspondently clean internal architecture in which independent vehicle subsystems such as sensing and data fusion, vehicle controlling, and path and task planning modules communicate with one another mostly via the world model. (Albeit there are usually a few time or safety critical mechanisms such as basic vehicle reflexes which must bypass the model). This modular, inherently parallelisable, model-based, data-driven architecture simplifies the introduction of new subsystems by providing a uniform inter-module interface; in particular it allows vehicle capabilities to be extended by the addition of more or less generic 'expertise modules' which need only interface with the relatively simple and vehicle-independent model database.

(iii) It decouples essentially robot-independent environmental information from the robot-specific software, allowing it to be handled by efficient specialist code such as the spatial database and shared between a number of robot and control processes. The result is an integrated model of the whole environment which is immediately extensible to large multi-robot systems and provides a solid foundation on which to build interacting-robot and factory-wide controllers.

Fig. 1 illustrates the model-based architecture we are developing for the AGV project. Before discussing this in detail we shall briefly list some of the 'expert modules' being

NATO ASI Series, Vol. F 71
Expert Systems and Robotics
Edited by T. Jordanides and B. Torby

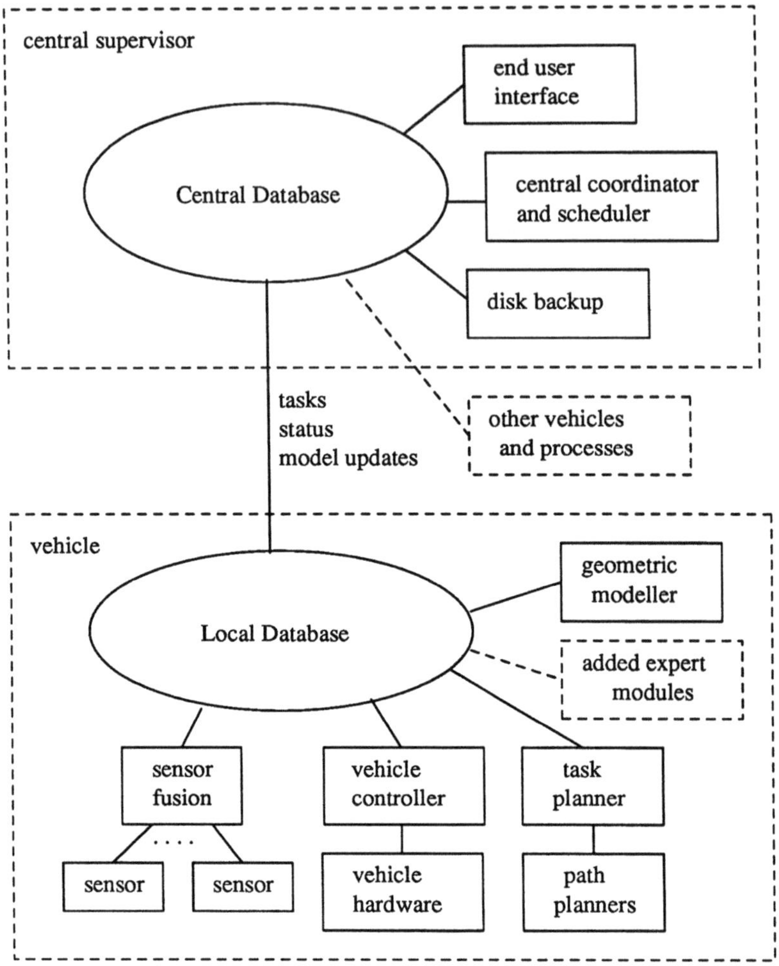

Figure 1: A modular architecture for a world-model based multi-robot system.

developed by other contributors to the project, which will eventually be interfaced to the world model:

(i) The Constructive Solid Geometry engine ROBMOD [3, 4], which was co-written by one of us, is being used for all of the solid modelling on the project. It provides a range of geometric expertise to its clients, including spatial bounds on objects for the indexing system, lists of visible edges for model-based vision, and very efficient three dimensional object interference detection for use by the path planning modules.

(ii) A range of model-based sensor systems are being developed, including a promising Kalman-filter-based sonar system [6], infra-red range and time-of-flight scanners, and single and multiple camera vision systems [7].

(iii) A number of different path planners are being evaluated, including generate-and-test, potential field and configuration space algorithms and several heuristics for common vehicle manoeuvres, with a view to incorporating some or all of them in a 'path planning expert' capable of choosing and invoking a suitable (combination

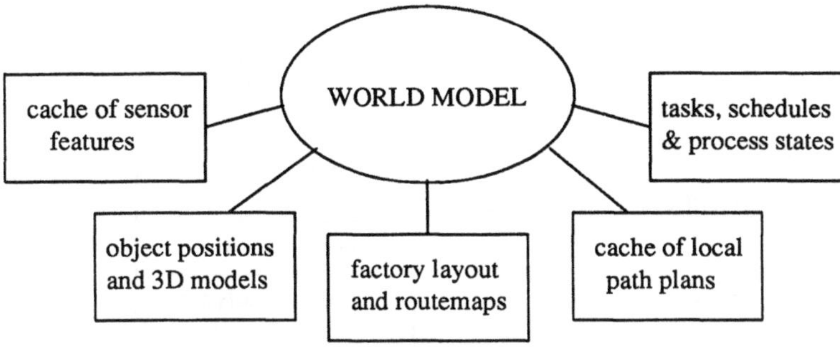

Figure 2: Types of information stored in the robot world model.

of) method(s) for any given problem [5]. This will probably take the form of a rule base driving a battery of hard-coded planning algorithms.

(iv) An 'expert palletiser' is being developed to plan the task of clearing a region cluttered with pallets and obstacles. This requires the integration of the above path planner, a rule-based system enhanced with several graph-based heuristics to select a suitable order for clearing the pallets, and a docking manoeuvre planner to supervise the vehicle as it approaches and picks up pallets.

It is hoped to make all of these modules sufficiently flexible to be used with other vehicles or systems.

A useful robot world model must contain a wide variety of information, for example (see fig. 2):

(i) Representations of the physical layout of the environment and the objects within it such as walls, pillars, pallets and robot vehicles.
(ii) More detailed descriptions of objects such as three dimensional solid or boundary models.
(iii) Maps of features important to model-based sensor systems, such as visible edges and sonar-reflecting walls and corners.
(iv) A decomposition of the workspace into functional regions such as roadways, processing sites and stores, and a corresponding route-map graph for large scale route planning.
(v) Caches of reusable local or temporary data such as details of paths planned around local obstacles and partial feature maps of unidentified objects.
(vi) State, task, and scheduling information for the modelled robots and manufacturing processes.

The critical component in this model-based architecture is the database which underlies the world model: it must be able to store and access at least the above types of information; it must be particularly efficient for the common geometrically or spatially organised data; it must interface easily both with the end user and with a potentially wide range of robot subsystems; and it must be sufficiently flexible to deal with unexpected types of information and subsystem as neither of these are known perfectly in advance.

Moreover, even for a single robot system such as the AGV we feel that the model database should be capable of supporting a distributed multi-robot multi-process world model because:

(i) This allows the robot software to be simply partitioned into a number of independent

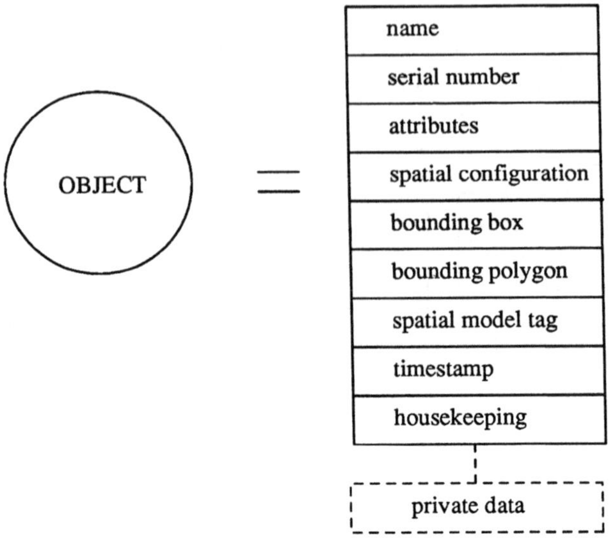

Figure 3: The Object header structure.

modules, each with access to the database, as above.

(ii) Even a robot as autonomous as our AGV must exchange a considerable amount of task, state and environmental information with a land-based supervising computer over a (low-bandwidth radio) link, so that both vehicle and supervisor world models of some sort are required, together with a reasonably efficient data-exchange protocol. It seems sensible to use the same format for both model databases and organise the communications around database records, in other words to build a primitive distributed database system: this distribution may as well be built in to the database at the outset.

(iii) A robot designed around a multi-robot database is more easily adapted to an integrated multi-robot environment. Such integrated systems are interesting in their own right and are likely to be of very considerable practical importance in the future.

The support of such a multi-process system requires some sophistication of the database, in particular the consistency of multiple distributed copies of database records must be maintained in real-time, preferably without undue complication to either the user interface or the implementation.

2. The Oxford Robot Database: Design

Since we are not aware of any existing database which meets the above requirements, we are developing our own modular object-oriented database system specifically for robot modelling in a distributed environment.

In this system generic header structures known as 'Objects' are attached to all data items; for the most part the database uses only these header fields, however one such field describes the type of data contained in the Object and if necessary the database can manipulate this private data with the type-specific Object routines supplied with each type. The Object header is intended to provide a succinct description of the enclosed data

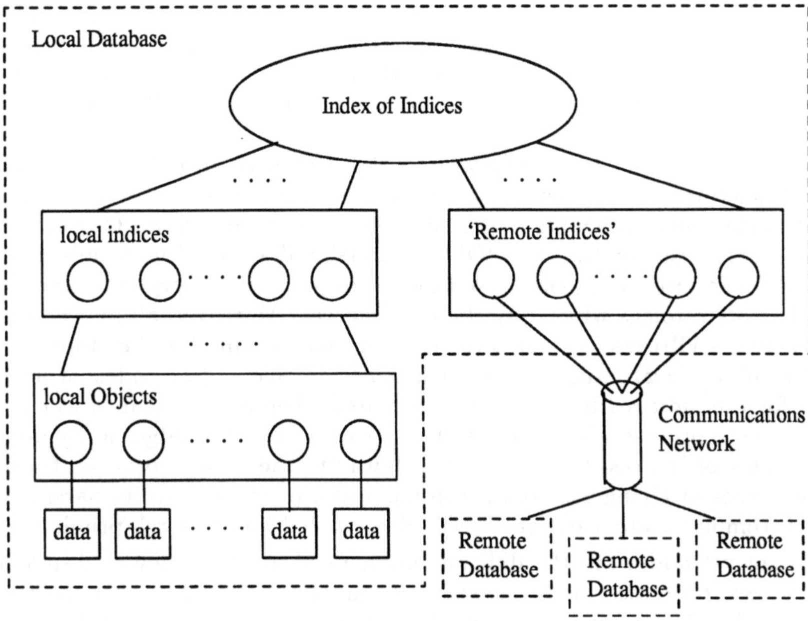

Figure 4: Architecture of the model database.

suitable for use with a range of indexing systems, however it actually contains enough information to describe simple physical objects in its own right (see fig. 3). Fields are provided for an object name string and a unique system-wide object serial number, for a vector of user-defined binary 'object attributes' and a time-of-last-modification of the object, and most importantly for a spatial configuration (position and orientation) and a description of the spatial region occupied by the object. The main types of index currently supported correspond to these fields — in particular it is possible to access an object by name or serial number and by two or three dimensional spatial position — however new types of index are very easily added.

The spatial parts of the system are 'two and a half dimensional': all positions are three dimensional but coordinate transformations always preserve the vertical axis so that object orientation is given by a single angle. The spatial extent of an object is specified by a three dimensional bounding box (aligned with the current coordinate system) and an optional two dimensional inscribed polygon; these allow the database to perform fast but fairly crude object localisation and interference detection, and efficient spatial indexing. The database kernel deliberately does not have solid modelling expertise, since if full three dimensional shape descriptions or interference tests are required (for example for visible edge prediction or path planning in a cluttered region) a specialist geometric modeller can easily be run as a database subprocess. (There is a tag in the Object header for a modeller-based shape description). At present this role of 'system geometry expert' is filled by the Constructive Solid Geometry modeller ROBMOD [3, 4], however the modeller is not tightly coupled to the database and could easily be changed or omitted if required.

In operation, the database kernel attempts to maintain appropriate multiple copies of objects and all indices without user intervention, so that when an object is locally created, modified or destroyed the changes are automatically propagated to all database processes which have expressed an interest in objects of that class, and are reflected in all indices, local or remote, which contain or should contain the object. Thus, each client process can behave as if its own copy of an object was the only one and use a variety of indexing systems without explicitly having to maintain them. This process, illustrated in fig. 4, is achieved as follows:

All indices present a uniform interface to the user with primitives such as object insertion and deletion and a generalised search which performs a user-defined action on every object in the index which fits a user-supplied 'Object Template'. Object Templates are data structures designed to act as 'Object filters' and are used uniformly throughout the system to select interesting classes of objects, in particular every index has a private Template which determines which objects it will accept. Moreover, all indices within the system are actually Objects in their own right and may themselves be stored in system indices. This allows local indices to be maintained by keeping an index of indices and searching it for the indices which should be updated when a particular object is altered. Similarly, a remote process with an interest in a class of objects lodges a 'Remote Index' in the local index of indices; this acts as a gateway to the remote process, transmitting appropriate classes of changes across a communications network to its parent. In both cases Object Templates are used for object selection and index search pruning.

We have not yet addressed the difficult problems of inconsistency or deadlock which arise when several processes simultaneously attempt to modify an object, however these are not expected to be severe in our domain as the few parts of the database which are frequently changed (such as robot states and positions) tend to be under the control of a single process.

3. The Oxford Robot Database: Implementation

At present the system is configured as a central database serving a number of sensor, controller and planning processes, each with embedded local database code. It is written in C and runs on (a network of) SUN workstations under UNIX, using SUN's Remote Procedure Call and External Data Representation mechanisms for network (INTERNET) communications in a machine-independent data format. The present implementation is only a prototype and the following changes are being considered: switching to an object-oriented language such as C++ would simplify the code considerably and allow a richer object hierarchy, although the loss of portability may be a problem; UNIX responds too slowly for realistic robot control and parts of the system will have to be ported to a different operating system, probably PARALLEL C or OCCAM as our vehicle controller is Transputer-based; and the Remote Procedure Call mechanism is not well adapted to our purposes and may have to be replaced with a lower level network protocol.

All of the basic types of indices are implemented using binary trees. The name and serial number indices are configured forms of a general purpose binary-tree index (with additional tree-balancing code) and the two and three dimensional spatial indices have a layered binary tree structure as follows:

Recall that spatial indices, being Objects, are provided with spatial configurations and bounding boxes; all indexing calculations are performed in index-configuration-aligned coordinates so that spatial indices with arbitrary spatial alignment are possible. For an object to be stored, its bounding box (in aligned coordinates) must overlap the aligned bounding box of the spatial index. If this is so, the index box is repeatedly bisected in the x-direction, the half containing the object being retained, until the object box is cut by a

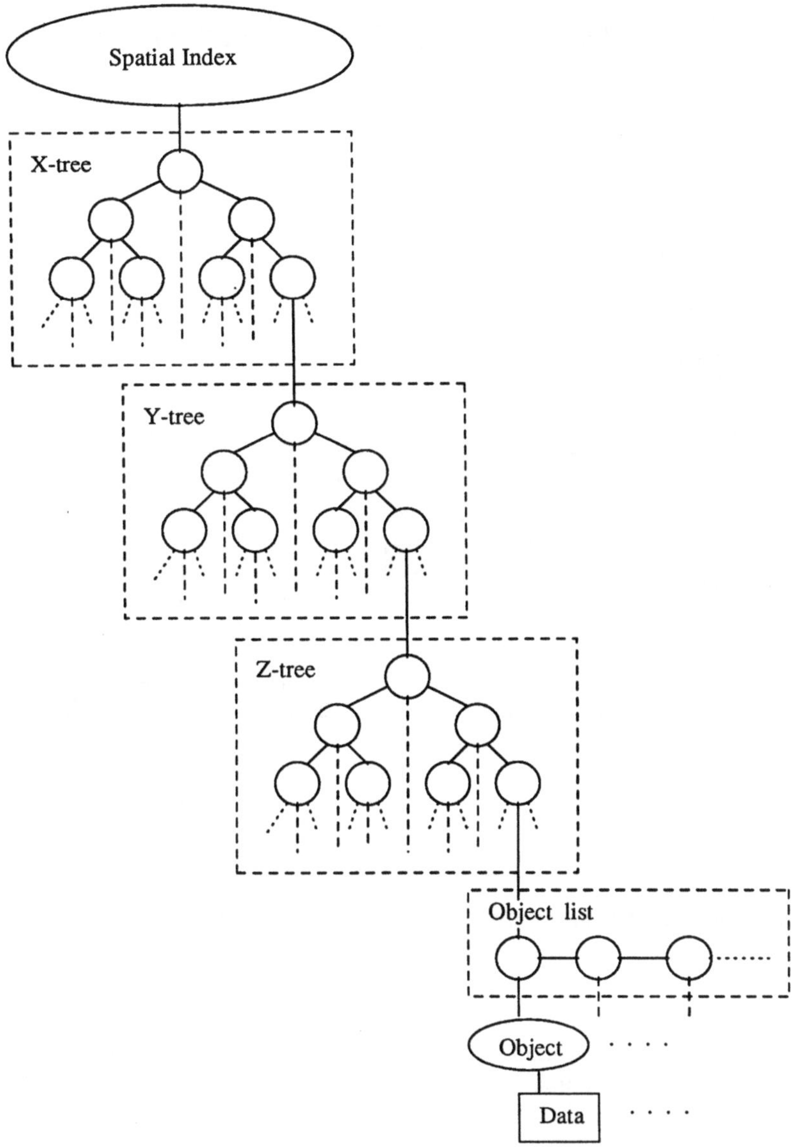

Figure 5: Spatial indices have a layered binary tree structure.

bisection rectangle; this rectangle is then bisected in the y-direction until the object box is cut by a bisection line; and the line is bisected in the z-direction until a bisection point falls within the object box; the object is then stored in a list of the objects associated with this bisection point. Thus, a three dimensional spatial index is a binary (x–) tree of binary (y–) trees of binary (z–) trees of lists of objects, as illustrated in fig. 5. Similarly, a two dimensional index is an x-tree of y-trees of object lists.

As is often the case with bisection-based spatial indices, each tree node is associated with a unique bounding box (the box which is bisected at the node), a unique spatial point (the centroid of the node box), and a unique set of three binary fractions (the coordinates

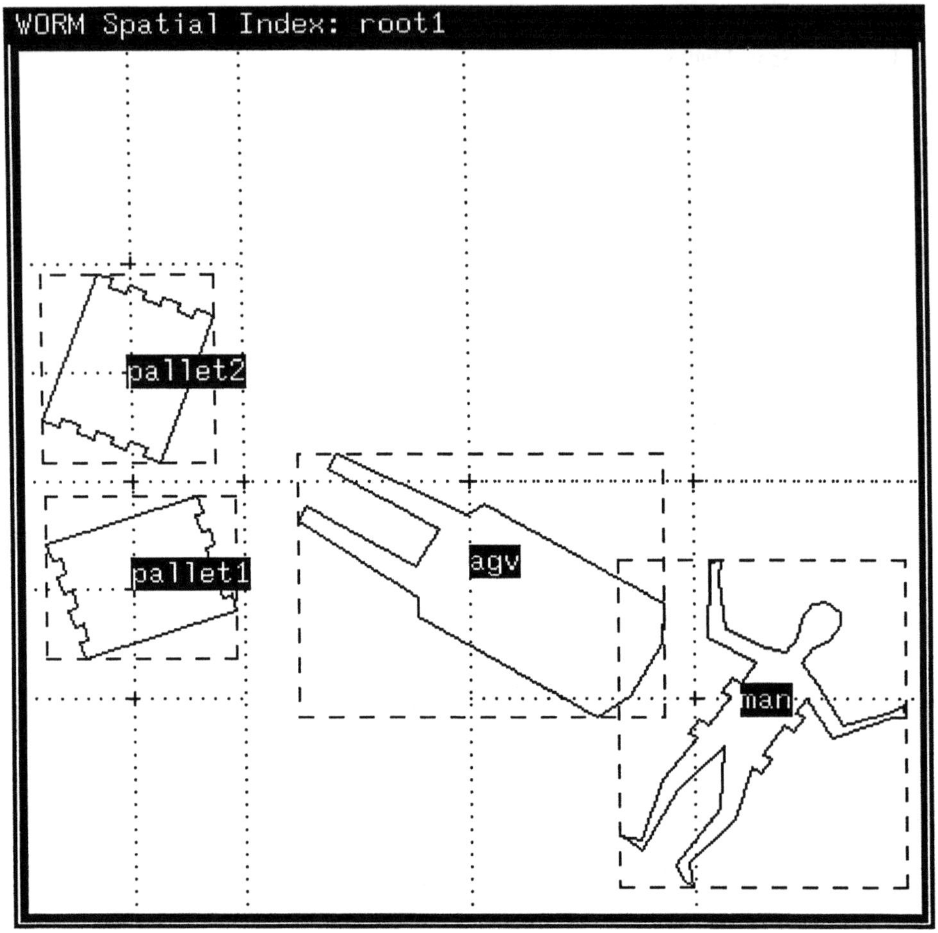

Figure 6: A two dimensional spatial index in action.

of the node centre in the system normalising the index box to $[0,1] \times [0,1] \times [0,1]$), which can conveniently be used for node identification and tree traversal. Objects are stored at the node corresponding to the smallest node box containing the object and the first node centre to fall inside the object box.

Fig. 6 illustrates a two dimensional spatial index in operation. The polygons and bounding boxes of the objects are shown, and the dotted lines are node bisectors; thus, 'pallet1' is cut by the third layer of bisection on the x-axis (which also cuts 'pallet2') and the third layer bisection on the y-axis, and its fractional node coordinates are $(.001, .011)$.

To prevent very small or thin objects falling too deep in the tree a resolution cut off is usually applied, objects being stored in the smallest remaining node box containing them. It is possible to alter the order of the node bisections without affecting the depth of any object in the tree.

Note that the node boxes repeatedly overlap one another: this means that finding the objects overlapping a given box requires a search of all of the parents (containing nodes) and many of the children (contained nodes) of the box node; however the search is easily implemented by recursively eliminating nodes (and hence subtrees) which do *not* overlap the given box and this requires only a single coordinate comparison at each node.

There are of course many alternative ways of organising such a bisection-based spatial index; in particular quad-tree or oct-tree based representations [8] (in which the node box is divided into quadrants or octants at each level) might be thought to be the obvious choice. The chosen bin-tree system has a number of advantages over these:

(i) The bin-tree representation has node-boxes of arbitrarily high aspect ratio and is therefore extremely efficient at circumscribing long thin objects such as walls, corridors and rows of storage racks, provided they are aligned with the index axes. Such objects are obviously very common in factory-like environments. By comparison, the quad– and oct-tree structures (at least in their simplest forms) force all of the tree nodes to have the same aspect ratio as the original index-box and may therefore be rather poor at object localisation, particularly if the region being indexed is far from square.

(ii) The bin-tree representation leads to slightly simpler and more compact code and is trivial to adapt to indices of any dimensionality; in fact the two and three dimensional indices share the same code in our system.

(iii) Although it is true that quad– or oct-trees tend to be shallower than their binary equivalents, it does not follow that they are much faster to traverse or more compact. Our experience is that access time depends more on the total amount of bisection required than on the tree depth because most of the time is spent doing bisection arithmetic, and that (at least in the robot modelling domain) the higher branching ratio trees are actually less compact because most of the node pointers are never used. One caveat is that because the bin-tree system uses a larger number of smaller nodes it is comparatively more dependent on the efficiency of the memory allocator for nodes.

In practice this layered binary tree system has proved to be highly efficient at object localisation, simple, fast and reasonably compact, despite the fact that quite deep trees are generated.

4. Summary

Many robots would be significantly more flexible, powerful and easy to use if they were fitted with a geometric model of their surroundings, and this is particularly true of autonomous robot vehicles. When such a world model is supported by a properly designed database the further advantages of a modular, data-driven, database-centred system architecture with a clean, consistent user interface are available, and if this database is also able to support distributed processes the resulting system forms the basis of a powerful integrated multi-robot environment. A particular advantage of this architecture is the ease with which it can be extended as new expertise becomes available.

Running such a model-based, data-driven system makes considerable demands of the underlying database, however we feel that these will largely be met by the modular object-oriented robot database system which we are developing as part of the Oxford Autonomous Guided Vehicle project.

Acknowledgements

The AGV project is supported by the SERC ACME Directorate, grant GR/E62776, and is based on a vehicle provided by GEC-FAST. Our accommodation at the ASI was supported by NATO.

References

[1] Brady, J. M.,Cameron, S. A., Durrant-Whyte, H. F., Fleck, M., Forsyth, D., Noble, A., and Page, I.: Progress towards a system that can acquire pallets and clean warehouses. In *Fourth Int. Symp. Robotics Research*, pages 359–374, Santa Cruz, August 1987.

[2] Cameron, S. A.: A geometric database for the Oxford autonomous guided vehicle. In B. Ravani, editor, *CAD Based Programming for Sensory Robots*, pages 511–526, Castelvecchio Pascoli, July 1988. Springer-Verlag. Ref. F-50.

[3] Cameron, S. A., and Aylett, J. C.: ROBMOD users guide. Software report, Department of Artificial Intelligence, University of Edinburgh (U.K.), 1987.

[4] Cameron, S. A., and Aylett, J. C.: ROBMOD: A geometry engine for robotics. In *IEEE Int. Conf. Robotics and Automation*, pages 880–885, Philadelphia, April 1988.

[5] Hague, T., and Cameron, S. A.: Path planning for the Oxford AGV. In *Proc. IEEE Int. Workshop on Intelligent Motion Control*, Istanbul, August 1990.

[6] Leonard, J. J., Cox, I. J., and Durrant-Whyte, H. F.: Mobile robot localization by tracking geometric beacons. *IEEE J. Robotics & Automation*, 1990. Submitted to.

[7] Reid, I., Brady, J. M., McIvor, A., Marshall, S., Knight, I. B., and Rixon, R. C.: Range vision and the Oxford AGV. In *Image Processing 89*, pages 51–72, Wembley, 1989. Online publications.

[8] Samet, H.: The quadtree and related hierachial data structures. *ACM Computing Surveys*, 16(2):187–260, June 1984.

The Blackboard Architecture in Knowledge-Based Robotic Systems

Spyros Tzafestas and Elpida Tzafestas

Intelligent Robotics and Control Unit (IRCU), Computer Engineering Division, National Technical University of Athens, Zographou 15773, Athens, Greece

Abstract : Integrated knowledge-based robotic systems that employ multi-sensor feedback information require the effective treatment of a large volume of complex knowledge. To this end, several methods of knowledge organisation and exploitation have been developed. One of them is the so-called **blackboard architecture**, which belongs to the class of distributed problem solving architectures, and employs more than one problem solving agents. The three basic elements of the blackboard architecture are : the blackboard, the knowledge sources and the control mechanism. The blackboard plays the role of a central working memory within the system and the controller stands for the conflict resolution mechanism. The knowledge about the system (problem) is split in a number of comparatively small knowledge bases called knowledge sources and controlled through the blackboard control (scheduling) mechanism which is usually a meta-knowledge control mechanism. This mechanism is not committed to any reasoning technique (e.g. forward or backward inference chaining), but has a rather opportunistic nature (i.e., the control action is determined by the optimal decision taken at each cycle of operation). The purpose of this paper is to provide an account of the main concepts and issues of the blackboard architecture and their use in designing integrated knowledge-based intelligent robotic systems. Some particular applications of this approach will also be discussed, one of which has been designed and tested by the authors. The paper is complemented by general design issues of intelligent robotic and flexible manufacturing systems.

Keywords : Blackboard architecture, robotic systems, flexible manufacturing systems, knowledge-based systems, intelligent control.

1. Introduction

Artificial Intelligence (AI) may be defined as the science of enabling computers to learn, reason and make judgements [23]. According to Elaine Rich [35], AI is the study of how to make

NATO ASI Series, Vol. F 71
Expert Systems and Robotics
Edited by T. Jordanides and B. Torby

computers do things at which, at the moment, people are better. Computer science and, in particular AI, does not have a long history, although the idea of creating an artificial intelligence is very old. Why is AI such a hot branch nowadays ? One might argue that AI has now started to build systems that are able to display human-level intelligence in certain domains. These include the "knowledge-based expert systems" which are appropriate for specialised tasks and formed the branching point away from the quest for generality that characterised early AI. Furthermore, the so-called "knowledge engineering" (i.e., knowledge extraction from human experts) slowly became refined, reducing development time for a first prototype. Another reason for AI's current domination might be traced to AI's undertaking practical issues.

The necessity for applying AI to engineering and control problems (in robotics, manufacturing systems, etc.) originates from the growing complexity of modern engineering and control systems, as well as from the traditional expense, time constraints and limited availability of human expertise. AI technology offers the tools that enable us to [6,10,13] :

(i) capture and retain expertise that was gained over many years of engineering,

(ii) amplify expertise that is needed to successfully deploy new methods and applications, and

(iii) design systems that reason intelligently about necessary actions to take in real time, thus freeing operational staff.

Intelligent robotic systems (IRS) are a specific type of intelligent machines that obey the principle of "**Increasing Intelligence with Decreasing Precision**" developed by Saridis [38]. They have decision-making capabilities and use multi-sensory information about their internal state and workspace environment to generate and execute plans for carrying out complex tasks. They can also monitor the execution of their plans, learn fron past experience, improve their behavior and communicate with a human operator in natural or almost natural language [21,39,57]. The key feature of an intelligent robot (which is not possessed by a non-intelligent one) is that it can perform a repertory of different tasks under conditions that may not be known a priori.

The principal components of any robotic system are : **effectors** (arms, hands, legs), **sensors** (vision, touch/force, range, acoustic), **computers** (local controllers, supervisors, coordinators) and **auxiliary equipment** (end-arm tools, fixtures, pallets, conveyors etc.). Through appropriate combinations of these components an intelligent robot can perform tasks that need flexibility and artificial intelligence (i.e., perception, interpretation, and planning capabilities).

The structure of any IRS is consistent with the general architecture of an intelligent controller, which is best illustrated by figure 1 [25]. The computer of an IRS communicates with the surrounding environment and performs the tasks of cognition, perception and planning/control. It also sends information to, and receives information from the plant under control which involves the effectors and sensors. According to Saridis [37], control of a process implies driving the process to effectively attain a prespecified goal, i.e., it only depends on the process under consideration and the goal this process must be driven forward. An intelligent controller can replace the human mind in making decisions, planning control strategies and learning new functions by training and performing other intelligent functions whenever the environment does not allow or does not justify the presence of a human operator.

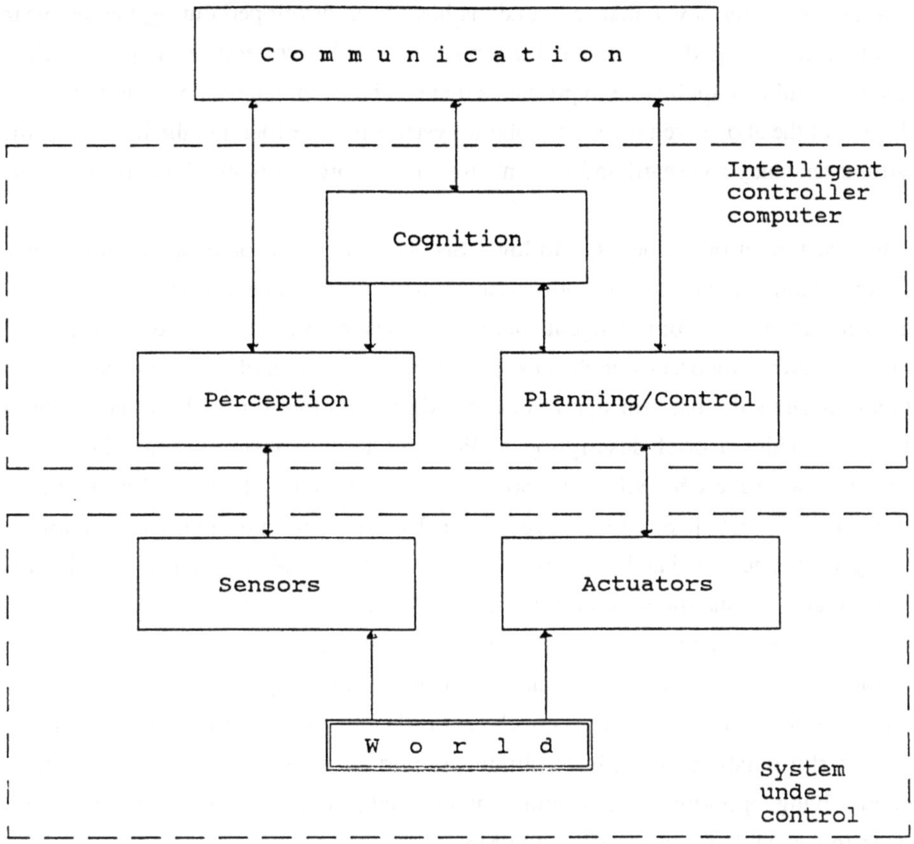

Fig.1 : General structure of an intelligent controller

Actually, intelligent control system (or machine) theory is the intersection of systems and control theory, artificial intelligence and operational research. Systems and control theory offers dynamics optimisation and feedback techniques, and artificial intelligence and computer science offer information processing, formal languages, learning, heuristics, inference techniques, knowledge representation, etc. Finally, operational research contributes with management, coordination, planning and scheduling techniques [38].

The blackboard model of problem solving is a distributed one and may integrate several inferencing techniques and several representation schemes, not only complementary, but also as different approaches of the same data. Furthermore, in most cases the blackboard data are organised hierarchically - the hierarchy may contain meta-knowledge levels [27]. Besides the original model, many new enhanced versions have been developed during the last years, the most important being the so-called blackboard control architecture [12], an architecture especially useful and applicable to problems that can be formulated as planning ones. Limited application of the above architecture in robotic systems may be found in the literature, still there is a strong tendency to regard and treat manufacturing problems in a blackboard framework [29].

The structure of this paper is as follows. Section 2 presents some background material on intelligent manufacturing and robotic systems, and in particular a brief account of Saridis' hierarchical architecture for intelligent control [38]. Section 3 describes in detail the blackboard architecture and its importance in the design and implementation of intelligent robotic systems. Section 4 presents the basic design issues of FMS in relation to the blackboard architecture. Section 5 provides a brief description of Purdue's 1-robot vision-based robotic assembly system and Karlsruhe's KAMRO autonomous robotic system which employ the blackboard architecture [22,36,58]. Section 6 presents a 2-robot (one stationary, one mobile robot) assembly system designed and implemented by the authors at NTUA. The system is flexible in the sense that it can handle several products in several volumes simultaneously using a rather simple blackboard structure, a sophisticated task planner and a combination of sensors. Finally, in section 7 some concluding remarks are provided about the capabilities and advantages of the blackboard architecture. Our general conclusion is that the blackboard architecture is one of the most powerful existing methods combining important features as parallel and distributed problem solving, opportunistic reasoning, incremental problem solving, hierarchical structure, and therefore suitable for use in IRS and FMS.

2. General Issues of Intelligent Robotic and Manufacturing Control Systems

Manufacturing involves many activities which can be monitored and controlled at several levels of abstraction. A manufacturing plant (shop-floor) involves a variety of work centers each one of which is composed of many manufacturing cells, and a manufacturing cell consists of a group of individual machines, robots and tools. Very broadly, activity planning in a manufacturing plant is the problem of selecting activities and allocating and scheduling resources at each level of abstraction so as to meet the desired production, quality and economic goals. By nature, the manufacturing environment is dynamic and so predictive planning does not lead to success. The manufacturing system has to be able to adapt to quick changes of its environment [31,34].

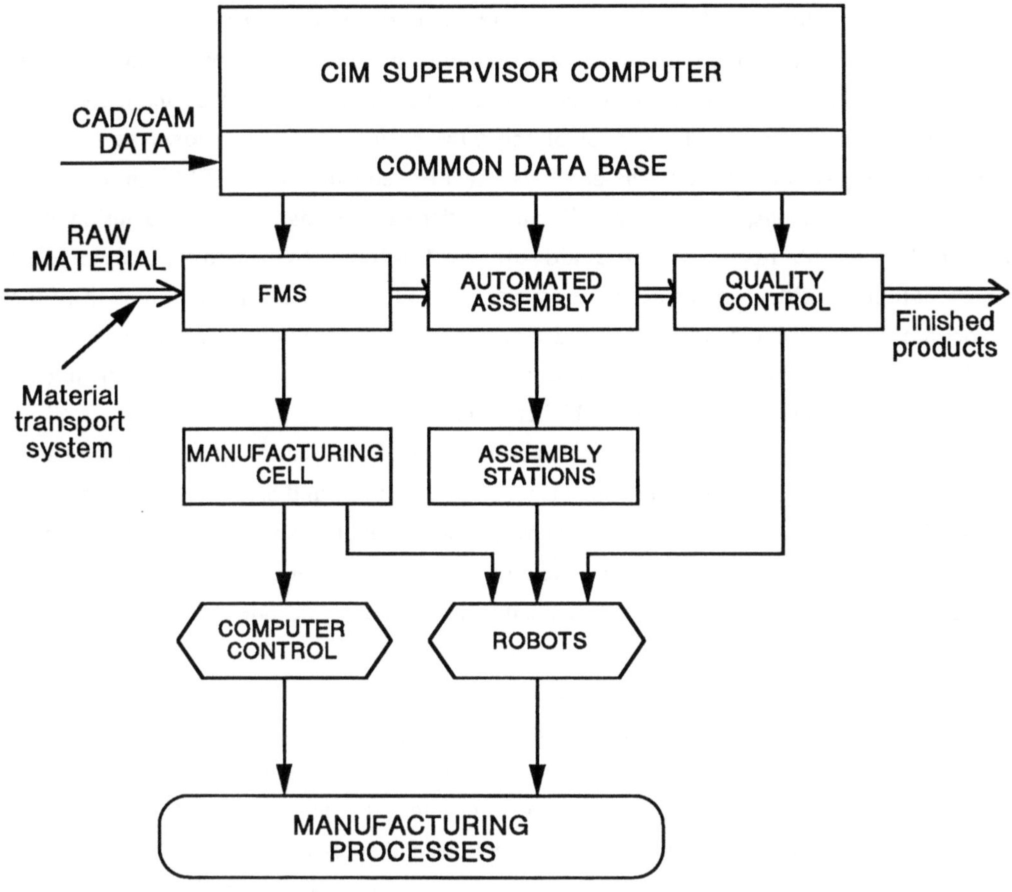

Fig. 2 : Hierarchical structure of computer integrated manufacturing shop floors.

Modern shop-floors have a hierarchical structure with at least three hierarchical levels (Fig. 2) which can have intelligence properties. At the highest level there is a computer which supervises the various manufacturing functions (activities or operations). In a system of this type, which is called computer integrated manufacturing (CIM) system, there is a coordinated cooperation of computers in all phases of the enterprise, viz. the design of a product, the planning of its manufacturing, the automatic production of parts, the automatic assembly, the automatic inspection, the control of flow of materials and parts through the plant, and so on. All these phases must be accomodated into a computer network, supervised by the CIM central computer which controls the various interrelated tasks on the basis of the overall management policy.

At the lowest level of this scheme there are stand-alone computer control systems of manufacturing processes and industrial robots (manufacturing cells). The operation of several manufacturing cells is coordinated by the central computer through a material handling system. This constitutes the intermediate level of the CIM system which is known as **flexible manufacturing system (FMS)**. The products (final or semifinal) of the FMS are tested by appropriate automatic inspection stations. The integration of the design and manufacturing tasks of a FMS can be made with the aid of the **CAD/CAM (computer-aided design/computer-aided manufacturing)** system. This leads to substantial reduction of the factory's production cycle. It is remarked that robots and computerised numerical control (CNC) machines can only replace the human power and skill. CIM can also replace the human intelligence and achieve an incomparably higher efficiency.

Manufacturing is one of the most attractive areas of application of knowledge-based techniques. However a considerable effort is required to capture and organise the accumulated knowledge of manufacturing engineers. There is a vast amount of knowledge here, extending over many processes, diverse situations, and an infinite array of parts and products. Particular attention should be given for representing this knowledge, and adequately representing explicitly the characteristics of machines and processes [4,7,18,19,24,32,41,43,45,59].

According to Nevins, Whitney and Endsall [26] two radically different approaches can be followed to provide intelligence to manufacturing systems, namely :

- By duplicating (imitating) the human's actions (e.g. in spray painting a robot is programmed to move a spray gun in exactly the same patterns as an expert painter).

- By studying the task's goal or expected functions and applying a convenient strategy that may bear little or no resemblence to the human's methods (e.g. automatic dishwashing or automatic sewing).

Both approaches have serious disadvantages. Today robotic systems do not have sufficient intelligence and sensory capabilities to address very complex manufacturing tasks. On the other hand, manufacturing (e.g. inspection) tasks can require great skill acquired through many years of experience and in many cases through formal engineering training. Usually the underlying physics is very complex and very poorly understood. The classical pure expert system approach to these problems would embody the skill aspects in heuristics which implicitly, but not explicitly, capture physics. The combined approach, called **engineering-based expert systems (EBES)** approach, explicitly models the physics and provides the proper solution. The engineering-based experts systems capture algorithmically the capabilities of human experts and trained process operators in an efficient and cost effective way. They use a blending of process modelling and identification, pattern recognition, hypothesis testing and control systems techniques, as well as conventional AI tools and expert systems rules and heuristics. The engineering-based expert systems approach to automated manufacturing is at the very beginning and provides a challenging area for further work.

A particular structure of intelligent control which can be used to design the control and operations of the various levels of a computer-aided manufacturing system is Saridis' "hierarchical architecture" which obeys the principle of "increasing intelligence with decreasing precision" mentioned in the introduction. This architecture involves three main levels, namely (from top to down) :

- Organisation level,
- Coordination level,
- Execution level,

with feedback connection from the lower to the middle level, and from the middle to the higher level. The organisational level, which is actually a decision-making level, carries out several sequential operations, and besides its control unit and input classifier, usually involves a knowledge base and an inference engine [40,58]. The coordination level consists of several coordinators (implemented by hardware and/or software) which provide to the execution level all necessary details for the successful execution of the selected best task sequence (plan), and send feedback information to the organisation level about the success or failure of this plan. In case of failure more attempts are made with alternative task sequences. In case of success the knowledge base is updated. The execution level consists of a number of devices each associated with a specific coordinator. The commands issued by the coordination level are executed at the execution level. In an intelligent robotic system the execution devices are the elements of the vision, arm, motion and sensory systems [38].

Organisation level : The robotic organiser preprocesses the user's command (i.e., it classifies this command into one of several existing types) and generates a suitable task plan of this command using forward reasoning or probability/possibility procedures. If the command is "new" then a new category is produced. The inference engine processes the information of the knowledge base through several steps. For example, in Valavanis' and Yuan's system [58] these steps are : initialisation, formulation, execution and finalisation. Once the task is known (retrieved or generated), the task plan is sent to the coordination level for execution and evaluation. If the desired task is performed successfully, the program returns to its main menu and updates the knowledge base. Otherwise the organiser tries to reformulate the next plan (by backtracking or by selecting the next probable plan).

Coordination level : The task is sent from the organiser to the coordination level which consists of several coordinators (each implemented, for example, by dedicated microprocessors). All necessary details must be provided so that the chosen plan is successfully executed. To execute plans simultaneously, the parallel processing task scheduler (PPTS)

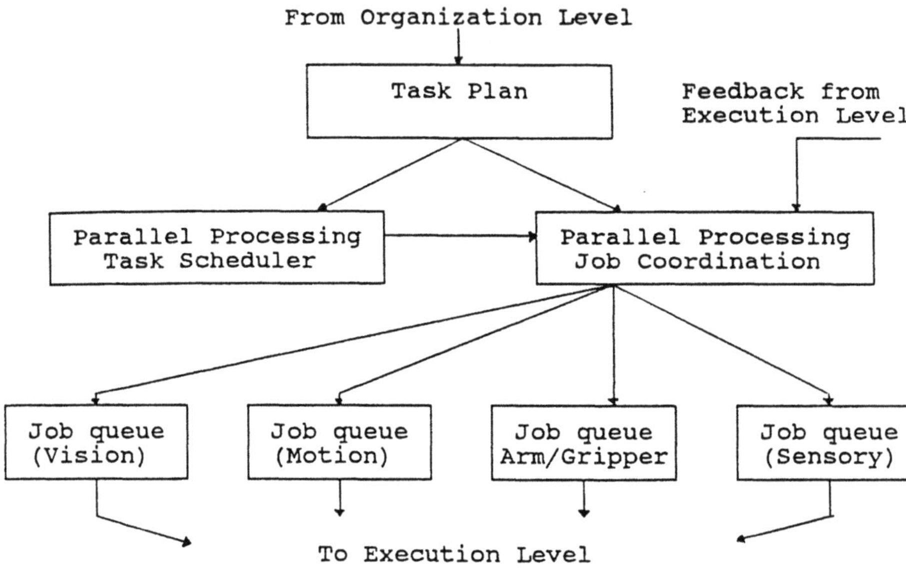

Fig. 3 : A particular implementation of the coordination level

assigns stamps to each process and synchronises these operations with the parallel processing job coordinator (PPJC) (see fig. 3). The PPTS determines whether two tasks in sequence can be executed in parallel, by searching through the parallel execution table. If the pair is found in the table, then these tasks could be executed in parallel and so they are assigned with the same time stamp [58]. An intelligent robotic system should include coordinators for the lower arm control (motion system coordinator), the upper arm control (gripper/end effector coordinator), the vision system and the sensory system. The sensory system must involve internal sensing for monitoring the operating state of the robotic system, and external sensing for providing information about the workspace environment to the processor for task planning and execution (proximity, force/torque, tactile, motion sensors).

Execution level : Actually, one cannot find a simple architectural model that embraces all possible situations and operations of the execution level. Each case is treated according to its special needs and specifications. However, to preserve the hierarchical structure, each device (agent) of this level is coordinated by a dedicated coordinator as indicated before. Although not precisely stated, the majority of intelligent robotic systems developed so far and documented in the open literature can be cast into the hierarchical structure described in the previous section (of course with differences in the actual implementation of each level).

3. The Blackboard Architecture

3.1 The Blackboard Model

The blackboard approach to computer science and engineering problems is by no way a new concept, but rather a refinement of the old and well-established notion of the common memory, and even more an attempt to integrate multiple and varying methods and techniques to a common field of action. The approach implies the partition of the knowledge about a problem into a number of agents, each one of which is used to solve a sub-problem of the initial one ; this behaviour permits the parallel and/or distributed problem solving, which is nowadays an area of enormous research interest, especially in real-time problems such as the industrial ones. A first observation may be that blackboard systems and computer systems show similar properties and needs - so that techniques used and tested in early computer systems might be useful to the design and development or modern blackboard systems. Secondly, the distributed character of blackboard systems permits their modular and hierarchical development, while, finally, the incorporation of techniques borrowed by different disciplines is straightforward. The above features make the blackboard approach a simple and powerful problem solving method that can be applied to complex, interdisciplinary, industrial problems.

Like the majority of the well-known problem solving models, the blackboard model is not a product of sudden inspiration, that proved useful later on, but derived from a common study of many examples and applications. Among the early blackboard systems, two have proved of great importance in the development of the primary model, namely the systems **Hearsay II** (a speech recognition system) and **HASP** (a system in the field of interpretation of continuous sonar signals).

The fundamental blackboard model (Fig. 4) consists of three parts :

- **Knowledge sources (KS)** : The knowledge about the problem (contains any empirical, statistical or other sort of data, including processing routines and programs) is divided into independent but complementary subsets, the "knowledge sources".

- **Blackboard data structure** : It is a central data structure, where all the knowledge sources have exclusive access for retrieval or storage/modification of information, and those "changes" in the state of the blackboard drive the system opportunistically and incrementally to a solution (not necessarily unique) of the problem at hand. It is interesting to observe that the communication and the interaction of the knowledge sources is possible only implicitly, through the blackboard. The stored data may contain partial solutions, constraints, properties and description of the environment of the problem, target limitations, etc. Usually, they are organised in a hierarchical/heterarchical structure, but they might instead belong to semantic nets or other representation structures.

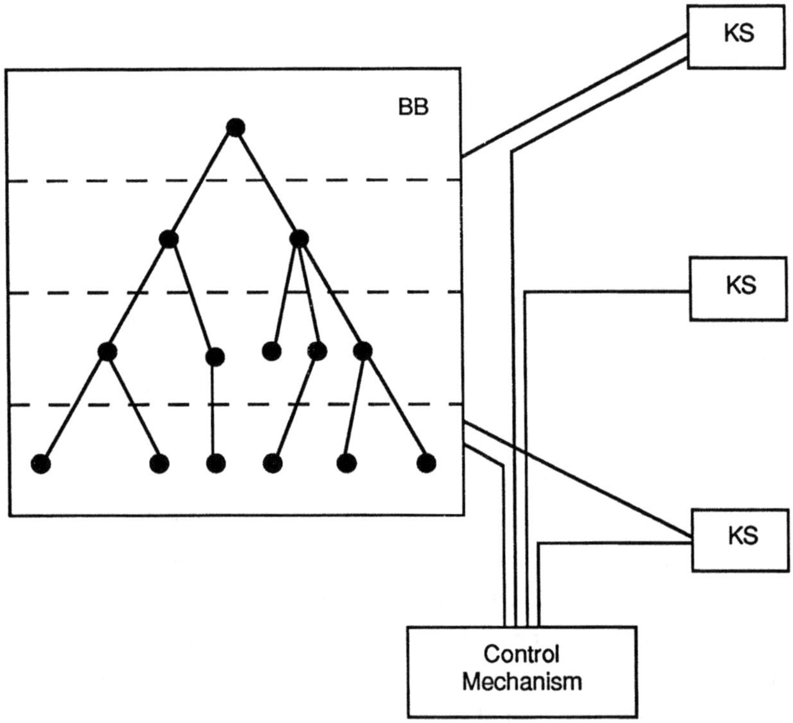

Fig. 4 : Fundamental blackboard architecture

- **Control mechanism** : There is no explicit control mechanism, but rather the KSs respond dynamically and opportunistically to the changes of the blackboard and are self-activating, in the sense that they know when and how to contribute to the solution of the problem. Actually, there need be a central monitor (or scheduler) of the execution process, so that some control data need be kept, upon which the monitor will reason and schedule the next action (execution of a KS), among the feasible ones. The focus of attention of the monitor might be a knowledge source or an object of the blackboard or a combination of the two. Based on the focus of attention (that may be static or change dynamically during the problem solving process), the monitor chooses a knowledge source to execute next, and the formula is repeated until a solution is found (or a deadlock traced). It is assumed that after each cycle of execution, all KSs examine their condition part, to see if they fire, i.e., if they are ready to execute. Since the monitor operates on the data of the higher hierarchical levels, it may be viewed as a meta-knowledge control mechanism.

Many blackboard systems have been developed based on the above model, each one in addition introducing its own features and properties, that enable them to treat properly various classes of problems, such as diagnostics [22], class and pattern recognition [27], operator intent inferencing [36], multiple task planning [12], intelligent control [30], etc. Even complete problem solving shells have been developed [12,14], and proper grammar models seeked [5]. Finally, the suitability of the model for various classes of problems has been examined and comparisons with other models attempted [3,11].

3.2 The Blackboard Control Architecture

A variation of the architecture, called the blackboard control architecture, is used in domains where the control of execution is the main concern, e.g. in manufacturing problems like the ones presented in later paragraphs. The aim of this architecture is to provide a field of effective solution of control problems - a control problem is a problem of what and how to do at each step of a process, so that it may constitute an abstract refinement of many existing and critical engineering problems. The control problem is regarded by the model as a **real-time planning problem** [12].

The architecture defines two independent blackboards, the **domain** and the **control blackboard**, and two corresponding classes of knowledge sources. The domain blackboard contains information about the problem at hand (including partial solutions, constraints, resources, etc.), while the control blackboard contains information about the strategy of solving the problem - a set of strategies (heuristics and algorithms) is used, both general purpose (e.g. graph search methods) and problem-dependent. Each triggering of a knowledge source (a KS is triggered when its conditions hold) produces a unique **knowledge source activation record (KSAR)**, among which the control mechanism chooses the one to be executed at each step of the process. The basic control loop, then, is : 1) Update KSARS, 2) Choose a KSAR, and 3) Execute the chosen KSAR. In that way, the control and domain knowledge sources are used in a simple and uniform manner, enabling also the extension of the system by introducing additional domain or control knowledge sources. It is also to observe here that the system has the power of interpreting and modifying its own knowledge representation and behaviour and applying this knowledge selectively and dynamically to changing situations of the problem-solving process.

Hence, the architecture has the following properties : 1) Decisions should be explicit, 2) The control mechanism should reconcile desirable and feasible actions, 3) It should adopt variable grain-size control heuristics and different degrees of detail, 4) It should adopt heuristics that may focus on whatever action attributes are regarded as critical at each point of execution, 5) It should adopt, retain and discard individual heuristics in response to dynamic problem-solving situations, 6) It should integrate heuristics of varying importance, 7) It should dynamically plan strategic sequences of actions, and 8) It should reason about the relative priorities of domain and control actions. [12]

4. Design Aspects of Flexible Manufacturing Systems

In general, the design of a manufacturing system consists of the following steps :

1) Qualitative specification of the production (definition of parts, products, tools, resources, etc.),

2) Partitioning of the knowledge to hierarchical levels, each one communicating if possible only with its adjacent ones - this partitioning is a non-trivial task and should be strictly deterministic,

3) Design of the control system of the production - the control system is a knowledge-based system that supervises and drives the process of manufacturing,

4) Hardware and resources selection - this step should be done in cooperation with the previous one, and

5) Operational study of the overall system.

The above hierarchical process of design offers the advantages of modularity, adaptivity to changing specifications, portability to different environments and possibility of correctness checks throughout the process.

A fully automated factory can thus be viewed as a set of production cells, to each one of which is assigned a specific job, e.g. production of a number of identical products - the cells are interdependent in some way, for example, the product of the first cell is treated as a part in the second cell.

A flexible manufacturing system differs from a non-flexible one, in that it is equipped with one or more reprogrammable machines/devices that can equally well treat materials and perform operations of comparison or join. Usually, these devices are nothing but robots explicitly programmed to perform one or more specific tasks.

Flexible manufacturing systems treat complex and ill-structured real-time problems that involve both numerical and symbolic data processing, so they need [1,16,30,52]:

- be easy to program,
- allow frequent program generation in order to adapt to changing production requirements,
- be easy to use and understand by people of varying abilities, education and needs,
- be easy to trim and tune,
- have high availability degree,
- employ a central computer system that supervises and controls the whole process, and through which all necessary communication between components of the system is performed,
- allow parallelism and/or distribution of single processes,
- be compatible with CAD or other systems - i.e., permit computer integration of the manufacturing process,
- allow control of the production in a higher level than that of implementation,
- observe quality and performance rules and requirements,
- possibly contain learning capabilities, and
- use only a small number of special-purpose devices and a larger number of reprogrammable devices.

Flexibility means adaptivity to changes on the geometry of the product and/or its parts, or even to changes on the type of the manufactured products. A measure of the flexibility of a system is its ability to alter its goals in the case of temporary malfunction of some component - this property is nothing but an ability to respond quickly and efficiently to situations that arise dynamically. Apparently, all flexible manufacturing systems should perform some kind of planning to avoid the computational explosion that may result from the need of exhaustive search of large amounts of data. Another reason for planning is that, although the production need be a priori programmed, some shop-floor level decisions cannot be taken in advance. Furthermore, different kinds of planning are requested in different steps of the manufacturing process. The most natural planning methods that are used are the hierarchical ones, not only because they are the most modern and the most efficient, but because they integrate the notions of production sub-systems and interdisciplinary approach to production. In this sense, the input to the first level of the hierarchy will be the production specification, while the output from the last level will be the product itself. The use of hierarchical planning methods is very natural in the case of nearly decomposable problems, and the problem of production control and coordination in a manufacturing system can be regarded as such.

Every flexible manufacturing system should employ and integrate both on-line and off-line planning components, in order to reconcile appropriately and effectively the corresponding flexibility and time (speed) requirements.

The need of manufacturing systems for extended automatisation and adaptivity to dynamic environments of production has imposed in the past the introduction of robots to the process of manufacturing. Naturally, however, the robots have no perfect mechanical adaptivity, whereas, on the other hand, through appropriate programming they can constitute powerful tools for manufacturing. Yet, robot programming is a very complex and time-consuming process, so that universities and enterprises keep developing new robot programming languages and shells, which tend to overlap existing difficulties, but still showing new drawbacks and making new aspects of systems appear.

The evolution of the manufacturing systems shows similarities and is driven by the evolution of the robot programming languages : far from the first purely mechanical manufacturing systems (corresponding to the joint or robot level programming languages), the specifications of which needed to be prescribed in advance and in detail, the modern trend is to design and develop interfaces between higher level software systems and robots, thus leading to object and task level systems (corresponding to object and task level robot programming languages). The input to those latter systems is a description of "what to do" rather than "how to do it" and is is obvious that such systems should employ **Artificial Intelligence** methods - research towards this direction needs to face problems like world modelling, automatic grasping, collision-free path planning and sensory programming.

No flexible manufacturing system can be regarded out of the environment of a larger computer system, including additional components, such as CAD systems, code generation components, simulation programs, optimisation filters, etc. This doesn't mean that there cannot exist a stand-alone FMS, but its power is only demonstrated through its integration to a larger system. In particular, the use of simulation methods, in the course of designing a FMS, may be critical due to the following reasons [2,52] : i) No sufficient analytic design methods exist that take into consideration all relevant interactions between sub-systems, ii) Large investments demand the use of some means of response control and evaluation as early as possible in the design process, iii) Only simulation can effectively handle larger and more complex systems, and iv) Dynamic response can only be predicted and estimated using simulation techniques. Yet, simulation constitutes no general remedy, since it involves high modelling and experiment cost, insufficiency of representation and result processing methods and, finally, it introduces inaccuracies in the resulting functions. To balance the high response time of pure expert systems that treat well-structured problems (usually, a manufacturing problem can be regarded as such, to some degree of abstraction), optimisation techniques can also be used [20], in order to reduce the knowledge base size, simplify the inference engine structure and integrate both qualitative and quantitative factors. However, the applicability scope of such techniques remains limited and the complexity of the corresponding problem rises forbiddingly.

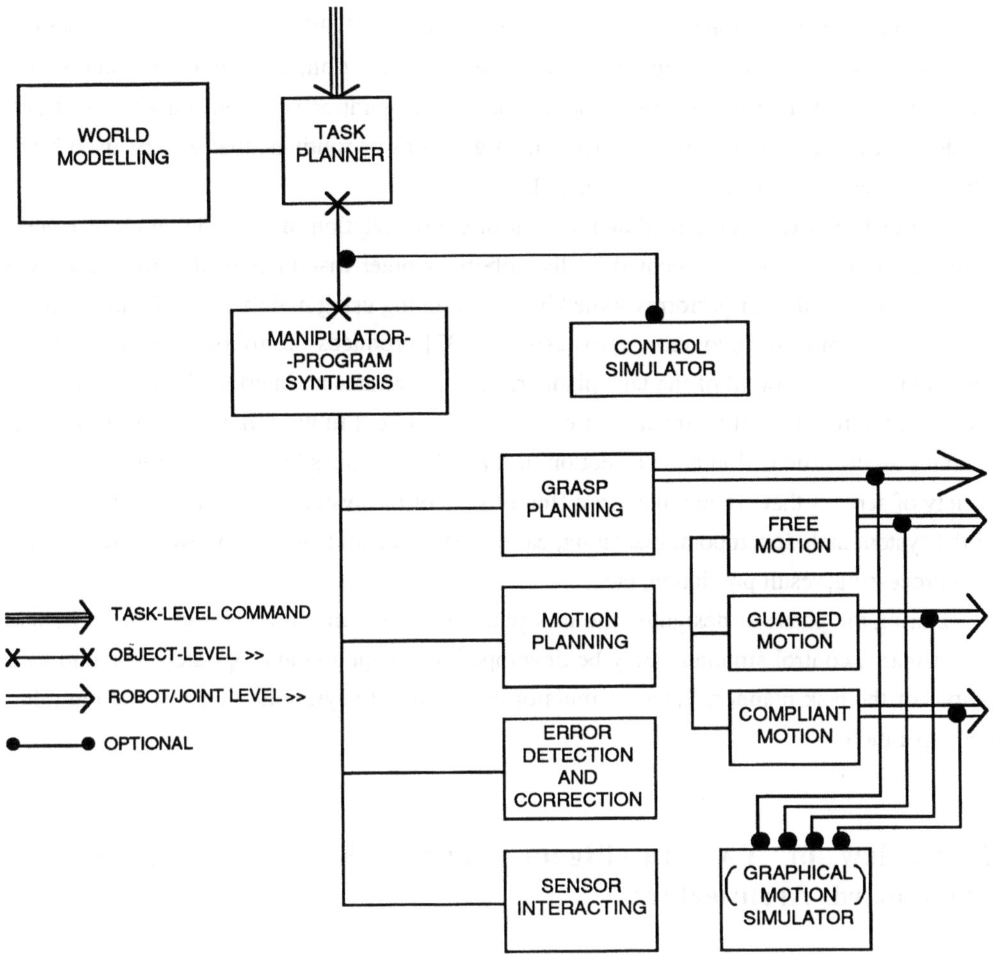

Fig. 5 : General architecture of a flexible manufacturing system

The general architecture of a FMS is presented in Fig. 5 - it is a knowledge-based system, that has in addition the right to interfere to lower level, e.g. the control of the gripper. It consists of several planning steps, still the methods used are not purely AI methods, as one might expect due to the use of the term "planning". Given the demand for cooperation and feedback between the different steps, a central computer system is needed to store the knowledge about the components of the system and serve as a means through which all communication and coordination will take place.

The **world model** can be created using several methods (e.g. using a CAD system or read data from a file) and must contain knowledge about the production environment and domain-specific knowledge : geometric and physical description of all objects, kinematic description of all robot links, information about the volume of production, any time or other sort of constraints, description of the absolute and relative configuration of all objects and robots, etc. In addition, it should contain any processing algorithms or routines, that will be used by the other modules of the system, e.g. the task planner.

The **task planner** is the module that, out of the description of an initial goal, produces a sequence of actions to be executed by the robots or other machines of the production cell. Usually, the task planner performs some kind of reasoning using mainly symbolic knowledge.

The **manipulator program synthesiser** [1,33] produces the robot language code that corresponds to the output of the task planner, so it constitutes an interface between the object level and the robot level commands. Besides the grasp and motion planning sub-modules, it contains an error detection and correction unit and has the possibility to communicate with a variety of sensors that are mounted to different sites of the system. The reasoning possibilities of the system include temporal reasoning, causal reasoning and abilities of event interpretation, data processing, result prediction, etc.

During the phase of designing the task planner, no real machines or robots need be used, but instead a control simulator may be developed to accept and appropriately respond to the output of the task planner. Such a simulator can be purely symbolic or allow also graphical demonstration.

5. Review of Two Intelligent Robotic Systems Employing the Blackboard Architecture

To obtain a better idea of how the blackboard can be used in intelligent robotic control, two particular systems will be briefly described here which were designed at Purdue University and Karlsruhe University respectively [15,42].

5.1 The Purdue 1-Robot Vision-Based System

This is an intelligent robotic system integrating vision feedback into a control system and trajectory planner for automated assembly [42]. It has been used in several applications, e.g. in performing an automated assembly operation on a carburator. The system design is modular. Each module has maximum flexibility and can be replaced by a complementary module without affecting the system as a whole. For example a program implementing a PID controller can be easily replaced by one implementing variable structure control without requiring any change to the other module. This is made by using a "**Star Network**" through a blackboard and multiprocessors as shown in figure 6. The communication code lies on the supervisory control module (motivator) in greater part, and partially distributed over the other modules. The code consists of a sequence of numbers identifying the transmitting module, the intended receiver module and the message type. The Star Network is centered around a blackboard, which stores information (such as the actual position of a manipulator, the status of a module, etc.) which can be accessed by any module at any time and is updated at regular intervals by an assigned module.

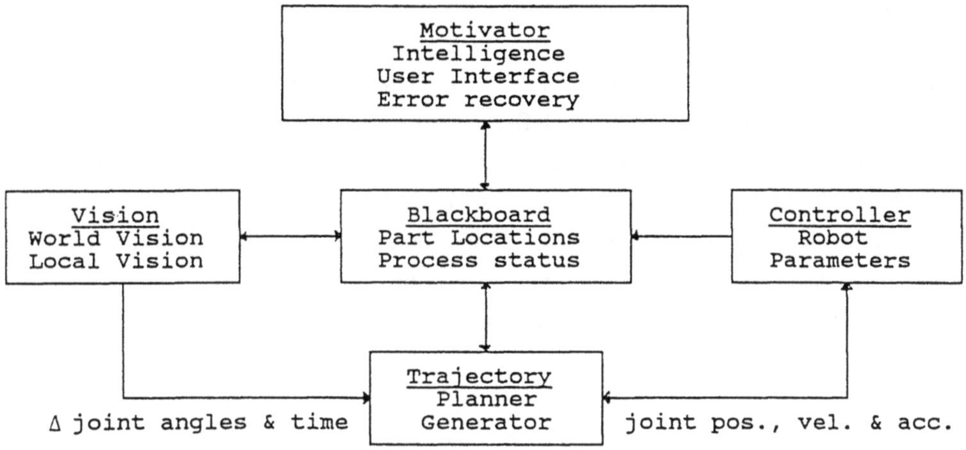

Fig. 6 : The Purdue system uses a star network based on a blackboard

The control structure involves a supervisor (organiser + coordinator) which is called **motivator** and a low level local controller. The local controller receives a desired position from the trajectory planner module and calculates the actuator inputs (here the current required to move the robot to the desired position). This control scheme allows to compensate for uncertainties by using the minimum and maximum values of the parameters of the model to add an uncertainty compensation factor to a proportional scheme.

Motivator : The high level control (coordination of internal activities and sensory information with what is happening externally) is performed by the motivator which is partitioned into three subprograms : the **user interface,** the **task manager** and the **error recovery** routines.

The task manager is implemented using STRIPS (a Prolog based tool). It involves four main procedures : Find (object), Goto (object/position), Get (object) and place-on (object1, object2). An unstack procedure can also be added for disassembly operations. There are three types of states, namely position, gripper and object, each one involving two or more states. For full interface of the Prolog task manager program with the C routines required to set-up and maintain the communication between the different modules, the general idea of STRIPS was translated into a C program and an alternative version of the system was constructed. However, this is not an optimal solution to the problem, since Prolog possesses many nice symbolic processing features that C doesn't.

Vision Module : The vision system has three modes of operation, namely supervisor, monitor and visual servoing. The vision hardware consists of a SUN 3/16 Workstation, Imaging Technology ITEX-151 image processing hardware and two Pulnix TM-540 CCD video cameras. The ITEX-151 has four 8-bit 512-by-480 image frames which are directly memory mapped into the SUN. Convolving and ALU operations can also be performed by the ITEX-151 system at video rates on user defined areas of interest in the image. One camera is mounted above the robot workspace and acts as a supervisor and a monitor of the workspace. A second camera is mounted on the last link of the robot and acts as a visual servoing camera. When the assembly begins, the supervisory vision mode identifies the carburator and gasket among other objects and transmits positions, orientations and time of image acquisition to the blackboard to be used by the trajectory planner. Once the object is identified and located, the stationary camera starts the monitoring of moving objects (monitoring mode). When the trajectory planner has positioned the parts close to the mating position, the end-effector camera starts operating and provides visual feedback to the fine motion control loop in real time. The end-effector camera software sends a small change in the last three joint angles directly to the trajectory generator, which in turn transmits the corrected change in joint angles to the robot controller.

Trajectory planner : Most industrial robots use a path that has been predetermined either by teaching points or by off-line programming. However, in uncertain work environment real time trajectory planning is necessary. The processed world data provided by the external sensors (vision in this case) provide a flexible object's position and orientation to the planner. The planner computes the best path to the object, obtains the gripping information from the data base, and causes a trajectory to be generated to pick up the object. The Purdue system uses a planner routine which is based on a set of "windows" to plan trajectories that can accomodate updated position data from the vision system. In this way accurate positioning of a flexible part with respect to a moving system is possible.

The Hardware Configuration : Three SUN workstations are used for the motivator, the trajectory and the vision modules. The motivator coordinates the operation of the three processes, which are carried out concurrently and ensures a smooth flow. The controller translates desired positions into motor currents for the drives of the PUMA 560 robot.

5.2 The Karlsruhe Autonomous Mobile Robot (KAMRO)

This is a 2-arm mobile robot consisting of two PUMA 260 robots in a hanging configuration, and arranged into three main subsystems : (i) the vehicle, (ii) the robots, and (iii) the sensory system [15]. The vehicle has four wheels of the MECANUM type allowing x-y motion and rotation on the spot. The navigation within the working space of the assembly cell is accomplished with the aid of ultrasonic range finders and a vision system. The camera is placed between the arms of the robot. The sensor system contains a digital image processing system for supervising the assembly task. The mobile robot KAMRO has two capabilities : (i) to transport parts between the work tables and storage systems, and (ii) to assemble parts making a product while the vehicle is docked stationary at a worktable.

The key feature of KAMRO is its ability to dynamically handle unexpected situations during real time operation. This is done with the combination of sensor guided actions and autonomous supervisory functions. The task sequence (action plan) is derived from a computer-aided planning system [9] by decomposing the task-oriented instructions into sequences of elementary instructions. Due to its complexity, planning is performed off-line, and so a number of unpredicted (erroneous) situations may appear during the on-line execution. These situations may occur for example because of modelling errors, or positioning errors of the mechanical motion devices, or due to the presence of unexpected objects within the workspace, etc. To cope with such erroneous situations one can improve the plan description and provide recovery routines for them. This technique however is not practical since one cannot determine beforehand all possible error situations.

The best approach is to use appropriate sensors and to adapt the execution of the plan to the actual state of the world through sensor controlled plans and sensor guided actions [15]. Each active component (agent) of an intelligent robotic cell (robot, NC machine, etc.) is provided a set of elementary operations (EO) (corresponding to the knowledge sources of a blackboard system as mentioned before), which cannot be decomposed any further and are executed by elementary operation modules (EOM). Each EOM is dedicated to a particular EO and comprises the expert knowledge required to execute this EO. The EO specifies the desired goal, and the EOM specifies how this goal should be actually achieved (through complex sensor guided tasks and suitable reaction to unpredicted events within a certain repertory). The corresponding control system has the hierarchical blackboard structure shown in fig. 7. The local planner handles the exceptions (errors) by modifying the initial plan. This is done on the basis of the information provided by the sensor system. The task monitor supervises and coordinates the activation of the entire system. The exception monitor is activated whenever an EOM cannot correctly execute an EO. The world model is actually a database (blackboard) in which all information about the agents and the assembly cell is stored, and through which the agents are communicating.

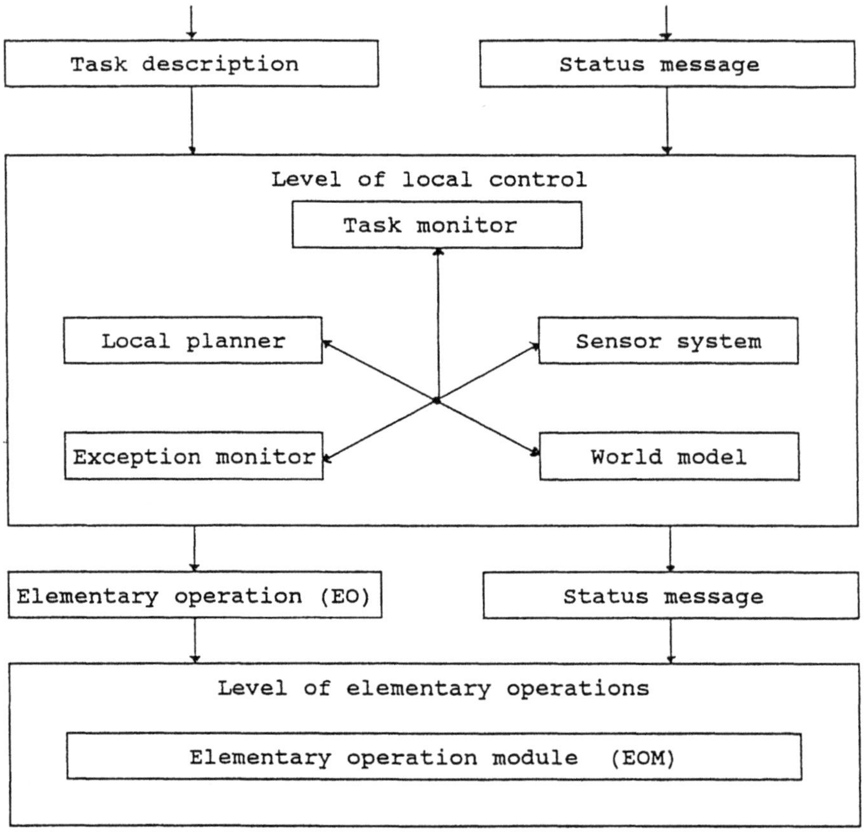

Fig. 7 : Hierarchical blackboard structure of the agent control system

The computer system used to realise the agent control system described above has the same hierarchical structure (see fig. 7). The elementary operations are performed by the computer which is on the mobile robot. Actually, a distributed multi-processing system is employed which is based on the MC 680X0 single board computers and connected through the VME bus. The local control is implemented using a SUN workstation (under UNIX) and the planning tools as well as the CAD world modelling system run on a VAX (under VMS). These computers are connected via a ETHERNET LAN with the TCP/IP communication protocols [9]. The high-level algorithms are coded in C and Prolog. The process and processor management is performed by the special type operating system HEROS (Hierarchically Extendable Real-time Operating System) which supports the process intercommunication by a message-passing mechanism.

6. Application of the Blackboard Architecture to a 2-Robot Robotic Assembly System

6.1 Introduction

The assembly problem may be defined as the problem of the incremental combination and join of parts to form a product in the optimum way, optimality having the sense that an optimum selection of one among the feasible next actions should be performed at each step of the process. This assembly problem can be easily formulated as a planning problem, and is usually treated as such - this approach enables the use of wellknown methods from other disciplines, e.g. computer science, for the development of assembly systems. Two issues should be stated here : the predictability of the environment of the manufacture and the need for combined on-line and off-line planning. To ensure correct and rational use of simulation methods, one must specify the environment of the manufacture as fully predictable - this assumption is not very far from reality. In fact, extreme and improbable events can be excluded from the specifications of the system and their treatment left to the operator (supervisor) - supervisory computer control is an approach that ensures both efficiency and correctness. To allow extended flexibility, an on-line planning module should be developed, while the need for satisfying time performance imposes the existence of an off-line planning module - an appropriate compromise between the two is required to achieve the specified goals of efficiency, cost, complexity, etc.

The system presented was developed in a SUN workstation under the operating system Unix 4.2BSD and it was implemented in C-Prolog, which constitutes a very powerful knowledge-based systems development tool. This approach has several benefits : portability of the system to other Unix environments, practically unlimited memory usage possibilities, support of various existing programs (simulation, graphics, etc.), real number processing possibilities, system easy to enhance and to connect to other devices, as well as programs, etc.

In its present version, the system is left open-ended, i.e. the robotic functions (grasp, move, etc.) are not implemented in the level of primitive commands program, and the external events as well as the response/reaction of the system are simulated. It is concise and relies on simple concepts, so that in the future it can be adapted to similar environments (where, for example, a larger number of robots is required and more complex tasks are specified) - the development of supporting programs, e.g. CAD programs, is also possible.

A variety of supporting programs (e.g. CAD programs) can be developed. In fact, the system constitutes the on-line module of a larger system to be developed, while the off-line designed module will consist of a repertory of routines that will be used for every single robot action (e.g. an assembly operation). These routines can be developed by any known methods; by instruction, i.e., "teach by showing, or by more sophisticated methods.

6.2 Knowledge Representation

The system has the ability to handle simultaneously several types of products in several volumes of production - a part type may be used by several product types and more than one time in the same product type.

For the representation of the state space of the system, the **Petri-net** model is used - this model corresponds to a single direction graph and involves the notions of state and transition. The benefit gained from this approach is the natural modelling of an assembly system, where several subassemblies of several products may be combined together in parallel - "merging solution islands" technique-, as well as the possibility to enhance the model to include additional information to the one assumed in the present system. In this way, no base part is required for any product; on the contrary, every transition fires opportunistically, without being committed to any prespecified order of assembly. For each product a different Petri-net is defined, the states corresponding to subassemblies and the transitions corresponding to legal assembly operations between two subassemblies. It is interesting to remark the special form of the Petri-net, each transition having two incoming arcs and only one outcoming. Parts and ready products are treated the same way as intermediate subassemblies, i.e., they are regarded as initial and final subassemblies respectively. Legality of operations has the sense that transitions should not violate any existing constraints (due to geometry, cost, product requirements, etc.), so that among the physically feasible subassemblies, some are not logically feasible - no assembly operations path will lead from the initial parts to them. So, the size of the state space, i.e., the Petri-net is substantially reduced, by-passing any combinatorial effects. In fact, in order to achieve better hierarchisation of the processes and knowledge-partitioning more effective to handle, the application engineers always tend to define products of small number of parts as strongly interconnected to each other as possible, so that the number of the actual existing subassemblies of any product doesn't practically exceed a threshold of 4 to 6, depending on the nature of the product. Many example products have been developed or borrowed from the literature. Such an example product [8] and the corresponding Petri-net are presented in figures 8 and 9.

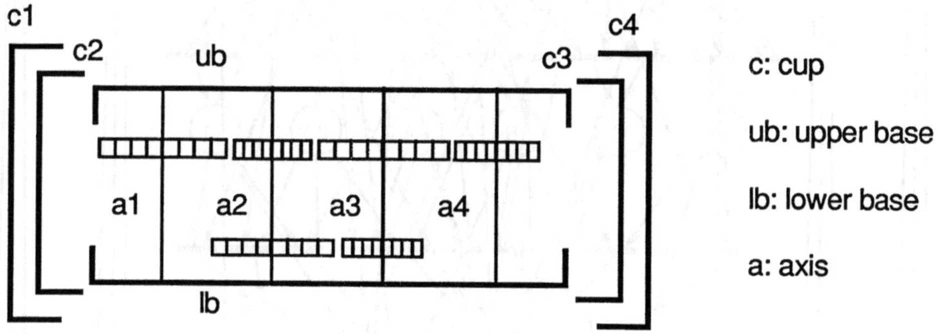

Fig. 8 : An example product : gear box

The current set of subassemblies at any moment in the process is described as a set of current states and current best transitions (since, for each state more than one transitions may be feasible at a time, only the best one is regarded and assigned to the current best transition - see below). Several attributes are associated with each state, namely the product type, a unique state number, the description of the subassembly as a list of parts and an integer that denotes the distance from the solution (ready product) - the distance is the maximum number of transitions that lead to the solution, rather than the number of remaining parts to assemble. Correspondingly, a transition is described by the numbers of the three involving states (two source states and a single target state), the product type and the list of tools required for the assembly. Current states are described by the state number, the product type, the product number and the current place of storage (e.g. pallet). Current transitions are described by the numbers of the three states involved, the product type, the product number and the distance.

No geometric or numeric information is integrated to the system; even the capacities of the storehouses are given as integers that denote the maximum number of objects to be stored.

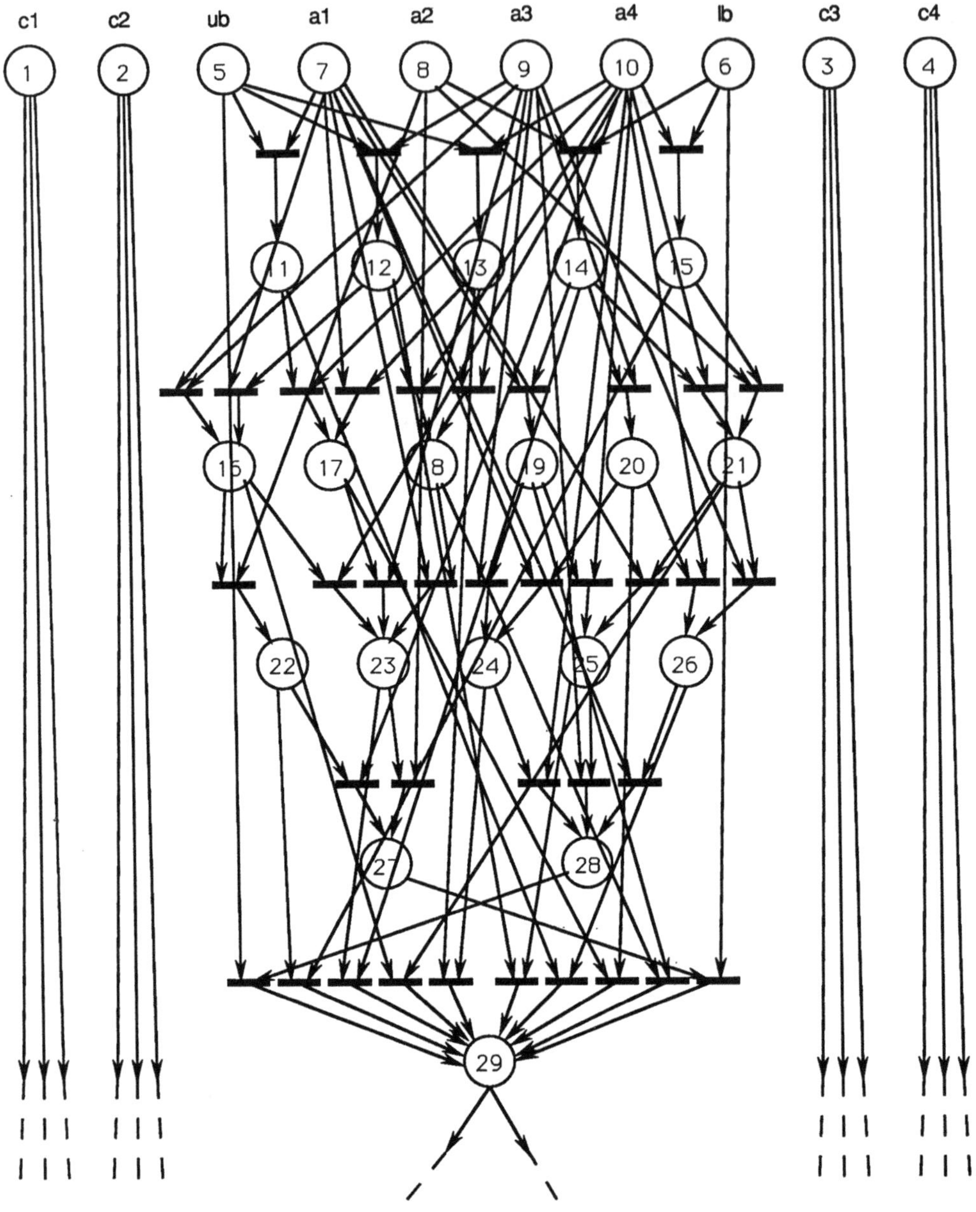

Fig. 9 : Part of the Petri-net corresponding to the example product of fig. 8

6.3 Assembly Environment

We assume that the assembly cell consists of the following : 1) Two single-armed robots, a stationary and a mobile one, 2) A pallet - parts and subassemblies temporary storage space, 3) A worktable to which the stationary robot is docked, 4) A platform docked to the mobile robot, 5) An input stream where the parts are supposed to arrive, 6) An output stream where the ready (assembled) products are put, 7) A dustbin where all faulty parts and subassemblies are thrown, 8) A toolkit, and 9) Sensors : any combination of sensors can be supported - at least a camera in the input stream and a tactile sensor at the worktable. In addition, two temporary storage places (one for each robot) are reserved. A top view of the assembly cell is given in figure 10.

6.4 Description of the System Operation

New parts arrive in any order and in any number (up to a certain limit) - the part arrival is detected in the input stream by an appropriate sensor (e.g. a weight sensor). The mobile robot moves from its safe position (abri) to a prespecified position where the input stream is within its reach. Using the sensors each new part is recognised - if it belongs to those needed to assemble the desired products, it is grasped and moved to the platform, otherwise it is ignored (but it may be useful to a subsequent cell !). The scanning of the input stream stops when either the platform is full or no more new parts to recognise exist. If any of the parts in the platform are found faulty, what involves new sensoring, the mobile robot moves to a position by the dustbin and throws them. If, after that, any new unrecognised parts remain, the previous scanning process is repeated, otherwise the mobile robot moves to abri and stores the platform contents to the pallet - if the pallet is not full. Any two objects (parts or subassemblies) in the pallet that fit are retrieved by the stationary robot and assembled together - the retrieval is possible if the worktable is not full. Assembly operation can also be performed if one or both of the objects to assemble are initially stored to the worktable, in which case retrieval of one or none object from the pallet is required. After an assembly operation new sensoring is performed, to find out if the subassembly is faulty or if we reached a ready product. The new object is then stored to the pallet, from where the mobile robot will carry it to the dustbin or the output stream respectively - it is assumed that the contents of the dustbin are flushed at specific moments in order to avoid overflow. All tools needed for the assembly operations are stored to the toolkit from where the stationary robot retrieves them when necessary, and it is possible that more than one tools will be required for some operations, in which case successive tool swapping will take place. There exists one single and rare case of "stuck" : when all storehouses are full, no assembly operation can be performed and new arrivals are detected. A special behaviour of the system is then required, e.g. to suspend all other operations and wait for the arrival of a part that fits with one of the current objects, in order to exit the deadlock. In reality, this critical situation doesn't arise, or, at least, only rarely, yet, it cannot be ignored, since the system should be effective in real-time and autonomous. In fact, the system is only semi-autonomous : extreme situations (e.g. loss of contact of a robotic arm with the grasped object) are not foreseen - instead, the operator/supervisor of the system can interfere to it (examining its current situation, changing parameters, etc.), or, in special cases, interrupt the normal flow of execution, control the system manually and, finally, resume automatic execution.

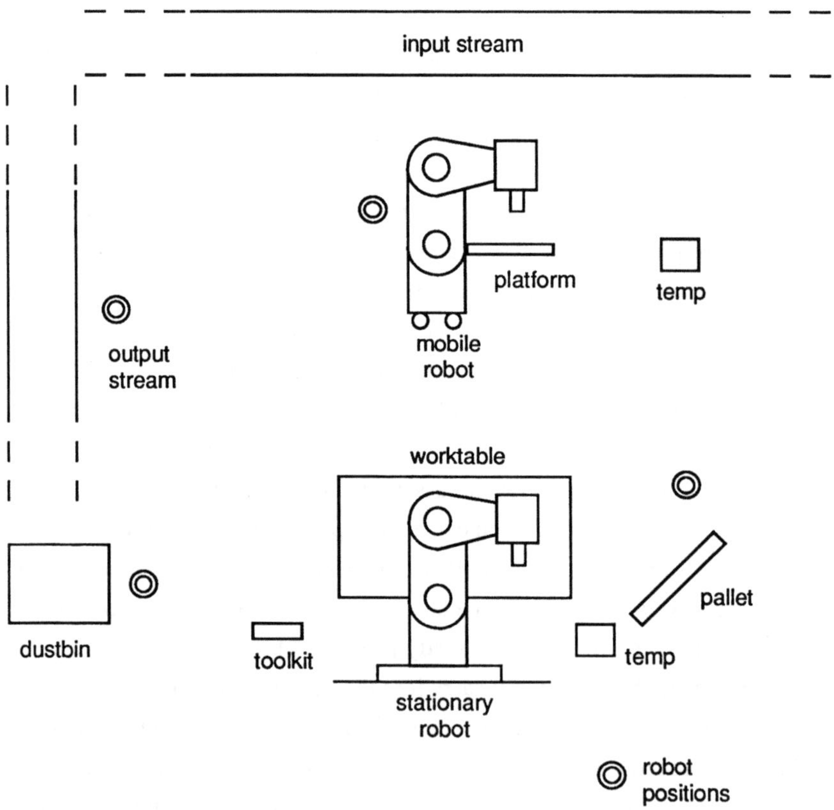

Fig.10 : Top view of the assembly cell

6.5 Blackboard structure

The blackboard is the central data storage section - various pieces of information are kept there, including current state description and control data. The domain knowledge is partitioned to four knowledge sources, namely : **arrival, access_pallet, assemble, user.** An additional domain-dependent module, the task planner, which serves as the control mechanism of the system, can be viewed as a special purpose knowledge source, that is executed at predetermined moments during the process. All knowledge sources communicate and interact implicitly through the blackboard data structure (Fig. 11). The task planner doesn't interact with the knowledge sources, but rather it schedules their execution.

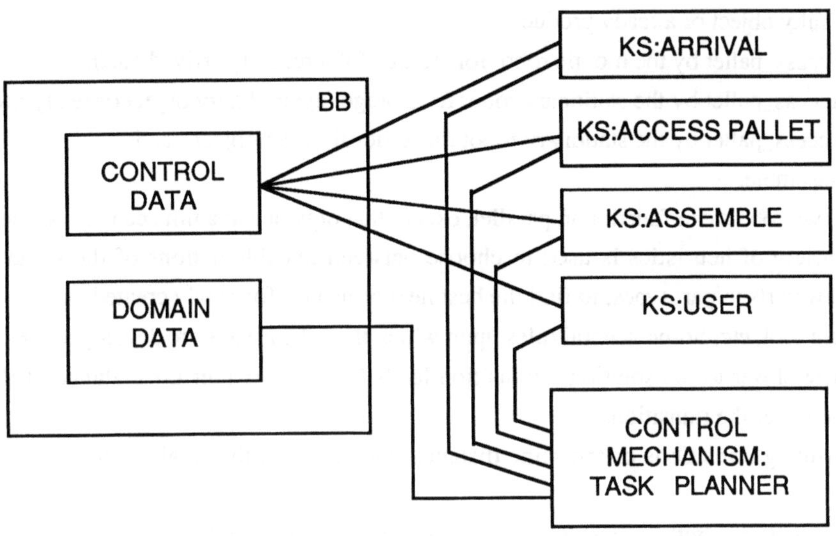

Fig. 11 : Blackboard implementation of the flexible manufacturing system

6.6 Task Planner

In order to support parallelism in execution and achieve efficiency, avoiding deadlock situations, the following rules are defined :

- A faulty object or a ready product must leave the system as soon as possible, in order to free some space in the storehouses.

- All robotic routines are symbolic predicates (always true, since the system is open-ended) and may be simple (e.g. move) or complex (e.g. fetch_from_pallet).

- Random variables are used to simulate the following external conditions : 1) User interference, 2) Part arrival, 3) Number of arrived parts, 4) Type of recognised part, 5) Number of faulty new parts, 6) Faulty assembly - the fault probability depends on the tools used during the assembly operation, with the lowest fault probability assigned to operations performed in absence of tools.

- The task planner chooses among simultaneously feasible actions using the following priority hierarchy :

1) User interference,

2) Part arrival,

3) Access_pallet by the mobile robot for retrieval (apparently, this cannot be but retrieval of either a faulty object or a ready product),

4) Access_pallet by the mobile robot for storage (of a recently arrived part),

5) Access_pallet by the stationary robot for storage (again, faulty object or ready product),

6) Access_pallet by the stationary robot for retrieval (of an object), and

7) Assemble.

The two robots can function in parallel, except that only one at a time can access the pallet.

A variety of heuristics is used to choose between feasible actions of the same type, to assign new parts to part types, to find the best next transition for newly created states, to update the current tool, etc. Some general rules upon which those heuristics are developed are :

1) The closer to the solution a transition leads (i.e., the smaller the distance of the target state), the better the transition.

2) Among transitions of the same distance, the one with the smaller tool list size is the best.

3) Transitions, both source subassemblies of which are stored to the worktable, are never inactivated, since they will lead to immediate execution of assembly operation within the next few process steps.

4) A newly arrived part is assigned to the product with the minimum number of remaining parts to arrive, provided that the part is legal for the given product.

5) In the case where swapping or removing of a subassembly is required (usually, due to limited storage space), a subassembly is chosen that belongs to a different product type than the one currently regarded. In absence of such a subassembly, one belonging to a different instance of the same product is chosen. If this criterion fails too, then one not belonging in the currently regarded transition is chosen, This last criterion never fails if the sizes of the storehouses are greater than 2 - and they usually are : the present version of the system was tested for sizes of worktable 3-8, platform 2-6 and pallet 20-35, which are the reasonable limits.

6.7 Evaluation

The system has been tested for integrity (correctness), consistency and robustness, precision and speed. It works as expected, producing the same results given the same data. It is robust in the sense that slightly different data lead to slightly different results, i.e., its performance doesn't deteriorate exponentially when the size of the data increases linearly. In a future, more complete version, explicit statistical data should be gathered to evaluate its performance mathematically rather than intuitively. Speaking about precision one point to take into consideration is whether the robotic functions correspond to reality - an attempt has been made to assume as realistic functions as possible, although they were left open-ended. It is remarkable that the accuracy of the random variables generators in this phase of the overall system design is of no interest. Finally, the speed of the system is satisfactory, at least in the present simulation stage. In general, the actual response of the system is expected to be slower than the simulated one. However, the time required for every single decision of the computer will not exceed a few seconds, while the execution of this decision, that will involve robot action, may require from half a minute to a quarter of an hour (depending on the complexity of the task) ! This demonstrates the need for extended parallelism and reduction of communication overheads that should be seeked in subsequent steps of the design process.

7. Conclusions

In this paper, an account of the blackboard approach to the design of knowledge-based robotic and flexible manufacturing systems was given. Two integrated system examples from the open literature and one robotic assembly system developed and tested through simulation by the authoes were described. Some other works dealing with several aspects of IRS and manufacturing systems can be found in [17,44][46-51][53-56]. The main advantages that make the blackboard approach very appropriate (and popular) in the area of robotic and manufacturing problem solving are :

- ability to solve large problems involving subproblems belonging to different domains,
- ability to call, use and integrate different heuristics, algorithms, etc., to arrive at the problem solution (opportunism),
- tendency to commit as less as possible to initial conditions, choices, etc. (similar to least commitment strategies) which leads to increased adaptivity and extensibility,
- ability to modify/enhance existing knowledge about the problem (especially the blackboard control architecture treats problem domain and control knowledge uniformly),

- the interface between the user/operator and the system is a knowledge source and so their interaction is very natural, and

- ability to integrate numeric and symbolic processing (some knowledge sources can handle numeric data, while others can process symbolic data).

8. References

[1] Almgren R.: *On Knowledge-Based Planning and Programming Systems for Flexible Automatic Assembly*, University of Linkoeping, Studies in Science and Technology, Thesis No. 176, LiU-Tek-Lic-1989:16, 1989.

[2] Binger G., V. Flemming: Graph Oriented Simulation of Assembly Process, *Systems Analysis, Modelling and Simulation*, 6(1989), 10, pp. 781-786.

[3] Collinot A., B. Hayes-Roth: Real-Time Control of Reasoning - Experiments with Two Control Models, Stanford University, Department of Computer Science, Knowledge Systems Laboratory, R*eport No. KSL 90-17*, March 1990.

[4] Connor D.E.: Using expert systems to manage and change complexity in manufacturing, in "*Artificial Intelligence Applications for Business*" by W. Reitman (Ed.), Ablex Publishing Corporation, pp. 149-158, 1984.

[5] Csuhaj-Varju E., J. Kelemen: Cooperating Grammar Systems : a Syntactical Framework for Blackboard Model of Problem Solving, in "*Artificial Intelligence and Information-Control Systems of Robots*" by I. Plander (Ed.), Elsevier Science Publishers B.V. (North-Holland), 1989.

[6] Faught W.S.: Applications of AI in Engineering, *Computer*, pp. 17-27, July 1986.

[7] Fisher E.L.: Expert systems can lay groundwork for intelligent CIM decision making, *Industrial Engineering*, Vol. 17, No. 3, pp. 78-83, 1985.

[8] Fox B.R.: The Implementation of Opportunistic Scheduling, *Intelligent Autonomous Systems Conference*, Amsterdam, Dec. 1986.

[9] Fromherz B., K. Hoermann: A Concept for a Robot Action Planning System, in "*Languages for Sensor-Based Control in Robotics*" by U. Rembold and K. Hoermann (Eds.), Springer Verlag, Berlin/New York, 1987.

[10] Gerencser M., R. Smetek : Artificial Intelligence: Technology and Applications, *Miltech : Military Technology,* No. 6, pp. 67-76, 1985.

[11] Hayes-Roth B. : The Blackboard Architecture : A General Framework for Problem Solving ?, Stanford University, Computer Science Department, Heuristic Programming Project, *Report No. HPP-83-30,* May 1983.

[12] Hayes-Roth B.: A Blackboard Architecture for Control, *Journal of Artificial Intelligence*, 26, 1985, pp. 251-321.

[13] Herrod R.A., B. Papas: Artificial Intelligence moves into industrial and process control, *The Industrial and Process Control Magazine*, pp. 45-52, March 1985.

[14] Hewett M.: BB1 User Manual - v2.1 Update (Common Lisp), Stanford University, Departments of Medical and Computer Science, Knowledge Systems Laboratory, *Working Report No. KSL 86-61a*, Feb. 1988.

[15] Hoermann A., T. Hugel, W. Meier: A Concept for an Intelligent and Fault-Tolerant Robot System, *Journal of Intelligent and Robotic Systems*, Vol. 1, pp. 259-286, 1988.

[16] Irizarry-Gaskins V.M., T.-C. Chang : Knowledge-Based Process Planning for Electronic Assembly, *Journal of Intelligent and Robotic Systems*, pp. 3-16, 3(1990).

[17] Kartsounis G., S.G. Tzafestas : Moving Object Displacement Estimation Using Second-Order Models : A R*obotic Vision Example,* Proceedings International Conference on Digital Signal Processing (Eds. : V. Capellini and A. Constantinides), pp. 703-710, Florence, September, 1987.

[18] King M.S., S.L. Brooks, R.M. Schaefer : Knowledge-based systems - how will they affect manufacturing in 80's ?, *Proceedings 1985 ASME International Computers in Engineering Conference*, Boston, MA, Vol. 2, pp. 383-390, 1985.

[19] Kusiak A. (Ed.) : *Artificial Intelligence Computer Integrated Manufacture*, Kempston, Bedford, UK : IFS Ltd., 1987.

[20] Kusiak A.: Manufacturing Systems : A Knowledge- and Optimisation-Based Approach, *Journal of Intelligent and Robotic Systems*, 3(1990), pp. 27-50.

[21] Lee G.S.G., R.C. Gonzalez, K.S. Fu : *Tutorial on Robotics*, IEEE Computer Society Press/North-Holland, NY/Amsterdam, 1983.

[22] Marik V., L. Lhotska : Some Aspects of the Blackboard Control in the Diagnostic FEL-Expert Shell, in "*Artificial Intelligence and Information-Control Systems of Robots*" by I. Plander (Ed.), Elsevier Science Publishers B.V. (North-Holland), 1989.

[23] McCalla G., N.Cercone : Techniques and Issues in the Design of Applied Artificial Intelligence Systems, *Computers and Mathematics with Applications*, Vol. 11, No. 5, pp. 421-430, 1985.

[24] Meyer R.J.: AI and expert systems - In pursuit of CIM, *Manufacturing Engineering*, pp. CT15-CT18, February 1987.

[25] Meystel A.: Intelligent Control : A Sketch of the Theory, *Journal of Intelligent and Robotic Systems*, Vol. 2, pp. 97-107, 1989.

[26] Nevins J.L., D.E. Whitney, A.C. Edsall : Intelligent Systems in Manufacturing, *Preprints of 10th IFAC World Congress on Automatic Control*, Vol. 4, pp. 130-139, Munich, July 1987.

[27] Nii H.P.: Blackboard Systems, Stanford University, Departments of Medical and Computer Science, Knowledge Systems Laboratory, Report No. KSL 86-18 & *AI Magazine*, Vols. 7-2 and 7-3.

[28] O' Grady P., K.H. Lee : A Hybrid Actor and Blackboard Approach to Manufacturing Cell Control, *Journal of Intelligent and Robotic Systems*, 3(1990), pp. 67-72.

[29] Pang G.K.H.: A Blackboard System for the Off-Line Programming of Robots, *Journal of Intelligent and Robotic Systems*, 2(1989), pp. 425-444.

[30] Pang G.K.H.: A Framework for Intelligent Control, under publication in *Journal of Intelligent and Robotic Systems*.

[31] Ranky P.: *Computer Integrated Manufacturing*, Prentice-Hall, Englewood Cliffs, NJ, 1986.

[32] Rayson P.T.: A review of expert systems principles and their role in manufacturing systems, *Robotica*, Vol. 3, pt. 4, pp. 279-287, 1985.

[33] Reeve M., S.E. Zenith (Eds.) : *Parallel Processing and Artificial Intelligence*, Wiley & Sons, 1989.

[34] Rembold U., R.Dillman (Eds.): *Computer-Aided Design and Manufacturing : Methods and Tools*, Springer Verlag, Berlin/New York, 1986.

[35] Rich E.: *Artificial Intelligence*, McGraw-Hill, Singapore, 1983.

[36] Rubin K., P. Jones, C.M. Mitchell : OFMspert : Inference of Operator Intentions In Supervisory Control Using a Blackboard Architecture, *IEEE Transactions on Systems, Man and Cybernetics*, Vol. 18, No. 4, July/August 1988.

[37] Saridis G.N.: *Self-Organising Control of Stochastic Systems*, Marcel Dekker, New York, 1977.

[38] Saridis G.N.: Intelligent Robotic Control, *IEEE Transactions on Automatic Control*, AC-28, pp. 547-557, 1983.

[39] Saridis G.N.: Foundations of the Theory of Intelligent Control, *Proceedings IEEE Symposium on Intelligent Control,* Troy, New York, 1985.

[40] Saridis G.N., K. Valavanis : Analytical Design of Intelligent Machines, *Automatica,* Jan./Feb. 1988.

[41] Schneider H.-J.: Systems of the 5th generation and their impact on the qualification, training and retraining of management, in T. Bernold (Ed.) "*Artificial Intelligence in Manufacturing*", Elsevier Science Publ. (North-Holland), pp. 85-96, 1987.

[42] Shoureshi R., M. Momot, O.R. Mitchell, J. Feddema : Vision-Based Intelligent Control for Automated Assembly, *Journal of Intelligent and Robotic Systems*, Vol. 2, pp. 153-170, 1989.

[43] Snyder R.A.: Artificial Intelligence in the future factory - Test program, *Test Measure World*, Vol. 4, No. 12, pp. 97-109, 1984.

[44] Tzafestas S.G.: Integrated Sensor-Based Intelligent Robot System, *IEEE Control Systems Magazine*, Vol. 8, pp. 61-72, 1988.

[45] Tzafestas S.G.: Expert Systems in CIM Operations : Key to productivity and quality, *Proceedings 3rd International Symposium on Systems Analysis and Simulation*, Berlin, GDR, September 1988.

[46] Tzafestas S.G., A. Ligeza : Expert Control Through Decision Making, *Found. Contr. Engrg.*, Vol. 13, No. 1, 1988.

[47] Tzafestas S.G. : *Knowledge-Based System Diagnosis, Supervision and Control*, Plenum Press, New York/London, 1989.

[48] Tzafestas S.G., A.M. Plouzennec : A Prolog-Based Expert System Prototype for Robot Task Planning, *Proceedings IFAC International Symposium on Advanced Information Processing in Automatic Control (AIPAC '89)*, Nancy, France, July 3-7, 1989.

[49] Tzafestas S.G.: Petri-Net and Knowledge-Based Methodologies in Manufacturing Systems, *Proceedings ESPRIT CIM Europe Conference*, Athens, May 17-19, 1989.

[50] Tzafestas S.G., A. Ligeza : A Framework for Knowledge-Based Control, *Journal of Intelligent and Robotic Systems*, Vol. 2, pp. 407-425, 1989.

[51] Tzafestas S.G.: System Fault Diagnosis Using the Knowledge-Based Methodology, in "*Fault Diagnosis in Dynamic Systems*" by R. Patton, P. Frank and R. Clark (Eds.), Ch. 15, Prentice Hall, New York/London, 1989.

[52] Tzafestas S.G.: AI Techniques in Computer-Aided Manufacturing, in "*Knowledge Engineering, Vol. II*" by H. Adeli (Ed.), McGraw-Hill, 1990.

[53] Tzafestas S.G.: Artificial Intelligence and Expert Systems in Control : An Overview, *Systems Analysis, Modelling and Simulation*, Vol. 7, No. 3, pp. 171-190, 1990.

[54] Tzafestas S.G., N. Papanikolopoulos : Incremental Fuzzy Expert PID Control, *IEEE Transactions on Industrial Electronics*, 1990.

[55] Tzafestas S.G., G. Kehriotis : A Numeric-Symbolic Expert System for 2-D and 3-D Object Recognition in Robotic Applications, *Math. Comp. Simul.*, Vol. 32, 1990.

[56] Tzafestas S.G.: Fuzzy Expert Control : Recent Results with Application to Robotic Systems, in "*Knowledge-Based Robot Control*" by G. Saridis and H. Stephanou (Eds.), Springer, Berlin/New York, 1991 (to appear).

[57] Tzafestas S.G.: *Intelligent Robotic Systems*, Marcel Dekker, 1991 (to appear).

[58] Valavanis K., P.H. Yuan : Hardware and Software for Intelligent Robotic Systems, *Journal of Intelligent and Robotic Systems*, Vol. 1, pp. 343-373, 1988.

[59] Warman E.: AI in manufacturing, An organic approach to manufacturing cells, *Data Processing*, Vol. 27, No. 4, pp.31-34, 1985.

IMPLEMENTATION OF SPECIFIC ROBOTIC FUNCTIONS DIRECTLY IN HARDWARE

John N. Lygouras, Basil G. Mertzios and Nicholas C. Voulgaris

Department of Electrical Engineering, Demokritos University of Thrace

Abstract: Motivated by the desire to add more intelligent functionality and enhanced real-time control capability to robotic systems, many authors suggest the implementation of specific robotic functions directly in hardware. This paper presents analog and digital techniques for the computation of the direct kinematic solution (DKS) on a single chip and the implementation of a velocity and acceleration estimator based on a position history. Analog and digital circuits for the DKS have been designed on the basis of the direct kinematic equations. A velocity and acceleration estimator based on a position history is implemented with two circuits dedicated to the computation of current joint velocities and accelerations of a robot arm. The first involves a microprocessor-based system and the second consists of clock-controlled analog circuitry. The above-described units have been constructed and tested under real conditions in an experimental computer-controlled, light-weight mechanical arm.

Keywords: Robotic functions, Hardware implementations, Direct kinematic solution, Velocity and acceleration estimators

Introduction

A. Forward kinematics

The control of robotic manipulators implies heavy computations, since they are highly nonlinear systems. It has been broadly realized that the computation burden is very critical in designing and controlling a robot. Because of this, the use of existing computationally complicated algorithms is usually prohibitive. Motivated by the desire to reduce the computational cost, a number of research efforts [2,4,7] have been directed to the efficient and fast calculation of the required computations. To relieve the computational burden, many authors suggest the implementation of specific robotic functions directly in hardware [3, 8].

NATO ASI Series, Vol. F 71
Expert Systems and Robotics
Edited by T. Jordanides and B. Torby

In this section two circuits are described for the implementation of a Cartesian-space sensor (CSS), both of which can be integrated in VLSI circuitry. In the first method, a hardware structure is described for deriving the trigonometric functions of the six joint angles directly from the joint potentiometers. The architecture of an analog circuit dedicated to the computation of the direct kinematic solution (DKS) of a robotic manipulator is also given. The accuracy of the solution obtained with the proposed circuit is lower than that achieved by solving the direct kinematic equations with a microproccesor, but its response is several orders of magnitude faster. The sensor implemented by this method can, therefore, be used in applications where high accuracy is not a prime factor in trajectory tracking (e.g., industrial painting).

In the second approach, the derivation of the trigonometric functions is based on the use of nonlinear A/D converters, while a hardware-implemented pipeline technique is used to perform all needed arithmetic operations. This method is more accurate compared with the first method, while its computational time is about two orders of magnitude faster than that achieved by using a microprocessor-based system.

For any robot kinematic structure with known geometric parameters, the direct kinematics can be written as:

$$\mathbf{n} = \mathbf{f}_n(\mathbf{q}), \quad \mathbf{s} = \mathbf{f}_s(\mathbf{q}), \quad \mathbf{a} = \mathbf{f}_a(\mathbf{q}), \quad \mathbf{p} = \mathbf{f}_p(\mathbf{q}) \tag{1}$$

The corresponding kinematic equations for the well known PUMA 600 arm are as follows [7]:

$$n_x = C_1\ [C_{23}(C_4C_5C_6 - S_4S_6) - S_{23}S_5C_6] - S_1[S_4C_5C_6 + C_4S_6]$$
$$n_y = S_1[C_{23}(C_4C_5C_6 - S_4S_6) - S_{23}S_5C_6] + C_1[S_4C_5C_6 + C_4S_6]$$
$$n_z = -S_{23}(C_4C_5C_6 - S_4S_6) - C_{23}S_5C_6$$
$$s_x = C_1\ [-C_{23}(C_4C_5C_6 - S_4C_6) + S_{23}S_5S_6] - S_1[-S_4C_5S_6 + C_4C_6]$$
$$s_y = S_1\ [-C_{23}(C_4C_5S_6 + S_4C_6) + S_{23}S_5S_6] + C_1[-S_4C_5S_6 + C_4C_6]$$
$$s_z = S_{23}(C_4C_5S_6 + S_4C_6) + C_{23}S_5C_6$$
$$a_x = C_1\ (C_{23}C_4S_5 + S_{23}C_5) - S_1S_4S_5$$
$$a_y = S_1\ (C_{23}C_4S_5 + S_{23}C_5) + C_1S_4S_5$$
$$a_z = -S_{23}C_4S_5 + C_{23}C_5$$
$$p_x = C_1(d_4S_{23} + a_3C_{23} + a_2C_2) - S_1d_3$$
$$p_y = S_1(d_4S_{23} + a_3C_{23} + a_2C_2) - C_1d_3$$
$$p_z = -(-d_4C_{23} + a_3S_{23} + a_2S_2)$$

B. Computational aspects

In order to compute the three right-hand columns of $^{0}T_{6}$, we require 12 transcendental function calls, 39 multiplications and 19 additions. The first column of $^{0}T_{6}$ can be obtained as the vector cross product of the second and third columns. For the derivation of the vector cross product, six more multiplications and three additions are needed giving a total of 12 transcendental function calls, 45 multiplications and 22 additions. If the joint coordinates are given, the position and orientation of the robot hand are obtained by evaluating these equations to obtain $^{0}T_{6}$.

The total time needed for a microproccesor to evaluate the kinematic equations (2) is typically about 8 msec. Taking into account that this evaluation is executed repeatedly once in each sampling interval, approximately every 45-60 msec, we can easily see that the demand for repetitive solution occupies a significant amount of the microproccesor work time. Programming language VAL-II, for example, provides some basic real-time processing power, such as trajectory calculation, but the time share of the trajectory and kinematic calculations is nearly half of the 28-msec control cycle [3].

For an intelligent robot, the capability of making real-time decisions based on the current kinematic information is essential. By feeding the current kinematic information back to a high-level controller, such as a trajectory planner, it can take appropriate actions according to the world model, its internal states, and the rules within its knowledge base.

Digital computers allow mathematical operations to be performed to a very high degree of accuracy. The quantities involved are often continuous signals (for instance, in the form of voltages), which, in turn, may correspond to some other measured quantities. In such cases, Analog-to-Digital (A/D) and Digital-to-Analog (D/A) converters are needed to interface with the digital computer. Such a high cost system is only justified, if the demand for accuracy cannot be met by analog computer circuitry. An upper limit for the accuracy of this kind of circuit is of the order of 99.9%.

C. Derivation of the trigonometric functions directly from the joint potentiometers

The derivation of the relevant position between two adjacent links is achieved mainly in two ways:

(i) by counting the number of revolutions of the joint actuating motor, using optical encoders, which corresponds directly to the displacement of the joint and

(ii) by using circular or linear potentiometers depending on whether the joint is rotational or prismatic.

The first method is more accurate and overcomes the need for A/D conversion, since the position is encoded directly in binary form. After the analog voltages from the potentiometers in the last method have been converted to binary form, the calculation of the trigonometric functions must follow. When these functions are available as part of a standard library, they are often computed from a series expansion at the cost of many multiplication and addition times. However, at the expense of the required memory, many manipulation systems employ look-up table implementations of the transcendental functions.

In addition to a linear rotation-vs-resistance characteristic, composition elements can have a wide range of nonlinear curves (tapers). Standard types are shown in Fig. 1. The usual rotation-vs-resistance characteristic is linear, but by tapering or otherwise shaping the resistance element, a limited amount of taper, logarithmic, sinewave, or some other functions can be generated.

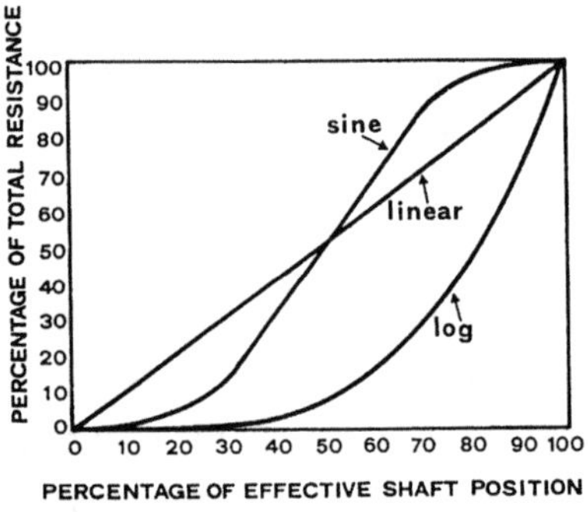

Fig. 1 Standard types of linear and nonlinear curves (tapers)

The Cartesian-space sensor

Recently developed CMOS multipliers achieved a better combination of linearity, bandwidth and dynamic range and faster response (in the order of nsec) than before. For example, Bult [1] reported a linearity better than

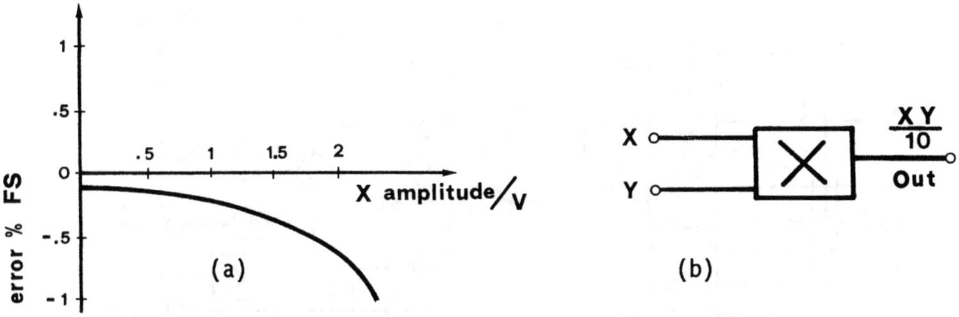

Fig. 2. (a) Typical error vs amplitude curve (b) Schematic symbol of an analog multiplier

0.14 percent. A typical error versus amplitude curve of a CMOS multiplier is shown in Fig. 2(a), while its schematic symbol is shown in Fig. 2(b).

A. Analog circuit for the DKS

The circuit shown in Fig. 3 is the hardware implementation of equations (2) based on the derivation of the trigonometric functions from the joint potentiometers and the analog multipliers. As an analog circuit, it has all the advantages and disadvantages of analog circuits, i.e.

- Very fast response (of the order of nanoseconds) and
- Reduced accuracy compared with digital computation.

The trigonometric functions of the six joint angles θ_1-θ_6 (assigned as c_i for cos θ_i and s_i for sin θ_i) are fed as inputs to the circuit in the form of analog voltages. After suitable multiplications and additions according to equation (2), the elements of matrix ${}^0\mathbf{T}_6$ are produced at the output in the form of analog voltages.

A major advantage of the above circuit is that it provides to the robot controller real-time knowledge of the current Cartesian-space position. The response time of this circuit is of the order of nanoseconds. An algorithm based on the information derived from the CSS, for applications requiring increased accuracy in trajectory tracking, could be the following:

The corner points of a trajectory are specified according to a specific task. If the demand for accuracy in trajectory tracking is no more than about 99%, the movement from point to point of the end-effector can be done in two stages; at first, the movement over the greater part between

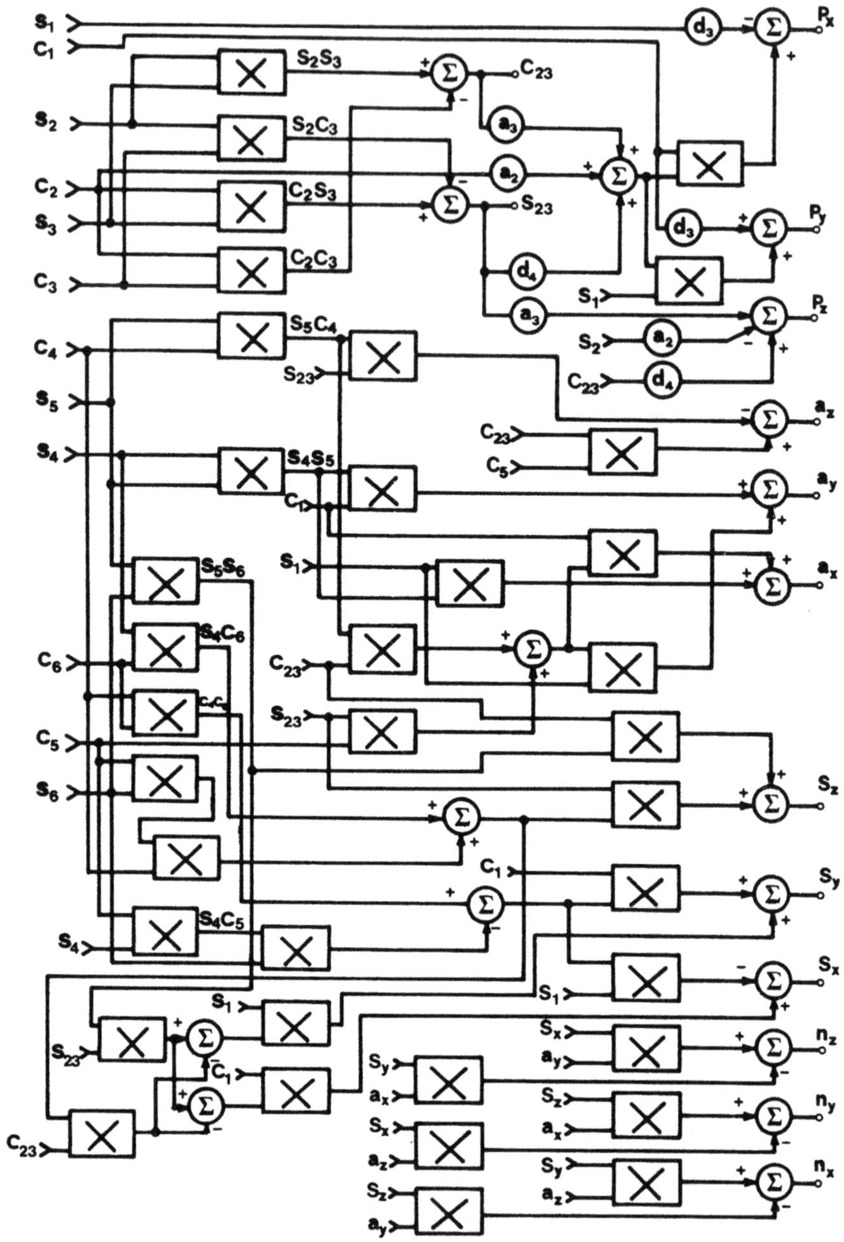

Fig. 3. The Cartesian space sensor implemented in analog circuitry

two successive corner points is done based on the information derived from the CSS, while for the remaining segment, in order to achieve the necessary correction, the trajectory can be followed based on the analytic solution of the direct kinematic equations. The distance of the second segment can be of the order of 5% of the total distance between the two corner points.

B. Derivation of the trigonometric functions in digital form

In this section, a method for entirely digital implementation of a Cartesian-space sensor is given. The method is based on the hardware implementation of the look-up table technique on an EPROM, and on linear interpolation to derive the trigonometric functions $\sin \theta_i$ and $\cos \theta_i$. The look-up table process in the A/D conversion and the whole construction constitute a nonlinear A/D converter [5,6].

As described earlier, the derivation of the relevant position between two adjacent links can be achieved in digital form through A/D converters or optical encoders. Both methods provide the position encoded in a binary n-bit word. The demand for resolution is at least 12 bits. These joint-angle set points are then read by a high-level controller, such as a trajectory planner.

To reduce the EPROM size as well as the cost, we can use linear interpolation (Fig. 4): The 12 most significant bits, out of the 16 bits, can be used as one sign bit and eleven magnitude bits which drive the 11 input lines of a decoder. The values of $\sin \theta$ have been stored in 11-bit words in an EPROM, organized as 1kx11 bits. The remaining 4 least significant bits of the A/D converter (or the position encoder) can be used directly to complete the value of $\sin \theta$. Thus, between two successive values of $\sin \theta$ stored in the EPROM, the linear interpolation of 16 additional values is achieved.

If we exclude for the moment the circuit in dashed lines, the remaining combinational circuit of 15 Exclusive OR (EXOR) gates is used as follows: the angular interval $0 \leq \theta < 2\pi$ is encoded into a 16-bit digital number. To obtain $\sin \theta$, we use the symmetry around the point $(\pi, 0)$. Thus, when $0 \leq \theta < \pi$, a digital number can represent the positive values of $\sin \theta$ and the negative values can be obtained by sign inversion. We notice that the MSB provides information concerning the polarity of $\sin \theta$.

Furthermore, because of the symmetry of $\sin \theta$ around the axis $\theta = \pi/2$, the 14 EXOR gates can be used to shift backwards the values on the address bus lines in order to derive the values of sin θ for the interval $\pi/2 \leq \theta < \pi$, using the values of sin θ for $0 \leq \theta < \pi/2$.

The values of cos θ can be obtained also from the same EPROM by shifting backwards the value on the address bus by $\pi/2$ or 1/4 FS. This can be achieved by the circuit shown in dashed lines in Fig. 4(a). When input line $\overline{S}/C=0$, then $F_1=A$ and $F_2=B$. Now, when $\overline{S}/C=1$, if B=1, then $F_1=A$ and $F_2=\overline{B}$, but if B=0, then $F_1=\overline{A}$ and $F_2=\overline{B}$. The above operations are shown in the truth table of Fig. 4(b) and the logic expressions for F_1 and F_2 can be written as:

$$F_1 = A \oplus [\overline{B} \cdot C] \quad \text{and} \quad F_2 = B \oplus C$$

where · represent logic AND and ⊕ EXOR operations.

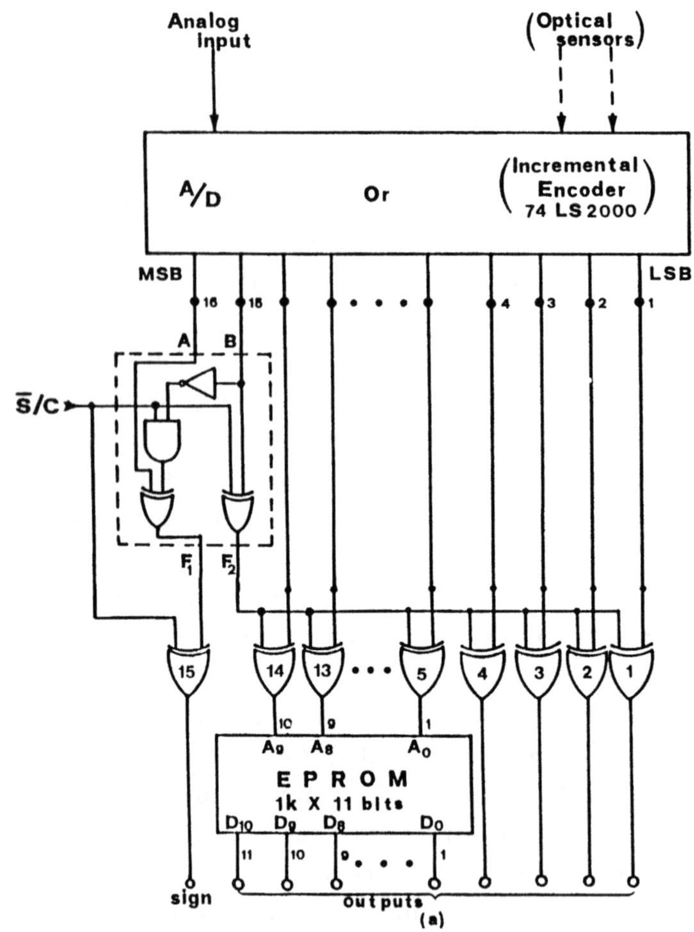

A	B	$\overline{S}/C$	F_1	F_2
0	0	0	0	0
0	0	1	1	1
0	1	0	0	1
0	1	1	0	0
1	0	0	1	0
1	0	1	0	1
1	1	0	1	1
1	1	1	1	0

(b)

Fig. 4 (a) Decoder for reading the value of sine and cosine from a 1K x 11-bit EPROM (b) its truth table

A digital circuit for the DKS

With the trigonometric functions of the joint angles available, the required arithmetic operations remain to be executed. Position-resolution requirements are of paramount importance in most industrial robotic applications. Thus the resolution achieved through the above method significantly affects the precision in the reachability of the final position by the end-effector. Recent surveys have shown that the position-resolution requirement for various robotic applications range from 0.1-10.00 mm for a working range of 1-2 m. Repeatability requirements of 0.01-1 mm are typical. These requirements have been achieved using 16-bit microprocessors.

Since the I/O values (i.e., joint-angle trigonometric functions/world coordinates) are readily transferable through a 16-bit bus, it is advantageous for the direct kinematic calculations to be computed using fixed-point numbers. The resulting calculation error depends not only on the word size used, but also on the accuracy of the trigonometric-function generation.

In the multiplications and the additions needed to be executed, the algorithm described in [3] can be followed. A two-stage multiplier with an adder in pipeline, both using two's complement, fixed-point arithmetic constitutes the basic computational structure. The method is based on the special structure of the rotation matrix, where the first column is always $[C_i \ \ S_i \ \ 0]^T$ and the other two columns are $[0 \ \ 0 \ \pm1]^T$ and $\pm[-S_i \ \ C_i \ \ 0]^T$ arranged in different order.

In Fig. 5, one multiplier (or two for higher speed-requirements) and one adder are used for the purpose of execution of the arithmetic operations. Fixed-point arithmetic is used throughout with 16-bit accuracy. R_1, R_2 and R_3 are double 16-bit temporary registers. The joint-angle trigonometric functions, derived from the circuit described previously in Fig. 4, are stored temporarily in trigonometric function registers. The constant parameters involved are stored in a PROM, and then in Constant-Parameter Registers (CPR).

Register files are used to temporarily store the intermediate results as well as the DKS. The control section of the circuit can be expressed in terms of a Register Transfer Language (RTL) program. The control signal table can then be used to drive a symbolic simulation program to verify the correctness of the design. The control signal can be stored in the same PROM with the constant parameters. This PROM will be personalized to a

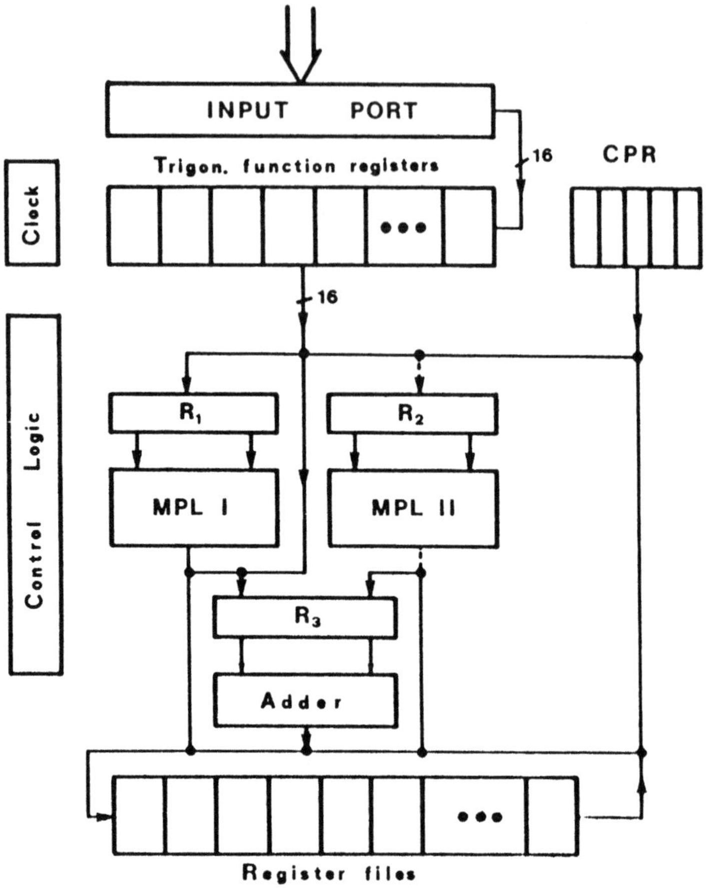

Fig. 5. Digital circuit dedicated to the DKS

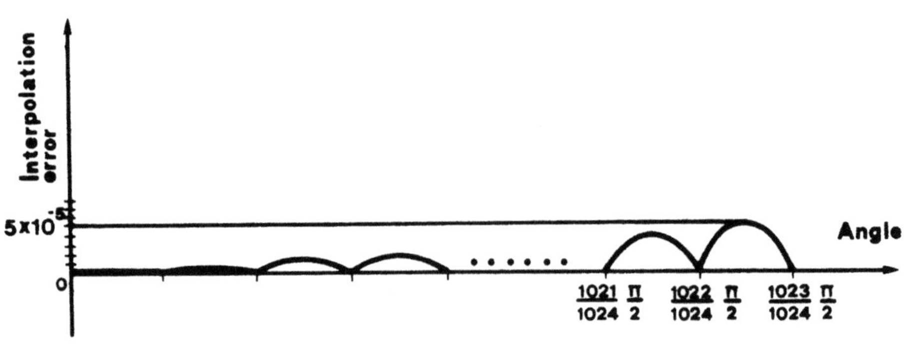

Fig. 6. Interpolation error vs. angular velocity

TABLE I
STEPS FOLLOWED FOR THE DKS

Involved functions	Register contents			Register file	Machine cycles
	R_1	R_2	R_3		
s_1, d_3, s_2, s_3	s_1, d_3	s_2, s_3	0 ,0	0, 0, 0....	4
			$s_1 d_3$	$s_2 s_3$,0,0,..	1
c_2, c_3, a_2	c_2, a_2	c_2, c_3		$s_1 d_3, s_2 s_3$,	4
				$c_2 a_2, c_2 c_3, s_1 d_3, s_2 s_3$,	2
$s_2 s_3, c_2 c_3$			$s_2 s_3, c_2 c_3$		2
				$c_{23}, s_1 d_3, c_2 a_2$,.	1
s_2, c_3, c_2, s_3	s_2, c_3	c_2, s_3			4
			$s_2 c_3, c_2 s_3$		1
. .	. .	. .	. .	. .	. .

specific robot arm configuration .

Since the trigonometric functions of the joint angles are now available, instead of the joint positions, the control program becomes much simpler than in reference [3], the speed of the DKS is higher. The increased number of EPROM entries (1024), as well as the high resolution in the value reading (16 bits) result in a resolution of 2^{-16} , while the maximum interpolation error in the value of sin θ and cos θ is found to remain less than 5×10^{-5} as shown in Fig. 6. The price paid for these improvements is an increase in complexity for each A/D converter because of the inclusion of the 1kx11-bit EPROM and the decoder. In TABLE I, some of the steps followed to derive the DKS are shown. Specifically, the contents of the registers are shown in the process of deriving the components of the position vector **p**, after a number of machine cycles.

Velocity and acceleration estimation

In research carried out in our laboratory, specific robotic functions have been implemented directly into hardware [3], and a light-weight robotic arm having six links and six joints has been constructed. Tubular straight links of aluminum with hollow rectangular cross sections have been used. This arm was used as a prototype for testing the proposed algorithms and methods under real conditions. This section describes the implementation in hardware of a velocity and acceleration estimator based on a position and velocity history, respectively. Two different circuits are given for the implementation, one digital and one clock-controlled analog, consisting of an arithmetic differentiator. This implementation may have more general applications.

The current joint velocities and accelerations are used by the arm-control unit in order to determine the Cartesian velocities and accelerations of the end-effector from the equations:

$$\dot{\mathbf{x}} = \frac{\partial \mathbf{f}}{\partial \Theta_a}\dot{\Theta}_a \quad \text{and} \quad \ddot{\mathbf{x}} = \frac{\partial \mathbf{f}}{\partial \Theta_a}\ddot{\Theta}_a + \frac{d}{dt}\left[\frac{\partial \mathbf{f}}{\partial \Theta_a}\right]\dot{\Theta}_a \tag{3}$$

where θ_a is the vector of actual joint angles and $\dot{\mathbf{x}}$ is the vector of Cartesian velocities. These quantities are commonly read directly from the joints using tachometers and accelerometers, or are estimated by the controlling computer using arithmetic methods. However, the first method is not convenient, since these instruments add to the inertia of the movable arm links, while the second method can take a significant percentage of the computer's run-time. Furthermore, in the second case, the joint-position sampling frequency, is restricted by the time taken by the arm-controlling computer to execute all required calculations. Whenever such a computation is done in present-day systems, usually just the values of θ_a are read and then the joint velocities and accelerations are computed numerically by first and second differences, as in [8]:

$$\dot{\Theta}_a(t) = \frac{\Theta_a(t) - \Theta_a(t-\Delta t)}{\Delta t} \quad \text{and} \quad \ddot{\Theta}(t) = \frac{\dot{\Theta}_a(t) - \dot{\Theta}_a(t-\Delta t)}{\Delta t} \tag{4}$$

However, such a numerical differentiation introduces noise and delay. To resolve this problem, one could either find a less expensive way of computing (3), or increase the computational speed, using a separate unit instead of the controlling computer. The method proposed in reference [1] for velocity estimation based on a position history is extended here for acceleration estimation in a similar manner and the results are extensively investigated. In this paper, the current velocity is estimated based on the present and the last *p* position values.

Estimation-algorithm description

The actual joint velocities and accelerations are read usually from joint tachometers and accelerometers. However, to avoid the use of such mechanical components in light-weight structures and to reduce further the link masses, a method for velocity estimation from a position history is commonly preferred. One such an estimation method is proposed in [1] where the current velocity is determined as a function of the current and the last *p* position values. For *p*=2, this method gives the following results:

If $q_j(i)$ denotes the generalized coordinate q at time i of joint j, the velocity of joint j at time (i-1) is approximated by:

$$E\left\{V_j(i-1)\right\} = \frac{q_j(i) - q_j(i-2)}{2} \tag{5}$$

and the expected value of the velocity at time i is:

$$E\left\{V_j(i)\right\} = \frac{3}{2}\, q_j(i) - 2\, q_j(i-1) + \frac{1}{2}\, q_j(i-2) \tag{6}$$

Since the time derivative of a function representing a velocity profile represents acceleration, an identical relation can be written for the acceleration of joint j at time i :

$$E\left\{A_j(i)\right\} = \frac{3}{2}\, V_j(i) - 2\, V_j(i-1) + \frac{1}{2}\, V_j(i-2) \tag{7}$$

However, this estimation method makes an unrealistic assumption in Eq. (5) which holds true only when the velocity profile is constant in the time interval [i,i-2]. For example, a common situation in robotics is shown in Fig. 7, where a joint is moved over a linear path with parabolic blends. In this example, we quickly accelerate, then coast at constant velocity, and then decelerate. Applying the above method for velocity estimation and then extending the results for an acceleration estimation for p = 1, 2 and 3, we obtain the graphs shown in Fig. 8. It is noted that the estimation based only on the present and the previous value (p = 1) gives better results than for larger values of *p* which tend to confuse the estimation.

Hardware implementation

Two methods for reading the values of the joint angles by the arm-control unit are now mainly used: with potentiometers in conjunction with A/D converters or with optical position encoders. Both methods provide the joint position encoded into a binary n-bit word. The demand for resolution is at least 12 bits in order to obtain the required precision. These joint-angle set points are then read by a high-level controller, such as a trajectory planner.

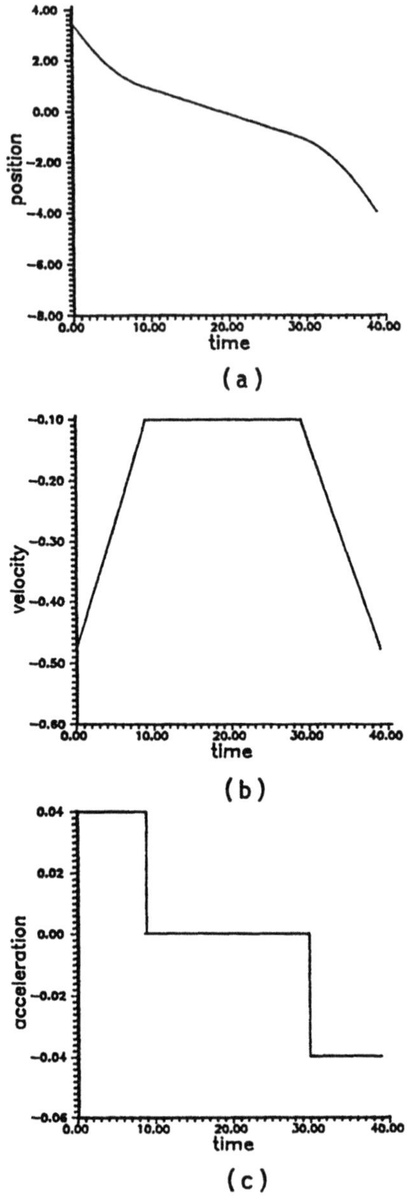

Fig. 7 (a) Position (b) Velocity and (c) Acceleration profiles for linear path with parabolic blends

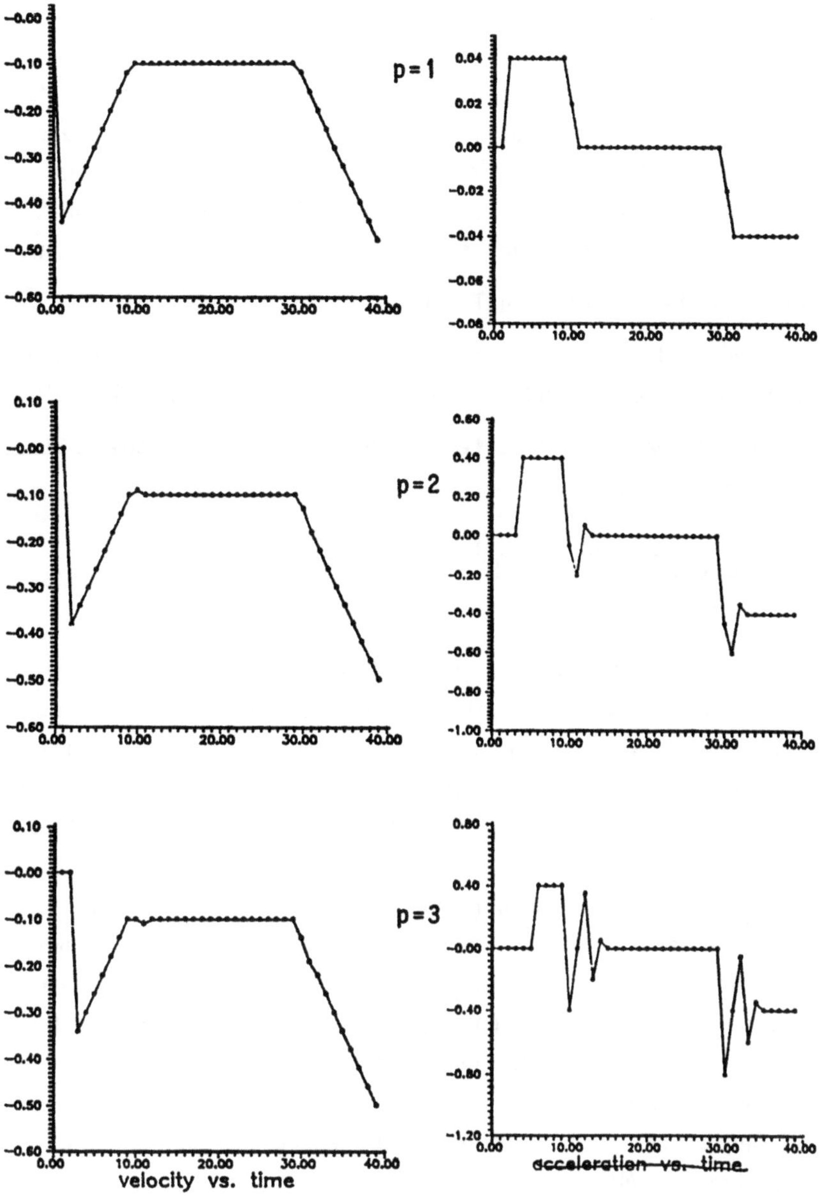

Fig. 8 Experimental results obtained by using the velocity estimation method based on a position history proposed in [1] for p = 1,2 and 3

A. Microprocessor-based circuit

A microprocessor-based circuit dedicated to the computation of the current joint velocities and accelerations may be used in the case where the joint angles are available encoded in digital form (e.g., by means of optical encoders). The construction of a separate arithmetic unit for the calculation of these parameters is deemed necessary mainly for two reasons:

- to relieve the computational burden of the arm-controlling computer and
- to increase the position-sampling frequency since, in this case, this quantity does not depend on the main computer's control cycle.

The second reason is dictated by the definition of the derivative of a function f(x) at a point x_0:

$$f'(x_0) = \lim_{\Delta x \to 0} \frac{f(x_0+\Delta x)-f(x_0)}{\Delta x} \tag{8}$$

and the calculated derivative value tends to the actual value, as the sampling interval Δx gets smaller. To illustrate the method for the implementation of a velocity estimator, a design example will be given. Here, the current and the previous position value are saved and the current velocity is determined as the difference between these two values, assuming that the normalized sampling interval Δt is equal to 1.

The estimation of the current joint velocities taking into account Eq.(5) is implemented using a microprocessor-based system (shown in Fig. 9) consisting of a Z80 CPU, a 2-Kbyte RAM, a 2-Kbyte ROM, and two I/O ports. It is assumed that the position $q_j(i)$ is available and encoded in a binary n-bit number. The microprocessor system can be programmed to execute the following operations:

- read sequentially the current joint angles from the input port and store them in RAM
- calculate the estimated velocity for joint j (j = 1, 2,..., N), as the difference between the current and the previous position values, which are read from the RAM. The result is then stored in the output latch L1
- calculate the estimated acceleration for joint j as the difference between the present and the previous velocity values multiplied by a factor of eight and store the result in the output latch L1.

In order to test the constructed arithmetic unit and to verify its response under various operation conditions, an A/D Converter (ADC) and a

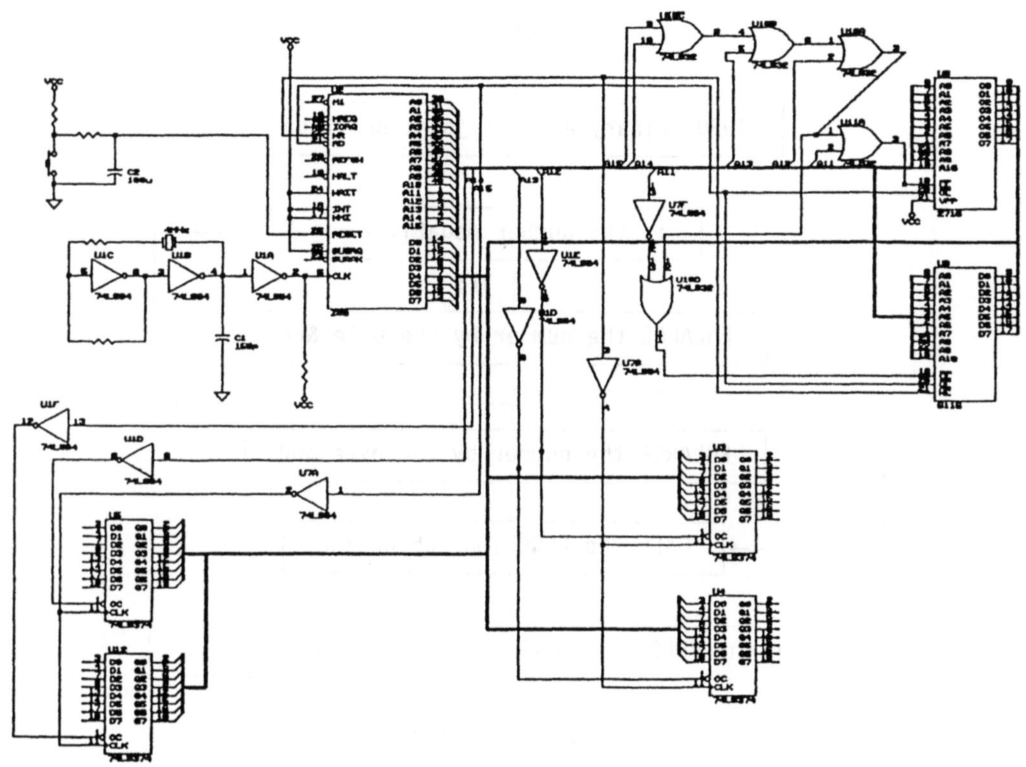

Fig. 9 Digital circuit for velocity and acceleration estimation using a Z-80 microprocessor

D/A Converter (DAC) were used in conjuction with the constructed arithmetic unit. The ADC was used to convert the current value of the position profile presented in the form of analog voltage at its input into a 12-bit digital number. The 12-bit DAC provides a stepped version of the calculated velocities and accelerations. Figure 10 shows the steps followed by the processing algorithm. Since the differences between two successive velocity values are quite small, these values are multiplied by a factor of eight. Fixed-point arithmetic and 2's complement representation are used for the quantities involved.

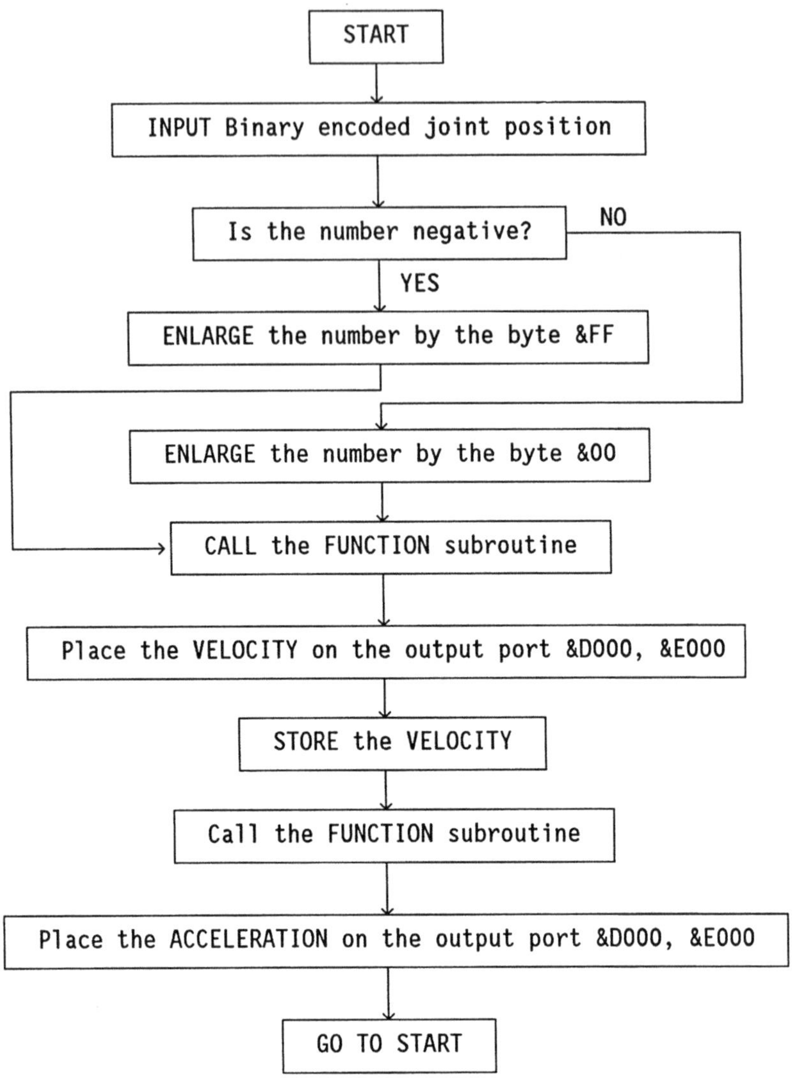

Fig. 10 Flowchart of the processing algorithm

B. The Clock-controlled analog circuit

The velocity and acceleration estimator based on a position history is also implemented using the analog circuit shown in Fig. 11(a). This circuit consists of two sample-and-hold circuits, an analog subtractor based on a low-noise operational amplifier, an 8-bit A/D Converter and the necessary timing circuitry. The two sample-and-hold circuits are controlled by the SAMPLE CONTROL inputs A and B whose waveforms are shown in the timing

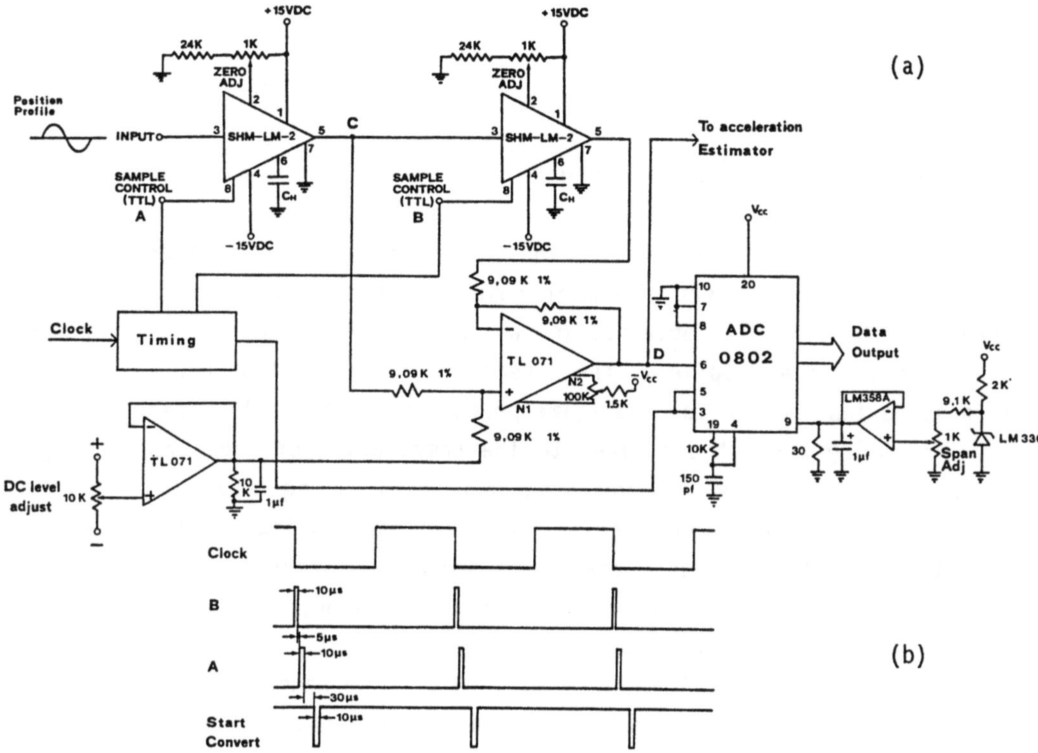

Fig. 11 Clock-controlled analog circuit for time-derivative calculation

diagram of Fig. 11(b). In this manner, the first sampler holds the current position value q(i), while the second sampler holds the previous value q(i-1). These two values are subtracted by an analog subtractor and the resulting difference corresponds to the time derivative of the waveform applied at the input. The difference of this technique compared with that used in Differential-Pulse-Code Modulation (DPCM) is due to the fact that the samples are subtracted in their analog form and not after they have been encoded. Thus the quantization error is avoided. The holding capacitors C_H are two 0.01 μF polystyrene type capacitors.

The analog voltage corresponding to the first derivative value at this point can be converted to a digital 8-bit number by means of an A/D converter. In an identical way, the velocity differences can be derived to calculate acceleration. Analog amplifiers can be used to amplify the weak velocity levels.

Experimental results and conclusions

The implementation of an analog and a digital circuit dedicated to the computation of the direct-kinematic solution of a robotic manipulator has been described. Results obtained in this work indicate that, with currently available analog multipliers, the direct-kinematic solution can be achieved in a speed fast enough for real-time processing of kinematic information.

An analog circuit dedicated to the computation of the components p_x, p_y and p_z of the position vector **p** according to equation (5) has been constructed in the laboratory. Four-quadrant analog multipliers and operational-amplifier-based adders were used. To investigate the linearity of analog multiplication as well as the precision, a sawtooth waveform (V_1). (where $-1V \leq V_1 \leq +1V$), was fed into the X input of one analog multiplier, while several DC voltages (V_2), in the range $-1V \leq V_2 \leq +1V$ were fed into the Y input. The circuit provided a very good linearity, as shown in the photograph of Fig. 12. In this photograph, the sawtooth (with peak values $-1V$ and $+1V$) was multiplied first by $V_2=-1V$ and then by $V_2=+0.5V$.

The decoder shown in Fig. 4(a) was also constructed in order to derive the trigonometric functions of the joint angles of a manipulator under construction in our laboratory. For our specific application, optical

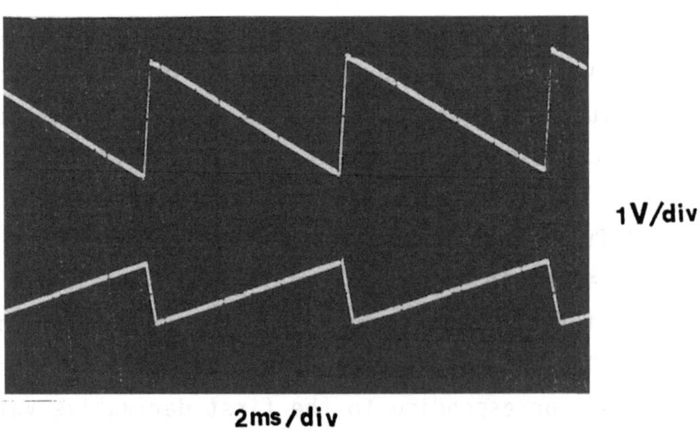

Fig. 12 Linearity of analog multiplier for a sawtooth input X multiplied by -1V and +0.5V through input Y

encoders in cooperation with a 10-bit up/down counter were used at the manipulator's joints to encode the joint angle in a 10-bit binary number. Ordinary TTL components were used for the decoder construction, while one EPROM 2708 organized as 1kx8 was used to store the values of sin θ for $0 \le \theta < \pi/2$ in 8-bit words. The two least-significant bits were derived by linear interpolation.

Figure 13 shows the waveforms corresponding to the quantized position and velocity profiles derived at points C and D of the analog estimator of Fig. 11(a). For a short time period, both samplers hold the same sample and the velocity becomes zero but it does not cause any problems.

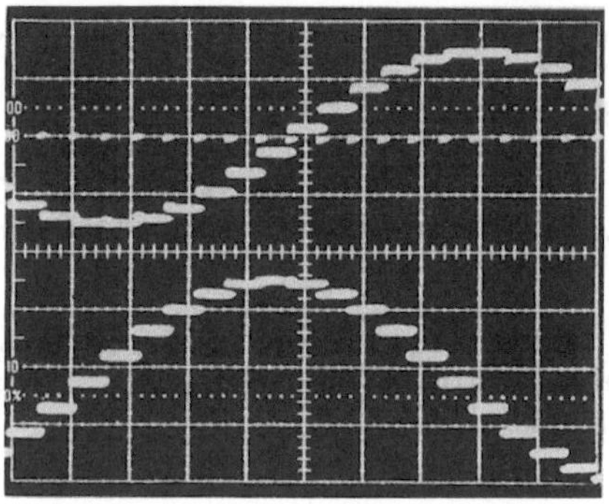

Fig. 13 Experimental results derived from the analog-parameter estimator

It must be noted that the circuit implementation of Fig. 11(a), which is attractive because of its simplicity and low cost, is more accurate than the circuit of Fig. 9, since no quantization error is introduced until the velocity and acceleration are encoded in digital words. Error in this case is introduced by the output offset voltage (which is less than 7 mV) and by the hold-mode drop which is less than 200 μV/msec.

Significant increase in the position-sampling rate has been achieved using both of the above-described dedicated hardware. Specifically, in order for the second circuit to achieve a maximum analog-velocity signal of 2V p-p, for an input signal 8V p-p, the sampling frequency must be thirty times the maximum frequency of the position signal.

References

1. Bult, K. and Wallinga, H.: A CMOS Four-Quadrant Analog Multiplier, IEEE J. Solid-State Circuits, Vol. SC-21, No.3, June, 19861.

2. Kim, B. and Shin, K.: Minimum-time path planning for robot arms and their dynamics, IEEE Trans. Syst., Man and Cyb., Vol. SMC-15, No.2, pp. 213-223, March/April 1985

3. Leung, S. S. and Shanblatt, M. A.: Real-Time DKS on a Single Chip, IEEE J. of Robotics and Autom. Vol. RA-3, No. 4, pp. 281-290, August 1987

4. Luh, J. Y. S. and Lin, C. S.: Optimum path planning for mechanical manipulators, Trans. ASME J. Dynamic Syst., Meas., Contr., Vol. 102, pp. 142-151, June 1981

5. Lygouras, J. N.: Nonlinear analogue-to-digital conversion through PWM, Int. J. Electronics, 62, pp. 347-352, March 1987

6. Lygouras, J. N.: Nonlinear ADC with digitally selectable quantizing characteristic, IEEE Trans. Nuclear Science, Vol. NS-35, No.5, pp. 1088-1091, Oct., 1988

7. Paul, R., Shimano, B. and Mayer, G.: Kinematic Control Equations for Simple Manipulators, IEEE Trans. Syst. Man and Cyb., Vol. SMC-11, No. 6, June 1981

8. Turney, J. L. and Mudge, T. N.: VLSI implementation of a numerical processor for robotics, in Proc. 27th Int. Instrumentation Symp., Indianapolis, IN, pp. 169-175, April 1981

The 3DP Real-Time Motion-Control Computer

Yulun Wang Amante Mangaser Steve Jordan
Partha Srinivasan Steven Butner

Computer Motion Inc.

Center for Robotic Systems
University of California, Santa Barbara

Abstract

Implementing advanced motion-control algorithms requires a computer which can satisfy the computational needs of the problem, a powerful development environment for programming complex and abstract ideas, and an open-architecture which gives the programmer full access to all of the system's capabilities. Conforming to industry standards is also important for system integration with other vendors' equipment.

We have developed a system - the 3DP controller - which satisfies these requirements. The 3DP has a unique processor architecture which exploits the intrinsic 3-dimensional nature of kinematic and dynamic equations. The majority of computations in advanced robot control algorithms are for solving kinematics and dynamics. By exploiting the innate 3-dimensional characteristic of the target problem the 3DP achieves execution speeds many times faster than any existing processor.

The 3DP is a second generation robot control computer which improves on the ideas developed in the RIPS architecture [1]. It is an attached processor which plugs into a workstation's VME bus. An optimizing C++ compiler lets the user program complex algorithms in a high-level object-oriented style, yet extremely efficient run-time code is produced. The user accesses the 3DP with familiar UNIX system calls, and the industry-standard VME bus allows the 3DP to communicate with other vendors' products. Even though the user views the 3DP through UNIX, real-time program execution and external communication is not degraded by UNIX's unpredictable response time.

Keywords: computer architecture, real-time dynamics, kinematics, robot control

1. Introduction

Implementing advanced motion-control algorithms requires a controller which can satisfy the computational needs of the problem, a powerful development environment for programming complicated and abstract ideas, and an open-architecture which gives the programmer full access to all of the system's capabilities. The experimental work of many roboticists has been limited by the capabilities of their laboratory set up.

The 3DP controller is designed to support the computational requirements of today's advanced motion-control methodologies. Model-based control, adaptive control, force control, and multi-manipulator cooperation are just a few examples of ideal application areas for the 3DP controller. The 3DP has a powerful object-oriented

NATO ASI Series, Vol. F 71
Expert Systems and Robotics
Edited by T. Jordanides and B. Torby

development environment which facilitates the implementation and modification of complex ideas and algorithms.

A wide array of processor architectures have been developed to handle the calculations required to solve today's problems. Most of these designs, however, employ "general-purpose" architectures so that they can be applied to a wide range of applications. The calculation requirements for data-base management and scientific computations are very different in structure and require a very flexible, or general, processor architecture to process both.

General purpose architectures sacrifice speed for flexibility. Their designs are not aimed at any particular class of problems. If a class of problems with a common structural characteristic is identified, and this class of problems encompasses enough applications, a special-purpose processor architecture may be justified. Digital signal processors (DSPs) are an example of an application area which has been identified and supported by special-purpose hardware. DSPs exploit the successive multiply/accumulate nature of signal processing applications with direct hardware support. Once DSPs were created, they were immediately found to be useful in many other application areas, for example in robotics [1-2].

We have identified a fundamental structural characteristic in motion-control problems and have designed a new processor architecture which exploits it. The result is a tremendous increase in the execution speed of equations which deal with the motion or representation of physical objects. We have encapsulated this processor design into a real-time motion-control system which provides unmatched real-time computational performance as well as a very powerful software development environment.

2. The Target Problem

Advanced robot control algorithms are derived from a few fundamental kinematic and dynamic properties. These equations explain relationships between physical quantities such as forces, moments, positions, velocities, and accelerations. In other words, they describe the behavior of physical objects. These equations are listed below.

$$V = \omega \times R$$
$$A_n = \omega \times (\omega \times R)$$
$$A_t = \alpha \times R$$
$$F = mA$$
$$M = I\alpha$$

where:

- V - linear velocity vector
- ω - angular velocity vector
- A_n - normal acceleration vector
- A_t - tangent acceleration vector
- R - radius vector
- F - force vector
- M - Moment vector
- α - angular acceleration vector

I - inertia tensor (3-by-3 matrix)
m - mass (scalar)

A homogeneous transformation is an important and frequently used operation. It is frequently used in robotics to rotate and translate a vector from one coordinate system to another.

$$\begin{bmatrix} s_i \\ s_j \\ s_k \\ 1 \end{bmatrix} = \begin{bmatrix} n_i & o_i & a_i & p_i \\ n_j & o_j & a_j & p_j \\ n_k & o_k & a_k & p_k \\ 0 & 0 & 0 & 1 \end{bmatrix} \begin{bmatrix} t_i \\ t_j \\ t_k \\ 1 \end{bmatrix}$$

By examining these equations it is apparent that there is an intrinsic 3-D character to them. An intuitive reason for this three dimensional structure is that kinematic and dynamic equations are mathematical expressions which explain the motion of 3-D rigid bodies in a 3-D space. Position, velocity, acceleration, force, and moment are all 3-D vector quantities. Consequently, this 3-D characteristic is fundamental to applications which require calculating the motion or representation of objects in the real-world.

Other researchers have tried to improve the calculation speed of kinematic and dynamic equations by first reformulating the problem for multiprocessor evaluation and then using multiple general purpose processors for real-time execution [5-6]. However, if the system to be controlled is dynamically coupled, such as a robot manipulator, this approach generates inadequate results. Kinematic and dynamic equations for a dynamic system are often very tightly coupled and do not lend themselves to parallel processing. By exploiting the parallelism of the problem at the *vector* level, much better results are obtained. Parallel processing techniques are applicable to dynamically uncoupled systems. In this case, multiple 3DPs could be used.

3. The 3DP Architecture

The 3DP's architecture accelerates three dimensional geometric computations. More importantly, it accomplishes this without sacrificing scalar performance. Even though the target problem is primarily 3-D, a significant number of scalar computations are also required. Reducing the scalar processing performance could easily offset any gain achieved by fast 3-D processing. The 3DP can be classified as a general-purpose processor with accelerated 3-D processing capabilities.

Other vector processors offer arbitrary vector lengths and use multi-staged pipelines, which have long latencies, to achieve high speed execution [7]. The short vector length of three-dimensional operations limits throughput with the multi-staged approach because of the large pipeline latency. Since the 3DP is built specifically for vectors of length 3, it exploits the intrinsic parallelism by providing three parallel execution units that can simultaneously operate on all three vector components. Scalar processing does not suffer with this parallel execution approach, whereas it would with a pipelined approach. The execution units are used individually for scalar operations. Another important capability of this architecture is that it supports vector-scalar operations, as illustrated in Figure 3.

A schematic block diagram of the 3DP is shown in Figure 1. This figure shows both the processor architecture and how we have encapculated it into a real-time motion controller. The data path can operate on two three-component vectors producing a three-component result. The data path originates (and ends) at the vector register file, where each vector register has an *i*, *j*, and *k* component. The instruction set is organized so that any vector register can be accessed as a single entity, such as **A**, or the components of the vector register can be accessed individually, such as **A.i** (the *ith* component of vector register **A**).

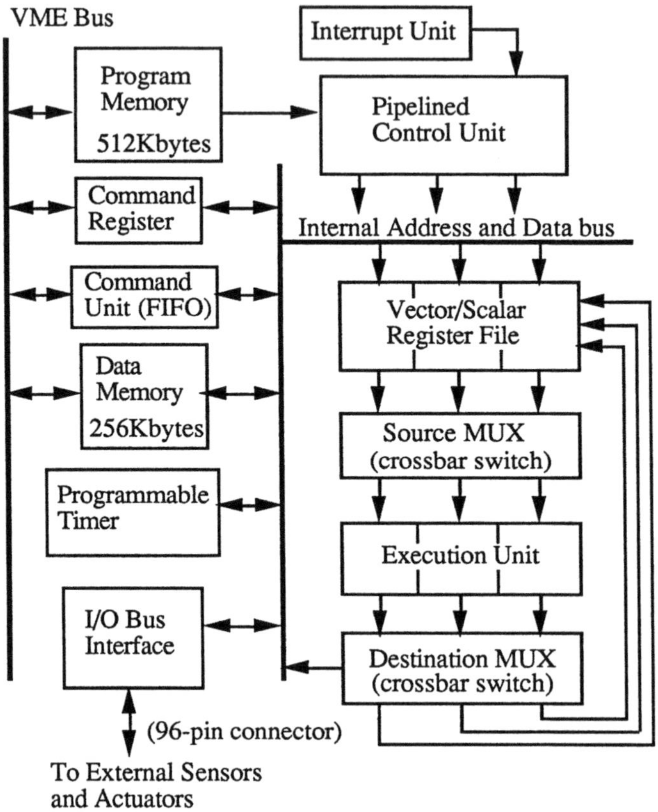

Figure 1. The 3DP

In the current implementation the register file has 64 vector registers, each having three 32-bit components. Therefore, the register can hold up to 64 vectors, 192 scalars, or any combination of the two. Each vector register can be dynamically replaced by 3 scalar registers. Consequently, the allocation of vector registers and scalar registers can be constantly changing throughout program execution. A large register file is provided to accomodate the large number of variables in complex motion-control algorithms.

The 3DP is a register-to-register architecture, which facilitates faster clock speeds and a simple instruction set [8]. The register file connections are shown in Figure 1. The output ports connect to the execution unit. One of the input ports comes from the external data bus, while the other one returns the operands from the execution unit via a multiplexer. The primary purpose of the register file is to store data operands. It also serves a secondary purpose, however. Since the processor has an internal vector arrangement, the register file is used to interface the processor's external single-width data bus to the internal three-wide execution bus. A single-width external bus is consistent with other general purpose microprocessor architectures, and standard interfacing techniques can be used to connect peripherals to the 3DP (e.g., memory and I/O ports).

The execution of every functional instruction begins by fetching the source operands from the register file. These operands enter the three-wide execution unit via six possible data paths. The execution unit consists of three ALUs and three FPUs connected pairwise in parallel, with each column consisting of one ALU and one FPU. Two data paths connect each column of the vector register file to the corresponding execution unit column.

A source multiplexer (shown in Figure 1) is used during inter-column operations. For example, when **A.i** is added with **B.k**. Inter-column operations are used in both vector and scalar operations. After the desired instruction is executed the destination multiplexer returns the operands to the register file. The destination multiplexer is capable of rearranging the resulting components before storing them into the register file. This unit is also used in scalar operations for intercolumn operations. The destination and source multiplexers give the 3DP its unique capability of supporting both fast and efficient 3-D geometric processing while maintaining good scalar performance. The best way to illustrate this capability is by example.

One of the most common three-dimensional geometric operations is the vector cross product. A cross product is necessary is to calculate the derivative of a vector in a rotating coordinate frame, for example. The 3DP performs these operations as follows:

```
RFMUR    R1, R2, R3     ; rotate vector R1 right 1 component,
                        ; floating-point vector multiply with R2,
                        ; rotate result right 1 component and store in
                        ; R3
LFMUL    R1, R2, R4     ; rotate vector R1 left 1 component, floating
                        ; point vector multiply with R2, rotate result
                        ; left 1 component and store in R4
NOP                     ; nop operation, pipeline break
FSU      R3, R4, R3     ; floating-point vector subtract of R4
                        ; from R3 with the result stored in R3
```

Note: the above rotates are vector component rotates
(i.e. for a right rotate $i \rightarrow j,\ j \rightarrow k,\ k \rightarrow i$)

Figure 2. Cross Product Operation (R3 = R1 $\times$ R2)

Since the 3DP completes one instruction every cycle, a floating-point vector cross product is executed in 4 cycles (or 400 ns). Another cycle is needed before the result of the cross product is available for further calculations.

Another common 3-dimensional operation is a vector rotation. Generally a rotation is accomplished by multiplying the vector with a rotation matrix. Figure 3 lists the instruction sequence that the 3DP performs to calculate a matrix-vector multiply. The rotational matrix columns are stored in vector registers R1, R2, and R3, and the vector to be rotated is stored in R4. The final result is placed in R5.

This operation completes in 6 cycles, or 600 ns. Note that the first 3 instructions are scalar-vector multiply instructions. This vector rotation sequence can be generalized to perform a homogenous transformation by adding a single vector addition. Hence, homogeneous transformations are performed in 700 ns on the 3DP.

```
FMU    R1, R4.i, R1    ; vector-scalar floating-point multiply
FMU    R2, R4.j, R2    ; vector-scalar floating-point multiply
FMU    R3, R4.k, R3    ; vector-scalar floating-point multiply
FAD    R1, R2, R2      ; vector addition
NOP                    ; no operation, pipeline dependency
FAD    R2, R3, R5      ; vector addition
```

Figure 3. Vector Rotation

A unique problem encountered during the design of the 3DP was the selection of the appropriate condition codes for program control instructions. Since there are 3 parallel execution units, there are 3 sets of condition flags. Many interesting usages of these conditions were considered. For example, it is possible to immediately determine whether or not a vector is zero by simultaneously testing all three zero flags. Another possibility is to provide vector quadrant tests, which determine whether or not a vector lies in a particular quadrant. We decided to provide the typical scalar test conditions, and also added a zero vector test. Other vector combinations can be included in future designs if particular ones seem useful.

The 3DP has a Harvard architecture [9], which separates program and data memory. The advantage of separate memories is that instruction and data fetches can occur simultaneously, thereby increasing execution speed. Another advantage is that program execution speed is more deterministic, an important property for real-time applications. Since program and data memory fetches do not interfere, one instruction is guaranteed to execute per clock cycle.

The 3DP is the second generation of an earlier design - the robotic processor. The robotic processor is the key computational element of a heterogeneous multi-processor system - RIPS [1]. The 3DP architecture was developed with the same 3-D philosophy as the robotic processor, but its design is improved. One of the major changes is a completely orthogonal datapath, which gives each register the same functionality. In other words, any register can be used for any of the assembly language instructions. Orthogonality is very important when developing a high-level language compiler for a processor architecture [10]. We discovered that the non-orthogonalities in the robotic processor design prevented our compiler from generating efficient assembly code.

4. 3DP Performance

The 3DP architecture has been incorporated into a real-time motion-control system - the 3DP Controller. Our current version implements the 3DP architecture in discrete logic and is clocked at 10Mhz. Since one instruction completes every cycle, and because there are 3 parallel execution units, the peak execution rate is 30 MFLOPS. Other processors capable of 30 MFLOPS performance exist. However, since the 3DP architecture is uniquely matched to 3-dimensional geometric operations, the *actual* execution speed for 3-D algorithms is close to 30 MFLOPS.

Table 1 shows the computational performance of this system. Note that the scalar addition, subtraction, and multiply take the same computational time as their 3-D counterparts. These times assume fully pipelined operation. Since the 3DP is a four-staged pipeline with a one cycle inter-dependency, one unrelated instruction (e.g., NOP) is required before any instruction which uses the result of an immediately previous instruction.

Table 1.
3DP Benchmarks

3-D addition/subtraction/multiply	100ns
3-D vector cross product	400ns
3-D vector rotation	600ns
homogeneous transformation	700ns
scalar addition/subtraction/multiply	100ns
scalar reciprocal approximation	500ns
vector divided by scalar	700ns

Table 2 shows some comparisons between the 3DP and other computer systems. The operations used to make the comparisons are funcions which are commonly used in advanced robot servoing techniques. These benchmarks were obtained from the same C++ program for a general revolute open-chained 6 degree-of-freedom robot

Table 2.
3DP Comparisons
Times are given in μicroseconds

Operation	3DP (10 Mhz)	SUN 4 (16 Mhz)	SGI 4D70 (12.5 Mhz)
Inv. Dynamics	438.8	1993.0	2491.0
Jacobian	339.6	896.0	1080.0
Kinematics & Dynamics	703.0	2828.0	3586.0
Forward Kinematics	301.4	783.0	966.0

Table 2 shows that the 3DP is significantly faster than the other systems. The current implementation uses discrete logic and is therefore clocked at a slower frequency. Once the 3DP is placed into VLSI, the performance improvement will be much more significant.

5. The 3DP Development Environment

A complete computer system requires not only a good hardware design but a powerful and easy-to-use software development environment. The 3DP controller is an attached processor integrated within the UNIX environment. It plugs directly into the VME bus of a SUN host computer. Program memory, data memory, and special communication registers are mapped into the VME address space. The dual-ported data memory supports data transfers with the host during program execution. The special communication registers include a bi-directional FIFO and a bi-directional register. Different applications require different methods of communication - our architecture provides the end user with several choices.

Since the 3DP is connected to a UNIX host via a device driver, all of the UNIX facilities are available for 3DP program development. Accessing the 3DP is performed using UNIX system calls. Math and remote procedure call libraries support program development. Figure 4 shows a block diagram of the 3DP software development environment.

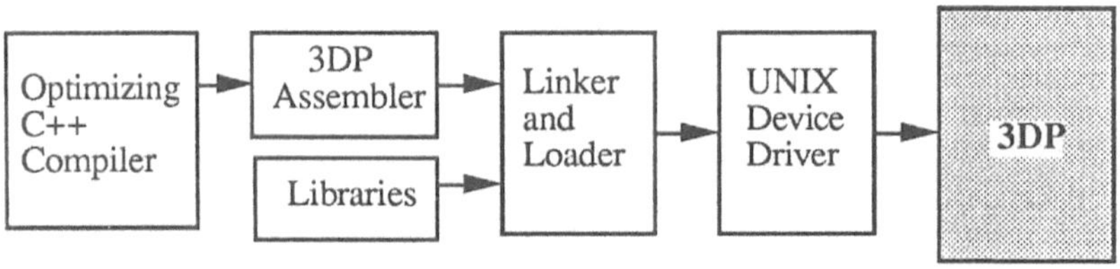

Figure 4. The 3DP Software Development Environment

6. Object-Oriented Programming on the 3DP

We have adapted the GNU C++ optimizing compiler [11] to the 3DP architecture to support high-level programming. There are several approaches to providing high level translation for a unique architecture. One choice is to define a custom language with its own syntax and semantics tailored to fit the 3DP architecture. While this seems advantageous for implementation, it has the significant disadvantage of requiring the users to learn a new custom language. Another approach is to extend an already existing standard language with specialized vector features. This approach extends both the syntax and semantics of the existing language to include a "machine" vector data type. The disadvantages of this approach are that the resulting language is no longer a standard language, and that major changes to an existing compiler must be made. A third approach is to develop a vectorizing compiler for a standard language. In this case, both the syntax and semantics of the language remain untouched. The compiler must include very sophisticated optimization techniques which automatically detect sections of code amenable to vectorization. Again, this approach requires major compiler modifications.

The last approach - the one we chose - is to begin with a user-extendible language like C++. In this case, the language allows the user to define new data types - e.g., 3-dimensional vector. The user also provides the compiler with an entire set of operators and functions which work on this data type. This solution is ideal since the standard language's syntax and semantics are left unchanged, and the user can write programs using the new data type which the underlying machine can efficiently manipulate.

C++ is a relatively new object-oriented programming language [12]. The object-oriented paradigm facilitates higher levels of abstraction than procedural languages. Object-oriented programming is especially well suited for manipulating real world objects. Even though C++ is a relatively new language, its following is rapidly increasing.

Instead of writing a compiler from scratch, we adapted the GNU compiler from the Free Software Foundation to the 3DP architecture. The GNU software has already developed a reputation for its high quality code. The GNU optimizing C and C++ compilers share the same retargettable back end, but use different front ends.

The object-oriented features of C++ allow us to generate efficient vector code for the 3DP. By introducing a VEC3 class (i.e. 3-dimensional vector), the programmer can directly manipulate 3-dimensional variables. A new VEC3 machine-mode is included within the compiler to enable the compiler to use the hardware support for this class. A 3-D vector/scalar register organization is also mapped into the compiler. Currently, the 64 vector registers are segmented into 48 scalar registers and 48 vector registers. This makes the register classes non-overlapping and simplifies the naming of the registers. The resulting compiler supports high-level object-oriented programming and exploits the parallel execution capabilities of the underlying machine.

7. Programming the 3DP in C++

Since motion-control problems are primarily derived from fundamental kinematic and dynamic relationships built from 3-D vector quantities, the VEC3 data type allows the programmer to implement motion-control algorithms in a very natural way. Furthermore, the VEC3 objects show the compiler where the intrinsic 3-D parallelism lies. This is what enables it to generate efficient vector code for execution on the 3DP.

C++ supports the concept of *operator overloading*, which lets the user define the semantics for operators operating on the new data type. We overloaded the arithmetic operators so that the C++ compiler understands how to manipulate VEC3 objects. Addition, subtraction, multiplication, and division are performed component-wise. For example, vector addition is defined such that the i components are added together, the j components are added together, and the k components are added together. Useful combinations of VEC3 objects and scalars are also overloaded. For example, the addition of a VEC3 object and a scalar is defined such that the scalar is added to each component of the vector.

The three components of a VEC3 object are also accessible as array components (e.g., s[0] refers to the *ith* component). These components can be accessed individually as well as collectively. Consequently, vector components can be individually used for scalar operations. This ability is useful in matrix-vector multiplies (such as the first 3 instructions of Figure 3), for example.

In order to illustrate the use of VEC3s, we can write a simple program which calculates the position of a point as a function of time. The equation which we will implement is the following:

pos = init_pos + init_vel * time + .5 * accel * time * time

where:

pos -	current position vector
init_pos -	initial position vector
init_vel -	initial velocity vector
accel -	acceleration vector
time -	time (scalar)

The corresponding program is as follows:

```
/* Class declaration */

class VEC3{
	public:
		float s[3];		// data members

	inline void init (float a, float b, float c) {s[0] = a; s[1] = b; s[2] = c;} // initialization

	friend inline VEC3 operator+( VEC3, VEC3);		// overloaded operator
	friend inline VEC3 operator*(VEC3, float);		// overloaded operator
}


/* VEC3_funcs defines the overloaded operators for VEC3s */
#include "VEC3_funcs"

main()
{
	VEC3 pos, init_pos, init_vel, accel;
	float time;

	init_pos.init (1.0, 2.0, 3.0);		// initialize position
	init_vel.init (5.0, 6.0, 7.0);		// initialize velocity
	init_accel.init (9.0, 8.0, 7.0);	// initialize acceleration

	for (time = 0; t < 10; t++){
		pos = init_pos + init_vel * time + .5 * accel * time * time;
	}
}
```

This program illustrates how simply equations involving 3-D vectors and scalars are programmed. The compiler is told how to add and multiply vectors, as well as combinations of vectors and scalars, by overloaded functions which are placed in the VEC3_funcs include file. This file is provided as part of the system. The end user need not be concerned with its implementation details. Once all of the necessary operators are overloaded, the user can work at a high-level of abstraction since the programmer no longer needs to be concerned with the details of the operation. The only consideration is the implementation of the algorithm.

VEC3s can also be used to build other classes. For example, we have created a class called ROT to support rotational transformations. A ROT is a 3-by-3 matrix made from 3 VEC3s whereby each VEC3 stores a different column of the matrix. These column can be individually accessed. The names of the vectors are n (normal), o (orientation), and a (approach), which is consistent with homogeneous transformation terminology. For example, if a ROT ttt was declared, ttt.n refers to column n of the matrix. Since each component of a vector can be accessed independently, ttt.n.i refers to the ith component of the nth column of matrix ttt.

8. The 3DP Real-time Controller

The 3DP controller is an attached processor for SUN workstations and it is accessed by the user through the UNIX operating system. Even though software development for the 3DP is supported by UNIX, real-time execution on the 3DP is not impeded by UNIX's non-real-time character. UNIX is a multi-tasking interrupt-driven system and was not designed to handle tasks under stringent real-time constraints. 3DP program execution is autonomous and can therefore operate under stringent real-time constraints. The UNIX host may communicate with the 3DP during run-time. However, if the 3DP is asked to wait for data from the host, the 3DP cannot maintain its deterministic response times. The host may sample data from the 3DP during execution without degrading its real-time performance.

An on-board programmable timer enables the 3DP to manage its own time, and a dedicated I/O bus lets the 3DP communicate with external sensors and actuators without using the UNIX-controlled VME bus. The dual-ported data memory and dedicated communication channels let the host computer communicate with the 3DP during program execution. However, since the host is assumed to operate under less stringent real-time requirements, the interface is designed to give the 3DP priority to its resources (e.g., data memory).

Figure 5 shows the software structure of the real-time execution environment. The remote procedure call (RPC) daemon supports UNIX system calls from the 3DP. For example, the 3DP can open, close, read, and write files on the UNIX file system. Furthermore, the RPC supports printf, scanf, and other important "service" function calls from the 3DP. With the RPC daemon, the 3DP has complete access to the UNIX file system as well as to any terminal or peripheral accessible by the host.

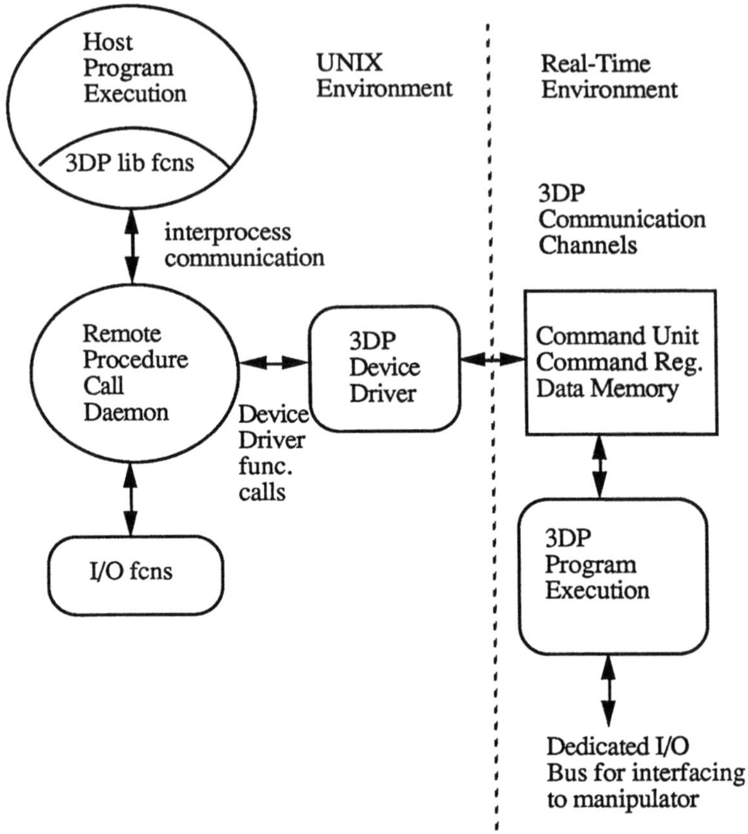

Figure 5. The 3DP Software Execution Environment

9. Conclusion

Today's robots are limited in their ability because their controllers have inadequate performance. Researchers have developed many new algorithms which would give robots more capability. However, current controllers are unable to satisfy the real-time computational demands of these algorithms. Furthermore, existing controllers offer poor programming interfaces, which reduces the effectiveness and speed with which complex algorithms can be implemented.

The 3DP real-time controller is an ideal platform for an advanced robotic system. It achieves superior performance by exploiting the underlying structure of the fundamental kinematic and dynamic relationships which describe object motion. These relationships are basic to the development of advanced motion-control algorithms.

An object-oriented programming environment provides a powerful and easy to use interface to the 3DP. Advanced algorithms and abstract concepts can be programmed directly at a high level. The object-oriented C++ compiler understands the 3DP's hardware architecture and is able to exploit its parallel processing capabilities. The 3DP controller is integrated within the UNIX environment and is accessible through a SUN host. Even though program development is performed under UNIX, the 3DP can operate in real-time because it can directly access sensors and actuators through a private I/O bus with an on-board programmable timer providing the notion of absolute time.

References

[1] Wang, Yulun and Steven E. Butner, "RIPS: A Platform for Experimental Real-Time Sensory-based Robot Control", *IEEE Transactions on Systems, Man, and Cybernetics*, vol 19, No. 4, July/August 1989.

[2] Takanashi, N., Ikeda, T., and N. Tagawa,"A High-Sampling Rate Robot Control System Using a DSP Based Numerical Calculation Engine," *Proc. of the IEEE Conf. on Robotics and Automation*, Scottsdale, Arizona, 1989.

[3] Luh, J.Y.S., M.W. Walker, and R.P.C. Paul,"On-Line Computational Scheme for Mechanical Manipulators," Trans. ASME, J. Dynam. Syst., Mea. Contr., vol. 120, pp. 69-76, June 1980.

[4] Goldstein, Herbert, *Classical Mechanics*, Addison-Wesley, 2nd ed., Reading Mass., 1980,

[5] Barhen, J., Halbert, E.C. and J.R. Einstein, "Advances in Concurrent Computation for Autonomous Robots," *Proc of the Robotics Research Conf*, Scottsdale, Arizona, August 1986.

[6] Lee, C.S.G and P.R. Chang,"Efficient Parallel Algorithm for Robot Inverse Dynamics Computation," *Proc of the IEEE Conf on Robotics and Automation*, San Francisco, CA 1986.

[7] Hwang, Kai, and Faye A. Briggs, *Computer Architecture and Parallel Processing*, McGraw-Hill, New York, 1984.

[8] Katevenis, Manolis G.H., *Reduced Instruction Set Computer Architectures for VLSI*, Ph.D. dissertation, University of California, Berkeley, Oct. 1983.

[9] Rosen, Saul,"Electronics Computers: A Historical Survey," Computing Surveys, v1, #1, March 1969, pp. 67-78.

[10] Wirth, N., "Hardware Architectures for Programming Languages and Programming Languages for Hardware Architectures," *Proc. of the ASPLOS II*, Palo Alto, CA., Oct., 1987.

[11] Stallman, R., "GNU Users Manual," Free Software Foundation, 1986.

[12] Stroustrop, Bjarne, "The C++ Programming Language," Addison-Wesley, Reading, Mass., 1986.

A KNOWLEDGE-BASED CONTROLLER FOR A PWR-TYPE NUCLEAR POWER PLANT

H. Levent Akin

Department of Computer Engineering, Bogazici University,
80815-Bebek, Istanbul, TURKEY

Abstract: In this study a rule-based, fuzzy-logic regulator was developed for a PWR type nuclear power plant based on knowledge acquisition through numerical simulations and making use of a validated mathematical model of the H.B. Robinson power plant. The performance of this controller with the single objective of minimizing weighted power was found to be better in this respect when tested against that of an optimal controller using several conflicting performance criteria. The behavior of this rule-based controller was investigated also under noisy operating conditions and in the presence of drift in process variables leading to negligible degradation.

Keywords: fuzzy-logic, rule-based controller, pwr-type nuclear power plant.

Introduction

As a means of control, the well developed analytic techniques have been used successfully for years. However, these systems, in addition to some drawbacks such as to having to accurately model, are sensitive to failures in sensors, and are inflexible. A large drift in process variables renders the controller almost useless, and in more dramatic cases dangerous. Recently, however, with the developments in the field of artificial intelligence (AI), a new type of controller, namely the rule-based or knowledge-based controllers has been developed and implemented. The advantage of rule-based controllers is that they are generally more robust than their analytic counterparts. As yet, there are no comprehensive guidelines for the design of rule-based controllers and such systems are quite difficult to calibrate.

In the field of nuclear engineering, rule-based controllers have been designed and implemented for research reactors [5], and for boiling water reactor recirculation flow-control system [4]. In this work, a rule-based fuzzy-logic controller was developed for a validated model of the PWR-type H.B. Robinson nuclear power plant and its performance is compared with that of an analytical controller also built for this plant [10]. The rule base was formed using numerical simulation as the main source of knowledge. The main design criteria were simplicity, ease of implementation and robustness.

NATO ASI Series, Vol. F 71
Expert Systems and Robotics
Edited by T. Jordanides and B. Torby

Mathematical Model of the Nuclear Power Plant

The development and testing stages of a knowledge based controller necessitates extensive simulations since its structure is not suited to mathematical techniques used for conceiving analytical controllers [5]. For this reason, the multi-input mathematical model formulated and validated against experimental results by Kerlin et al. [7] for predicting the dynamic response of the H.B. Robinson nuclear power plant(NPP) during full-power operation was used. It is a simple model and suitable for fast simulations. H. B. Robinson is a pressurized-water reactor with slightly enriched fuel. It has three vertical U-tube type steam generators.

The original model [7,3] is based on the basic conservation laws for neutrons, mass and energy. It includes the representation for point kinetics, core heat transfer, pressurizer, piping and the steam generator components. However, for controller design, the model is further simplified by making the following assumptions: 1) a single delayed neutron group is employed based on the weighted harmonic mean of the original six delayed neutron group constants; 2) the pressurizer is assumed to be large enough to accommodate the steam generator disturbances, therefore its dynamics is neglected in the model formulation; and 3) since the purpose of this controller is power regulation, it is assumed that no power level change occurs [1,9,10].

Reactor core:

Neutronics:

$$\frac{d\delta P}{dt} = -\frac{\beta}{\Lambda}\,\delta P + \sum_i \lambda_i \delta C_i + \frac{\alpha_f P_o}{\Lambda} \sum_{\substack{fuel\\nodes}} F_{fi}\,\delta T_{fi} + \frac{\alpha_p P_o}{\Lambda}\,\delta P_p + \frac{\alpha_c P_o}{\Lambda} \sum_{\substack{coolant\\nodes}} F_{ci}\,\delta T_{ci} \tag{1}$$

$$\frac{d\delta C_i}{dt} = \frac{\beta_i}{\Lambda}\,\delta P - \lambda_i\,\delta C_i \quad , i=1,\ldots,6 \tag{2}$$

Core heat transfer:

$$\frac{d\delta T_{c1i}}{dt} = \left(\frac{UA_f}{MC_p}\right)_{ci} (\delta T_{fi} - \delta T_{c1i}) - \frac{2}{\tau}\,(\delta T_{c1i} - \delta T_{cin}) \tag{3}$$

$$\frac{d\delta T_{c2i}}{dt} = \left(\frac{UA_f}{MC_p}\right)_{ci} (\delta T_{fi} - \delta T_{c1i}) - \frac{2}{\tau} (\delta T_{c2i} - \delta T_{c1i}) \tag{4}$$

$$\frac{d\delta T_{fi}}{dt} = \frac{Q_{fi}}{(MC_p)_{fi}} \delta P - \left(\frac{UA_f}{MC_p}\right)_{fi} (\delta T_{fi} - \delta T_{c1i}) \tag{5}$$

Steam generator:

$$\frac{d\delta P_s}{dt} = D_1 \delta P_s + D_2 \delta T_m + D_3 \delta T_{FW} + D_4 \delta W_{FW} + D_5 \delta W_{so} \tag{6}$$

$$\frac{d\delta T_p}{dt} = \frac{1}{\tau_{SG}} \delta T_{IP} - \frac{(hA)_{pm}}{M_p C_p} (\delta T_p - \delta T_m) - \frac{1}{\tau_{SG}} \delta T_p \tag{7}$$

$$\frac{d\delta T_m}{dt} = \frac{(hA)_{pm}}{M_m C_m} (\delta T_p - \delta T_m) - \frac{(hA)_{ms}}{M_m C_m} \left[\delta T_m - \left(\frac{\partial T_{sat}}{\partial P_s}\right) \delta P_s\right] \tag{8}$$

Piping and plena:

$$\frac{d\delta T}{dt} = \frac{1}{\tau} \delta T_{in} - \frac{1}{\tau} \delta T \tag{9}$$

Numerical values for the coefficients are obtained using the H. B. Robinson nuclear plant data presented in Tables I, II, III and IV.

TABLE I. Reactor Design Data.

Core Thermal and Hydraulic Characteristics	
Total primary heat output, MW (th)	2200
Nominal primary system pressure, psi (MPa)	2250 (15.82)
Total coolant flow rate, lb/h (kg/s)	101.5×10^6 (1.279×10^4)
Average coolant velocity along fuel rods, ft/sec (m/s)	14.3 (4.358)
Total mass of coolant in primary loop, lb (kg)	406 050 (184 347)
Nominal coolant inlet temperature, ^oF (C)	546.2 (285.66)
Nominal coolant outlet temperature, ^oF (C)	602.1 (316.72)
Active heat transfer surface area, ft^2 (m^2)	42460 (3944.7)
Average heat flux, Btu/(h ft^2) (W/m^2)	171600 (5.44×10^4)
Fuel-to-coolant heat transfer coefficient (includes resistance in fuel) Btu/(h ft^2-F) (W/m2-C)	176 (100.4)
Kinetic Characteristics	
Doppler coefficient, $(\Delta k/k)/^oF$ $((\Delta k/k)/^oC)$	-1.3×10^{-5} (-2.34×10^{-5})
Moderator temperature coefficient, $(\Delta k/k)/^oF$ $((\Delta k/k)/^oC)$	-2.0×10^{-4} (-3.6×10^{-4})
Moderator pressure coefficient, $(\Delta k/k)$/psi $((\Delta k/k)$/MPa)	$+3.0 \times 10^{-6}$ (4.27×10^{-4})
Prompt neutron lifetime, sec	1.6×10^{-5}
Delayed neutron fraction	0.0064

TABLE II. Delayed Neutron Constants.

Mean Life (sec)	Decay Constant (λ_i, sec^{-1})	Fraction
80.4	0.0124	0.00021
32.8	0.0305	0.00140
8.98	0.111	0.00125
3.32	0.301	0.00253
0.88	1.14	0.00074
0.332	3.01	0.00027

TABLE III Pressurizer Design Data.

Water volume, full-power, ft^3 (m^3)	780 (22.09)
Steam volume, full-power, ft^3 (m^3)	520 (14.72)
Electric heater capacity, kW (total)	1300

TABLE IV. Steam Generator Data (for each unit)

Steam Generator	
Number of U-tubes	3260
U-tube diameter, In. (cm)	0.875 (2.22)
Average tube wall thickness, in. (cm)	0.050 (0.13)
Mass of U-tube metal, lb (kg)	91 800 (41 677)
Total heat transfer area, ft^2 (m2)	44 430 (4127.7)
Steam Conditions at Full Load	
Steam flow, lb/h (kg/s)	3.169x 10^6 (400)
Steam temperature, F (C)	516 (268.89)
Steam pressure, psig (MPa)	770 (5.3)
Primary Side Coolant	
Reactor coolant flow, lb/h (kg/s)	33.93x 10^6 (4279)
Reactor coolant water volume. ft^3 (m^3)	928 (26.28)
Secondary Side Fluid	
Feedwater temperature, F (C)	435 (223.89)
Secondary side water volume, full-power, ft^3 (m^3)	1526 (43.21)
Secondary side steam volume, full-power, ft^3 (m^3)	3203 (90.70)

Controller Development

Knowledge Acquisition

The primary task in building a rule-based fuzzy-logic controller is acquisition of the knowledge necessary for forming the rules that make up the knowledge base. The usual procedure in knowledge acquisition is based on the cooperation of the knowledge engineer with an identified expert in the field of concern. However, the regulation function to be performed by this controller is too fast a process for a human being to cope with, and therefore, beyond operator experience. For this reason the knowledge base had to be built without the cooperation of an appropriate expert and the only suitable method of forming a set of exhaustive, non-redundant, error and conflict-free set of rules making up the base proved to be numerical simulation.

In order to aid the development and implementation stages, an interactive software that allows the user to stop the simulation and observe each and all of the variables as time-domain plots has been developed. The programming language chosen was Turbo Pascal 4.0 [3] due to its ease in programming and ample graphic capabilities which were essential for simulation. The information about plant behavior necessary for rule formation and proper selection of fuzzy variables was acquired through simulations using this software.

Formation of the Knowledge Base

The information thus gathered had to be structured in order to: (a) to determine the manipulated and controlled variables; (b) to deduce the linguistic labels that can be used to describe the measured variables and determine the range of each label which will form the basis of the fuzzy sets to be used; (c) to determine the rules that will relate the linguistic labels to specific control actions.

In the original H.B. Robinson model [7,1] there are two variables that can be manipulated for implementing the desired control action, namely rod movement and steam flow rate. However, as the control rods are generally used for power-level change, only steam flow was manipulated for power regulation in accordance with previous work [9,10].

Of the measurable variables available to the operator, such as power (neutron flux), hot and cold-leg primary-water temperature, primary-loop pressure, feedwater flow rate, steam pressure, core outlet temperature, steam generator level, steam flow rate and rod position, power is the variable chosen to be controlled. During the aforementioned simulation runs it was seen that a sufficiently effective manual control of NPP was possible by acting only on the deviation of power from its steady-state value. The same set of simulation runs suggested the rate and sign of the deviation as other important factors that would reasonably be considered by a human controller. Based on these observations and simulation runs, it was decided to choose power error (PE) corresponding to δP in the model and the rate of change in power error (RCPE) as the linguistic variables. This choice also follows from a desire to make the rules as simple as possible both for purposes of development and implementation.

The allowable range for power variation that would be the subject of regulation was chosen to be ± 6 percent since the steady-state operation point is full-power at 2200 MW. In order to decrease chattering, a ± 0.25 percent band was taken as the dead band where no control action would be applied.

After the selection of the regulation range, the number of necessary fuzzy sets, their ranges and corresponding membership functions were determined. The number of fuzzy sets and their ranges depends on the fineness of the control rules and the desired control action. In this work, the initial functional representation for fuzzy sets was obtained by dividing the allowable positive and negative ranges into intervals of 2 percent of full-power. Positive-small, positive-medium and positive-large fuzzy sets have maximum membership values at 2, 4 and 6 percent of full-power and negative-small and negative-medium and negative large fuzzy sets have maximum membership values at -2, -4 and -6 percent of full-power,respectively. Adjustments were made to the membership values of intermediate values. Negative and positive-zero fuzzy sets correspond to the above mentioned dead band. Thus the selected allowable ranges were spanned with a set of eight fuzzy sets, each comprising the range of values of the linguistic variable PE, and two other sets of seven fuzzy sets for RCPE and change in steam flow rate(CSFR). The fuzzy sets are "negative-large", "negative-medium", "negative-small", "negative-zero", "zero", "positive-zero", "positive-large", "positive-medium", and "positive-small" as presented in Tables V and VI. The membership functions for these fuzzy sets and their parameters were adjusted through the process of calibrating the controller. The final fuzzy sets were decided upon after test runs, and have broken-line type membership functions.

TABLE V. Broken-Line Fuzzy Subsets Used in this Study.

Subset	Range	DOF	Subset	Range	DOF
Power Error [MW] :					
positive-zero	0.0 ≤ x < 5.5	1.0	negative-zero	-5.5 < x ≤ 0	1.0
	5.5 ≤ x < 44	1.1 - x/55		-44 < x ≤ -5.5	1.1 + x/55
	44 ≤ x < 66	0.9 - 3x/220		-66 < x ≤ -44	0.9 + 3x/220
	otherwise	0.0		otherwise	0.0
Rate of Change in Power Error(RCPE) [MW/s] :					
zero	0.0 ≤ x < 22	1.0 - 3x/220			
	22 ≤ x < 44	1.1 - x/55			
	44 ≤ x < 66	0.9 - 3x/220			
	-22 < x ≤ 0	1.0 + 3x/220			
	-44 < x ≤ -22	1.1 + x/55			
	-66 < x ≤ -44	0.9 + 3x/220			
	otherwise	0.0			
Power Error [MW] and Rate of Change in Power Error(RCPE) [MW/s] :					
positive-small	0.0 ≤ x[a] < 5.5	3x/55	negative-small	-5.5 < x ≤ 0	3x/55
	5.5 ≤ x < 22	1/6 + 4x/165		-22 < x ≤ -5.5	1/6 - 4x/165
	22 ≤ x < 44	0.4 + 3x/220		-44 < x ≤ -22	0.4 - 3x/220
	44 ≤ x < 66	1.6 - 3x/220		-66 < x ≤ -44	1.6 + 3x/220
	66 ≤ x < 88	1.9 - x/55		-88 < x ≤ -66	1.9 + x/55
	88 ≤ x < 110	1.5 - 3x/220		-110< x ≤ -88	1.5 + 3x/220
	otherwise	0.0		otherwise	0.0
positive-medium	22 ≤ x < 44	-0.3 + 3x/220	negative-medium	-44 < x ≤ -22	-0.3 - 3x/220
	44 ≤ x < 66	-0.5 + x/55		-66 < x ≤ -44	-0.5 - x/55
	66 ≤ x < 88	-0.2 + 3x/220		-88 < x ≤ -66	-0.2 - 3x/220
	88 ≤ x < 110	2.2 - 3x/220		-110 < x ≤ -88	2.2 + 3x/220
	110 ≤ x < 132	2.7 - x/55		-132 < x ≤ -110	2.7 + x/55
	132 ≤ x < 154	2.1 - 3x/220		-154 < x ≤ -132	2.1 + 3x/220
	otherwise	0.0		otherwise	0.0
positive-big	66 ≤ x < 88	-0.9 + 3x/220	negative-big	-44 < x ≤ -22	-0.3 - 3x/220
	88 ≤ x < 110	-1.3 + x/55		-66 < x ≤ -44	-0.5 - x/55
	110 ≤ x < 132	-0.8 + x/220		-88 < x ≤ -66	-0.2 - 3x/220
	132 ≤ x	1.0		-110 < x ≤ -88	2.2 + 3x/220
	otherwise	0.0		otherwise	0.0
Change in Steam Flow Rate					
positive-small	0 ≤ x < 16	x/16	negative-small	-16 < x ≤ 0	-x/16
	otherwise	0		otherwise	0
positive-medium	16 ≤ x < 32	-1 + x/16	negative-medium	-32 < x ≤ -16	-1 - x/16
	otherwise	0		otherwise	0
positive-big	32 ≤ x < 48	-2 + x/16	negative-big	-48 < x ≤ -32	-2 - x/16
zero	∀x	0			

[a] x is either PE or RCPE.

TABLE VI. S-Shaped Fuzzy Sets Used in this Study.

Subset	Range	DOF	Subset	Range	DOF
Power Error [MW] :					
positive-zero	$x < 0$	0	negative-zero	$0 < x$	0
	$0 \le x < 5.5$	1		$-5.5 < x \le 0$	1
	otherwise	$(1+(2/55(x-5.5))^2)^{-1}$		otherwise	$(1+(-2/55(x+5.5))^2)^{-1}$
Rate of Change in Power Error(RCPE) [MW/s] :					
zero	$\forall x$	$(1+(x/33)^2)^{-1}$			
Power Error [MW] and Rate of Change in Power Error(RCPE) [MW/s] :					
positive-small	$x^a \le 5.5$	0	negative-small	$-5.5 \le x$	0
	otherwise	$(1+(-4/121(x-44))^2)^{-1}$		otherwise	$(1+(4/121(x+44))^2)^{-1}$
positive-medium	$x \le 5.5$	0	negative-medium	$-5.5 \le x$	0
	otherwise	$(1+(-1/33(x-88))^2)^{-1}$		otherwise	$(1+(1/33(x+88))^2)^{-1}$
positive-big	$x \le 5.5$	0	negative-big	$-5.5 \le x$	0
	$5.5 < x \le 132$	$(1+(-1/33(x-132))^2)^{-1}$		$-132 \le x < -5.5$	$(1+(1/33(x-132))^2)^{-1}$
	$132 < x$	1		$x \le -132$	1

[a] x is either PE or RCPE.

The conditional rules to be assessed during the manual control trials were of the following type

$$IF\ Condition(PE)\ \wedge\ Condition(RCPE)\ THEN\ Action. \qquad (10)$$

where Condition(PE) and Condition(RCPE) together with the conjunction operator form the antecedent part of the rule, and Action which is the consequent, represents the change to be made in the steam flow rate. These rules were written down and investigated as to their physical correctness and the rule base was formed after an iterative process in which new rules were added, previous ones deleted, or changes were made in the existing ones. This process is by no means a trivial task since the manipulated variable is one of the secondary loop while the controlled variable is that of the primary one. Due to the heat capacity of the steam generators, any action in one of the loops is felt in the other after a considerable time delay thus hindering the credit assignment to the applied rules. After numerous simulation runs the twenty two rules presented in Table VII were identified. All of these are based on the inherent negative temperature feedback in PWR type nuclear reactors and each rule is an expert attempt at reversing the adverse developments back to the operating level at an appropriate rate.

Note that the same set of rules would be applicable to larger perturbations leading to power variations greater than ± 6 % were it not for the fact that the mathematical model is a linear one, describing the behaviour of the actually nonlinear system in the vicinity of a point of interest, in this case the full operating power. Furthermore, these rules, having been chosen as a simple and a small set are in a sense similar to a PID controller in what and how they perform. This is because in this case the rules themselves are partly inspired by the analytical model. However, in general this does not have to be the case, and the rule set can be made much more sophisticated so as to be able to deal with models that would not be amenable to the analytical techniques of optimal control, or even with systems for which an analytical model is not available. For this reason, although the rule base resembles a conventional PI(D) controller, the RBC is expected to perform better under degraded conditions, e.g., the presence of noise and drift in the process variables.

TABLE VII. The Rule Base of the Controller

Rule No.	IF Power Error	AND Change in Power Error	THEN Change in Steam Flow Rate
1	negative-big or negative-medium	negative-small	positive-medium
2	negative-small	positive-small	positive-medium
3	negative-zero	positive-big or positive-medium	positive-medium
4	negative-zero	negative-big or negative-medium	negative-medium
5	negative-zero or positive-zero	zero	zero
6	positive-zero	negative-big or negative-medium	positive-medium
7	positive-zero	positive-big or positive-medium	negative-medium
8	positive-small	positive-small or zero	negative-medium
9	positive-big or positive-medium	negative-small	negative-medium
10	negative-small	zero	positive-medium
11	positive-medium or positive-big	zero or positive-small	negative-big
12	negative-big or negative-medium	zero or positive-small	positive-big
13	negative-big or negative-medium	positive-big	positive-big
14	positive-big or positive-medium	positive-big	negative-big
15	negative-small	positive-big or positive-medium	positive-big
16	positive-small	positive-medium or positive-big	negative-big

TABLE VII. The Rule Base of the Controller (Contd.)

Rule No.	IF Power Error	AND Change in Power Error	THEN Change in Steam Flow Rate
17	negative-small	negative-big or negative-medium	negative-small
18	positive-small	negative-big or negative-medium	positive-small
19	negative-small	negative-small	zero
20	negative-zero	negative-small	negative-small
21	positive-zero	negative-small	positive-small
22	positive-small	negative-small	zero

The rules thus selected must be expressed in a form suitable for digital processing. Initially Prolog was chosen as the programming language for knowledge representation since knowledge representation and inferencing can be done separately. But this approach was later abandoned for the following reasons:(a) the required inferencing mechanism is forward chaining while that of Prolog is essentially backward chaining and this would necessitate special programming in violation of the separability of representation and inferencing; (b) the controller tests require numerical simulation, however Prolog is not suited to numerical calculations and thus interfacing to another language such as C is necessary. This is generally an awkward process with many pitfalls, and which also means recoding the numerical simulation software which was already coded in Turbo Pascal 4.0 since it is not possible to interface Prolog to Turbo Pascal; (c) execution of a compiled Prolog program is essentially slower while regulation requires faster than one second responses; the number of test cases that must be run which are necessary for forming the rule base dictates a fast response; (d) the necessary forward-chaining inferencing is essentially simple because any, or all rules may be fired at the same time without resorting to any additional conflict-resolution strategy. This approach is in line with the general trend in the industrial AI applications where there is an increasing use of more conventional languages such as C and Pascal.

Inferencing

The procedure adopted in the next step of generating the controller action (which requires the application of an inferencing mechanism to combine the rules) is as follows: At every sampling interval for which a suboptimal value is determined as explained below, δP (PE) and its rate of change(RCPE) are calculated which is actually a simulation of the sensor input. Using these the degrees of membership of PE and RCPE in every labeled group is determined, because a calculated value of PE and RCPE can be a member of more than one fuzzy set. The degree of fulfillment (DOF) of the antecedent part of each and every rule i is calculated by

$$DOF(antecedent_i) = DOF(Condition_i(PE)) \wedge DOF(Condition_i(CPE)), \quad i=1,...,n \tag{11}$$

where n is the number of rules. In fuzzy-logic conjunction is defined as

$$P \wedge Q \triangleq \min(P,Q) \tag{12}$$

and disjunction as

$$P \vee Q \triangleq \max(P,Q) \tag{13}$$

[1]. Using (11) together with (12) and (13), the Action of each rule is calculated from the corresponding fuzzy set for CSFR given in the consequent part of the rule. The final control action is calculated by weighing the Action of each rule by its DOF as given in (11):

$$CSFR^* = \frac{\sum_i ACTION_i \, DOF_i}{\sum_i DOF_i} \tag{14}$$

This process can be best illustrated with an example. Let us assume that we measure PE to be 39.5 MW (% 1.8 FP) and calculate RCPE to be -36.72 MW/s. We proceed as follows: first, the degree of fulfillment of the fuzzy sets are calculated. PE is then classified as "positive-zero" to degree 0.382, "positive-small" to degree 0.939, and "positive-medium" to degree 0.239. RCPE is "zero" to degree 0.432, "negative-small" to degree 0.901, and "negative-medium" to degree 0.201. Degrees of membership of all other sets are zero. Using these values all the antecedents of the rules are calculated. For example, rule 6 states that "If power error is positive-zero, and change in power error is negative-big or negative-medium, then change in steam flow rate is positive-medium." Applying the OR operation to "negative-big" and "negative-medium" for RCPE gives us the union of these sets which is the maximum of the degrees of membership, i.e, 0.201. Since the clauses for PE and RCPE are connected by AND, the intersection of these or the minimum of the degrees of membership, must be calculated which is 0.201 for 0.382 is greater. Therefore the DOF of rule #6 is 0.201. The recommended action by this rule is, in this case, CSFR so as to obtain "positive-medium" to degree 0.201. Using the inverse function, we calculate the change in steam flow rate that satisfies this condition as 19.21 lb/s (8.72 kg/s). Repeating this process for all the rules, we find that only rules 5, 6, 8, 9, 11, 18, 21, and 22 have DOFs greater than 0. They are 0.382, 0.201, 0.432, 0.239, 0.239, 0.201, 0.382, and 0.901 respectively. The recommended changes in steam flow rate by these rules are similarly calculated by using the inverse functions of the fuzzy-set descriptions for CSFR to be found as 0, 19.21, -22.92, -19.82, -35.82, 3.21, 6.11, and 0 respectively. The net control action is calculated by weighing the action of each rule by its DOF. Thus, substituting these values to (14),

$$\begin{aligned} CFSR^* = & [(0.382)(0.0)+(0.201)(19.21)+(0.432)(-22.92) \\ & +(0.239)(-19.82)+(0.239)(-35.82)+(0.201)(3.21) \\ & +(0.382)(6.11)+(0.901)(0.0)] \\ & \div[0.382+0.201+0.432+0.239+0.239+0.201+0.382+0.901] \\ & = -5.50\ lb/s \end{aligned} \tag{15}$$

is obtained. Therefore, there has to be a decrease in steam flow rate by 5.5 lb/s (2.50 kg/s). It is evident that this represents a compromise between differing rules.

Calibration and Testing of Controller

The only presently known method to determine and improve the performance of a rule-based controller (RBC) is testing and simulation.

An ITAE (integral of time multiplied by absolute error) type performance index was selected as the criterion to be used for comparing the performance of the controller, since the errors that persist are more heavily weighted than the initial ones [6]:

$$PI = \int_0^{\infty} t\ |PE|\ dt \tag{16}$$

An ISE(integral of square of error) type criterion was found to be poorly discriminating.

First a calibration for the sampling interval was made. In this study, the responses of the sensors and actuators are assumed to be instantaneous or a negligible function of the sampling interval. The results of the simulation

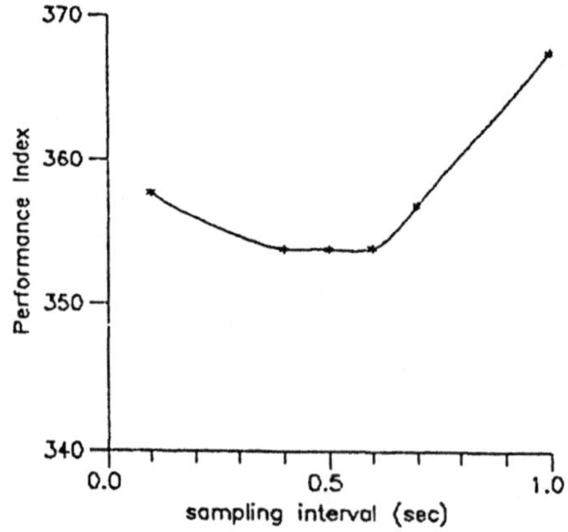

FIGURE 1. Variation of PI with control interval.

for different values of sampling intervals are shown in Fig.1 where it is evident that the minimum value of PI lies in the vicinity of t=0.5 which is chosen as the sampling interval.

The gain calibration is important for the RBC, as with all other controllers, since it has a marked effect on the overshoot, oscillation and steady-state error characteristics of the controller. After determining the suboptimal control interval, repeated simulations are performed for different values of control gain and a calibration curve as in Fig.2 is obtained. The gain where PI takes its minimum value is selected.

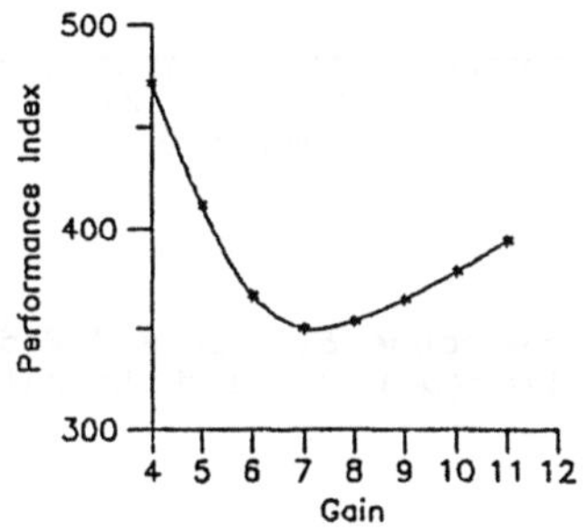

FIGURE 2. Controller gain calibration curve.

Usually, the fuzzy-logic controllers are compared with PI(D) controllers because the two are similar [4]. It was decided to compare the performance of RBC with an optimal controller, also developed for the H. B. Robinson plant [10], whose weights were adjusted through the placement of poles at desired locations together with the goal of minimum-energy expenditure. The selected test case is an initial impulse disturbance of $\delta T_{LP}(0)$=2 F (1.1 C) reported in previous work [10]

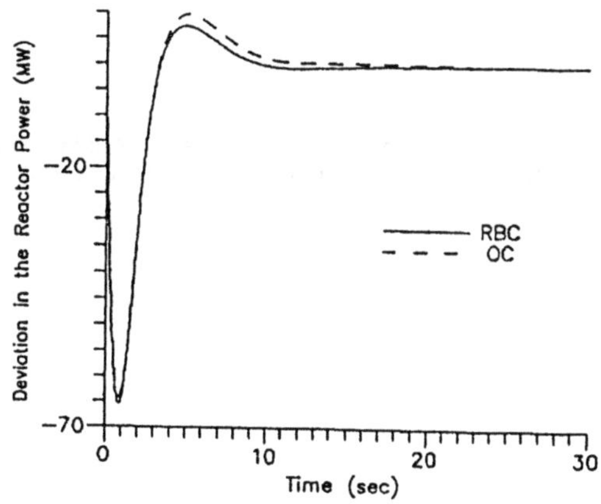

FIGURE 3. Comparison of the response δP for a 2 F disturbance in $\delta T_{LP}(0)$ for the optimal controller(OC) [10] and the rule-based controller (RBC).

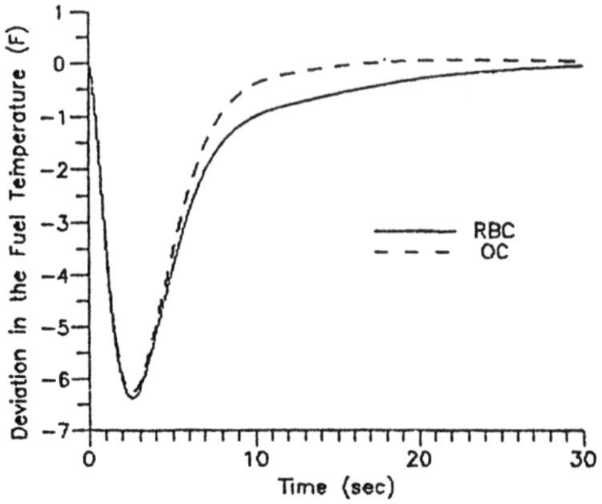

FIGURE 4. Comparison of the response δT_f for a 2 F disturbance in $\delta T_{LP}(0)$ for the optimal controller(OC) [10] and the rule-based controller (RBC).

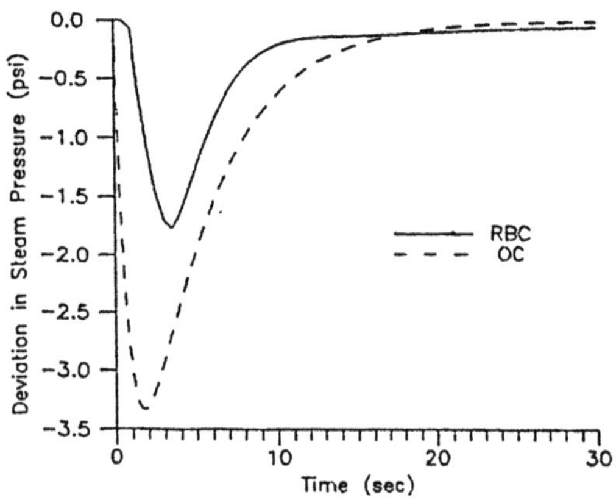

FIGURE 5. Comparison of the response δP_s for a 2 F disturbance in $\delta T_{LP}(0)$ for the optimal controller(OC) [10] and the rule-based controller (RBC).

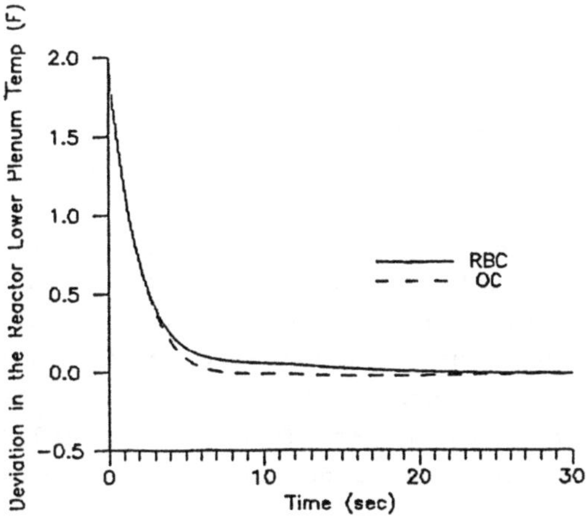

FIGURE 6. Comparison of the response δT_{LP} for a 2 F disturbance in $\delta T_{LP}(0)$ for the optimal controller(OC) [10] and the rule-based controller (RBC).

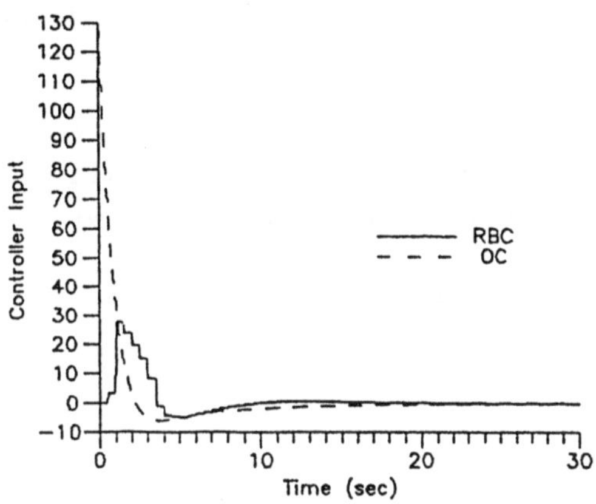

FIGURE 7. Comparison of the controller input for a 2 F disturbance in $\delta T_{LP}(0)$ for the optimal controller(OC) [10] and the rule-based controller (RBC).

(see Figs. 3-7). The performance index of the optimal controller is 474.24, while that of rule-based controller (RBC) turns out to be 353.80. An examination of Fig.3 reveals that the overshoot of the rule-based controller is about 2 MW less than that of the optimal controller at 2200 MW (full-power) and this is a significant amount. It can also be seen from Fig.5 that the decrease in steam pressure is less. Furthermore, the range of values of the input is much narrower for the RBC (see Fig. 7). In these respects, it can be concluded that the performance of rule-based controller is better than that of optimal controller. However, it should be kept in mind that our objective function deals with only power variation as opposed to the performance index of reference [10] where the placement of some poles at predetermined locations implies incorporation of a set of conflicting criteria.

Since the success of the actual implementation of any controller depends on the dynamical performance of the sensors supplying the necessary feedback information, and since, unfortunately, sometimes either the sensors fail or their signal is smeared with noise, a successful controller implementation is expected to perform even under the degraded conditions of noise or failing sensors, or at least degrade gracefully without causing dangerous oscillatory or divergent behavior in the system. Therefore, in order to investigate the behavior of RBC under noisy conditions, the following noise model is used [8]:

$$\delta P_n(t) = \delta P(t) \; [1 + b.r(t)] \qquad \textbf{(17)}$$

where $\delta P_n(t)$ is the variation in power with noise, $\delta P(t)$ the true variation in power, b half-width of noise band, r(t) random number between -1 and 1. The result of the simulations are shown in Fig.8. It can easily be seen that RBC is extremely robust under these conditions and the degradation in its performance is negligible.

Generally assumed time invariance of the process to be controlled, although valid during the interval the controller is effective, is not so throughout the plant life during which drifts or changes will occur in the process parameters thus decreasing the validity of the model. This is generally compensated for by tuning the controllers, as the effect becomes noticeable.

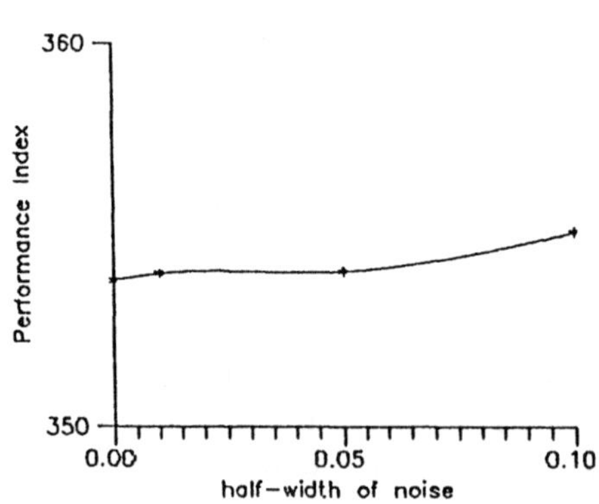

FIGURE 8. Effect of noise on controller performance.

In the case of an NPP with a PWR core, the moderator coefficient of reactivity becomes more negative with burnup, primarily as a result of boric acid dilution, but also to a significant extent from the effects of the fission-product build-up [2]. In this work, for a parameter variation of 2% from its nominal value, the moderator temperature coefficient of reactivity is considered [9]. The variation in performance is almost unnoticeable as is reflected in the small variation in PI: 360.52 for the case of off-nominal operation and 353.80 for the nominal case.

Discussion and Conclusions

Although, the usual approach in developing such a knowledge-based controller starts with knowledge acquisition through interaction with an identified expert in the field, in this work, knowledge acquisition was accomplished through numerical simulation, analogy with similar systems and an examination of the mathematical model of the power plant. This somewhat inferior technique is necessitated by the lack of an appropriate human expert because the time scales involved in the implementation of the controller are far beyond the limits of human response based on observation. Such a knowledge acquisition process, though long and tedious and with inherent pitfalls, proved to be successful in hindsight as seen from the results produced and can be resorted to in similar cases. The knowledge thus acquired is represented as production rules expressed in terms of two linguistic variables which are the deviation of power from its steady-state value and its time rate of change. The values of these linguistic variables can be represented by using the fuzzy sets such as "negative-small", "positive-big", etc.

The structure of the controller is determined by its rule base whereas its performance can be tuned by adjusting the fuzzy sets. Two of the different membership functions for the fuzzy sets that can be used for this purpose, the broken-line and the s-shaped, were considered in this work. After tests, the broken-line membership function was decided upon, which can easily be modified and is computationally inexpensive. The persistent residual behavior of the s-shaped membership functions in the neighborhood of 0 degree of membership is one of the fundamental factors that influence the tuning process and is intuitively hard to account for.

The most important factor that must be considered in the development of a fuzzy-logic controller is the completeness of the rule base because the allowable and conceivable test cases are limited. Thus, unanticipated transients may be overlooked and it is not possible to guarantee the exhaustiveness of the rules and the stability of the controlled system. In this work, twenty two rules were identified through knowledge acquisition with simulation. All of them are based on the inherent negative temperature feedback of the PWR type nuclear power plants.

In the implementation, instead of an artificial-intelligence programming language such as Prolog which was initially considered, the rule-base and the inferencing scheme was coded in Pascal mainly due to problems in interfacing to simulation software. Although, in this case, modification in rules requires some major changes in coding, the final implementation is considerably faster, being also facilitated by the fact that inferencing is data driven or forward chaining only. This will help in actual implementations, since the control interval in this work is less than a second, and the decision process must take considerably less time.

A performance index was used in order to compare different controllers and the different implementations of rule-based controllers alike. Of the two indices considered, the ITAE(integral of time multiplied with absolute error) type was found to be more discriminating than the ISE(integral of square of error) type. This is mainly due to the considerable time delay between the manipulated and the controlled variables which reside in the secondary and primary loops respectively. The ISE type punishes the initial errors severely, where the possible latter fluctuations are shadowed by this effect.

The control interval was determined through a calibration process at the end of which the value giving the minimum performance index was taken. Time delay was observed to be the main factor influencing the sampling interval because for very short sampling intervals the control action has less time to become effective due to time delay. As the sampling interval is increased, again due to time delay, the nature of the transient changes faster than the control action can accommodate. However, it is apparent that there is not a marked degradation in the controller response. Therefore, in implementing a RBC of this type, the practical necessities determine the sampling interval.

It was also possible to tune the gain of the controller by varying the gain, constructing a calibration curve, and taking the value corresponding to the minimum value of the performance index. It was noted that this process requires almost no iteration, because interaction of the gain and sampling interval was weak for the range of values considered.

In the comparison of the RBC with an optimal controller, it was found that the RBC had a better performance in the sense that overshoots were less, but this is because the optimal controller incorporates conflicting objectives while RBC has the single objective of minimizing power variations. It should be stated at this point that the main advantage of the RBC controller lies in its speed and applicability, even in cases where an analytical model is not available for a system, or a nonlinear one is available but is not amenable to analytical optimal control techniques. Otherwise the performance of an RBC controller with a thoroughly comprehensive rule base can at best equal that of an optimal controller with the same performance criterion if the optimal solution exists for the latter.

Also investigated were the behavior of the rule-based controller under noisy operating conditions and drift in process variables. Using the noise model of MacDonald et al. [8] the effect was observed to be negligible up to 10 percent noise level. For the case of 2 % variation in moderator temperature-reactivity coefficient [9], the degradation is also negligible. Thus it can be concluded that the RBC is sufficiently robust which is an extremely important asset for nuclear power plant operation.

List of Symbols

A	heat-transfer area, m^2
A_c	flow area of the channel, m^2
A_f	heat-transfer area, m^2
δC_i	deviation of normalized precursor concentration from its steady-state value
C_m	specific-heat of the tube metal,
C_p	specific-heat capacity of the fluid,
C_{pFW}	specific-heat of the feedwater,
CSFR	fuzzy variable for change in steam flow rate
D_i	coefficients
DOF	degree of fullfilment
F_{ci}	reactivity importance for temperature changes in the i th coolant node.
F_{pci}	the fraction of total power released in coolant node i
F_{fi}	reactivity importance for temperature changes in the i th fuel node.
M	mass of coolant in node i, kg
M_m	mass of tube metal, kg
M_p	mass of primary-coolant in the steam generator, kg
P	the total power, MW
PE	fuzzy variable for power error
PI	performance index
P_o	initial steady-state power level, MW
δP	deviation of reactor power from initial steady-state value, MW
PI	performance index, MJ-s
P_p	primary loop pressure, MPa
P_s	steam pressure, MPa

RCPE	fuzzy variable for the rate of change of power error
Q_{fi}	fraction of the total reactor power generated in fuel node i
δT_{ci}	deviation of coolant temperature in the i th coolant node from its initial steady-state value, C
δT_{cin}	deviation in inlet temperature of the first coolant node from its initial steady-state value, C
δT_{fi}	deviation of fuel temperature in the i th fuel node from its initial steady-state value, C
T_{in}	fluid temperature at entrance,C
T_m	steam generator tube metal temperature, C
δT_p	primary-coolant temperature in the steam generator, C
U	overall fuel-to-coolant heat-transfer coefficient
W_{FW}	feedwater flow rate, kg/s
W_{wi}	mass flow of water into(or out of) the pressurizer, kg/s
$W_{s,o}$	steam flow rate to the turbine, kg/s
c	specific heat capacity,
h	film heat-transfer coefficient,
h_{ms}	heat-transfer coefficient for metal to secondary coolant
h_{pm}	heat-transfer coefficient for primary-coolant to metal
Λ	neutron generation time
α_c	coolant temperature coefficient of reactivity, 1/C
α_f	fuel temperature coefficient of reactivity, 1/C
α_p	coolant pressure coefficient of reactivity, 1/C.
β	total delayed neutron fraction
β_i	delayed neutron fraction for the i th delayed neutron group
λ	delayed neutron decay constant for i th delayed neutron group
ρ	fluid density, density, kg/m_3

τ	residence time of fluid in node i, s
τ_{SG}	residence time of coolant in the steam generator, s

References

1. Akin, H. L.:Development of a Knowledge-Based Regulator for A PWR-Type Nuclear Power Plant, Ph.D. Dissertation, Bogazici University, 1989

2. Anonymous:Preliminary Safety Analysis Report. Wisconsin Utilities Project, Wisconsin Electric Power Company, 1975.

3. Anonymous: TURBO PASCAL Owner's Handbook Version 4.0. Scotts Valley: Borland International 1987

4. Arakawa, A., Sekimizu, K., Sumida, S.:Fuzzy Logic Control Application for BWR Recirculation Flow Control System. J. Nucl. Sci. Tech. 25, 263-273 (1988)

5. Bernard, J. A.:Use of a Rule-Based System for Process Control. IEEE Cont. Sys., 3-13 (1988).

6. Doebelin, E.:Control System Principles and Design. John Wiley 1985.

7. Kerlin, T. W.:Theoretical and Experimental Dynamic Analysis of the H.B. Robinson Nuclear Plant. Nucl. Tech. 30, 299-316 (1976)

8. MacDonald, J. L., Koen, B. V.:Application of Artificial Intelligence Techniques to Digital Computer Control of Nuclear Reactors. Nucl. Sci. Eng. 56, 142-151 (1975)

9. Nair, P. P., Gopal, M.:Sensitivity-Reduced Design for A Nuclear Pressurized Water Reactor. IEEE Trans. Nucl. Sci. 34, 1834-1842, (1987)

10. Saif, M.:A Novel Approach for Optimal Control of a Pressurized Water Reactor. IEEE Trans. Nucl. Sci. 36, 1317-1325 (1989)

TELEPRESENCE AND INTELLIGENT CONTROL FOR A LEGGED LOCOMOTION ROBOT

Manuel Armada
Instituto de Automatica Industrial
Consejo Superior de Investigaciones Cientificas
C.N.III, Km.22,800 LA POVEDA, ARGANDA DEL REY
28500 MADRID SPAIN

Abstract : The aim of this paper is to present the actual research that is carried out in order to provide a fully operative sense of telepresence, and on-board intelligent control for a four legged locomotion robot. The paper describes the whole project showing mechanics and kinematics of the walking machine, and the teleoperation station and on-board software architecture. Special attention is given to those levels at which intelligence is necessary to improve performance or to achieve desired goals. Also, from the control point of view, the paper discusses the need and the necessary means to achieve robust and/or intelligent control under uncertain conditions posed by the operation of the legged locomotion robot in hostile and/or unstructured environments.

Keywords : Telepresence, robotics, computer aided teleoperation, hostile environments, intelligent control, robust control, legged robots, walking machines.

1. INTRODUCTION

The aim of this paper is to present the actual research that is being carried out in order to obtain a fully operative sense of telepresence and on-board intelligent control for a four-legged locomotion robot. The robot is under development by the joint effort and close cooperation of the Instituto de Automatica Industrial (IAI) and the Centro de Investigaciones Medioambientales y Tecnologicas (CIEMAT) robotic laboratories.

Within this project, we are trying to investigate in a very practical way a number of topics that shall permit us to obtain a telerobot with legged displacement capabilities over irregular terrain, and that can be used to succesfully perform tasks in a remotely located hostile environment. The hostile environments over which we focused our attention at first instance were those related to nuclear power plants, although recently we became also

NATO ASI Series, Vol. F 71
Expert Systems and Robotics
Edited by T. Jordanides and B. Torby

interested in fusion reactors. In this kind of environment it is necessary to carry out a variety of tasks (inspection, manipulation of radiactive parts, welding, grinding, and so on), which implies that human operators could be exposed to high radiation doses. Also, there are a great number of potential applications that cannot be performed by human operators. These considerations lead in a natural way to the use of robotic devices. They provide two main advantages : drastic reduction of operator exposure to radiactive doses, and the possibility of taking into account new or classical applications with a higher degree of security and fiability [3,6-8,12-13].

The main reason for the work which is to be depicted here is based on the research that was realized in the IAI in the last years relating to the design and the developement of a telerobot. This robot designed for performing inspection and maintenance tasks in the steam generator's channel head of PWR nuclear power plants. These investigations has yielded very practical and high performance results in the field of Telepresence, Computer Aided Teleoperation, complex kinematics, trajectory planning, collision avoidance, and robust robot control. Such a telerobot is actually in the first stages of industrialization, and it is expected that the experience and know-how acquired are to be applied and extended in our new researches.

One of the most important results achieved with earlier investigations was the design and development of a "general purpose" Teleoperation Station. Although the results are not "so general", the experience and knowledge gained from such an effort is of major relevance for our actual work.

Also, it should be noticed that the work we are going to present in this paper is related to another experimental prototype, a self-propelling robot that has the possiblility of reaching remote locations by means of special tools (i.e., electromagnets) attached to both ends of its kinematic chain [1].

The content of this paper is as follows. First the RIMHO, a four-legged locomotion robot will be presented and details about its mechanical configuration will be shown along with its kinematics. After Telepresence is discussed, a number of other questions are dealt with : software organization, real-time 3D-animated simulation, environmental modelling and multisensor integration. The last part of the paper is devoted to the control problem, where robust and intelligent controllers are discussed. Finally, conclusions, remarks and future research will be presented.

2. RIMHO : A FOUR-LEGGED LOCOMOTION ROBOT

The frontier between autonomous and semi-autonomous (or tele-autonomous) systems is not well defined, and practical experience shows that for the time being if we want to obtain real working prototypes, it is necessary to implement in some way supervisory control. This is the main reason why the Telerobotic system which we are building is composed by two main subsystems : the four legged locomotion robot which we call RIMHO, and a Teleoperation Station which can be located in a safe, remote position.

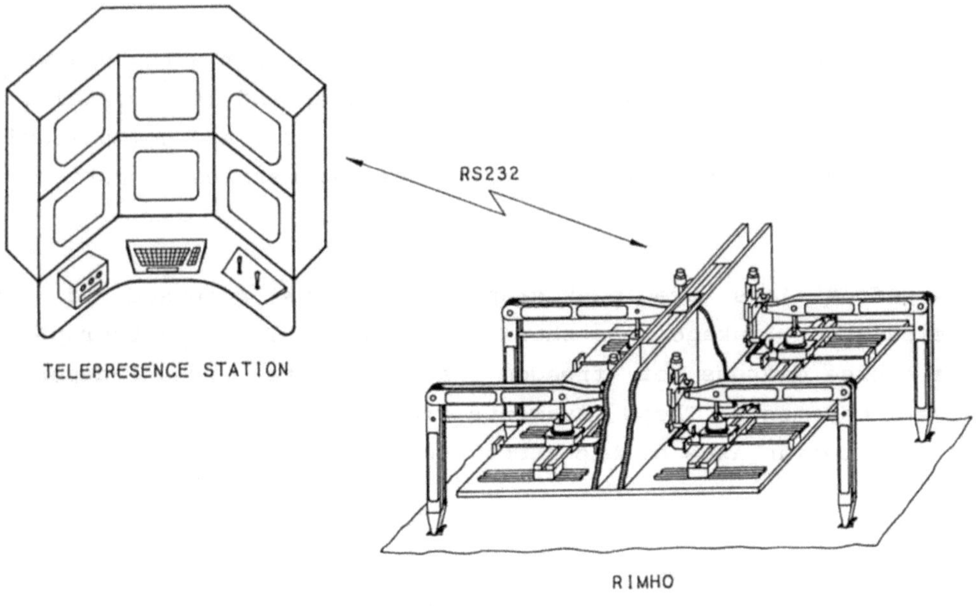

FIG.1 RIMHO : A four legged locomotion robot

Although we are not looking for complete autonomy, we are looking for obtaining a certain degree of "local" intelligence.

The increasing number of degrees of freedom that is present with legged locomotion systems, as compared to other kinds of wheeled or tracked vehicles, leads to a unique terrain adaptability that constitutes the major feature of this devices [11]. But it is to be recognized, that this relevant feature of legged robots places significant difficulties on its design and control.

The aim of this research on legged locomotion, therefore is to obtain practical experience in both electromechanics and control by using laboratory prototypes that permit us to attack and solve some of the problems posed by this kind of vehicle.

2.1 MECHANICS

The RIMHO (see Fig. 1) is a four-legged locomotion robot that was designed to be able to carry a 20 kg. payload, and pass over significant obstacles while moving on irregular terrain with the possibility of adaptive suspension.

The body of the RIMHO is build up of an aluminium structure whose dimensions are : 736 mm. long, 710 mm. wide, and 344 mm. tall. In the center of this structure is the control and communication electronics. It si placed there to get symmetrical location of masses.

The four legs are pantograph mechanisms composed of four articulated linkages, as shown in Fig. 1 . Each pantograph leg has three degrees of freedom that originate three independent linear movements. These movements are, by design, along the three main Cartesian axes of the robot body.

By means of the 3 dof, the end of the leg can be positioned following the movement of each dof in the same direction. The position has an amplification factor of five as will be shown later.

With this configuration, it is possible to reach a rectangle of 1125 mm X 800 mm in the horizontal plane, and to raise up/down the end of the leg about 450 mm.

2.2 KINEMATICS

As was just mentioned above, the four RIMHO legs are pantographs with three dof for each one, so the robot has twelve dof. Each dof is obtained through linear movements in the **X, Y, Z** Cartesian directions, and results in decoupled amplified movements along the same directions for the leg's foot. The transmision factor, **TF**, is defined as follows,

$$\mathbf{TF} = \frac{\mathbf{L + A}}{\mathbf{A}}$$

For the RIMHO we have chosen **L = 400 mm., A = 100 mm.**, so **TF = 5**. This selection of the transmision factor is of major relevance for the displacement capabilities of the robot and has important influence on the possible gaits, strides, etc..

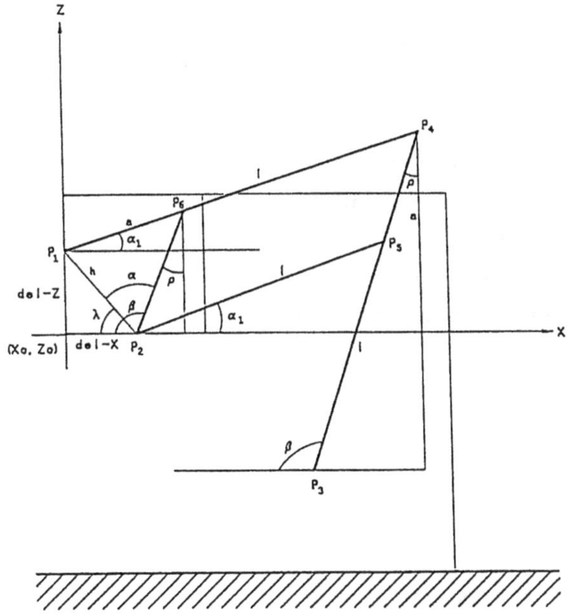

FIG.2 One leg kinematics

The direct and inverse kinematic models for one leg in the **X-Z** plane for illustrative purposes are as follows (see Fig. 2).

There are six points of interest (P_1, P_2, ..., P_6) that can be used to describe completely the leg configuration in the **X-Z** plane. The center of the reference frame (X_0, Z_0) is located in the crossing point of the leg's three d.o.f., and is moving along with the robot frame. The direct geometrical model for the right legs is given by the following algorithm :

```
DEL_XD : displacement of X dof
DEL_ZD : displacement of Z dof

H = sqrt(DEL_XD**2 + DEL_ZD**2)
α = atan2(sqrt(4*A**2-H**2),H)
λ = atan2(DEL_ZD,DEL_XD)
β = α + λ
ρ = - β + π/2

PX_1 = X0
PZ_1 = Z0 + DEL_ZD
PX_2 = X0 + DEL_XD
PZ_2 = Z0
PX_6 = PX_2 - A*sin(ρ)
PZ_6 = PZ_2 + A*cos(ρ)
α1 = atan2(PZ_6-PZ_1,PX_6-PX_1)
PX_4 = PX_1 + (A+L)*cos(α1)
PZ_4 = PZ_1 + (A+L)*sin(α1)
PX_5 = PX_2 + L*cos(α1)
PZ_5 = PZ_2 + L*sin(α1)
α2 = atan2(PX_5-PX_4,PZ_5-PZ_4)
PX_3 = PX_4 + (L+A)*sin(α2)
PZ_3 = PZ_4 + (L+A)*cos(α2)
```

where, as it is shown in Fig. 2 , (**PX_3,PZ_3**) are the Cartesian coordinates of the foot. It is easy to demonstrate that,

```
PX_3 = X0 + TF*DEL_XD
```

PZ_3 = Z0 + (1-TF)*DEL_ZD

The inverse kinematics are given by,

DEL_XD = (PX_3 - X0)/TF
DEL_ZD = (PZ_3 - Z0)/(1-TF)

With these equations it is possible to know the foot position for any displacement of the dof's and vice versa (what is very important for obstacle avoidance and space intrusion). The above kinematics are easily extended to the third dof and are used for the whole vehicle kinematics. They are extensively employed in the simulations as we will see later.

The leg's kinematic solution permits positioning the foot at any place inside its reachable volume or to know the leg internal configuration for a given foot position. By using the kinematics from each leg, we can modify the chassis attitude and position. For example, it is very easy to move the platform up and down along the **Z**-axis by giving the same **DEL_Z** to the four legs. If we give different **DEL_Z** for each leg, we will obtain a change in the α, β, λ (pitch, roll, yaw) angles of the center of gravity. When using the **Y**-axis or **X**-axis movements, we are changing the position of the center of gravity. By lifting, transferring and placing the legs in a specific pattern (which is called a gait), we can move the RIMHO along a desired trajectory.

3. TELEPRESENCE

The RIMHO on board control system is in full-duplex communication with a Teleoperation Station (see Fig.1) from where it is possible to program and/or supervise the robot displacements. This Teleoperation Station is based on a workstation with 3D graphic capabilities, along with CCTV monitors and a device with two joysticks.

We undertand telepresence as the sense of "being there" experienced by the operators of remote systems. Such a sense is obtained throughout a series of means provided by recent advances of technology, such as CCTV (including, or not including computer vision), interactive graphic systems, and innovative computer architecture, and by the major contributions of automatic control and robotics.

Our basic idea for advanced teleoperation is based upon implementing in some way the concept of telepresence. This means that we must give to the operator(s) of the system the necessary information for the accomplishment of the tasks to be performed. This kind of information must be relevant to the task in progress, and we must avoid giving the operator an excess of information that he or she cannot process adequately. One of our goals is to implement intelligent or advisory telepresence in the sense of furnishing aids to the operators that releases them from tedious, repetitive, and time consuming tasks. In doing so, we will be walking towards more autonomous systems.

We obtain telepresence in the following way. First, we present to the operator(s) 3D-animated graphics showing the movement of the legged robot from several viewpoints : Frontal, Plant and Lateral. These graphics are continuously updated using position feedback from the RIMHO joint encoders, and the operator(s) can select at any time the particular preferred view and zoom or pan. Secondly, we have the CCTV visual feedback, and the operator has the possibility of selecting, moving and controlling the cameras to visualize the most relevant part of the scenery. Another feedback is provided by the microphones located on board the robot which brings workplace sounds. Finally, a number of lights and indicators show the operator(s) the state of the system (alarms, etc).. The operator(s) can drive the RIMHO by several means, such as a pair of joysticks unit.

This unit has its own local processor that permits, among other things, changing of sampling rates or sensitivities (simple gains or high-order digital filter parameters). It is possible to perform a number of experiments dealing with human operator tracking capabilities to decide speed, direction of travel and heading of the vehicle, when using this unit.

Because of the great number of degrees of freedom, the transmission-execution time delays, and other factors like operator(s) psychology, there are serious difficulties in obtaining good performance when using this system. We propose to use AI in the form of an Expert System so that the man-machine interface can be improved. This ES will be used as an aid to the operator for facilitating trajectory planning/following, obstacle avoidande and navigation. Several partial examples have been simulated and the obtained results are very promising.

3.1 SOFTWARE ORGANIZATION

The modelling for the software organization of the Teleoperation Station is based on "evaluation nets" (see Fig. 3) derived from Petri networks. An evaluation network consists of an oriented graph whose nodes are of three kinds :

- States (represented by circles).
- Transitions (represented by horizontal lines).
- Triggering of transitions (represented by rectangles).

When the system is in the state S, the reception of a requirement R triggers a transition that yields to a new state T. Some actions can be associated with each transition :

- The sending of a trigger condition by means of a rectangle labeled with the name of the requirement.

- The execution of a complex action. This can be described, in each case, by a number of instructions, or by indication of a separate descriptive procedure.

- The final state, or associated action with one transition, can depend on certain condition represented by a variable C with several possible values (c1, c2, ...,cn).

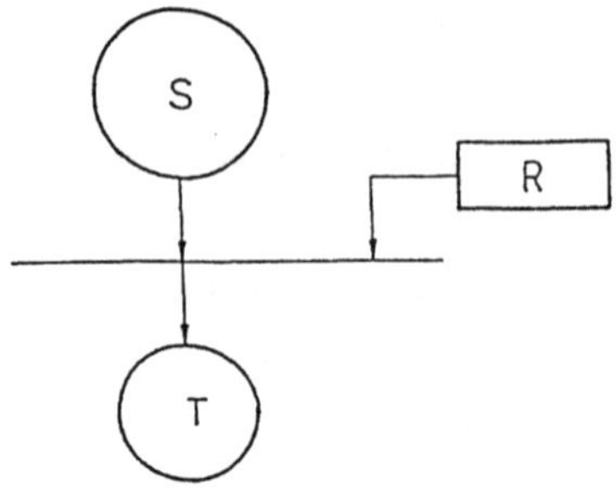

FIG.3 Evaluation networks

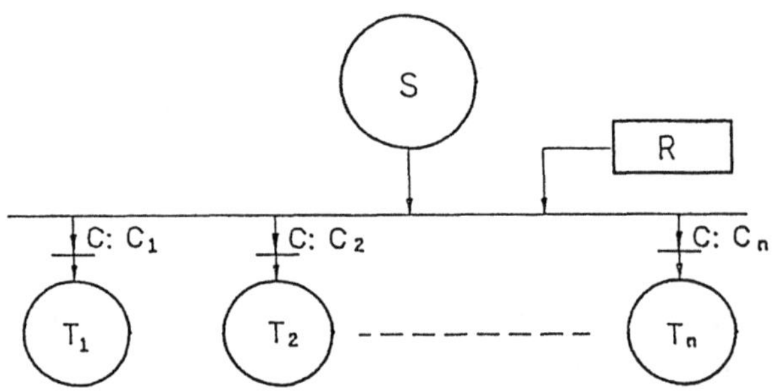

FIG.4 Actions associated with each transition

The corresponding diagram is shown in Fig. 4, and has the following meaning,

Case C in
C1 : <final state T1>
....
Cn : <final state Tn>
End Case.

Using the above notation we have established the general diagram of Fig. 5 that describes the whole behaviour of the Teleoperation Station, where,

INSTRUCTIONS	**STATES**
PO : Operator Request	**A : Start**
	SE : Session
	C : Create
	P : Program
	SI : Simulate
	O : Operation
	E : Autonomous Run
	CM : Hand control
	F : End of session

START : When we turn on the system it enters in a transition state in which it will perform a series of initializations and verifications of the various system components performance :

* Initialization of the variables and parameters of the Teleoperation Station.

* Checking of communications and of the on-board controller state.

* Checking of auxiliary units (joysticks, CCTV, etc.).

SESSION : State that is reached after the functions defined by the START procedure are realized. It constitutes the operating state of the system and it can require the operator to :

* Create data bases for the description of the environment.

* Define kinematics of the legged robot.

* Define/modify the environment model.

* Edit RIMHO programs.

* Simulate and debug programs.

* Control the system by hand.

* Run programmed tasks in autonomous mode, and supervise its performance by using the telepresence feedback.

CREATE : This module facilitates to creation, to modification and to visualization the data bases that have the information for the legged-locomotion robot, its tools and of the environment arrangement.

PROGRAM : This mode is oriented to give the necessary tools for the creation of the inspection and maintenance programs. It is made up of two big interlaced modules that permit definition of the plant operation environment. The first step is to build the real-world model over which the legged robot is to work, and once this is accomplished, it will be possible to create the RIMHO program, based on the previously fixed environment model.

To facilitate the operator(s) work there are a series of windows with specific functions :

- Graphics editing window : intended for graphical representation of the environment with the RIMHO and to create the environmental model.

- RIMHO programming window.

- Position table editing window.

The programming philosophy for the RIMHO is based on the concept of off-line programming, that is, creating and debugging application programs without using the robot. We are using the C programming language as a basic tool, and we have extended its capability by adding new elements specific to the robotic applications.

The application program consists of two modules that can be edited in a separate way :

- Program module. It has all the instructions (including those specific for the robot) and the explicit data (constants) that are necessary for the task definition.

- Positions table. It has the positions to be reached by the legged machine when it is in autonomous mode. Each position is stored under a symbolic name that will be used in the programming file. There are three methods for position specification : Textual entering by operator; teaching by manually driving the robot; and position obtention from the graphical window.

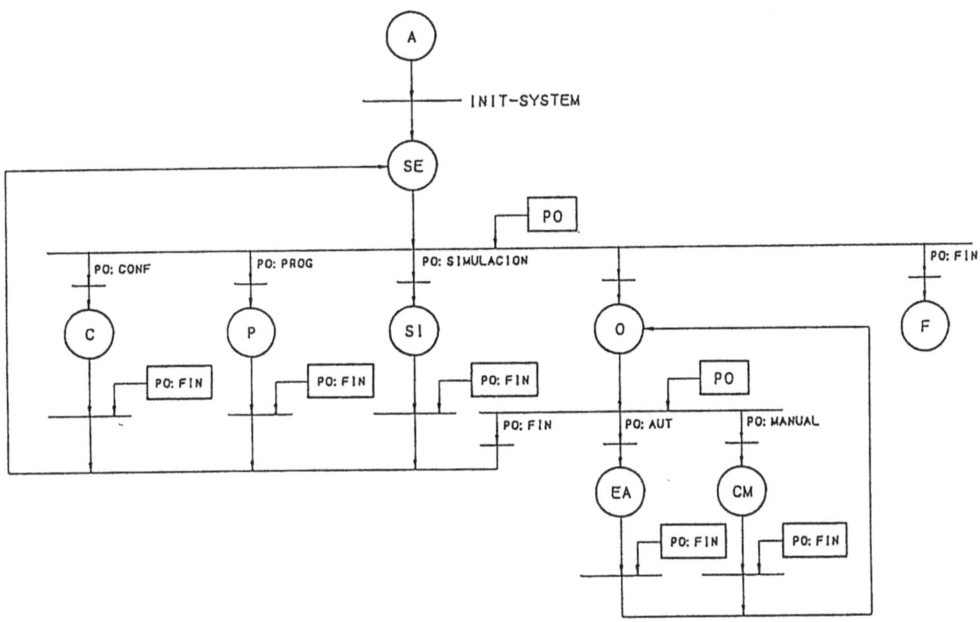

FIG.5 Telepresence Station software organization.

The Teleoperation Station permits the operator(s) of the system two main ways of working : autonomous program running and direct manual control.

When the operator commands the autonomous running of a desired task, the following facts are to take place :

- The RIMHO animated drawing is updated in the graphical screen at some rate so that its evolution can be followed by the operator(s).

- The operator(s) can choose their preferred view according to their actual necessities, and can request a graphical representation of the working environment in which, by means of special purpose symbols, certain aspects related to the state of the work in progress will be displayed.

- The operator(s) will have information at all time about the real-legged robot configuration, its state, the state of the on-board controller unit, messages, errors, etc..

- The operator(s) are enabled to abort the program.

- The operator(s) can suspend the execution of a program to solve unexpected problems, or to analyze any kind of information of particular interest, and after that command, continue the program.

- The operator(s) can, at their will, command the manual operation of the system.

The system is driven by menus. The basic functions are executed selecting the corresponding items from the menus. At the same time, some parameters (for example the robot positions) can be introduced graphically by moving the mouse pointer over a specially defined valuator scale. Several windows are possible simultaneously, with specific functions for each one (menu window, message window, ...). These windows are the tool for man/machine communication.

Under manual operation, the RIMHO is to be driven by means of a pair of three-axis joysticks each of one has several switches. By using these switches, the operator can select, among other things, the coordinate reference system to be used. Besides this there are available a number of additional commands that permit performing of special tasks (homing, etc.).

3.2 REAL-TIME 3D-ANIMATED SIMULATION

The simulation capability of our system is oriented to try and debug an application program without needing to drive the RIMHO. We have both an off-line programming aid, and the possibility of training operators. These simulation capabilities permit us to obtain application programs in a short time with a high degree of performance, and without either risk of collision nor operator errors. The 3D simulations are programmed using PHIGS.

We can use the simulation in two ways : Step-by-step or continuous simulation. In both cases, these simulation modules provide to the operator(s) the necessary tools to debug the tasks to be performed.

One of the most interesting experiments that can be realized are those related with forward simulation [2], that is, moving the robot in another time scale (faster) different from the real-time position data acquisition and display.

Fig. 6, shows simulations (presented here in the vertical plane for simplicity) of the RIMHO walking on an inclined surface, and changing its configuration.

3.3 ENVIRONMENTAL MODELLING AND MULTISENSOR INTEGRATION

When the environment is known, it can be modelled by 3D graphics (wire and/or solid modelling) and this environmental model is joined with the RIMHO model, so that the robot evolution can be followed inside this environment. We are now focused on this problem. But we are also trying to conduct experiments that yield to a step-by-step environmental modelling by mixing graphics with video, and using CAD to draw partial scenery.

We need to use proximity sensors to obtain terrain maps and to detect obstacles. We will use mainly ultrasonic range sensors located on the RIMHO legs and chassis, because we have long experience with these kind of sensors when attached to robot end effectors or when they are located around the robot body. Other kinds of sensors to be used are simple on-off and force (strain gages) sensors. Because there are many sensors it is necessary to integrate all of the information and use it in the best way. Some information will be conflicting. To accomplish this task we are starting to use Expert Systems.

4. ROBUST AND INTELLIGENT CONTROL

We are going now to discuss the control problem for the legged-locomotion robot. Robustness is a major goal to be looked for in

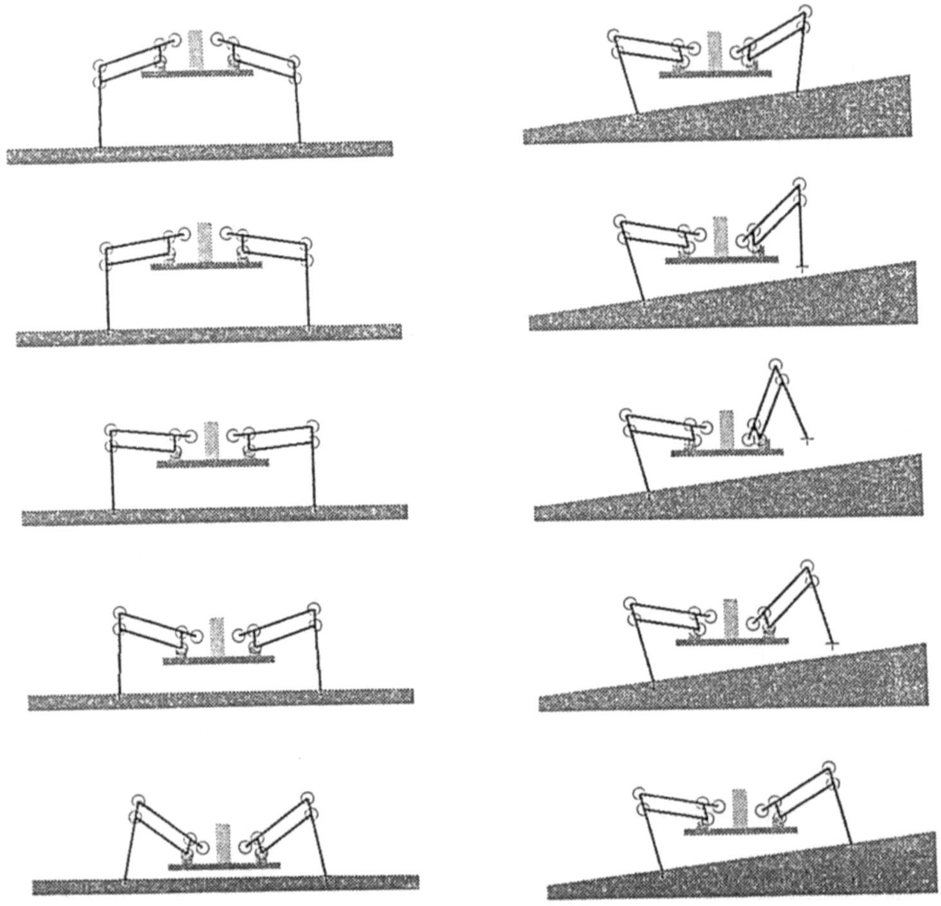

FIG.6 Animated simulation of RIMHO displacements.

this case because of parameter variations, and because of the external disturbances acting upon the system caused by the hostile and/or poorly structured or partially known environment. The prime movers of the RIMHO joints are low inertia dc motors

(15 watts) with incremental optical encoders (500 ppr). No tachometer feedback is used in order to save space and weight, and also because the digital controller we employ does not need it. The power to the dc motors is supplied by an integrated PWM system which is directly driven by the digital controller. This power stage was realized by using chopper integrated circuits. So we have a high performance system that occupies a reduced volume, that is low energy in consumption and the power supply is on board dc (batteries).

With this electromechanical arrangement, the maximum speed of the RIMHO is as follows. For horizontal displacements the speed is 250 mm/s, and for vertical displacements the velocity is 25 mm/s. The RIMHO has a total weight, including electronics, of 25 kg..

On board the RIMHO there is a powerful computer architecture distributed over three hierarchical levels. The lower level is dedicated to the digital control of each joint and uses specialized dc motor microcontrollers.

The intermediate level uses 16 bit microprocessors to control the leg's displacement. We use one microprocessor for two legs.

The highest level uses a 32 bit microprocessor to realize the most complex calculus tasks (stability, gaits, ...).

The digital controller for each joint is based upon a dc motor microcontroller integrated circuit that offers, among other things the following functions : 24 bits absolute counter for incremental encoder feedback, digital filter with programmable parameters, outputs for driving speed regulator or PWM H-bridge. This control system at joint level was successfully tried and its performance is excellent.

To design the digital filters (that are different for each joint) we have employed classical control theory approaches like root locus, Bode, Nyquist, and pole placement algorithms. One method that was very useful was the Mitrovic multiparameter approach because it gave very good results and yields to robust designs in a simpler way.

We are using the RIMHO and another laboratory prototype to get insight and to obtain a firmly practical experience-based knowledge about the control theory methods that are more suitable for obtaining high performance real-time robust controllers for robots. In this way, we are using the RIMHO as a testbed for evaluating and comparing the behaviour of different control algorithms that arise from using a variety of control approaches.

Also, with this intention in mind, we are using and developing a series of means (hardware and software) to facilitate our work. We are employing a powerful electronic device that performs FFT (and DFFT) to identify the transfer functions of the joints servomechanism, and to investigate the influence of the robot space configuration on the transfer function parameters. This constitutes an invaluable aid, and a correcting information source to the direct mathematical modelling and simplification that is classically used. What is more important, it gives us the information needed for getting robust control.

We have developed software aids that permit us to obtain not only the siso transfer functions, but also the mimo transfer function matrices, so we are designing the compensators in the multivariable frequency domain, mainly using the Inverse Nyquist Array.

But we are also attacking the robust control problem in the time domain by using singular perturbations and linear and bilinear dynamical models. This means that we decompose the time response into two time scales (faster, legs movement; slow, body center of gravity movement). We are starting now with experiments, and the simultion results obtained are very good.

In our opinion the use of intelligent controllers at joint level is not very clear and several experiments and simulations on fuzzy controllers have resulted in no advantage over robust algorithms. But for the whole vehicle, when moving over soft terrain (or terrain that changes suddenly from hard to soft), this may be different, because in this case a certain degree of intelligence seems desirable and promising. So we think that at the higher control levels, the use of intelligent control that uses expert systems to integrate multisensor information, and that takes into account the previous robot experience (learning), should be explored. One of the open posibilities for application of ES resides in the gait selection which is recognized as a major problem.

5. CONCLUSIONS

We have presented in this paper the actual research that is being performed in order to provide a fully operative sense of telepresence and on-board intelligent control for a four-legged locomotion robot, RIMHO, that is being developed by the IAI and CIEMAT robotic laboratories. Starting from our early experience in the design of teleoperation stations and control systems for robots, we are extending the available results to this twelve dof legged robot, which is an important research effort.

The concept of telepresence and the means we are using to implement it, as well as the software organization of the teleoperation station are presented, and are actually operative. Special attention was paid to the 3D simulation capabilities that are required for a good telepresence feedback to the operator(s).

One of the main reasons for this reseach resides in the fact that we are trying to disminish the gap between control theory and its practical application to robotics by using experimental prototypes. This idea has resulted in a number of practical realizations that range from multi-joint, position-speed monitoring software to compare trajectories achieved with different algorithms under variable load and/or space robot configurations, to modular software packages that permit us to identify the tranfer function matrices of the robot dynamical model.

With the information experimentally obtained, it was possible to use the most suitable mathematical model in each case for designing the robust digital compensators for the RIMHO joints.

Also, it signaled the opened research directions dealing with both issues of telepresence and robust/intelligent control.

6. REFERENCES

1. Armada, M., Gonzalez, P., Nieto, J. & Araujo, D. : On the design and control of a Self-Propelling robot for hazardous environments. Proc. 21st ISIR, 1990.
2. Conway, L.,Volz, R., & Walker, M. : Teleautonomous systems : Methods and Architectures for Intermingling Autonomous and Telerobotic Technology. Proc. 1987 IEEE Intern. Conf. on Robotics and Automation, vol. II, pp. 1121-1130.
3. Da Costa, L., De Buck, F., Gortz, R., Kirchner, E., & Leconte, M. : Systems for Remotely Controlled Decommissioning Operations. Graham&Trotman Ltd. for the CEC, 1986.
4. Gorinevsky, D.M. & Shneider, A. Yu. : Force control in Locomotion of Legged Vehicles over Rigid and Soft Surfaces". Robotics Research, vol. 9, num. 2, april 1990.
5. Gupta, S.C. & Hasdorff, L. : Fundamentals of automatic control. John Wiley & Sons, Inc., 1970.
6. Larcombe, M.H.E. & Halsall, J.R. : Robotics in nuclear Engineering. Graham&Trotman Ltd. for the CEC, 1984.
7. Martin, H.L. & Kuban, D.P. (Editors) : Teleoperated robotics in hostile environments ". SME 1985.
8. Mason, C.A. (Ed.) : Teleoperation and control". Proc. of the Intern. Symposium, Bristol, England, 1988.
9. Phillips, C.l. & Nagle Jr.,H.T. : Digital Control Systems Analysis and Design". Prentice Hall.
10. Sugeno, M. : An introductory survey of fuzzy control. Information Sciences 36, 59-83, Elsevier Science P. Co., 1985.
11. Todd, D.J. : Walking Machines. An Introduction to Legged Robots. Kogan Page, 1985.
12. Vertut, J. & Coiffet, P. : Teleoperation and Robotics. Evolution and developement. Hermes Publishing, 1984.
13. Vertut, J. & Coiffet, P. : Key note on Teleoperation. Proc. ICAR'85.
14. Vukobratovic, M., Borovac, B., Surla, D., & Stokic, D. : Biped Locomotion. Springer Verlag : Scientific Fundamentals of Robotics 7, 1990.
15. Winston, P.H. : Artificial Intelligence. Adison Wesley,1984.

Robust Adaptive Control Of Robotic Manipulators

Anastassios G. Chassiakos[1] and Petros A. Ioannou[2]

[1] California State University, Long Beach, CA. 90840-5602

[2] University of Southern California, Los Angeles, CA. 90089-0781

Abstract: An adaptive control scheme for the control of robotic manipulators is presented. The control scheme is shown to be robust with respect to bounded disturbances and parameter variations. The adaptive controller is compared to a fixed-gain computed-torque scheme and is seen to have superior performance in the case of unknown dynamical parameters or in the case of changes in the manipulator payload.

Keywords: Adaptive Control, Robust Adaptive Control, Robot Control, Manipulator Control.

1. Introduction

The dynamic control of robotic manipulators has been of interest to researchers for a long time. Advanced control algorithms have been proposed in the literature, ranging from adaptive control to variable structure controllers, nonlinear robust controllers and repetitive control. A comprehensive survey of adaptive algorithms for robot control is given in ref. [4].

Adaptive control is a natural choice as an alternative to fixed controllers: the adaptation mechanism will adjust to the changing operating environment (payload changes, workpiece variations, etc.), hence achieving better performance and precise positioning and tracking.

The use of direct-drive arms for robotic manipulators has several advantages over arms using mechanical gearing: it eliminates gearing backlash, reduces friction and wear and increases the joint's mechanical stiffness. On the other hand, a direct-drive arm reflects all the dynamic interactions between links to the actuators in an unattenuated fashion.

The production of this publication was partially funded by the California State University, Long Beach, through a Summer Fellowship Award

NATO ASI Series, Vol. F 71
Expert Systems and Robotics
Edited by T. Jordanides and B. Torby

As a result, any parameter changes (such as mass or inertia changes due to the unknown payload variations) will have a severe effect on the controller performance, hence adaptive control strategies become even more necessary.

In what follows we present the dynamic manipulator model (Section 2), the adaptive controller with a discussion of robustness issues and stability proof (Section 3), simulations and conclusions (Sections 4 and 5).

2. Manipulator Model

Assume an n-degree-of-freedom *rigid link* robotic manipulator. The model of the manipulator dynamics is written in the form

$$u(t) = M(\theta(t), p^*)\ddot{\theta}(t) + C(\theta(t), \dot{\theta}(t), p^*)\dot{\theta}(t) - d(t) \tag{1}$$

where

- $u(t)$ is a $n \times 1$ vector of joint torques
- $\theta(t)$ is a $n \times 1$ vector containing the joint variables
- $M(\theta(t), p^*)\ddot{\theta}(t)$ represents the contribution of the inertial forces to the dynamical equation, hence the matrix $M(\theta(t), p^*)$ represents the inertia matrix of the manipulator
- $C(\theta(t), \dot{\theta}(t), p^*)\dot{\theta}(t)$ represents the contribution of the Coriolis forces
- $d(t)$ is a $n \times 1$ bounded disturbance vector, with upper bound $\|d(t)\| \leq \bar{d}$
- p^* is a parameter vector, whose elements are functions of the geometric and inertial characteristics of the manipulator links and of the payload, i.e., p^* depends on the lengths and moments of inertia of each individual link and the payload.

Remark 2.a The parameter vector p^* can be constant in time (for example in case of constant payload) or it can be varying as a function of time, $p^* = p^*(t)$, as in the case of changing payload.

Remark 2.b If equation (1) is derived via a Lagrangian formulation, it is seen that the total contribution of the Coriolis forces will be given by a term of the form $C_0(\theta(t), \dot{\theta}(t), p^*)$. By appropriate algebraic manipulations, this can be rewritten as

$$C_0(\theta(t), \dot{\theta}(t), p^*) = C(\theta(t), \dot{\theta}(t), p^*)\dot{\theta}(t)$$

It should be noted that the choice of $C(\theta(t), \dot{\theta}(t), p^*)$ is not unique. There are several ways of extracting $C(\theta(t), \dot{\theta}(t), p^*)$ from the Coriolis related terms. A particular choice is the one used in [12] for which the matrix

$$[2C(\theta(t), \dot{\theta}(t), p^*) - \dot{M}(\theta(t), p^*)]$$

turns out to be *skew-symmetric*: $(2C-\dot{M})^T+(2C-\dot{M})=0$ and concequently the quadratic form $x^T(2C-\dot{M})x=0\ \forall x\in\Re^n$. This last equation will be of significance in the stability proof of the adaptive control scheme.

Remark 2.c The only assumption on the signal $d(t)$ is that it is bounded. Hence it can model, among other disturbances, the Coulomb friction type of disturbance.

Remark 2.d From now on, for notational convenience, the dependence on t will be dropped from all signals.

3. Adaptive Control of a Robotic Manipulator

The adaptive control scheme presented in this section is a modification of the algorithms given in [10], [11], [12] and has been presented in [1], [8] and [9].

3.1 Controller Structure

The following notation is used for the adaptive controller:

$\hat{p}=$ the estimated value of the parameter vector p^*.

$\tilde{p}=\hat{p}-p^*=$ parameter error

$p_0=$ an upper bound on the (unknown) norm $\|p^*\|$: $\|p^*\|\leq p_0$

$M^*=M(\theta(t),p^*)=$ the exact value of the inertia matrix

$C^*=C(\theta(t),\dot{\theta}(t),p^*)=$ the exact value of the matrix C

$\hat{M}=M(\theta(t),\hat{p})=$ the estimated value of the inertia matrix

$\hat{C}=C(\theta(t),\dot{\theta}(t),\hat{p})=$ the estimated value of the matrix C

$\theta_m,\ \dot{\theta}_m,\ \ddot{\theta}_m=$ desired manipulator trajectory

$\dot{\theta}_r,\ \ddot{\theta}_r=$ auxiliary signals generated on-line

$e=\theta-\theta_m=$ tracking error signal

$\epsilon=$ auxilliary tracking error signal

$\Gamma=\Gamma^T>0,\quad Q=Q^T>0\quad$ and $\quad\Lambda=\Lambda^T>0$, are positive-definite matrices, to be chosen as design parameters

$\sigma_0>0$, a positive design parameter.

The structure of the adaptive controller is shown in Fig.1 and in equations (2)-(12).

Manipulator model:

$$M^*\ddot{\theta}+C^*\dot{\theta}-d=u \tag{2}$$

Control law:

$$u=\hat{M}\ddot{\theta}_r+\hat{C}\dot{\theta}_r-Q\epsilon \tag{3}$$

Adaptation law:

$$\dot{\hat{p}} = -\Gamma Y^T \epsilon - \sigma \Gamma \hat{p} \tag{4}$$

$$Y p^* = M^* \ddot{\theta}_r + C^* \dot{\theta}_r \tag{5}$$

$$\sigma = \begin{cases} 0, & \text{if} \quad \|\hat{p}\| < p_0 \\ \sigma_0 (\|\hat{p}\|/p_0 - 1), & \text{if} \quad p_0 \le \|\hat{p}\| \le 2p_0 \\ \sigma_0 > 0, & \text{if} \quad \|\hat{p}\| > 2p_0 . \end{cases} \tag{6}$$

Tracking errors:

$$e = \theta - \theta_m \tag{7}$$

$$\dot{e} = \dot{\theta} - \dot{\theta}_m \tag{8}$$

$$\epsilon = \dot{e} + \Lambda e \tag{9}$$

Auxiliary signals:

$$\dot{\theta}_r = \dot{\theta}_m - \Lambda e \tag{10}$$

$$\ddot{\theta}_r = \ddot{\theta}_m - \Lambda \dot{e} \tag{11}$$

Parameter error:

$$\tilde{p} = \hat{p} - p^* \tag{12}$$

Remark 3.a The control law

$$u = \hat{M}\ddot{\theta}_r + \hat{C}\dot{\theta}_r - Q\epsilon = \hat{M}\ddot{\theta}_r + \hat{C}\dot{\theta}_r - Q(\dot{e} + \Lambda e)$$

is of the "computed-torque" type, containing a *nonlinear part*: $(\hat{M}\ddot{\theta}_r + \hat{C}\dot{\theta}_r)$ and a *linear, proportional-plus-derivative* control action: $Q(\dot{e} + \Lambda e)$. This control law will be compared to the standard computed-torque law

$$u = \hat{M}\ddot{\theta}_m + \hat{C}\dot{\theta} - \hat{M}(\Lambda_1 \dot{e} + \Lambda_2 e) \tag{13}$$

Remark 3.b: Error equation By substituting (3) into (2) and using equations (7)-(12), the error ϵ is shown to satisfy the following error equation

$$M^* \dot{\epsilon} + (C^* + Q)\epsilon = Y\tilde{p} + d \tag{14}$$

3.2 Robustness Considerations

This adaptive controller is based on a Lyapunov function design. Adaptation laws of the form (5) without the switching σ term, i.e., $\dot{\hat{p}} = -\Gamma Y^T \epsilon$, can lead to instability even when the disturbance $d(t)$ is bounded, due to *parameter drift*, [6]. In reference [8] it is shown that even for $d(t) \to 0$, the adaptation law (5) with $\sigma = 0$ may yield $\hat{p}(t) \to \infty$ as

$t \to \infty$ for certain choices of the disturbance $d(t)$. In the next sections it will be proven that the switching σ term *improves the robustness* of the system with respect to *bounded disturbances and parameter variations* (e.g., changing payload conditions). Moreover, the switching σ can yield a robust system in the case of *unmodeled dynamics* (e.g., fast actuator dynamics), [8].

3.3 Known Parameters and No Disturbances

In this case, $d(t) = 0$ and the parameter vector is assumed known, $\hat{p} = p^*$. Then no adaptation is necessary and the tracking error ϵ goes asymptotically to zero as $t \to \infty$. To see this, consider the positive-definite function $V(t) = \frac{1}{2}\epsilon^T M^* \epsilon$. Then the time derivative of $V(t)$ for the solutions of (14) will be

$$\begin{aligned} \dot{V}(t) &= \epsilon^T M^* \dot{\epsilon} + \frac{1}{2}\epsilon^T \dot{M}^* \epsilon \\ &= \epsilon^T \left[Y\tilde{p} - (C^* + Q)\epsilon\right] + \frac{1}{2}\epsilon^T \dot{M}^* \epsilon \\ &= \epsilon^T Y\tilde{p} + \frac{1}{2}\epsilon^T (\dot{M}^* - 2C^*)\epsilon - \epsilon^T Q\epsilon \\ &= -\epsilon^T Q\epsilon \leq 0 \end{aligned}$$

since $\tilde{p} = 0$ (parameters assumed known) and the quadratic form $\epsilon^T (\dot{M}^* - 2C^*)\epsilon = 0$. It follows that all the signals in the loop are bounded and that the tracking errors $\epsilon \to 0$, and $e \to 0$ as $t \to \infty$, hence perfect tracking is achieved.

3.4 Unknown Parameters and Nonzero Disturbances

When $d(t) \neq 0$ and the parameter vector p^* is unknown, the following theorem holds:

Theorem 1: *Using the adaptation law (4)-(6), all the signals in the closed-loop system (2)-(12) remain bounded in the presence of a bounded disturbance signal $\|d(t)\| \leq \bar{d}$. Moreover the tracking error ϵ converges to the residual set*

$$E = \left\{\epsilon : \limsup_{T \to \infty} \frac{1}{T} \int_{t_0}^{t_0+T} \|\epsilon\|^2 \, dt = \Delta \bar{d}^2 \right\}$$

for some constant Δ.

Proof: Consider the positive-definite function

$$V(t) = \frac{1}{2}\epsilon^T M^* \epsilon + \frac{1}{2}\tilde{p}^T \Gamma^{-1} \tilde{p} \tag{15}$$

The derivative of $V(t)$ along the solutions of (2)-(12), (14), will be

$$\begin{aligned}\dot{V}(t) &= \epsilon^T M^* \dot{\epsilon} + \tilde{p}^T \Gamma^{-1} \dot{\tilde{p}} \\ &= \epsilon^T Y \tilde{p} - \epsilon^T Q \epsilon - (\Gamma Y^T \epsilon + \sigma \Gamma \hat{p})^T \Gamma^{-1} \tilde{p} + \epsilon^T d \\ &= -\frac{1}{2}(\epsilon - Q^{-1} d)^T Q (\epsilon - Q^{-1} d) + \frac{1}{2} d^T Q^{-1} d - \sigma \hat{p}^T \tilde{p} - \frac{1}{2} \epsilon^T Q \epsilon\end{aligned} \tag{16}$$

after completing the squares in the last equality.

Now letting $\lambda_0 = \min \lambda(Q)$, (16) can be rewritten as

$$\dot{V}(t) \le \frac{1}{2\lambda_0} \|\bar{d}\|^2 - \frac{1}{2}\lambda_0 \|\epsilon\|^2 - \sigma \hat{p}^T \tilde{p} \tag{17}$$

For $\|\hat{p}\| > 2p_0$ we have

$$\begin{aligned}\sigma \hat{p}^T \tilde{p} &= \sigma_0 \hat{p}^T \tilde{p} \\ &= \sigma_0 \tilde{p}^T \tilde{p} + \sigma_0 p^{*T} \tilde{p} \\ &= \frac{1}{2}\sigma_0 (\tilde{p} + p^*)^T (\tilde{p} + p^*) - \frac{1}{2}\sigma_0 \|p^*\|^2 + \frac{1}{2}\sigma_0 \|\tilde{p}\|^2\end{aligned}$$

after completing the squares.

Then (17) is written as

$$\dot{V}(t) \le -\frac{1}{2}\sigma_0 \|\tilde{p}\|^2 - \frac{1}{2}\lambda_0 \|\epsilon\|^2 + \frac{1}{2\lambda_0} \|\bar{d}\|^2 + \frac{1}{2}\sigma_0 \|p^*\|^2 \tag{18}$$

from which it follows that $\dot{V}(t) < 0$ whenever $V(t) \ge V_0$ for some finite constant V_0. Hence $V(t)$ is bounded and $\tilde{p}$, ϵ are uniformly bounded and all the signals in the closed-loop system (2)-(12) are bounded.

Now due to the particular choice of the switching σ, the following holds

$$\sigma \hat{p}^T(t) \tilde{p}(t) \ge 0 \quad \forall \ t > 0 \tag{19}$$

To obtain the residual set E, integrate both sides of (17) given (19) and the boundedness of $V(t)$:

$$\limsup_{T \to \infty} \frac{1}{T}\{V(t_0 + T) - V(t_0)\} \le \limsup_{T \to \infty}\{\frac{1}{T}\int_{t_0}^{t_0+T} \frac{1}{2}\lambda_0 \|\epsilon\|^2 dt + \frac{1}{2\lambda_0}\|\bar{d}\|^2\}$$

from which

$$\limsup_{T \to \infty} \frac{1}{T}\int_{t_0}^{t_0+T} \|\epsilon\|^2 dt = \Delta \|\bar{d}\|^2$$

for some constant $\Delta > 0$.

3.5 Nonzero Disturbances and Parameter Variations

In this case the (unknown) parameter vector p^* is a function of time $p^* = p^*(t)$. The assumptions about the system and the parameter variations follow those in reference [5], where a general approach is presented. The variations of $p^*(t)$ are assumed to consist

(i) of a smooth part $p_s^*(t)$ such that $\sup_t \|\dot{p}_s^*(t)\| = m_1 < \infty$

(ii) of a finite number of parameter jumps Δp_j^* occuring at times t_j, $j = 1, \ldots, N$ such that $max_j \|\Delta p_j^*\| = m_2 < \infty$.

The time derivative of $p^*(t)$ is given by

$$\frac{d}{dt} p^*(t) = \dot{p}_s^*(t) + \sum_{j=1}^{N} \Delta p_j^* \delta(t - t_j) \tag{20}$$

Let t_{min} be the minimum time between two consequtive jumps. Then the following theorem holds:

Theorem 2: *There exists a time t^* such that if $t_{min} > t^*$ all the signals in the closed loop are uniformly bounded, in the presence of a bounded disturbance $\|d(t)\| \leq \bar{d}$ and parameter variations modeled as in (20). Moreover the tracking error ϵ converges to the residual set*

$$E = \left\{ \epsilon : \limsup_{T \to \infty} \frac{1}{T} \int_{t_0}^{t_0+T} \|\epsilon\|^2 dt = \Delta(\bar{d}^2 + \bar{m}) \right\}$$

where $\bar{m} = \max(m_1, m_2)$.

Proof: Consider the positive-definite function

$$V(t) = \frac{1}{2} \epsilon^T M^* \epsilon + \frac{1}{2} \tilde{p}^T \Gamma^{-1} \tilde{p}$$

Let $\gamma = \|\Gamma^{-1}\|$. The time derivative of $V(t)$ for the solutions of (2)-(12), (14), (20) is given as

$$\begin{aligned} \dot{V}(t) &= \epsilon^T M^* \dot{\epsilon} + \tilde{p}^T \Gamma^{-1} \dot{\tilde{p}} \\ &\leq -\frac{1}{2}\lambda_0 \|\epsilon\|^2 - \frac{1}{2}\sigma_0 \|\tilde{p}\|^2 + \frac{1}{2\lambda_0} \|\bar{d}\|^2 + \frac{1}{2}\sigma \|p^*\|^2 + \dot{p}_s^*(t) \Gamma^{-1} \tilde{p} \\ &\quad + \Big(\sum_{j=1}^{N} \Delta p_j \delta(t - t_j) \Big) \Gamma^{-1} \tilde{p} \end{aligned} \tag{21}$$

Let t_1 be the time when the first jump occurs. Then:

(i) For $t < t_1$, whenever $\|\hat{p}\| > 2p_0$ we have

$$\begin{aligned}\dot{V}(t) &\leq -\frac{1}{2}\lambda_0\|\epsilon\|^2 - \frac{1}{2}\sigma_0\|\tilde{p}\|^2 + \frac{1}{2\lambda_0}\|\bar{d}\|^2 + \frac{1}{2}\sigma_0\|p^*\|^2 + \gamma\|\tilde{p}\|m_1 \\ &\leq -\frac{1}{2}\lambda_0\|\epsilon\|^2 - \frac{1}{4}\sigma_0\|\tilde{p}\|^2 - \sigma_0\left(\|\tilde{p}\|/2 - \gamma m_1/\sigma_0\right)^2 \\ &\qquad + (\gamma m_1)^2/\sigma_0 + \frac{1}{2\lambda_0}\|\bar{d}\|^2 + \frac{1}{2}\sigma_0\|p^*\|^2 \\ &\leq -\frac{1}{2}\lambda_0\|\epsilon\|^2 - \frac{1}{4}\sigma_0\|\tilde{p}\|^2 + (\gamma m_1)^2/\sigma_0 + \frac{1}{2\lambda_0}\|\bar{d}\|^2 + \frac{1}{2}\sigma_0\|p^*\|^2\end{aligned} \tag{22}$$

from which it follows that $\dot{V}(t) < 0$ whenever $V(t) \geq V_0$ for some finite constant V_0, hence $V(t)$ and $\epsilon, \tilde{p}$ are bounded $\forall\, t < t_1$.

(ii) At time $t = t_1^+$, after the first parameter jump, we obtain from (21)

$$V(t_1^+) - V(t_1^-) \leq \gamma\|\tilde{p}(t_1)\|m_2 \leq b_0 V_0$$

for some constant b_0, since $\|\tilde{p}(t_1)\|$ is bounded. Hence

$$V(t_1^+) \leq V(t_1^-) + b_1\|\tilde{p}(t_1)\| \leq b_2 V_0$$

for some constants b_1, b_2 with $b_2 > 1$.

(iii) For the time interval $t \in (t_1, t_2)$ we have $V(t) \leq b_2 V_0 e^{-b_3(t-t_1)}$ for some constant b_3 since $\dot{V}(t) < 0$ for $V(t) \geq V_0$. In particular, for $t = t_2^-$ we have $t_2^- - t_1^+ \geq t_{min}$ and $V(t_2^-) \leq b_2 V_0 e^{-b_3 t_{min}}$. Now if t_{min} is chosen as

$$t_{min} > \frac{1}{b_3}\ln b_2 = t^* \tag{23}$$

then $V(t_2^-) < V_0$.

(iv) It follows by induction that if (23) is satisfied then $V(t)$ is uniformly bounded, hence $\tilde{p}$, ϵ are uniformly bounded and all the signals in the closed loop remain bounded. Integrating both sides of (21) and using the boundedness of $V(t)$ and $\|pt\|$, and the fact that $\sigma\tilde{p}^T\hat{p} \geq 0\ \forall\, t > 0$, we get the residual set

$$\limsup_{T\to\infty}\frac{1}{T}\int_{t_0}^{t_0+T}\lambda_0\|\epsilon\|^2 dt \leq k_1 m_1 + k_2 m_2 + k_3\|\bar{d}\|^2$$

for some constants k_1, k_2, k_3 > 0, from which follows

$$\limsup_{T\to\infty}\frac{1}{T}\int_{t_0}^{t_0+T}\lambda_0\|\epsilon\|^2 dt \leq \Delta(\bar{m} + \bar{d}^2)$$

for some constant Δ.

Remark If no parameter variations and no disturbances are present, then $m_1 = m_2 = 0$, $\bar{d} = 0$ and E, $\epsilon \to 0$ as $t \to \infty$.

4. Simulations

4.1 Simulation Model

The computer model used, simulates the Berkeley/NSK robotic arm. This is a direct-drive, two-degree-of-freedom SCARA arm. The computer package that simulates the response of the arm was developed by R. Horowitz and his group at UC Berkeley, [3]. The objective of developing this package was to provide a *"benchmark"* robotic system for testing several control algorithms under realistic conditions [7].

The model parameters (lengths of links, link masses and inertias, sensor and actuator characteristics, etc.) were measured on the actual Berkeley/NSK arm and the identified values were used in the simulation package. The package has capabilities of simulating realistic conditions as closely as possible, by including:

- sensor parasitic dynamics
- actuator parasitic dynamics
- actuator saturation nonlinearities
- Coulomb friction
- quantization effects

The program is written in C and can run on a MS-DOS machine with the Microsoft C compiler.

4.2 Manipulator Parameters and System Matrices

The parameter values used in the simulation are shown in the following table:

Symbol	SI Value	Definition
I_1	0.2675	Motor 1 rotor inertia
I_2	0.360	Arm 1 inertia about c.g.
I_3	0.0077	Motor 2 rotor inertia
I_{3c}	0.040	Motor 2 stator inertia
I_4	0.051	Arm 2 inertia about c.g.
I_p	0.046	Payload inertia
M_1	73.0	Motor 1 mass
M_2	10.6	Arm 1 mass
M_3	12.0	Motor 2 mass
M_4	4.85	Arm 2 mass
M_p	6.81	Payload mass
L_1	0.36	Arm 1 length
L_2	0.24	Arm 2 length
L_3	0.139	Arm 1 radius of gyration
L_4	0.099	Arm 2 radius of gyration
TH_1	5.5	1st axis Coulomb friction magnitude
TH_2	0.9	2nd axis Coulomb friction magnitude

Table 1: Geometric and Inertial Parameter Values

If the parameter vector $p = [p_1, p_2, p_3]^T$ is chosen as

$$p_1 = I_1 + I_2 + I_{3c} + L_3^2 M_2 + L_1^2 (M_3 + M_4 + M_p) + p_2$$

$$p_2 = I_3 + I_4 + I_p + L_4^2 M_4 + L_2^2 M_p$$

$$p_3 = L_1 L_4 M_4 + L_1 L_2 M_p$$

then the system matrices $M(\theta, p)$ and $C(\theta, \dot{\theta}, p)$ are written as:

$$M(\theta(t), p) = \begin{pmatrix} [\ 1,\ 0,\ 2\cos\theta_2\ ]p, & [\ 0,\ 1,\ \cos\theta_2\ ]p \\ [\ 0,\ 1,\ \cos\theta_2\ ]p, & [\ 0,\ 1,\ 0\ ]p \end{pmatrix}$$

and

$$C(\theta(t), \dot{\theta}(t), p) = \begin{pmatrix} [\ 0,\ 0,\ -\dot{\theta}_2 \sin\theta_2\ ]p, & [\ 0,\ 0,\ -(\dot{\theta}_1 + \dot{\theta}_2)\sin\theta_2\ ]p \\ [\ 0,\ 0,\ \dot{\theta}_1 \sin\theta_2\ ]p, & [\ 0,\ 0,\ 0\ ]p \end{pmatrix}$$

One choice for the matrix Y used in the adaptive law and satisfying eq. (5) is given by

$$Y = \begin{pmatrix} \ddot{\theta}_{r1}, & \ddot{\theta}_{r2}, & 2\ddot{\theta}_{r1}\cos\theta_2 + \ddot{\theta}_{r2}\cos\theta_2 - \dot{\theta}_{r1}\dot{\theta}_2 \sin\theta_2 - \dot{\theta}_{r2}(\dot{\theta}_1 + \dot{\theta}_2)\sin\theta_2 \\ 0, & \ddot{\theta}_{r1} + \ddot{\theta}_{r2}, & \ddot{\theta}_{r1}\cos\theta_2 + \dot{\theta}_{r1}\dot{\theta}_1 \sin\theta_2 \end{pmatrix}$$

where θ_{r1} and θ_{r2} are the components of the θ_r vector: $\theta_r = [\theta_{r1},\ \theta_{r2}]^T$.

4.3 Fixed-Gain Controller

The performance of the adaptive controller will be compared to that of a fixed-gain proportional-plus-derivative computed-torque controller given by

$$u(t) = u_N(t) - u_L(t)$$

where the nonlinear computed-torque term is

$$u_N(t) = \hat{M}\ddot{\theta}_m + \hat{C}\dot{\theta}$$

and the proportional-plus-derivative term is

$$u_L(t) = \hat{M}(\Lambda_1(\dot{\theta} - \dot{\theta}_m) + \Lambda_2(\theta - \theta_m)).$$

4.4 Task Specification

The manipulator is to execute a pick-and-place task of duration 8.0 *sec* as shown in Fig. 2. The task is decomposed into two subtasks:

- *Subtask 1:* High-speed motion. Each joint will move from 0° to 180° and back to 0° within 4.0 *sec*. At time $t = 2.0$ *sec*, the arm picks up an *unknown* payload (of unkown mass and inertia).
- *Subtask 2:* Low-speed motion. Each joint will move from 0° to 45° and back to 0° within 4.0 *sec*. At time $t = 6.0$ *sec*, the arm drops the payload off.

The desired position, velocity and acceleration profiles, as well as the payload history, are shown in Figs. 2a-2d.

4.5 Simulation Results

The performance of the adaptive scheme of Section 3 as applied to the Berkeley/NSK arm model is shown in Figs.3-9.

In Figs. 3 and 4, the position and velocity responses of the adaptive algorithm are presented and compared to the desired trajectories. The tracking performance is seen to be extremely good and robust, despite the system uncertainties (unmodeled dynamics, etc.), disturbances (friction) and payload variations (unknown inertial characteristics). The control inputs generated by the adaptive controller are shown in Fig. 5.

In Fig. 6 the parameter estimates $\hat{p} = [\hat{p}_1, \hat{p}_2, \hat{p}_3]^T$ are presented. The parameters are initialized close to their true values without a payload, since we already have a rough idea of the geometric and inertial characteristics of the manipulator. After the payload is picked up at $t = 2.0$ *sec* the controller is readjusting the parameters to account for the new conditions. The same readjustment process occurs after the payload has been dropped off at $t = 6.0$ *sec*.

In Figs. 7-9, the performance of the controller is compared to that of a fixed-gain PD computed-torque scheme, as described in Section 4.3. The fixed-gain scheme is seen to produce much higher errors after the payload has been picked up, whereas the adaptive scheme compensates for the unknown payload characteristics. Figs. 7 and 8 present the position and velocity errors in joint-space, and Fig. 9 presents the tip position errors in Cartesian-space coordinates.

5. Conclusions

The adaptive control scheme presented herein is seen to have robust performance in the face of unmodeled dynamics, bounded disturbances, unknown geometric and inertial manipulator characteristics and unkown payload variations. It is also seen to improve trajectory tracking as compared to a fixed-gain PD computed-torque scheme.

Extensions of this robust adaptive scheme will be made for the case of flexible manipulators (including joint or link flexibility as in [2]). Robust-adaptive-control algorithms for robot control can be applied to situations with imprecise or uncertain knowledge of the environment and tasks for which the controller must compensate for the uncertainties. These applications include among others: applications in space, in nuclear plants and in medicine (e.g., dentistry, neurosurgery, orthopedic surgery for hip/elbow prostheses).

6. References

1. Chassiakos A.G., Pak A., Ioannou P., "A Nonlinear Robust Adaptive Controller for Manipulator Tracking", *Proceedings of the American Control Conference*, pp.1469–1475, June 1989.
2. Chassiakos A.G., and Bekey G.A., "On the Modelling and Control of a Flexible Manipulator Arm by Point Actuators", *Proceedings of the IEEE 25th Conference on Decision and Control*, pp.1145–1150, Dec. 1986.
3. Horowitz A., Kao W.W., Kang C.G., "A Computer Simulation Program for a Two Link SCARA Manipulator", *Mechanical Eng. Dept, UC Berkeley*, 1988.
4. Hsia T.C., "Adaptive Control of Robotic Manipulators: A Review", *IEEE Intern. Conf. on Robotics and Automation*, April 1986.
5. Ioannou P.A., Tsakalis K.S., "A Robust Adaptive Controller", *IEEE Trans. on Automatic Control*, AC-31, Nov. 1986.

6. Ioannou P.A., Kokotovic P.V., "Instability Analysis and Improvement of Robustness of Adaptive Control", *Automatica*, Vol. 20, pp.583–594, Sept. 1984.
7. Kang C.G., Kao W.W., Boals M., Horowitz R., "Modeling, Identification and Simulation of a two Link SCARA Manipulator", *Proc. of Winter Annual Meeting of ASME*, pp.393–407, Nov. 1988.
8. Reed J.S., "Instability Analysis and Robustness Improvement of Adaptive Control of Robot Manipulators", *PhD Thesis, University of Southern California*, 1988.
9. Reed J.S., Ioannou P.A., "Instability Analysis and Robust Adaptive Control of Robot Manipulators", *IEEE Trans. on Robotics and Automation*, Vol. 5, pp. 381–386, 1989.
10. Sadegh N., Horowitz R., "Stability Analysis of an Adaptive Controller for Robotic Manipulators", *IEEE Intern. Conf. on Robotics and Automation*, April 1987.
11. Sadegh N., Horowitz R., "Stability and Robustness Analysis of a Class of Adaptive Controllers for Robotic Manipulators", *Int. Journal of Robotics Research*, Vol.9, pp.74–92, 1990.
12. Slotine J.E., Li W., "On the Adaptive Control of Robot Manipulators", *Int. Journal of Robotics Research*, Vol.6, pp.49–59, 1987.

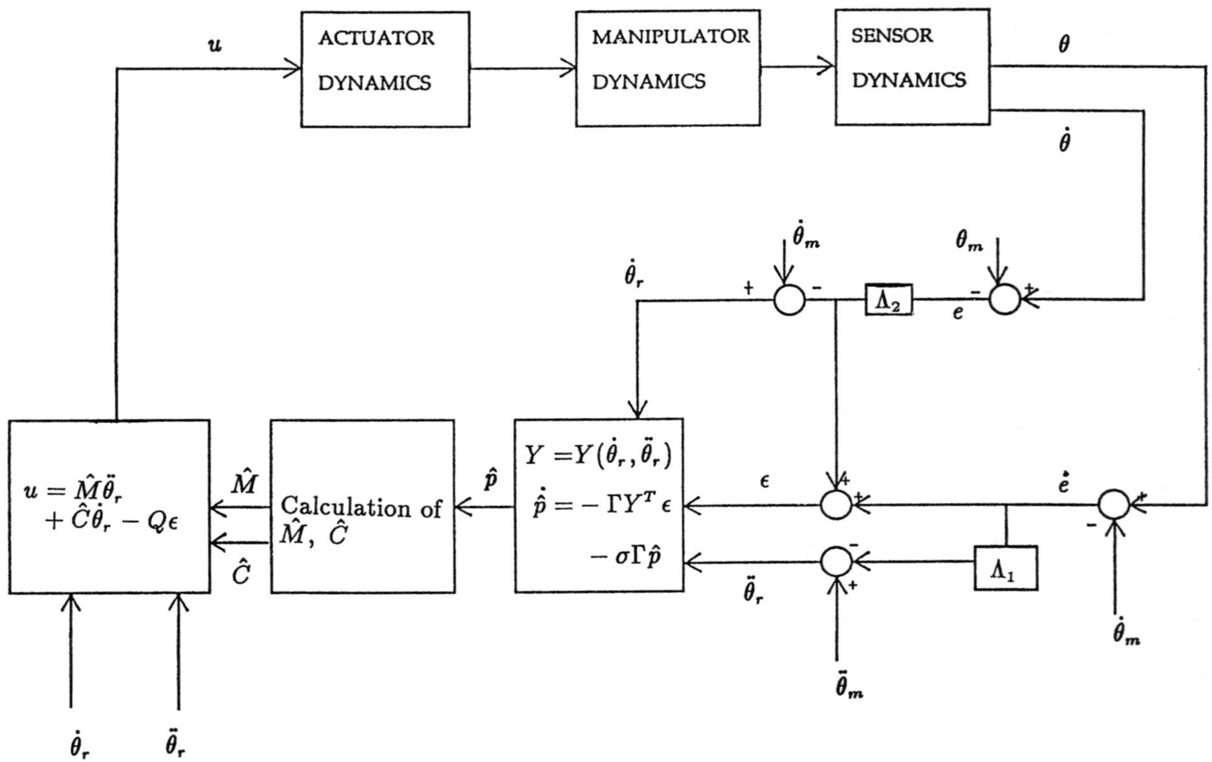

Fig. 1: Structure of the Adaptive Controller

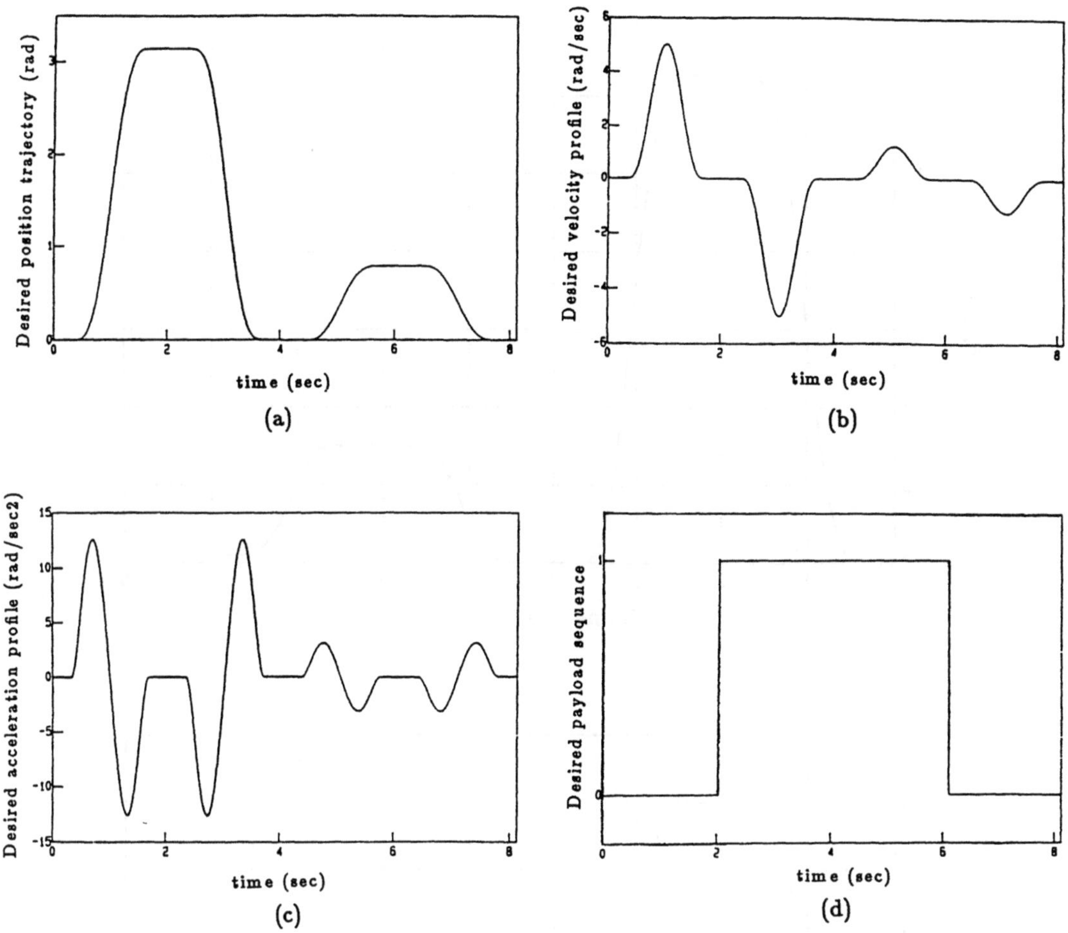

Fig. 2: Task Specification

(a) Desired position trajectory (rad)

(b) Desired velocity profile (rad/sec)

(c) Desired acceleration profile (rad/sec^2)

(d) Desired payload sequence

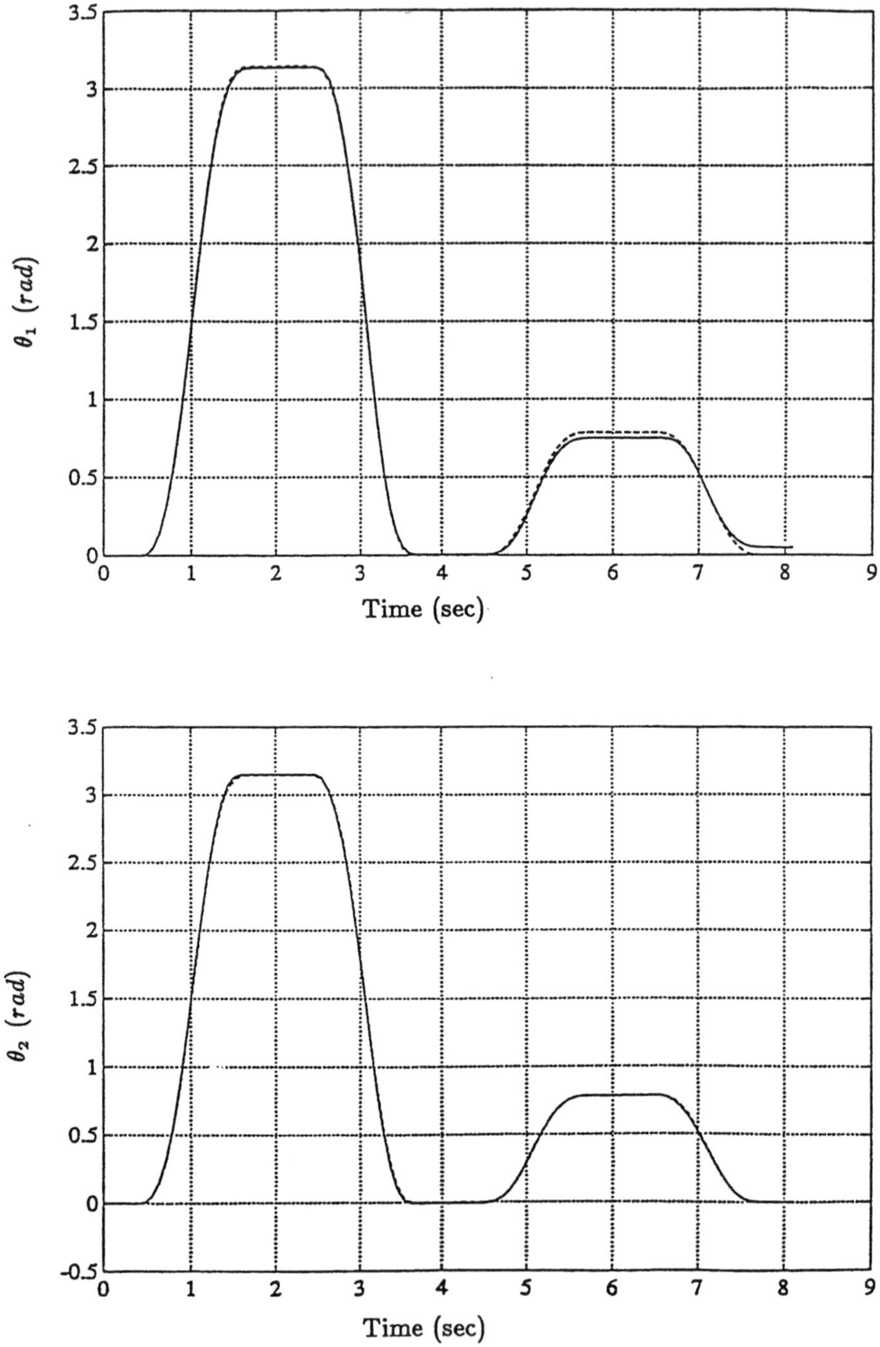

Fig. 3: Position Response Under Adaptive Control

Dashed line: desired position

Solid line: response under adaptive control

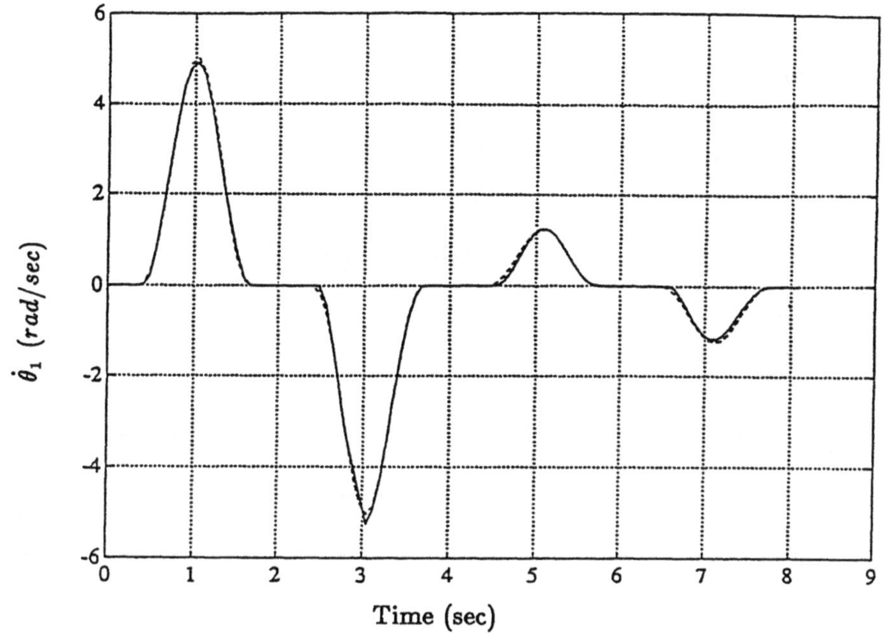

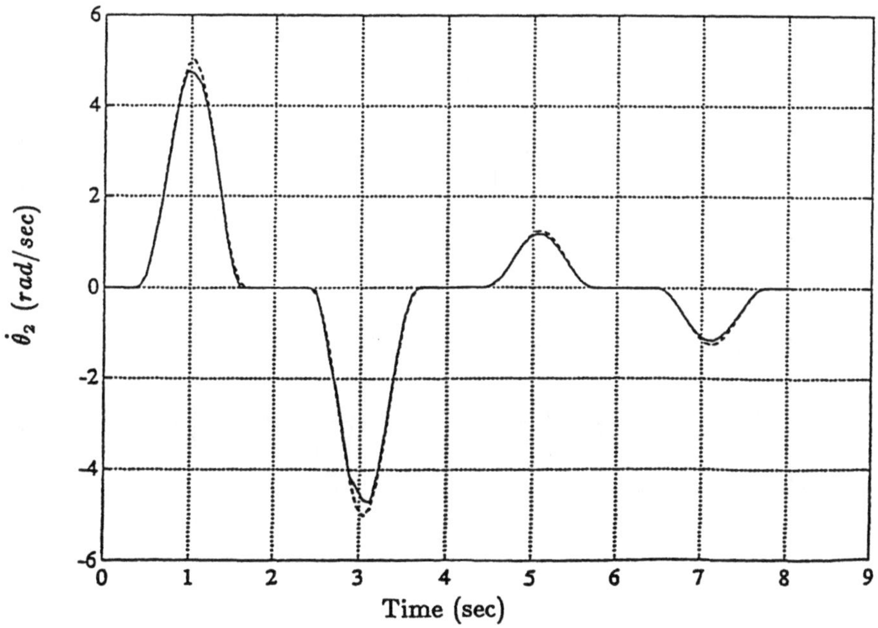

Fig. 4: Velocity Response Under Adaptive Control

Dashed line: desired velocity

Solid line: response under adaptive control

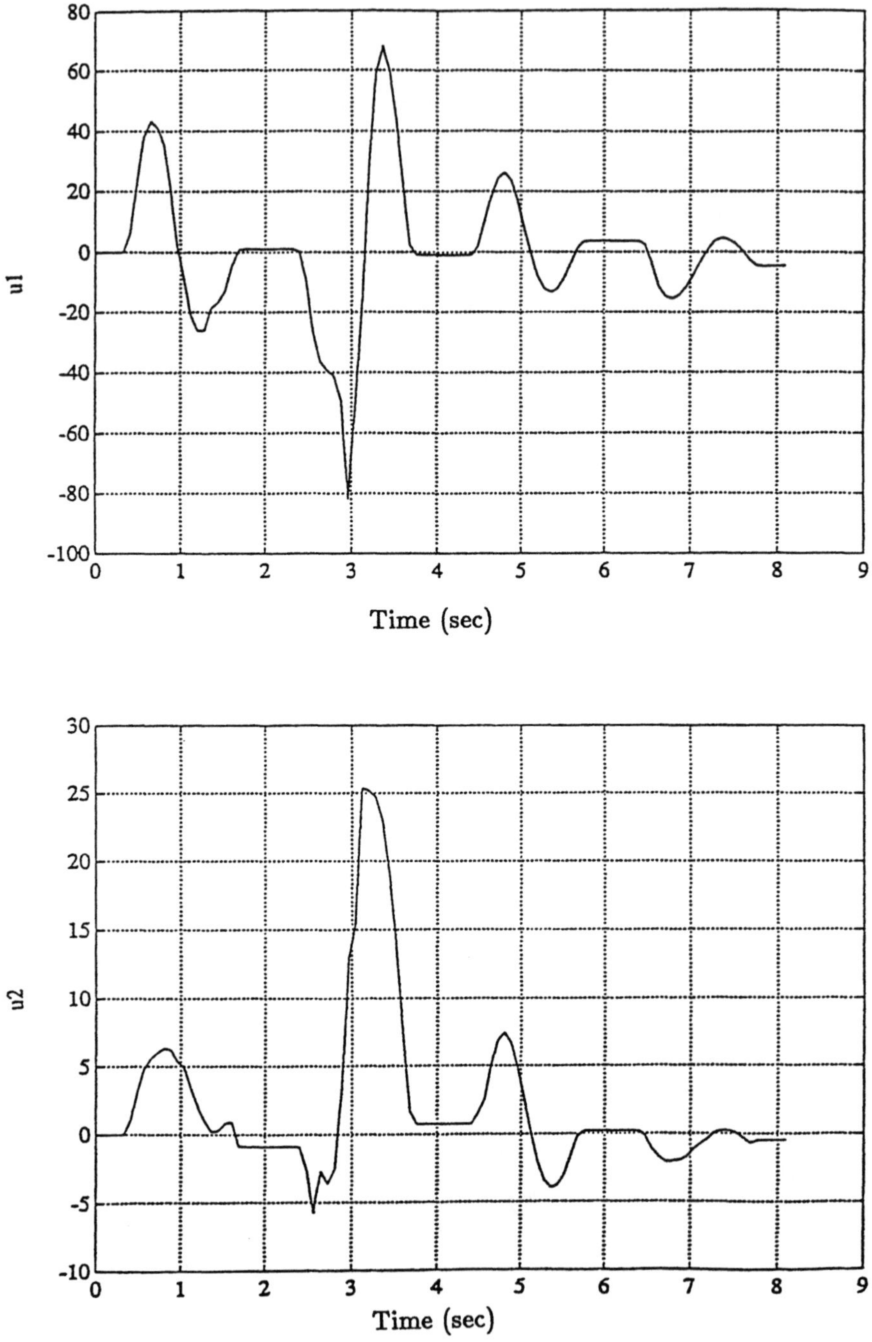

Fig. 5: Control Inputs Generated by the Adaptive Controller

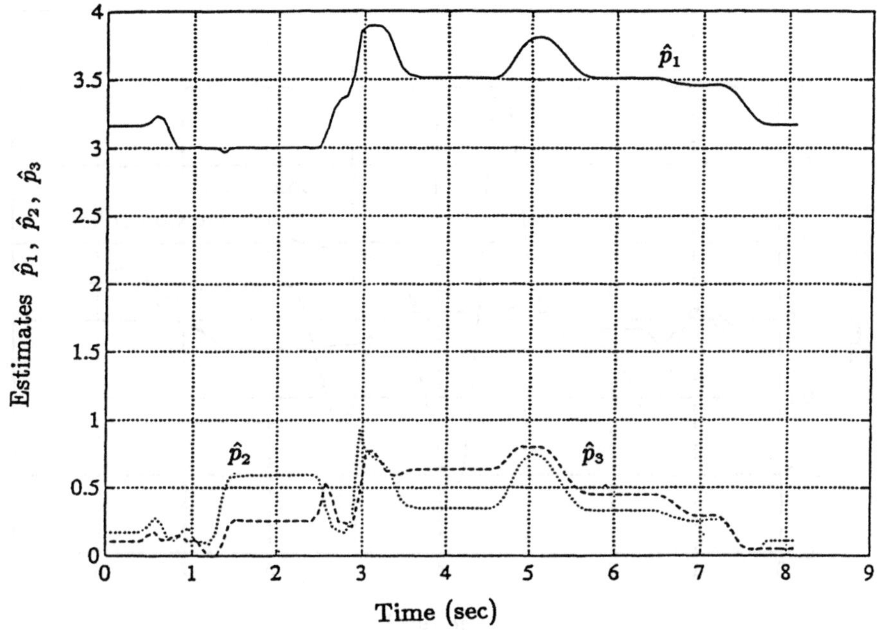

Fig. 6: Parameter Estimates $\hat{p}_1$, $\hat{p}_2$, $\hat{p}_3$

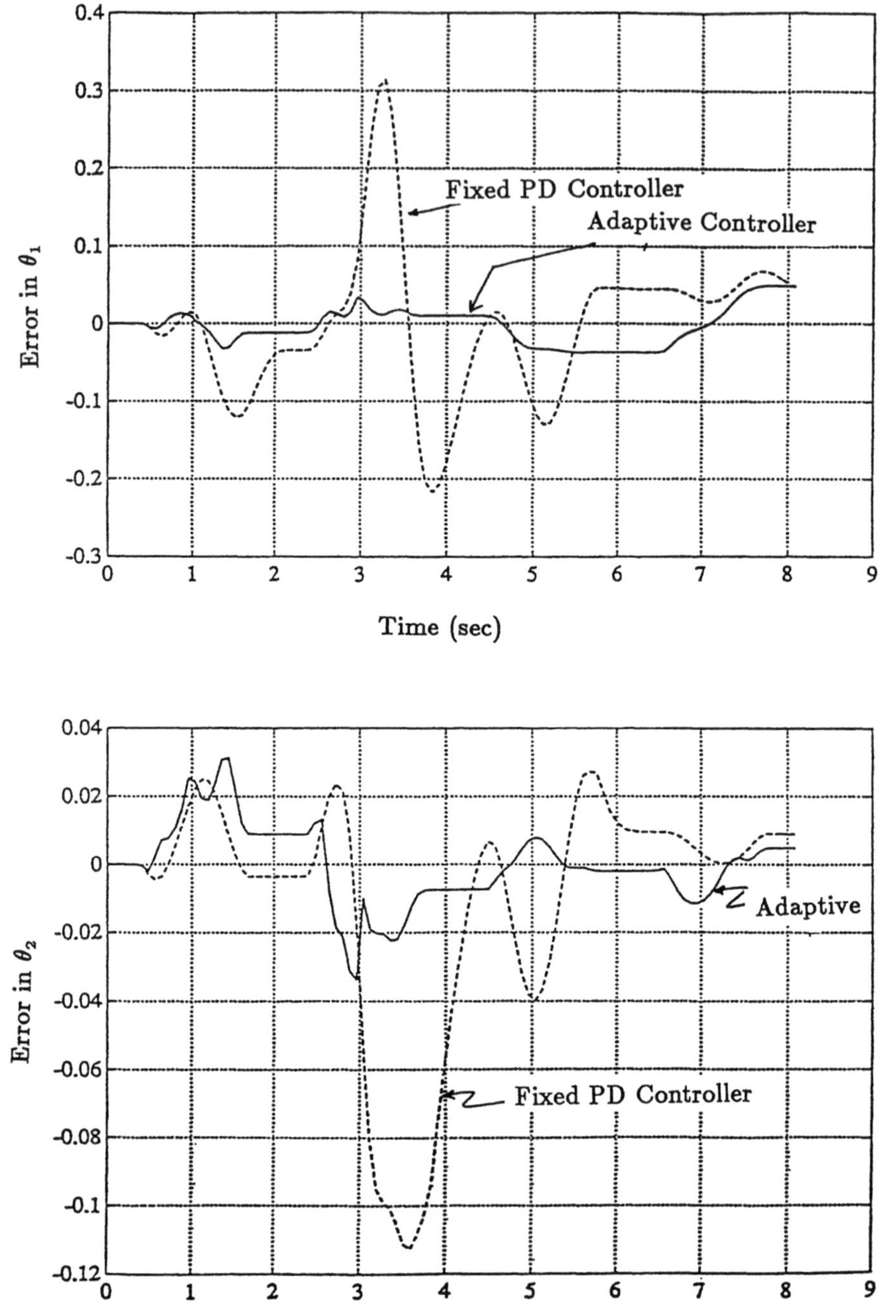

Fig. 7: Position Errors

Dashed line: Fixed PD computed-torque controller

Solid line: Adaptive controller

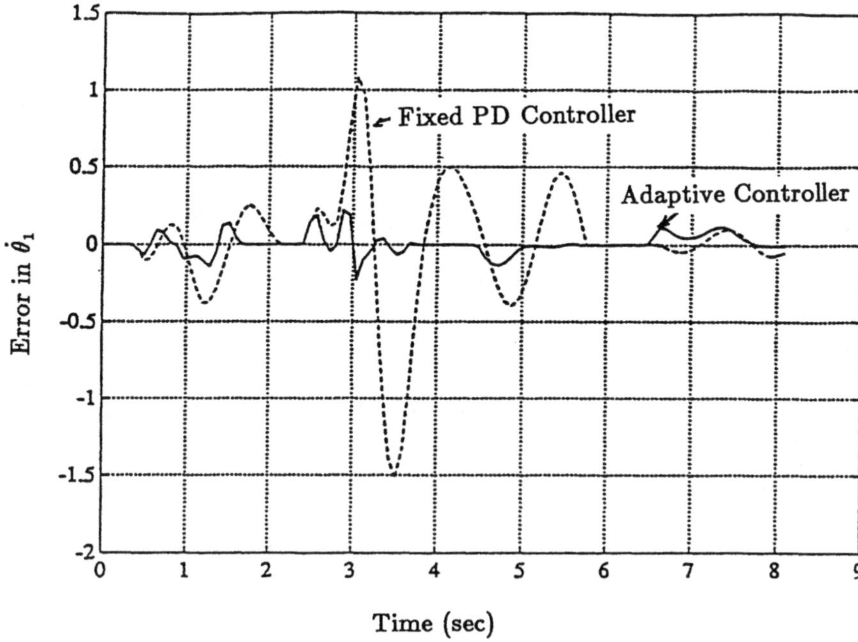

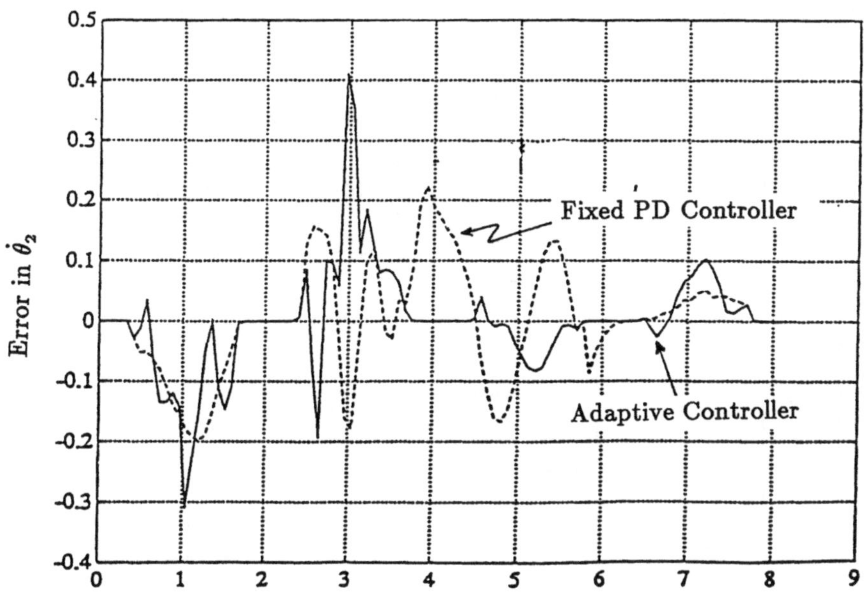

Fig. 8: Velocity Errors

Dashed line: Fixed PD computed-torque controller

Solid line: Adaptive controller

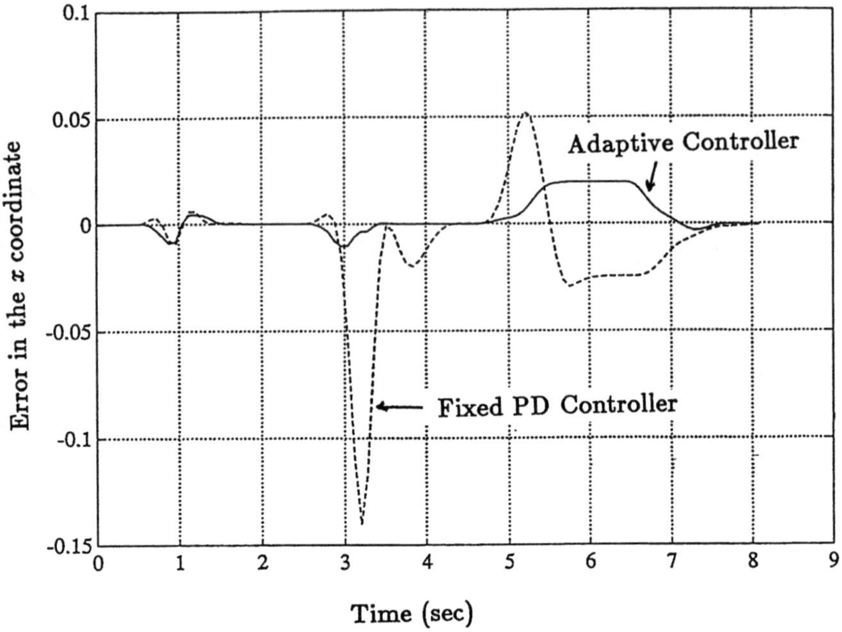

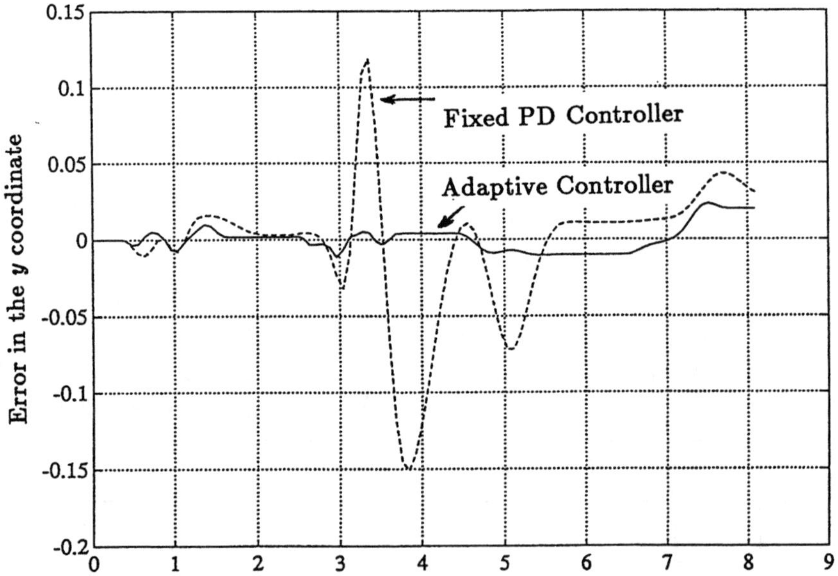

Fig. 9: Position Errors in Cartesian Coordinate Space

***Dashed line*: Fixed PD computed-torque controller**

***Solid line*: Adaptive controller**

About Qualitative Robot Control

Ph. Coiffet[1], J. Zhao[1], J. Zhou[1], F. Wolinski[2], P. Novikoff[1], D. Schmit[3].

1 Laboratoire de Robotique de Paris, Université Pierre et Marie Curie,
BP 340-75229-Paris Cedex 05-France.

2 EDF/DER,
Quai Watier-78400-Chatou-France.

3 INSTN Centre d'Etudes Nucléaires de Saclay
BP 2-91191-Gif sur Yvette Cedex-France

Abstract

Among the obstacles to be overcome for reaching a full mastering of robot control, three of them seem of great importance :

1) To be able to find the optimal bridges linking the qualitative and quantitative control approaches. This point exhibits the problem of precision during a robot physical action.
2) When a robot "brain" has to make a decision, to be sure that this decision is not a mistake. The reliability problem is raised with this question.
3) All computations modelled into the percéption and reasoning phenomena must be performed in real time. Consequently, the computational efficiency problem is brought up.

The three questions are not totally independant and, up to now, no one seems able to propose satisfactory solutions. The first step must be attached to explore all qualitatives approches, knowing that quantitative approches, have been extensively studied.

We present new results in four types of qualitative control. The two first ones are devoted to trajectory generation using the viability theory and fuzzy sets. The third is dedicated to an adaptative navigation control process and he last is concerned with the planning generator problem.

1-Introduction

No living being seems to use precise computations when it performs acts in its life time. However it is able to succeed in very delicate physical operations such as in the case of man.

But the final purpose of a robot is surely to have in a particular field the same dexterity and the same reasoning power that a man has. We have to conclude with the idea that a basically qualitative control is better adapted to robot control than a quantitative one.

Nevertheless, almost all industrial robots have control algorithms based upon computations of supposed exact values deduced from quantitative modelling [Craig & al, Dubowsky & al, Featherstone, Liégeois, Low & Duby, Koivo & Paul, Morgan & Ozguner, Nakamura Hanafusa, Samson, Slotine, Meung & Chen, Yoshikawa]. Can we replace the algorithms with qualitative ones and achieve improvement? Many researchers have worked on the general decision process which is a part of the problem. But the more we approach the final execution step of the task, the more the qualitative processes is difficult to use with efficiency (time and precision). At the opposite end, the more we are far from this final step, the more a quantitative control is important to the qualitative solution.

NATO ASI Series, Vol. F 71
Expert Systems and Robotics
Edited by T. Jordanides and B. Torby

Somewhere an optimal zone probably exists in which we should have to change the control mode. We want to illustrate the phenomenon with several instances of qualitative approaches.

2 Robot control from a qualitative point of view

A qualitative point of view information obviously describes a property that can be used if context conditions are satisfied.

In the matter of robot control, we have a goal : the performance of a task (wich is a modification of the environment) via only one kind of action: the motion of the robot (coordinated motions off all its joints). We understand easily that the more we are near the final execution step, the more the motion must be accurate.

So, at any time during a task performance (or a mission execution), we can define in the space of the robot operational variables (task variables $\underline{X}$) a volume (an hypervolume) inside of which these variables must lie in order to get a correct progression of the task performance.

The idea of qualitative control is directly linked to this volume. When the volume is large, qualitative information allows the robot to move correctly. When the volume becomes very thin, the accuracy demands a quantitative computation (see instance in figure 1).

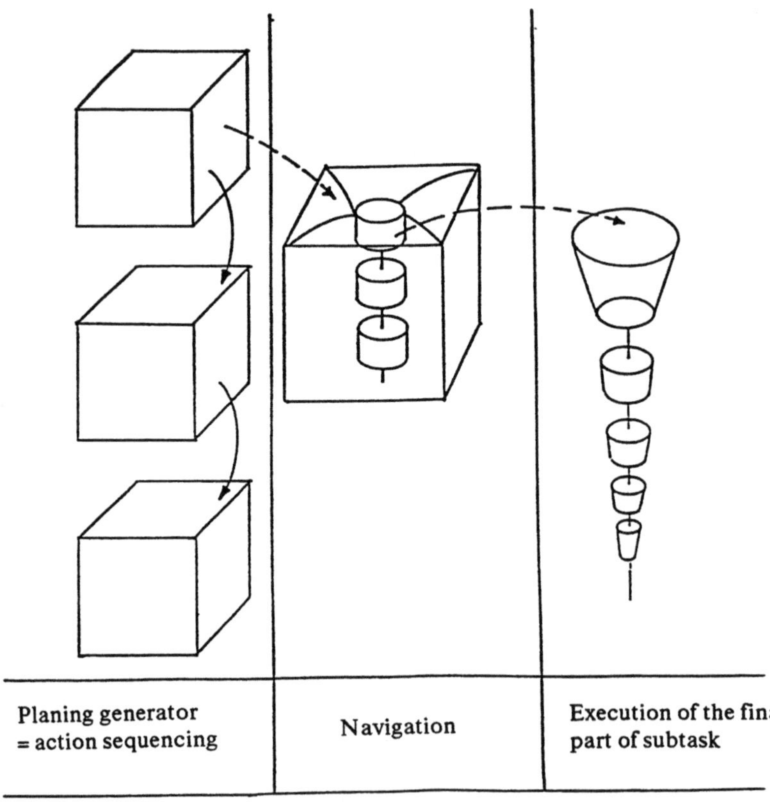

FIGURE 1

Each volume represents the volume in which the robot can be in terms of task variable for a subtask implied into the task.

Illustration of the link of qualitative and quantitative control in the robot control problem.

In fact, the problem is a bit more complex if we include the decision process which can always be described by :

if (conditions) then (action)

or

if (domain) then (domain)

which means : if constraints are inside a domain, then the operational coordinates must be inside a domain. When the domains are reduced to a set of simple points, a qualitative control is no longer possible.

But the number of possibilities for the shapes of these domains is very large. We can have, for instance, some strict contraints of equality and some contraints with a larger range. We can have innaccurate conditions and accurate actions and etc...

Finally, the conditions of shifting from a qualitative control to a quantitative control are very difficult to formalize. Here we want only to describe four instances of original qualitative controls : two about trajectory generation, one in navigation and one in planning generation.

3-Viability theory and trajectory generation

3.1 Theory Recall

With general control theory, the dynamic evolution of a system is described in terms of differential equations such as :

$$(t) = f(x(t), u(t))$$
$$x(0) = x_0 \qquad (1)$$

Where $x(t)$ is the state vector ;
$u(t)$ is the system control vector ;
$f(x(t), u(t))$ is the state description function.

In this case, we assume the following :

- there is no uncertainty in the system and its environment ;
- the structure of the system is invariant relative to the mathematic model ;
- the model is deterministic ;
- there is no formal consideration about the system constraints.

These conditions rarely hold true when we use the system model (1) for any practical problem.

To overcome the above difficulties, the viability theory can be efficient thanks to its fundamental properties.

Through viability theory [Aubin & al, Aubin], a system state can be expressed in terms of differential inclusions :

$$\mathbf{x}'(t) \in F(x(t), u(t))$$
$$\mathbf{x}(0) = \mathbf{x}_0 \qquad (2)$$

where F is a set-valued map from a Banach space $x \in X$, to itself $x' \in X$, which delivers a set of velocity solutions at every time ***t***.

The system (2) is interesting because the set-value map ***F*** allows the state evolution to be chosen among a set of propositions Image(***F***).

However, we cannot confirm, up to here, whether the system has a viable solution for a given set of propositions. So, we need to define another set-valued map $\mathbf{T_K(x)}$ called a contingent cone on a set $\mathbf{x} \in K \subset X$:

$$T_K(x) := \{v \in X | \liminf_{h \to 0+} \frac{d_K(x+hv)}{h}$$

Where $d_K(y) := \inf \{ \| y - z \|, z \in K \}$ is the distance from y to the set $K \subset X$

REMARKS

- The set K is defined as a viable domain since for $\forall x \in K$, the system (2) is assumed *viable*. This implies that the system states are limited within some bounded space and that there is no viable solution in the outer of K.

- A contingent cone $T_K(x)$ to K is a kind of tangent space of K. It characterizes a state velocity space in which the system evolution guarantees the state viability.

Viability domain definition : *Let X be a Banach space and* $\Omega \subset X$ *is a set valued map. A subset* $K \subset \Omega$ *is a viablility domain if and only if :*

$$Vx \in K, F(x(t)) \cap Tk(x) = \varnothing \quad (4)$$

The system (2) is said viable when for $\forall x_0$, K and $t > 0$, there exists at least one solution such that $x(t) \in K$ starting from an initial x_0.

Viability theorem : *Let us consider a nontrivial closed set-valued map* $F : x \Rightarrow x$ *with convex images and a closed subset* $K \subset Dom(F)$ *of finite dimensional vector space X. There exists a constant c such that :*

$$\sup_{x \in K} \frac{d(0,\ F(\mathbf{x}) \cap Tk(\mathbf{x}))}{1+\|\mathbf{x}\|} \leq c \leq +\infty$$

Then for all initial states $\forall x_0 \in K$*, there exists a viable solution in* $[0, \infty[$ *to differential inclusions (2), which belongs to the space* x.

3.2 Application to a 6 d.o.f robot

Basically the control consists in defining a field of constraints for all variables and to verify at any time that the robot state is located in the viability domain.

The robot evolution can be represented with the differential inclusions :

$$\dot{\theta}(t) = \underline{f}(\theta(t), u(t))$$

$$\underline{u} \in (t)\ \underline{H}(\theta(t))$$

$$\underline{\theta}(0) = \theta_o$$

$\underline{\theta}(t)$ = joint angles

$\underline{u}(t)$: system inputs

The viability domain K is described in terms of several sets of robot constraint interpreted in the state space (via the inverse geometric model of the robot)

These contraints are :

* physical constraints

- $\underline{\theta}_{i\,min} < \underline{\theta} < \underline{\theta}_{i\,max}$: The range of each angular joint is limited (and depending on the value of other joints)

- $\underline{X}(\underline{\theta}) \in E_r$: The task variables (position and orientation of the end effector) must stay inside the physical reachable zone.

* model constraints
- the system has to be prevented from passing through singular configurations

* task constraints
- the reachable space may be reduced with presence of obstacles

The viability domain K can be describe with the intersection of sets in joint space determined by the previous contraints, i.e.,

C_j : set of ranges of joints angles
C_r : set of instantaneous reachable points in task space

$$C_r = \{\underline{\theta} \mid \underline{X} = T(\underline{\theta}) \in E_r\}$$

C_s : the complement of the set of neighbourhoods of the singular configurations
C_g : set of constraints imposed by the tasks

So the viability domain can be written as :

$$K = C_j \cap C_r \cap C_s \cap C_g \tag{7}$$

We must add the frontier conditions defined with the contingent cone :

$$T_K(\underline{\theta}) = T_{C_j} \cap T_{C_r} \cap T_{C_s} \cap T_{C_g} \tag{8}$$

the controls are composed of two parts :
- comparison of input values (desired state deduced from theoretical computation) with output values (actual state given by the system)

- verification of velocity values $\dot{\theta}$, in order to be sure that they are inside the admissible range, and verification of the state system to see that it stays within the viability domain K.

These principles were used in the teaching of a six d.o.f welding robot. The robot has a set of collision sensors and the robot supervisor is described with logical equations in the controller.

The system learns off-line a point-to-point trajectory. Then we try to link these points. During the simulation, at every instant the controller verifies that the robot state is inside the viability domain. As soon as a border is reached, the simulation is stopped and a diagnostic on what constraint is violated is given. So we can try either the same trajectory with another type of interpolation a (for instance : linear interpolation, circular interpolation, splines....), or another slightly modified trajectory. In automatic mode, the system proposes a trajectory which satisfies the constraints, passes through the first and the last point, and passes as close as possible to the others points. This controller is running in an application of industrial welding and seems to be satisfactory.

In conclusion, we have here an instance of a kind of hybrid qualitative control through the choice of a non-empty space of constraints, and quantitative control in the final choice of trajectory.

4 Fuzzy control and trajectory generation

The application is devoted to a redundant robot. The joint trajectory generation can be considered as a control process in which the task coordinates are the inputs, and the joint variables are the outputs.

This problem almost always has been considered as a problem of inversion of the Jacobian matrix [Yoshikawa & al., Aubin & al., Nenchev, Vaccaro & al., Hsia &al., Pedrycz, Sugeno.]. Using the qualitative approach, a classical scheme [Wolovich & Elliot] based on the work of Wolovich and Elliot is given in figure 2 where notations are indicated.

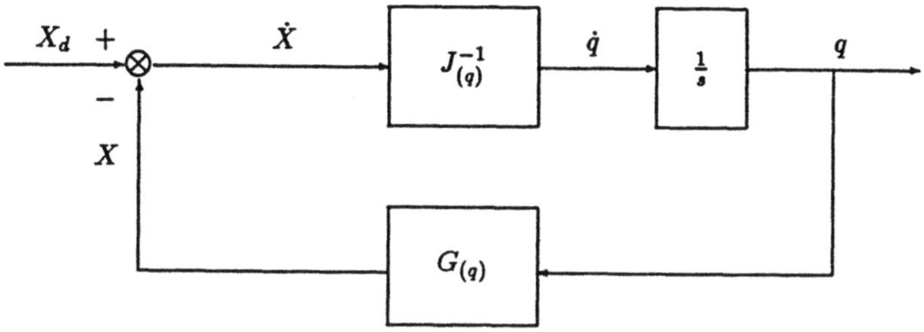

where : X_d - desired task coordinates, X - actual task coordinates, $\dot{X}$ - dX/dt, q - joint varibles, $\dot{q}$ - velocity of joint variables, $J_{(q)}$ - Jacobian matrix, $1/s$ - integration relative to time, $G_(q)$ - direct geometrical model endflushleft

Figure 2 : Deterministic scheme of a trajectory generation in joint space from a desired trajectory in task space

This deterministic scheme can be transformed into a qualitative scheme using fuzzy sets for determining the matrix J(q) as depicted in figure 3.

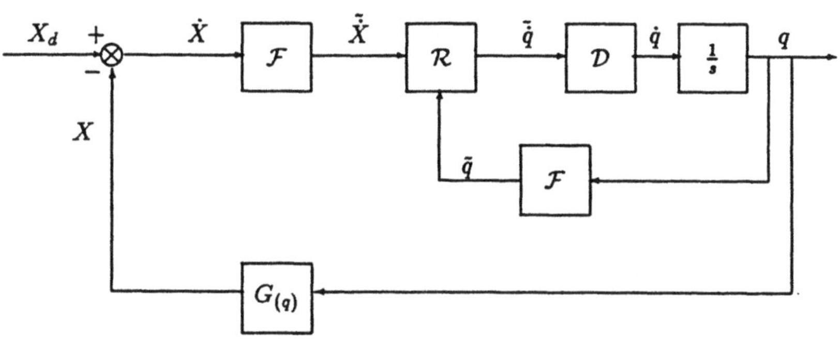

where : $\tilde{\dot{X}}$ - fuzzy set for $\dot{X}$, $\tilde{\dot{q}}$ - fuzzy set for $\dot{q}$, $\tilde{q}$ - fuzzy set for q, $\mathcal{F}$ fuzzification process, $\mathcal{D}$ - defuzzification process, $\mathcal{R}$ - fuzzy relation

Figure 3 : Fuzzy scheme of a trajectory generation in joint space from a desired trajectory in task space

4.1 Fuzzy control design

In fact, we have to find a fuzzy relation of q with X and q replacing J(q).
We have the well know relation

$$|\dot{X}| = J\,\dot{q}$$

Let us write

$$\dot{q} = [\dot{q}_1 , \dot{q}_2 , \dot{q}_3, \ldots\ldots\ldots\ldots \dot{q}_n]$$

j_i = column number i of J

We have

$$\dot{X} = \Sigma\, j_i\, \dot{q}_i$$

Locking all the joint variables except the variable number i, we have a series of decoupled equations :

$$\dot{X} = j_i\, \dot{q}_i + O_i \;;\; i = 1, \ldots n \tag{9}$$

O_i is an error vector.

For each equation of (9) we are going to choose a value of q_i which minimizes O_i. For that, we define the angle separating the vector $\dot{X}$ from the vector j_i with

$$1)\quad \alpha_i = \frac{j_i\dot{X}}{|j_i|\ |\dot{X}|} \tag{10}$$

$$1.\quad \dot{q} = |\dot{X}|\, \alpha_i \tag{11}$$

α_i represents the contribution of the ith joint and has the following properties :

if	$\lvert \alpha_i \rvert = 0$	then	$MIN(O_i) = \lvert \dot{X} \rvert$
if	$\lvert \alpha_i \rvert = 1$	then	$MIN(O_i) = 0$
if	$\lvert \alpha_j \rvert > \lvert \alpha_k \rvert$	then	$MIN(O_j) < MIN(O_k)$

The more $\lvert \alpha_i \rvert$ is close to 1, the more O_i is close to O.

So a fuzzy control can be described by :

$$\dot{q}_i = \lvert \dot{X} \rvert \alpha_i \quad \text{if} \quad \alpha_i = MAX \lvert \alpha_j \rvert$$

$$\dot{q}_i = 0 \quad \text{elsewhere} \qquad (13)$$

$MAX \lvert \alpha_j \rvert$ = maximum value of α_j

The scheme of figure 3 can be modified as shown in the figure 4.

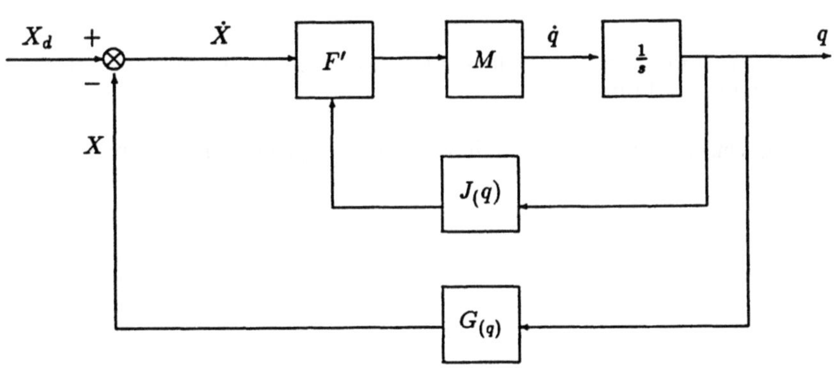

where : F' - computation, M - using of equation [13]

Figure 4 : Fuzzy control design

4.2 Joint limits and singularities

The system depicted in figure 4 can have problems when a joint reaches its limit, or when the robot passes through a singular configuration.

As far as the limit problem is concerned, we can use the previous result [Aubin & Cellina] concerning q_i and combine it with the contraint of maintaining q_i as far as possible from its positive and negative limit.

For that we express the distance from qi to its limits with a value Li included between 0 and 1. The sign of ai indicates which limit we have to consider in the computation.

Therefore, we have a set composed of :

(α_j, L_i^+) if $\alpha_j > 0_i$

(α_j, L_i^-) if $\alpha_j < 0_i$

Considering the suitable set, we select first for each couple (α_j, L_i) the minimum value (which can be α_j or L_i). We keep the set of these minimum values. Then we select in this set the $L_{i\ min\ max}$. Finally the chosen couple will be $(\alpha_j, L_{i\ min\ max})$.

We can characterize also the singular configurations or their neighbourhood with the condition :

1) $\min [(\dot{X} \in \tilde{k}_v), \min(q \in \tilde{k}_v)] = \tilde{1}$

2) $\tilde{k}_v$ and $\tilde{k}$ being fuzzy sets

3) $\tilde{1}$ being the fuzzy value for 1

When this condition is satisfied, the control uses the computation $\dot{q} = f(q, \dot{X})$, f being a known mathematical function.

Finally, the whole fuzzy control will have the form represented in figure 5.

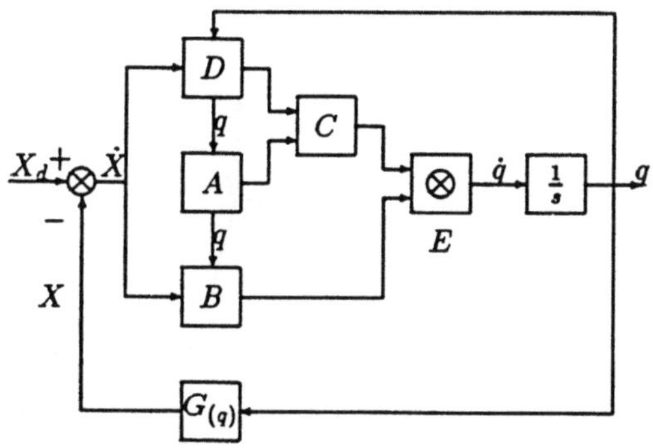

where : D - computation of L_i, A - computation of joint limit·, C - choice of L_i, B - computation of singular configuration condition, E - exclusing "OR"

Figure 5

4.3 Illustration

Let us consider the redundant robot depicted in figure 6 which has as joint limits

1) q_{max} = [0,35 ; 3,14 ; 1,86 ; 1,48 ; 0,76]
2) q_{min} = [-0,35 ; - 3,14 ; 1,46 ; - 1,05 ; 0,60]
3) (lengths in meters and angles in radians)

and two singular points at

1) $\theta_1 = 0$ with $\dot{X} = [-1, 0,0]^T$

2) $\theta_2 =$ with $\dot{X} = [-\cos\theta_1, -\sin\theta_1, 0]^T$

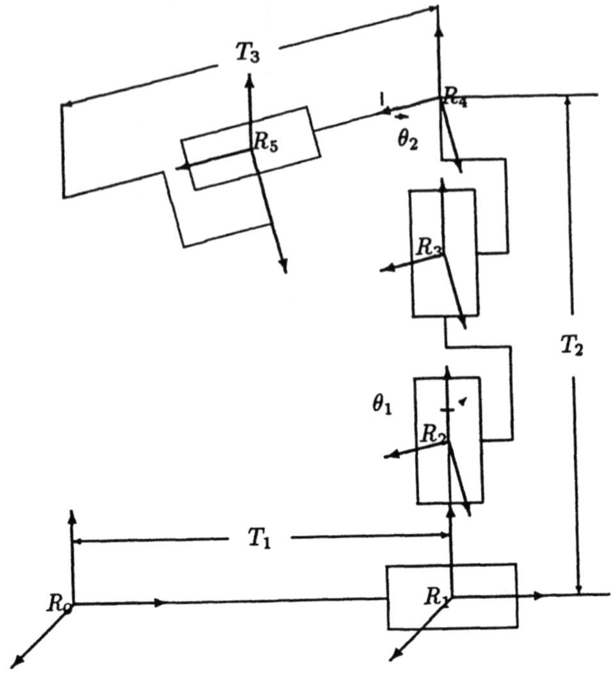

Figure 6 : 5 DOF robot
(2 rotations θ_1, θ_2 and 3 translations T_1, T_2, T_3)

The fuzzy controller can be written as following :

- if $\alpha_1 = \tilde{1}$ and $L_1^+ = \tilde{0}$, then $\dot{T}_1 = |\dot{X}|\alpha_1$
- if $\alpha_1 = -\tilde{1}$ and $L_1^- = \tilde{0}$, then $\dot{T}_1 = |\dot{X}|\alpha_1$
- if $\alpha_2 = \tilde{1}$ and $L_2^+ = \tilde{0}$, then $\dot{\theta}_1 = |\dot{X}|\alpha_2/T_3 \cos\theta_2$
- if $\alpha_1 = \tilde{1}$ and $L_2^+ = \tilde{0}$, then $\dot{\theta}_1 = |\dot{X}|\alpha_2/T_3 \cos\theta_2$
- if $\alpha_1 = \tilde{1}$ and $L_3^+ = \tilde{0}$, then $\dot{T}_2 = |\dot{X}|\alpha_3$
- if $\alpha_1 = \tilde{1}$ and $L_3^- = \tilde{0}$, then $\dot{T}_2 = |\dot{X}|\alpha_3$

- if $\alpha_1 = \tilde{1}$ and $L_4^+ = \tilde{0}$, then $\dot{\theta}_1 = |\dot{X}| \alpha_4 / T_3$

- if $\alpha_1 = \tilde{1}$ and $L_4^- = \tilde{0}$, then $\dot{\theta}_1 = |\dot{X}| \alpha_4 / T_3$

- if $\alpha_1 = \tilde{1}$ and $L_5^+ = \tilde{0}$, then $\dot{T}_3 = |\dot{X}| \alpha_5$

- if $\alpha_1 = \tilde{1}$ and $L_5^- = \tilde{0}$, then $\dot{T}_3 = |\dot{X}| \alpha_5$

- if $\theta_1 = \tilde{0}$ and $L_1^+ = \tilde{0}$, and $\dot{X}_1 = [-1,0\,0]^T$

  then $\theta_1 = -a\mathrm{Cos}\,((R - |\dot{X}| / R)$

  $\dot{T}_1 = R\ \mathrm{Sin}\ \theta_1$

  $R = T_3\ \mathrm{Cos}\ \theta_2$

- if $\theta_1 = \tilde{0}$ and $L_1^- = \tilde{0}$ and $\dot{X} = [-1, 0,0]$

  then $\theta_1 = a\mathrm{Cos}\ (R - |\dot{X}| / R)$

  $\dot{T}_1 = R\ \mathrm{Sin}\ \dot{\theta}_1$

  $R = T_3\ \mathrm{Cos}\ \theta_2$

  if $\theta_2 = \tilde{0}$ and $L_3^+ = \tilde{0}$ and $\dot{X} = [\ -\mathrm{Cos}\theta_1, -\mathrm{Sin}\theta_1, 0]^T$

  then $\dot{\theta}_2 = a\ \mathrm{Cos}\ ((T_3 - |\dot{X}| / T_3$

  $\dot{T}_2 = T_3\ \mathrm{Sin}\dot{\theta}_2$

- if $\theta_2 = \tilde{0}$ and $L_3^- = \tilde{0}$ and $\dot{X} = [\ -\mathrm{Cos}\theta_1, -\mathrm{Sin}\theta_1, 0]^T$

  then $\dot{\theta}_2 = a\ \mathrm{Cos}\ ((T_3 - |\dot{X}| / T_3$

  $\dot{T}_2 = -T_3\ \mathrm{Sin}\dot{\theta}_2$

In figure 7 is shown an instance of evolution of joint variables when the desired trajectory is defined by

$$164.\quad Xd = \begin{pmatrix} 0.68-0.272t \\ 0.4t \\ 1.66+0.136t \end{pmatrix} \qquad 0 < t < 2.5\ s$$

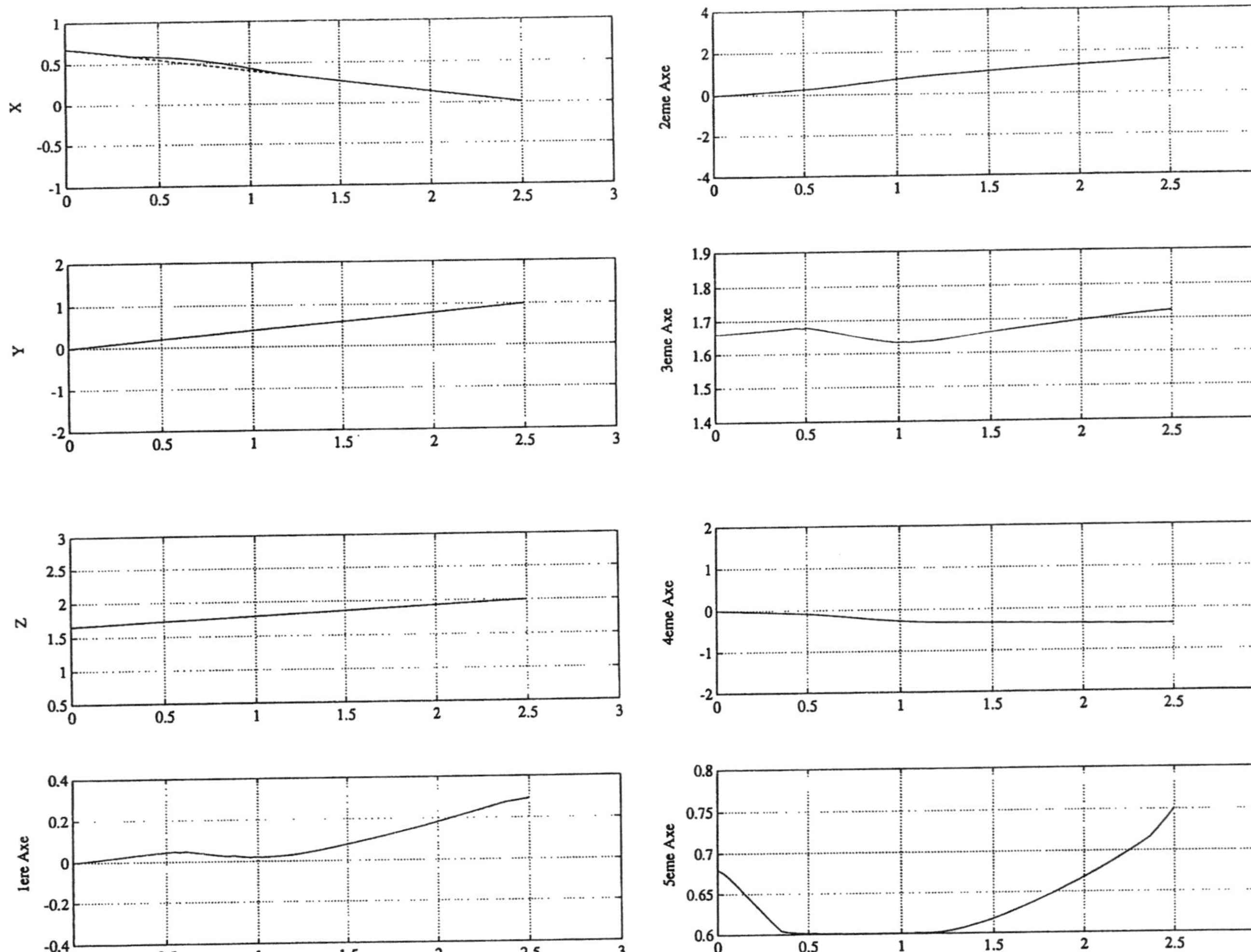

FIGURE 7. Trajectory generation using the fuzzy controller.

Conclusion

The fuzzy controller based upon a qualitative approach for determining the joint evolution exhibits good behaviour. Computations are easier than with the quantitative method consisting of the computation of the Jacobian matrix.

5 Navigation

5-1 Decision zones

The mission of a mobile robot is to move from one point to another in order to perform a task. We are interested in the displacement of the robot. A displacement order that the robot receives is a pure qualitative order like, "pass between the tap and the bench and then follow the left wall of the corridor until you reach a door". A path is made of multiple small subpaths (e.g. "pass between the tap and the bench") each connected to the other by a zone in which the subgoal changes. We call the zone a *decision zone* [Novikoff & al.].

The decision zone is very important because the robot must pass through it. The robot then has to make a decision to find the next subpath. A decision zone is not clearly defined in geometrical terms. For instance, what is the intersection of two corridors, or the place between a tap and a bench, or the front of a door? Humans have an idea about these zones but the machine does not. The robot has to give geometrical orders to its actuators (a distance and a direction). So these purely qualitative zones are unsuitable for the robot.

The decision zone is an abstraction. The main question is: how can a robot recognize an abstraction? It needs some references, something that can tell it that it is in the decision zone. We therefore need another concept. This concept must be geometrical, and there must be an easy link with the decision zone. We call this concept a *characteristic zone*, because it is referenced to landmarks (a tap, a bench...) and the way the robot sees these landmarks is characteristic of this
zone.

5.2 Characteristic zones

The characteristic zone is referenced to *landmarks*. A landmark is a point, or a set of connected points that the robot can recognize quickly, and in which the robot can easily make two kinds of operations : <u>compute the distance</u> from the robot to the landmark and compute the <u>heading</u> the robot must have to reach the landmark. For example, a door, a wall, or an artificial landmark like a panel or a flag may be landmarks. The two elementary operations that can be done on landmarks (<u>distance and orientation</u>) can be linearly combined to form a *measure*. For instance, the difference between the orientations of two landmarks forms a measure.

A use of measures is to tell whether the robot has reached the end of its displacement or not. Each measure performed must be in some given interval. For instance, "the difference in orientation between the tap and the bench must be within [175° ,185°]. Boolean values representing the ownership of a measure to an interval are derived. We call a *position test* some logical combination of these boolean values. For example, "the end of the corridor must be within [-5° ,5°] **and** the door must be within [85°, 95°]. To this position test, a set of points on the ground corresponds to where the robot can be. We call these points a characteristic zone. For instance, if the position test is : "the difference between the distances to two landmarks is 1.5m", then the characteristic zone is an hyperbola.

A landmark cannot be seen from everywhere. As each characteristic zone is defined by landmarks, a direction of progression must be linked with each of them. This is the other use of measures. The robot is under the control of some measures. For example, an order can be "go in the direction of the bench until you cross a door". The direction is the first part of this order and is relative to the landmark "bench". The robot needs for each displacement a measure to follow, and a control law. In our example, the measure is an orientation near zero, the control law can be: "if M is the measure, rotate -M before moving one step forward".

5.3 The link between qualitative and quantitative control

To sum up, a decision zone (DZ) is the end of an elementary displacement for the human being. It is not well defined in geometrical terms and is direction independent. A characteristic zone (CZ) is the end of an elementary displacement for the robot. A CZ is defined by a position test, a measure that the robot must follow and the explanation of how it can be followed (control law). It corresponds to a set of physical points on the ground. So, it is well defined in geometrical terms; but it is orientation dependent. The first notion is qualitative. The second is quantitative.

The link between the quantitative and the qualitative notions is evocated by the distance from CZ to DZ. Since a CZ is direction dependant and not DZ, multiple CZ are needed to represent a DZ. Let us take an example : a crossing of two corridors is a DZ (figure 8). We will try to find the corresponding CZ. If the robot progresses from A towards C, the landmark indicating the end of the displacement can be B or D. Let us say that D is more visible. The CZ is then defined by C for the direction and D specifies the end of the displacement. In the opposite way (from C towards A), the CZ would be defined by A for the direction and maybe B for the end of the displacement. For the robot, the four CZ corresponding to the four possible directions are completely different. They are referenced to different landmarks. The parameters are not the same. The physical points on the ground will be, in the four cases, about the same. We say that the union of these points on the ground forms the quantitative representation of a qualitative notion : DZ.

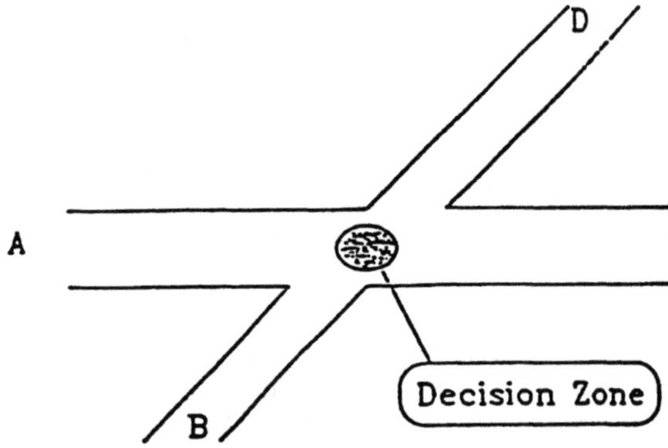

Figure 8 A crossing of two corridors.

5.4 The Grafcet of the robot

After the presentation of the principles that guide our qualitative approach, we present the different states that are required. Our aim is to introduce the Grafcet that the robot must follow to manage the qualitative control.

FIGURE 9

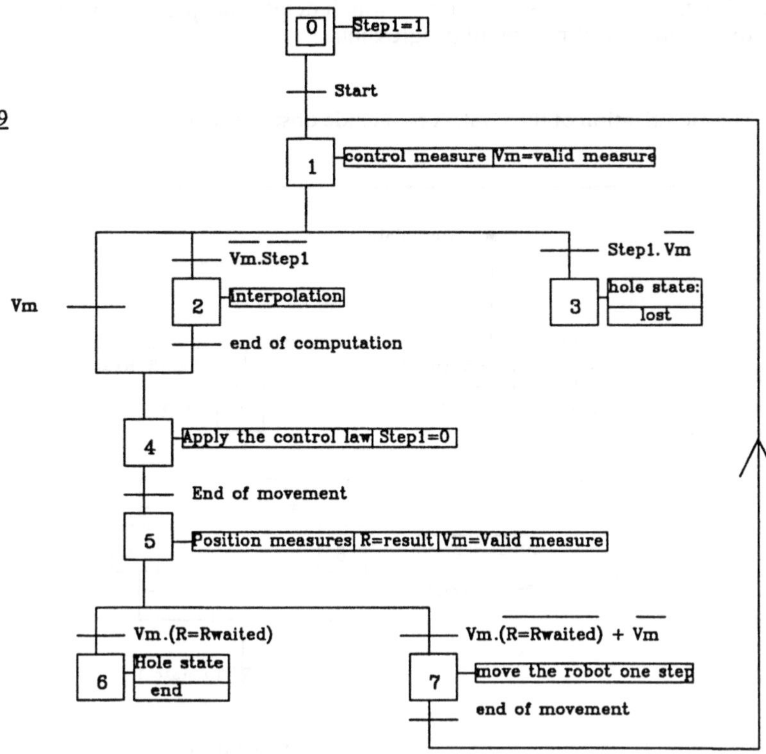

The Grafcet of the robot control without the obstacle avoidance

---

The robot has two mains actions to perform : to progress in the correct direction and to verify that it has not reached in the CZ. We represent these actions by the states 4 and 7 in figure 9. To manage these actions, measures on landmarks are required and they are done in states 1 and 5. However, the measure is not always available and we need to add some action when the landmarks are invisible. This is achieved by the transitions controlled by the variable Vm (1-2,1-3,5-7). State 3 is reached when the robot is at its first step and does not see the landmarks. It does not know where to go. It is lost. If it is not the first step of the robot, it can make an interpolation of the measure and this is the aim of state 2. Transitions 5-6 and 5-7 are controlled by the equation (R=Rwaited). This equation is true if the position measurement R is one of the values of the characteristic zone. In that case, the robot falls into a hole state which represents the end of the displacement. In the other case the robot moves and returns to state 1.

Actually, obstacle avoidance must be a part of the control of the robot. That is why we must modify the Grafcet. First, we must choose a way to avoid obstacles that is compatible with the control. It must be a local modification of the path. The method we kept was inspired by the Lumelsky's Bug2 algorithm [Lumelsky]. It tells that one can move forward in a straight line until an obstacle is reached. In that case, one has to follow the obstacle until the straight line is crossed again at a point nearer to the goal. This requires the following modification of the Grafcet (cf. figure 10). A state of obstacle avoidance has been added (state 8). This state is reached if Fw is false. Fw is set by state 7 which detects the presence of an obstacle in an area near the robot and in its direction of progression.

The modification of the Grafcet to include obstacle avoidance

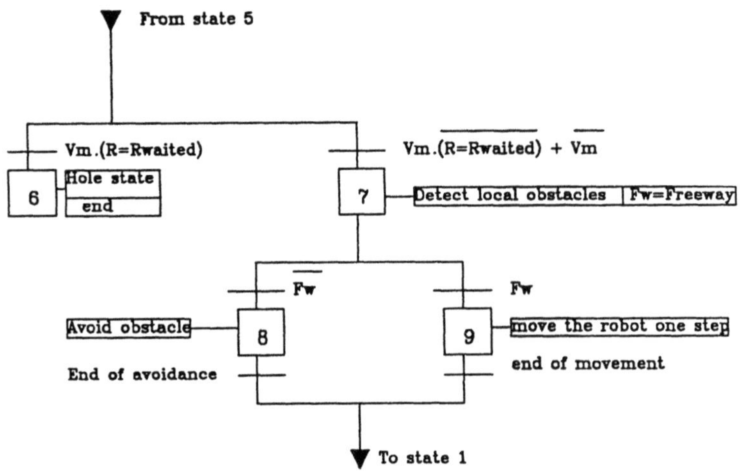

FIGURE 10

5.5 An experiment to validate the theory with Systalk

Presentation of Systalk

Systalk is a knowledge representation tool specialized in modelling and simulation; it is written in the Smalltalk environment [Goldberg & Robson]. The choice of Smalltalk as a good environment to develop robotics applications, and especially mobile robotic applications, has already been discussed [Oppenheim & al.]. In particular, it provides a programming environment based on the object paradigm which is simultaneously a high-level knowledge representation frame and an efficient software engineering tool. Nevertheless, it has been shown that the pure object paradigm is not sufficient on its own for modelling and simulation of robotics systems [Wolinsky 1989, 1990] and Systalk is an example of an object-oriented modelling environment specialized in robotics.

Systalk provides a uniform, modular, and extensible formalism which empowers the user to build, to modify ,and to re-use his models. He can then initialize them in order to run simulations. The knowledge representation tool is based on the system paradigm. In Systalk, a system is an object that has several points of view and that defines a part-whole hierarchy that is made of sub-systems. Each point of view, which is called activity, is a domain field that the user gives to its model. It is composed of two parts : a functional one, which expresses the way the model acts in each perspective, and a structural one, which expresses the way the model is bounded to other models in each field. Systalk supplies a library of about ten functional activities that cover the main robotic fields (e.g. geometric referential, geometric primitive, kinematics, planification, control, measurement,...). All theses activities are made of a small hierarchy of classes that expresses variety of behaviour in each field. Modelling is in pure algebraic form : activities assigning, systems structuring and systems bounding.

Presentation of the #Automaton activity

Modelling mobile robots involves the use of a formalism adapted to the Grafcet that has been choosen for driving the robot. Procedures are too rigid to implement some kind of intelligent systems. On the contrary, rule-based systems, which are not supplied in Smalltalk but can be added, are not sufficiently structured for such a purpose. So, we decided to implement concurrent finite states automatons in Smalltalk. When it is active, such an automaton executes a method where it can change its state according to the answer resulting from any question. If the state is not changed, the automaton will re-do the same method the next time it will be active; if the state is changed then, it will execute another method, and so on. We get a kind of modular second-order formalism, that as we will see, is rich enough to implement generic and extensible automatons.

Presentation of the experiment

The Grafcet that has been designed can be implemented with a five-state automaton (figure 11) :

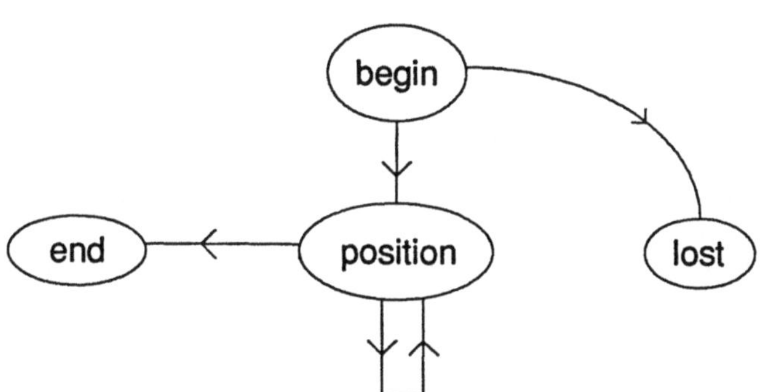

FIGURE 11

begin , initial state, tries to orientate itself; if it is able, goes to *position* state, if not goes to *lost* state,
position tests if the mission is accomplished; if it is, goes to *end* state, if it is not, goes to *measure* state,
measure tries to orientate itself; then goes to *position* state,
end stops the automaton; the mission is over,
lost stops the automaton; the mission can't be executed.

This automaton is generic and can be used for any mission. The mission just defines two procedures: measurement that brings the robot into control and position that tests if the mission is accomplished. So, adding a new mission consists in writing these two methods. One implements an acquisition, then makes a linear combination of distances or orientations and answers an angle. The other implements a position test by making a Boolean combination of elementary tests (a distance or an angle inside an interval) and answers a Boolean. Another kind of generic is supplied. As a mission consists of two procedures, those procedures use some parameters (e.g., number of each landmark, arguments of the test). So we have defined generic missions where one mission is fixed by giving the values of its parameters.

This automaton is extendable and it is able to modify or to add any state. For instance, inside the main loop *measure-control/position-test* it is possible to add an obstacle-test state that can be completed by an *obstacle-avoidance* one. This facility is very useful in order to implement and to test new navigation strategies.

Since automatons that have been implemented are concurrent, it is possible to have several mobile robots in the same scene. It is therefore possible to consider one mobile robot made of three independant automatons : one for navigation strategy, one for measure acquisition and one for control. Also it is possible, in order to make some comparisons, to execute at the same time two (or more) independant scenes implementing different navigation strategies.

description of one experiment

Systalk provides a man-machine interface that shows the part-whole hierarchy of the scene (figure 12). It is composed of a mobile robot, four landmarks and a cross corridor made of eight walls. Special functions are available providing geometric, automation and perception representation (figure 13). We can then ask the robot to start a mission. Since user keeps control of the system during the execution, he can move a landmark, for instance; this change will be taken into account at the next measure state of the robot.

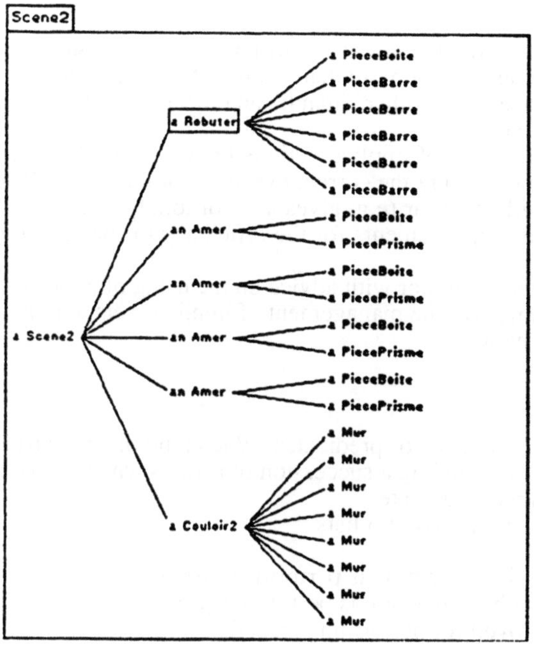

FIGURE 12

Part - Whole hierachy of the scene provided by systalk.

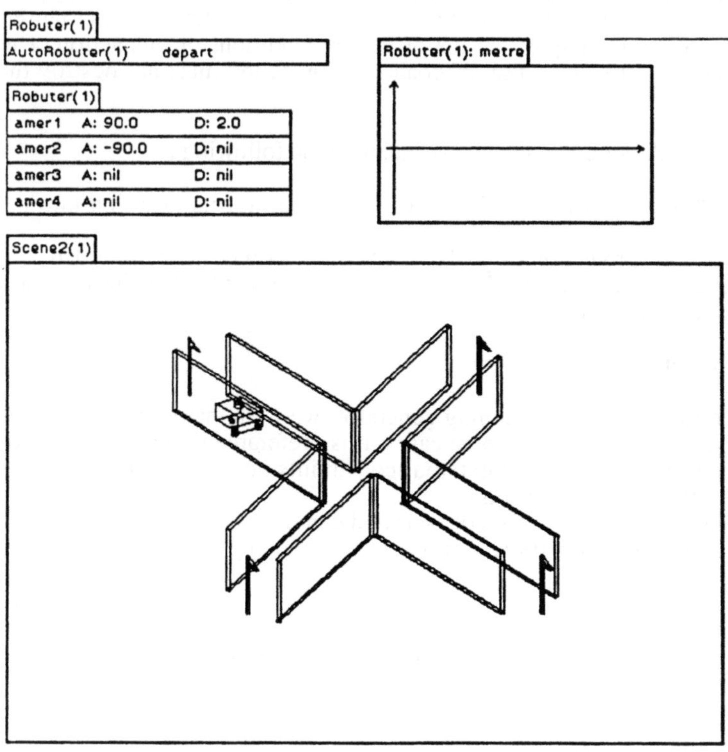

FIGURE 13

View of robot and environnement (corridors).

6 On-line planning generation

In our examples, the plannings to be producted are at "short term" that is to say they have to be generated in 1 to 10 seconds and concern not more than a tenth of simple actions.

A simple action can be : "go from a place to another, "handle an object", etc. They can be auto-adaptive such as "go up to a target", etc.

Our generator will give only a sequentiality of implied actions knowing that for each action a navigation and trajectory generator will ensure the correct execution of the action. The planning generator is not hierarchical and will be linear (e.g. gives a set of totally ordered). It is based on RSTRIPS but has two interesting improvements for the efficiency of the research planning :

1. first, we want to help the generator with advice on the using of actions
1) secondly, we want to improve the management of implied facts with a distinction about their nature.

6-1 Recall on RSTRIPS

The state of the world is represented with a set of predicates. When a predicate is true, a fact is proven generating a planning consists in finding a succession of actions which transfer the world from a known initial state to a desired final state.

The model of an action in STRIPS is composed of 3 lists :

1. - the fact which become true when an action is performed (list 1)
1) - the fact which become false when an action is performed (list 2)
2) - the facts which have to be proven for performing an action (preconditions list)

RSTRIPS starts from the final desired world state. It has to find a set of facts named goals. If a goal is not proven (in the initial world state), it looks for an action executing the goal with help of the first previous list. As soon as it has found a correct action, it tries to insert this action into the planning with the help of the list of the preconditions of this action. The only restriction is concerned with the protection of goals : a new action must not destroy the previously reached goals.

RSTRIPS is not perfect. Among the defects we can note the following :

1. once RSTRIPS has chosen an action, it can only look for its insertion into the planning
1) consequently to the protection of goals, a necessary action can be discarded
2) almost all the time used by RSTRIPS is devoted to verify that the facts are proven and the goals are not deleted.

6-2 New representation of actions

For improving the efficiency of the planning generator, we wish that it is able to think to the "correct" action as soon as possible. In some cases, it is preferable to not use an action which leads to the goal because we can have external constraints not directly linked to the planning problem (for example safety constraints).

Our action modelling consists in five lists of facts and conditions :

1. the list n 1 and 2 of RSTRIPS which represent the <u>consequences</u> of an action performance
1) the preconditions list of RSTRIPS, facts to be proved for starting the performance of an action
2) a list of effects : facts brought by an action performance which justify its using

3) a list of contextual preconditions : conditions which must be verified when the generator wants to use an action.

The list of effects

The modeller indicates the results of the action for which it is possible to use this action. These effects are a subset of the consequences of the action described in list 1 and 2.

When the generator looks for an action leading to a goal, it look at lists of effects for finding the action (and not at the list n°1 as in RSTRIPS). This distinction of "effects" and "consequences" allows as to reduce the numbers of choices of the generator facing a goal to be reached.

Instance :

Let us assume a mobile robot pushes a load. Functioning in the presence of the load in front of the robot, an action of displacement will have as a consequence the displacement of the load, or no displacement of the load. We model this fact with two different actions "go from A to B" and "push the load from A to B".

The consequences of "push the load C from A to B" are :

1. - C is in B
1) - Robot is in B
2) - C is not in A
3) - Robot is not in A

The effect of the action is

1. - C is in B

So, if the desired goal is to lead a load in a place, this action will be retainned.

But if the goal is to lead the robot in this place, this action will not be retained (althougth the goal is a consequence of this action).

The list of contextual preconditions

In RSTRIPS, once an action has been chosen, the generator looks for a planning satisfying its execution preconditions. This action will not be accepted if such a planning cannot be found because the generation has proven impossible to use it. This proof spends a lot of time. We try to avoid it.

For that, we have introduced other kinds of preconditions : the contextual preconditions which have to be verified when the action is under consideration. The preconditions do not demand a research of a partial planning for their performance. They allow a second filtering of possible actions or/and perform the action under consideration.

An action is then chosen for function of a desired effect (execution of goal) and for the context in which it is under consideration. Once an action has been chosen, the research process goes on as in RSTRIPS.

If we come back to our previous instance with a mobile robot and a load, thanks to what we have defined, we can say that the action "push" has as the effect to place the robot in B, but only after the use of the action "go".(We take into account the case for which the robot cannot have a motion without pushing the load).

6.3 Conclusion

The planning generation approach is clearly a qualitative approach for which the quantitative part has disappeared. However, we can see it raises the problem of efficiency which can be translated in terms of time computation (for on-line planning generator) and to terms of completeness, that is to say, of escaping to a rigid reasoning mode which does not allow one to take under consideration all realistic cases.

7 General Conclusion

In this paper we have presented new results in qualitative robot control for the three main levels of action control : generation planning, navigation inside an action, and execution of a precise action via trajectory generation.

In addition to these new results, our main purpose is to contribute to the understanding of the qualitative approach of robot control.

In the future, the questions to which we would like to answer are the following : can we make a whole robot control using only qualitative notions, this control satisfying the properties of precision, reliability and efficiency? If the answer is no, what is the best trade-off of qualitative and quantitative control? For what criteria, and following what formalisms must we shift from the qualitative part to the deterministic part of the control process. At this time we have only intuitive thougtht but no methodology for the propose.

REFERENCES

Aubin, J.P. & Cellina, A.: *Differential inclusions*, Université Paris IX, 1982.

Aubin, J.P.: *Viability Theory*, Université Paris IX, 1982.

Craig, J.J, Hsu, P. & Sastry, S.S.: *Adaptative Control of Mechanical Manipulator*, Int. J. Robotics Res., vol 6 N° 2, 1987.

Dubowsky, S. & Desforges, D.T.: *The Application of Model Referenced Adaptative Control to Robotic Manipulator*, J. of Dyn and Contr., vol. 101, 1979.

Featherstone, R., *Position and Velocity Transformation between Robot End-effector Coordinates and joint Angles*, Int. J. Robotics Res., vol. 2 N° 2, 1983.

Goldberg, A., Robson, D.: *The language and its implementation*, Smalltalk-80, Addison-Wesley Publishing Compagny,1983.

Hill, S.D., & Vaccaro, J.: *A Commande Generator for 6 DOF Robotic Manipulator*, Proc. 1987 Int. Conf. robotics and Automation, Raleigh, pp. 1104-1109, 1987.

Hill, S.D., & Vaccaro, R.J.: *A discrete Time Robotic Command Generator*, Proc. 1988 IEEE Conf. Robotics and Automation, Philadelphian, pp. 675-680, 1988.

Hsia, T.C., Gua, Z.Y. & Li, L.: *An Adaptative Joint Trajectory Generation Scheme for Cartesian Robot Control*, Proc. Eleventh IASTED Int. Sym. on Robotics and Automation, Santa Barbara, pp. 120-124, May 1988.

Hsia, T.C & Guo, Z.Y.: *Joint Trajectory Generation for Redundant Robot*, Proc. 1989 IEEE Int. Conf. on Robotics and Automation, pp. 277-282, 1989.

Koivo, A.J., & Paul, R.P.: *Manipulator with Self-turning Controller*, IEEE Conference on Cybernetics and Society, Massachussets, 1980.

Liegeois, A.: *Automatic Supervisory Control of the Configuration and Behaviour of Multibody Mechanisms*, IEEE Trans. Sys., Man. Cyber., SMC-7, N° 12, 1977.

Low, K.H. & Dubey, R.N.: *A Comparative Study of Generalised Coordinates for Solving the inverse - Kinematics Problem of a 6R Robot Manipulator*, Int. J. Robotics Res., vol. 5, N° 4, 1986.

Lumelsky, V.: *Algorithmic and Complexity Issues of Robot Motion in an Uncertain Environment*, journal of complexity 3, pp 146-182, 1987.

Morgan, R.G. & Ozguner, U.O.: *A Decentralized Variable Structure System Algorythm for Robot Manipulator*, IEEE J. Robotics Automat., RA-1, Mar.85

Nakamura,Y. & Hanafusa, H.: *Optimal Redundancy Control of Robot Manipulators*, Int. J. Robotics, Res, vol. 6, N° 1, 1987.

Nenchev, D.N.: *Redundancy Resolution through Local Optimation : A Review*, Journal of Robotic System 6 (6), 769-798, 1989.

Novikoff, P., Chellali, R. & Milgram, M.: *Navigation of a Mobile Robot with Obstacle and Landmark Recognition*, ICARCV'90, Singapore 1990.

Oppenheim, I., Rehak, D., Keirouz, W. & Woodbury, R.: *Robotic task planning - Domain modelling and geometric reasonning*, Final Report, MIT 1987.

Pedrycz, W.: *Fuzzy control Systems*, Research Studies Press LTD., 1989.

Samson, C.: *Une Approche pour la Synthèse et l'Analyse de la Commande des Robots Manipulateurs Rigides*, Publication interne IRISA, N° 356, Avril 1987.

Slotine, J. E.:*The Robust Control of Robot Manipulators*, Int. Robotics Res., vol. 4, 1985.

Sugeno, M.: *An Introductory Survey of Fuzzy Control*, Information Sciences 36, 59-83, 1985.

Vaccaro, R.J. & Hill, S.D.: *A Joint Space Command Generator for Cartesian Control or Robotics Manipulators*, IEEE Journal of Robotics an Atomation, vol. RA-4, N° 1, PP. 70-76, February 1988.

Wolinski, F.: *Modeling and simulation of robotic systems using the Smalltalk-80 environment*, TOOLS, Paris, pp.141-149, 1989.

Wolinski, F.: *Représentation de systèmes robotiques en Smalltalk*, Convention IA, Paris, pp. 685-699, 1990.

Wolovich, W.A & Elliot, H.: *A Computational Technique for Inverse Kinematics*, Proc. 23rd conf. Decision and Control, Las Vegas, pp. 1359-1363, 1984.

Yeung, K. & Chen, Y.P.: *A New Controller Design for Manipulators Using the Theory of Variable Structure Systems*, IEEE Trans. on Auto, vol. 33, N° 2, 1988.

Yoshikawa, T.: *Manipulating of Robotic Mechanism*, Int. J. Robotics Res., vol. 4, N° 2, 1985.

MOTION ESTIMATION FROM TARGET TRACKING

Juan Frau[1], Vicenç Llario[2], Joan Codina[1]

[1] Dept. ESAII. Facultat d'Informàtica de Barcelona. Universitat Politècnica de Catalunya.c/ Pau Gargallo, 5. 08028-Barcelona. SPAIN. Phone:(34-3) 401 69 74. Fax:(34-3) 401 70 40.
[2] RALUX,S.A.c/ Cai Celi 11, 08014-Barcelona. SPAIN. Phone: (34-3) 325 85 00. Fax: (34-3) 325 82 54.

Abstract: This paper describes a tracking vision system under development at the Automatic Control and Computer Engineering department (ESAII) in the Polytechnical University of Catalonia. The system consists of two recognition modules which allow the tracking of a predefined target on a multi-target environment. Binocular tracking provides, at the same time, an estimate of the relative distance which decreases the uncertainty of geometric positioning allowing a visual feedback for control purposes as well as navigation.

Data association which models target motion defined by the two tracking modules, allows one to perform an oriented search based on the prediction of the motion of the target. Thus, the system is able to restrict the searching window to a defined zone of the image where foreseeable the target will be detected.

The system implemented classifies the objects according to a feature vector associated with each object in the scene (area, perimeter, coordinates of the c. g, main axis of inertia with respect to the absolute reference,. . .). Simultaneously, the system stores a polar coding of every detected object, taking the relative reference formed by the main axis of inertia, and the perpendicular straight line through the c. g of the object.

Once the user has chosen one of the targets among several which has to be tracked, its geometric features as well as its rotation invariant description (contour shape) make up its dynamic model. We call it "dynamic model" since it is updated from image to image. As a matter of fact, we can distinguish two different filtering stages. A first stage performs a rough filtering based on geometric features, while a second stage refines the classification of possible candidate targets based on the dynamic shape model with respect to their polar coding.

The comparison criterion is based on a discrimination function which supplies a quantitative measurement of the divergence between the model and the potential candidates. This function implements a summation of least-square differences of 360 vectors for each candidate, coding their shapes from their relative references.

This discrimination function is calculated every 40ms for all the probable targets. Thus, it is possible to define an upper threshold for this function to guarantee the recognition whenever a value below this threshold is obtained. The candidate which in every case shows the lowest function value, below the threshold, is assumed to be the target. In the mean time, the updating process is validated, and the system stores the new parameter values of both the geometric model and the shape model. This detection threshold allows a definite robustness of the system in front of sharp changes in lighting conditions, low contrast effects, low visibility, occlusions, or other unpredictable noise sources.

Keywords: Target tracking/ Motion estimation/ Nonlinear systems/ Mobile robot control

NATO ASI Series, Vol. F 71
Expert Systems and Robotics
Edited by T. Jordanides and B. Torby

INTRODUCTION

Motion analysis has been one of the topics which has attracted a great deal of researchers all over the world during the last two decades. The usefulness of motion prediction to deal with tasks such as detection, recognition and tracking of moving targets has powered the research in this field. Since the main goal of a robust tracking system is to orient the search for the target to a defined zone of the image, it seems clear that defining the most probable zone in the most accurate way will lead to an increase in the efficiency of the tracking system.

The main features which constrain the general tracking problem we would like to face are the answers to two different questions. On one hand, what are the feature points or feature vectors we will take as a reference to be able to deal with the tracking of the target?. In other words "what are we going to track", and on the other hand, where are we going to look for the target? Or , where are we going to find the target in the next image?.

Trying to answer both questions we tested various alternatives and finally reached an adequate trade-off between robustness and performance.

The way we model the target is a combination of geometric plus cinematic features which leads to a dynamic model updated from frame to frame. The reference point we are going to track is a virtual point corresponding to the center of gravity of the target projection on the image plane. Calculation of coordinates of this reference point from image to image will be disturbed by two main sources of noise: acquisition noise and pre-processing noise.

As we mentioned before, the second step, after recognizing the target on a defined image, consists of orienting the search in order to guarantee the recognition in the next image. We have tested a tracking window with the main feature of being self-adjusting depending on the appearance of the target and its orientation on the image plane.

Finally , we have set up a predictive model of target motion with two main objectives: to provide the sequence of positions of the tracked feature point and as a consequence restrict positioning of the tracking window. We propose to use a dynamic model optimally fitting the trajectory followed by the target.

This motion prediction allows on one hand, one to orient the detection process by constraining the searching window to a defined zone of the whole image, and on the other hand, to associate a tracking window of variable size depending on the smoothness of the motion. The size of this tracking window will depend as well on the superimposed noise intrinsic to the observations and the target itself, considered as a system.

Different approaches have been used to implement the extrapolation of motion in target - tracking tasks. The simplest criterion based on constant speed, without taking into account any source of noise in the predictive model, up to the extended Kalman filter (typically 1st or 2nd order) have been used to predict the state of the system and/or the equations related to the observations.

The system presented has been developed trying to attain a reasonable trade-off between system robustness, noise level of the data to be processed and real time performance. Some polynomial regression algorithms(least squares estimation) have been tested with quite satisfactory results when tracking targets with an inherently smooth motion and small bandwidth.

APPROACHES TO GET A PREDICTIVE MODEL OF TARGET MOTION

There are two principal approaches to take with the target-tracking problem [9]. The methods used are based either on optimal filtering theory: trying to find a model of the trajectory, or heuristic approaches: taking advantage of the correspondence problem over multiple images. There are a great deal of studies which do not involve filtering theory, although we would like to concentrate on those researchers who use more than two frames to analyze motion. This is the case of the studies developed by Yen and Huang in 1983 [28], where they determined the three-dimensional motion and structure of a rigid body by establishing correspondence of straight lines over three image frames. The application of an iterative Gauss-Newton method allows one to determine the general motion given seven correspondence lines, while a linear method allows one to determine the screw motion from eight correspondence lines.

Tsai and Huang made an extension to the studies involving only two consecutive frames. They derived the three-dimensional motion parameters of translation and rotation using the method of singular value decomposition of a 3x3 matrix, given four correspondence points on three consecutive image frames [27].

Shariat and Price proposed a specific method to estimate motion parameters in a sequence of five consecutive images. They solved a set of five nonlinear equations with four unknowns applying the Gauss-Newton nonlinear least-square method. [24].

The filtering approach to the tracking problem consists of finding the model of the target trajectory on the image plane. Consequently, there must be some criterion of goodness which will allow an optimal estimate of motion parameters. The problem is to find a stochastic model able to estimate motion parameters from statistics, taking into account the variations or changes in the real world.

One of the techniques of widespread use is the linear filter introduced by Kalman and Bucy. This filter has been extensively applied in the processing of radar signals while only a few authors have used it in image processing.

Futrelle and Speckert combined the Kalman filter with optical flow processing to estimate position and velocity. [13]. Legters and Young tried to solve the occlusion problem by using the Kalman filter and a two-dimensional mathematical motion model to deal with prediction. [18]. Stuller and Krishnamurti faced the problem of estimating the television image motion by means of the Kalman filter [25] while Broida and Chellappa studied the motion parameter estimation for a target with unforced motion. [7]. Some recent studies have the aim of decentralizing the Kalman filtering algorithm into several nodes that they associated with a matrix of cameras. It is possible to explore the intrinsic parallelism of the algorithm by substituting a central processor by an array of processors (i. e. , an array of transputers), allowing a faster response time, . [11] and [23].

Another recent work already mentioned is the one developed by Chang, Lee and Chung where they proposed an adaptive nonlinear filtering approach to estimate the motion parameters for tracking a maneuvering target by means of an Extended Kalman Filter (EKF), augmented with semi-Markov process and Bayesian theory. [9].

The system we propose is based on regression analysis and could be included inside what we have called optimal filtering approaches.

UNCOUPLED POLYNOMIAL REGRESSION ANALYSIS

Regression analysis has been used to estimate parameters in some computer vision systems as reported in the Annual Report 1988/89 of the University of Maryland. The methods used fall into three main categories: least-squares method, least-median squares and weighted-least squares.

We have applied polynomial regression analysis to find the best estimates for the uncoupled motion along the x and y axis on the image plane. Since what we are tracking is just the projection of a 3D rigid body maneuvering with the most general motion in 3D, we must assume that the laws of motion for each axis on the image plane are different. We can write for x and y, coordinates of the virtual center of gravity:

$$x_t=a_0+a_1t+a_2t^2+\ldots +a_pt^p+e$$

$$y_t=b_0+b_1t+b_2t^2+\ldots +b_rt^r+e'$$

The problems we must solve given these expressions are as follows:

We must find the optimal values for p an r, let´s say, the order of each polynomial.

We must analyze the weight of the coefficient corresponding to the highest order term in the polynomial and accurately estimate its value.

We must be able to predict the next location x_{t+1}, y_{t+1} based on the optimal fitting of regression polynomials.

Finally, we must optimize the size of the tracking window in order to decrease false detections and at the same time speed up the response time.

The method we propose to face the problems mentioned above is based on the application of a recursive algorithm given the order of the polynomials and the number of observations N.

If the number of observations N is constant then, in the case of a parabolic regression, the solution of a system of three equations with three unknowns leads to the model for each coordinate just by recalculating a vector from image to image since the matrix T (time depending) is constant by the fact that it is time dependent and we assume a sampling period either of 20 or 40ms that is constant. Anyway, the order of the polynomial associated to each coordinate can be found by means of statistics.

Finally, the optimal dimensions of the tracking window can be obtained from the analysis of the variance of regression equations together with the projections over the two axis of the whole boundary image and this will guarantee a self-adjusting tracking window.

CONTROL FEEDBACK THROUGH 3D-MOTION ESTIMATION

The 3-D tracking module has been used as a motion feedback in a mobile robot. The X and Z coordinates of the target are supplied to a controller that tries to maintain the target's c. o. g inside an area around the geometric center of both image planes and to keep a prudential distance between the target and the mobile platform. As a result the mobile robot has to follow the target (fig. 1,2).

The control problem

The mobile robot with the in-built tracking system must follow the target with some conditions:

- -Keep a prudential distance between the vision system and the target. If the target is too far it will be difficult to distinguish it, but if the distance is shorter than a lower limit, then the risk of collision could be high.
- -Try to follow the target, not to catch it. The tracking system must be predictive, and the robot has to be always after the target and "looking" at it.
- -Filtering the results coming from the 3-D tracking system

One of the main problems when controlling the robot is that the direction it takes is the direction it looks at, and the camera's aperture angle (a) puts an upper limit on the maximum error allowed between the direction of the robot's speed vector and the angle it makes with the target.

To solve this problem let's think about the way a person solves this. When we have to go after someone to an unknown place, what we try is just to track the same path they are following. This implies a delay between the moment we see them turning and the moment we begin to turn ourselves. This delay is a function of the distance and the speed that we are moving. Besides that, if what we want is to catch someone, then we run focussed on the direction that they appear from.

Now let's look at our mobile robot. It must try to follow the target and noι go for it. As a matter of fact, there is a delay, but during this period of time the object might move out from the scene, being lost by the tracking system. To avoid this, we need a system with adaptive reactions depending on the motion´s smoothness. Under normal conditions it has to react slowly, but when the object goes out of a security angle, the system must be fast.

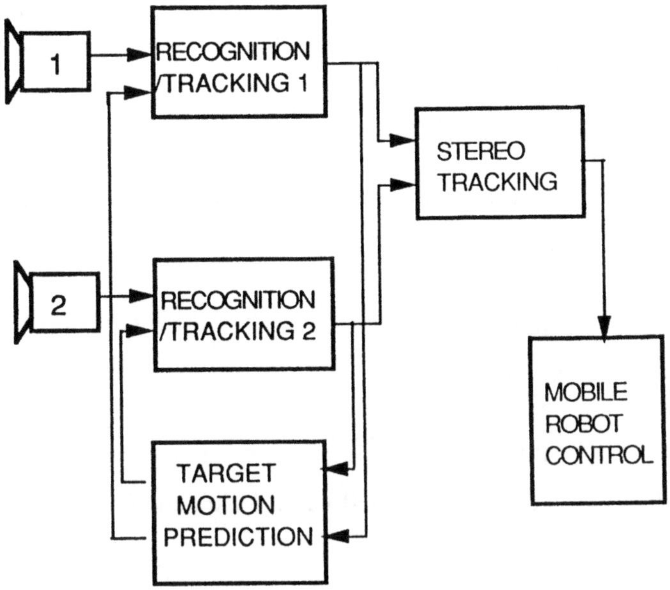

Fig. 1 Binocular Tracking system structure

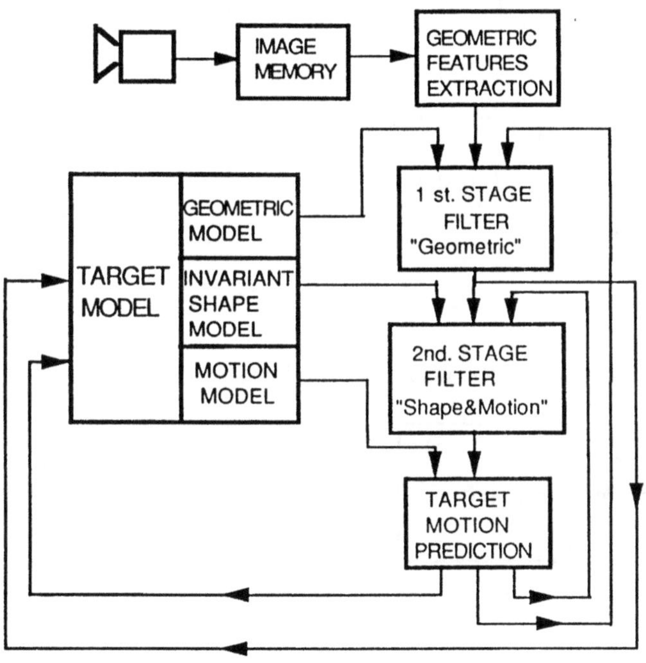

Fig. 2 2D-tracking block diagram with "geometric-shape-motion" model of a target.

If we are able to achieve this trade-off, then while the error will be inside the security angle our system will have a smooth response, filtering the errors coming from the vision system. But if the target goes out from this security angle, then it will react, even if the input was just noise.

So we must implement a function with a gain near zero in the security area and a gain equal to one outside this area. This function must be smooth, otherwise we'll get a pulse generator.
The function we propose is, as shown in (fig. 3,4,5)

$$F_{\mu,n}(x)= x^{2n+1}/(x^{2n}+\mu^{2n})$$

where μ is the radius of the dead zone (Security angle). This function has values:

$$F_{\mu,n}(\mu) = 0.5\mu$$

$$F_{\mu,n}(-\mu)= -0.5\mu$$

$$F_{\mu,n}(0) = 0$$

$$F_{\mu,n}(x) = x \text{ if } x >> \mu$$

If n is big enough, we can consider $F_{\mu,n}(x)=0$ x if in $(-\mu,\mu)$, and $F_{\mu,n}(x)=x$ if x in (μ,∞) or x in $(-\infty,-\mu)$

The derivative of $F_{\mu,n}(x)$ is:

$$F_{\mu,n}'(x)= 1 + ((2n-1)x^{2n}\mu^{2n}-\mu^{4n})/(x^{2n}+\mu^{2n})^2$$

where for n=1

$$F_{\mu,1}'(x) <1 \text{ if } x<\mu$$

$$F_{\mu,1}'(x) >1 \text{ if } x>\mu$$

and if n>1

$$F_{\mu,1}'(x) <1 \text{ if } x<\mu/(2n-1)^{1/2n}$$

The proposed control system block diagram is as follows:

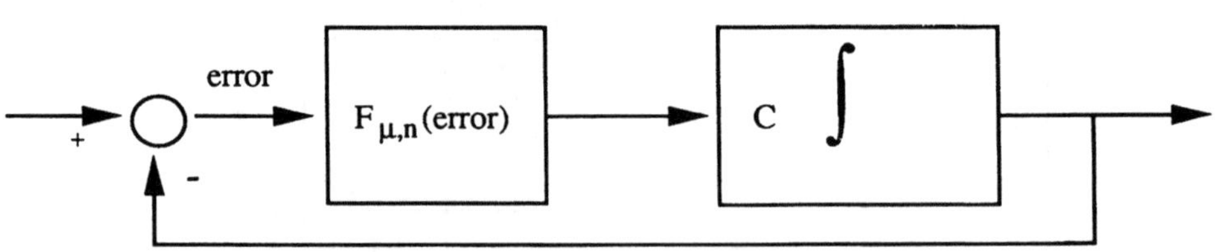

This is a nonlinear system that can be linearized as shown:

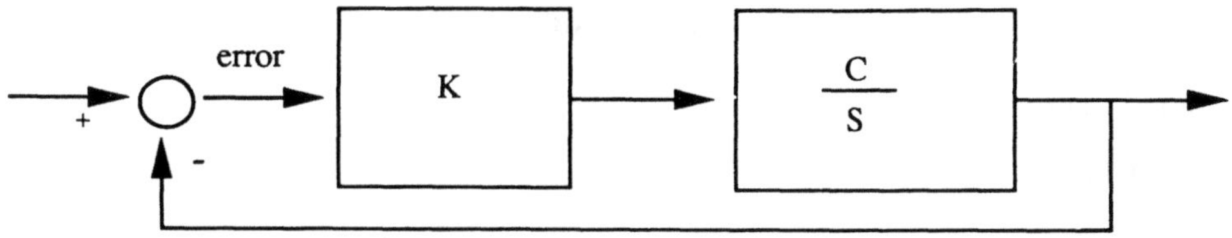

And having a transfer function

$$H(s) = KC /(s+KC).$$

This is a first-order, low pass filter with cut-off frequency KC. K is the gain of $F_{\mu,n}(e)$, and it can vary from 0 to 1 (fig. 6) so when the error is 0 the cut-off frequency is zero. And as the error increases, K tends to 1 and the system will be a low-pass filter with frequency KC.

We calculate the robot radius of curvature as $F_{\mu,n}$(error), with the appropriate μ and n=1, and error being the angle we see the target. The integrator is the robot itself. Applying this, we have a system that has a smooth response to the noise and to small changes, with some kind of dead zone around zero. This is important, because the error always tends slowly to zero when the target goes straight ahead.

Such a function is also used to calculate the module of the robot's speed. In this case the error is the difference of speeds, and as before, we need fast reactions in some situations, but we have to filter the noise.

To avoid collisions, we have introduced a security factor in the distance, so the distance between two objects will be a default distance Z_0 plus a distance proportional to the mobile platform speed ($Z_0+Z_1*|V|$).

Being X(k),Z(k) the coordinates of tthe estimated relative position of the target with respect to the robot in the plane of robot's movement. we can calculate

$$\text{Distance}(k)=(X(k)^2+Z(k)^2)^{1/2}$$

$$D(k)= \frac{d\,\text{Distance}(k)}{dt} = \frac{\text{Distance}(k)-\text{Distance}(k-1)}{S}$$

where S is the sample time.

The angle we see the object is more easily calculated from the position of the object's c. o. g. in the image plane, and is denoted by ß.

Defining

- Sp(k) The theoretical linear speed of the robot.
- V2(k) The theoretical linear speed of the robot's left wheel.
- V1(k) The theoretical linear speed of the robot's right wheel.
- Ω(k) The theoretical angular speed.
- R(k) The theoretical radius of curvature of the robot movement.
- V(k) The real linear speed of the mobile robot
- C_0,C_1,C_2,Z_0,Z_1. Constants
- d Distance between the driving wheels

we can write the control equations as follows:

$$Sp(k+1)=V(k)+C_0*F_{\mu 1,n1}(SpError(k))$$

$$SpError(k)=D(k)+C_1*(Distance(k)-(Z_0+Z_1*|V(k)|))$$

$$\Omega(k+1)=C_2*F_{\mu 2,n2}(\beta(k))$$

IF $\Omega(k+1) <> 0$

THEN

$$R(k+1)=Distance(k)/2*\sin(\Omega(k+1))$$

$$V1(K+1)= \frac{Sp(k+1)*(R(k+1)-d/2)}{R(k+1'}$$

$$V2(K+1)= \frac{Sp(k+1)*(R(k+1)+d/2)}{R(k+1)}$$

ELSE

$$V2(k+1)=V1(k+1)=Sp(k+1)$$

The controller block diagram is:

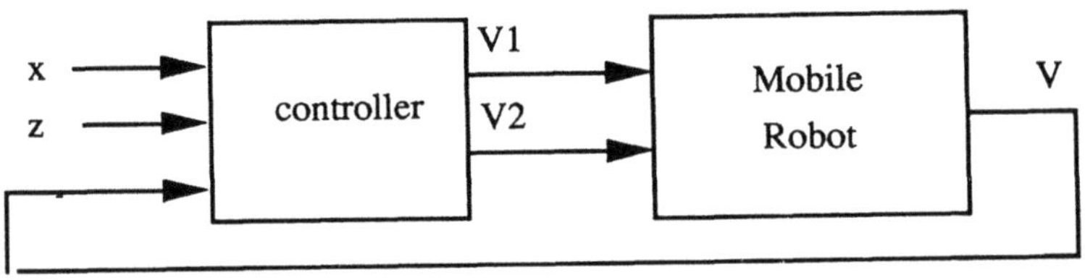

To have an appropriate response of the system it is necessary to choose the right values for the constants, and those depend on the system inertia. Z_1, C_2 must be small, μ_1 less than $Z_0/4$, if μ_1 is similar to Z_0 then, we loose the security margin. The limit of μ_2 depends on a (the camera aperture angle). The choice of n depends on how strict we want the dead zone.

Features of the control system

Using $F_{\mu,n}$ we get a nonlinear system that has a special kind of dead zone with the following main features:

- Filters the noise
- The error tends to zero
- Dead zone
- A delayed response

All this is done in a specific way due to the fact that the system will not respond till the error is $\pm\mu$. So if the error is zero, and there is noise, then it will be filtered. If the error is near μ then the controller will only filter the negative part of the noise but not the positive one. This is normal, being at the edge of the permissible error. If there is an increase in the error the system can not wait two or three samples to be sure that it is not noise.

Due to the shape of $F_{\mu,n}$, in the dead zone (the function has low values but not zero), the error tends to zero.

The system has three areas of convergence: $-\mu$, 0, μ. When the derivative of the system input is zero, the error decreases to zero. If the derivative is positive, then the error tends to be μ and the time the error needs to reach this value is the system delay. If the derivative is bigger than μ, then the system can be considered just a linear integrator. For a negative derivative the quiescent point is $-\mu$.

Some simulations have been done with the following and achieved satisfactory results:

$C_0 = C_1 = C_2 = 1$
$Z_0 = 4$
$Z_1 = 0.1$
$n_2 = 2$
$\mu_2 = 1$
$n_1 = 1$
$\mu_1 = 0.5$

Some simulations have been done for different trajectories of the target, comparing the performance of both the nonlinear and a simple linear system. The graphics are labeled T: for target, NL: for the nonlinear system L: for the linear system.

To see the nonlinear filter response we have given a constant acceleration plus a sinusoid and it can be seen than it filters the highest frequencies, and the lowest are filtered while the error is bounded by μ (fig 7,8,9).

The same has been done for circular trajectories with some sinusoidal noise added to it and compared with an equivalent linear system. (fig. 10,11,12).

To see the effect of the dead zone, three simulations have been done each of them with three different initial conditions. We can see that in a system where the target has constant speed the error decreases quickly until it enters into the dead zone, and then slowly tends to zero (fig. 9). If the target has acceleration then the system tends to keep the error equal to $\pm \mu_1$ (fig. 13,14,15) depending on the sign of the acceleration.

Photograph of the mobile robot "EIXERIT"

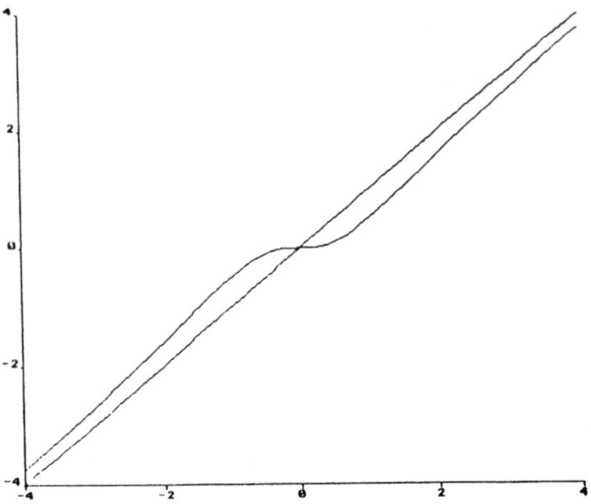

Fig. 3 $F_{\mu,n}$ with n= 1 and $\mu = 1$.

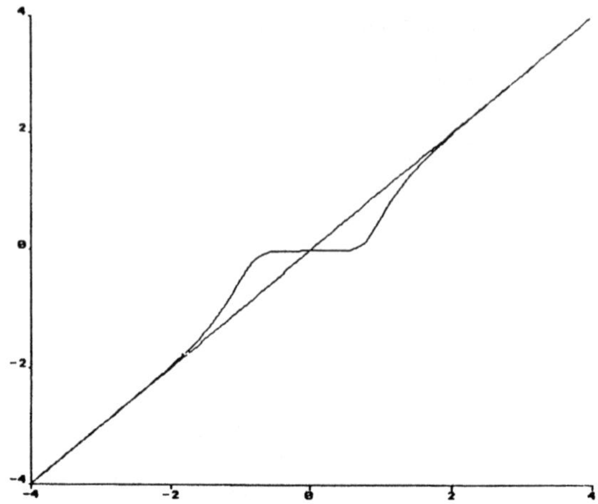

Fig. 4 $F_{\mu,n}$ with n= 2 and $\mu = 1$.

Fig. 5 $F_{\mu,n}$ with n= 3 and $\mu = 1$

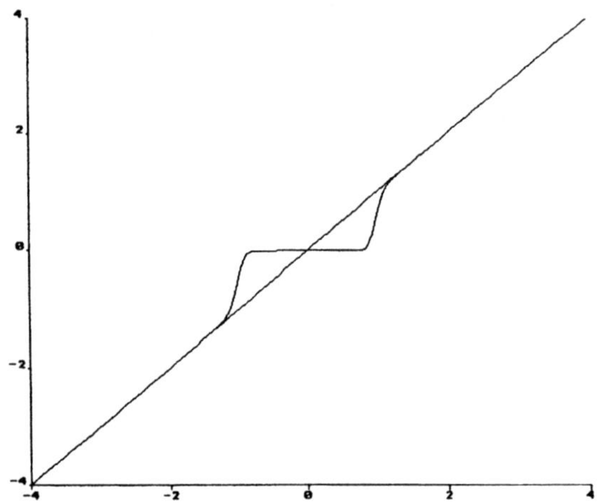

Fig. 6 Gain of $F_{\mu,n}$ with n= 1,2,5 and $\mu = 1$

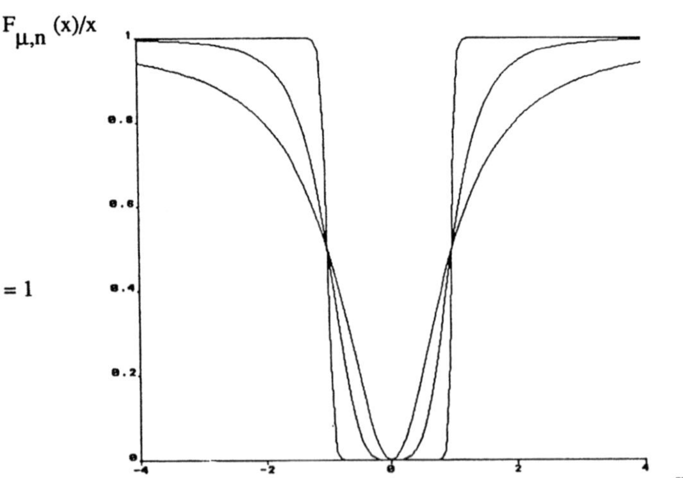

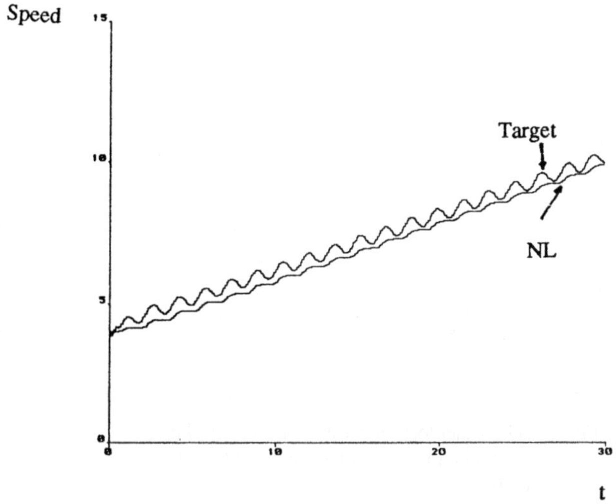

Fig. 7 Target, and nonlinear system speed .

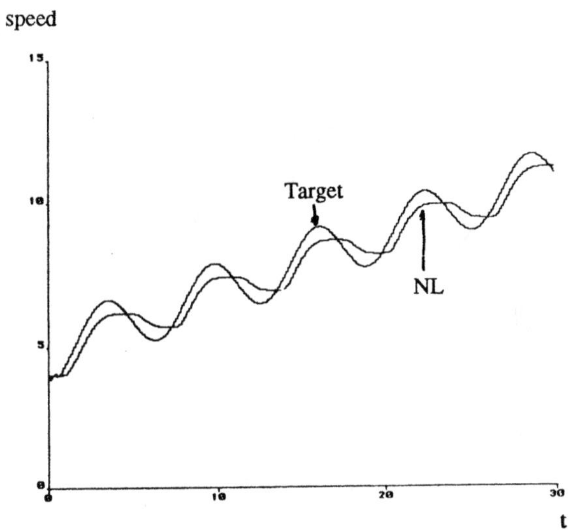

Fig. 8 Target, and nonlinear system speed .

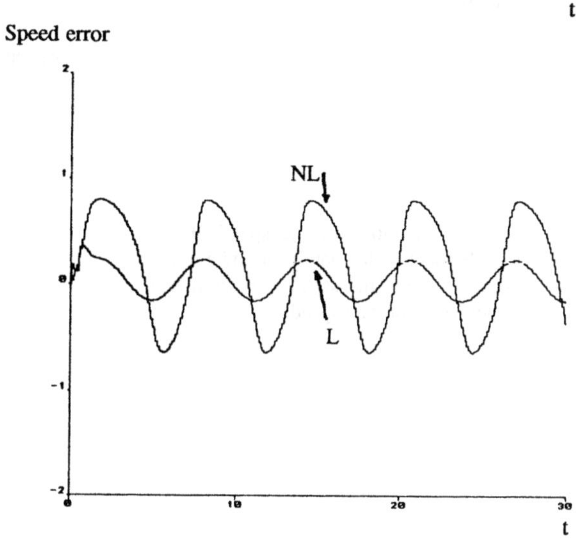

Fig. 9 Speed error produced compared with a linear system with the target speed of fig 8.

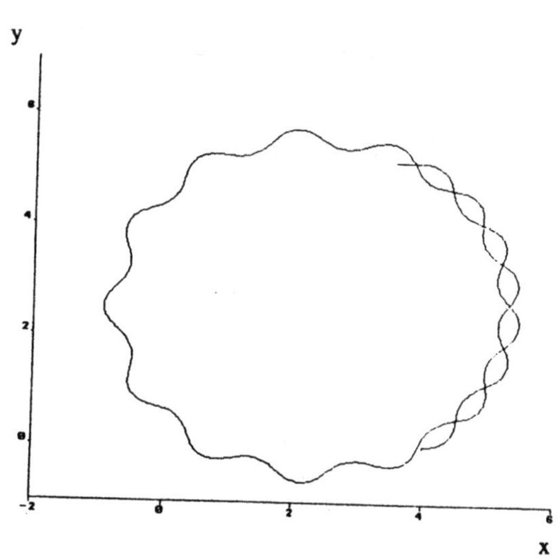

Fig. 10 A trajectory in the plane of the target.

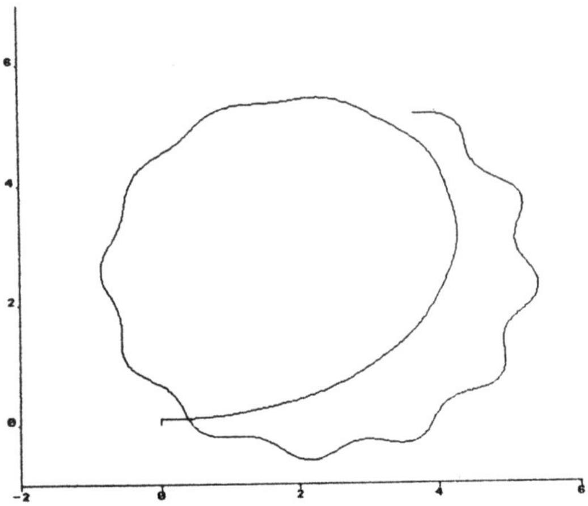

Fig. 11 The trajectory of a linear system in the plane when the target has the trajectory of fig 10

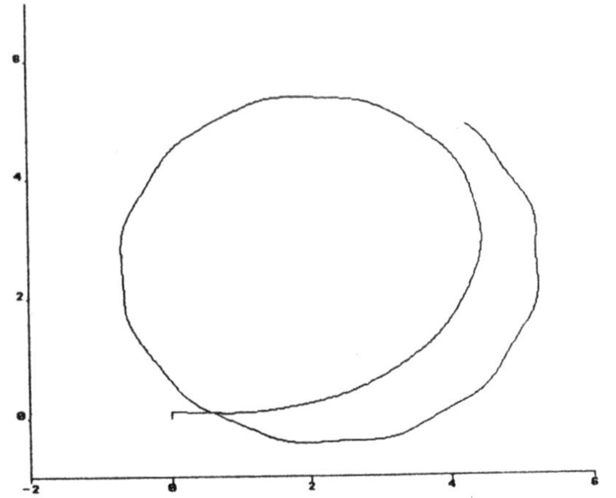

Fig. 12 The trajectory of a nonlinear system in the plane when the target has the trajectory of fig 10

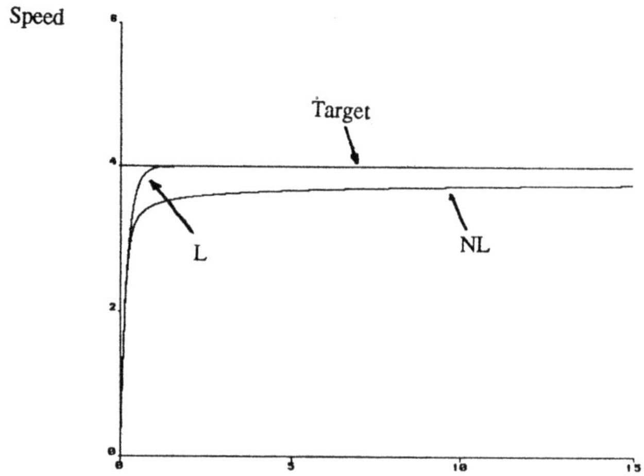

Fig. 13 Time evolution of the speed of the mobile robot with a nonlinear and a linear control. The initial speed is 0 while the target has a constant speed 4.

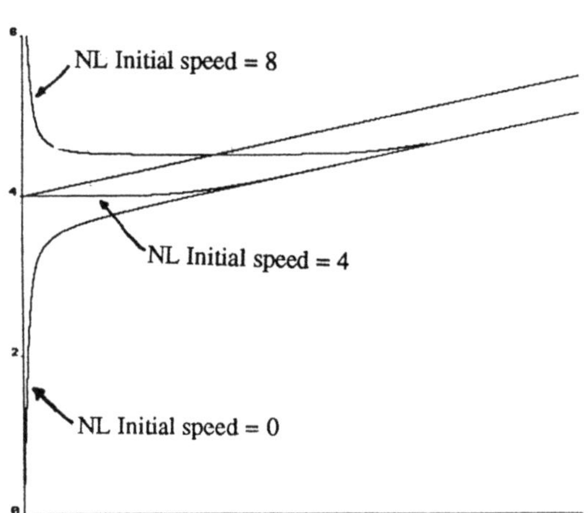

Fig. 14 Time evolution of the speed of the mobile robot. The target has positive acceleration, begining from different initial conditions.

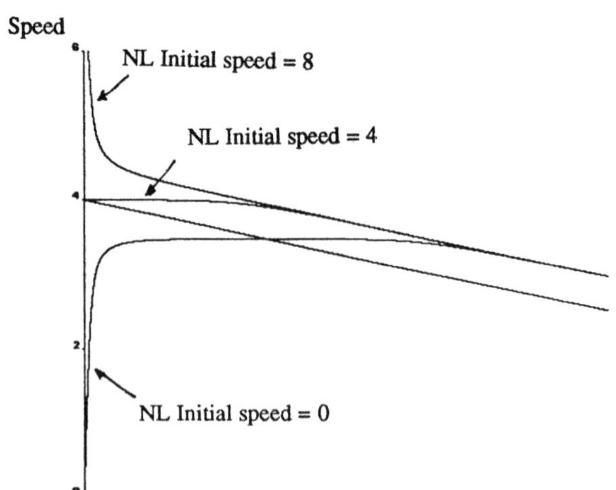

Fig. 15 Time evolution of the speed of the mobile robot. The target has negative acceleration, begining from different initial conditions.

CONCLUSIONS

In this paper we have proposed a polynomial regression analysis to deal with real-time target tracking. The main features of the method presented are the ability to model the trajectory of the target, the need for less computation effort than heuristic approaches, and the ease of estimation due to its stochastic nature and adaptability to real-world variations.

To use the results coming from the tracking system, a special control algorithm has to be designed to filter the noise but avoid losing the target from the scene.

A nonlinear system with a kind of dead zone, has been shown to have competence and good performance in these kind of systems.

REFERENCES

1. AMAT. J, CASALS. A, "Real-Time Tracking of Targets From a Mobile Robot", Proceedings of an International Conference on Intelligent Autonomous Systems-2, vol. 1,pp. 361-367, 1989.
2. ASCHWANDEN. P, "Methods for Real-Time Tracking of Uncooperative Targets", Proceedings of the IEEE International Conference on Acoustics, Speech and Signal Processing, vol. 3, pp. 1560-63, Glasgow, UK. 23-26 May, 1989.
3. AYACHE. N, FAUGERAS. O. D. , "Maintaining Representations of the Environment of a Mobile Robot", Robotics research, the 4th. International Symposium. Cambridge, MIT Press, pp. 337-350, 1988.
4. BAR-SHALOM. Y, FORTMANN. T. E. , "Tracking and Data Association", Academic Press, vol. 179, 1988.
5. BHANU. B. , BURGER. W. , "Qualitative Motion Detection and Tracking of Targets From a Mobile Platform", Image Understanding Workshop, pp. 289-318, April 1988.
6. BHANU. B, PANDA. D. , "Qualitative reasoning and modeling for robust target tracking and recognition from a mobile platform", Image Understanding Workshop, pp. 96-102, April 1988.
7. BROIDA. T. J. , CHELLAPPA. R. , "Estimation of Object Motion Parameters From Noisy Images", IEEE Trans. on Pattern Analysis and Machine Intelligence PAMI-8 (1):90-99.
8. BROWN. C. , "Kalman Filter Utilities for Tracking and Control", Skids document DT-SKI-2A-010016, Oxford University. November 1988.
9. CHANG. K-C. , LEE. H-J. , CHUNG. C-G. , "Adaptively Estimating Motion Parameters for Tracking a Maneuvering Target Using Images", International Journal of Robotics and Automation, vol. 4, No. 1, 1989, pp. 43-52.
10. COULON. P. Y. , CELLARD. D. , "A Dual Video Tracking System", Proceedings of the 6th International Conference on Robot Vision and Sensory Controls, pp. 179-186, France, June 1986.
11. DURRANT-WHYTE. H. F. , RAO. B. Y. S. , HU. H, "Toward a Fully Decentralized Architecture for Multi-sensor Data Fusion", Technical report SKIDS (ESPRIT 1560), Oxford 1989.
12. FRAU. J. I. , BALCELLS. LL. , "Traslational Stabilization of Images in Real-Time (in Spanish)", 1er. Congreso de Automatización y Robótica Industrial (AER), pp. 71-82, Zaragoza 1989.
13. FUTRELLE. R. P. , SPECKERT. G. C. , "Extraction of Motion Data by Interactive Image Processing", 1978 IEEE Conference on Pattern Recognition and Image Processing, 1978, pp. 405-408.
14. GENNERY. D. B. , "Modelling the Environment of an Exploring Vehicle by Stereo Vision", Ph. D. Thesis, Stanford A. I. Laboratory, June 1980.
15. HILDRETH. E. C. , "The Measurement of Visual Motion", Ph. D. Thesis. Depart. of Electrical Engineering and Computer Sciences, MIT, Cambridge, 1983.

16. JENKIN. M. , "The Stereopsis of Time-Variyng Images", RBVC Technical Report 84-3. Depart. of Computer Science. University of Toronto, 1984.
17. LAWSON. D. T. , "Processing Dynamic Image Sequences From a Moving Sensor". Ph. D. Thesis. COINS Depart. , University of Massachussets, 1984.
18. LEGTERS. G. R. , YOUNG. T. Y. , "A Mathematical Model for Computer Image Tracking", IEEE Trans. on Pattern Analysis and Machine Intelligence PAMI-4 (6): 583-594.
19. LUO. R. C. , MULLEN. R. E. , WESSELL. D. E. , "An Adaptive Robotic Tracking System Using Optical Flow", Proceedings of IEEE International Conference on Robotics and Automation, vol. 2, pp. 568-573, April 1988.
20. LLARIO. V. , MARTINEZ. A. B. , "Obtaining Depth Maps: Active Methods", in Computer Vision: Theory and Industrial Applications. Carme Torras Editor (in press). 1990.
21. LLARIO. V. , "Hardware-Software Trade-offs in Robot Vision", Nato Advanced Research Workshop on Real Time object and environment measurement and classification. Maratea, Italy, 1987.
22. MATTHIES. L. , SZELISKI. R. , KANADE. T. , "Kalman Filter-based Algorithms for Estimating Depth From Image Sequences",Image Understanding Workshop, pp. 199-211, April 1988.
23. RHODES. I. B. , "A Parallel Decomposition for Kalman Filters", IEEE Transactions on Automatic Control, vol. 35, No. 3, March 1990, pp. 322-324.
24. SHARIAT. H. , PRICE. K. E. , "How to Use More Than Two Frames to Estimate Motion", IEEE Workshop on Motion: Representation and analysis, 1986, pp. 119-124.
25. STULLER. J. , KRISHNAMURTY. G. , "Kalman Filter Formulation of Low-Level Television Image Motion Estimation", Computer vision, Graphics and Image Processing 21 (1983): pp. 169-204.
26. TAN. J. , KYRIAKOPOULOS. N. , "Implementation of a Tracking Kalman Filter on a Digital Signal Processor", IEEE Transactions on Industrial Electronics, vol. 35, No. 1, pp. 126-134, February 1988.
27. TSAI. R. Y. , HUANG. T. S. , "Estimating Three-Dimensional Motion Parameters of a Rigid Planar Patch, III:Finite point correspondences and the three-view problem", IEEE Transactions on Acoustics, Speech ans Signal Processing ASSP-32 (2): pp. 213-219.
28. YEN. B. L. , HUANG. T. S. , "Determining 3-D Motion/Structure of a Rigid Body over 3 Frames Using Straight Line Correspondences", IEEE Conference on Computer Vision and Pattern Recognition, 1983, 267-272.

Robust Control of Robotic Manipulators

F. Hamano
Department of Electrical Engineering, California State University, Long Beach, 1250 Bellflower Boulevard, Long Beach, California 90840-8303, USA

Abstract: A robotic manipulator must often be operated in an uncertain environment, e.g., picking up different payloads of different or unknown weights, accidental drop of a payload, imprecise knowledge of the robot itself, disturbances, etc. Therefore, the controller must be designed so that the manipulator will perform its tasks despite the existence of some uncertainties. In this article methods of designing stabilizing state feedback control laws are discussed based on the worst case scenario. The results are applied to the tracking problem of a robot manipulator. The second method of Lyapunov and its extensions are extensively used. Our focus is given to the controller design motivated by the game theory with conflicting agents.

Keywords: Robust control, uncertain dynamical systems, robot, manipulator, tracking, exponential stability, practical stability, feedback linearization

1. Introduction

In many control applications such as robotics, aerospace, etc., available mathematical models may be imprecise or some parameters may vary. In particular, in controlling robotic manipulators the inertial parameters may change significantly when they pick up payloads; the model parameters may be slightly inaccurate or some variations may occur; a part of dynamics may be ignored; friction may be different from what is expected; there may be excessive external reaction forces applied to the hands, etc.. Therefore control laws must be capable of handling the above uncertain situations, i.e., they must be robust against modeling uncertainties and disturbances. If a priori statistical characterizations are available, controller design may be based on these data, but in many cases they may not be available. In such situations, if the bounds on the uncertainties are known, the controller can be designed by using deterministic theory. In this paper our focus will be placed on the method based on

The production of this publication was partially funded by California State University, Long Beach.

NATO ASI Series, Vol. F 71
Expert Systems and Robotics
Edited by T. Jordanides and B. Torby

Lyapunov's second method [11] for which the form of the controller originates in a min-max solution of differential game theory. See [1-3, 5, 6, 8, 9, 11-13, 14, 15.17]. The method will be applied to robust tracking problem of robotic manipulators. See [7, 16, 18, 19] also.

In Section 2 a single degree-of-freedom robot is discussed to see why robust control methods are necessary. Sections 3 and 4 discuss modeling of robotic manipulators. In particular, Section 3 presents the most simplified model, and Section 4 discusses a modeling that incorporates elasticity in the drive trains. In Section 5 the feedback linearization technique is discussed for the models developed in the previous sections, which will be used for the controller design to be discussed in the next section. Section 6 formulates the trajectory tracking problem in terms of error dynamics equations, and it also presents a method of controlling a robot under uncertain environment. The type of controller used here originates in game theory. See [5, 6, 12]. Section 7 provides the theory to prove that the controller adopted in section 6 in fact achieves the tracking, and Section 8 provides the simulation results for the single degree-of-freedom manipulator.

The following notations will be used throughout the article: R and R+ respectively denote the set of real numbers and the set of non-negative numbers. Also, R^n stands for the set of real n-(column) vectors. Let $x_1, \cdots, x_n \in R$ and also let $h_1, \cdots, h_m: R^n \rightarrow R$ be differentiable functions. Define column vectors $x:=[x_1, \cdots, x_n]'$ and $h:=[h_1, \cdots, h_m]'$. Then, $\frac{\partial h(x)}{\partial x}$ denotes (the mxn matrix whose (i,j)-th element is defined by $\frac{\partial h_i(x)}{\partial x_j}$. For simplicity of notation the arguments of functions are omitted in most cases (with the exceptions where the meanings becomes unclear by omission).

The existence and uniqueness of the solution of differential equations are assumed throughout the article. For the discussions of the subjects and the proofs see [13] for instance.

2. Motivating Example

To see the technical issues arising from uncertainty, let us consider a single degree of freedom robotic manipulator as in [19]. See Fig.2.1. The equation of motion for the mechanism is given by

$$I\ddot{\theta} + MgL\sin\theta = u \tag{2.1}$$

where θ is the joint angle, u is the input torque, I is the total moment of inertia about the joint axis, M is the total mass, L is the distance from the joint axis to the center of mass of the system and g is the gravitational acceleration. Let $x_1 := \theta$ and $x_2 := \dot{\theta}$. It is desired that $\theta(t)$ follow the reference trajectory $\theta^d(t)$. Then the above equation of motion can be written as

$$\dot{x}_1 = x_2, \tag{2.2a}$$

$$\dot{x}_2 = -(MgL/I)\sin\theta + (1/I)u. \tag{2.2b}$$

Therefore, if we use the "linearizing" (or "inverse dynamics") control law of the form

$$u = MgL\sin\theta + Iv \tag{2.3}$$

we obtain

$$\dot{x}_1 = x_2, \tag{2.4a}$$

$$\dot{x}_2 = v. \tag{2.4b}$$

The reference (or desired) trajectory can be rewritten as

$$\dot{x}_1^d = x_2^d, \tag{2.5a}$$

$$\dot{x}_2^d = \ddot{\theta}^d. \tag{2.5b}$$

Defining the tracking error $e_i := x_i - x_i^d$, $i = 1, 2$, we obtain the error dynamics

$$\dot{e}_1 = e_2, \tag{2.6a}$$

$$\dot{e}_2 = v - \ddot{\theta}^d, \tag{2.6b}$$

which is a simple second order linear system. Clearly the application of

$$v = k_1 e_1 + k_2 e_2 + \ddot{\theta}^d = k_1(x_1 - x_1^d) + k_2(x_2 - x_2^d) + \ddot{\theta}^d \tag{2.7}$$

yields $e_1 \to 0$, $e_2 \to 0$ as $t \to \infty$ for $k_1, k_2 < 0$.

However, a simple linear equation, such as (2.6), can not be obtained if some of the parameters, say, I and L are not precisely known, e.g., we may only know that $5 \leq I \leq 10$ and $5 \leq MgL \leq 10$. In this case the above linearizing control law can not be determined precisely, instead we may be forced to use its estimate

$$u = \widehat{Mg}\hat{L}\sin\theta + \hat{I}v \tag{2.8}$$

where $\widehat{M}$, $\hat{I}$ and $\hat{L}$ are respectively the estimates of M, I and L. This results in the nonlinear system

$$\dot{x}_1 = x_2, \tag{2.9a}$$

$$\dot{x}_2 = (\hat{I}/I)v + (g/I)(\widehat{M}\hat{L}-ML)\sin x_1, \tag{2.9b}$$

and accordingly,

$$\dot{e}_1 = e_2, \tag{2.10a}$$

$$\begin{aligned}\dot{e}_2 &= (\hat{I}/I)v + (g/I)(\widehat{M}\hat{L}-ML)\sin(e_1+\theta^d) - \ddot{\theta}^d \\ &= v - \ddot{\theta}^d + \{(\hat{I}/I) -1\}v + (g/I)(\widehat{M}\hat{L}-ML)\sin(e_1+\theta^d).\end{aligned} \tag{2.10b}$$

Therefore, it is not clear whether the choice of v via (2.7) still allows tracking.

3. Rigid-Body Manipulator Model

In this section we will discuss how to derive the equations of motions of manipulators under the rigid-body assumption. Let q_i, i=1, ... , m be the i-th generalized coordinate of a robotic manipulator, e.g., the relative angle between the two adjacent links if the i-th joint is revolute, or the relative displacement of one link with respect to the adjacent link if the i-th joint is prismatic. Define an m-column vector $q := [q_1, \dots, q_m]'$. Then the kinetic energy of the system is given by

$$\boldsymbol{K}_{energy} = \frac{1}{2}\dot{q}'D(q)\dot{q} \tag{3.1}$$

where D(q) is the mxm "inertia matrix", which is symmetric and positive definite for each $q \in R^m$. See [19] for the details. If we denote the potential energy by $V(q)$, the Lagrangian L is given by

$$L = \boldsymbol{K}_{energy} - V(q) = \frac{1}{2}\dot{q}'D(q)\dot{q} - V(q). \tag{3.2}$$

Using the Euler-Lagrange equation

$$\frac{d}{dt}\frac{\partial L}{\partial \dot{q}_i} - \frac{\partial L}{\partial q_i} = \tau_i \tag{3.3}$$

where τ_i is the i-th generalized force, e.g., torque for a revolute joint and force for a prismatic joint, we obtain the equation of motion

$$D(q)\ddot{q} + h(q,\dot{q}) = u \tag{3.4}$$

where $u := [\tau_1, \dots, \tau_m]'$. Here, $h(q,\dot{q})$ is defined by

$$h(q,\dot{q}) := C(q,\dot{q})\dot{q} + (\frac{\partial V(q)}{\partial q})' + h_{fr}(q,\dot{q}) \tag{3.5}$$

where $C(q,\dot{q})$ is the mxm matrix whose (i,i)-element is given by

$$\sum_{k=1}^{m} \frac{1}{2}\{\frac{\partial d_{ij}}{\partial q_k} + \frac{\partial d_{ki}}{\partial q_j} - \frac{\partial d_{jk}}{\partial q_i}\}\dot{q}_k \tag{3.6}$$

and $h_{fr}(q,\dot{q})$ is an m-column vector representing the terms due to nonconservative forces such as friction. Let $x_1 := q$ and $x_2 := \dot{q}$. Then, we have the equations of motion in terms of state variables:

$$\dot{x}_1 = x_2, \tag{3.7a}$$

$$\dot{x}_2 = -D^{-1}(x_1)h(x_1,x_2) + D^{-1}(x_1)u. \tag{3.7b}$$

4. Elastic Joint Manipulator

The previous equations of motion has been obtained by ignoring elasticity in each joint (i.e., drive train dynamics) and actuator dynamics. In this section the equations of motions are derived, which incorporate joint elasticity. A more detailed discussion of the subject can be found in [18].

For simplicity we will limit ourselves to the revolute joint case where the rotor is directly coupled to the link. We assume the following

Assumption 4.1: The kinetic energy of the rotor is due only to its own rotation, i.e., the motion of the rotor is a pure rotation with respect to an inertial frame.

Assumption 4.2: The rotor/gear inertia is symmetric about the rotor axis of rotation so that the gravitational potential of the system and the velocity of the rotor center of mass are both independent of the rotor position.

The elastic joint under consideration is depicted in Fig.4.1. Let $q_1 := [q_1, q_3, \dots, q_{2m-1}]'$ and also let $q_2 := [q_2, q_4, \dots, q_{2m}]'$. Here,

q_{2i-1} := the angle of the i-th link with respect to the (i-1)-th link, i=1, 2, ... , m

$q_{2i} := (1/r_i)\theta_i$, i=1, 2, ... , m (4.1)

where θ_i is the angular displacement of rotor i and r_i is the corresponding gear ratio. Under Assumption 4.1 the kinetic energy of the system is described by

$$\boldsymbol{K}_{energy} = \frac{1}{2}\dot{q}_1{}'D(q_1)\dot{q}_1 + \frac{1}{2}\dot{q}_2{}'J\dot{q}_2 \quad (4.2)$$

Here, $D(q_1)$ is the inertia matrix of the "rigid robot" as in (3.1) and J is given by $J := \mathrm{diag}[r_1^2 I_{zz1}, \ldots, r_m^2 I_{zzm}]$ where I_{zzi} is the moment of inertia around the principal axis for the i-th motor, i=1, ... , m. Due to Assumption 4.2, the gravitational potential is a function only of q_1. Therefore, the potential energy $V(q_1,q_2)$ of the system is given by

$$V(q_1,q_2) = V(q_1) + \frac{1}{2}(q_1 - q_2)'K(q_1 - q_2) \quad (4.3)$$

where $V(q_1)$ is the potential energy for the rigid part of the manipulator as in (3.2), the second term is the elastic potential energy of the "spring", and $K := \mathrm{diag}[k_1, \ldots, k_m]$ where k_i is the spring constant for the i-th joint, i=1, ..., m. Using the Lagrangian L defined by

$$L := \boldsymbol{K}_{energ} - V(q_1,q_2) \quad (4.4)$$

and also using the Euler-Lagrange equations (3.3), we obtain

$$D(q_1)\ddot{q}_1 + h(q_1,\dot{q}_1) + K(q_1 - q_2) = 0 \quad (4.5a)$$

$$J\ddot{q}_2 + B\dot{q}_2 - K(q_1 - q_2) = u \quad (4.5b)$$

where B is the diagonal matrix defined by $B := \mathrm{diag}[r_1^2 b_1, \ldots, r_m^2 b_m]$, b_i is the friction coefficient corresponding to the i-th motor and gears, $h(q_1,\dot{q}_1)$ is as in (3.5) and $u := [\tau_1, \ldots, \tau_m]'$. Defining $x_1 := q_1$, $x_2 := \dot{q}_1$, $x_3 := q_2$, $x_4 := \dot{q}_2$, we get

$$\dot{x}_1 = x_2, \quad (4.6a)$$

$$\dot{x}_2 = -D^{-1}(x_1)\{h(x_1,x_2) + K(x_1 - x_3)\} \quad (4.6b)$$

$$\dot{x}_3 = x_4, \quad (4.6c)$$

$$\dot{x}_4 = J^{-1}(x_1)\{K(x_1 - x_3) - Bx_4\} + J^{-1}u. \quad (4.6d)$$

5. Feedback Linearization

As we have observed in Section 2 a nonlinear system with certain structure can be transformed into a linear system by applying some nonlinear transformation. More generally, let $f : R^n \to R^n$, $g_i : R^n \to R^n$, $u_i : R_+ \to R$ where $i=1, \ldots , m$, $G(x) := [g_1(x), \ldots , g_m(x)]$, $u := [u_1 \ldots u_m]'$, $\alpha : R^n \to R^m$ and $\beta : R^n \to R^{mxm}$. Then we have (see [18] for instance.) the

Definition 5.1: A nonlinear system

$$\dot{x} = f(x) + \sum_{i=1}^{m} g_i(x)u_i$$
$$= f(x) + G(x)u \tag{5.1}$$

is said to be <u>feedback linearizable</u> in a neighborhood U_o of the origin of the state space R^n if there is a diffeomorphism (i.e., differentiable transformation for which the inverse transformation is also differentiable) $T : R^n \to R^n$ and a nonlinear feedback

$$u = \alpha(x) + \beta(x)v \tag{5.2}$$

such that the transformed state $\xi = T(x)$ satisfies the linear system equation

$$\dot{\xi} = A\xi + Bv \tag{5.3}$$

where A and B are, respectively, nxn and nxm matrices for which the pair (A, B) is reachable (or controllable).

We will now see that the manipulator state equations are feedback linearizable. The necessary and sufficient condition under which a system is feedback lenearizable is known in general in terms of the tools from Differential Geometry, which can be found in [10]. However, we will avoid the use of Differential Geometry by focusing on the specific models described in the previous sections.

Consider the rigid-body manipulator state equations (3.7). Let

$$u = h(x_1, x_2) + D(x_1)v \tag{5.4}$$

where $v : R_+ \to R^m$ is the "new" control. Then, by substitution it is trivial to verify that

$$\dot{x}_1 = x_2, \tag{5.5a}$$

$$\dot{x}_2 = v. \tag{5.5b}$$

Now consider the elastic robot state equations (4.6). Unlike the previous case, the control does not enter the same channel as the nonlinearity. Therefore it is not as obvious as in the previous case whether the system is feedback linearizable. But, as given in [18], it can be shown that the following choice of coordinate transformation can be used to feedback linearize the system. We choose the coordinate transformation so that the new state $\xi := [\xi_1 \dots \xi_n]'$ consists of the position, velocity, acceleration and jerk, i.e., $\xi_1 = x_1$, $\xi_2 = x_2 = \dot{x}_1$, $\xi_3 = \dot{x}_2$, $\xi_4 = \ddot{x}_2$. More precisely, let $(d/dt)D^{-1}(x_1)$ denote the time derivative of $D^{-1}(x_1)$ along the state trajectory, i.e., if the (i,j)-th element of $D^{-1}(x_1)$ is given by $\bar{d}_{ij}(x_1)$, then the (i,j)-th element of $(d/dt)D^{-1}(x_1)$ is defined by $\frac{\partial}{\partial x_1}\bar{d}_{ij}(x_1)\cdot x_2$. We define the new state ξ by

$$\xi_1 := x_1, \tag{5.6a}$$

$$\xi_2 = x_2, \tag{5.6b}$$

$$\xi_3 := -D^{-1}(x_1)\{h(x_1,x_2) + K(x_1 - x_3)\}, \tag{5.6c}$$

$$\xi_4 := f_4(x_1,x_2,x_3) + D^{-1}(x_1)Kx_4 \tag{5.6d}$$

where

$$f_4(x_1,x_2,x_3) := -\{(d/dt)D^{-1}(x_1)\}\{h(x_1,x_2) + K(x_1 - x_3)\} - D^{-1}(x_1)[\frac{\partial h(x_1,x_2)}{\partial x_1} x_2 +$$

$$\frac{\partial h(x_1,x_2)}{\partial x_2}\{-D^{-1}(x_1)\{h(x_1,x_2) + K(x_1 - x_3)\} + K(x_2 - x_4)].$$

The above transformation defines a global diffeomorphism, and its inverse is given by

$$x_1 = \xi_1, \tag{5.7a}$$

$$x_2 = \xi_2, \tag{5.7b}$$

$$x_3 = \xi_1 + K^{-1}\{D(\xi_1)\xi_3 + h(\xi_1,\xi_2)\}, \tag{5.7c}$$

$$x_4 = K^{-1}D(\xi_1)\{\xi_4 - f_4(\xi_1, \xi_2, \xi_1 + K^{-1}\{D(\xi_1)\xi_3 + h(\xi_1,\xi_2)\}). \tag{5.7d}$$

By taking the derivative of (5.6d) we obtain

$$\dot{\xi}_4 = f_5(x) + D^{-1}(x_1)KJ^{-1}u \tag{5.8}$$

where

$$f_5(x) := \frac{\partial f_4(x_1,x_2,x_3)}{\partial x_1} x_2 - \frac{\partial f_4(x_1,x_2,x_3)}{\partial x_2} D^{-1}(x_1)\{h(x_1,x_2) + K(x_1 - x_3)\} + \frac{\partial f_4(x_1,x_2,x_3)}{\partial x_3} x_4$$
$$+ \{(d/dt)D^{-1}(x_1)\}Kx_4 + D^{-1}(x_1)KJ^{-1}K(x_1 - x_3).$$

Therefore, choosing u by

$$u = JK^{-1}D(x_1)\{v - f_5(x)\}$$
$$= -JK^{-1}D(x_1)f_5(x) + JK^{-1}D(x_1)v \quad (5.9)$$

clearly leads to

$$\dot{\xi}_4 = v.$$

Thus, we have

$$\dot{\xi}_1 = \xi_2, \quad (5.10a)$$
$$\dot{\xi}_2 = \xi_3, \quad (5.10b)$$
$$\dot{\xi}_3 = \xi_4, \quad (5.10c)$$
$$\dot{\xi}_4 = v. \quad (5.10d)$$

6. Tracking of a Robotic Manipulator and the Control Law

In this section we will discuss the problem of finding a feedback control law with which the robot state x tracks a prescribed reference trajectory x^d (in the joint space).

In the previous section it has been shown that , if the equations of motion of the robotic manipulator is described by (5.1), then it can be globally linearized to the form

$$\dot{\xi} = A\xi + Bv \,. \quad (6.1)$$

by application of a suitable coordinate transformation $\xi=T(x)$ and an appropriate control law of the form

$$u = \alpha(x) + \beta(x)v \quad (6.2)$$

In practice, however, the functions T, α and β are not exactly known and the measurement of the state x may have some error. Therefore, instead of (6.2) we can only apply the control law

$$u = \hat{\alpha}(y) + \hat{\beta}(y)v \quad (6.3)$$

$$y = x + w \quad (6.4)$$

where $\hat{\alpha} : R^n \to R^m$ and $\hat{\beta} : R^n \to R^{m\times m}$ represent our "models" of α and β, $y : R_+ \to R^n$ is the measurement of x and $w : R_+ \to R^n$ is the measurement noise. Let $P \subset R^n$ be a compact set, and we make the following assumptions:

Assumption 6.1: The function $w : R_+ \to R^n$ is an (unknown) Lebesgue measurable function and there is a positive number ρ_w satisfying

$$\|w(t)\| \leq \rho_w \tag{6.5}$$

for all $t \in R_+$.

Assumption 6.2: $p : R_+ \to R^{\bar{n}}$ is an (unknown) Lebesgue measurable function whose value belongs to $P \subset R^n$, i.e., $p(t) \in P$ for all $t \geq 0$.

Assumption 6.3: $\beta : R^n x P \to R^{m\times m}$ and $\hat{\beta} : R^n \to R^{m\times m}$ are continuous functions satisfying the following two conditions: There exist positive constants β_{min} and β_{max} satisfying

$$\beta_{min} \leq \|\beta^{-1}(x,p)\| \leq \beta_{max} \tag{6.6}$$

for all $x \in R^n$ and for all $p \in P$. There is a positive constant μ such that

$$\| \beta^{-1}(y-w,p)\hat{\beta}(y) - I \| \leq \mu < 1 \tag{6.7}$$

for all $y \in R^n$, for all $p \in P$ and for all $w \in R^n$ satisfying $\|w\| \leq \rho_w$.

Assumption 6.4: $T : R^n x P \to R^n$ and $\hat{T} : R^n \to R^n$ are continuous functions.

Assumption 6.5: $\alpha : R^n x P \to R^m$, $\hat{\alpha} : R^n \to R^m$ are continuous functions.

Remark. The equation (6.7) of Assumption 6.3 can always be satisfied by the following choice of $\hat{\beta}$. Let

$$\hat{\beta} := \hat{\beta}_o := \frac{2}{\beta_{max}+\beta_{min}} I. \tag{6.8}$$

Then, we have (See [18, 19].)

$$\| \beta^{-1}(y-w)\hat{\beta}_o(y) - I \| \leq \frac{\beta_{max}-\beta_{min}}{\beta_{max}+\beta_{min}} < 1 \tag{6.9}$$

for all $y \in R^n$, for all $p \in P$ and for all $w \in R^n$ satisfying $\|w\| \leq \rho_w$.

Notice that without the feedback law applied, the system is described by

$$\dot{\xi} = A\xi + B\beta^{-1}(x)(u - \alpha(x)). \tag{6.10}$$

Therefore, when the feedback law defined by (6.3) and (6.4) is used, we have

$$\begin{aligned}\dot{\xi} &= A\xi + B\beta^{-1}(x)(\widehat{\alpha}(y) + \widehat{\beta}(y)v - \alpha(x)) \\ &= A\xi + Bv + B\{(\beta^{-1}(x)\widehat{\beta}(y) - I)v + \beta^{-1}(x)(\widehat{\alpha}(y) - \alpha(x))\} \\ &= A\xi + Bv + B\{Ev + \beta^{-1}(x)\Delta\alpha\}\end{aligned} \tag{6.11}$$

where $E(x,y) := \beta^{-1}(x)\widehat{\beta}(y) - I$ and $\Delta\alpha(x,y) := \widehat{\alpha}(y) - \alpha(x)$.

The reference (or desired) state trajectory used here is described by

$$\dot{\xi}^d = A\xi^d + Bv^d \tag{6.12}$$

where $v^d : R_+ \to R^m$. Define the tracking error by $e := \xi - \xi^d$. Then, we have

$$\begin{aligned}\dot{e} &= \dot{\xi} - \dot{\xi}^d \\ &= A\xi + Bv + B\{Ev + \beta^{-1}(x)\Delta\alpha\} - (A\xi^d + Bv^d) \\ &= Ae + B(v - v^d) + B\{Ev + \beta^{-1}(x)\Delta\alpha\}.\end{aligned} \tag{6.13}$$

To state the problem precisely, we introduce the following notion of stability. Consider in general a system described by

$$\dot{e}(t) = f(e(t), t) \tag{6.14}$$

where $e(t) \in R^n$ and $t \in R_+$ and $f : R^n x R_+ \to R^n$, and let $N(r) := \{e \in R^n : \|e\| \le r\}$, i.e., the closed ball with radius $r \ge 0$ centered at the origin.

Definition 6.1. ([4, 8]) The system (2.1) is <u>(uniformly) exponentially stable to within</u> $N(r)$ <u>with the exponential rate of convergence</u> $\sigma \ge 0$ if the following two conditions are satisfied: (i) the system is <u>(uniformly) exponentially convergent to</u> $N(r)$ <u>with the exponential rate of convergence</u> σ, i.e., there exists a region $N(\bar{r}_o) \subset R^n$ with the property that, for any initial conditions $t_o \ge 0$ and $e_o \in N(\bar{r}_o)$, there exists a number $c(e_o) \ge 0$ such that if $e : R_+ \to R^n$ is any solution of (6.14) with $e(t_o) = e_o$ then

$$\|e(t)\| \le r + c(e_o)\exp[-\sigma(t - t_o)] \tag{6.15}$$

for all $t \geq t_o$, and (ii) the state trajectory remains in N(r) if e_o is sufficiently close to the origin, i.e.,

$$\lim_{e_o \to 0} c(e_o) = 0. \tag{6.16}$$

Using the above definition, we can now state the

Problem. Given σ and $r \geq 0$, find the control law v for (6.13) (therefore for (6.11)) so that the resulting system is (uniformly) exponentially stable to within N(r) with the exponential rate of convergence $\sigma \geq 0$

Denote by $\lambda_i(A+BF)$ the i-th eigen-value of A+BF where $i = 1, \ldots, n$, and let F be an mxn real matrix such that $Re[\lambda_i(A+BF)] < -\sigma < 0$ for all $i = 1, \ldots, n$. Further define the computed (or estimated) error $\hat{e}$ and computed (or estimated) state $\hat{\xi}$ by $\hat{e} := \hat{\xi} - \xi^d$ and $\hat{\xi} :=$ $\hat{T}(y)$ where the diffeomorphism $\hat{T} : R^n \to R^n$ is the estimated (therefore, known) coordinate transformation. The control law for v will be of the form

$$v = F\hat{e} + v^d + \Delta v \tag{6.17}$$

and Δv is a nonlinear control law which will be defined in the sequel.

Remark. If there is no (or sufficiently small) uncertainties, Δv is not necessary. This term is used to compensate for the uncertainty.

Substituting (6.17) into (6.13), we get

$$\begin{aligned}\dot{e} &= Ae + B(F\hat{e} + \Delta v) + B\{E(F\hat{e} + v^d + \Delta v) + \beta^{-1}(x)\Delta\alpha\}\\ &= Ae + B(F(e + \hat{\xi} - \xi) + \Delta v) + B\{E(F(e + \hat{\xi} - \xi) + v^d + \Delta v) + \beta^{-1}(x)\Delta\alpha\}.\end{aligned}$$

Writing $\Delta\xi := \hat{\xi} - \xi$ and $\bar{A} := A+BF$, we obtain

$$\begin{aligned}\dot{e} &= Ae + B(F(e + \Delta\xi) + \Delta v) + B\{E(F(e + \Delta\xi) + v^d + \Delta v) + \beta^{-1}(x)\Delta\alpha\}\\ &= \bar{A}e + B\Delta v + B\{E(Fe + v^d + \Delta v) + \beta^{-1}(x)(\Delta\alpha + \hat{\beta}(y)F\Delta\xi)\}.\end{aligned}$$

Let $\eta := E(Fe + v^d + \Delta v + \beta^{-1}(x)(\Delta\alpha + \hat{\beta}(y)F\Delta\xi)$. Then,

$$\dot{e} = \bar{A}e + B\Delta v + B\eta. \tag{6.18}$$

The control law Δv will be chosen so that the following inequalities are satisfied for some scalar valued function $\rho: R^n x R_+ \to R_+$ (to be chosen in the sequel):

$$\|\Delta v\| \leq \rho(y,t) \tag{6.19}$$

$$\|\eta\| \leq \rho(y,t). \tag{6.20}$$

Note

$$\begin{aligned}\|\eta\| &= \|E(F(\hat{e} - \Delta\xi) + v^d + \Delta v) + \beta^{-1}(x)(\Delta\alpha + \hat{\beta}(y)F\Delta\xi)\| \\ &\leq \|E\Delta v\| + \|E(F(\hat{e} - \Delta\xi) + v^d) + \beta^{-1}(x)(\Delta\alpha + \hat{\beta}(y)F\Delta\xi)\| \\ &\leq \mu\|\Delta v\| + \|E(F(\hat{e} - \Delta\xi) + v^d) + \beta^{-1}(x)(\Delta\alpha + \hat{\beta}(y)F\Delta\xi)\|.\end{aligned} \tag{6.21}$$

To chose the above $\rho(y,t)$, we make the following

Assumption 6.6: There exist functions $\hat{\phi}: R^n x R_+ \to R_+$ and $\phi: R^n x R_+ \to R_+$ such that, for each $y \in R^n$,

$$\|E(F(\hat{e} - \Delta\xi) + v^d) + \beta^{-1}(x)(\Delta\alpha + \hat{\beta}(y)F\Delta\xi)\| \leq \hat{\phi}(y,t) \tag{6.22}$$

for all $p \in P$ and for all $w \in R^n$ satisfying $\|w\| \leq \rho_w$ where $\hat{e} = \hat{T}(y) - \xi^d$ and $\Delta\xi = \hat{T}(y) - T(y-w,p)$, and for each $e \in R^n$,

$$\hat{\phi}(y,t) \leq \phi(e,t) \tag{6.23}$$

for all $p \in P$ and for all $w \in R^n$ satisfying $\|w\| \leq \rho_w$ where $y = T^{-1}(e+\xi^d(t),p) + w$.

Remark. This assumption is not restrictive since one can always choose $\hat{\phi}(y,t)$ as follows.

$$\begin{aligned}&\|E(F(\hat{e} - \Delta\xi) + v^d) + \beta^{-1}(x)(\Delta\alpha + \hat{\beta}(y)F\Delta\xi)\| \\ &\leq \|E\| \, \|F(\hat{e} - \Delta\xi) + v^d\| + \|\beta^{-1}(x)\| \, \|\Delta\alpha + \hat{\beta}(y)F\Delta\xi\| \\ &\leq \mu(\|F(\hat{e} + v^d)\| + \|F\| \, \|\Delta\xi\|) + \beta_{max}(\|\Delta\alpha\| + \|\hat{\beta}(y)\| \, \|F\| \, \|\Delta\xi\|) \\ &\leq \mu\|F(\hat{e} + v^d)\| + \beta_{max}\|\Delta\alpha\| + (\mu + \beta_{max}\beta_o)\|F\| \, \|\Delta\xi\|) \\ &\leq \mu\|F(\hat{T}(y) - \xi^d(t)) + v^d(t)\| + \beta_{max}\sup\|\Delta\alpha\| + (\mu + \beta_{max}\beta_o)\|F\|\sup\|\Delta\xi\| \\ &=: \hat{\phi}(y,t).\end{aligned} \tag{6.24}$$

where "sup" is taken over all the uncertain parameter variations.

Then,

$$\|\eta\| \leq \mu\|\Delta v\| + \hat{\phi}(y,t). \tag{6.25}$$

The function $\rho(y,t)$ is now defined by

$$\mu\rho(y,t) + \hat{\phi}(y,t) = \rho(y,t),$$

i.e., $\rho(y,t)$ is the function satisfying

$$\rho(y,t) := \frac{1}{1-\mu}\hat{\phi}(y,t). \tag{6.26}$$

Note that the equation (6.20) is clearly satisfied for the above choice of $\rho(y,t)$.

Now let Q be an nxn positive-definite matrix, and let P be a unique, symmetric, positive-definite solution of the Lyapunov equation:

$$(\overline{A} + \sigma I)'P + P(\overline{A} + \sigma I) = -Q. \tag{6.27}$$

Using the above quantities, the control law Δv is defined, for a given positive number ε, by the equations:

$$\Delta v := -\frac{B'P\hat{e}}{\|B'P\hat{e}\|}\rho(y,t) \tag{6.28a}$$

if $\|B'P\hat{e}\| \geq \varepsilon$, and

$$\Delta v := -\frac{B'P\hat{e}}{\varepsilon}\rho(y,t) \tag{6.28b}$$

if $\|B'P\hat{e}\| < \varepsilon$ where $\hat{e} = \widehat{T}(y) - \xi^d$. Note that (6.19) clearly hold since $\|\Delta v\| = \rho(y,t)$.

Remark. The above choice of control law Δv is motivated by the min-max solution of a differential game with conflicting agents. See [5, 6, 12, 13].

7. Exponential Stability of Error Dynamics

In the previous section we formulated the tracking problem as a problem of finding a control law so that exponential stability to within a neighborhood is achieved in the error dynamics.

We have also discussed the control law to be used. In this section we show that the exponential stability is achieved if such a control law is used.

Let us consider the equation (6.18) describing the dynamics of the error $e(t)$ and the control law given by (6.28) where $\rho(y,t)$ and $\hat{\phi}(y,t)$ are defined by (6.26) and (6.25) (or (6.22)). The system described by these equations will be referred to as Σ_e. Recalling (6.23) and (6.26), write $\bar{\rho}(e,t) := \frac{1}{1-\mu}\,\phi(e,t)$, so that $\rho(y,t) \leq \bar{\rho}(e,t)$ where $y = T^{-1}(e+\xi^d(t),p) + w$. We assume

Assumption 7.1: There are non-negative numbers $\tilde{r}_o$, a_o, a_1, b_o and b_1 such that

$$\bar{\rho}(e,t) \leq a_o + a_1\|e\| \tag{7.1}$$

for all $e\in R^n$ satisfying $\|e\| \leq \tilde{r}_o$ and $t \geq 0$, and such that

$$4\|B'P\|\,\|\Delta\xi\| \leq b_o + b_1\|e\| \tag{7.2}$$

for all $e\in R^n$ satisfying $\|e\| \leq \tilde{r}_o$, for all $t \geq 0$, for all $y\in R^n$, for all $p\in P$ and for all $w\in R^n$ satisfying $\|w\| \leq \rho_w$ where $\Delta\xi := \hat{\xi} - \xi = \hat{T}(T^{-1}(e+\xi^d(t),p)+w) - (e+\xi^d(t)))$.
Define a number r_1 by

$$r_1 := (1/c_2)(c_1 + \sqrt{c_1^2+c_oc_2}\,) \tag{7.3}$$

where $c_o := a_ob_o+(\varepsilon/2)a_o$, $c_1 := \dfrac{a_ob_1+a_1b_o+(\varepsilon/2)a_1}{2}$ and $c_2 := \lambda_{min}(Q) - a_1b_1$. Then, we have

Theorem 7.1. Under the assumptions (A6.1)-(A6.6) and (A7.1), if

$$r := \sqrt{\frac{\lambda_{max}(P)}{\lambda_{min}(P)}}\,r_1 \leq \sqrt{\frac{\lambda_{min}(P)}{\lambda_{max}(P)}}\,\tilde{r}_o, \tag{7.4}$$

then the system Σ_e is (uniformly) exponentially stable to within $N(r)$ with exponential rate of convergence σ. Furthermore, if $e(t_o) = e_o$, then

$$\|e(t)\| \leq r + \sqrt{\frac{\lambda_{min}(P)}{\lambda_{max}(P)}}\,e_o\exp[-\sigma(t-t_o)]. \tag{7.5}$$

Remark. If the state measurement error is zero, i.e., $w(t,p) = 0$ for all $t \geq 0$ and for all $p\in P$, and if the coordinate transformation is exact, i.e., $\hat{T}(x) = T(x,p)$ for all $x\in R^n$ and for all $p \in P$, then $r \to 0$ as $\varepsilon \to 0$, i.e., r can be made arbitrarily small by choosing ε to be arbitrarily small.

The right hand side of (7.5) provides an upper bound for $\|e(t)\|$. The actual $e(t)$ enters the ball $N(r)$ after some time, and we have

Corollary 7.1. Suppose that all the conditions in Theorem 7.1 hold. Let $\|e_o\|$, $r_o \leq \sqrt{\frac{\lambda_{min}(P)}{\lambda_{max}(P)}}\tilde{r}_o$, and define $t_{D1} := \frac{1}{2\sigma}\ln\frac{e_o' P e_o}{\lambda_{max}(P) r_1^2}$ and $\bar{t}_{D1} := \frac{1}{2\sigma}\ln\frac{r_o^2}{r_1^2}$. Then, $e(t_o) = e_o$ implies $\|e(t)\| \leq r$ for all $t \geq t_o + t_{D1}$, and $\|e(t_o)\| \leq r_o$ implies $\|e(t)\| \leq r$ for all $t \geq t_o + \bar{t}_{D1}$.

Proof of Theorem 7.1: (Outline) Define a Lyapunov candidate function by $V(e) := e'Pe$ for all $e \in R^n$. (See [11] for the Lyapunov function.) Using (6.18), the derivative of V along the state trajectory can be shown to be, for each $e \in R^n$,

$$\dot{V}(e,t) := \left(\frac{\partial V(e)}{\partial e}\right)(\bar{A}e + B\Delta v + B\eta)$$

$$= -2\sigma V(e) - e'Qe + 2e'PB(\Delta v + \eta)$$

where $\frac{\partial V(e)}{\partial e}$ denotes the gradient of V with respect to e. Noting that $\|\eta(x,t,p,w)\| \leq \bar{\rho}(e,t)$ for all $e \in R^n$, for all $p \in P$, $w \in R^n$ satisfying $\|w\| \leq \rho_w$ where x is related to t, p, w and e through the transformations defined in Section 6, and using a similar technique to the one given in [13], it can be shown that

$$\dot{V}(e,t) \leq -2\sigma V(e) - e'Qe + \{(\varepsilon/2) + 4\|B'P\Delta\xi\|\}\bar{\rho}(e,t)$$

$$\leq -2\sigma V(e) - e'Qe + \{(\varepsilon/2) + 4\|B'P\|\,\|\Delta\xi\|\}\bar{\rho}(e,t),$$

which holds for any uncertainties satisfying the required conditions. Applying (7.1) and (7.2), we obtain

$$\dot{V}(e,t) \leq -2\sigma V(e) + L_\varepsilon(e,t).$$

where $L_\varepsilon(e,t) := -c_2\|e\|^2 + 2c_1\|e\| + c_o$. Since $L_\varepsilon(e,t) < 0$ if $\|e\| > r_1$, $\dot{V}(e,t) < -2\sigma V(e)$ if $\|e\| > r_1$, from which the results follow.

Proof of Corollary 7.1: (Outline) The results follows from the fact that any state trajectory $e : R_+ \to R^n$ with $\|e(t_o)\| = e_o$ satisfies

$$V(e(t)) \leq V(e_o)\exp\{-2\sigma(t-t_o)\}.$$

8. Simulation

In this section simulation results are given for the one-degree-of-freedom robot discussed in Section 2 to which the control law developed in Section 6 is applied.

For simplicity we first assume that the state can be measured accurately, i.e., $w \equiv 0$. Then, from (2.10) we have

$$\dot{e}_1 = e_2, \tag{8.1a}$$

$$\dot{e}_2 = v - \ddot{\theta}^d + \{(\hat{I}/I) - 1\}v + (g/I)(\widehat{ML} - ML)\sin x_1. \tag{8.1b}$$

As in [19] if we apply $v = -100e_1 - 20e_2 + \ddot{\theta}^d + \Delta v$, we obtain

$$\dot{e} = \overline{A}e + B\Delta v + B\eta. \tag{8.2}$$

where $\overline{A} = \begin{bmatrix} 0 & 1 \\ -100 & -20 \end{bmatrix}$, $B = \begin{bmatrix} 0 \\ 1 \end{bmatrix}$ and $\eta = \{(\hat{I}/I) - 1\}v + (g/I)(\widehat{ML} - ML)\sin x_1$.
Now assume as in Section 2 that I and MgL are uncertain to the extent $5 \le I \le 10$ and $5 \le MgL \le 10$, and suppose we choose $\hat{I} = 5$ and $\widehat{MgL} = 5$. Then, it can be shown that

$$\|\eta\| \le \rho(x,t) = \|-100e_1 - 20e_2 + \ddot{\theta}^d\| + (4/5). \tag{8.3}$$

Let $Q = I$. Then the solution of the Lyapunov equation

$$(\overline{A} + 9I)'P + P(\overline{A} + 9I) = -Q \tag{8.4}$$

is given by $P = \begin{bmatrix} 2.5305 & 0.2278 \\ 0.2278 & 0.0208 \end{bmatrix}$.

The reference trajectory chosen is described by $\theta^d(t) = \{1 - \exp(-8t)\}^3$, and therefore, $\theta(0) = \dot{\theta}(0) = 0$ in particular, and the plot is given in Fig.8.1. It is also superimposed in Fig's 8.2(a), 8.3(a), 8.4(a), 8.5 and 8.6 for easy comparison.

The simulation results using the controller described by (6.17), (6.27) and (6.28) with the above values are given in the graphs in Fig's 8.2 and 3 as well as the plots when Δv is set to zero for comparison purposes. The plot of θ (or x_1) in Fig.8.2 is for the case where $\theta(0) = \dot{\theta}(0) = 0$, and the plot of θ in Fig.8.3 is for the case where $\theta(0) = 0.2$ (rad) and $\dot{\theta}(0) = 0.2$(rad/sec). The results show that the method presented here is promising. The results using the controller given in [19] (which does not guarantee specified exponential convergence) is also depicted in Fig.8.4 for comparison for the initial conditions $\theta(0) = 0.2$ (rad) and $\dot{\theta}(0) = 0.2$(rad/sec).

In addition, the simulation results are given for the case where the state measurement noise is non-zero. The equations for w are chosen to be $w(t) = 0.01\sin 376.8t$ (for Fig.8.5) and $w(t) = 0.1\sin 376.8t$ (for Fig.8.6). The initial conditions are set to be $\theta(0) = 0.2$ (rad) and $\dot{\theta}(0) = 0.2$(rad/sec). In either case the tracking error still converges to within a small neighbourhood of the zero error.

9. Conclusion

In this article a method of robust control has been presented, which is applicable to controlling a robotic manipulator, and simulation results have been given for a simple manipulator to demonstrate the performance of the controller. For further information on the methods along this line see [7, 16, 18, 19].

10. Acknowledgement

The author is grateful to Mr. Stephen Clingenpeel, Logicon for performing simulations.

References

1. Barmish, B. R.,: Necessary and sufficient conditions for quadratic stabilizability of an uncertain system. J. Optimization Theory and Applications, vol. 46, No. 4, pp. 399-408, Aug. 1985
2. Barmish, B. R., Corless, M. and Leitmann, G.: A new class of stabilizing controllers for uncertain dynamical systems. SIAM J. Control and optimization,Vol. 21, No2, pp.246 - 255, March 1983
3. Barmish, B. R. and Leitmann, G.: On ultimate boundedness control of uncertain systems in the absence of matching assumptions. EEE Trans. Automat. Contr., vol. AC-27, No. 1, pp.153-158, 1982
4. Corless, Garofalo and Leitmann,: Guaranteeing exponential convergence for uncertain systems. Proceedings of the International Workshop on Robustness in Identification and Control, Torino, Italy, 1988
5. Corless, M. and Leitmann, G.: Continuous state feedback guaranteeing uniform ultimate boundedness for uncertain dynamic systems. IEEE Trans. Automat. Contr., vol.AC-26, No.5, pp.1139-1144, 1981
6. Gutman, S. and Leitmann, G.: Stabilizing feedback control for dynamic systems with bounded uncertainty. Proceedings of the IEEE Conference on Decision and Control, 1976
7. Ha, I. J. and Gilbert, E. G.: Robust tracking in nonlinear systems. IEEE Trans. Automat. Contr., vol. AC-32, No. 9, pp. 763-771, Sept. 1987
8. Hamano, F.: Exponential stabilization to a neighborhood of zero state for nonlinear uncertain dynamical systems by linear feedback. Proc. of 28th SICE (Society of Instrumentation and Control Engineers) Annual Conference, Matsuyama, Japan, July, 1989
9. Hamano, F.: Practical stabilization of uncertain dynamical systems by continuous state feedback based on Riccati equation and a sufficient condition for robust practical stability. Proc. of the 28th Conference on Decision and Control, Tampa, Dec. 1989

10. Isidori, A.: Nonlinear Control Systems: An Introduction. Lecture Notes in Control and Information Sciences, vol 17. Berlin: Springer 1985
11. Kalman, R. E. and Bertram, J. E.: Control system analysis and design via the "Second Method" of Lyapunov I, Continuous-time systems. J. Basic Engineering, Trans. ASME, vol. 82, No. 2, pp. 371-393, 1960
12. Leitmann, G.: Stabilization of dynamical systems under bounded input and parameter uncertainty. Proc. 2nd Kingston Conf. on Differential Games and control Theory, pp. 47-64. New York: Marcel Dekker 1976
13. Leitmann, G.: On the efficacy of nonlinear control in uncertain linear systems. ASME Journal of Dynamic Systems, Measurement and Control, vol. 102, pp.95-102, June 1981
14. Leitmann, G. and Pandey, S.: A controller for a class of mismatched uncertain systems. Proc. of the 28th Conference on Decision and Control, Tampa, Dec. 1989
15 Petersen, I. R. and Hollot, C. V.: A Riccati equation approach to the stabilization of uncertain linear systems. Automatica, vol. 22, No. 4, pp. 397-411, 1986
16. Ryan, E. P., Leitmann, G. and Corless, M.: Practical stabilizability of Uncertain dynamical systems: Application to robotic tracking. J. Optimization Theory and Applications, vol. 47, No. 2, pp. 235-252, Oct. 1985
17. Schmitendorf, W. E.: Designing stabilizing controllers for uncertain systems using the Riccati equation approach. IEEE Trans. Automat. Contr., vol. 33, No. 4, pp. 376-379, April 1988
18. Spong, M. W.: Modeling and control of elastic joint robots. ASME J. of Dyn. Sys. Meas. and Cont., vol 109, pp. 310-319, Dec. 1987
19. Spong, M. W. and Vidyasagar, M.: Robot Dynamics and Control. New York: Wiley 1989

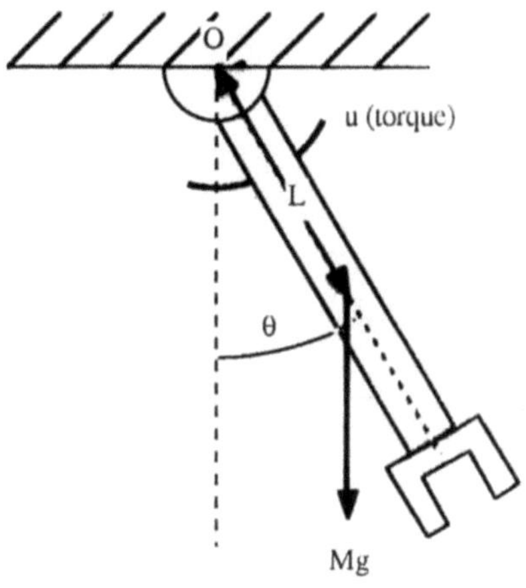

Fig.2.1. Single degree-of-freedom robot. The moment of inertia arround O is I.

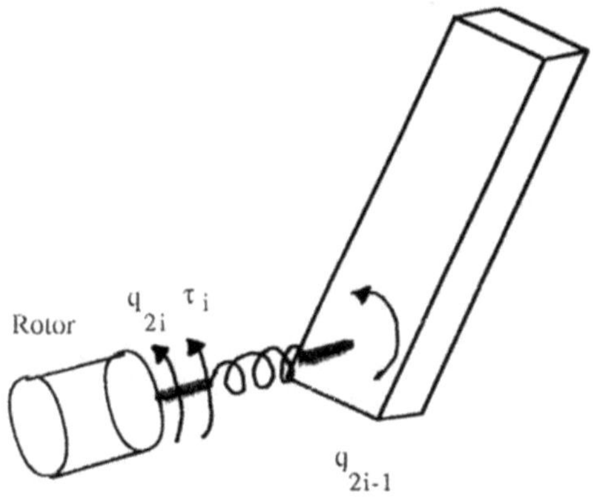

Fig.4.1. Elastic model of the i-th joint.

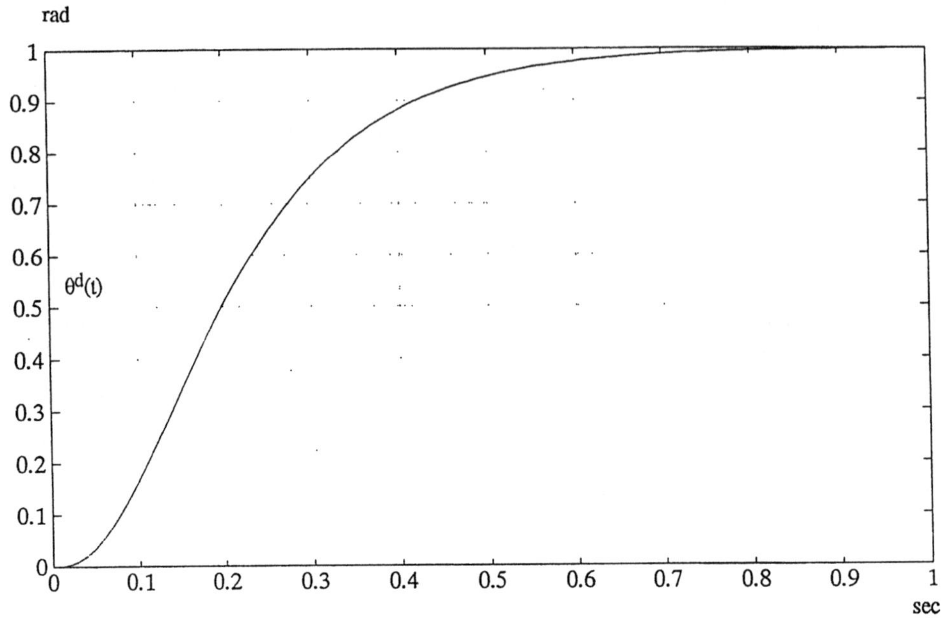

Fig.8.1. Reference trajectory $\theta^d(t)$

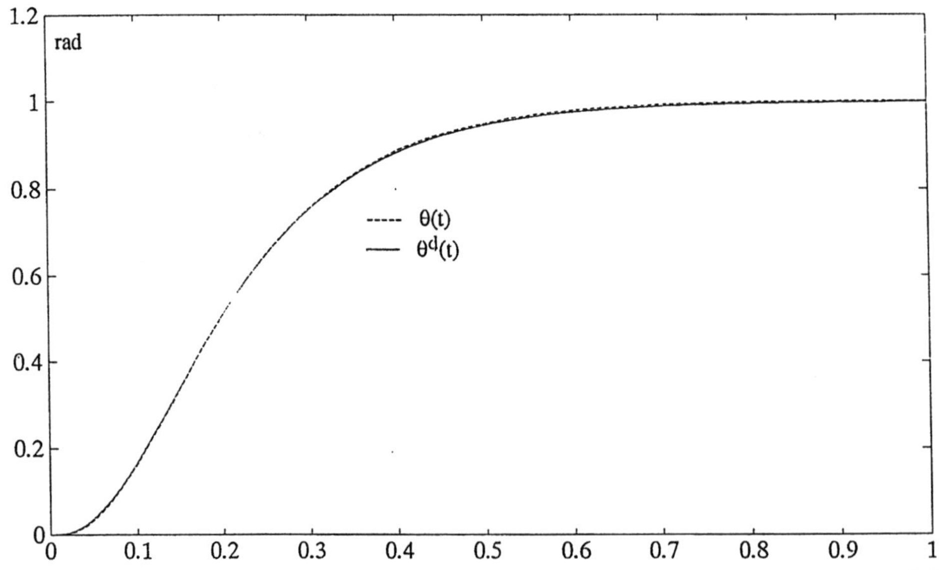

Fig.8.2(a). Plot of $\theta(t)$ for the zero initial conditions

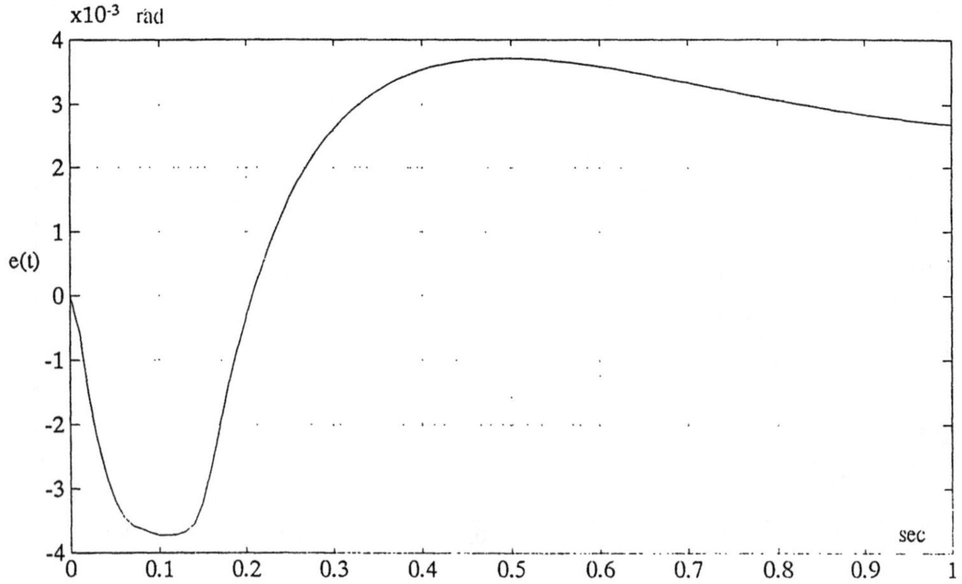

Fig.8.2(b). Plot of $e(t) := \theta(t) - \theta^d(t)$ for the zero initial conditions

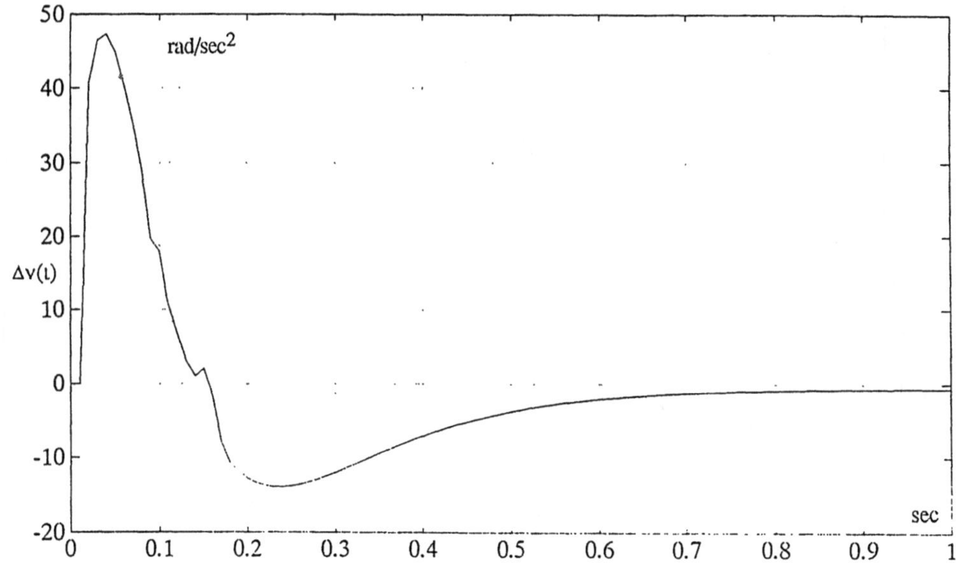

Fig.8.2(c). Plot of $\Delta v(t)$ for the zero initial conditions

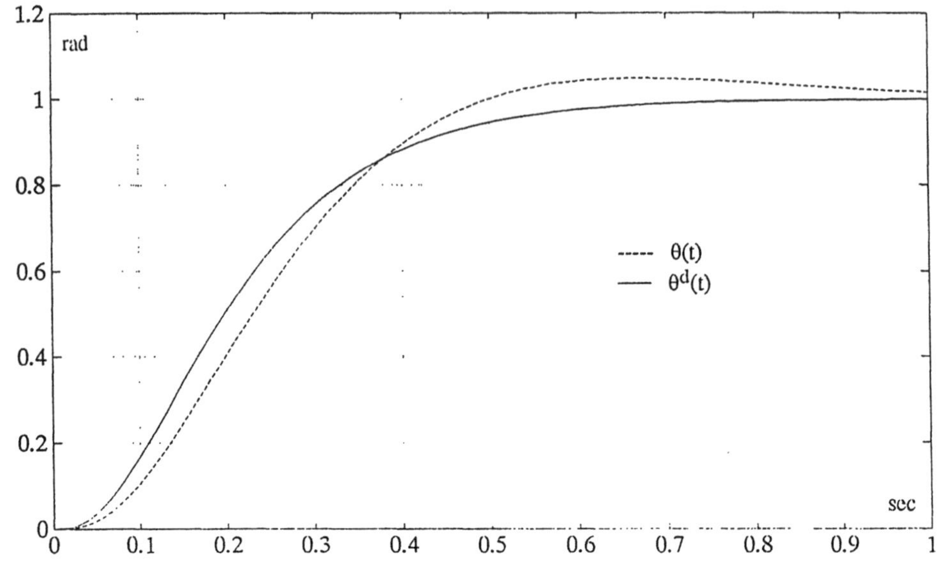

Fig.8.2(d). Plot of $\theta(t)$ for the zero initial conditions when $\Delta v(t)$ is not used, i e., set zero for all time

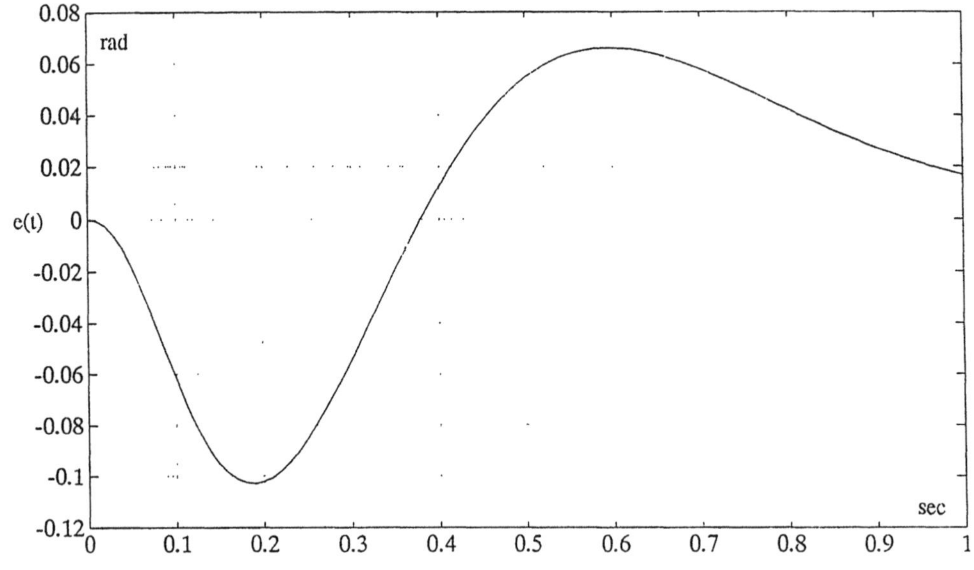

Fig.8.2(e). Plot of $e(t)$ for the zero initial conditions when $\Delta v(t)$ is not used, i e., set zero for all time

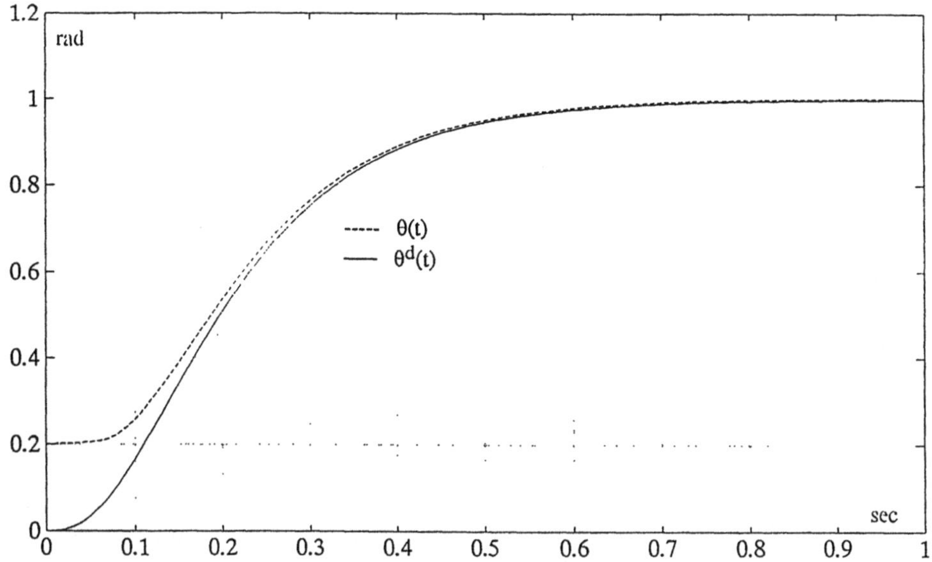

Fig.8.3(a). Plot of $\theta(t)$ for $\theta(0) = 0.2$ (rad) and $\dot{\theta}(0) = 0.2$(rad/sec)

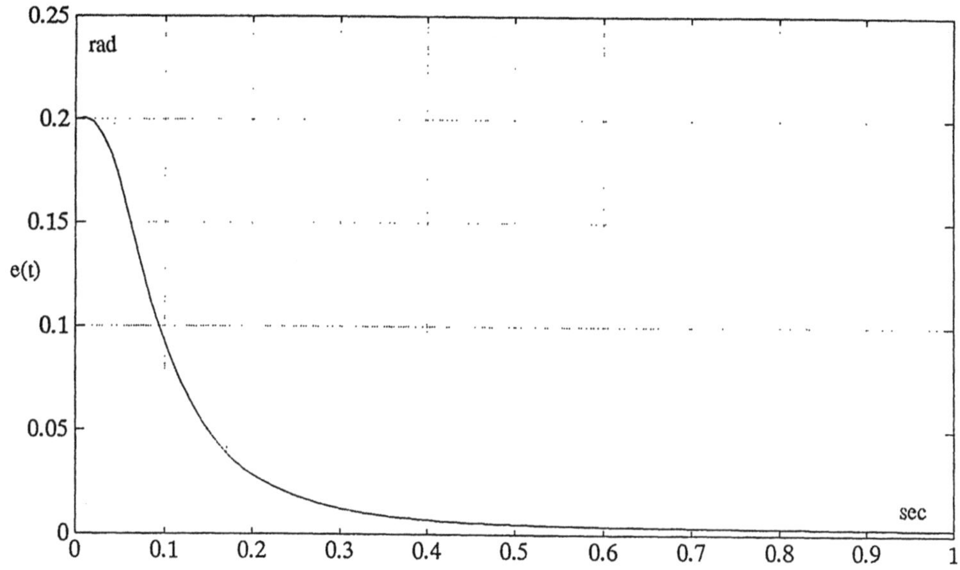

Fig.8.3(b). Plot of $e(t):=\theta(t)-\theta^d(t)$ for $\theta(0) = 0.2$ (rad) and $\dot{\theta}(0) = 0.2$(rad/sec)

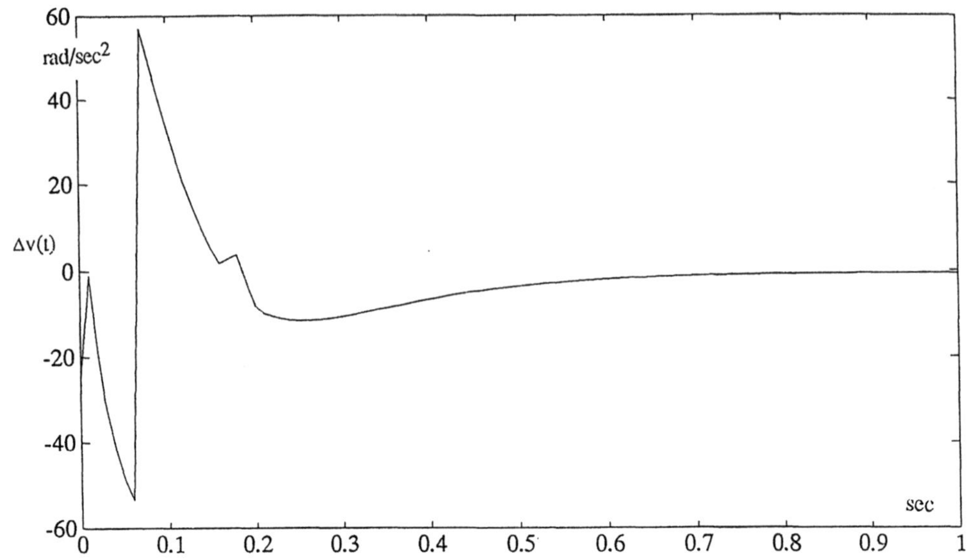

Fig.8.3(c). Plot of $\Delta v(t)$ for $\theta(0) = 0.2$ (rad) and $\theta(0) = 0.2$(rad/sec)

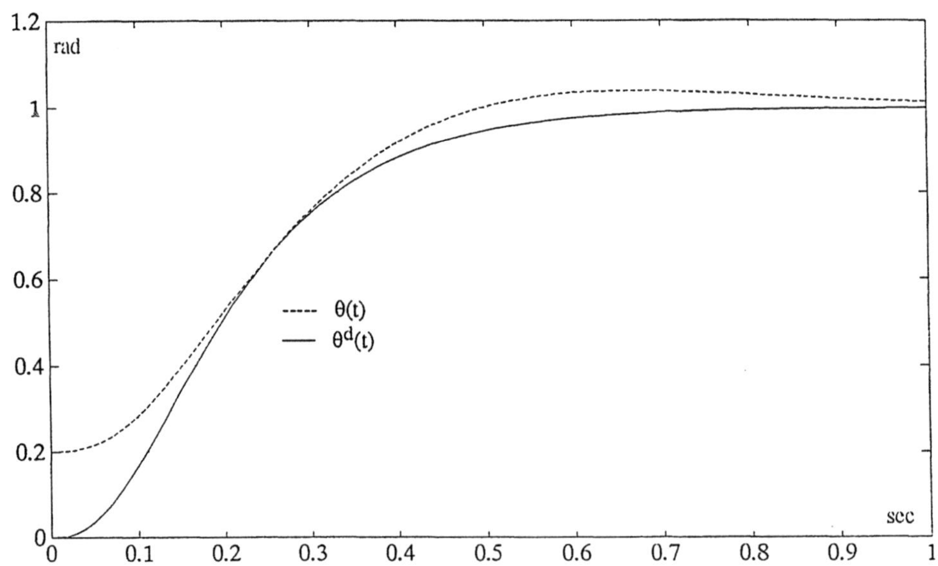

Fig.8.3(d). Plot of $\theta(t)$ for $\theta(0) = 0.2$ (rad) and $\theta(0) = 0.2$(rad/sec) when $\Delta v(t)$ is not used, i e., set zero for all time

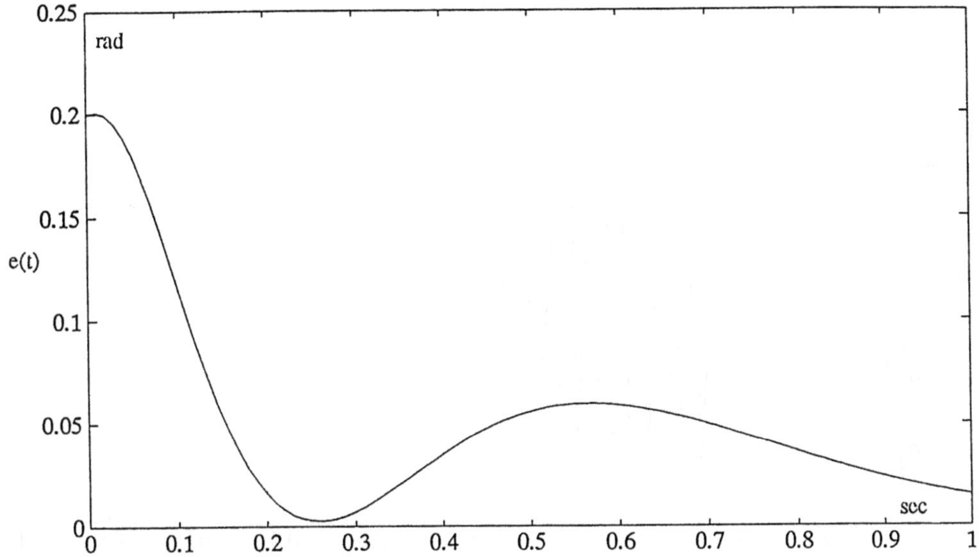

Fig.8.3(e). Plot of e(t) for $\theta(0) = 0.2$ (rad) and $\dot{\theta}(0) = 0.2$(rad/sec) when $\Delta v(t)$ is not used, i.e., set zero for all time

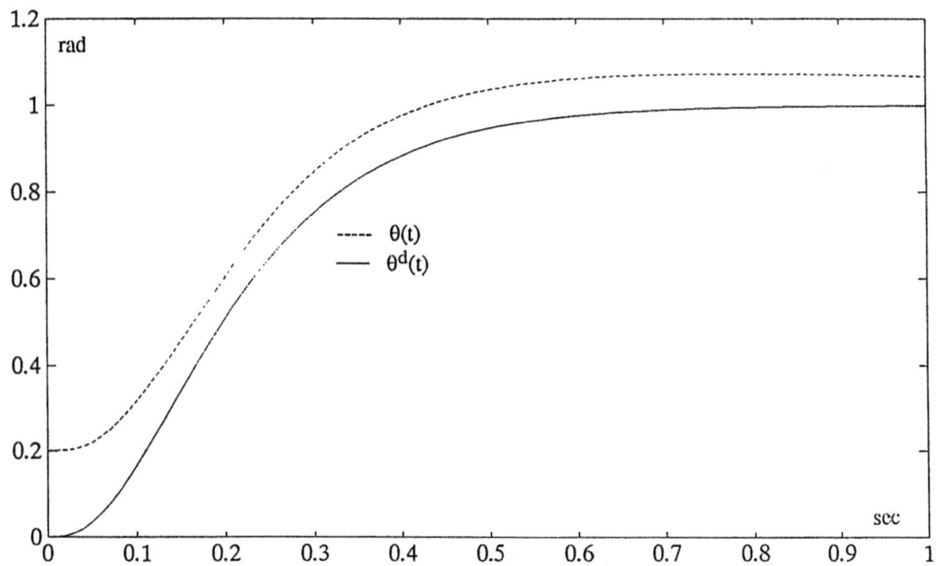

Fig.8.4(a). Plot of $\theta(t)$ for $\theta(0) = 0.2$ (rad) and $\dot{\theta}(0) = 0.2$(rad/sec) when the controller in [19] is used

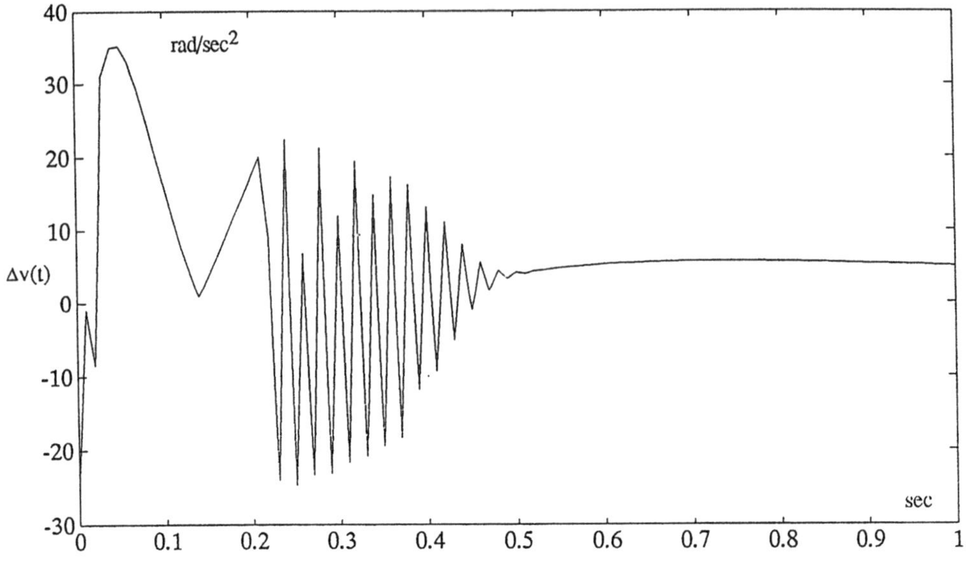

Fig.8.4(b). Plot of $\Delta v(t)$ for $\theta(0) = 0.2$ (rad) and $\theta(0) = 0.2$(rad/sec) when the controller in [19] is used

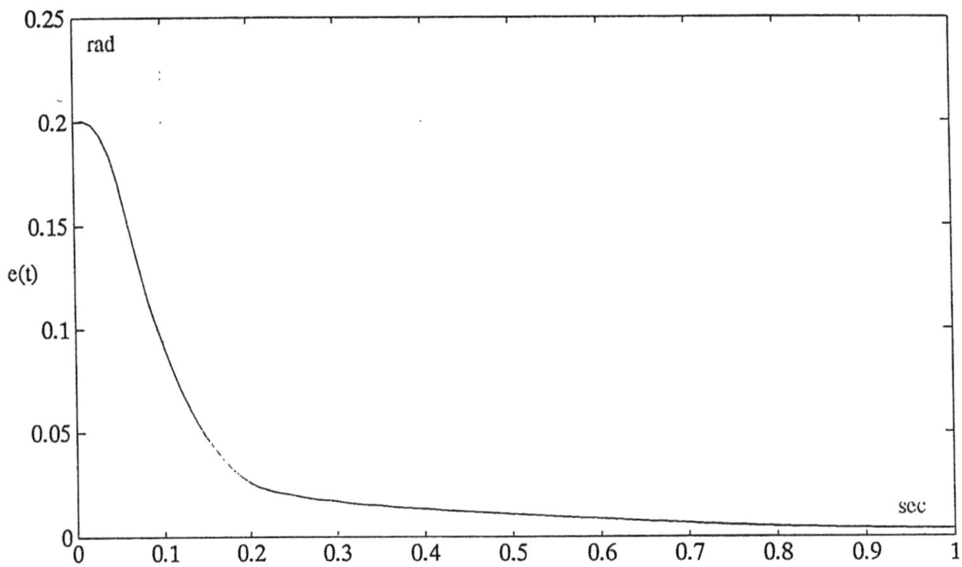

Fig.8.5. Plot of $e(t){:=}\theta(t)-\theta^d(t)$ with $w(t) = 0.01\sin 376.8t$ for $\theta(0) = 0.2$ (rad) and $\theta(0) = 0.2$(rad/sec)

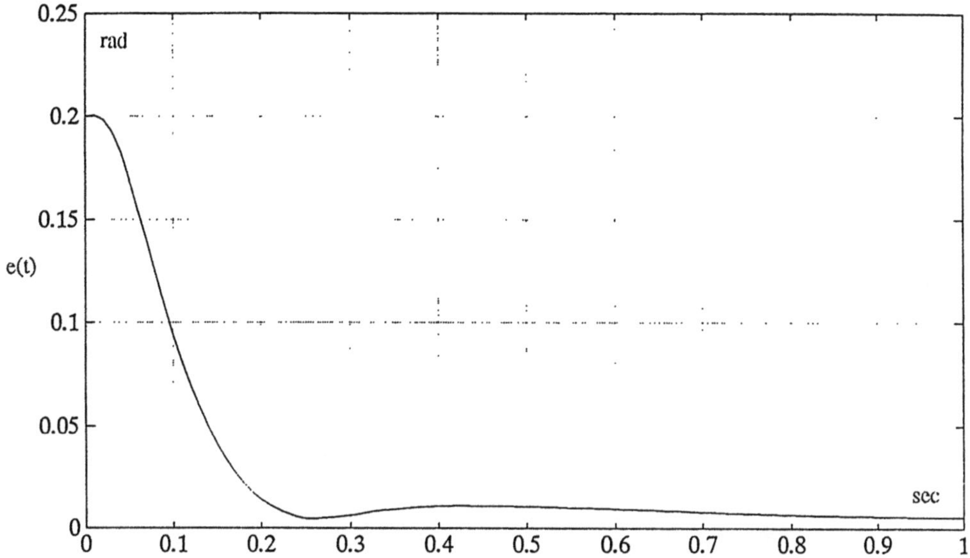

Fig.8.6. Plot of $e(t):=\theta(t)-\theta^d(t)$ with $w(t) = 0.1\sin 376.8t$ for $\theta(0) = 0.2$ (rad) and $\dot{\theta}(0) = 0.2$(rad/sec)

Automatic Determination of Quasi-Time Optimal Joint Trajectories and Adaptive Control for a Robotic Arm

Yorgo Istefanopulos, Luciano Casagrande

Electrical and Electronics Engineering Department, Boğaziçi University, Istanbul, Turkey

Abstract: In this work an original algorithm is developed to generate quasi-time optimal trajectories for each joint of a planar robot arm on the basis of the desired trajectory of its end effector, expressed in Cartesian coordinates, along with various time specifications. The algorithm takes into account the following three cases: free time for motion completion; fixed time for motion completion; motion with specified times at intermediate points. The generated reference trajectories are used in two different robot control algorithms based on the computed-torque method and the proportional-derivative feedback with feedforward control, where both methods include load adaptation.

Keywords: end-effector trajectory, joint trajectories, redundant planar robot arm, computed-torque control, proportional double-derivative control, load adaptation.

1. Introduction

In most applications of an industrial robot the objective is to control the movement of its gripper (or end-effector) to perform industrial jobs such as material handling, assembly, painting, welding, or pick-and-place operations. Since a robot is an expensive machine, in most cases the use of a robot must be justified by the fact that the robot can complete a task faster and better than a human operator.

Generally all algorithms developed for robot control assume that the desired trajectory of motion for each joint is available as a piece-wise continuous function of time.

Often, however, trajectory specifications for robot end-effector motion are given in Cartesian coordinates and therefore they must be converted to the corresponding joint positions.

The aim of this work is to develop an algorithm that obtains the desired joint trajectories as a function of time when the desired trajectory of the end-effector is specified in Cartesian coordinates.

The first problem that has to be solved when the desired trajectory of the end-effector is specified either as a sequence of points or as a continuous function, is an inverse kinematic problem. The solution of this problem results in a sequence of values for joint position coordinates from which piece-wise continuous first or second-order polynomials can be generated to describe the joint trajectories between any pair of consecutive points.

A second problem to consider is the accommodation of the time requirements. If no time is specified for the completion of the entire end-effector trajectory the algorithm finds a time optimal trajectory for each joint and then rearranges and modifies the generated polynomials

NATO ASI Series, Vol. F 71
Expert Systems and Robotics
Edited by T. Jordanides and B. Torby

according to the greatest joint time needed between two consecutive points. Obviously the trajectories determined in this way are quasi-time optimal.

Time specifications on the other hand can be given in two ways:

1) total motion completion time for the end-effector trajectory,

2) time instants that each point on the trajectory should be visited.

The developed polynomials are used as reference joint trajectories in a control algorithm. In this work two control algorithms are used for the end-effector trajectory tracking for a loaded robot arm.

The first control scheme is the computed-torque method which includes also mass adaptation for the load.

The second control scheme consists of a multivariable proportional-double-derivative (PDD) feedforward controller to ensure trajectory tracking and a multivariable proportional-derivative (PD) feedback controller to achieve pole placement.

2. Robot Model

The robot model used in this work is the model developed by H. Hemami [1],[3]. A planar manipulator can be modeled as an n rigid-body linkage each link with mass m_i, moment of inertia I_i, length l_i, and mass center at k_i for i=1,2,...,n. The angle of link i with respect to the vertical is θ_i. The input torque at the ith joint is u_i.

The dynamic model in matrix form for the manipulator deacribed above can be written as :

$$J_o \theta'' = J_{co} \theta'^2 + G_o + EU \qquad (2.1)$$

In this model, J_o represents the matrix of system parameters, J_{co} contains terms accounting for centrifugal and Coriolis forces, G_o represents gravity loading, and EU represents the torque applied at the joints.

The outermost link of the robot has an end-effector that can hold a load with mass M, radius r and moment of inertia I.

The load affects the outermost link of an n-link planar robot as follows [2]:

$$m_n \rightarrow m_n + M \qquad (2.2)$$

$$k_n \rightarrow k_n + M (l_n - k_n) / (m_n + M) \qquad (2.3)$$

$$I_n \rightarrow I_n + I + M\, m_n (l_n - k_n)^2 / (m_n + M) \qquad (2.4)$$

Replacing the new values of m,k,I for the outermost link in the equation (2.1), the dynamic equation for the loaded manipulator is obtained:

$$J \theta'' = J_c \theta'^2 + G + EU \qquad (2.5)$$

In this model now, J , J_c , G are the new matrices for the robot dynamics in the case of a loaded manipulator.

Defining

$$J - J_o = M\, J_1 \tag{2.6}$$

$$J_c - J_{co} = M\, J_{c1} \tag{2.7}$$

$$G - G_o = M\, G_1 \tag{2.8}$$

where M is the mass of the load, the dynamic equation for a planar loaded n-link manipulator becomes [2]:

$$(J_o + M\, J_1)\, \theta'' = (J_{co} + M\, J_{c1})\, \theta'^2 + (G_o + M\, G_1) + EU \tag{2.9}$$

The matrices and vectors in the dynamic equations are presented in appendix A.

3. Inverse Kinematics

The kinematic equations must be solved for joint angle values, given the desired end-effector position in Cartesian coordinates.

The kinematic equations to be solved are nonlinear and therefore if solutions exist, which means that the end-effector position is in the workspace of the manipulator, multiple solutions may exist for the same end-effector location. A unique solution is not obtained unless additional constraints are imposed on the variables.

In this work the inverse kinematics problem is solved for the two and three link planar manipulators by imposing suitable additional constraints.

3.1 Two-Link Planar Robot Arm

For a two link manipulator the kinematic equations can be written as:

$$X = l_1 \cos \alpha + l_2 \cos (\alpha+\beta) \tag{3.1}$$

$$Y = l_1 \sin \alpha + l_2 \sin (\alpha+\beta) \tag{3.2}$$

where (X,Y) are the Cartesian coordinates of the desired end-effector position, α is the angle made by link 1 with the horizontal and b is the angle made by link 2 with respect to link 1.

Squaring (3.1) and (3.2), summing and using trigonometric identities we obtain:

$$X^2 + Y^2 = l_1{}^2 + l_2{}^2 + 2\, l_1\, l_2 \cos \beta \tag{3.3}$$

Solving for β we obtain two angle values and choose the one that corresponds to the elbow up case.Using this value of β, α is obtained from the kinematic equations.

The angles made by the links with the vertical are then evaluated as:

$$\theta_1 = \pi / 2 - \alpha \tag{3.4a}$$

$$\theta_2 = \pi / 2 - \alpha - \beta \tag{3.4b}$$

3.2 Three-Link Planar Robot Arm

For a three-link manipulator the kinematic equations can be written as :

$$X = l_1 \cos \alpha + l_2 \cos (\alpha+\beta) + l_3 \cos (\alpha+\beta+\tau) \tag{3.5}$$

$$Y = l_1 \sin \alpha + l_2 \sin (\alpha+\beta) + l_3 \sin (\alpha+\beta+\tau) \tag{3.6}$$

where α, β are as defined above and τ is the angle made by link 3 with link 2.

Squaring (3.5) and (3.6), summing and using trigonometric identities we obtain:

$$X^2 + Y^2 = l_1^2 + l_2^2 + l_3^2 + 2\, l_2\, l_3 \cos\beta + 2\, l_1\, l_3 \cos(\beta+\tau) + 2\, l_2\, l_3 \cos\tau \tag{3.7}$$

In order to obtain a unique solution to equation (3.7), a somehow restrictive constraint is chosen as:

$$\beta = \tau \tag{3.8}$$

which assumes that the incremental angles of the two outer links are equal.

Solving equation (3.7) after the substitution of the constraint two values for $\cos\beta$ are obtained.Using the acceptable value of $\cos\beta$, two values for β are also obtained.Among those two values a choice is made in order to avoid collision with the horizontal plane. Substituting the value of β in the kinematic equations the value of α is evaluated.

The angles made by the links with the vertical are then evaluated as:

$$\theta_1 = \pi / 2 - \alpha \tag{3.9a}$$

$$\theta_2 = \pi / 2 - \alpha - \beta \tag{3.9b}$$

$$\theta_3 = \pi/2 - \alpha - \beta - \tau \tag{3.9c}$$

4. Quasi-Time Optimization

The solution of the inverse kinematics problem for the end-effector desired trajectory points leads to a set of joint angle values for each link.

For each link moving between two successive points, first or second-order time polynomials will be computed to describe the link trajectory between those points [4].

The first step of the algorithm is the computation of those polynomials for each individual joint.

4.1 Quasi-Time Optimal Polynomial Calculation

In this section the procedure to be followed in the calculation of the quasi-time optimal polynomials for each individual joint is explained. This procedure is repeated for each link of the manipulator.

In this procedure the following variables are represented by:

x_i : the joint angle value of point i for the current link

v_i : the joint angular velocity at point i

v_m : the maximum attainable velocity of a link; it must be specified by the user

a_m : the maximum attainable acceleration of a link; it must be specified by the user.

Step I: The first step of the procedure consists in the initialization of variables.

1) Calculation of the velocity of point i. This velocity is computed using the time derivative of the polynomials calculated to describe the trajectory between point i-1 and point i. The velocity of point i can be one of the following:

$v_i = 0$

$0 < | v_i | < v_m$

$| v_i | = v_m$

If i=1, then the initial velocity is taken as zero.

2) Calculation of the distances D between point i and point i+1 and the next distance D_n between point i+1 and point i+2:

$$D = x_{i+1} - x_i \tag{4.1}$$

$$D_n = x_{i+2} - x_{i+1} \tag{4.2}$$

Step II: The second step consists in the qualitative estimation of the velocity of point i+1.

Three cases are possible according to the velocity of point i and the sign of D, D_n:

1)	$vi = 0$	$D > 0$ and $D_n < 0$	$v_{i+1} = 0$
		$D < 0$ and $D_n > 0$	$v_{i+1} = 0$

		$D > 0$ and $D_n > 0$	$0 < v_{i+1} < v_m$
		$D < 0$ and $D_n < 0$	$-v_m < v_{i+1} < 0$
2)	$0 < v_i < v_m$	$D > 0$ and $D_n < 0$	$v_{i+1} = 0$
		$D > 0$ and $D_n > 0$	$0 < v_{i+1} \leq v_m$
	$-v_m < v_i < 0$	$D < 0$ and $D_n < 0$	$-v_m \leq v_{i+1} < 0$
		$D < 0$ and $D_n > 0$	$v_{i+1} = 0$
3)	$v = v_m$	$D > 0$ and $D_n > 0$	$v_{i+1} = v_m$
		$D > 0$ and $D_n < 0$	$v_{i+1} = 0$
	$v_i = -v_m$	$D < 0$ and $D_n < 0$	$v_{i+1} = -v_m$
		$D < 0$ and $D_n > 0$	$v_{i+1} = 0$

Step III: The third step of the procedure consists in the computation of the polynomials using the data obtained in the first two steps of the procedure.According to the velocity of point i and point i+1, four main cases will be considered in the initial computation of the polynomials.

CASE I.1. Acceleration from Rest, Deceleration to Rest.

In this case the velocity of point i and point i+1 are equal to zero. Here according to the distance to cover between point i and i+1 two possibilities may occur.

If the distance is not sufficient to reach the maximum velocity and then decelerate to rest,

$$D < v_m^2 / a_m \tag{4.3}$$

the trajectory between these points is described by two second-order time polynomials in terms of the following parameters:

$$t_1 = \sqrt{D / a_m} \tag{4.4}$$

$$v_1 = a_m t_1 \tag{4.5}$$

The polynomials are then written as:

$$x(t) = x_i + 1/2\, a_m t^2 \qquad 0 \leq t \leq t_1 \tag{4.6}$$

$$x(t) = x(t_1) + v_1 (t - t_1) - 1/2\, a_m (t - t_1)^2 \qquad t_1 \leq t \leq 2\, t_1 \tag{4.7}$$

If the distance between the points is large enough to reach maximum velocity or

$$D \leq v_m^2 / a_m$$

the trajectory between the two points is described by two second-order and one first-order time polynomial in terms of the parameters:

$$t_1 = v_m / a_m \tag{4.8}$$

$$t_2 = (D - v_m^2 / a_m) / v_m + t_1 \tag{4.9}$$

$$t_3 = t_2 + t_1 \tag{4.10}$$

The polynomials are then written as:

$$x(t) = x_i + 1/2\, a_m\, t^2 \qquad 0 \le t < t_1 \tag{4.11}$$

$$x(t) = x(t_1) + v_m\, (t - t_1) \qquad t_1 \le t < t_2 \tag{4.12}$$

$$x(t) = x(t_2) + v_m\, (t - t_2) - 1/2\, a_m\, (t - t_2)^2 \qquad t_2 \le t \le t_3 \tag{4.13}$$

CASE I.2. Acceleration from an Arbitrary Velocity

In this case, the velocities of point i and point i+1 are in the following ranges:

$$0 \le | v_i | < v_m$$

$$0 < | v_{i+1} | \le v_m$$

According to the distance between the point the trajectory will be described by one second-order time polynomial or by one second-order and one first-order time polynomial.

If the distance between the points is not sufficient to reach maximum velocity, the trajectory is described by a single second-order time polynomial.

The time needed to reach point i+1 is obtained from the solution of the following second-order equation :

$$1/2\, a_m\, t_1^2 + v_i\, t_1 - D = 0 \tag{4.15}$$

The polynomial is then written as :

$$x(t) = x_i + v_i\, t + 1/2\, a_m\, t^2 \qquad 0 \le t \le t_1 \tag{4.16}$$

If the distance between the points is sufficient to reach maximum velocity, the trajectory is described by one second-order and one first-order time polynomial using the parameters

$$t_1 = (v_m - v_i) / a_m \tag{4.17}$$

$$t_2 = (D - (v_i\, t_1 + 1/2\, a_m\, t_1^2)) / v_m + t_1 \tag{4.18}$$

The polynomials are written as:

$$x(t) = x_i + v_i\, t + 1/2\, a_m\, t^2 \qquad 0 \le t \le t_1 \qquad (4.19)$$

$$x(t) = x(t_1) + v_m\, (t - t_1) \qquad t_1 \le t \le t_2 \qquad (4.20)$$

CASE I.3. Maintain Maximum Velocity

In this case, the link has reached maximum velocity at point i and it will stay the same point i+1.The trajectory between the two points will be described by a single first-order time polynomial.

The time needed to reach point i+1 is:

$$t_1 = D / v_m \qquad (4.21)$$

The polynomial for this case is:

$$x(t) = x_i + v_m\, t \qquad 0 \le t \le t1 \qquad (4.22)$$

CASE I.4. Deceleration from Arbitrary Velocity to Rest

In this case, the velocity of point i is different from zero while the end-effector has to stop its motion at point i+1.The trajectory between the two points is described by one first-order and one second-order time polynomial.

Defining the intermediate time t_1 and final time t_2:

$$t_1 = (D - v_i^2 / 2\, a_m) / v_i \qquad (4.23)$$

$$t_2 = v_i / a_m + t_1 \qquad (4.24)$$

The polynomials are written as:

$$x(t) = x_i + v_i\, t \qquad 0 \le t \le t_1 \qquad (4.25)$$

$$x(t) = x(t_1) + v_i\, (t - t_1) - 1/2\, a_m\, (t - t_1)^2 \qquad t_1 \le t \le t_2 \qquad (4.26)$$

In this case where braking is involved, the distance between the points is important. If this distance is smaller than a certain value (which depends on the physical parameters of the link) the end point will be passed. This possibility has been considered and a check is performed when the points of the trajectory are specified in order to avoid such problems during the evaluation of the quasi-time, optimal trajectory polynomials.

4.2 Modification of the Time Polynomials to Satisfy Time Requirements

The time polynomials describing the joint trajectories need to be modified at different steps. These modifications are needed in order to arrange the polynomials according to time requirements and to synchronize the motion of the links. In the modification procedure solutions have been found for the different cases discussed in section 4.1.

In the following, two new variables are defined:

T_n :time requirement to be met for the interval;

T : actual time value for which the polynomials are calculated for the interval;

In this procedure four different modification cases are discussed.

CASE M.1. Acceleration from Rest - Deceleration to Rest

In this section the case in which initial and final velocities are equal to zero is treated. If the distance between the points is described by two second-order time polynomials

$$x(t) = x_i + 1/2\, a_m\, t^2 \qquad 0 \le t < t_1 \qquad (4.27)$$

$$x(t) = x(t_1) + v_1\, (t - t_1) - 1/2\, a_m\, (t - t_1)^2 \qquad t_1 \le t \le T \qquad (4.28)$$

which means that maximum velocity is not reached,the following modifications are brought

$$t_{1n} = T_n / 2 \qquad (4.29)$$

$$a = D / (t_{1n}^2) \qquad (4.30)$$

and the new peak velocity v_n can be calculated as the product of the acceleration and time given above.

The new polynomials are

$$x(t) = x_i + 1/2\, a\, t^2 \qquad 0 \le t < t_{1n} \qquad (4.31)$$

$$x(t) = x(t_{1n}) + v_n\, (t - t_{1n}) - 1/2\, a\, (t - t_{1n})^2 \qquad t_{1n} \le t \le T_n \qquad (4.32)$$

If the trajectory for the interval is described by two second-order and one first-order time polynomial

$$x(t) = x_i + 1/2\, a_m\, t^2 \qquad 0 \le t < t_1 \qquad (4.33)$$

$$x(t) = x(t_1) + v_m (t - t_1) \qquad t_1 \le t \le t_2 \qquad (4.34)$$

$$x(t) = x(t_2) + v_m (t - t_2) - 1/2\, a_m (t - t_2)^2 \qquad t_2 \le t \le T \qquad (4.35)$$

the following changes are done

$$t_{1n} = (T_n - (t_2 - t_1)) / 2 \qquad (4.36)$$

$$a = (D - v_m (t_2 - t_1)) / t_{1n}^2 \qquad (4.37)$$

$$t_{2n} = T_n - t_{1n} \qquad (4.38)$$

The polynomials are then written as

$$x(t) = x_i + 1/2\, a\, t^2 \qquad 0 \le t < t_{1n} \qquad (4.39)$$

$$x(t) = x(t_{1n}) + v_m (t - t_{1n}) \qquad t_{1n} \le t < t_{2n} \qquad (4.40)$$

$$x(t) = x(t_{2n}) + v_m (t - t_{2n}) - 1/2\, a\, (t - t_{2n})^2 \qquad t_{2n} \le t < T_n \qquad (4.41)$$

CASE M.2. Acceleration from Zero Initial Velocity

The second case considered by the modification procedure is the one in which the velocity of point i is zero, but the velocity of point i+1 is different from zero. In such a case, as already mentioned in section 4.1., the trajectory between the points can at most be described by one second-order and one first-order time polynomials such as

$$x(t) = x_i + 1/2\, a_m t^2 \qquad 0 \le t < t_1 \qquad (4.42)$$

$$x(t) = x(t_1) + v_m (t - t_1) \qquad t_1 \le t < T \qquad (4.43)$$

If the distance between the points is not sufficient to reach the maximum velocity, the interval is only described by the second-order polynomial.

In order to meet the new requirements, the following modifications are done

$$t_{1n} = T_n - T \qquad (4.44)$$

$$t_{2n} = t_{1n} + t_1 \qquad (4.45)$$

The polynomials are then written as

$$x(t) = x_i \qquad 0 \le t < t_{1n} \qquad (4.46)$$

$$x(t) = x_i + 1/2\, a_m (t - t_{1n})^2 \qquad t_{1n} \le t < t_{2n} \qquad (4.47)$$

$$x(t) = x(t_{2n}) + v_m (t - t_{2n})) \qquad t_{2n} \le t < T_n \qquad (4.48)$$

CASE M.3. Acceleration from Nonzero Initial Velocity

In this case both the velocity at point i and the velocity at point i+1 are different from zero. Three possibilities have to be considered

i) $| v_i | < v_m$ and $| v_{i+1} | < v_m$

ii) $| v_i | < v_m$ and $| v_{i+1} | = v_m$

iii) $| v_i | = v_m$ and $| v_{i+1} | = v_m$

a) For the pair of velocities indicated in (i) above, the trajectory between the points is described by a single second-order polynomial function

$$x(t) = x_i + 1/2\, a_m t^2 \qquad 0 \le t \le T \qquad (4.49)$$

Since the link is already moving when it is at point i+1, it is not possible to delay the motion as done in the case above in order to meet the following requirements:

$$t_{1n} = (T_n - T) / 2 \qquad (4.50)$$

$$a = 2\, v_i / t_{1n} \qquad (4.51)$$

$$t_{2n} = 2\, t_{1n} \qquad (4.52)$$

The polynomials are then written as:

$$x(t) = x_i + v_i\, t - 1/2\, a\, t^2 \qquad 0 \le t \le t_{1n} \qquad (4.53)$$

$$x(t)= x(t_{1n}) - v_i (t - t_{1n}) + 1/2\, a\, (t - t_{1n})^2 \qquad t_{1n} \le t \le t_{2n} \qquad (4.54)$$

$$x(t)= x_i + v_i (t - t_{2n}) + 1/2\, a_m (t - t_{2n})^2 \qquad t_{2n} \le t \le T_n \qquad (4.55)$$

As can be seen, the aim of the first two polynomials is to delay motion in order to meet the time requirement. In this modification, a problem that can be encountered is that the acceleration needed to meet the time requirements may be greater than the maximum acceleration of the link. In such a case, maximum acceleration is used, and the error is compensated by the third polynomial which is computed using as the initial value point i.

b) Here, the velocity of point i is less than the maximum link velocity, while maximum velocity is reached at point i+1. For the pair of velocities indicated in (ii) above, the trajectory is described by one second-order and one first-order time polynomial

$$x(t) = x_i + v_i\, t + 1/2\, a_m\, t^2 \qquad 0 \le t \le t_1 \qquad (4.56)$$

$$x(t) = x(t_1) + v_m\, (t - t_1) \qquad t_1 \le t \le T \qquad (4.57)$$

and the following arrangement is made

$$t_c = T_n - t_1 \qquad (4.58)$$

$$d = D - (v_i\, t_1 + 1/2\, a_m\, t_1^2) \qquad (4.59)$$

where t_c represents the time interval during which the acceleration must be zero and d represents the distance to cover in the interval when the velocity is constant. In this modification, two sets of polynomials are possible according to the value of d.

If $d \ge v_i\, t_c$, then

$$t_{1n} = t_c - |(d - v_i\, t_c)| / (v_m - |v_i|) \qquad (4.60)$$

$$t_{2n} = t_{1n} + t_1 \qquad (4.61)$$

The polynomials are written as

$$x(t) = x_i + v_i\, t \qquad 0 \le t \le t_{1n} \qquad (4.62)$$

$$x(t)=x(t_{1n}) + v_i\, (t - t_{1n}) + 1/2\, a_m\, (t - t_{1n})^2 \qquad t_{1n} \le t \le t_{2n} \qquad (4.63)$$

$$x(t)=x(t_{2n}) + v_m\, (t - t_{2n}) \qquad t_{2n} \le t \le T_n \qquad (4.64)$$

If $d < v_i\, t_c$, then

$$v = 2\, (d - v_i\, t_c) / t_c \qquad (4.65)$$

$$a = 2\, (v - v_i) / t_c \qquad (4.66)$$

$$t_{1n} = t_c / 2 \qquad (4.67)$$

$$t_{2n} = t_{1n} + t_c / 2 \qquad (4.68)$$

The polynomials are then written as

$$x(t) = x_i + v_i\, t - 1/2\, a\, t^2 \qquad 0 \le t \le t_{1n} \qquad (4.69)$$

$$x(t)=x(t_{1n}) + v\, (t - t_{1n}) + 1/2\, a\, (t - t_{1n})^2 \qquad t_{1n} \le t \le t_{2n} \qquad (4.70)$$

$$x(t)=x(t_{2n}) + vi\, (t - t_{2n}) + 1/2\, a_m(t - t_{2n})^2 \qquad t_{2n} \le t \le T_n \qquad (4.71)$$

c) If the velocity both at point i and point i+1 is equal to the maximum velocity, the trajectory is defined as

$$x(t) = x_i + v_m\, t \qquad 0 \le t \le T \qquad (4.72)$$

The following modification is applied to meet the requirements

$$a = 4\, (v_m\, T_n - D) / T_n^2 \qquad (4.73)$$

$$t_{1n} = T_n / 2 \qquad (4.74)$$

$$v = v_m - a\, t_{1n} \qquad (4.75)$$

In the case that the acceleration a of (4.73) results in a value greater than the maximum attainable acceleration, the maximum value is used.

The polynomials are

$$x(t) = x_i + v_m\, t - 1/2\, a\, t^2 \qquad 0 \le t \le t_{1n} \qquad (4.76)$$

$$x(t) = x(t_{1n}) + v\, (t - t_{1n}) + 1/2\, a\, (t - t_{1n})^2 \qquad t_{1n} \le t \le T_n \qquad (4.77)$$

CASE M.4. Decelerate from Arbitrary Velocity to Rest

In the case that velocity of point i is different from zero, and it is zero at point i+1, as described in section 4.1., the trajectory between the two points can at most be defined by two polynomials such as

$$x(t) = x_i + v_i\, t \qquad 0 \le t \le t_1 \qquad (4.78)$$

$$x(t) = x(t_1) + v_i (t-t_1) - 1/2\, a_m (t - t_1)^2 \qquad t_1 \leq t \leq T \qquad (4.79)$$

The requirement is met defining

$$t_{1n} = t_1 \qquad (4.80)$$

$$t_{2n} = T \qquad (4.81)$$

The polynomials are written as

$$x(t) = x_i + v_i\, t \qquad 0 \leq t \leq t_{1n} \qquad (4.82)$$

$$x(t)=x(t_{1n}) + v_i (t - t_{1n}) - 1/2\, a_m (t - t_{1n})^2 \qquad t_{1n} \leq t \leq t_{2n} \qquad (4.83)$$

$$x(t) = x_{i+1} \qquad t_{2n} \leq t \leq T_n \qquad (4.84)$$

The polynomials that have been modified to meet all the requirements can be used as joint reference trajectories in a control algorithm.

5. Control Algorithms

In this work, in order to achieve tracking of the desired trajectory of motion by the end-effector, two alternative control algorithms will be used:
i) Computed-torque control, including adaptation to the load
ii) Proportional-double-derivative control (PDD) ,including adaptation to the load

5.1. Computed-Torque Method

In this section two kind of controllers will be considered:
i) Computed-torque control algorithm for known robot and load parameters
ii) Adaptive computed-torque control algorithm.

5.1.1. Computed-Torque Controller with Known Parameters

The aim of the control is to make each joint track the desired trajectory, that is, the control must force the error

$$e = \theta - \theta_d \qquad (5.1)$$

to zero as t approaches infinity, where e, θ, θ_d are n dimensional vectors. The vectors represents:

θ_d : desired joint positions

θ : actual joint positions

e : error between desired and actual joint positions.

The dynamic model, as computed in (2.9) is nonlinear and computationally complex. The controller for trajectory control will consist of two stages [2]:
i) A primary nonlinear controller for known M and I, which includes the inverse dynamics of the manipulator :

$$U_1(t) = E^{-1} (J \theta_d'' - J_c \theta'^2 - G) \quad (5.2)$$

ii) A secondary controller that compensates for undesirable deviations of the motion from the desired trajectory.One of most commonly used secondary controllers is the proportional-derivative controller:

$$U_2(t) = E^{-1} J (B (\theta' - \theta_d') - K (\theta - \theta_d)) \quad (5.3)$$

K and B are n x n matrices to be determined. The total control strategy is evaluated as [2]:

$$U(t) = U_1(t) - U_2(t) \quad (5.4)$$

Substitution into (2.5) yields:

$$J [(\theta'' - \theta_d'') - B (\theta' - \theta_d') - K (\theta - \theta_d)] = 0 \quad (5.5)$$

or ,in terms of the error ,

$$J (e'' - \tilde{B} e' - K e) = 0. \quad (5.6)$$

By suitable choice of B and K as negative-definite matrices the error equation will be stable.

5.1.2. Adaptive Computed-Torque Method

The adaptive control algorithm is used to achieve tracking of the desired trajectory by the end-effector with an unknown load attached to the outermost link.

The load parameters M and I are unknown, but based on the assumptions of load size and configuration ,I is computed from M and r. Therefore, the only parameter to be estimated is the mass M of the load.

5.1.2.1 Estimation of the Unknown Load Mass

Suppose that the initial state of the manipulator is on the desired trajectory and that the computed-torque algorithm with M=0 is used. The unknown load not only increases the error in q_n of the outermost link but also increases the errors of all other links. From a mechanical point of view, the geometric error in the vertical direction

$$\Delta y = -\sum_{i=1}^{n} l_i \left(\cos\theta_i - \cos\theta_{d\,i} \right) \tag{5.7}$$

in some sense, represents the potential energy difference between the actual motion and the desired motion. This error can be used to estimate the mass of the load [2]. Let m(t) be an estimate of the unknown mass M. The estimate mass m(t) must have memory such that zero error does not make m(t) zero. For $t \le T$, where T is a large positive number, m(t) must satisfy

$$| m(t) - M | < \varepsilon_m \tag{5.8}$$

where ε_m is a small positive number.

The estimation algorithm used here is

$$m(t) = -K_m \int_0^t \sum_{i=1}^{n} l_i \left(\cos\theta_i(z) - \cos\theta_{d\,i}(z)\right) dz \tag{5.9}$$

where K_m is a suitably chosen positive constant. In actual applications, K_m must be selected such that the position, velocity, and mass estimate error converge roughly at the same rate. Alternatively,

$$m'(t) = -K_m \sum_{i=1}^{n} l_i \left(\cos\theta_i - \cos\theta_{d\,i}\right) \tag{5.10}$$

5.1.2.2. Adaptive Controller

The adaptive control strategy is similar to the strategy discussed in section 5.1.1. except that the value of the load mass used in the control equations is continuously updated by the adaptive mechanism.

The adaptive control strategy is defined by:

$$U(t) = E^{-1} (-J_{co}\, \theta'^2 - G_o + J_o\, \theta_d'') + E^{-1}\, m(t)\, (-J_{c1}\, \theta'^2 - G_1 + J_1\, \theta_d'')$$
$$- E^{-1} ((J_o + m(t)\, J_1) \left(B (\theta' - \theta_d') + K (\theta - \theta_d)\right) \tag{5.11}$$

After adaptation

$$\theta_i(t) = \theta_{di}(t) \tag{5.12}$$

$$\theta_i'(t) = \theta_{di}'(t) \tag{5.13}$$

$$m'(t) = 0 \tag{5.14}$$

equation (5.11) reduces to equation (5.4).

Substitution of equation (5.11) into equation (2.9) gives :

$$(J_o + J_1\, m(t))\, ((\theta'' - \theta_d'') - B\, (\theta' - \theta'_d) - K\, (\theta - \theta_d)) =$$
$$(M - m(t))\, (-J_1\, \theta'' + J_{c1}\, \theta'^2 + G_1) \qquad (5.15)$$

If the estimate m(t) is equal to the actual M, then the right side of (5.15) is zero and the equation becomes similar to equation (5.6).

5.2. Proportional-Double-Derivative Method

This control algorithm consists of a multivariable proportional-double-derivative (PDD) feedforward controller to ensure trajectory tracking and a multivariable proportional-derivative (PD) feedback controller to achieve pole placement [5].

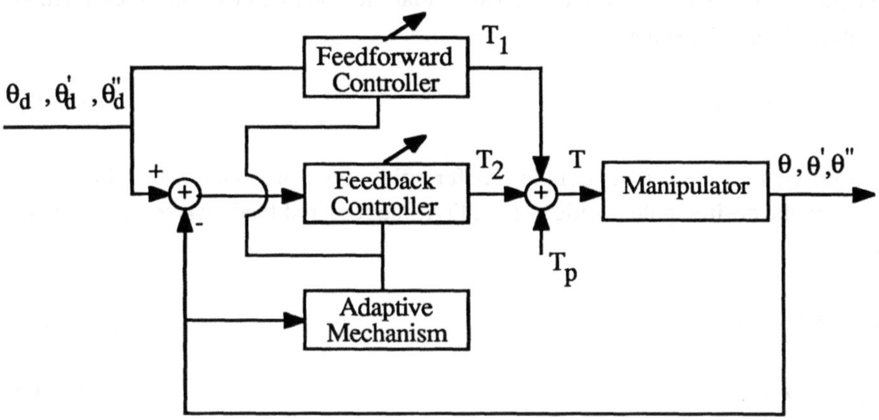

Figure 1. Block diagram of the PDD control algorithm with load estimation

5.2.1 Feedforward Controller

The feedforward controller is the minimal order inverse of the linearized robot model around an operating point P $(T_p, \theta_p, \theta_p')$ [5].

$$T_1(t) = A\, \theta_d'' + B\, \theta_d' + C\, \theta_d \qquad (5.16)$$

where

$$A = [\, J\,]\, p \qquad (5.17a)$$

$$B = -\frac{\partial\,[J_c\,\dot{\theta}^2]\,p}{\partial\dot{\theta}} \tag{5.17b}$$

$$C = -\frac{\partial\,[J_c\,\dot{\theta}^2 + G]\,p}{\partial\theta} \tag{5.17c}$$

5.2.2. Feedback Controller

The state feedback controller is used to place the poles of the linear model at any desired location in the complex plane.

$$T_2(t) = K_p\,(\theta_d - \theta) + K_v\,(\theta_d' - \theta') \tag{5.18}$$

The gain matrices K_p, K_v are chosen such that the tracking error vanishes asymptotically and the error differential equation

$$e'' + D_1\,e' + D_2\,e = 0 \tag{5.19}$$

is decomposed into a set of uncoupled differential equations. Given the linearized model {A,B,C} and the desired pole positions $\{D_1, D_2\}$, the required feedback controller gains can be computed as

$$K_p = A\,D_1 - B \tag{5.20a}$$

$$K_v = A\,D_2 - C \tag{5.20b}$$

5.2.3 Total Control Law

The total control law can be written as

$$T(t) = T_p + T_1(t) + T_2(t) \tag{5.21}$$

where

$$T_p = T_l - C\,\theta_d - B\,\theta_d' - A\,\theta_d'' \tag{5.22}$$

and $\{T_l, \theta_d, \theta_d', \theta_d''\}$ are the operating point torque, desired position, velocity and acceleration vectors, respectively.

During gross motion of the manipulator, the operating point P changes with time, and the controllers are updated to cope with the variations in the model matrices {A,B,C}.

5.2.4 Proportional Double-Derivative Method with Mass Adaptation

The proportional derivative feedback with feedforward control can also be used to achieve tracking of the desired trajectory by the end effector with an unknown load attached to the outermost link.

The estimation of the unknown load mass is performed using the estimation algorithm explained in section 5.1.2.1.

The control law used is (5.21). The controller gains A, B, C, K_p, K_v are used in this equation are updated using the equations (5.17a), (5.17b), (5.17c), (5.20a), (5.20b) respectively. In updating the controller gains, the current load mass estimate is used.

6. Results and Conclusions

In this work quasi-time optimal trajectories for each joint of a robot arm are generated on the basis of the desired end-effector trajectory along with various time specifications. The developed algorithm generates the joint trajectories for two-link and three planar-link robot arms. The polynomials computed initially for each joint are exact. After the polynomials are modified to meet time requirements, small variations may occur, especially when the link is moving at maximum speed. The computed polynomials are used as reference trajectories for each joint of the robotic arm in the control algorithms.

6.1. Simulation Example

In this simulation example, joint reference trajectories are calculated for a three-link planar robot arm that has to visit the following points:

1.(50;0) 2.(40;10) 3.(50;20) 4.(40;25) 5.(30;10)

The motion completion time is specified as 6.5 seconds.

The generated polynomials are used as joint reference trajectories for a three link robot arm carrying a load of 5 Kg. The mass of the load is assumed to be unknown.

Two different control algorithms are used to achieve gross motion control and load mass estimation:

i) PDD control with load estimation.

The simulation results for this control strategy are shown from Figure 2. through Figure 13.

ii) Computed-torque control with load estimation.

The simulation results for this control strategy are shown from Figure 14. through Figure 25.

In both simulations, the initial estimate of the mass is zero. The estimates of the load mass for both controllers are shown in Figure 26. and 27. respectively.

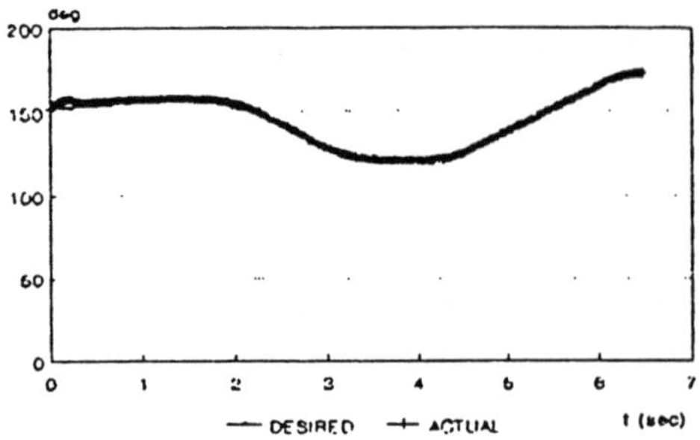

Figure 2. Position of joint 1
PDD Control & Load Estimation

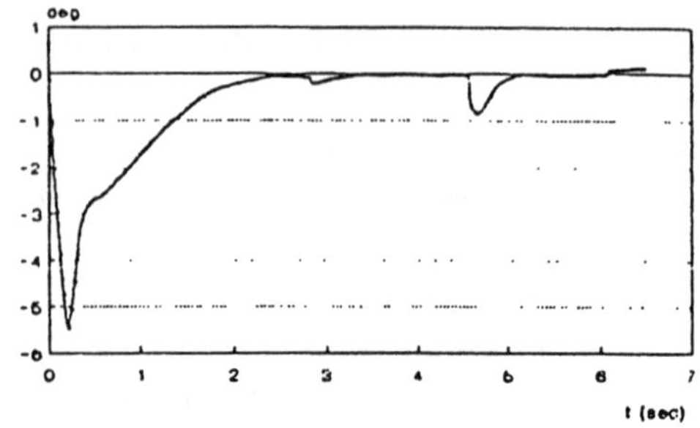

Figure 3. Position error of joint 1
PDD Control & Load Estimation

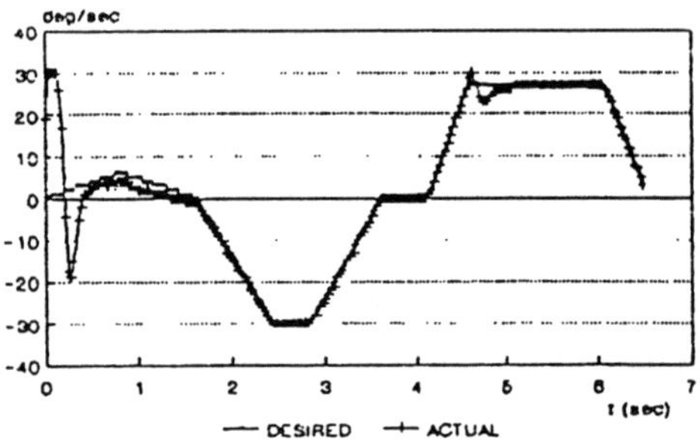

Figure 4. Velocity of joint 1
PDD Control & Load Estimation

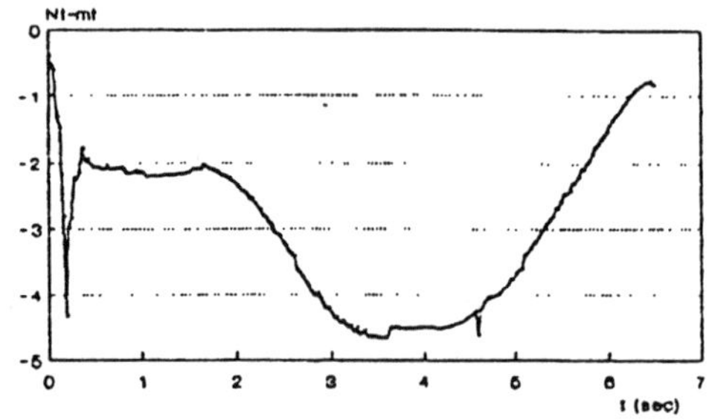

Figure 5. Torque applied to joint 1
PDD Control & Load Estimation

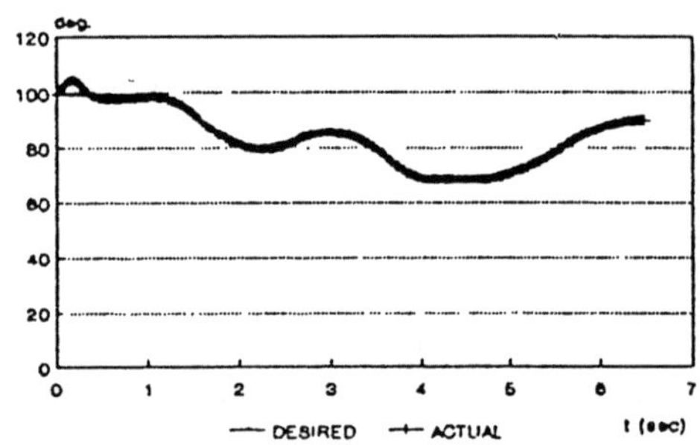

Figure 6. Position of joint 2
PDD Control & Load Estimation

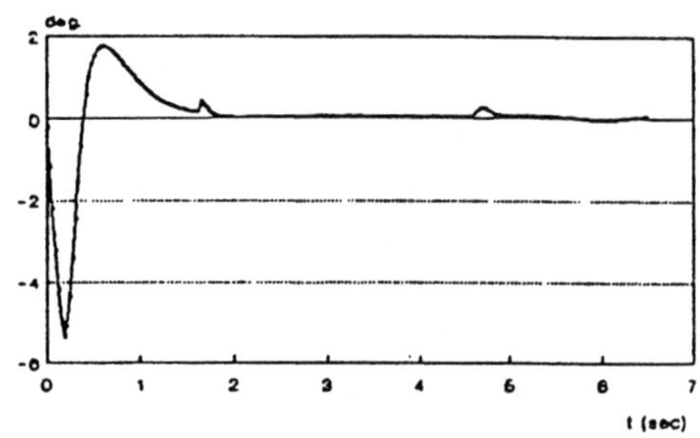

Figure 7. Position error of joint 2
PDD Control & Load Estimation

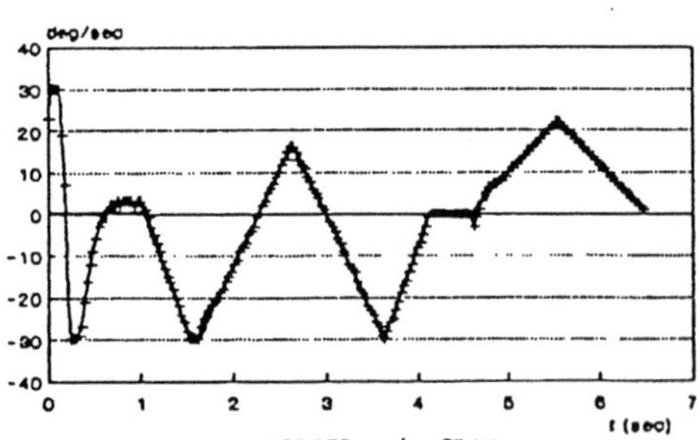

Figure 8. Velocity of joint 2
PDD Control & Load Estimation

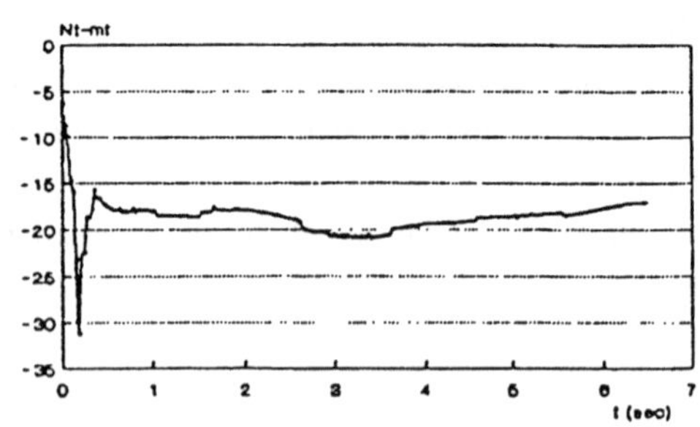

Figure 9. Torque applied to joint 2
PDD Control & Load Estimation

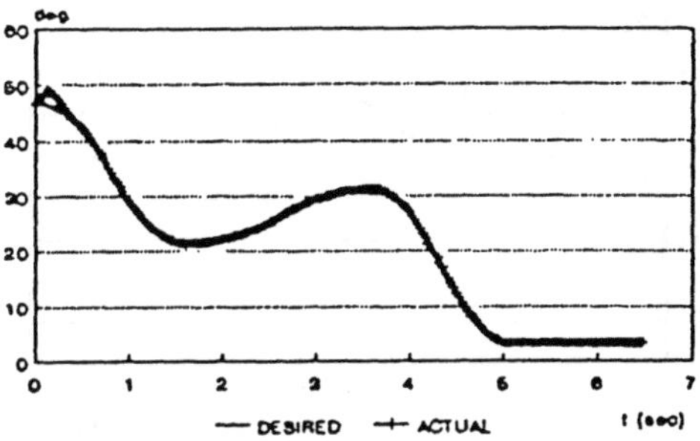

Figure 10. Position of joint 3
PDD Control & Load Estimation

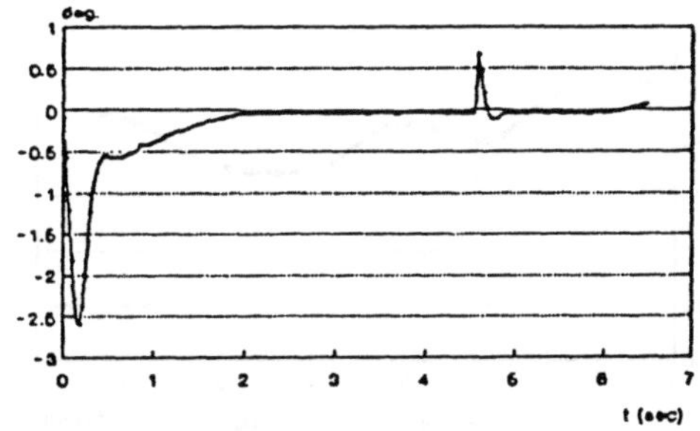

Figure 11. Position error of joint 3
PDD Control & Load Estimation

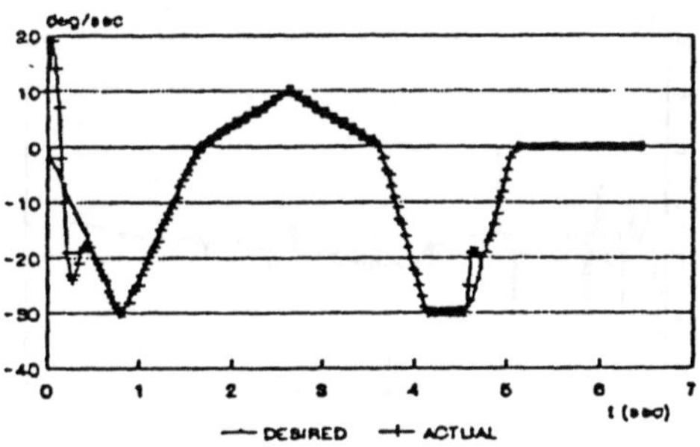

Figure 12. Velocity of joint 3
PDD Control & Load Estimation

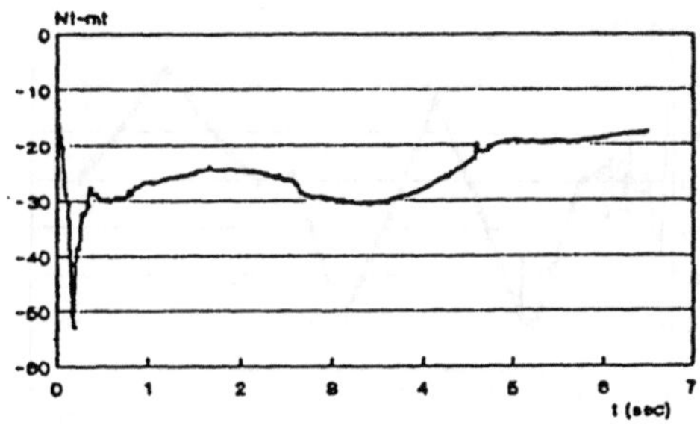

Figure 13. Torque applied to joint 3
PDD Control & Load Estimation

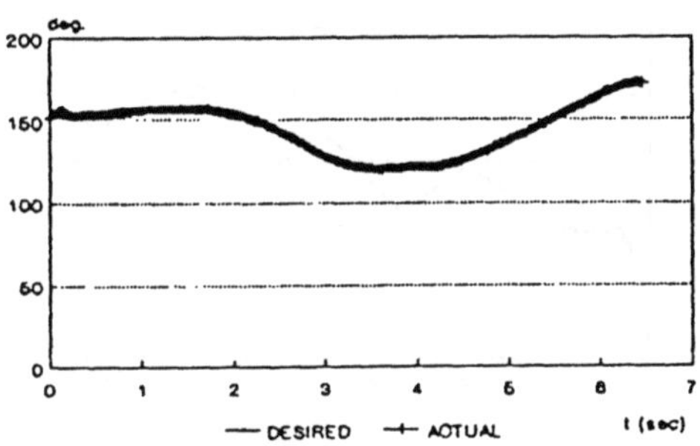

Figure 14. Position of joint 1
Computed Torque & Load Estimation

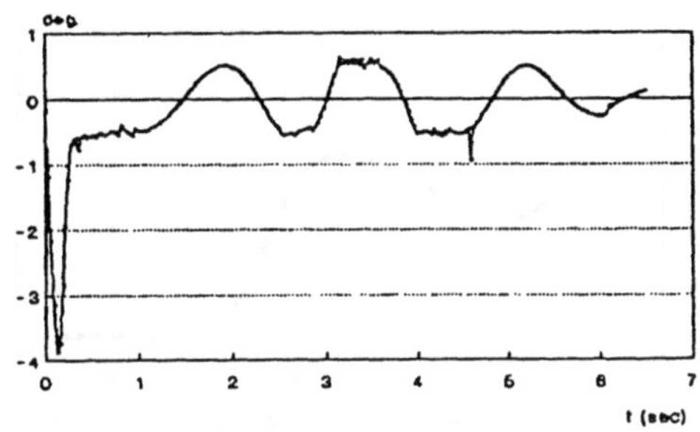

Figure 15. Position error of joint 1
Computed Torque & Load Estimation

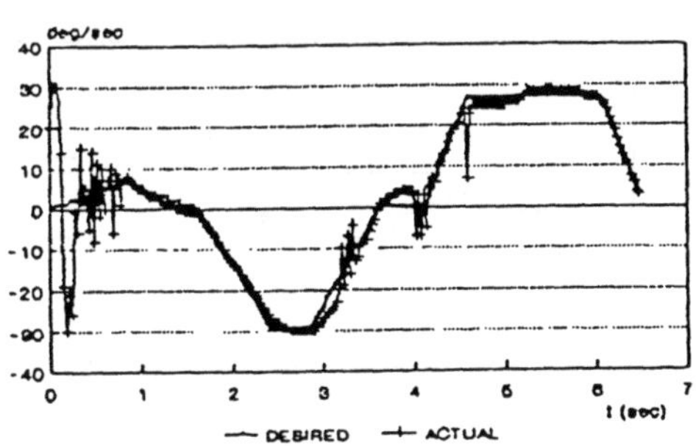

Figure 16. Velocity of joint 1
Computed Torque & Load Estimation

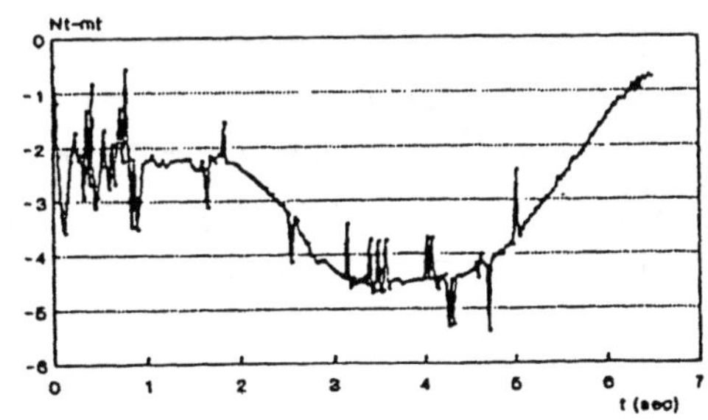

Figure 17. Torque applied to joint 1
Computed Torque & Load Estimation

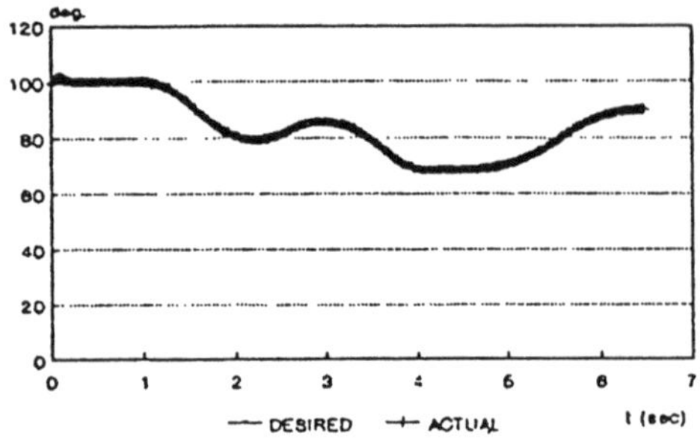

Figure 18. Position of joint 2
Computed Torque & Load Estimation

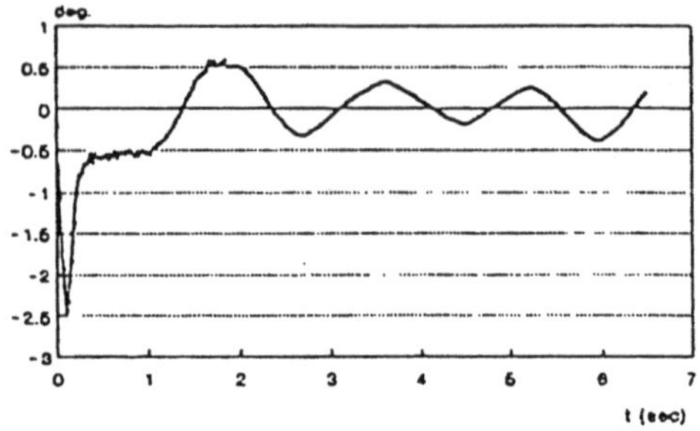

Figure 19. Position error of joint 2
Computed Torque & Load Estimation

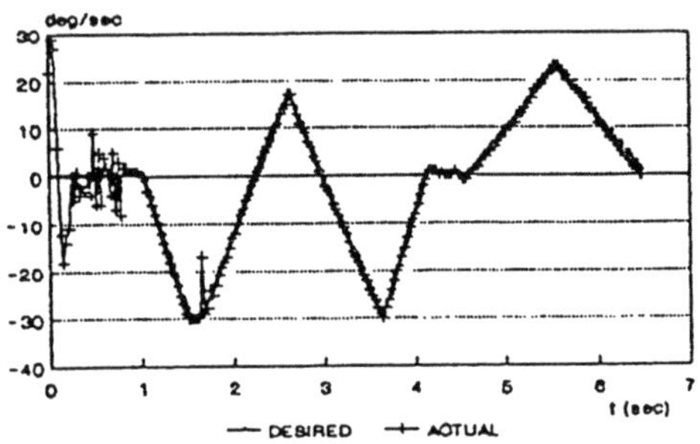

Figure 20. Velocity of joint 2
Computed Torque & Load Estimation

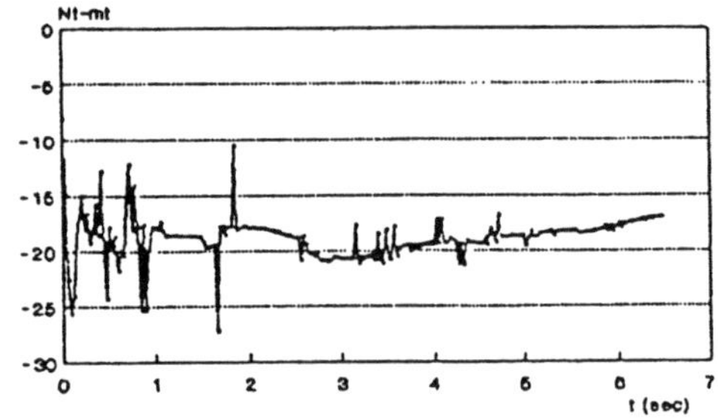

Figure 21. Torque applied to joint 2
Computed Torque & Load Estimation

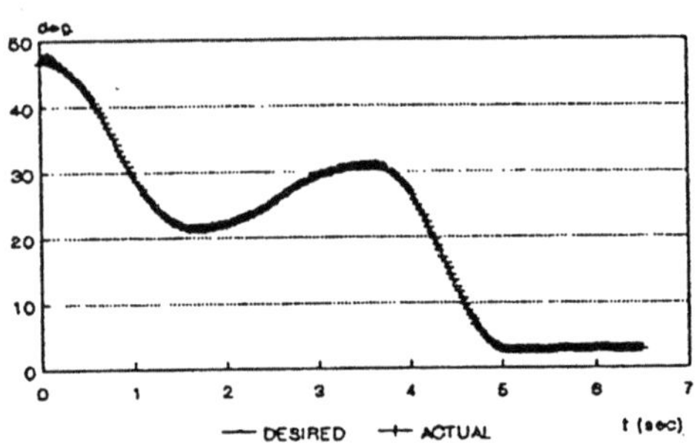

Figure 22. Position of joint 3
Computed Torque & Load Estimation

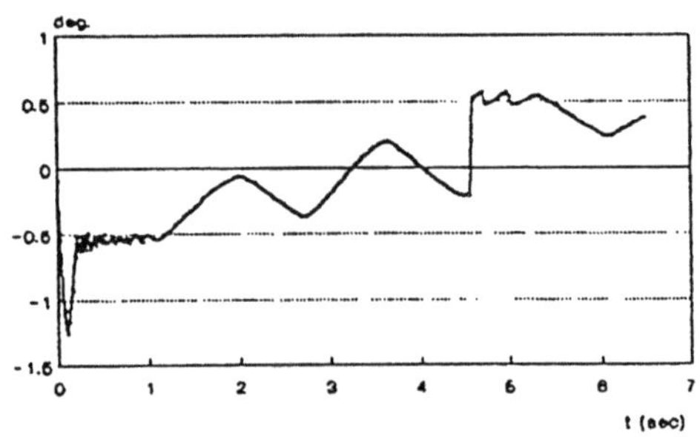

Figure 23. Position error of joint 3
Computed Torque & Load Estimation

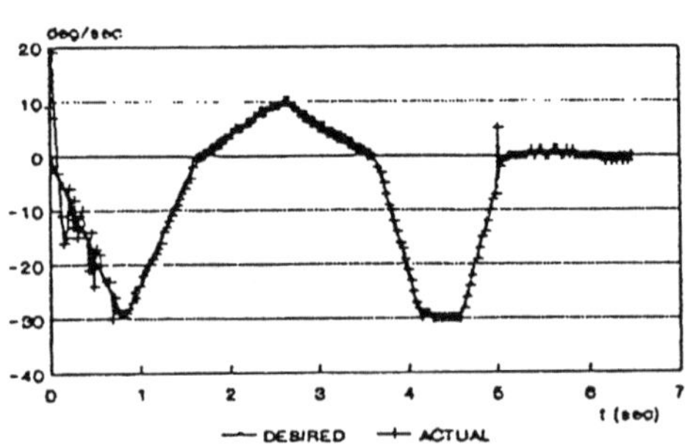

Figure 24. Velocity of joint 3
Computed Torque & Load Estimation

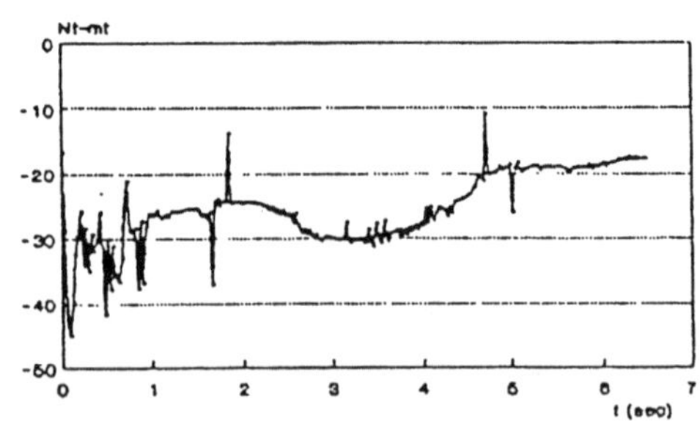

Figure 25. Torque applied to joint 3
Computed Torque & Load Estimation

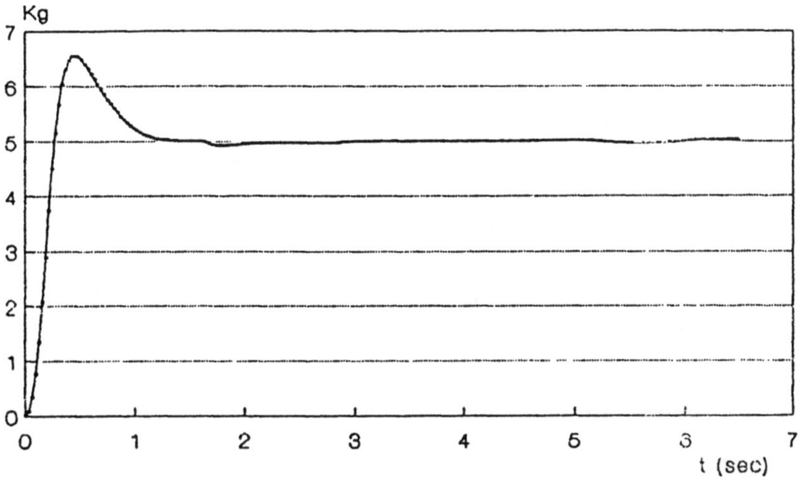

Figure 26. Mass estimate
PDD Control & Load Estimation

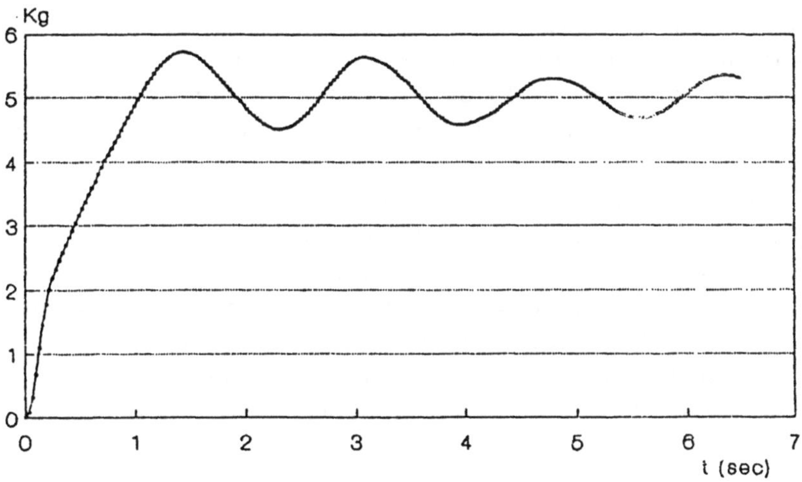

Figure 27. Mass estimate
Computed Torque & Load Estimation

In the simulations the sampling time is 0.01 sec, and the parameters of the planar robot arm are given in the table below:

Link i	Mass m_i (kg)	Inertia I_i (kg m^2)	Length l_i (cm)	Center of Mass k_i (cm)
1	1.0	0.001	10	5
2	1.0	0.005	25	15
3	1.0	0.005	25	15

6.2. Conclusions

The computed-torque method gives satisfactory results ,even when the initial conditions of the manipulator are not known exactly. Since in this technique the nonlinearities of the robot dynamics are cancelled out using a nonlinear feedback controller, the controller is of the same order of complexity as the robot dynamics itself.

The second control algorithm used is the proportional-derivative feedback controller with feedforward control (PDD). In this method, the controller gains are computed from simple explicit expressions given the linearized model of the robot dynamics. The performance in terms of position error depends on how often the controller gains are updated during gross motion control of the robot. In the simulations done using this controller, the gains have been updated every ten sampling instants,and the results obtained are satisfactory.

Both mentioned control algorithms are model-based control systems and therefore require the dynamic model and the parameter values of the robot in order to generate the control action.

In the computed-torque control algorithm with load adaptation,load mass estimation is performed during gross motion control of the robot arm. The algorithm requires the exact knowledge of the initial conditions for efficient mass estimation and gross motion control.

In the PDD method with load adaptation,the load mass estimation is performed during gross motion control of the robot arm and by updating the gains every five sampling instants. Satisfactory results are obtained both in motion control and mass estimation.

Comparing the two algorithms which perform both gross motion control and mass estimation, the PDD control with load-adaptation provides a better mass estimation compared to the load-adaptive computed torque. Although the modifications performed in the load-adaptive computed torque algorithm have improved its performance in tracking the desired position and velocity profiles, the performance in estimating the mass has not reached the performance level of the load-adaptive PDD method.

The original proportional-derivative feedback controller with feedforward control (PDD) and load adaptation has shown a high level of performance not only in tracking the desired position and velocity profiles, but in estimating the load mass. In this method, the unknown mass value is estimated in a short interval of time, and after the true value has been found, the variations that occur in the estimate value are very small.

Acknowledgement

This research has been supported by the Research Fund of Boğaziçi University.

Bibliography

[1] Brown, D. B. , H. Hemami and Y. Istefanopulos,"On-Line Exploration of an Unknown Surface by a Robotic Probe", Journal of Robotic Systems 6 (5),1989, pp 521-543.

[2] Han, J. Y., H. Hemami,S. Yurkovic," Nonlinear Control of an N-Link Robot with Unknown Load ", The International Journal of Robotics Research, Vol 6, No.3, Fall 1987, pp 71-85.

[3] Hemami, H., V. C. Jaswa and R.B. McGhee, " Some Alternative Formulations of Manipulator Dynamics for Computer Simulation Studies," in Proc. 13th Allerton Conf. on Circuit and System Theory, Univ. of Illinois, Urbana, Oct. 1975.

[4] Koivo, A.J., Fundamentals for Control of Robotic Manipulator, New York: John Wiley & Sons,Inc.,1989.

[5] Seraji, H. ," An Approach to Multivariable Control of Manipulators", Transactions of the ASME Vol 109, June 1987, pp 146-154.

APPENDIX A

VECTORS AND MATRICES USED IN THE ROBOT DYNAMIC EQUATIONS

The vectors and matrices used in robot dynamic equation (2.1) are defined as follows [2]:

$$\dot{\theta}^2 = \left[\dot{\theta}_1^2 \ \dot{\theta}_2^2 \ldots \dot{\theta}_n^2\right]^T \tag{A.1}$$

$$G_0 = \begin{bmatrix} m_1 k_1 g \sin\theta_1 \\ (m_1 l_2 + m_2 k_2)\, g \sin\theta_2 \\ \vdots \\ \left(\left(\sum_{i=1}^{n-1} m_i\right) l_n + m_n k_n\right) g \sin\theta_n \end{bmatrix} \tag{A.2}$$

$$E = \begin{bmatrix} 1 & 0 & 0 & 0 & \cdots & 0 \\ -1 & 1 & 0 & 0 & \cdots & 0 \\ 0 & -1 & 1 & 0 & \cdots & 0 \\ \vdots & \vdots & \vdots & \vdots & 1 & \vdots \\ 0 & 0 & 0 & 0 & -1 & 1 \end{bmatrix} \tag{A.3}$$

$$E^{-1} = \begin{bmatrix} 1 & 0 & 0 & 0 & \cdots & 0 \\ 1 & 1 & 0 & 0 & \cdots & 0 \\ \vdots & \vdots & & & & \vdots \\ 1 & 1 & 1 & 1 & \cdots & 1 \end{bmatrix} \tag{A.4}$$

$$J_0 = \begin{bmatrix} a_{11} & a_{12}\cos(\theta_1-\theta_2) & \cdots & a_{1n}\cos(\theta_1-\theta_n) \\ a_{21}\cos(\theta_1-\theta_2) & a_{22} & \cdots & a_{2n}\cos(\theta_2-\theta_n) \\ \vdots & \vdots & & \\ a_{n1}\cos(\theta_n-\theta_1) & a_{n2}\cos(\theta_n-\theta_2) & \cdots & a_{nn} \end{bmatrix} \tag{A.5}$$

where

$$a_{ii} = I_i + \sum_{j=1}^{i-1} m_j l_i^2 + m_i k_i^2 \qquad i = 1, 2, \ldots, n \tag{A.6}$$

$$a_{ij} = (m_i k_i + l_i \sum_{t=1}^{i-1} m_t) l_j \qquad i < j, \quad i, j = 1,2, \ldots, n \tag{A.7}$$

$$a_{ji} = a_{ij} \qquad i > j \tag{A.8}$$

$$J_{co} = \begin{bmatrix} 0 & -a_{12} \sin(\theta_1 - \theta_2) & \cdots & -a_{1n} \sin(\theta_1 - \theta_n) \\ a_{21} \sin(\theta_1 - \theta_2) & 0 & \cdots & -a_{2n} \sin(\theta_2 - \theta_n) \\ \vdots & \vdots & \ddots & \vdots \\ a_{n1} \sin(\theta_1 - \theta_n) & a_{n2} \sin(\theta_2 - \theta_n) & \cdots & 0 \end{bmatrix} \tag{A.9}$$

where a_{ij} is defined as in (A8). Obviously, J_o is a symmetric matrix, and J_{co} is an asymmetric matrix.

The matrices used in equations (2.6) through (2.9) are defined as follows:

$$J-J_o = \begin{bmatrix} M\,(l_1^2 + \frac{1}{2} r^2) & M\, l_1 l_2 \cos(\theta_1-\theta_2) & \cdots & M\, l_1 l_n \cos(\theta_1-\theta_n) \\ M\, l_1 l_2 \cos(\theta_1-\theta_2) & M\; l_2^2 & \cdots & M\, l_2 l_n \cos(\theta_2-\theta_n) \\ \vdots & \vdots & \ddots & \vdots \\ M\, l_1 l_n \cos(\theta_1-\theta_n) & M\, l_2 l_n \cos(\theta_2-\theta_n) & \cdots & M\; l_n^2 \end{bmatrix} \tag{A.10}$$

$$J - J_o = M\, J_1 \tag{A.11}$$

where

$$J_1 = \begin{bmatrix} (l_1^2 + \frac{1}{2} r^2) & l_1 l_2 \cos(\theta_1-\theta_2) & \cdots & l_1 l_n \cos(\theta_1-\theta_n) \\ l_1 l_2 \cos(\theta_1-\theta_2) & l_2^2 & \cdots & l_2 l_n \cos(\theta_2-\theta_n) \\ \vdots & \vdots & \ddots & \vdots \\ l_1 l_n \cos(\theta_1-\theta_n) & l_2 l_n \cos(\theta_2-\theta_n) & \cdots & l_n^2 \end{bmatrix} \tag{A.12}$$

$$J_c - J_{c0} = \begin{bmatrix} 0 & -Ml_1l_2\sin(\theta_1-\theta_2) & \cdots & -Ml_1l_n\sin(\theta_1-\theta_n) \\ Ml_1l_2\sin(\theta_1-\theta_2) & 0 & \cdots & -Ml_2l_n\sin(\theta_2-\theta_n) \\ \vdots & \vdots & \cdots & \vdots \\ Ml_1l_n\sin(\theta_1-\theta_n) & Ml_2l_n\sin(\theta_2-\theta_n) & \cdots & 0 \end{bmatrix} \quad (A.13)$$

$$J_c - J_{co} = M\, J_{c1} \quad (A.14)$$

where

$$J_{c1} = \begin{bmatrix} 0 & -l_1l_2\sin(\theta_1-\theta_2) & \cdots & -l_1l_n\sin(\theta_1-\theta_n) \\ l_1l_2\sin(\theta_1-\theta_2) & 0 & \cdots & -l_2l_n\sin(\theta_2-\theta_n) \\ \vdots & \vdots & & \vdots \\ l_1l_n\sin(\theta_1-\theta_n) & l_2l_n\sin(\theta_2-\theta_n) & \cdots & 0 \end{bmatrix} \quad (A.15)$$

$$G - G_0 = \begin{bmatrix} M\, l_1\, g \sin\theta_1 \\ M\, l_2\, g \sin\theta_2 \\ \vdots \\ M\, l_n\, g \sin\theta_n \end{bmatrix} \quad (A.16)$$

$$G - G_o = M\, G_1 \quad (A.17)$$

where

$$G_1 = \begin{bmatrix} l_1\, g \sin\theta_1 \\ l_2\, g \sin\theta_2 \\ \vdots \\ l_n\, g \sin\theta_n \end{bmatrix} \quad (A.18)$$

J_1 is a symmetric matrix and J_{c1} in an asymmetric matrix.

A Robust Approach for the Control of Robotic Manipulators: Predictive Control

Okyay Kaynak
Michigan Technological University
Houghton, Michigan 49931
(on leave from Bogazici University,
Bebek, 80815 Istanbul, TURKEY)

Abstract: In this paper a novel scheme, named Generalized Predictive Control (GPC) is formulated for trajectory tracking control of a two degree-of-freedom manipulator. The simulation results are presented and compared with those obtained by two other approaches, namely the conventional computed-torque method and a pole-placement algorithm. It is seen that a GPC algorithm results in a much better performance, resulting in an accurate tracking of the reference trajectory in spite of the discrepancies simulated between the robot and its model. The prediction scheme employed adds a high degree of intelligence to the system, enabling it to cope with the system-model mismatches and the nonlinearities. The paper concludes with the presentation of some results obtained from a DSP based, one degree-of-freedom experimental set up.

Keywords: Robotics/Control of Manipulators/Trajectory Control/ Predictive Control

1. Introduction

Servo motors are the most widely used electrical machines in industrial control applications. Over the past hundred years, the theory and the practice of their control have been well developed and matured and, with the application of modern control algorithms, a high degree of accuracy can now be ensured and various engineering specifications can be met. However, in the case of robotic manipulators, the situation is somewhat different. The highly nonlinear and coupled nature of the equations describing the dynamics of a robotic manipulator makes it a difficult candidate for accurate trajectory control. The difficulties are more pronounced when a fast motion is required. To overcome the problem, many different schemes have been proposed in the literature. Some of these, generally referred to as "computed-torque technique", use a nonlinear model derived on the basis of Newtonian mechanics to compute the joint-input generalized forces [14,17] while some others resort to nonlinear compensation schemes to eliminate interactions and propose nonlinear control laws [5,15]. These schemes, like many others, necessitate high computational power and an accurate model of the manipulators. With the recent developments in ULSI (Ultra Large Scale Integration) technology, the former does not pose as much of a problem as before, but the latter problem is not easy to alleviate. Even when an accurate model is available, there very often are

NATO ASI Series, Vol. F 71
Expert Systems and Robotics
Edited by T. Jordanides and B. Torby

uncertainties in the load that the gripper carries. Therefore, the recent tendency in the control of robotic manipulators has been the use of techniques that are insensitive to parameter variations. The use of adaptive techniques [2,4,12] or approaches based on variable structure systems theory [10,16] are examples of this among the many studies that have appeared in the literature. Another approach is to treat all the uncertainties and the nonlinearities as a disturbance and to use an observer for it so as to enable its compensation in a feedforward manner in an acceleration control scheme [6,13].

The main features of a controller for an industrial manipulator should be as following:

1. Desired trajectory should be tracked smoothly and in a fast manner (servo behavior).
2. The manipulator is to be kept at its set point in steady-state (regulatory behavior).
3. Design of the controller should need minimal knowledge of the manipulator and not be effected by changes in the conditions (robustness).
4. Controller algorithm should be computationaly fast enough to be applicable in real-time.

A control scheme, which has recently been widely reported in the literature [3], is a likely candidate to meet the above specifications. It is based on a discrete-time prediction scheme and therefore named Generalized Predictive Control (GPC). Some results on the application of a version of this approach to an actual industrial manipulator have already been reported [7]. In its formulation for trajectory control, a CARIMA (Controlled Auto-Regressive Integrated Moving Average) model is assumed for each joint of the manipulator, instead of the highly coupled and nonlinear dynamic equations. A prediction scheme based on such a model generates a set of predictive outputs. The difference between the model output and the actual output is the prediction error, which is a direct outcome of the unmeasurable disturbances and system-model mismatches. This error is used to improve the prediction.

The GPC algorithm assumes that the future set point sequence is known and minimizes a quadratic function of the future errors. In doing this, a control horizon is chosen and the increments in control further to this are assumed to be zero. In the implementation of the algorithm, a receding horizon approach is used, i.e., the first element of the future controls obtained as described is applied to the servo motor of a joint, and the process is repeated at the next sampling interval.

In this paper, the GPC approach described above is formulated for trajectory tracking control of a two degree-of-freedom manipulator. The simulation results are presented and compared with those obtained by two other approaches, namely the conventional computed-torque method and a pole-placement algorithm. It is seen that the GPC algorithm results in a much better performance, resulting in an accurate tracking of the reference trajectory in spite of the discrepancies simulated between the robot and its model. The prediction scheme employed adds a high degree of intelligence to the system, enabling it to cope with the system-model mismatches and the nonlinearities. The paper concludes with the presentation of some results obtained from a DSP based, one degree-of-freedom experimental set up. The practicality of the approach is thus verified, the system being capable of tracking a given trajectory intelligently in spite of the changes in the system parameters.

2. CARIMA Model for Joint Motion

The dynamics of an N joint manipulator is modelled by the well known nonlinear matrix differential equation.

$$\tau = M(\theta)\ddot{\theta} + V(\theta,\dot{\theta}) \tag{1}$$

where θ, $\dot{\theta}$, *and* $\ddot{\theta}$ are N-dimensional vectors of joint positions, velocities, and accelerations respectively, τ is the N-dimensional vector of applied joint torques or forces, $M(\theta)$ is the positive-definite (NxN) inertia matrix including the acceleration related coefficients of the joints, and finally $V(\theta,\dot{\theta})$ is an N-dimensional vector signifying Coriolis, centrifugal, Coulomb, viscous-friction torques or forces, and the gravitational terms. It can be seen that the model is very complex and has a highly cross-coupled and nonlinear nature. In order to compensate for these effects, let us use a compensator of the following form:

$$\tau = \hat{M}(\theta)\,u + \hat{V}(\theta,\dot{\theta}) \tag{2}$$

where the caret signifies the estimated values and u is the input vector to the compensated system. The joint dynamics will then be described by

$$\ddot{\theta} = u - [M(\theta)]^{-1}([M(\theta) - \hat{M}(\theta)]\ddot{\theta} + V(\theta,\dot{\theta}) - \hat{V}(\theta,\dot{\theta})) \tag{3}$$

Let us assume that the compensated system described by Eqn.(3) can be modelled by N difference equations of the following form.

$$A_i(q^{-1})\,y_i(t) = B_i(q^{-1})\,u_i(t-1) + x_i(t)\,; \quad i = 1,2,\ldots N \tag{4}$$

where $u_i(t)$ and $y_i(t)$ are the system input and output at the sampling instant t and $A_i(q^{-1})$ and $B_i(q^{-1})$ are polynomials of degrees n and m respectively in the backward shift operator q^{-1}, $A_i(q^{-1})$ being monic; i.e.,

$$\begin{aligned} A_i(q^{-1}) &= 1 + a_{i1}q^{-1} + a_{i2}q^{-2} + \ldots + a_{in}q^{-n} \\ B_i(q^{-1}) &= b_{io} + b_{i1}q^{-1} + b_{i2}q^{-2} + \ldots + b_{im}q^{-m} \end{aligned} \tag{5}$$

The signal $x_i(t)$ is a disturbance acting on the i<u>th</u> subsystem, signifying the errors introduced by linearization, discretization and decoupling as well as the effects of the system-model mismatches and other environmental stochastic effects. It is generally assumed to be of the form

$$x_i(t) = C_i(q^{-1})\xi_i(t) \tag{6}$$

where $C_i(q^{-1})$ is a further monic polynomial in the backward shift operator q^{-1}. Equations (4) and (6) thus result in a CARMA (Controlled Auto-Regressive Moving Average) model. Although in self-tuning regulator theory such a model is generally assumed, in many industrial applications the disturbances are not stationary and a more realistic model for the disturbance is [3]

$$x_i(t) = C_i(q^{-1})\xi_i(t)/\Delta \tag{7}$$

where Δ is the differencing operator, i.e., $\Delta=1-q^{-1}$. Coupling Eqns.(4) and (7) results in a CARIMA model. In what follows, the subscripts and the arguments will be dropped for clarity so that each joint of the compensated manipulator can be represented by a difference equation of the following form.

$$y(t+1) = q(1-A)y(t) + Bu(t) + C\xi(t+1)/\Delta \tag{8}$$

2.1. Prediction Scheme

The GPC and the pole-placement control algorithms that will be described in the subsequent chapters necessitate a prediction of the output variable of the joint at the future sampling instants. For the formulation of an optimal predictor, the approach proposed in [3] can be followed. Below, however, a simpler approach is described which results in the same predicted values.

With a slight abuse of the terminology, let us name $\hat{y}(t+1)$ obtained from

$$\hat{y}(t+1)=q(1-A)y(t) + Bu(t) \tag{9}$$

as the estimation of y(t+1). For the prediction of y(t+1), denoted by $y^*(t+1)$, let us complement the estimated value by the estimation error of the previous sampling instant, i.e.,

$$y^*(t+1) = \hat{y}(t+1) + \tilde{y}(t) \tag{10}$$

where

$$\tilde{y}(t) = y(t) - \hat{y}(t) \tag{11}$$

Similarly, a j-step-ahead predictor is given by

$$y^*(t+j) = \hat{y}(t+j) + \tilde{y}(t+j-1) \tag{12.a}$$

$$= q(1-A)y^*(t+j-1)+Bu(t+j-1)+y^*(t+j-1)-q(1-A)y^*(t+j-2)-Bu(t+j-2) \tag{12.b}$$

where, as indicated in Eqn.(12.b), the predicted values are used when measured values are not available, i.e., $y(t+i)=y^*(t+i)$ for $i>0$.

To show that the predictor of Eqn.(12.a), arrived in such an ad hoc manner, is the same as the optimal predictor described by Clarke et al [3], consider Eqn. (10):

$$\begin{aligned} y^*(t+1) &= q(1-A)y(t) + Bu(t) + y(t) - q(1-A)y(t-1) - Bu(t-1) \\ &= q(1-A)y(t) + y(t) - (1-A)_y(t) + B\Delta u(t) \\ &+ q(1-\tilde{A})y(t) + B\Delta u(t) \end{aligned} \tag{13}$$

where, in accordance with the nomenclature of the referred work $\tilde{A} = A\Delta$. It can be seen that Eqn.(13) is the same as one given in [12].

Now, for the j-step-ahead predictor, let us assume the form given in [3], i.e.,

$$y^*(t+j) = E_jB\Delta u(t+j-1) + F_jy(t) \tag{14}$$

where E_j and F_j satisfy the identity

$$1 = E_j(q^{-1})A\Delta + q^{-j}F_j(q^{-1}) \tag{15}$$

If we use Eqn.(12.a), i.e., the predictor described in this paper to predict the next output, it is found that

$$\begin{aligned} y^*(t+j+1) &= q(1-A)y^*(t+j) + Bu(t+j) + \tilde{y}(t+j) \\ &= q(1-\tilde{A})y^*(t+j) + B\Delta u(t+j) \\ &= q(1-\tilde{A})\,[E_jB\Delta u(t+j-1) + F_jy(t)] + B\Delta u(t+j) \end{aligned} \tag{16}$$

which after some manipulations using Eqns.(9) and (15) reduces to

$$y^*(t+j+1) = (E_j+q^{-j}f_0)B\Delta u(t+j) + q(F_j-\tilde{A}f_0)y(t) \tag{17}$$

where f_0 is the first element of F_j. A comparison of Eqn.(17) with the recursion formulas given in [12] results in

$$y^*(t+j+1) = E_{j+1} + B\Delta u(t+j) + F_{j+1}y(t) \tag{18}$$

and this completes the proof that the predictor of Eqn.(12.a) is the same as the optimal predictor described by Clarke et al.

3. Generalized Predictive Control (GPC) Algorithm

In trajectory control of robotic manipulators, the desired trajectory is generally known beforehand. Let us assume that the set $w(t+j)$; $j=1,2,\ldots$ that describe it is available. Our aim is to ensure that future plant outputs $y(t+j)$ are as close to the desired ones as possible. To achieve this objective a receding horizon approach is used as follows. Firstly, by means of the method described above, the predicted values of the output are calculated. Then a quadratic function of the future predicted errors and controls is minimized assuming that after some control horizon NU the increments in control are zero. The first element of the control sequence thus calculated is applied to the system and the process is repeated at the next sampling interval.

Using Clarke's terminology [3], the most general quadratic function that can be used is

$$J(N1,N2) = E\left(\sum_{j=N1}^{N2} [y^*(t+j)-w(t+j)]^2 + \sum_{j=1}^{N2} \lambda(j)\,[\Delta u(t+j-1)]^2 \right) \quad \textbf{(19)}$$

where N1 and N2 are the minimum and the maximum costing horizons respectively, and $\lambda(j)$ is the control weighting sequence. In this work, N1 and NU are both chosen as one sampling interval and $\lambda(j)$ is set to zero. NU=1 means that the future controls are the same as the present one, in other words, an infinite costing is placed on the future control increment. The proposed algorithm is therefore applicable to non-minimum phase plants. This is important in trajectory control of manipulators because the equation that models a joint, Eqn.(8), generally contains a zero very near to the unit circle. It is due to this reason that most adaptive controllers for robotic manipulators are based on speed outputs of the joints.

Summarizing what is stated above, the algorithm described in this paper is realized as follows:

1. Setting $u(t+j)=u(t)$; $j=1,2,\ldots N2-1$, evaluate $y^*(t+j)$; $j=1,2,\ldots N2$ using Eqn.(12).
2. By minimizing the quadratic costing function, evaluate $u(t)$.

$$J = \sum_{j=1}^{N2} [y^*(t+j) - w(t+j)]^2 \quad \textbf{(20)}$$

3. Use $u(t)$ thus found in Eqn.(2) to find the torque to be applied to the joint.
4. Go back to Step 1 to find the torque to be applied at the next sampling interval.

Let us assume that the degrees of the polynomials A and B are 2 and 1 respectively. If, as stated above, we set $u(t+j)=u(t)$; $j=1,2,\ldots N2-1$, it can easily be shown that

$$y^*(t+j) = g_j u(t) + k(j) \quad \textbf{(21)}$$

where

$$\begin{aligned} g_1 &= b_0 \\ g_2 &= (b_0+b_1) - a_1 g_1 \\ g_3 &= (b_0+b_1) - a_1 g_2 - a_2 g_1 \\ g_4 &= (b_0+b_1) - a_1 g_3 - a_2 g_2 \\ g_5 &= (b_0+b_1) - a_1 g_4 - a_2 g_3 \\ &\vdots \end{aligned} \tag{22}$$

etc. and

$$\begin{aligned} k_1 &= b_0 u(t-1) - a_1 y(t) - a_2 y(t-1) + \ldots\ldots \\ k_2 &= -a_1 k_1 - a_2 y(t) + \ldots\ldots \\ k_3 &= -a_1 k_2 - a_2 k_1 + \ldots\ldots \\ k_4 &= -a_1 k_3 - a_2 k_2 + \ldots\ldots \\ k_5 &= -a_1 k_4 - a_2 k_3 + \ldots\ldots \\ &\vdots \end{aligned} \tag{23}$$

and so on. If we take the derivative of J as given in Eqn.(20) and equate it to zero, u(k) is found as

$$u(k) = \left(\sum_{j=1}^{N2} g_j^2 \right)^{-1} (g_j [w(t+j) - k_j]) \tag{24}$$

4. Predictive Pole-Placement Algorithm

The objective of the control algorithm [8] presented in this section is for the tracking error to decay with a prespecified dynamics, i.e., it is desired that

$$P(q^{-1})\, e(t+1) = 0 \tag{25}$$

where $P(q^{-1})$ is a monic polynomial of degree r, given by

$$P(q^{-1}) = 1 + p_1 q^{-1} + p_2 q^{-2} + \ldots. + p_r q^{-r},\ r \le n \tag{26}$$

and e(t+1) is the error between the reference value for the output and the actual value, i.e.,

$$e(t+1) = y(t+1) - w(t+1) \tag{27}$$

Using the predicted value of y(t+1) in Eqn.(25), the following equation can be obtained:

$$Pw(t+1) - y^*(t+1) - q(P-1)y(t) = 0 \tag{28}$$

By substituting Eqn.(10) in the above, the control law can be derived as

$$u(t) = -\frac{1}{b_0}[Fy(t) + B_1u(t) - Pw(t+1) + \tilde{y}(t)] \tag{29}$$

where $F=A+q(P-1)$ and $B_1=B-b_0$.

It should be stated here, as was mentioned above, that in the case of robotic manipulators, B contains a zero very close to the unit circle. In order not to cause a pole-zero cancellation by the controller, which would result in an undesirable ringing effect in the torque or force inputs, the pole-placement algorithm should either be used for the inner velocity control loop or it should be ensured that the final closed-loop transfer function contains this zero. The former is the case usually followed, the speed reference $\dot{\theta}_r$ being defined as

$$\dot{\theta}_r(t+1) = \dot{\theta}_d(t+1) + K[\theta_d(t) - \theta(t)] \tag{30}$$

where θ_d and $\dot{\theta}_d$ are the desired values for the joint position and velocity respectively and K is a design constant. The dynamics of the positional error $e(t+1) = \theta_d(t+1) - \theta(t+1)$ is therefore governed by Eqn.(30) and the following equation:

$$P[\dot{\theta}_r(t+1) - \dot{\theta}(t+1)] = 0 \tag{31}$$

If we use backward-difference approximations for the velocities, i.e., if we assume that

$$\begin{aligned} T\dot{\theta}_d(t+1) &\cong \theta_d(t+1) - \theta_d(t) \\ T\dot{\theta}(t+1) &\cong \theta(t+1) - \theta(t) \end{aligned} \tag{32}$$

the following equation which governs the error dynamics is easily obtained from Eqns.(30), (31) and (32).

$$P(q^{-1})\ [1-(1-KT)q^{-1}]\ e(t+1) = 0 \tag{33}$$

5. Computed-Torque Method

In the computed-torque method, the forces and torques are evaluated from

$$\tau = \hat{M}(\theta)\,u + \hat{V}(\theta,\dot{\theta}) \tag{34}$$

and the control law is given by

$$u = \ddot{\theta}_d + K_v(\dot{\theta}_d - \dot{\theta}) + K_p(\theta_d - \theta) \tag{35}$$

where K_v and K_p are constant diagonal feedback matrices. The application of Eqn.(35) to Eqn.(1) results in the following uncoupled error equation

$$\ddot{e} + K_v\dot{e} + K_p e = w \tag{36}$$

where w represents the discrepancies between the actual and the estimated system matrices.

6. Simulation Studies

In order to demonstrate the performance of the proposed GPC algorithm, a number of simulation studies were carried out on an industrial manipulator [9] the configuration of which is shown in Fig.1. It has a rotational joint and a translational joint in the x-y plane. The arm can be lifted along the vertical z-axis, which constitutes a third degree of freedom. However, the dynamics of this axis is completely decoupled from the other two axes and is therefore not taken into consideration.

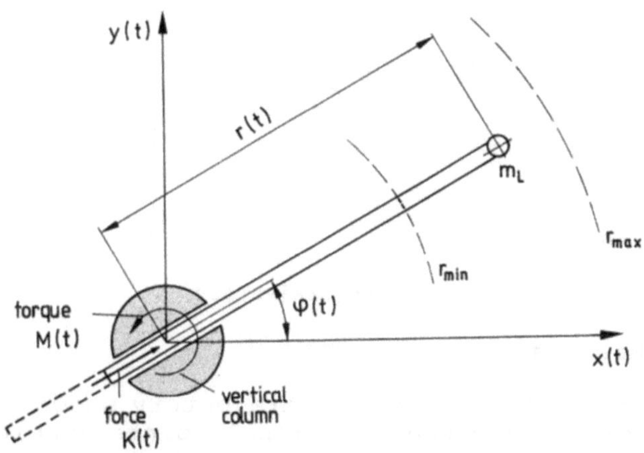

Fig.1. The industrial manipulator considered.

6.1. Dynamic Model of the Manipulator

Using Lagrange's equations, the dynamic model of the manipulator can be obtained as

$$\begin{bmatrix} K(t) \\ M(t) \end{bmatrix} = \begin{bmatrix} m+m_L & 0 \\ 0 & \theta_f+ml^2/3+(m+m_L)r^2(t)-mlr(t) \end{bmatrix} \begin{bmatrix} \ddot{r}(t) \\ \ddot{\phi}(t) \end{bmatrix}$$

$$+ \begin{bmatrix} -\left((m+m_L)r(t)-\frac{1}{2}ml\right)\dot{\phi}^2(t) \\ (2(m+m_L)r(t)-ml)r(t)\dot{\phi}(t) \end{bmatrix} \qquad (37)$$

where K(t) and M(t) are the force and the torque applied to the translational and the rotational joints, m and m_L are the masses of the arm and the load, and θ_f and l are the moment of inertia of the vertical column and the length of the arm respectively. The values of the parameters are m=20 kg, m_L=15 kg, θ_f=0.8 kgm and l=2m.

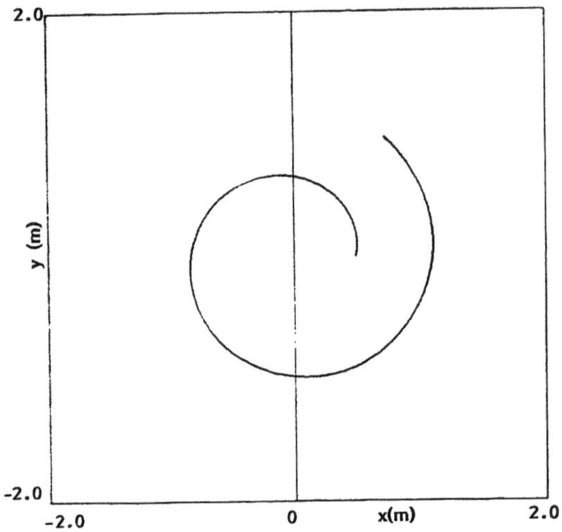

Fig.2. The desired trajectory.

6.2. Simulation Results

For simulation studies, a desired trajectory shown in Fig.2 is used. The translational and the rotational joints accelerate from rest having the initial values 0.5 m and 0 rad at the rates of 0.5 m/s^2 and 5 $rads/s^2$ respectively, and reach constant speeds of 0.2 m/s and 2 rad/s. Towards the end of the trajectory, they decelerate at the same rates to come to a rest. The sampling interval is chosen as 10 ms. In order to demonstrate the effectiveness of the predictive-control algorithm even under extreme conditions, a minimum amount of compensation is attempted. It is assumed that

$$\hat{M}(\theta) = \begin{bmatrix} m+\hat{m}_L & 0 \\ 0 & \theta_f+\dfrac{ml^2}{3} \end{bmatrix}, \hat{V}(\theta,\dot{\theta}) = 0 \tag{38}$$

where m_L is the estimated value of the load mass. It is taken as 10 kg instead of the actual value of 15 kg.

A comparison of Eqns.(37) and (38) indicates the amount of simplification. Coriolis and centrifugal torques and forces are all neglected together with the speed dependent terms of the inertial matrix. Additionally, it is assumed that the system so compensated can be considered as a double integrator plant. This is to say that each joint can be represented for positional output as

$$y(k+1) = 2y(k) - y(k-1) + (T^2/2)[u(t)+u(t-1)] \tag{39}$$

where $y(\cdot)$ is $r(\cdot)$ for the translational joint and $\phi(\cdot)$ for the rotational joint. The parameters of the A and B polynomials are therefore $a_1=-2$, $a_2=+1$ and $b_0=b_1=0.00005$. By use of Eqn.(24), u(t) can easily be calculated.

Figure 3 is an example to the results of the simulation studies. This figure shows the errors in x and y coordinates for a maximum costing horizon of N2=5. As can be seen, the tracking performance is exceptionally good in spite of the amount of simplifications explained above.

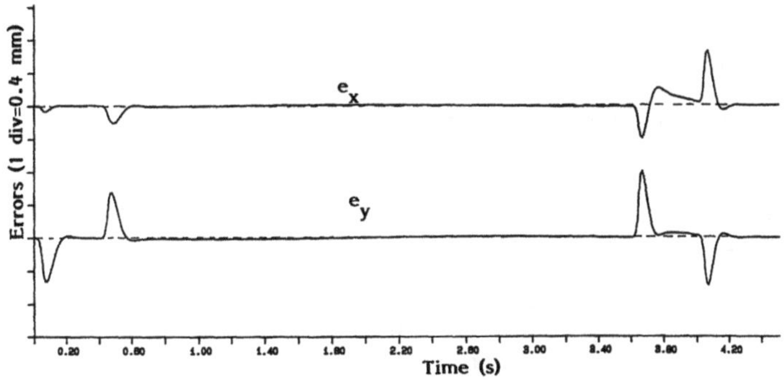

Fig.3. Error waveform for predictive-control algorithm (N2=5).

The effect of the maximum costing horizon on the system performance is also studied. It is seen that the error waveforms for N2=3 are not significantly different than those for N2=5. However, as shown in Fig.4, N2=3 results in more control activity.

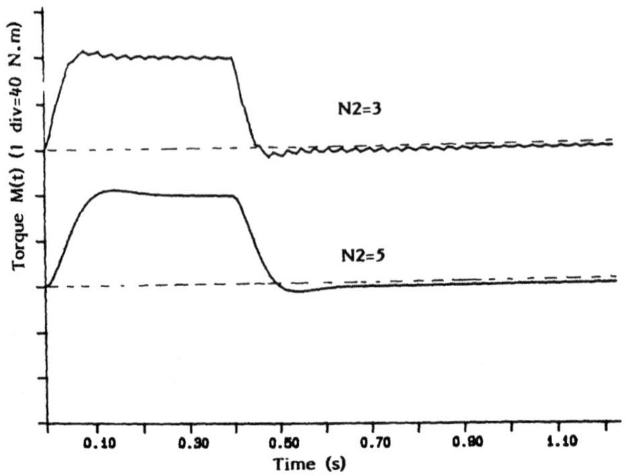

Fig.4. Torque to be applied to the rotational joint for N2=3 and N2=5.

The proposed algorithm is also realized in an adaptive manner by use of a RLS identification routine for the identification of the coefficients of A and B polynomials. For this, the well known equations

$$\hat{\theta}(k+1) = \hat{\theta}(k) + K(k)\,[y(k+1) - \phi(k+1)\theta(k)]$$

$$K(k) = P(k)\phi^T(k+1)\,[\lambda + \phi(k+1)P(k)\phi^T(k+1)]^{-1} \qquad \textbf{(40)}$$

$$P(k+1) = [I - K(k)\phi(k+1)]\,P(k)/\lambda$$

of the literature [1] are used where the forgetting factor λ is chosen as 0.95. The algorithm is initiated with P taken as 1000 times an identity matrix and the initial values of the parameter vector θ are approximately the same as the values obtained with double-integrator plant assumption. Figure 5 shows that the identified b parameters of the translational (b_{0r} and b_{1r}) and the rotational ($b_{0\phi}$ and $b_{1\phi}$) axes vary appreciably during the trajectory. The coefficients of the A polynomials on the other hand were seen to stay almost constant around $a_{1r}=a_{1\phi}=-2$ and $a_{2r}=a_{2\phi}=1$ and these are not shown.

The error waveforms obtained with the RLS routine running are shown in Fig.6. As can be seen, the performance is not very different than that shown in Fig.3. It is seen that the correction term used in the prediction scheme ensures that the performance with constant assumed parameters is as good as those obtained with an adaptive approach.

For simulation studies with the pole-placement algorithm described in Sec.4, a second-order response

$$s^2 + 2\delta w_n s + w_n^2 = 0 \qquad \textbf{(41)}$$

is aimed at, the damping constant δ and the natural frequency w_n being chosen as unity and 10 rads/s^2, respectively. For T=0.01 s, the equivalent of Eqn.(41) in the z-plane is

$$z^2 - 1.81z + 0.82 = 0 \tag{42}$$

In order to ensure that the dynamics of the positional error is governed by Eqn.(41), it can easily be found from Eqn.(33) that the P polynomial should be $P(q^{-1}) = 1 + 0.905q^{-1}$ and K = 9.5.

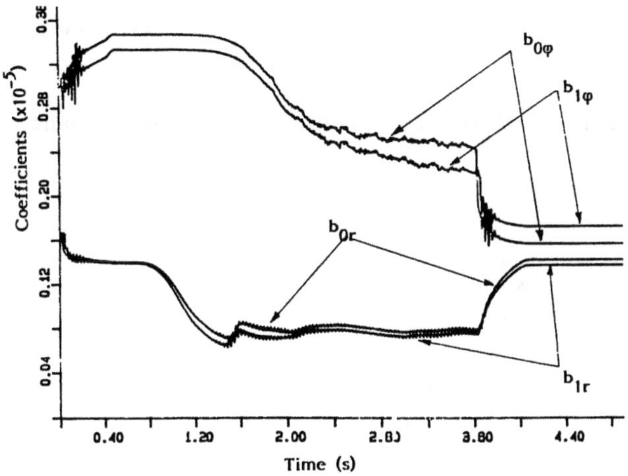

Fig.5. Identified coefficients with RLS routine.

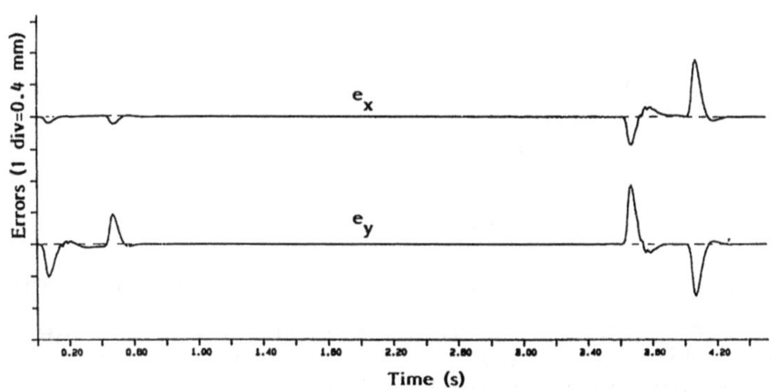

Fig.6. Error waveforms with RLS identification.

Figure 7 shows the error waveforms for the pole-placement algorithm. For this simulation study, the acceleration and deceleration rates, and the maximum speed of the desired trajectory are halved. The completion of the trajectory therefore takes almost 8.5 s. In spite of this slower trajectory, the error is much larger than that seen for the GPC algorithm.

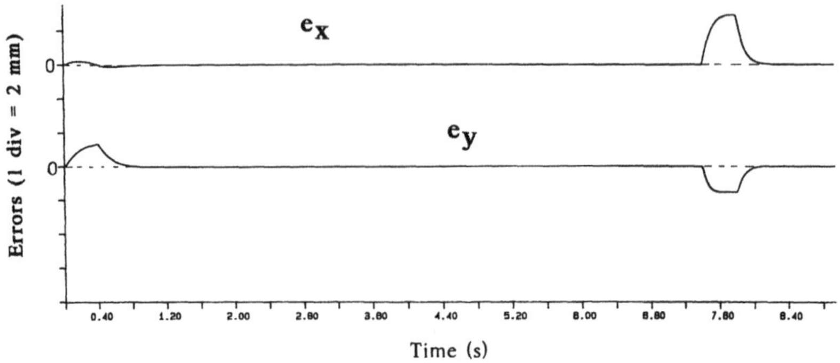

Fig.7. Error waveforms for predictive pole-placement algorithm.

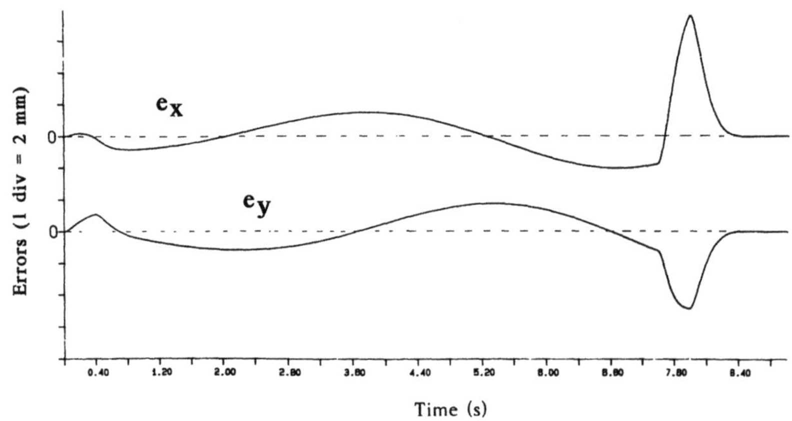

Fig.8. Error waveforms for computed-torque algorithm.

Finally, to provide another basis for comparison, the computed-torque method is also simulated. For this, full compensation is used with no mismatch, i.e., $\hat{M}(\theta) = M(\theta)$ and $\hat{V}(\theta,\dot{\theta}) = V(\theta,\dot{\theta})$. The values for K_v and K_p were chosen so as to result in the same dynamics as for the case of the pole-placement algorithm. Figure 8 shows that the errors are much higher than both the GPC and the pole-placement algorithms.

7. Experimental Studies

In order to demonstrate the practicality of the GPC algorithm, some experimental studies were carried out [11] on the two degree-of-freedom experimental manipulator shown in Fig.9. The actuator of the first link is a servo motor with a coupled 1000 pulse/rev encoder and a tachogenerator. It has a harmonic gear reduction ratio of 88:1. The second link is activated by a stepper motor. For the purpose of the experimental investigation this joint was not activated, the stepping motor being dismantled.

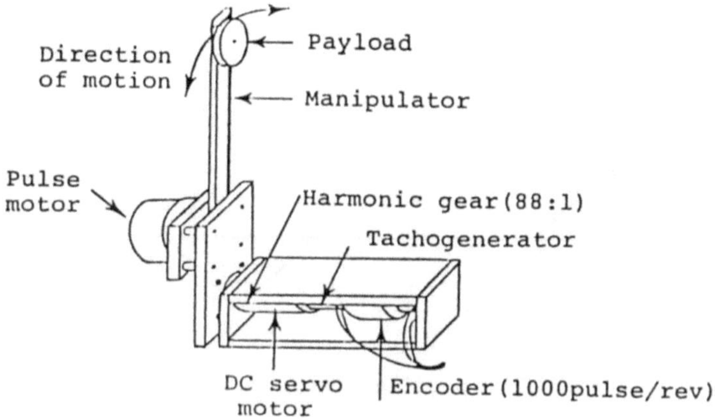

Fig.9. The structure of the experimental manipulator.

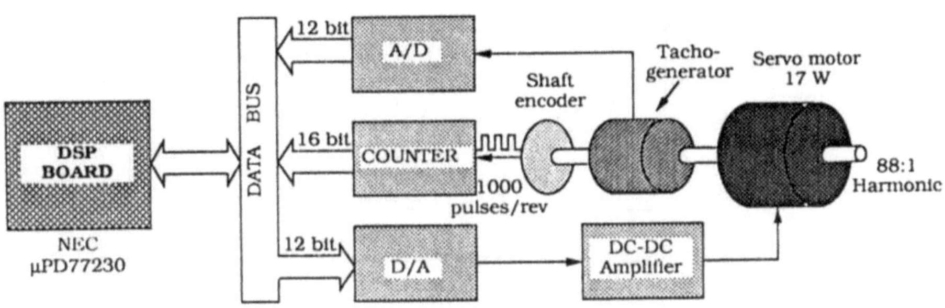

Fig.10. Experimental set-up for data acquisition and processing.

Figure 10 shows how the data is handled. For the implementation of the GPC algorithm, a DSP (Digital Signal Processor) based system was used with a 12-bit A/D converter and a 16-bit counter circuit to process the tachogenerator and the encoder signals, respectively. The value of the control signal that is calculated by the DSP is converted into analog form by a 12-bit D/A converter and then amplified by a dc amplifier to drive the servo motor. The transfer function of the system was assumed to be $G(s) = 30/s^2$ neglecting the payload which was 0.5 kg placed 30 cm away from the center of the shaft, and the weight of the arm (1.32kg). These assumptions meant that $\hat{M}(\theta) = 30$ *and* $\hat{V}(\theta, \dot{\theta}) = 0$.

The control algorithm was realized for maximum costing horizons (N2) of 3, 5 and 10 sampling intervals; one sampling interval being 0.005 s. For experimental investigations, the acceleration profile shown in Fig.11 was used; the total distance covered by the arm thus being about 2.5 rad. The initial point of trajectory was with the arm in the up position.

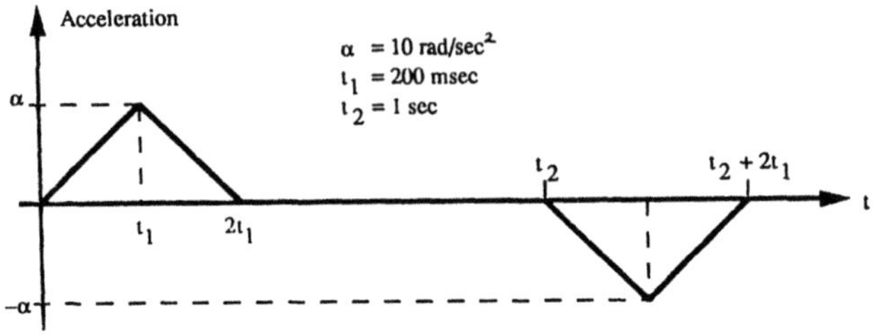

Fig.11. Acceleration profile of the reference trajectory.

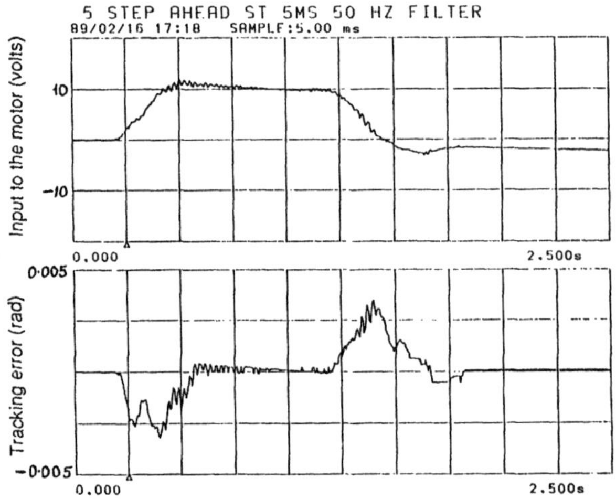

Fig.12. Input to the motor and the tracking error.

Fig.12 shows a typical result for N2=5. As can be seen, the tracking performance is exceptionally good; the maximum tracking error being around 0.003 radians. In order to smooth the control input to the power amplifier, a low-pass filter with a cut-off-frequency of 50 Hz was used. This ensured minimization of the control activity. Further experimental investigations which are not presented here showed, as would be expected, that the control signal was smoother for N2=10, with a slight degradation of the tracking performance. Decreasing the control horizon had the opposite effect. A full discussion of the "tuning knobs" (N1, N2 and NU) can be found in [3].

8. Conclusions

The simulation and the experimental results presented in this paper have demonstrated that predictive control is an important contender for high-performance applications in robotics. The algorithm is simple and easily implementable. It requires minimal a priori knowledge of the manipulator, only a rough estimate of the diagonal terms of the inertial matrix is sufficient. The correction term used in the prediction scheme adds a high degree of intelligence to the controller, enabling it to cope with uncertainties and nonlinearities.

Another contribution of the paper is the prediction scheme. It is simple and has a physical meaning, and it results in the same predicted values as the optimal predictor popularly referred to in the literature. The latter necessitates the solution and the subsequent recursion of two Diophantine equations and is therefore harder to implement.

9. Acknowledgements

The author would like to acknowledge the fellowships of the Alexander von Humboldt-Stiftung of Fed. Rep. of Germany and Institute of Industrial Science (University of Tokyo) which enabled the accomplishment of some parts of the work reported. The support of Bogazici University Research Fund (Project No:88A0215) is also acknowledged.

10. References

1. Aström, K. J. and B. Wittenmark: Computer Controlled Systems, Theory and Design. Prentice Hall Int. Inc. (1984)

2. Choi, Y. K., M. J. Chung and Z. Bien: An adaptive control scheme for robotic manipulators. Int. J. Control, 44, 1185-1191 (1987)

3. Clarke, D. W., C. Mohtadi and P. S. Tuffs: Generalized predictive control, Part I and II. Automatica, 23, 137-160(1987)

4. Dobowsky, S. and D. T. DesForges: The application of model reference adaptive control to robotic manipulators. J. Dynm. Syst. Mea. and Contr., 101, 193-200(1969)

5. Freund, E.: Fast non-linear control with arbitrary pole-placement for industrial robots and manipulators. Int. J. Robotics Research, 1, 65-78(1982)

6. Hoshinaka, T. and K. Ohnishi: Fast motion control of mechanical manipulator for path tracking. Proc. IECON'88, 163-168(1988)

7. Jacubasch, A. H., B. Kuntze, Ch. Arber and J. Richalet: Anwendung eines neuen Verfahrens zur schnellen und robusten Positionregelung von Industrierobotern. Robotersysteme, 3, 129-138(1987)

8. Kaynak, O. and H. Hoyer: Adaptive pole placement controllers for robotic manipulators with predictive aciton (Invited Paper). Proc. IECON'87, 149-157(1987)

9. Kaynak, O. and H. Hoyer: Predictive control of a robotic arm. Robot Control 1988 (Syroco '88), Ed. U. Rembold, Pergamon Press, 213-218(1988)

10. Kaynak, O. and S. Z. Sarptürk: An adaptive control algorithm for manipulators based on sliding mode control concept. Int. J. Robotics and Automation, 3, 105-114(1988)

11. Kaynak, O., H. Hashimoto, H. Kuroyanagi and F. Harashima: Robust tracking control based on long range prediction. Proc. IECON'89, 296-300(1989)

12. Koivo, A. J. and T. G. Guo: Adaptive linear controller for robotic manipulators. IEEE Trans. Autom. Contrl., AC-28, 162-171(1983)

13. Komada, S. and K. Ohnishi: Force feedback control of robot manipulator by the acceleration tracing orientation method. IEEE Trans. Indust. Electron. 37, 6-12(1990)

14. Luh, T.Y.S., M.W. Walker and R.P. Paul: Resolved acceleration control of mechanical manipulators. IEEE Trans. Autom. Contrl., AC-25, 468-474(1980)

15. Tourrasis, V.D. and C.P. Neuman: Robust nonlinear feedback control for robotic manipulator. IEE Proc., 132, Pt.D, 134-143(1985)

16. Xu, J.X., H. Hashimoto, J.J.E. Slotine, Y. Arai and F. Harashima: Implementation of VSS control to robotic manipulators. IEEE Trans. Indust. Electron. 36, 321-329(1989)

17. Whitney, D.E.: Resolved motion rate control of manipulators and human prosthese. IEEE Trans. Man-Machine Systems, MMS-1, 47-53(1969)

Analysis of Cooperating Robot Manipulators on a Mobile Platform

Steve H. Murphy John T. Wen George N. Saridis

Center for Intelligent Robotic Systems for Space Exploration
Rensselaer Polytechnic Institute
Troy, NY 12180–3590

Abstract

When robot manipulators are to be used in a space environment, the interactions between the manipulators and the platform upon which they are mounted increases the complexity of the dynamic model. The situation is further complicated when the manipulators must cooperate with each other to move objects or to perform assembly and disassembly operations. In this work, the dynamic equations of motion for two or more cooperating manipulators on a freely-moving mobile platform are presented. The formulation includes the full dynamic interactions from arms-to-platform and arm-tip to arm-tip and the possible translation and rotation of the platform. The equations of motion are shown to be identical in structure to the fixed-platform, cooperative manipulator dynamics. The system of cooperating robot manipulators forms a closed kinematic chain where the forces of interaction must be included in the simulation of robot and platform dynamics. The structure of the closed-chain dynamics allows the use of any solution for the open topological tree of base and manipulator links. A MATLAB simulation of two cooperating PUMA manipulators on a mobile platform is presented and displayed through SILMA Cimstation.

KEYWORDS: simulation; cooperating robots; mobile platform; space manipulation; robot dynamics

1 Introduction

The effective implementation of complex robot-control systems requires extensive simulation and experimentation prior to implementation in the field. It is expected, that through simulation, one can freely evaluate different control schemes without damage to property or extensive setup times. The inherent complexity of multiple-cooperating robot manipulators requires an effective evaluation in a simulated environment before implementation. This is especially true for space and undersea applications of multiple manipulators where human intervention is not available.

For the difficult physical environments of space, simulations offer many other uses. The simulation can model the dynamics of the entire space environment with all the complexity of three

NATO ASI Series, Vol. F 71
Expert Systems and Robotics
Edited by T. Jordanides and B. Torby

dimensional motion, unlike simplified laboratory experiments. When combined with a graphical display, the simulation can provide response information that would otherwise be difficult to discern from data plots. For environments with large time delays, such as space-telerobotic applications, it may be advantageous to use the simulation to evaluate the effect of a control input ahead of the real-time feedback.

Simulation of robot cooperation in a space environment consists of two issues, the simulation of robot cooperation, and the simulation of robot motion on a mobile platform. Previous work has concentrated on either of the two issues but not as a whole, which is necessary whenever multiple manipulators work together on a mobile platform. In the space environment, the motion of the robot manipulators will affect the platform motion and the platform motion will affect the cooperative manipulation.

The simulation solution requires the calculation of the forward dynamics of the closed-chain, mobile-platform system: given the joint torques, and the torques and forces on the platform, calculate the joint and platform accelerations. The inverse dynamics provides the joint torques and platform forces for given accelerations.

The work on both the forward and inverse dynamics for cooperating manipulators has been concerned with the fixed-platform system. Anderson and Pittlekau [1][2] presented a method for the forward dynamics of two cooperating manipulators with an expression for the interactive force and a load that was included as a part of the manipulators. Unseren and Koivo [32] simulated a closed kinematic chain of two manipulators and showed the position and interactive force output for two 2 DOF manipulators under constant torque input. Tarn, et. al. [27][28] [29] developed two formulations for the inverse and forward dynamics of two cooperating manipulators on a fixed base and simulated two dimensional motions with the first three joints of simulated PUMA 560's. Hayati [6] derived the inverse dynamics of a multi-arm robot system but without regards towards the forward dynamics calculation. Inverse dynamics of systems where the load has its own degrees of freedom have been presented by Zheng and Luh, [40][41][42], and Zheng [39]. A fixed base simulation for two manipulators grasping a floating object was presented by Walker and Dionise [34], and two fixed cooperating PUMA 560s with an adaptive controller were simulated by Walker et al. [35].

Oh and Orin [19] develop a dynamic simulation method for multiple cooperative manipulators on a fixed base and legged vehicles in contact with a fixed earth, no simulation results are reported. Rodriguez and Kreutz [22] use a similar notation but are able to develop an algorithm for calculating the fixed platform forward dynamics which grow linearly with the total number of links in the closed kinematic chain. Both methods may use any recursive or closed-form solution to the forward dynamics of a single manipulator in the simulation of the fixed-base closed chain.

Much of the work in the area of mobile platforms has concentrated on single manipulator systems. Simulation analysis of manipulators with base disturbances have been developed by Joshi and Desrochers [9], and Tanner [26]. Umetani and Yoshida [31][30] report experimental results for a single-arm manipulator on a free-floating, 2-DOF platform. Motion of the platform due to arm motion is included. Simulation results for a similar system when subject to actuator saturation limits have been presented by Dubowsky et al. [4]. Vafa and Dubowsky [33], Papadopoulos and Dubowsky [21] derived the equations of motion of a single arm on a space platform through the conservation of momentum and the concept of a virtual manipulator. The analysis used velocity rather than torques as the input variable. Nakamura and Mukherjee [17] also used conservation of momentum for determining the motion of a single manipulator on a space platform for path planning purposes. Longman et al. [14] examined both the kinematics and dynamics of a single manipulator on a space platform and derived expressions for the torques and forces acting between the base of the robot and the platform. Wang [36] presented the symbolic equations for two planar robots on a planar space platform. Arm–load dynamic interactions are not considered.

The solution to the motion of such systems of rigid links attached to a space platform complements the area of dynamic multibody simulation. There is a large amount of work dealing with the dynamic simulation of flexible, and rigid multibody systems. A good survey of this area was presented by Lips and Singh [13]. Methods, such as developed by Kane and Levinson [11][10] may also be used. However, the goal for simulation of multiple cooperating robotic manipulators is to determine not only the motion of the closed-chain system, but also the forces exerted on one segment of the closed chain, the load. As the load may change between simulations, the approach must be flexible.

This work presents the dynamic equations of motion for multiple cooperating robot manipulators on a mobile platform. The following problems are addressed by this work:

- Modeling of mobile-base manipulators in a form suitable for control development and analysis.
- Modeling a complex combination of manipulators and platform for the application of efficient simulation techniques.
- Simulation of cooperating manipulators on a space platform.

While previous work on space manipulators has concentrated on models based upon angular and linear velocities, or single manipulator systems, this work produces a model using complete information of forces and torques. The full dynamic interactions between the arms and platform and between the arms are included in the development.

A system of two, six degree-of-freedom manipulators on a freely moving platform was used in the development of the equations of motion. The attitude and position of the platform will respond to arm motion and may also be controlled through external inputs such as reaction wheels or left to move freely. The system has a total of 18 degrees of freedom and the model is directly extendable to more than 2 manipulators. The model allows for the use of any solution for the forward dynamics of a topological tree of rigid links [24][10] [18][8]. Due to the large number of degrees of freedom, the recursive solutions which are linear in the number of links become much more efficient. Simulation of the system is accomplished through a MATLAB program and the response is visualized through Silma Cimstation [25].

The paper is organized as follows: Section 2 presents the closed form solution for a topological tree of two manipulators and base using the Newton-Euler method. In Section 3 this result is used to develop the equations for the forward dynamics of the closed chain of manipulators. Section 4 describes the MATLAB implementation, the results of the simulation, some of the simulation issues, and the graphical presentation. Conclusions are given in Section 5.

2 Closed-chain Inverse Dynamics

In this section the closed form solution for the inverse dynamics of two manipulators on a mobile platform is presented. The solution is derived from the Newton–Euler equations of motion. While a Lagrange-Euler solution could be used, the Newton-Euler formulation provides greater insight into the structure of the multiple-robot dynamics. This insight allows for the development of recursive solutions to the forward-dynamics problem. The approach is to first develop the equations of motion for the open kinematic chain on a mobile platform, and then close the chain through the forces on the load.

2.1 Notation

The notation used is the coordinate-free spatial notation presented in [7][23], with similar, but coordinate-dependent notation in [19]. All vectors and inertias are written coordinate independent and are represented in specific coordinate frames only at the final stages of implementation. Links are numbered from the base to the tip of each manipulator. Link 0 is the base, common to both manipulators. Subscripts are used to specify individual robots. For example, $M_1(2)$ represents the mass matrix of link 2 on robot 1. Each robot is assumed to have the same number of links with a coordinate frame attached to each link. Thus $O(2)$ represents the origin of the frame attached to link 2. Modified Denavit-Hartenburg labeling is used [3]. The coordinate frame at the tips of the manipulators is $O(N+1)$ where N is the number of links in one manipulator. Some of the key notations are: $\mathcal{V}$ is a Euclidean vector space with dimension 3, ^{T} is the operator adjoint or transpose, $\cdot$ is the vector dot product operator, $\times$ is the vector cross product operator.

$O(k)$	The point located on the axis of rotation of link k.
$l(k,k+1)$	The vector from the point $O(k)$ to the point $O(k+1)$.
$\omega(k)$	The vector of angular velocity of link k with respect to the inertial frame.
$v(k)$	The vector of linear velocity of link k at point $O(k)$ with respect to the inertial frame.
$V(k)$	Spatial velocity of $O(k)$, $V(k) \in \mathcal{V}\times\mathcal{V}$, $V(k) = \begin{bmatrix} \omega(k) \\ v(k) \end{bmatrix}$
$\alpha(k)$	Spatial acceleration of point $O(k)$, $\alpha(k) \in \mathcal{V}\times\mathcal{V}$, $\alpha(k) = \begin{bmatrix} \dot{\omega}(k) \\ \dot{v}(k) \end{bmatrix}$
$F(k)$	Vector of force acting upon link k at point $O(k)$.
$N(k)$	Vector of torque or moments acting upon link k at point $O(k)$.
$f(k)$	Spatial force acting upon link k at point $O(k)$, $f(k) \in \mathcal{V}\times\mathcal{V}$,

$$f(k) = \begin{bmatrix} N(k) \\ F(k) \end{bmatrix}.$$

$p(k)$	The vector from the point $O(k)$ to the center of mass of link k.
$\phi(k+1,k)$	Linear operator which translates a spatial velocity at point $O(k)$ to point $O(k+1)$, $\phi(k+1,k) : \mathcal{V} \times \mathcal{V} \rightarrow \mathcal{V} \times \mathcal{V}$ $$\phi(k+1,k) = \begin{bmatrix} I & 0 \\ -l(k,k+1)\times & I \end{bmatrix},$$ where I is the identity operator, $I : \mathcal{V} \rightarrow \mathcal{V}$.
$\phi^T(k+1,k)$	Linear operator which translates a spatial force acting at point $O(k+1)$ to a spatial force acting at $O(k)$. $\phi^T(k+1,k) : \mathcal{V} \times \mathcal{V} \rightarrow \mathcal{V} \times \mathcal{V}$ $$\phi^T(k+1,k) = \begin{bmatrix} I & l(k,k+1)\times \\ 0 & I \end{bmatrix}.$$
$m(k)$	The mass of link k.
$I(k)$	The inertia tensor for link k at point $O(k)$.
$M(k)$	The spatial mass matrix of link k at point $O(k)$ which incorporates the inertia tensor, $I(k)$, link mass, and location of center of mass, $p(k)$. $$M(k) = \begin{bmatrix} I(k) & m(k)p(k)\times \\ -m(k)p(k)\times & m(k)I \end{bmatrix}, \; M(k) : \mathcal{V} \times \mathcal{V} \rightarrow \mathcal{V} \times \mathcal{V}.$$
$a(k)$	The spatial bias accelerations of link k, $a(k) \in \mathcal{V} \times \mathcal{V}$. For a rotational joint, $a(k) = \begin{bmatrix} w(k-1) \times w(k) \\ w(k-1) \times [w(k-1) \times l(k-1,k)] \end{bmatrix}$
$b(k)$	The spatial bias forces acting on link k at point $O(k)$, $b(k) \in \mathcal{V} \times \mathcal{V}$. For a rotational joint, $b(k) = \begin{bmatrix} w(k) \times I(k)w(k) \\ m(k)w(k) \times [w(k) \times p(k)] \end{bmatrix}$.
$h(k)$	The vector aligned with the axis of rotation for link k and defining the direction of positive rotation, $h(k) \in \mathcal{V}$.
$H(k)$	The spatial axis of motion for link k, $H(k) \in \mathcal{V} \times \mathcal{V}$. For a rotational joint, $H(k) = \begin{bmatrix} h(k) \\ \vec{0} \end{bmatrix}$, where $\vec{0}$ is the null vector.
$H^T(k)$	Projection operator onto spatial axis of motion, $H^T(k) = \begin{bmatrix} h(k)\cdot & \vec{0}\cdot \end{bmatrix}$ $H^T(k) : \mathcal{V} \times \mathcal{V} \rightarrow \mathcal{R}$

2.2 Recursive Inverse Dynamics

Let link 0 represent the common base. The equations for the propagation of velocities and accelerations from the common base to the tips of the manipulators may be written:

$$V_1(0) = V_2(0) = V(0), \quad \text{Velocity of the Base}$$

$$\alpha_1(0) = \alpha_2(0) = \alpha(0), \quad \text{Acceleration of the Base} \tag{1}$$

for $k = 0, ..., N$

$$V_1(k) = \phi_1(k, k-1)V_1(k-1) + H_1(k)\dot{\theta}_1(k) \tag{2}$$
$$V_2(k) = \phi_2(k, k-1)V_2(k-1) + H_2(k)\dot{\theta}_2(k) \tag{3}$$

$$\alpha_1(k) = \phi_1(k, k-1)\alpha_1(k-1) + H_1\ddot{\theta}_1(k) + a_1(k) \tag{4}$$
$$\alpha_2(k) = \phi_2(k, k-1)\alpha_2(k-1) + H_2\ddot{\theta}_2(k) + a_2(k) \tag{5}$$

At the tips of the manipulators:

$$V_1(N+1) = \phi_1(N+1, N)V_1(N) \tag{6}$$
$$\alpha_1(N+1) = \phi_1(N+1, N)\alpha_1(N) + a_1(N+1) \tag{7}$$
$$V_2(N+1) = \phi_2(N+1, N)V_2(N) \tag{8}$$
$$\alpha_2(N+1) = \phi_2(N+1, N)\alpha_2(N) + a_2(N+1) \tag{9}$$

where the subscripts refer to robots 1 and 2, respectively, and $V_1(0) = V_2(0)$ and $\alpha_1(0) = \alpha_2(0)$ because the robots share a common base.

The propagation of torques/forces from the tips of the manipulators to the common base and the calculation of joint torques may be written in terms of the accelerations found above:

$$f_1(N+1) = \text{tip forces exerted on the load from robot 1} \tag{10}$$
$$f_2(N+1) = \text{tip forces exerted on the load from robot 2} \tag{11}$$

for $k = N, ..., 1$

$$f_1(k) = \phi_1^T(k+1, k)f_1(k+1) + M_1(k)\alpha_1(k) + b_1(k) \tag{12}$$
$$f_2(k) = \phi_2^T(k+1, k)f_2(k+1) + M_2(k)\alpha_2(k) + b_2(k) \tag{13}$$
$$T_1(k) = H_1^T(k)f_1(k) \tag{14}$$
$$T_2(k) = H_2^T(k)f_2(k) \tag{15}$$

On the base, $f_1(1)$ and $f_2(1)$ are the forces acting on the base of each robot. The forces on the base may be written;

$$f(0) = \phi_1^T(1,0)f_1(1) + \phi_2^T(1,0)f_2(1) + M(0)\alpha(0) + b(0) \tag{16}$$

where $f(0)$ is the 6x1 vector of torques and forces acting on the platform at the platform frame, $O(0)$. This frame may be placed at any point on the platform. The forces $f(0)$ may be created, for example, by reaction jets under the control of a position and orientation control system. If the torque/force $f(0)$ was zero, then the base will move in reaction to the manipulator forces.

2.3 Closed-form Inverse Dynamics

Define the vectors and matrices for robot 1:

$$\phi_1 = \begin{bmatrix} \phi_1(1,1) & 0 & \dots & 0 \\ \phi_1(2,1) & \phi_1(2,2) & \dots & \vdots \\ \vdots & \vdots & \ddots & 0 \\ \phi_1(N,1) & \phi_1(N,2) & \dots & \phi_1(N,N) \end{bmatrix} \tag{17}$$

$$H_1 = \text{diag}[H_1(1),\dots,H_1(N)]; \quad M_1 = \text{diag}[M_1(1),\dots,M_1(N)] \tag{18}$$

$$B_1 = [0,\dots,0,\phi_1(N+1,N)]; \quad E_1^T = \left[\phi_1^T(1,0),0,\dots,0\right] \tag{19}$$

$$a_1 = \begin{bmatrix} a_1(1) \\ \vdots \\ a_1(N) \end{bmatrix}; \quad b_1 = \begin{bmatrix} b_1(1) \\ \vdots \\ b_1(N) \end{bmatrix}; \quad \alpha_1 = \begin{bmatrix} \alpha_1(1) \\ \vdots \\ \alpha_1(N) \end{bmatrix}; \quad f_1 = \begin{bmatrix} f_1(1) \\ \vdots \\ f_1(N) \end{bmatrix}; \quad V_1 = \begin{bmatrix} V_1(1) \\ \vdots \\ V_1(N) \end{bmatrix} \tag{20}$$

Here we made use of the operator properties of $\phi_1(i,j)$:

$$\phi_1(i,i) = I; \quad \phi_1^{-1}(i,j) = \phi_1(j,i); \quad \phi_1(i,k)\phi_1(k,j) = \phi_1(i,j) \tag{21}$$

The matrices for the second robot are defined similarly with only the subscript changing.

Stacking up all the vectors with the above definitions allows the equations for the force propagation (12–13) to be written:

$$f_1 = \phi_1^T \left(M_1\alpha_1 + b_1 + B_1^T f_1(N+1)\right) \tag{22}$$

$$f_2 = \phi_2^T \left(M_2\alpha_2 + b_2 + B_2^T f_2(N+1)\right) \tag{23}$$

Now continue the stacking operation for both the robots and the base. Define

$$\phi = \begin{bmatrix} I_{6x6} & 0 & 0 \\ \phi_1 E_1 & \phi_1 & 0 \\ \phi_2 E_2 & 0 & \phi_2 \end{bmatrix} \tag{24}$$

where

$$\phi_1 E_1 = [\phi_1^T(1,0), \phi_1^T(2,0), \dots, \phi_1^T(N,0)]^T \tag{25}$$

$$\phi_2 E_2 = [\phi_2^T(1,0), \phi_2^T(2,0), \dots, \phi_2^T(N,0)]^T \tag{26}$$

Also define

$$M = \begin{bmatrix} M(0) & 0 & 0 \\ 0 & M_1 & 0 \\ 0 & 0 & M_2 \end{bmatrix}; \quad \alpha = \begin{bmatrix} \alpha(0) \\ \alpha_1 \\ \alpha_2 \end{bmatrix}; \quad b = \begin{bmatrix} b(0) \\ b_1 \\ b_2 \end{bmatrix} \tag{27}$$

$$B = \begin{bmatrix} 0 & B_1 & 0 \\ 0 & 0 & B_2 \end{bmatrix}; \quad f = \begin{bmatrix} f(0) \\ f_1 \\ f_2 \end{bmatrix}; \quad f(N+1) = \begin{bmatrix} f_1(N+1) \\ f_2(N+1) \end{bmatrix}; \tag{28}$$

Then the force propagation equations, (22), (23), and (16), may be written:

$$f = \phi^T \left(M\alpha + b + B^T f(N+1) \right) \tag{29}$$

In order to write the velocity and acceleration propagation in the same manner, we make the following definitions:

$$V = \begin{bmatrix} V(0) \\ V_1 \\ V_2 \end{bmatrix} \quad ; \quad a = \begin{bmatrix} 0 \\ a_1 \\ a_2 \end{bmatrix} \tag{30}$$

$$W = \begin{bmatrix} V(0) \\ \dot{\theta}_1(1) \\ \vdots \\ \dot{\theta}_1(N) \\ \dot{\theta}_2(1) \\ \vdots \\ \dot{\theta}_2(N) \end{bmatrix} \; ; \; T = \begin{bmatrix} f(0) \\ T_1(1) \\ \vdots \\ T_1(N) \\ T_2(1) \\ \vdots \\ T_2(N) \end{bmatrix} \; ; \; H = \begin{bmatrix} I_{6x6} & 0 & 0 \\ 0 & H_1 & 0 \\ 0 & 0 & H_2 \end{bmatrix} \tag{31}$$

where $V(0)$ is the combined angular and linear velocity of the base, and I_{6x6} is a 6x6 identity matrix. W is the velocity vector for the platform and both manipulators, T is the force/torque vector for forces and torques acting on the platform and around each joint. Note that $a(0) = 0$, as there are no bias accelerations of the base.

With these definitions, the equations (2–5) may be written:

$$V = \phi H W \tag{32}$$

$$\alpha = \phi H \dot{W} + \phi a \tag{33}$$

with the same boundary conditions as before, equation (1). Define,

$$\alpha(N+1) = \begin{bmatrix} \alpha_1(N+1) \\ \alpha_2(N+1) \end{bmatrix} \quad ; \quad a(N+1) = \begin{bmatrix} a_1(N+1) \\ a_2(N+1) \end{bmatrix} \tag{34}$$

The angular and linear accelerations at the tips of the manipulators may be determined:

$$\alpha(N+1) = B\phi H \dot{W} + B\phi a + a(N+1). \tag{35}$$

Combining equations (29) and (33) gives the equation for the forces on each link of the manipulators and the forces on the base:

$$f = \phi^T M \phi H \dot{W} + \phi^T M \phi a + \phi^T b + \phi^T B^T f(N+1) \tag{36}$$

The torques on the links are found by projecting the link torque/forces onto the axis of rotation. Using the definition of H:

$$T = H^T f \tag{37}$$

$$= H^T \phi^T M \phi H \dot{W} + H^T \phi^T M \phi a + H^T \phi^T b + H^T \phi^T B^T f(N+1) \tag{38}$$

$$= \mathcal{M} \dot{W} + \mathcal{C} + J^T f(N+1) \tag{39}$$

where the mass matrix, $\mathcal{M} = H^T\phi^T M\phi H$, the Coriolis and Centrifugal forces, $\mathcal{C} = H^T\phi M\phi a + H^T\phi^T b$, and the combined Jacobian for both robots and the base, $J^T = H^T\phi^T B^T$.

Equation (39) gives the closed form solution to the inverse dynamics of an open chain of two robot manipulators on a mobile platform. The form of the equations is identical to a single manipulator system and could be derived equally as well by the Lagrange–Euler method. However, the advantage of the above approach is the insight it gives into the structure of $\mathcal{M}$.

Expanding $\mathcal{M}, \mathcal{C}$, and J^T and defining the submatrices gives:

$$\mathcal{M} = \begin{bmatrix} M(0) + E_1^T\phi_1^T M_1\phi_1 E_1 + E_2^T\phi_2^T M_2\phi_2 E_2 & E_1^T\phi_1^T M_1\phi_1 H_1 & E_2^T\phi_2^T M_2\phi_2 H_2 \\ H_1^T\phi_1^T M_1\phi_1 E_1 & H_1^T\phi_1^T M_1\phi_1 H_1 & 0 \\ H_2^T\phi_2^T M_2\phi_2 E_2 & 0 & H_2^T\phi_2^T M_2\phi_2 H_2 \end{bmatrix} \tag{40}$$

$$\equiv \begin{bmatrix} \mathcal{M}_B & \mathcal{M}_{1B}^T & \mathcal{M}_{2B}^T \\ \mathcal{M}_{1B} & \mathcal{M}_1 & 0 \\ \mathcal{M}_{2B} & 0 & \mathcal{M}_2 \end{bmatrix} \tag{41}$$

$$\mathcal{C} = \begin{bmatrix} b(0) + E_1^T\phi_1^T(M_1\phi_1 a_1 + b_1) + E_2^T\phi_2^T(M_2\phi_2 a_2 + b_2) \\ H_1^T\phi_1^T(M_1\phi_1 a_1 + b_1) \\ H_2^T\phi_2^T(M_2\phi_2 a_2 + b_2) \end{bmatrix} \tag{42}$$

$$\equiv \begin{bmatrix} \mathcal{C}_B \\ \mathcal{C}_1 \\ \mathcal{C}_2 \end{bmatrix} \tag{43}$$

$$J^T = \begin{bmatrix} E_1^T\phi_1^T B_1^T & E_2^T\phi_2^T B_2^T \\ H_1^T\phi_1^T B_1^T & 0 \\ 0 & H_2^T\phi_2^T B_2^T \end{bmatrix} \tag{44}$$

$$\equiv \begin{bmatrix} J_{1B}^T & J_{2B}^T \\ J_1^T & 0 \\ 0 & J_2^T \end{bmatrix} \tag{45}$$

The mass matrix for the combined robot and platform in equation (41) can be expressed as the mass matrix of each robot, as if it were on a fixed platform, $\mathcal{M}_1$, $\mathcal{M}_2$, plus additional elements due to the effect of platform motion on the robots and robot motion on the platform. The C vector can also be seen as the Coriolis terms for each individual arm where now the motion of the base is included. The J_1^T, and J_2^T terms are exactly the Jacobians for the manipulators on a fixed base platform. Thus, in the calculation of the inverse dynamics, any previously developed solution for the mass matrix and Jacobian matrix of fixed-base manipulators could be used in place of the $\mathcal{M}_1$, $\mathcal{M}_2$ matrices and the J_1^T, and J_2^T matrices.

2.4 Interaction with a Common Load

When the two manipulators carry a load, the forces exerted from the tips of the manipulators act to accelerate and/or compress the load. The motion of a point C on the load may be written [22]:

$$M(C)\alpha(C) + b(C) = A^T f(N+1) \tag{46}$$

where $M(C)$ is the 6x6 mass matrix of the load, $b(C)$ is the 6x1 vector of bias forces due to the angular velocity of the load, and A^T is the grasp matrix:

$$A^T = [\phi_1^T(N+1, C), \phi_2^T(N+1, C)] \tag{47}$$

which relates torques and forces at the manipulator tips to net torque/force at the point C.

The acceleration at point C can be related to the accelerations at the tips of the manipulator:

$$\alpha(N+1) = J\dot{W} + \dot{J}W \tag{48}$$

$$= A\alpha(C) + a(C) \tag{49}$$

where $a(C)$ are the bias accelerations induced by the angular velocity and differences in coordinate frames.

2.5 Summary

In summary, equations (39), (46), (48), and (49) give the complete closed-form solution to the inverse dynamics of two cooperating robot manipulators on a common, mobile platform, or:

$$T = \mathcal{M}\dot{W} + \mathcal{C} + J^T f(N+1) \tag{50}$$

$$\alpha(N+1) = J\dot{W} + \dot{J}W = A\alpha(C) + a(C) \tag{51}$$

$$M(C)\alpha(C) + b(C) = A^T f(N+1). \tag{52}$$

3 Mobile-Platform Closed-chain Forward Dynamics

The above equations present the solution to the inverse dynamics of the mobile closed-chain system. The forward dynamics problem consists of calculating the system acceleration, $\dot{W}$, when given the current state, W, θ_1, θ_2, and the input torques and forces on the joints and base, T.

Because we have placed the equations of motion for the mobile base in the same form as the equations for a fixed-base system, we may follow the same approach as in [22]. The approach consists of finding the forces exerted by each manipulator on the load, $f(N+1)$. With this force, one can then calculate the motion of the system from equation (50).

Using equations (50) and (51),

$$J\dot{W} = J\mathcal{M}^{-1}T' - J\mathcal{M}^{-1}J^T f(N+1) \tag{53}$$

where $T' = T - \mathcal{C}$. $\mathcal{M}$ is the combined mass matrix of the system and is always positive definite.

From equations (51) and (52),

$$J\dot{W} = A\alpha(C) + a(C) - \dot{J}W \quad (54)$$
$$= AM^{-1}(C)A^T f(N+1) + a' \quad (55)$$

where $a' = -AM^{-1}(C)b(C) + a(C) - \dot{J}W$.

Combining (53), (55)

$$\Omega f(N+1) = J\mathcal{M}^{-1}T' - a' \quad (56)$$

where

$$\Omega = J\mathcal{M}^{-1}J^T + AM^{-1}(C)A^T\,. \quad (57)$$

If it is assumed that the manipulators are not at a singular configuration then Ω may be inverted and the forces exerted at the tips of the manipulators can be uniquely determined. A stronger result for the invertibility of Ω for a fixed-base system was developed by Milman et. al. [15]. This result states that the probability that the system is at a configuration where Ω is uninvertible is zero. For the mobile-platform case we will assume that the manipulators are not at a singular configuration. The forces exerted at the tips may be calculated,

$$f(N+1) = \Omega^{-1}\left(J\mathcal{M}^{-1}T' - a'\right). \quad (58)$$

From the known tip forces, the motion of the system may be calculated.

However, the above equations for the calculation of the tip forces and resultant motion are not advantageous for the efficient calculation of manipulator dynamics in simulation. The forward dynamics can be rewritten using the concept of free dynamics. Define the free dynamics of the system as the dynamics with no tip forces: [22]

$$\mathcal{M}\dot{W}_f + \mathcal{C} = T \quad (59)$$

$$\alpha_f(N+1) = J\dot{W}_f + \dot{J}W \quad (60)$$
$$= J\dot{W}_f + B^T\phi^T a + a(N+1) \quad (61)$$

where the subscript f will be used to denote the free dynamics and equation (35) was used to derive the expression for $\dot{J}W$.

From (59),

$$\mathcal{M}\dot{W}_f = T - \mathcal{C} = T' \quad (62)$$

substituting into (56) and using the definitions of $\alpha_f(N+1)$, and a' gives:

$$\Omega f(N+1) = \alpha_f(N+1) + AM^{-1}(C)b(C) - a(C) \quad (63)$$

which gives the tip force in terms of the free accelerations of the tips. Thus for simulation purposes, the forward dynamics of the system are first calculated as if there was no load, or the manipulators are free from constraints, providing the free motion $\alpha_f(N+1)$ and $\dot{W}_f$. Knowledge of the free motion may then be used to calculate the actual tip forces when the manipulator tips are constrained to the common load. From the actual tip forces and free motion, the true system motion may be easily expressed. Let

$$\Delta\dot{W} = \dot{W} - \dot{W}_f \tag{64}$$

then (50) and (59) yield

$$\Delta\dot{W} = -\mathcal{M}^{-1}J^T f(N+1) \tag{65}$$

The result is a solution for the forward dynamics which parallels Rodriguez and Kreutz [22], where here the system includes the motion of the base.

1. Find the free accelerations:

$$\mathcal{M}\dot{W}_f + C = T \tag{66}$$

$$\alpha_f(N+1) = J\dot{W}_f + \dot{J}W \tag{67}$$

2. Find the tip forces, $f(N+1)$

$$\Omega = \left(J\mathcal{M}^{-1}J^T + AM^{-1}(C)A^T\right) \tag{68}$$

$$\Omega f(N+1) = \alpha_f(N+1) + AM^{-1}b(C) - a(C) \tag{69}$$

3. Find the correction acceleration, $\Delta\dot{W}$,

$$\mathcal{M}\Delta\dot{W} = -J^T f(N+1) \tag{70}$$

4. Find the true accelerations of robots and platform,

$$\dot{W} = \dot{W}_f + \Delta\dot{W} \tag{71}$$

With this approach, any solution to the forward dynamics of an open topological tree of rigid links may be used to solve equations (66), (70). For two, six DOF manipulators on a freely mobile platform in space, the mass matrix, $\mathcal{M}$, is 18x18. Instead of inverting this mass matrix at every simulation time step, is becomes effective to examine more efficient solutions to the forward-dynamics equations. Recursive approaches such as Featherstone, [5], Rodriguez [24] or Nasser [18], become increasingly efficient as the size of the mass matrix increases with the degrees of freedom.

4 Computer Simulation

Using the above equations, we simulated the cooperative motion of two PUMA 560 manipulators on a free space platform with a load of 3 kg. The solution to equations (66–71) produces the vector $\dot{W}$. This is an 18x1 vector of angular and linear accelerations of the platform, and joint accelerations for robot 1 and robot 2. The joint angles for both robots were found by numerically integrating the joint accelerations. The linear position of the platform was calculated by numerically integrating the linear acceleration of the platform. The orientation of the platform was determined by integrating the angular acceleration vector and using the resultant angular velocity to calculate the rate of change of the rotation matrix.

Figure 1 shows the initial positions of of the manipulators, platform, and load. The simulation of the dynamics was written in MATLAB [16], and a 4th-5th order Runge-Kutta routine was used to integrate the equations of motion.

4.1 Control

As an initial investigation of the methods for cooperative control, we chose to implement an object level controller for the positioning of the frame C. This is just one of many possible methods for the control of object position and internal force in a cooperative multi-manipulator system [38] [12][20]. The position and orientation of the frame C on the load was specified relative to the platform and not the inertial frame. The controller consisted of a PD tracking control on the Cartesian position of the load and a PD - quaternion-based control on the orientation of the load [37]. The forces necessary to move the load were evenly divided between the manipulators and converted to joint torques through the Jacobian transpose. No load-balancing or squeeze-force controllers were simulated at this time.

The platform was left uncontrolled in order to demonstrate the effects of the robot motion on the position and orientation of the platform during motion.

4.2 Results

The motion of the system is shown in Figures 1–3. The desired motion was to move the load downward 45 cm in 1 second, wait 0.5 seconds and then rotate the load 45 degrees about the load's z axis in 1 second. After another wait of 0.5 seconds, the motion was reversed. Figure 4 shows the Cartesian position errors relative to the platform frame. The system exhibits the expected large overshoot as the dynamics of the manipulators are not used in the control calculation. Other multiple-arm controllers have been described above which may be examined in future simulations.

4.3 Visualization

Because of the vast amounts of data present in the motion of an 18 DOF multi-robot system, it was found necessary to visualize the simulated motion on a graphics screen. The MATLAB dynamics simulation outputs a sequence of joint angles for both manipulators and a sequence of platform positions and orientations separated in time. Animation of the motion was presented through Silma Cimstation, a graphical robot modeling package [25]. Code was written in Cimstation to read the kinematic data provided from MATLAB and move the manipulators and base accordingly. This allowed for direct visual inspection of the manipulator and base motion.

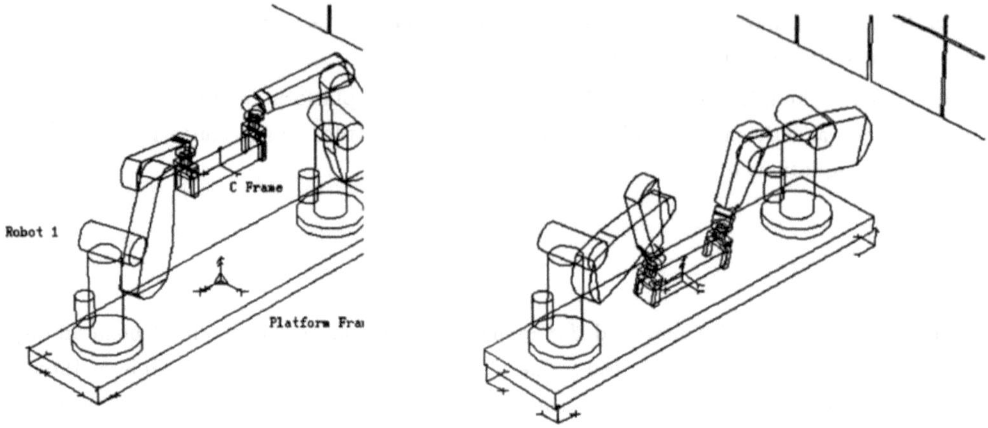

Initial Configuration

After 1.5 Seconds

Figure 1

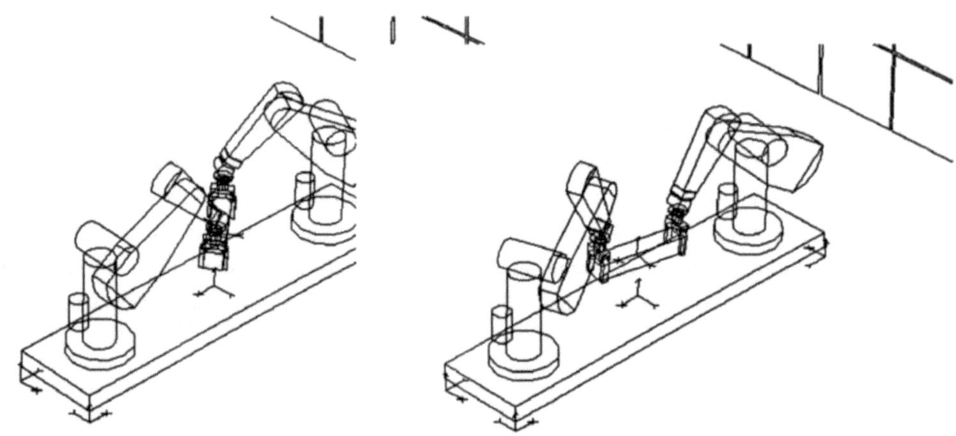

After 3.0 Seconds

After 4.5 Seconds

Figure 2

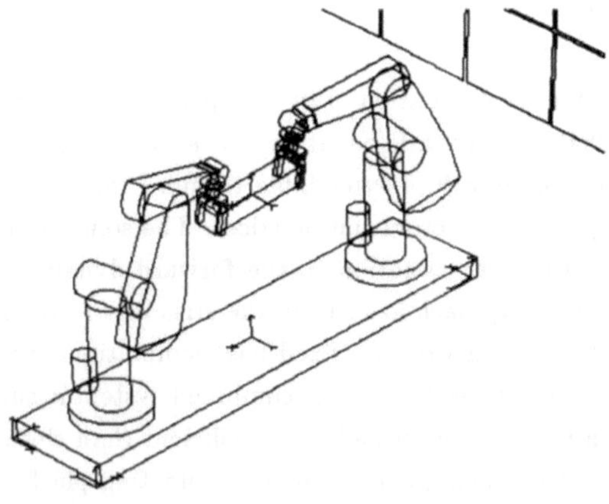

Figure 3 After 6 Seconds

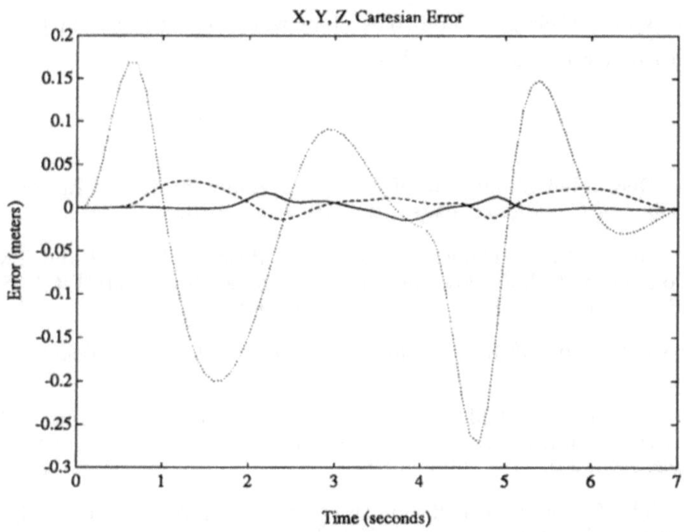

Figure 4 Cartesian Error of Load in Platform Frame. Solid = X, Dashed = Y, Dotted = Z

5 Conclusion

This paper has presented a method for the simulation of two cooperating robot manipulators on a mobile platform. We have shown that the dynamic equations for such a system have an identical form as the fixed-platform case. The approach is extendable to more than two manipulators by increasing the dimensions of the appropriate vectors and matrices. The solution to the cooperative motion is able to incorporate any form of the solution to the forward dynamics of a topological tree of manipulators and platform. The approach has shown the structure in the inverse dynamics of the system where portions of the existing work on fixed-base simulations may be included in the calculations. The number of degrees of freedom of the combined system is sufficiently large to make recursive, dynamic calculation methods potentially more efficient than closed form solutions. A complete simulation with two, six DOF manipulators on a free-floating platform was presented along with a multiple-arm controller to position the common load.

Acknowledgment

The authors wish to express their gratitude to A. Jain who provided the starting point for the simulation code and to K. Kreutz for his insights.

References

[1] Anderson, J. N. Decoupled Simulation of Two-Arm Robot Systems. In *Proceedings of the 1986 American Control Conference*, pages 127–129, June 1986.

[2] Anderson, J. N. and M. E. Pittelkau. Reflex-Action Position Control and Two-Arm Load-Sharing Force Control for Robotic Manipulators. Technical Report MCTR-1-88-1, Center for Manufacturing Research, Tenn. Tech. Univ, January 1988.

[3] Craig, J. J. *Introduction to Robotics, Mechanics & Control.* Addison–Wesley, Reading, Mass., 1986.

[4] Dubowsky, S., E. E. Vance, and M. A. Torres. The Control of Space Manipulators Subject to Spacecraft Attitude Control Saturation Limits. In *Proceedings of the NASA Conference on Space Telerobotics.* JPL, Jan 31 – Feb 2 1989.

[5] Featherstone, R. The Calculation of Robot Dynamics Using Articulated-Body Inertias. *The International Journal of Robotics Research*, Vol. 2, pp 13–29, 1983.

[6] Hayati, S. Hybrid Position/Force Control of Multi-Arm Cooperating Robots. In *Proceedings of the 1986 IEEE International Conference on Robotics and Automation*, pages 82–89, April 1986.

[7] Jain, A. Unified Formulation of Dynamics for Serial Rigid Multibody Systems. Technical Report 347–88–236, Jet Propulsion Laboratory Engineering Memorandum, April 1989.

[8] Jerkovsky, W. The Structure of Multibody Dynamics Equations. *Journal of Guidance and Control*, Vol. 1, No. 3, pp 173–182, 1978.

[9] Joshi, J. and A. A. Desrochers. Modeling and Control of a Mobile Robot Subject to Disturbances. In *Proceedings of the 1986 IEEE International Conference on Robotics and Automation*, pages 1508–1513, 1986.

[10] Kane, T. R. and D. A. Levinson. Formulation of Equations of Motion for Complex Spacecraft. *Journal of Guidance and Control*, Vol. 3, No. 2,, 1980.

[11] Kane, T. R. and D. A. Levinson. The Use of Kane's Dynamical Equations in Robotics. *The International Journal of Robotics Research*, Vol. 2, No. 3, pp 3–18, 1983.

[12] Kreutz, K. and A. Lokshin. Load Balancing and Closed Chain Multiple Arm Control. In *Proceedings of the 1988 American Control Conference*, pages 2148–2155, 1988.

[13] Lips, K. W. and R. P. Singh. Obstacles to High Fidelity Multibody Dynamics Simulation. In *Proceedings of the 1988 American Control Conference*, pages 587–594, 1988.

[14] Longman, R. W., R. E. Lindberg, and M. F. Zedd. Satellite-Mounted Robot Manipulators – New Kinematics and Reaction Moment Compensation. *The International Journal of Robotics Research*, Vol. 6, No. 3, pp 87–103, 1987.

[15] Milman, M., K. Kreutz, and A. Lokshin. Coordination and Load Balancing for Multi–Arm Systems. In *Int. Symp. on Robotics*, November 1988. Albuquerque, NM.

[16] Moler, C., J. Little, and S. Bangert. *Pro-Matlab User's Guide*. The MathWorks, Inc., 1987.

[17] Nakamura, Y. and R. Mukherjee. Nonholonomic Path Planning of Space Robots. In *Proceedings of the 1989 IEEE International Conference on Robotics and Automation*, pages 1050–1054, 1989.

[18] Nasser, M. G. Recursive Newton Euler Formulation of Manipulator Dynamics. In *Proceedings of the NASA Conference on Space Telerobotics*. JPL, Jan 31 – Feb 2 1989.

[19] Oh, S. Y. and D. E. Orin. Dynamic Computer Simulation of Multiple Closed–Chain Robotic Mechanisms. In *Proceedings of the 1986 IEEE International Conference on Robotics and Automation*, pages 15–20, 1986.

[20] Orin, D. E. and S. Y. Oh. Control of Force Distribution in Robotic Mechanisms Containing Closed Kinematic Chains. *Journal of Dynamic Systems, Measurement and Control*, Vol. 102, pp 134–141, June 1981.

[21] Papadopoulos, E. and S. Dubowsky. On the Nature of Control Algorithms for Space Manipulators. In *Proceedings of the 1990 IEEE International Conference on Robotics and Automation*, May 1990.

[22] Rodriguez, G. and K. Kreutz. An Operator Approach Formulated Dynamics for Multiple Arms Grasping a Commonly Held Object. Technical Report 347–88–237, Jet Propulsion Laboratory Engineering Memorandum, April 1988.

[23] Rodriguez, G. and K. Kreutz. An Operator Formulated Manipulator Dynamics. Technical Report 347–88–236, Jet Propulsion Laboratory Engineering Memorandum, April 1988.

[24] Rodriquez, G. Recursive Dynamics for Topological Trees of Rigid Bodies by Kalman Filtering and Bryson-Frazier Smoothing. In *VPI/SU Symposium on Dynamics and Control of Structures*, 1987. Blacksburg, VA.

[25] SILMA, Inc., 1601 Saratora-Sunnyvale Road, Suite 155, Cupertino, CA 95014. *Cimstation User's Manual.*

[26] Tanner, A. B. Study of Robotic Manipulators Subject to Base Disturbances. Master's thesis, Massachusetts Institute of Technology, January 1987.

[27] Tarn, T. J., A. K. Bejczy, and X. Yun. Coordinated Control of Two Robot Arms. In *Proceedings of the 1986 IEEE International Conference on Robotics and Automation*, pages 1193–1202, April 1986.

[28] Tarn, T. J., A. K. Bejczy, and X. Yun. Dynamic Coordination of Two Robot Arms. In *Proceedings of the 25th Conference on Decision and Control*, pages 1267–1270, December 1986.

[29] Tarn, T. J., A. K. Bejczy, and X. Yun. Design of Dynamic Control of Two Cooperating Robot Arms: Closed Chain Formulation. In *Proceedings of the 1987 IEEE International Conference on Robotics and Automation*, pages 7–13, 1987.

[30] Umetani, Y. and K. Yoshida. Continuous Path Control of Space Manipulators Mounted on OMV. *ACTA Astronautica*, Vol. 15, No. 12, pp 981–986, 1987.

[31] Umetani, Y. and K Yoshida. Experimental Study on Two Dimensional Free–Flying Robot Satellite Model. In *Proceedings of the NASA Conference on Space Telerobotics*. JPL, Jan 31 – Feb 2 1989.

[32] Unseren, M. A. and A. J. Koivo. Kinematic Relations and Dynamic Modeling for Two Cooperating Manipulators in Assembly. In *Proceedings of the 1987 International Conference on Systems, Man, and Cybernetics*, pages 798–802, October 1987.

[33] Vafa, Z. and S. Dubowsky. On the Dynamics of Manipulators in Space Using the Virtual Manipulator Approach. In *Proceedings of the 1987 IEEE International Conference on Robotics and Automation*, pages 579–585, 1987.

[34] Walker, M. W. and J. Dionise. On the Simulation of Space Based Manipulators with Contact. In *Proceedings of the NASA Conference on Space Telerobotics*. JPL, Jan 31 – Feb 2 1989.

[35] Walker, M. W., D. Kim, and J. Dionise. Adaptive Coordinated Motion of Two Manipulator Arms. In *Proceedings of the 1989 IEEE International Conference on Robotics and Automation*, pages 1084–1090, 1989.

[36] Wang, P. K. C. Control Strategy for a Dual–Arm Maneuverable Space Robot. In Rodriguez, G., editor, *Proceedings of the Workshop on Space Telerobotics*, pages 257–266. JPL Publication 87–13, 1987. Volume 2.

[37] Wen, J. T. and K. Kreutz. Attitude Control of an Object Commonly Held by Multiple Robot Arms: A Lyapunov Approach. In *Proceedings of the 1988 American Control Conference*, pages 1790–1794, 1988.

[38] Wen, J. T. and K. Kreutz. Stability Analysis of PD Control for Multiple Robot Manipulators Holding a Common Object. Technical Report 347–88–229, Jet Propulsion Laboratory Engineering Memorandum, February 1988.

[39] Zheng, Y. F. Kinematics and Dynamics of Two Industrial Robots in Assembly. In *Proceedings of the 1989 IEEE International Conference on Robotics and Automation*, pages 1360–1365, 1989.

[40] Zheng, Y. F. and J. Y. S. Luh. Control of Two Coordinated Robots in Motion. In *Proceedings of the 24th Conference on Decision and Control*, pages 1761–1765, 1985.

[41] Zheng, Y. F. and J. Y. S. Luh. Joint Torques for Control of Two Coordinated Moving Robots. In *Proceedings of the 1986 IEEE International Conference on Robotics and Automation*, pages 1375–1380, April 1986.

[42] Zheng, Y. F. and J. Y. S. Luh. Optimal Load Distribution for Two Industrial Robots Handling a Single Object. In *Proceedings of the 1988 IEEE International Conference on Robotics and Automation*, pages 344–349, 1988.

ON THE THEORY OF INTELLIGENT MACHINES

George N. Saridis
Electrical, Computer and Systems Engineering Department
Rensselaer Polytechnic Institute
Troy, New York 12180-3590

ABSTRACT

The Theory of Intelligent Machines has been formulated analytically. The functions of an Intelligent Machine are executed by Intelligent Controls. The Principle of Increasing Precision with Decreasing Intelligence is used to form a hierarchical structure of the control systems. The three levels of the Intelligent Control, e.g., the Organization, Coordination and Execution Levels are described as originally conceived. New designs as Neural-nets for the organization level and Petri-nets for the coordination level are included. Application to Intelligent Robots for space exploration has been used as examples in this work.

NATO ASI Series, Vol. F 71
Expert Systems and Robotics
Edited by T. Jordanides and B. Torby

1 INTRODUCTION

Several researchers have developed viable theories for Intelligent Machines and created working models to implement such theories (Albus 1985, Meystel I986, Saridis 1985, Zames 1979, etc.), in the past fifteen years. Such machines were designated to perform anthropomorphic tasks with minimum interaction with a human with potential applications on robotic systems designed to operate in remote, inaccessible, hazardous, unfamiliar or other environments as need appeared.

Since the task was enormous and the available technologies rather limited, the results of such an effort have been meager. The theoretic efforts that have come into the picture to reinforce the development of Intelligent Machines have taken two distinct directions: the logic-based approach (Nielson 1969) and the analytic approach (Saridis 1988, Meystel 1986).

The results on the analytic approach, which concerns this particular paper, have been reported regularly by the author (Saridis 1977, Saridis 1979, Saridis 1983, Saridis 1985c, Saridis 1988) and have reached a level of maturity both theoretically and experimentally. A summary of the work produced is given in the next section.

In the past few years, new methodologies and new techniques like Neural-nets, Petri-nets, Boltzmann machines etc., have appeared in the literature and have provided new tools for the analytic formulation of the theory of Intelligent Machines. The further development and adaptation of such tools, along with the better understanding of the process led to modifications and refinements of the theory aimed to strengthen, simplify and integrate the proposed design of Intelligent Machines.

The refinements introduced herein are due to the better understanding of some of the basic concepts of the Intelligent Machines, e.g., the Principle of Increasing Precision with Decreasing Intelligence (IPDI), the ability to develop Boltzmann Machines and Petri-Nets as realizations of Inference Engines and Linguistic Decision Schemata, respectively, and the use of entropy measures for the evaluation of the performance at every level of the machine.

A review of the analytic formulation of the Intelligent Machines is given in the next section, followed by a set of pertinent definitions and a discussion on the principle of IPDI. A brief explanation of the development of the Boltzmann machine as an inference engine for the organization level is next. The following section presents details on the three levels of the Intelligent Machine with introduction of Neural-nets to the organization level and

Petri-nets to the coordination level. The next section places the Intelligent Machine in a Distributed Intelligence environment, followed by applications, discussions and conclusions.

2 REVIEW OF THE ANALYTIC FORMULATION OF INTELLIGENT CONTROLS

Intelligent Machines require the use of "generalized" control efforts in order to perform intelligent functions such as simultaneous utilization of a memory, learning, or multilevel decision making in response to simple qualitative commands. Intelligent Controls have been developed by Saridis (1977, 1983) to implement such functions.

Intelligent Control can be considered as a fusion between the mathematical and linguistic methods and algorithms applied to systems and processes. They utilize the results of cognitive systems research effectively with various mathematical programming control techniques.

The control intelligence is hierarchically distributed according to the Principle of Precision with Decreasing Intelligence (IPDI), evident in all hierarchical management systems, and it is further discussed in a future section (Saridis 1988b). The resulting structure is composed of three basic levels of controls, each level of which may contain more than one layer of tree-structured functions (Saridis 1979) (See Figure 1):

1. The organization level.
2. The coordination level.
3. The execution level.

The functions involved in the upper levels of an intelligent machine are imitating functions of human behavior and may be treated as elements of knowledge-based systems. Actually, the activities of planning, decision making, learning, data storage and retrieval, task coordination, etc., may be thought of as knowledge handling and management. Therefore, the flow of knowledge in an intelligent machine may be considered as the key variable of such a system.

Knowledge flow in an intelligent machine represents respectively:

1. Data Handling and Management.

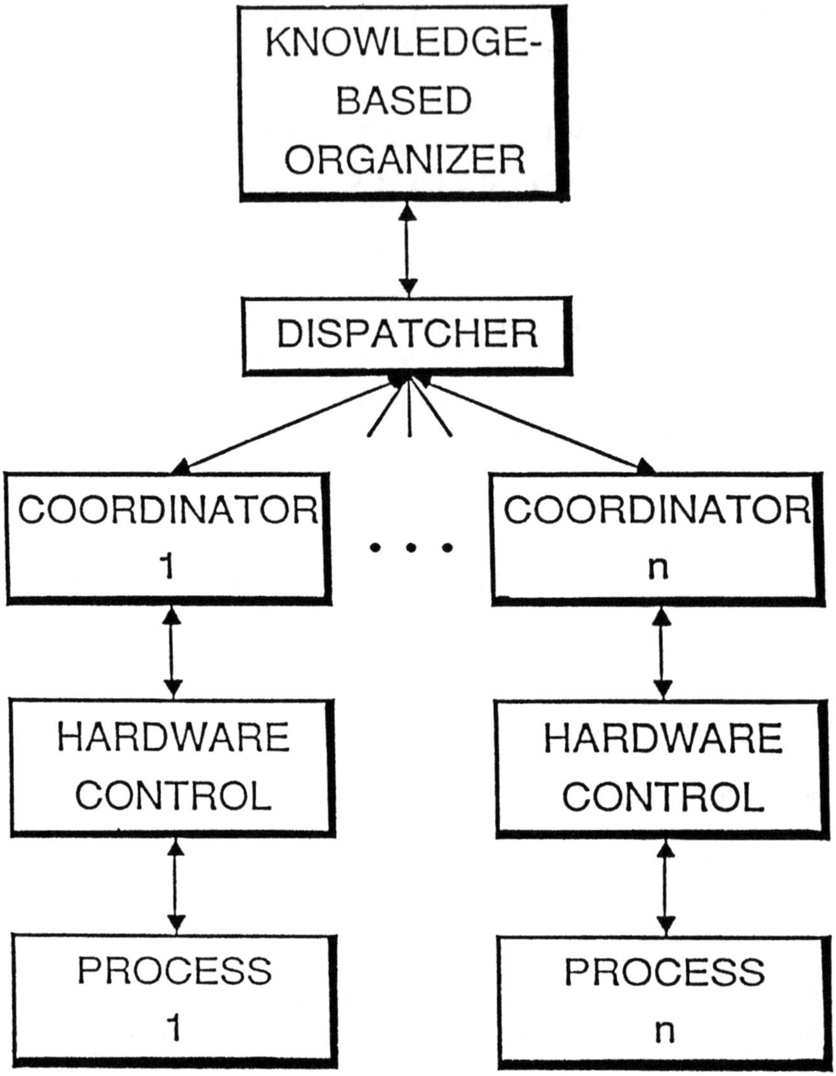

Fig. 1. Hierarchical Intelligent Control System.

2. Planning and Decision performed by the central processing units.

3. Sensing and Data Acquisition obtained through peripheral devices.

4. Formal Languages which define the software.

Subjective probabilistic models are assigned to the individual functions. Their entropies may be evaluated for every task executed. This provides an analytical measure of the total activity.

Artificial Intelligence methods, using among other logic functions have been used to implement Intelligent Machines (Albus 1975, Meystel 1985). However, they lack that rigor and precision that analytic techniques provide. Nevertheless new methodologies have been adapted to analytic models to perform tasks at the various levels of an Intelligent Machine.

Moed and Saridis (1990), proposed a neural net approach to perform reasoning, planning and decision making in the organization level of an Intelligent Machine. A Boltzmann machine, suitable for the discrete binary state model of this particular level, is a natural device for organizing actions and rules necessary for the execution of a given command, regardless of the particular world model the machine is inhabiting.

Wang and Saridis (1988) proposed a Petri-net transducer to implement the Linguistic Decision Schemata (Saridis and Graham 1984), which serve as model coordinators and decision makers at the machine's coordination level. These devices set up the communication protocols, with the help of small real-time memories, and serve apply in real time the rules generated by the organization level to properly generate and sequence the subtasks in the particular environment of the machine, in order to execute the given original command.

Finally, Saridis (1988a) was able to reformulate the system control problem to use entropy as a control measure and therefore integrate all the hardware activities associated with the Intelligent Machine with the other levels regardless of the discipline they belong to. Thus, vision coordination, motion control, path planning, force sensing, etc., in a robot paradigm, may be integrated into the pertinent actions of the machine and evaluated by common entropy functions.

Since all levels of a hierarchical intelligent control can be measured by entropies and their rates, then the optimal operation of an "intelligent machine" can be obtained through the solution of mathematical programming problems.

Another development of this theory the structure of the "nested hierarchical" systems (Meystel, 1986). Even when the hierarchy is not tree-like, still using hierarchy is beneficial since the hierarchy of resolutions (errors per level) helps to increase the effectiveness of the system under limited computing power which is important to mobile systems.

The various aspects of the theory of hierarchically intelligent controls may be summarized as follows:

> The theory of intelligent machines may be postulated as the mathematical problem of finding the right sequence of decisions and controls for a system structured according to the principle of increasing precision with decreasing intelligence (constraint) such that it minimizes its total entropy.

3 DEFINITIONS AND THE IPDI

3.1 Definitions

It remains to investigate the general concepts of Intelligent Control Systems which pertain to the fundamental functions Intelligent Machines. Such are the notions of Machine Knowledge, its Rate and Precision.

> Definition 1. Machine Knowledge is defined to be the structured information acquired and applied to remove ignorance or uncertainty about a specific task pertaining to the Intelligent Machine.

Knowledge is a cumulative quantity accrued by the machine and cannot be used as a variable to execute a task. Insteady, the Rate of Machine Knowledge is a suitable variable.

> Definition 2. Rate of Machine Knowledge is the flow of knowledge through an Intelligent Machine.

Intelligence is defined by the American Heritage Dictionary of the English Language (1969) as the capacity to acquire and apply knowledge.

In terms of Machine Intelligence, this definition may be modified to yield:

Definition 3. Machine Intelligence (MI) is the set of actions or rules which operates on a data-based (DB) of events to produce flow of knowledge (R).

On the other hand, one may define Precision as follows:

Definition 4. Imprecision is the uncertainty of execution of the various tasks of the Intelligent Machine.

and

Definition 5. Precision is the complement of Imprecision, and represents the complexity of a process.

Analytically, the above relations may be summarized as follows:

Knowledge (K) representing a type of information may be represented as

$$K = -\alpha - lnp(K) \tag{1}$$

where $p(K)$ is the probability density of Knowledge.

From equation (1) the probability density function $p(K)$ satisfies the following expression in agreement with Jaynes' Principle of Maximum Entropy (1957):

$$p(K) = e^{-\alpha-K}; \quad \alpha = ln \int_X e^{-K} dx \tag{2}$$

The Rate of Knowledge R which is the main variable of an intelligent machine with discrete states is defined over a fixed interval of time T:

$$R = \frac{K}{T}$$

It was intuitively thought (Saridis 1983), that the Rate of Knowledge must satisfy the following relation which may be thought of expressing the principle of Increasing Precision with Decreasing Intelligence.

$$(MI) : (DB) to (R) \tag{3}$$

An analytic formulation of the above principle derived from simple probabilistic relation among the Rate of Knowledge, Machine Intelligence and the Data Base of Knowledge, is presented in the next section. The entropies of the various functions come naturally into the picture as a measure of their activities.

3.2 The Analytic Formulation of IPDI

In order to formulate mathematically the concepts of knowledge-based systems, one must consider the state space of knowledge Ω_s with states $s_i, i = 1, 2, ...n$. They represent the state of events at the nodes of a network defining the stages of a task to be executed.

Then knowledge between two states is considered as the association of the state s_i with another state s_j and is expressed as

$$K_{ij} = \frac{1}{2} w'_{ij} s_i s_j \tag{4}$$

where w_{ij} are state transition coefficients, which are zero in case of inactive transmission.

Knowledge at the state of s_i is the association of that stat with all the other active states s_j and is expressed as

$$K_i = \frac{1}{2} \sum_j w'_{ij} s_i s_j \tag{5}$$

Finally, the total knowledge of a system is considered as

$$K = \frac{1}{2} \sum_i \sum_j w'_{ij} s_i s_j \tag{6}$$

and has the form of energy of the underlying events. The rate (flow) of knowledge is the derivative of knowledge and for the discrete state space Ω_s is defined respectively

$$R_{ij} = \frac{K_{ij}}{T} , \quad R_i = \frac{K_i}{T} , \quad R = \frac{K}{T} \tag{7}$$

where T is a fixed time interval.

The principle of Increasing Precision with Decreasing Intelligence is expressed probabilistically by

$$PR(MI, DB) = PR(R) \tag{8}$$

where MI is the machine intelligence and DB is the data base associated with the task to be executed and represents the complexity of the task which is also proportional to the precision of execution. The following relation is obtained by conditioning and taking the natural logarithms:

$$lnp(MI/DB) + lnp(DB) = lnp(R) \tag{9}$$

Taking the expected value on both sides

$$H(MI/DB) + H(DB) = H(R) \tag{10}$$

where $H(x)$ is the entropy associated with x. For a constant rate of knowledge which is expected during the conception and execution of a task increase of the entropy of DB requires a decrease of the entropy of MI for the particular data base, which manifests the IPDI. IF MI is independent of DB then

$$H(MI) + H(DB) = H(R) \tag{11}$$

A detailed proof of the Principle is given in Saridis (1989).

This Principle is applicable both across one level of the Intelligent Hierarchy as well as throught the levels of the Hierarchy, in which case the flow R represents the throughput of the system in an information theoretic manner. The partition law of information rate applies naturally to such a system (Saridis 1985c).

As case study demonstrating the validity of the above is given in Saridis and Valavanis (1988).

4 THE ANALYTIC STRUCTURE OF THE INTELLIGENT MACHINE

In order to implement an Intelligent Machine on analytic foundations, the theory of Intelligent Control has been developed by Saridis (1979), and briefly discussed in Section 2. This theory assigns analytic models to the various levels of the machine and improve them through a generalized concept of selective feedback.

The Intelligent Control System is composed of three levels in decreasing order of intelligence and increasing order of precision as stipulated by the IPDI. However, with the better understanding of the basics, new methodologies are proposed to analytically implement the various functions, without significantly changing the models at each level.

The Organization Level is designed to organize a sequence of abstract actions or rules from a set of primitives stored in a long-term memory regardless of the present world model. In other words it serves as the generator of the rules of an Inference Engine by processing (intelligence) high level of information, for reasoning, planning and decision making. This can be accomplished by a two level Neural-net, analytically derived as a Boltzmann machine by Saridis and Moed (1988 and 1990).

The Coordination Level is an intermediate structure serving as an interface between the organization and execution levels. It deals with real- time information of the world by generating a proper sequence of subtasks pertinent to the execution of the original command.

It involves coordination of decision making and learning on a short term memory, e.g., a buffer. It utilizes Linguistic Decision Schemata with learning capabilities defined in Saridis and Graham (1984), assigned subjective probabilities for each action. The respective entropies may be obtained directly from these subjective probabilities. Petri Net Transducers have been investigated by Wang and Saridis (1988), to implement such decision schemata. In addition, Petri-nets provide the necessary protocols to communicate among the various coordinators, in order to integrate the activities of the Machine. Complexity functions may be used for real-time evaluation.

The Execution Level performs the appropriate control functions on the processes involved. Their performance measure can also be expressed as an entropy, thus unifying the functions of an Intelligent Machine.

Optimal control theory utilizes a non-negative functional of the states of a system in the state space, and a specific control from the set of all admissible controls, to define the performance measure for some initial conditions, representing a generalized energy function. Minimization of the energy functional yields the desired control law for the system.

The Principle of IPDI is applicable at every level of the Machine, reaffirming its universal validity. However, the coordination may serve as a salient example of its application where the intelligence provided by the organization level as a set of rules is applied to the database provided by the execution level to produce flow of knowledge.

A more detailed description of the analytic functions of each level is given in the sequel.

4.1 The Neural-Net Based Organization Level

The function of the organizer, the highest level of the hierarchy of Intelligent Controls, is based on several AI (knowledge based) concepts forming the foundations of Machine Intelligence. These concepts translated into probabilistic models form the functions of representation and reasoning, planning, decision making, long-term memory exchange and learning through feedback to set up a task in response to some outside command. The probabilistic model generated provides the mechanism to select the appropriate task for the appropriate command. The principle followed here is that instead of task decomposition a collection of tasks is generated from a list of primitives stored in the memory and matched against the input command applied.

To specify analytically the model of the organizer, it is essential to derive the domain of the operation of the machine for a particular class of problems (Saridis and Valavanis 1988). Assuming that the environment is known, one may define the following functions on the organization level:

a) Machine Representation and Abstract Reasoning, (RR) is the association of the compiled command to a number of activities and/or rules. A probability function is assigned to each activity and/or rule and the Entropy associated with it is calculated. When rules are included one has active reasoning (inference engine).

   In order to generate the required analytic model of this function the following sets are defined:

   The set of commands $C = \{c_1, c_2, ..., c_m\}$ in natural language, is received by the machine as inputs. Each command is compiled to yield an equivalent machine code explained in the next section.

   The task command of the machine contains a number n of independent events.

   The events $E = \{e_1, e_2, ..., e_n\}$ are individual primitive objects or actions e_i stored in the long-term memory and repesenting primitive tasks to be executed. The task domain indicates the capabilities of the machine. Events represent the nodes of the Neural-net.

   Activities A, are groups of events concatenated to define a complex task; e.g., $A_{234} = \{e_2, e_3, e_4\}$. If the events are ordered then we have an ordered activity.

   A set of random variables $X = \{x_1, \ldots x_n\}$ representing the state of events is associated with each individual event e_i. If the random variable x_i is binary (either 0 or 1), it indicates whether an event e_i is inactive or active, in a particular activity and for a particular command. If the random variables x_i are continuous (or discrete but not binary) over [0,1], they reflect a membership function in a fuzzy decision making problem. At this point, the x_i's are considered to be binary.

A set of probabilities P associated with the random variables X is defined as follows.

$$P = \{p_i = Prob[x_i = 1]\}$$

The probabilities P are known at the beginning of the representation stage. In order to reduce the problem of dimensionality a subset of events is defined for the given command c_k.

$$S_k = \{e_i; p_i > a\} \subset E$$

b) Machine Planning, (P), is ordering of the activities.

The ordering is obtained by properly concatenating the appropriate abstract primitive events $e_i \subset S_k$ for the particular command c_k, in order to form the right abstract activities (sentences or text).

The ordering is generated by a Boltzmann machine which measures the average flow of knowledge from node j to node i on the Neural-net by

$$R_{ij} = \frac{1}{2}E\{w_{ij}x_ix_j\} = \frac{1}{2}w_{ij}p_ip_j \geq 0 \tag{12}$$

The probability due to the uncertainty of knowledge flow into node i, is calculated by:

$$p(R_i) = e^{\alpha_i - \frac{1}{2}\sum_j w_{ij}p_ip_j} \tag{13}$$

where

$w_{ij} \geq 0$ is the interconnection weight between nodes i and j

$w_{ij} = 0$

$\alpha_i > 0$ is a probability normalizing factor.

The average Flow of Knowledge R_i into node i, is:

$$R_i = \alpha_i + \frac{1}{2}E\{\sum_j w_{ij}x_ix_j\} = \alpha_i + \frac{1}{2}\sum_j w_{ij}p_ip_j$$

with probability $P(R_i)$, (Jaynes' Principle):

$$P(R_i) = exp[-\alpha_i - \frac{1}{2}\sum_j w_{ij}p_ip_j].$$

The Entropy of Knowledge Flow in the machine is

$$H(R) = -\sum_i P(R_i) \text{ in } [P(R_i)] =$$

$$\sum_i(\alpha_i + \frac{1}{2}\sum_j w_{ij}p_ip_j)exp[-\alpha_i - 1/2\sum_j w_{ij}p_ip_j.] \qquad (14)$$

The normalizing factor α_i is such that $1/2^n \leq P(R_i) \leq 1$.

The entropy is maximum when the associated probabilities are equal, $p(R_i) = \frac{1}{2^n}$ with n the number of nodes of the network. By bounding $p(Ri)$ from below by $\frac{1}{2^n}$ one may obtain a unique minimization of the entropy corresponding to the most like sequence of events to be selected.

Unlike the regular Boltzmann machines, this formulation does not remove α_i when $p_i = 0$. Instead, the machine operates from a base entropy level $\propto_i \epsilon^{-\alpha_i}$ defined as the Threshold Node Entropy which it tries to reduce (Saridis and Moed, 1988).

c. Machine Decision Making, (DM) is the function of selecting the sequence with the largest probability of success.

This is accomplished through a search to connect a node ahead that will minimize the Entropy of Knowledge Flow at that node:

$$H(R_i) = (\alpha_i + \frac{1}{2}\sum_j w_{ij}p_ip_j)\ exp[-\alpha_i - \frac{1}{2}\sum_j w_{ij}p_ip_j]$$

A modified genetic algorithm, involving a global random search, has been proposed by Moed and Saridis (1990) as a means of generating the best sequence of events that minimized the uncertainty of connections of the network expressed by the entropy (14).

This algorthim, proven to converge globally compared favorably with other algorithms like the Simulated Annealing and the Random Search. The DM process is illustrated in Figure 2.

d. Machine Learning, (ML) (Feedback). Machine Learning is obtained by feedback devices that upgrade the probabilities p_i and the weights w_{ij} by evaluating the performance of the lower levels after a successful iteration.

For y_k representing either p_i or w_{ij}, corresponding to the command c_k, the upgrading algorithms are:

$$\begin{aligned} y_k(t_k+1) &= y_k(t_k) + \beta_k(t_k+1)[\xi(t_k+1) - y_k(t_k)] \\ J_k(t_k+1) &= J_k(t_k) + \gamma_k(t_k+1)[V^k_{obs}(t_k+1) - J_k(t_k)] \end{aligned} \tag{15}$$

where J_k is the performance estimate, V^k_{obs} its observed value and

$$\begin{aligned} p_i &: \xi_k(t_k+1) = x(t_k) \\ w_{ij} &: \xi_k(t_k+1) = \begin{cases} 1 & \text{if } J = {}^{min}_{\;e} J_e \\ 0 & \text{otherwise} \end{cases} \end{aligned} \tag{16}$$

β_k and γ_k are members of the harmonic sequence.

e. Memory Exchange (ME), is the retrieval and storage of information from the long-term memory, based on selected feedback data from the lower levels after the completion of the complex task.

The above functions may be implemented by a two level Neural-net, of which the nodes of the upper level represent the primitive objects e_{oi} and the lower level of primitive actions relating the objects e_{ai} of a certain task. The purpose of the organizer may be realized by a search in the Neural-net to connect objects and actions in the most likely sequence for an executable task.

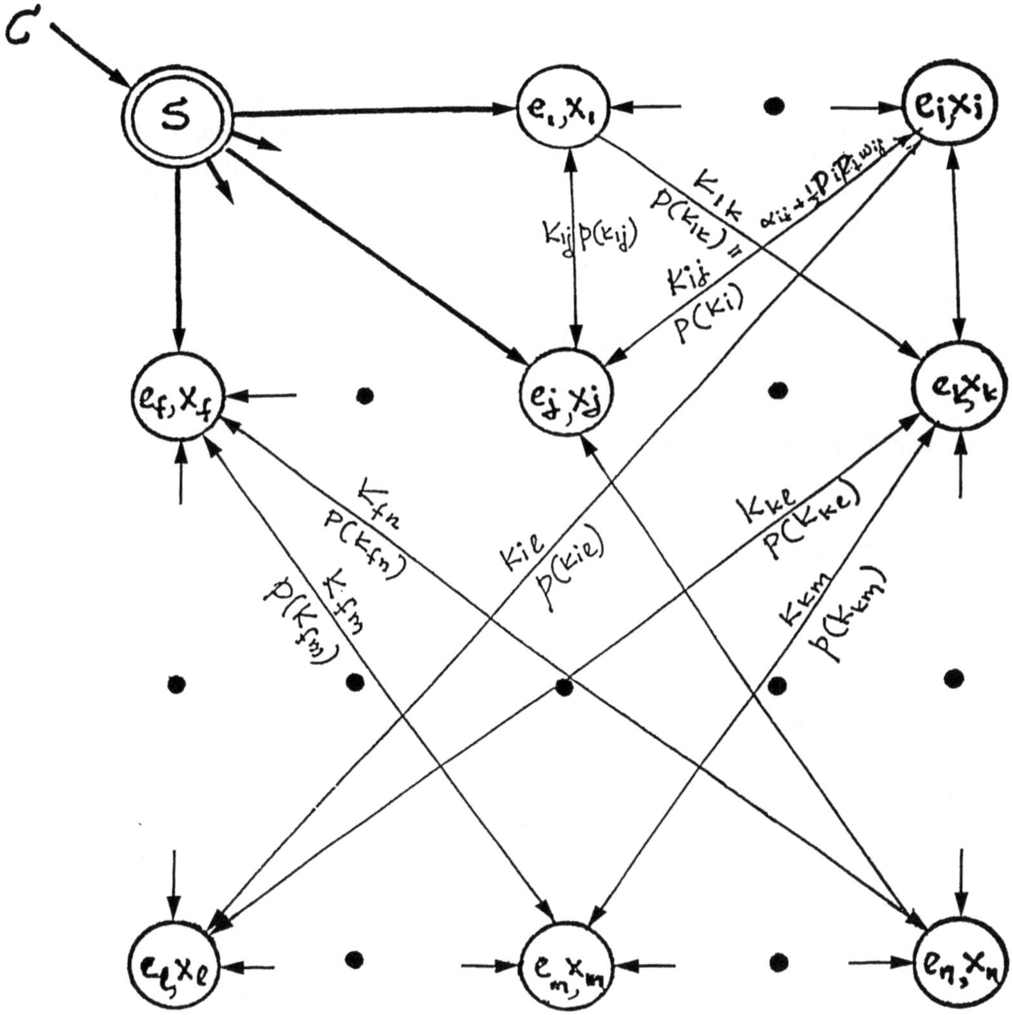

Fig. 2. The Boltzmann Machine for the Organization Level.

e_i = primitive event

x_i = state of event e_i, $c\{1,0\}$; with prob p_i

K_i = energy at node i, $= \alpha_i + \frac{1}{2} \sum_j p_i\, p_j\, w_{ij}$

w_{ij} = learned weights

$p(K_i)$ = probability of connection i-j.

4.2 The Coordination Level and Petri Net Transducers

The coordination level is an intermediate structure serving as an interface between the organization and the execution level. It is essential for dispatching and communicating organizational information to the execution level. Its objective is the actual formulation of the control problem associated with the most probable complete and compatible plan generated by the organization level and based on real-time acquired information about the world model.

The functions of the coordination level are summarized as follows:

a. Dispatching of tasks requested by the organization level.

b. Identification and of the current environment.

c. Real-time decision-making.

d. Data transfer and communication.

e. Use of formal languages.

f. Learning (Feedback)

g. Interfacing.

The coordination level is composed of a dispatcher and a number of specialized coordinators (Figure 3). Specific hardware (execution devices) from the execution level is associated with each coordinator. These devices execute well defined tasks when a command is issued to them by their corresponding coordinator (Saridis and Valavanis 1988). The dispatcher serves as both the communicator of information from the organization level to the coordinators and on-line exchange of data among the coordinators. A Petri-net formulation of these activities has been recently proposed by Wang and Saridis (1988).

Petri-nets have been proposed as devices to communicate and control complex heterogenous processes. These nets provide a communication protocol among stations of the process as well as the control sequence for each one of them (Peterson 1977).

Abstract task plans, suitable for many environments are generated at the organization level by a grammar:

$$G = (N, \Sigma_o, P, S)$$

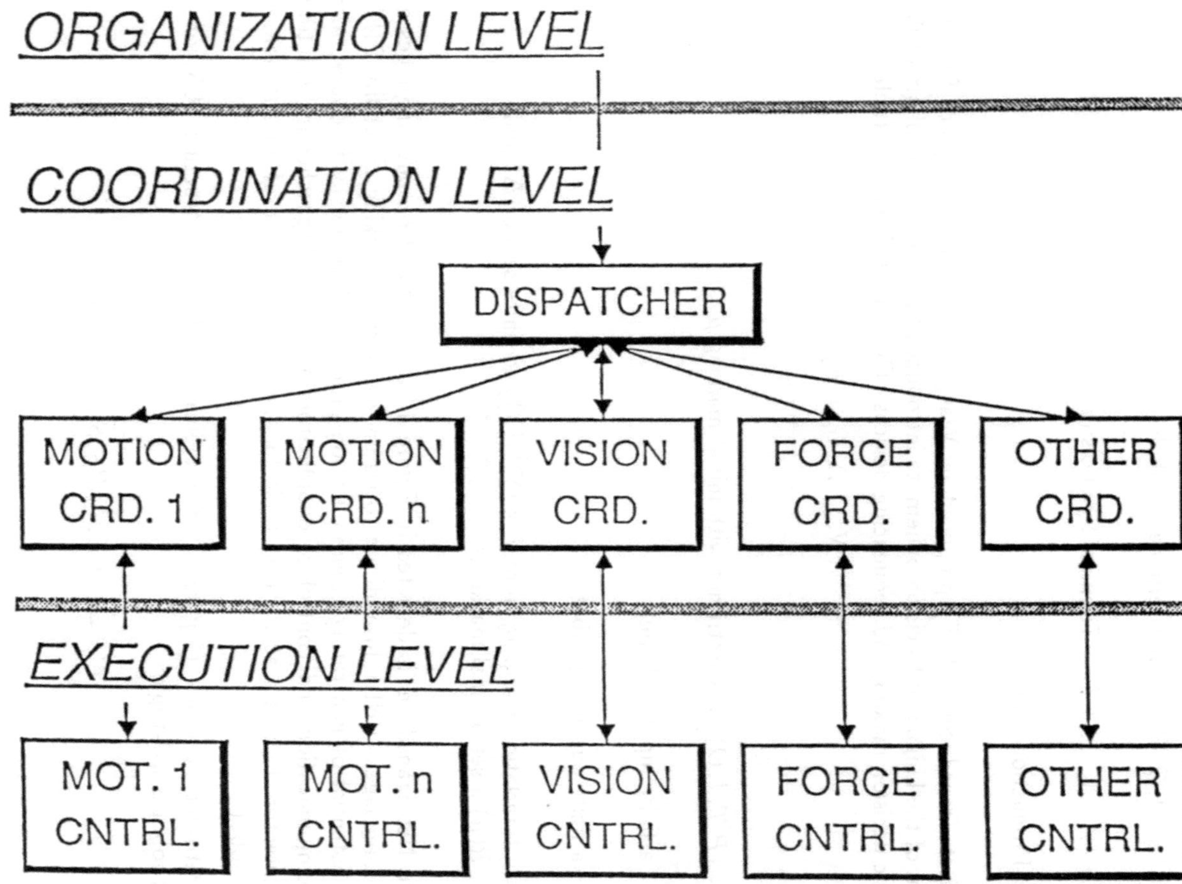

Fig. 3. Hierarchical Intelligent Manipulator Control System.

where

$N = \{S, M, Q, H\}$ = Non-terminal symbols

$\sum_o = \{A_1, A_2 \dots A_n\}$ = Terminal Symbols (activities)

P = Production rules

Petri Net Transducers (PNT) proposed by Wang and Saridis (1988) are Petri-net realizations of the Linguistic Decision Schemata introduced by Saridis and Graham (1984) as linguistic decision making and sequencing devices. They are defined as 6-tuples.

$$M = (N, \sum, \Delta, G, \mu. F)$$

where

$N = (P, T, I, O)$ = A Petri-net with initial marking μ

$\sum$ = a finite input alphabet

Δ = a finite output alphabet

σ = a translation mapping from T x $(\sum \cup \{\lambda\})$ to finite sets of Δ^* and $F \subset R(\mu)$ a set of final markings.

A Petri Net Transducer is depicted in Figure 4. Its input and output languages are Petri Net Languages (PNL). In addition to its on-line decision making capability PNT's have the potential of generating communication protocols, learning by feedback, ideal for the communication and control of coordinators and their dispatcher in real time. Their architecture follows a scenario suitable for the implementation of an autonomous intelligent robot.

The Petri-net Structure (Figure 5) of a typical coordination structure (CS) of an intelligent robot is a 7-tuple:

$$CS = (D, C, F, R_D, S_D, R_C, S_C)$$

where

$D = (N_d, \sum_o, \Delta_o, G_d, \mu_d, F_d)$ = The PNT dispatcher

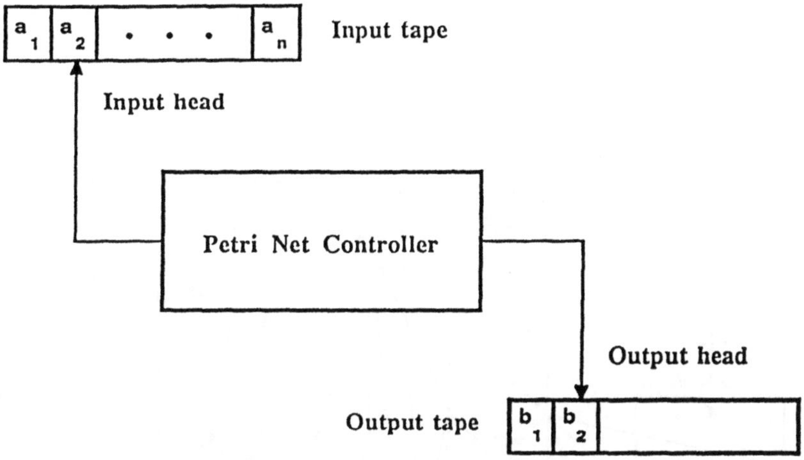

Figure 4. The Petri Net Transducer (PNT)

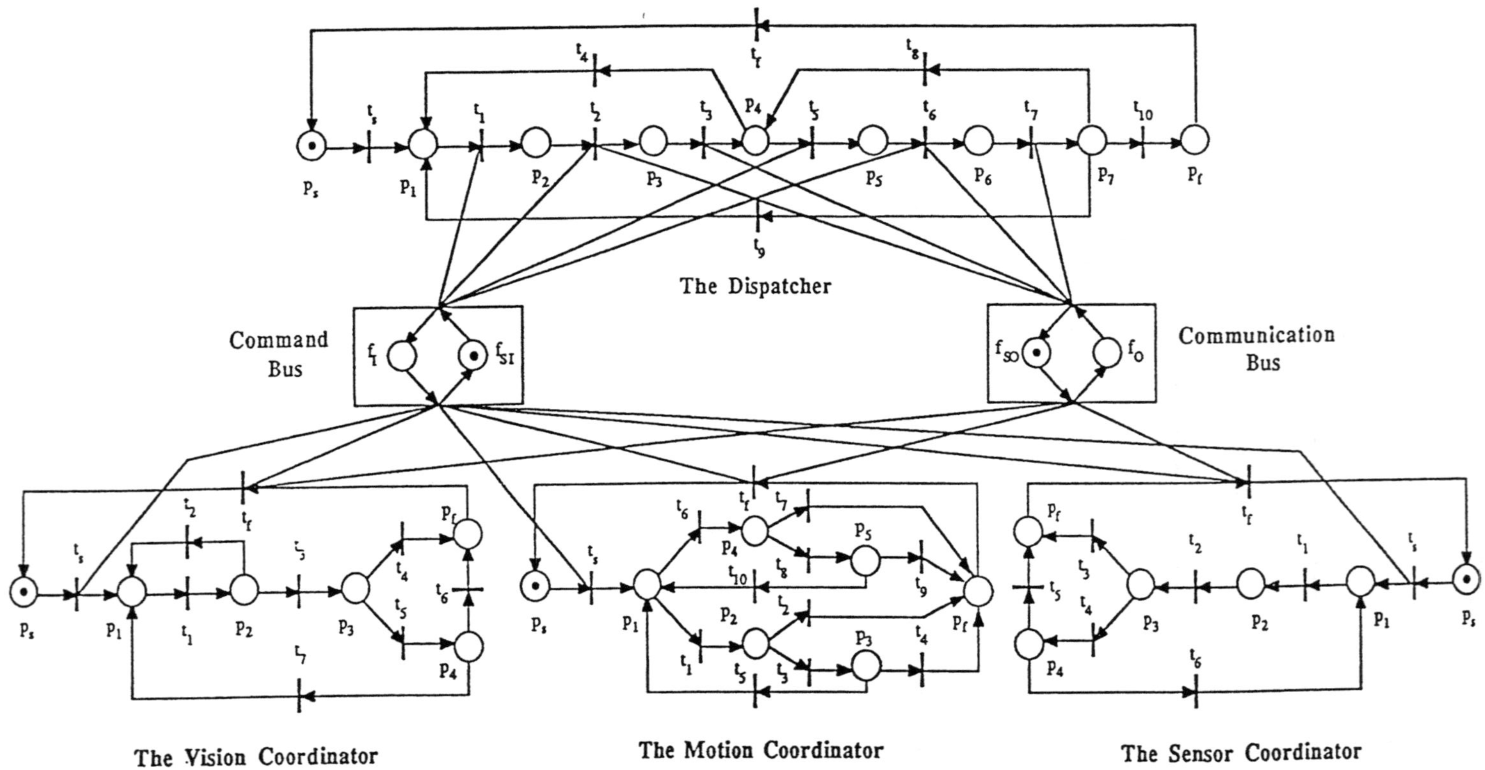

Figure 5: The Coordination Structure of An Intelligent Manipulator System

$C = \{C_1, \ldots C_n\}$ = set of coordinators

$C_i = (N_c^i, \Sigma_c^i, \Delta_c^i, G_c^i, F_c^i)$ = the ith PNT coordinator

$F = \bigcup_{i=1}^{n} \{f_I^i, f_{SI}^i, f_o^i, f_{SO}^i\}$ = set of connection points

R_D, R_C = Receiving maps for the dispatcher and coordinators

S_D, S_C = Sending maps for the dispatcher and coordinators

Decision making in the coordination structure is accomplished by Task Scheduling and Task Translation, e.g., for a given task find σ an enabled t such that $\sigma(t, a)$, is defined and then select the right translation string from $\sigma(t, a)$ for the transition t.

The sequence of events transmitted from the organization level is received by the dispatcher which requests a world model with coordinates from a vision coordinator. The vision coordinator generates appropriate database and upon the dispatcher's command communicates it to the planning coordinator which set a path for the arm manipulator. A new command from the dispatcher sends path information to the motion controller in terms of end points, constraint surface and performance criteria. It also initializes the force sensor and proximity sensor control for grasp activities. The vision coordinator is then switched to a monitoring mode for navigation control, and so on.

The PNT can be evaluated in real-time by testing the computational complexity of their operation which may be expressed uniformly in terms of entropy. Feedback information is communicated to the coordination level from the execution level during the execution of the applied command. Each coordinator, when accessed, issues a number of commands to its associated execution devices (at the execution level). Upon completion of the issued commands feedback information is received by the coordinator and is stored in the short-term memory of the coordination level. This information is stored in the short-term memory of the coordination level. This information is used by other coordinators if necessary, and also to calculate the individual, accrued and overall accrued costs related to the coordination level. Therefore, the feedback information from the execution to the coordination level will be called on-line, real-time feedback information.

The performance estimate and the associated subjective probabilities are updated after the k_{ij}-th execution of a task $[(u_t, x_t)_i, S_j]$ and the measurement of the estimate of the observed cost J_{ij}:

$$J_{ij}(k_{ij} + 1) = J_{ij}(k_{ij}) + \beta(k_{ij} + 1)[J_{obs}(K_{ij} + 1) - J_{ij}(k_{ij})] \tag{17}$$

$$p_{ij}(k_{ij}+1) = p_{ij}(k_{ij}) + \gamma(k_{ij}+1)[\xi_{ij}(k_{ij}+1) - p_{ij}(k_{ij})]$$

where

$$\xi_{ij} = \begin{cases} 1 & J_{ij} = \text{min} \\ 0 & \text{elsewhere} \end{cases}$$

and β and γ are harmonic sequences. Convergence of this algorithm is proven in (Saridis and Graham 1984).

The learning process is measured by the entropy associated to the subjective probabilities. If

$$H(M) = H(E) + H(T/E) \tag{18}$$

where $H(E)$ is the environmental uncertainty and $H(T/E)$ is the pure translation uncertainty. Only the latter may be reduced by learning.

4.3 The Execution Level With Entropy Measures

The cost of control at the hardware level can be expressed as an entropy which measures the uncertainty of selecting an appropriate control to execute a task. By selecting an optimal control, one minimizes the entropy, e.g., the uncertainty of execution. The entropy may be viewed in the respect as an energy in the original sense of Boltzmann, as in Saridis (1988).

Optimal control theory utilizes a non-negative functional of the state of the system $x(t)\epsilon\Omega_x$ the state space, and a specific control $u(x,t)\epsilon\Omega_u xT; \Omega_u \subset \Omega_x$ the set of all admissible feedback controls, to define the performance measure for some initial conditions $x_0(t_0)$, representing a generalized energy function, of the form

$$V(x_0, t_0) = E\{\int_{t_0}^{t_f} L(x, t, u(x,t))dt\} \tag{19}$$

where $L(x, t, u(x,t)) > 0$, subject to differential constraints dictated by the underlying process

$$\begin{array}{l} \frac{dx}{dt} = f(x, u(x,t), w, t) \; x(t_o) = x_o \\ z = g(x, v, t) \; x(t_f) \epsilon M_f \end{array} \tag{20}$$

where $x_o, w(t), v(t)$ are random variables with associated probability densities $p(x_o)$, $p(w(t))$, $p(v(t))$ and M_f a maninfold in Ω_x. The trajectories of the system (20) are defined for a fixed but arbitrarily selected control $u(x,t)$ from the set of admissible feedback controls Ω_u.

In order to express the control problem in terms of an entropy function, one may assume that the performance measure $V(x_0, t_0, u(x,t))$ is distributed in Ω_u according to the probability density $p(u(x,t))$ of the controls $u(x,t)\epsilon\Omega_u$. The differential entropy $H(u)$ corresponding to the density is defined as

$$H(u) = -\int_{\Omega_u} p(u(x,t)) ln p(u(x,t)) dx$$

and represents the uncertainty of selecting a control $u(x,t)$ from all possible admissible feedback controls Ω_u. The optimal performance should correspond to the maximum value of the associated density $p(u(x,t))$. Equivalently, the optimal control $u^*(x,t)$ should minimize the entropy function $H(u)$.

This is satisfied if the density function is selected to satify Jaynes' Principle of Maximum Entropy (1956), e.g.,

$$p(u(x,t)) = \exp\{-\lambda - \mu V(x_o, t_o, u(x,t))\} \tag{21}$$

where λ and μ are normalizing constants.

It was shown by Saridis (1985b) that the expression $H(u)$ representing the entropy for a particular control action $u(x,t)$ is given by

$$\begin{aligned} H(u) &= \int_{\Omega_u} p(x, u(x,t)) V(x_o, t_o, u(x,t)) dx \\ &= \lambda + \mu V(x_0, t_0, u(x,t))\} \end{aligned} \tag{22}$$

This implies that the average performance measure of a feedback control problem corresponding to a specifically selected control, is an entropy function. The optimal control $u^*(x,t)$ that minimizes $V(x,t,u(x,t))$, maximizes $p(x, u(x,t))$, and consequently minimizes the entropy $H(u)$.

$$u^*(x,t) \quad : \quad E\{V(x_o,t_o,u^*(x,t))\}$$
$$= \min_u \int_{\Omega_u} V(x_o,t_o,u(x,t))p(u(x,t))dx \qquad (23)$$

This statement is the generalization of a theorem proven in (Saridis 1988) and establishes equivalent measures between information theoretic and optimal control problem and provides the information and feedback control theories with a common measure of performance.

This optimal control theory designed mainly for motion control, can be implemented for vision control, path planning and other sensory system pertinent to an Intelligent Machine by slightly modifying the system equations and cost functions. After all one is dealing with real-time dynamic systems which may be modeled by a dynamic set of equations.

5 APPLICATION TO ROBOTIC SYSTEMS

The theory of Intelligent Controls has direct application to the design of Intelligent Robots. The IPDI provides a means of structuring hierarchically the levels of the machine. Since for a passive task the flow of knowledge through the machine must be constant, it assigns the highest level with the highest machine intelligence and smallest complexity (size of data base), and the lowest level with the lowest machine intelligence and largest complexity. Such a structure agrees with the concept of most organizational structures encountered in human societies. Application to machine structures is straight forward (Figure 6).

Even at the present time there is a large variety of applications for intelligent machines. Automated material handling and assembly in an automated factory, automation inspection, sentries in a nuclear containment are some of the areas where intelligent machines have and will find a great use. However, the most important application for the author's group is the application of Intelligent Machinesto unmanned space exploration where, because of the distance involved, autonomous anthropomorphis tasks must be executed and only general commands and reports of executions may be communicated.

Such tasks are suitable for intelligent robots capable of executing anthropomorphic tasks in unstructured uncertain environments. They are structured uncertain environment. They are structured usually in a human-like shape and are equipped with vision

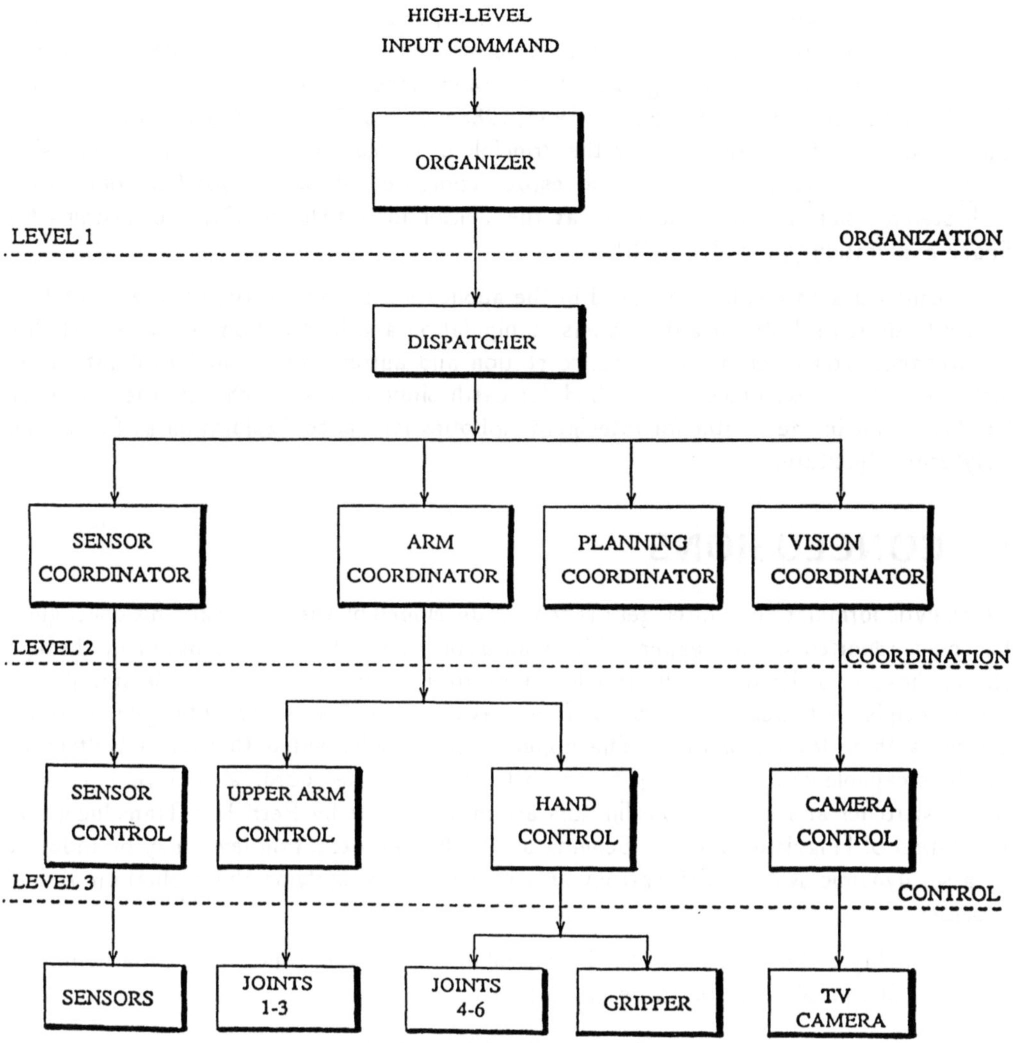

Fig. 6. Hierarchically Intelligent Control for a Manipulator with Sensory Feedback

and other tactile sensors to sense the environment, two areas to execute tasks and locomotion for appropriate mobility in the unstructured environment. The controls of such a machine are performed according to the theory of Intelligent Machines previously discussed (Saridis and Stephanou 1977), (Saridis 1983, 1985a, 1985b, 1988a), (Meystel 1985, 1986). The three levels of controls, obeying the Principle of Increasing Precision with Decreasing Intelligence, are presently tested on a testbed composed of two PUMA 600 robot arms with stereo vision and force sensing, at the Center for Intelligent Robotic Systems for Space Exploration (CIRSSE) at RPI.

Recent research has been focused in the application of the Theory of Intelligent Machines to design robots for autonomous manipulation and locomotion in space. Satellite maintenance, construction of the space station and autonomous planet exploration vehicles are typical examples. A testbed for earth simulation of such activities in space has been built in the Center for Intelligent Robotics for Space Exploration at Rensselaer Polytechnic Institute.

6 CONCLUSIONS

An analytic formulation of Intelligent Machines developed by the author and his colleagues, has been reported in this paper. The realization of the Machine is obtained through the application of Hierarchically Intelligent Control based on a better understanding of the Principle of Increasing Precision (complexity) with Decreasing Intelligence, which utilizes a three level structure. The upper level is implemented through a Boltzmann machine capable of task planning at an abstract level. The coordination level composed of a dispatcher and several coordinators are implemented by Petri Net Transducers, as realization of Linguistic Decision Schemata. Finally, the execution level may be modeled by a set dynamic system with entropy as a cost function, unifying the evaluation various processes.

Optimality is still searched at the individual levels. Total optimization, a mathematical programming problem is still to be investigated.

ACKNOWLEDGEMENTS

This work was supported by NASA Grant NAGW-1333.

REFERENCES

Albus, J.S. (1975), "A New Approach to Manipulation Control: The Cerebellar Model Articulation Controller", *Transactions of ASME, J. Dynamics Systems, Measurement and Control, 97*, pp. 220-227.

Fu, K.S. (1971), "Learning Control Systems and Intelligent Control Systems: An Intersection of Artificial Intelligence and Automatic Control", *IEEE Trans on Automatic Control*, Vol. AC-16, No. 1, pp. 70-72.

Hayes-Roth, F., et al. (1983), *Building Expert Systems*, Addison-Wesley, New York.

Hinton, G.E. and Sejnowski, T.J. (1986), "Learning and Relearning in Boltzmann Machines", pp. 282-317, in *Parallel Distributed Processing*, ed D.E. Rumelhart and J.L. McClellan, MIT Press.

Jaynes, E.T. (1957), "Information Theory and Statistical Mechanics", *Physical Review*, pp. 106, 4.

Meystel, A. (1986), "Cognitive Controller for Autonomous Systems", *IEEE Workshop on Intelligent Control 1985*, p. 222, RPI, Troy, New York.

Minsky, M.L. (1972), *Artificial Intelligence*, McGraw-Hill, New York.

Moed, M.C. and Saridis, G.N. (1990), "A Boltzmann Machine for the Organization of Intelligent Machines", *IEEE Transactions on Systems Man and Cybernetics*, Vol. 20, No. 5, Sept.

Nilsson, N.J. (1969), "A Mobile Automation: An Application of Artificial Intelligence Techniques", *Proc. Int. Joint Conf. on AI*, Washington, D.C.

Peterson, J.L. (1977), "Petri-Nets", *Computing Survey, 9*, No. 3, pp. 223-252, September.

Saridis, G.N. (1977), *Self-Organizing Controls of Stochastic Systems*, Marcel Dekker, New York, New York.

Saridis, G.N. (1979), "Toward the Realization of Intelligent Controls", *IEEE Proceedings*, Vol. 67, No. 8.

Saridis, G. N. (1983), "Intelligent Robotic Control", *IEEE Trans. on AC-29*, 4.

Saridis, G.N. (1985a), "Intelligent Control-Operating Systems in Uncertain Environments", Chapter 7, in *Uncertainty and Control*, J. Ackermann Editor, Springer-Verlag, Berlin, pp. 215-233.

Saridis, G.N. (1985b), "Control Performance as an Entropy", *Control Theory and Advanced Technology*, 1, 2.

Saridis, G.N. (1985c), "Foundations of Intelligent Controls", *Proceedings of IEEE Workshop on Intelligent Controls*, p. 23, RPI, Troy, New York.

Saridis, G.N. (1988), "Entropy Formulation for Optimal and Adaptive Control", *IEEE Transactions on AC*, Vol. 33, No. 8, pp. 713-721.

Saridis, G.N. (1989), "Analytic Formulation of the IPDI for Intelligent Machines", *Automatic the IFAC Journal*, Vol. 25, No. 3, pp. 461-467.

Saridis, G.N. and Graham, J.H. (1984), "Linguistic Decision Schemata for Intelligent Robots", *Automatica*, Vol. 20, No. 1, 121-126.

Saridis, G.N. and Lee, C.S.G. (1979), "Approximation of Optimal Control for Trainable Manipulators", *IEEE Trans. on SMC*, Vol. SMC-8, No. 3.

Saridis, G.N. and Moed, M.C. (1988), "Analytic Formulation of Intelligent Machines as Neural Nets", Symposium on Intelligent Control, Washington, D.C., August.

Saridis, G.N. and Stephanou, H.E. (1977), "A Hierarchical Approach to the Control of a Prosthetic Arm", *IEEE Trans. on SMC*, Vol. SMC-7, No. 6, pp. 407-420.

Saridis, G.N. and Valavanis, K.P. (1988), "Analytical Design of Intelligent Machines", *Automatica the IFAC Journal*, Vol. 24, No. 2, pp. 123-133.

Sannon, C. and Weaver, W. (1963), *The Mathematical Theory of Communications*, Illini Books.

Wang, F., Kyriakopoulos, K., Tsolkas T., Saridis, G.N., (1990) "A Petri-Net Coordination Model of Intelligent Mobile robots" CIRSSE Technical Report #50, Jan. 1990.

Wang, F. and Saridis, G.N. (1988), "A Model for Coordination of Intelligent Machines Using Petri Nets", Symposium on Intelligent Control, Washington, D.C., August.

Wang, F., Saridis, G.N. (1990) "A Coordination Theory for Intelligent Machines" Automatica the IFAC Journal, Vol. 35, No. 5, Sept.

Zames, G. (1979), "On the Metric Complexity of Casual Linear Systems, ε-entropy and ε-dimension for Continous Time", *IEEE Trans. Automatic Control*, Vol. AC-124, pp. 220-230, April.

SYMBOLIC COMPUTATION FOR ROBOT DESIGN

C. Vibet

Université Paris XII, France

Abstract : This report presents the basic derivations needed to develop a CAD robot system devoted to automatic analysis and design of controlled mechanisms. The feedback laws required by sophisticated robot controllers are directly derived from the Lagrangian of mechanical systems by means of *generating rules*. The dynamical model is not explicitly needed in this derivation which only requires partial derivatives and matrix calculus. In addition it is shown how the dynamical model can be computed from the control laws so that dynamics of controlled and uncontrolled mechanisms can be derived in the same framework. Finally, the sampling rate to use in computerized servo-mechanisms is defined from a simple rule.

Keywords : *CAD robot system, controlled robot designer, generating rules, nonlinear Decoupling Method.*

INTRODUCTION

The advent of expert systems has opened new facilities to solve the robotic problems. For example, consultation programs have been developed to help production engineers in selecting the industrial robot architecture. Among them, the X-ARS program of Acaccia, et al. [1-6] which has a knowledge-based architecture with access governed by a rule-based expert-module works with and data-base of commercially available robots. At the same level, the inverse kinematics problems are now symbolically solved by means of expert systems without support of algebra systems as REDUCE or MACSYMA. In such systems, a knowledge-base of mathematical solution rules, or a set of prototype equations with known a priori solutions, are used as a knowledge base to solve iteratively the global data base consisting of kinematics relationships. For example, the Symbolic Kinematics Inversion

NATO ASI Series, Vol. F 71
Expert Systems and Robotics
Edited by T. Jordanides and B. Torby

Program SKIP of Rieseler and Wahl [7] performs the inverse kinematics of the PUMA 562 in 14s on a PC-AT equipped with the INTEL 80386 processor and 80387 coprocessor, for a clock rate of 16MHz.

At the robot design level, there is also a need to solve as a whole control and path-modelling problems [8]. The goal is to obtain static and dynamical accuracy of the tool trajectories. This represents another application field of expert systems, because numerous design alternatives have to be solved in a convenient manner as, for example, how accurate the dynamics modelling and feedback laws must be to design a precise manipulator. For the moment, these are the traditional approaches which are applied to develop, in connection with the trajectory generation problem, smart servo-controlled robots. However, these two aspects of a single problem are analyzed in two independent steps. For example, the path generation problem is frequently addressed by workers in the literature without any link with the control problem [9], while generally, the design of sophisticated robot controllers does not consider whether the desired path of a manipulator is specified in terms of location, speed, acceleration, or jerk of the end-effector. The most consistent approach to solve these problems is probably due to Khatib and Le Maitre [10,11]. They have introduced the operational-space formulation which enables (i) the construction of the manipulator end-effector dynamic model in Cartesian space (ii) the development of an efficient control scheme [12,13] which includes an active force control scheme and (iii) the trajectory generation for the manipulator's path in particular in terms of collison avoidance of links. This last approach is based on the artificial potential field concept [13] in which the position to be reached is an attractive pole for the end-effector, and obstacles are repulsive surfaces for the manipulator parts. Collison avoidance for moving objects is obtained by means of a continuously time-varying potential field. However, the operational approach cannot be easily generalized to the case of dynamically driven mechanisms nor to compliant and sensored systems. Such a limitation does not exist when using the static feedback linearization method.

This report presents a methodology to solve dynamics modelling and robot control problems [14] in connection with a spline representation of the path. *Generating rules* are established from a direct application of the nonlinear decoupling method to Lagrange's equations [15-

19] which leads to the definition of the control laws in an explicit form. In addition, a basic parameter as the sampling rate to be used in robot controllers [20], is also specified from a simple rule [21]. So, the main elements required to build a solver-program in the context of expert systems are provided.

I-USUAL APPROACHES IN CONTROLLER DESIGN

Most approaches used in controller design consist of starting from a *dynamical model of robots* which involves ordinary differential equations. This is the case of the Cartesian or operational approach [12,13] in which a Jacobian transformation permits the mapping between the joint dynamical model of a robot arm and its counterpart in work space Fig 1. The resulting dynamical equations are then looked at as a generalization of Newton's second law from

NEWTON/EULER   LAGRANGE's EQUATIONS   KANE's METHOD

etc.

↘ $\Downarrow$ ↙

DYNAMICAL MODEL

CONTROL LAWS

SIMULATION

Fig 1 : Traditional methods to derive feedback laws for robot controllers from dynamical equations.

which the control algorithms are synthetized in a simple and heuristic manner [22]. Though more systematic than the previous method, application of the Nonlinear Decoupling Method (NLDM) on the robot model [23-31] requires a state-space representation of the joint equations. Because this approach is fundamentally a synthesis method, the nonlinear feedback laws involved by controllers can be therefore systematically derived. Moreover, this method permits the design of both joint or Cartesian robot controllers. As shown in Fig 1, the operational approach and the NLDM lead to the same control scheme [32]. However, note that the NLDM is a synthesis procedure which can be applied to any Lagrangian or other system, and in particular no reference has to be made to Newton's laws to derive the required feedback laws.

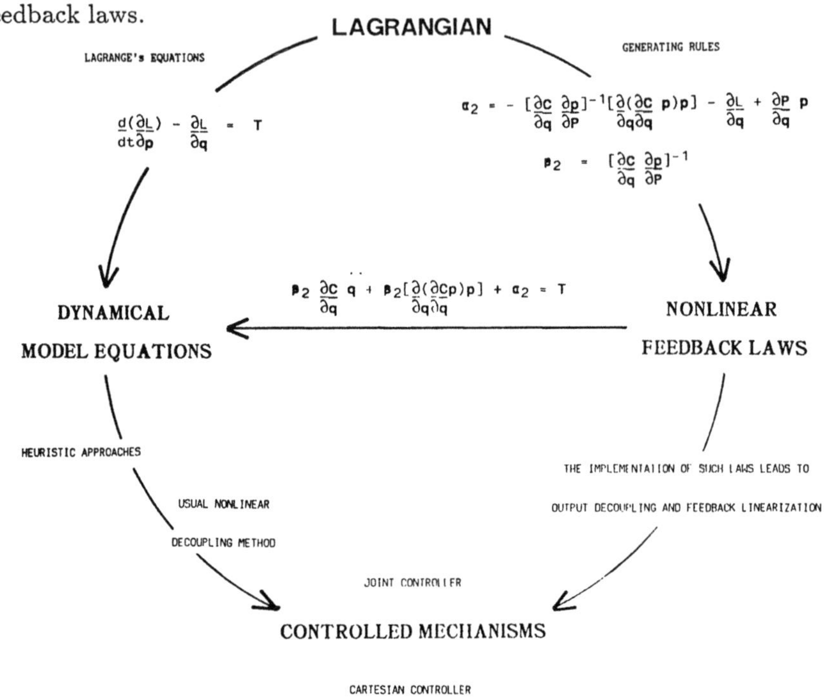

Fig 2 : Two alternate ways to derive control schemes

(i) from the dynamical equations, and (ii) from the Lagrangian itself.

In an alternate approach, the dynamical model of robots *is not explictely needed* to design robot control schemes. In particular, sophisticated robot controllers can be directly built from the Lagrangian itself. This is depicted in Fig 2, where a comparison between this new method and the traditional approaches is given.

II-CONTROL LAWS DERIVATED FROM THE LAGRANGIAN

This section shows how the feedback laws involved in controlled mechanisms can be directly derived from the Lagrangian. In addition, the new proposed technique yields a way to compute the dynamical model both in joint, or in operational spaces, from the control laws themselves, see Fig 2. One advantage of this approach is that the dynamical equations of mechanisms are not explicitly needed in designing robot controllers so that a very simple formulation of the problem results. Due to Cotsaftis and Vibet [15], the method uses a canonical form of Euler-Lagrange's equations on which the nonlinear decoupling method is directly applied. For non-conservative systems, these equations are defined by

$$\frac{d}{dt}\left(\frac{\partial L}{\partial \dot{\mathbf{q}}}\right) - \frac{\partial L}{\partial \mathbf{q}} + \mathbf{D} = \mathbf{T}, \tag{1}$$

where L represents the Lagrangian which is function of n generalized coordinates $\mathbf{q}$, and $\mathbf{p}$ represents the generalized velocities $\mathbf{p} = d\mathbf{q}/dt$. In addition, the dissipation function is $\mathbf{D}$ while $\mathbf{T}$ is the generalized force driving the joints. The state space representation of eqn(1) is

$$\begin{aligned} \dot{\mathbf{q}} &= \mathbf{p} \\ \dot{\mathbf{P}} &= \frac{\partial L}{\partial \dot{\mathbf{q}}}(\mathbf{q}, \dot{\mathbf{q}}) - \mathbf{D}(\mathbf{q}, \dot{\mathbf{q}}). \end{aligned} \tag{2}$$

In eqns(2), the canonical momentum or the conjugate momentum with respect to the generalized coordinate $\mathbf{q}$ is defined by

$$\mathbf{P} = \frac{\partial L}{\partial \dot{\mathbf{q}}}. \tag{3}$$

To derive a phase plane representation of Euler-Lagrange's equations from its state space representation, use is now made of the following identity

$$\frac{d\mathbf{P}}{dt} = \frac{\partial \mathbf{P}}{\partial \mathbf{p}}\,\dot{\mathbf{p}} + \frac{\partial \mathbf{P}}{\partial \mathbf{q}}\,\mathbf{p}. \tag{4}$$

With the help of eqn(4), and because $\dot{\mathbf{p}} = \ddot{\mathbf{q}}$, a phase-plane representation of Euler-Lagrange eqns(2) is

$$\begin{aligned} \dot{\mathbf{q}} &= \mathbf{p} \\ \dot{\mathbf{p}} &= \left(\frac{\partial \mathbf{P}}{\partial \mathbf{p}}\right)^{-1}\left(\frac{\partial L}{\partial \mathbf{q}} + \mathbf{D} - \frac{\partial \mathbf{P}}{\partial \mathbf{q}}\,\mathbf{p} + \mathbf{T}\right). \end{aligned} \tag{5}$$

The synthesis of control algorithms are now derived through the use of the nonlinear decoupling method. To do that, eqns(5) are written in a canonical form defined by

$$\dot{\mathbf{Z}} = \mathbf{F}(\mathbf{Z}) + \mathbf{H}(\mathbf{Z})\mathbf{U}(t), \tag{6}$$

where $\mathbf{Z}=$ col($\mathbf{q}$,$\mathbf{p}$), and

$$\mathbf{F}(\mathbf{Z}) = \begin{pmatrix} \mathbf{p} \\ (\frac{\partial \mathbf{P}}{\partial \mathbf{p}})^{-1}(\frac{\partial L}{\partial \mathbf{q}} - \mathbf{D} - \frac{\partial L}{\partial \mathbf{q}}\mathbf{p}) \end{pmatrix}, \tag{7}$$

with

$$\mathbf{H}(\mathbf{Z}) = \begin{pmatrix} 0 & 0 \\ 0 & (\frac{\partial \mathbf{P}}{\partial \mathbf{p}})^{-1} \end{pmatrix}, \text{ and } \mathbf{U}(t) = \begin{pmatrix} 0 \\ \mathbf{T} \end{pmatrix}. \tag{8}$$

When considering the design of a Cartesian controller, the output of the system $\mathbf{Y}$ is therefore expressed from the Jacobian matrix as

$$\dot{\mathbf{Y}} = \mathbf{J}(\mathbf{q})\dot{\mathbf{q}}. \tag{9}$$

By applying the nonlinear decoupling method, it can be shown that the torque law which gives decoupled controls and Cartesian output linearization for the dynamical system is given, with $\mathbf{p} = \dot{\mathbf{q}}$, by

$$\mathbf{T} = \alpha(\mathbf{q}, \dot{\mathbf{q}}) + \beta(\mathbf{q}, \dot{\mathbf{q}})\mathbf{w}(t), \tag{10}$$

where $\mathbf{w}(t)$ represents the input of the decoupled system, and

$$\alpha(\mathbf{q}, \dot{\mathbf{q}}) = -\frac{\partial \mathbf{P}}{\partial \mathbf{p}}\mathbf{J}(\mathbf{q})^{-1}\frac{d}{dt}[\mathbf{J}(\mathbf{q})\mathbf{p}] - \frac{\partial L}{\partial \mathbf{q}} + \mathbf{D} + \frac{\partial \mathbf{P}}{\partial \mathbf{q}}\ \mathbf{p}. \tag{11}$$

$$\beta(\mathbf{q}, \dot{\mathbf{q}}) = \frac{\partial \mathbf{P}}{\partial \mathbf{p}}\mathbf{J}(\mathbf{q})^{-1}. \tag{12}$$

These generating rules allow the derivation of nonlinear control laws directly from the Lagrangian and Jacobian equations of the mechanical system while (i) the dynamical equations are not needed in this analysis (ii) knowledge of the nonlinear decoupling method is not required to use the generating rules, and though not shown here, (iii) the pole placement problem can be realized [18-24]. Finally, it can be verified that the control algorithms obtained by the application of the generating rules are the feedback laws resulting from the application of the usual nonlinear decoupling method to the dynamical equations of the system. Note that this last technique requires a state-space representation of the differential equations. This is not required when applying the generating rules.

III-CONTROLLED DYNAMICALLY-DRIVEN MECHANISMS

When fast robot manipulators are to be designed, the motor's characteristics have to be taken into account to model the resulting dynamically-driven mechanism. This leads non-Lagrangian systems, i.e., systems for which the operational approach fails because there is no "counterpart" in Newton's second law which allow the feedback laws construction. In fact, the usual application of the nonlinear decoupling method offers a way to solve this problem, and this section introduces *generating rules* [18] to systematically derive such nonlinear feedback laws from the Lagrangian and outputs equations, i.e., from a scalar and a vector. If D.C. electric motors are used to drive a mechanism, the system equations is

$$\begin{aligned}
\dot{\mathbf{q}} &= \mathbf{p} \\
\dot{\mathbf{p}} &= (\frac{\partial \mathbf{P}}{\partial \mathbf{p}})^{-1}(\frac{\partial L}{\partial \mathbf{q}} - \mathbf{D} - \frac{\partial \mathbf{P}}{\partial \mathbf{q}}\,\mathbf{p} + \mathbf{K}_m \mathbf{I}) \\
\dot{\mathbf{I}} &= \mathbf{L}_m^{-1}(-\mathbf{R}_m \mathbf{I} - \mathbf{K}_m \mathbf{p}) + \mathbf{L}_m^{-1}\mathbf{V},
\end{aligned} \tag{13}$$

where $\mathbf{V}$ =col $(V_1, V_2, \ldots, V_n)$ represents the voltage vector applied to motor armatures, $\mathbf{L}_m$ =diag $(L_1, L_2 \ldots, L_n)$ is the matrix of the armature inductance, $\mathbf{K}_m$ = diag $(k_1, k_2, \ldots, k_n)$ is the torque constant matrix, $\mathbf{R}_m$ =diag $(R_1, R_2, \ldots, R_n)$ is the armature resistance matrix and $\mathbf{I}$ =col $(I_1, I_2, \ldots, I_n)$ is the armature current vector. The state-space form of eqn(13) is

$$\dot{\mathbf{Z}} = \mathbf{F}(\mathbf{Z}) + \mathbf{H}(\mathbf{Z})\mathbf{U}(t), \tag{14}$$

where

$$\mathbf{F}(\mathbf{Z}) = \begin{pmatrix} \mathbf{p} \\ (\frac{\partial \mathbf{P}}{\partial \mathbf{p}})^{-1}(\frac{\partial L}{\partial \mathbf{q}} - \mathbf{D} - \frac{\partial L}{\partial \mathbf{q}}\mathbf{p}) \\ \mathbf{L}_m^{-1}(-\mathbf{R}_m \mathbf{I} - \mathbf{K}_m \mathbf{p}) \end{pmatrix}, \tag{15}$$

with $\mathbf{Z}$=col($\mathbf{q}$,$\mathbf{p}$,$\mathbf{I}$), and

$$\mathbf{H}(\mathbf{Z}) = \begin{pmatrix} 0 & 0 & 0 \\ 0 & 0 & 0 \\ 0 & 0 & \mathbf{L}_m^{-1} \end{pmatrix}, \text{ and } \mathbf{U}(t) = \begin{pmatrix} 0 \\ 0 \\ \mathbf{V} \end{pmatrix}. \tag{16}$$

To design a robot controller, the outputs are written as

$$\mathbf{Y} = \mathbf{C}(\mathbf{Z}), \tag{17}$$

where in fact, $\mathbf{C}(\mathbf{Z}) = \mathbf{C}(\mathbf{q})$. Decoupling and linearization are obtained by using a control law defined by

$$\mathbf{U}(t) = \boldsymbol{\alpha}(\mathbf{Z}) + \boldsymbol{\beta}(\mathbf{Z})\mathbf{W}(t), \tag{18}$$

with $\mathbf{W}(t) = \text{col }(0, 0, \mathbf{w}(t))$, and

$$\boldsymbol{\alpha}(\mathbf{z}) = col\ [0, 0, \boldsymbol{\alpha}_3(\mathbf{z})],\ \boldsymbol{\beta}(\mathbf{z}) = diag\ [0, 0, \boldsymbol{\beta}_3(\mathbf{z})]. \tag{19}$$

By taking the third derivative of outputs defined by eqn(17) with respect to t, one gets with $\mathbf{D} = \mathbf{D}(\mathbf{q}, \dot{\mathbf{q}})$

$$\dddot{\mathbf{Y}} = \frac{\partial}{\partial \mathbf{q}}[\frac{\partial}{\partial \mathbf{q}}(\frac{\partial \mathbf{C}}{\partial \mathbf{q}}\mathbf{p})\mathbf{p}]\mathbf{p} + \frac{\partial}{\partial \mathbf{q}}[\frac{\partial \mathbf{C}}{\partial \mathbf{q}}(\frac{\partial \mathbf{P}}{\partial \mathbf{p}})^{-1}(\frac{\partial L}{\partial \mathbf{q}} - \mathbf{D} - \frac{\partial \mathbf{P}}{\partial \mathbf{q}}\mathbf{p} + \mathbf{K}_m\mathbf{I})]\mathbf{p}$$

$$+\left\{\frac{\partial}{\partial \mathbf{q}}(\frac{\partial \mathbf{C}}{\partial \mathbf{q}}\mathbf{p}) + \frac{\partial}{\partial \mathbf{p}}(\frac{\partial \mathbf{C}}{\partial \mathbf{q}}(\frac{\partial \mathbf{P}}{\partial \mathbf{p}})^{-1}(\frac{\partial L}{\partial \mathbf{q}} - \mathbf{D} - \frac{\partial \mathbf{P}}{\partial \mathbf{q}}\mathbf{p}))\right\}.(\frac{\partial \mathbf{P}}{\partial \mathbf{p}})^{-1}(\frac{\partial L}{\partial \mathbf{q}} - \mathbf{D} - \frac{\partial \mathbf{P}}{\partial \mathbf{q}}\mathbf{p} + \mathbf{K}_m\mathbf{I})$$

$$+\frac{\partial \mathbf{C}}{\partial \mathbf{q}}(\frac{\partial \mathbf{P}}{\partial \mathbf{p}})^{-1}\mathbf{K}_m\mathbf{L}_m^{-1}(-\mathbf{R}_m\mathbf{I} - \mathbf{K}_m\mathbf{p} + \mathbf{V}). \tag{20}$$

By introducing eqns(51-52) into eqn(53) and solving to obtain output-decoupling and feedback-linearization, the generating rules valid for dynamically-driven mechanisms are defined by

$$\boldsymbol{\beta}_3(\mathbf{Z}) = \mathbf{L}_m\mathbf{K}_m^{-1}\frac{\partial \mathbf{P}}{\partial \mathbf{p}}(\frac{\partial \mathbf{C}}{\partial \mathbf{q}})^{-1}, \tag{21}$$

$$\alpha_3(\mathbf{Z}) = \mathbf{R}_m\mathbf{I} + \mathbf{K}_m\mathbf{p} \tag{22}$$

$$-\beta_3(\mathbf{Z})\left\{\frac{\partial}{\partial\mathbf{q}}[\frac{\partial}{\partial\mathbf{q}}(\frac{\partial\mathbf{C}}{\partial\mathbf{q}}\mathbf{p})\mathbf{p})]\mathbf{p} + \frac{\partial}{\partial\mathbf{q}}[\frac{\partial\mathbf{C}}{\partial\mathbf{q}}(\frac{\partial\mathbf{P}}{\partial\mathbf{p}})^{-1}(\frac{\partial L}{\partial\mathbf{q}} - \mathbf{D} - \frac{\partial\mathbf{P}}{\partial\mathbf{q}}\mathbf{p} + \mathbf{K}_m\mathbf{I})]\mathbf{p}\right.$$

$$\left. +\{\frac{\partial}{\partial\mathbf{q}}(\frac{\partial\mathbf{C}}{\partial\mathbf{q}}\mathbf{p}) + \frac{\partial}{\partial\mathbf{p}}(\frac{\partial\mathbf{C}}{\partial\mathbf{q}}(\frac{\partial\mathbf{P}}{\partial\mathbf{p}})^{-1}(\frac{\partial L}{\partial\mathbf{q}} - \mathbf{D} - \frac{\partial\mathbf{P}}{\partial\mathbf{q}}\mathbf{p}))\}.(\frac{\partial\mathbf{P}}{\partial\mathbf{p}})^{-1}(\frac{\partial L}{\partial\mathbf{q}} - \mathbf{D} - \frac{\partial\mathbf{P}}{\partial\mathbf{q}}\mathbf{p} + \mathbf{K}_m\mathbf{I})\right\}. \tag{23}$$

These relations which are defined from a scalar (i.e. the Lagrangian) and from a vector (i.e., the controller ouput equation), yield the nonlinear feedback-laws, the implementation of which leads to decoupled controls. It can be observed that these equations do not involve any derivative with respect to t. On the other hand, to give a physical meaning to terms involved in eqns(54,55), it is useful to reconsider the previous derivations by introducing the mapping relationship between the joint and Cartesian spaces (or operational space). So, the controller output is now defined by

$$\dot{\mathbf{Y}} = \mathbf{J}(\mathbf{z})\,\dot{\mathbf{Z}}, \tag{24}$$

where $\mathbf{J}(\mathbf{q})$ is the Jacobian matrix of the mechanism. In addition, because its components are very well understood, the differential equations of the driven mechanisms can be written from the usual joint dynamical model as

$$\begin{aligned} \dot{\mathbf{q}} &= \mathbf{p} \\ \dot{\mathbf{p}} &= \mathbf{A}^{-1}(-\mathbf{B}(\mathbf{q},\dot{\mathbf{q}}) - \mathbf{g}(\mathbf{q}) + \mathbf{K}_m\mathbf{I}) \\ \dot{\mathbf{I}} &= \mathbf{L}_m^{-1}(-\mathbf{R}_m\mathbf{I} - \mathbf{K}_m\mathbf{p}) + \mathbf{L}_m^{-1}\mathbf{V}, \end{aligned} \tag{25}$$

where $\mathbf{A}(\mathbf{q})$ represents the inertia matrix of the mechanism, $\mathbf{B}(\mathbf{q},\dot{\mathbf{q}})$ is the Coriolis plus centrifugal torque, while $\mathbf{g}(\mathbf{q})$ is the gravity torque. The third derivative with respect to t of the controller output defined by eqn(56) is

$$\dddot{\mathbf{Y}} = \frac{\partial}{\partial \mathbf{q}}(\frac{\partial}{\partial \mathbf{q}}[\mathbf{J}(\mathbf{q})\mathbf{p}]\mathbf{p})\mathbf{p} + \frac{\partial}{\partial \mathbf{q}}\{\mathbf{J}(\mathbf{q})\mathbf{A}^{-1}(\mathbf{B}(\mathbf{q},\dot{\mathbf{q}}) - \mathbf{g}(\mathbf{q}) - \mathbf{K}_m\mathbf{I})\}\mathbf{p}$$

$$+\left\{\frac{\partial}{\partial \mathbf{q}}[\mathbf{J}(\mathbf{q})\mathbf{p}]\mathbf{p} + \mathbf{J}(\mathbf{q})\mathbf{A}^{-1}\frac{\partial}{\partial \mathbf{p}}(-\mathbf{B}(\mathbf{q},\dot{\mathbf{q}}) - \mathbf{g}(\mathbf{q}) - \mathbf{K}_m\mathbf{I})\right\}.\mathbf{A}^{-1}(\mathbf{B}(\mathbf{q},\dot{\mathbf{q}}) - \mathbf{g}(\mathbf{q}) - \mathbf{K}_m\mathbf{I})$$

$$+\mathbf{J}(\mathbf{q})\mathbf{A}^{-1}\mathbf{K}_m\mathbf{L}_m^{-1}(-\mathbf{R}_m\mathbf{I} - \mathbf{K}_m\mathbf{p} + \mathbf{V}). \tag{26}$$

The comparison of these results with the previous ones leads to the following relationships

$$\frac{\partial \mathbf{C}}{\partial \mathbf{q}} = \mathbf{J}(\mathbf{q}), \qquad \frac{\partial \mathbf{P}}{\partial \mathbf{p}} = \mathbf{A}(\mathbf{q}), \tag{27}$$

$$\frac{\partial L}{\partial \mathbf{q}} - \frac{\partial \mathbf{P}}{\partial \mathbf{q}}\mathbf{p} - \mathbf{D} = -\mathbf{B}(\mathbf{q},\dot{\mathbf{q}}) - \mathbf{g}(\mathbf{q}), \qquad \mathbf{P} = \frac{\partial L}{\partial \mathbf{p}} \tag{28}$$

$$\frac{\partial}{\partial \mathbf{q}}(\frac{\partial}{\partial \mathbf{q}}[\mathbf{J}(\mathbf{q})\ \mathbf{p}]\ \mathbf{p})\ \mathbf{p} = \frac{d^2}{dt^2}(\mathbf{J}(\mathbf{q})\mathbf{p}). \tag{29}$$

These relations which are called *generating rules* yield the robot nonlinear feedback laws either (i) directly from the Lagrangian and output equation, or (ii) from the Jacobian matrix and the usual dynamical joint model of the dynamically-driven mechanisms. In comparison with the amount of manipulations associated with the usual application of the nonlinear decoupling method to system equations, the use of the generating rules represents probably the shortest way known to derive the control laws to be implemented in robot control schemes. In particular, it must be pointed out that since no derivative with respect to t is needed, the generating rules enable the symbolic computation of robot controller feedback laws with less memory space needed.

IV-POLE ASSIGNMENT BY NONLINEAR FEEDBACK

Previous paragraphs have shown how the nonlinear laws leading to decoupled controls and feedback linearization of controlled mechanisms can be derived. In fact, the implementation of these feedback laws produces a closed loop system in which each input w_i influences one and only one output y_i by means of (1) a double integrator, for mechanisms; (2) a triple integrator for dynamically driven mechanical systems (i.e., driven by hydraulic or D.C. motors for example), and so on. To obtain both decoupling and arbitrarily pole assignment by nonlinear feedback, a general control law is to be used in place of eqn(43). For mechanisms, the the form is the following

$$\mathbf{U}(t) = \boldsymbol{\alpha}(\mathbf{X}) + \boldsymbol{\beta}(\mathbf{X}) \begin{pmatrix} 0 \\ \mathbf{KQ}(\mathbf{X}) + \boldsymbol{\Gamma}\mathbf{z} \end{pmatrix}, \tag{30}$$

Taking $\mathbf{K}$ =diag $(\mathbf{K}_1, \mathbf{K}_2, \ldots, \mathbf{K}_n)$, with $\mathbf{K_i} = (-k_0^i, -k_1^i)$, and $\boldsymbol{\Gamma}$ =diag (σ_i), the dynamics of the decoupled system becomes

$$\ddot{\mathbf{Y}}_i = \ -k_1^i \dot{\mathbf{Y}}_i - k_0^i \mathbf{Y}_i + \sigma_i z_i, \tag{31}$$

provided the control law is defined from

$$\mathbf{K}_i \mathbf{Q}_i(\mathbf{X}) = -k_1^i \dot{\mathbf{Y}}_i - k_0^i \mathbf{Y}_i, \tag{32}$$

and $\boldsymbol{\alpha}(\mathbf{X})$ and $\boldsymbol{\beta}(\mathbf{X})$ are defined by eqns(44,45). In the case of dynamically driven mechanisms, the control law to be used is

$$\mathbf{U}(t) = \boldsymbol{\alpha}(\mathbf{X}) + \boldsymbol{\beta}(\mathbf{X}) \begin{pmatrix} 0 \\ 0 \\ \mathbf{KQ}(\mathbf{X}) + \boldsymbol{\Gamma}\mathbf{z} \end{pmatrix}, \tag{33}$$

but now $\mathbf{K_i} = (-k_0^i, -k_1^i, -k_2^i)$. So, taking

$$\mathbf{K}_i \mathbf{Q}_i(\mathbf{X}) = -k_2^i \ddot{\mathbf{Y}}_i - k_1^i \dot{\mathbf{Y}}_i - k_0^i \mathbf{Y}_i, \tag{34}$$

the dynamics of the resulting decoupled system becomes

$$\dddot{\mathbf{Y}}_i = \ -k_2^i \ddot{\mathbf{Y}}_i - k_1^i \dot{\mathbf{Y}}_i - k_0^i \mathbf{Y}_i + \sigma_i z_i. \tag{35}$$

With these expressions, it is possible to design pole location by choosing the appropriate parameters of $\mathbf{K}$ and $\boldsymbol{\Gamma}\mathbf{z}$. In fact, the poles of the system can be placed to satisfy any performance criteria [33]. For a second-order system, the pole location can be selected as $p_{1,2} = 0.809 \pm i0.5878$ so that the overshoot D=1.3%. Such a choice leads to $k_0^i = \omega_o^2$, $k_1^i = 1.618\omega_o^2$. For a third-order system the pole location could be selected as $p_{1,2} = 0.6515 \pm i0.7823$ and $p_3 = 0.964$. This choice yields an overshoot D=2.4%, when $k_0^i = \omega_o^3$,

$k_1^i = 2.294\omega_o^2$ and $k_2^i = 2.268\omega_o$. In comparison with these design parameters, the ITAE criteria leads to $k_0^i = \omega_o^3$, $k_1^i = 1.75\omega_o^2$, $k_2^i = 2.15\omega_o$ and the overshoot for the step response is D=5%.

V-CONTROLLER SAMPLING RATE

One important parameter which is to be selected in order to implement the computerized feedback laws in controlled systems is the sampling rate [20]. Though for the moment, there is no convenient solution to select the sampling period magnitude to be used in nonlinear controllers, an engineering approach consists of the Madwed's rule [21].
Experiments with nonlinear computer-controlled systems show that decreasing the sampling period T (i.e., the inverse of the sampling frequency) improves the controllers performance. In addition, it appears that for a given computer a compromise must be made between the maximum number of mathematical operations required by the control algorithms and the allowed maximum value of the sampling period. This section introduces a guide, i.e., a rule, to choosing the sampling rate to be used in systems working with computer-controlled loops. This rule comes from the number series transformation method, a technique based on the concept of representing a function by means of a triangular number series transformation. Following such an approach, it is shown that if the characteristic equation of a differential equation is written as

$$a_n\sigma^n + a_{n-1}\sigma^{n-1} + \ldots + a_1\sigma + 1 = 0, \tag{36}$$

then, the number series method becomes as accurate as the analytic method if the sampling period is taken as

$$T \leq \frac{1}{5}\,(a_n)^{\frac{1}{n}}. \tag{37}$$

This last expression, which appears as the limit of Hermite's polynomials, can be used as a guide for deriving the sampling period to be used in computerized-robot controllers. To see that, recall first that the previous sections of this paper have shown how the nonlinear decoupling method leads to linearized and decoupled systems once the appropriate feedback laws have been implemented. In particular for mechanisms, the closed-loop behavior of the related controller is then described by a linear second-order differential equation, see eqn(63). Naturally this result is obtained by assuming that the nonlinear feedback laws are true continuous signals. In fact, Madwed's rule indicates rightly which limit of the sampling rate is to be used so that the dynamical behavior of a computerized controlled system appears as a continuous one. For example, application of eqn(69) to the case of a controlled mechanism leads us to take the following sampling rate

$$T \leq \frac{1}{5\sqrt{k_0^i}}. \tag{38}$$

For a third-order system, i.e., a dynamically-driven robot, the sampling rate must be such that

$$T \leq \frac{1}{5}(k_0^i)^{-\frac{1}{3}}, \tag{39}$$

and so on for higher-order decoupled and linearized systems. Now practical cases are investigated to illustrate the use of Madwed's rule. Reports about the effects of the sampling rate variation on the performance of robot controllers have been analyzed on true robots [20,32]. This is the case with the CMU Direct-Drive Arm II which reveals that the convenient sampling period to be used in a six degree-of-freedom robot is lower than $5ms$, while increasing the sampling rate from 2 to $5ms$ results in a noteworthy degradation of the controller performances [34]. In addition these experiments show that stiffer controller systems are obtained by using high sampling rate. More precisely the control schemes which have been implemented yields a stable closed loop characteristic equation

$$s^2 + k_{vj}s + k_{pj} = 0, \tag{40}$$

where $k_{vj} = 2\sqrt{k_{pj}}$ so that a small overshoot results. The real-time results show that the maximum permissible gains depend on the sampling rate of the control system. The following table gives the identified transfer function for the six joints of the CMU Direct-Drive Arm as well as the allowed gains. The evaluation comes from experiments on the dynamic tracking accuracy when the sampling rate used was $5ms$.

Joint (j)	Transfer Function	k_{pj}	k_{vj}	Sampling Rate
1	$\frac{1}{12.3s^2}$	2.7	3.3	120
2	$\frac{1}{2s^2}$	15	7.5	51.6
3	$\frac{1}{0.25s^2}$	256	32	12.5
4	$\frac{1}{0.007s^2}$	1285	71.5	5.6
5	$\frac{1}{0.006s^2}$	625	50	8
6	$\frac{1}{0.0003s^2}$	1110	50	$6(ms)$

Here, these results have been used to apply Madwed's rule and the sampling-rate evaluation is given in last column of the previous table. It is observed that the lower sampling rate to be used in joint controllers is $T \leq 5.6ms$, while the measurements are realized with an effective sampling rate of $5ms$. Other experiments at $2ms$ leads to

Joint (j)	Transfer Function	k_{pj}	k_{vj}	Sampling Rate
1	$\frac{1}{12.3s^2}$	40	12.6	31
2	$\frac{1}{2s^2}$	58	15.2	26
3	$\frac{1}{0.25s^2}$	400	40	10
4	$\frac{1}{0.007s^2}$	2800	106	3.7
5	$\frac{1}{0.006s^2}$	1200	69.3	5.7
6	$\frac{1}{0.0003s^2}$	3000	110	$3.6(ms)$

With these results, it is noted that the maximum permissible sampling rate allowed when applying Madwed's rule is $T \leq 3.6ms$, while measurements were taken on the CMU Direct-Drive robot when the sampling rate was $2ms$. So, Madwed's rule, as translated into the robotic field, is a convenient guide in selecting the maximum sampling rate to be used for robot controllers.

VI-DYNAMICAL MODEL DERIVATION FROM CONTROL LAWS

At the opposite of traditional methods, the dynamical model of a mechanism is derived here from the control laws [35,36] which have been computed by means of the generating rules.

In the simple case of conservative systems, the Cartesian control laws defined by eqns(44,45) are introduced into the phase-plane representation of Euler-Lagrange eqns(38). This leads to

$$\ddot{\mathbf{q}} = (\beta\frac{\partial \mathbf{C}}{\partial \mathbf{q}})^{-1}(-\alpha - \beta\frac{d}{dt}(\mathbf{J}\mathbf{p}) + \mathbf{T}). \tag{41}$$

Eqn(73) defines the joint space dynamical model of a mechanism in terms of the control laws which define the corresponding work-space controller. In the same way the Cartesian dynamical model can defined from

$$\mathbf{J}^{-T}(\beta\ddot{\mathbf{Y}} + \alpha) = \mathbf{F}. \tag{42}$$

In the case of dynamically-driven systems, i.e., when motor characteristics are taken into account, the work space dynamical model can be derived from eqns(21-23) as

$$\beta_3\dddot{\mathbf{Y}} + \alpha_3 = \mathbf{V}, \tag{43}$$

and the armature current appearing in α_3, is computed from

$$\mathbf{I} = \mathbf{K}_m^{-1}(\frac{\partial \mathbf{P}}{\partial \mathbf{p}}\frac{d}{dt}(\mathbf{J}^{-1}\dot{\mathbf{Y}}) - \frac{\partial L}{\partial \mathbf{q}} + \frac{\partial \mathbf{P}}{\partial \mathbf{q}}\mathbf{p}), \tag{44}$$

where $\mathbf{J}$ represents the Jacobian matrix of the system, and $\dot{\mathbf{Y}}$ is the Cartesian speed. Eqns(41-43) reveal that either the joint space, or the work-space dynamical-mechanism model can be computed from the nonlinear feedback laws, the implementation of which leads to decoupled controls and output linearization.

VII-ANALYSIS OF MECHANICAL SHOCKS

One of the advantages of the formulation of Euler-Lagrange equations as a canonical form is to permit the construction of an equivalent block scheme from which consistent initial conditions [37] can be found. To investigate a shock effect on a mechanism [38], the work-space model required is

$$\mathbf{J}(\mathbf{q})^{-1}\frac{\partial \mathbf{P}}{\partial \mathbf{p}}\mathbf{J}(\mathbf{q})^{-1}(\ddot{\mathbf{Y}} - \dot{\mathbf{J}}(\mathbf{q})\mathbf{p}) = \mathbf{J}(\mathbf{q})^{-T}(\frac{\partial L}{\partial \mathbf{q}} + \frac{\partial \mathbf{P}}{\partial \mathbf{q}}\mathbf{p}) + \mathbf{F}. \tag{45}$$

where $\mathbf{F}$ represents the Cartesian view of the generalized torque which drives the joints of the mechanisms, and $\mathbf{Y}$ is the Cartesian-coordinates vector of the end-effector. In the following, a mechanical shock is viewed as resulting from the application of an impulse force $\mathbf{F}_o\delta(t)$ on the mechanism, which can instantaneously stop the system. This concept is useful to simulate shocks on the computer model of mechanical systems. For a robot, the previous impulse force induces a speed-jump of the end-effector Cartesian speed, the magnitude of which is found by opening some loops of the block-scheme representation of the work-space dynamical equations at the instant of the shock [36]. Thus, the jump of the Cartesian speed is given by

$$\mathbf{J}(\mathbf{q})^{-T}\frac{\partial \mathbf{P}}{\partial \mathbf{p}}\mathbf{J}(\mathbf{q})^{-1}\delta\dot{\mathbf{Y}} = \mathbf{F}_o, \tag{46}$$

where $\delta\dot{\mathbf{Y}} = \mathbf{Y}^+ - \mathbf{Y}^-$ is the speed jump, while $\dot{\mathbf{Y}}^+$ is the speed after the shock, and $\dot{\mathbf{Y}}^-$ is the speed before the shock. In terms of the generalized coordinates of the system, the corresponding jump of the joint-speed is given by

$$\frac{\partial \mathbf{P}}{\partial \mathbf{p}}\delta\dot{\mathbf{q}} = \mathbf{J}(\mathbf{q})^T\ \mathbf{F}_o. \tag{47}$$

Such a method can be applied to investigate electric and mechanical shocks and generalized to the case of higher-order systems. Finally, the equations given in this paper can be used for simulations of mechanical shocks or used in pulse control schemes, the strategy of which remains to be defined.

CONCLUSION

Main basic elements needed by a solver-program devoted to the analysis and simulation of controlled stiff robots have been given in this paper. In particular *generating rules* have been introduced to derive directly from the Lagrangian of mechanisms the explicit form of precise feedback laws without reference to the dynamical model. These laws are valid for mechanisms or for driven mechanisms, i.e., systems in which the motor's characteristics have to be taken into account. Moreover, simple expressions have been introduced to derive systematically, from the feedback laws, the dynamical equations of uncontrolled systems. In addition, a simple rule or a guide has been proposed to define the sampling rate to be used in robot controllers. To prove the feasibility of code based on these generating rules, most of needed algebraic manipulations have been realized with a program based on REDUCE, which checks also the results by applying the usual nonlinear decoupling method. The advantage of this approach comes from the fact that for a n degree-of-freedom robot, the amount of trace of matrices to be computed increases as $\sum_{n=1}^{N} \frac{n+n^2}{2}$ when applying *generating rules*, while the dynamic modeling based on Uicker's method requires the computation of $\sum_{n=1}^{N}(n+1)\frac{n+n^2}{2}$ traces. For $n = 6$, this represents 56 and 322 traces of matrices respectively.

In the next work, a knowledge base of rules will be included to handle procedures for symplifying, in a job context, the nonlinear robot models. In conclusion, the direct and inverse protocol features of the *generating rules* represent a key property to design CAD systems devoted to the selection of components, modelling and simulation of controlled mechanical systems.

ACKNOWLEDGMENTS

The author wishes to acknowledge the contribution of Prof. Cotsaftis from UGRA, CEA-FAR, France, for early discussions and mathematical developments. The author is also indebted to Prof. Torby, from California State University, Long Beach, for reading the manuscript.

REFERENCES

1- Acaccia, G.; Callegari, M.; Michelini, R. and Molfino, R. : Architectural analysis of robotic industrial manipulators. Int. AFCET Conf. on Robotics and Artificial Intelligence, Toulouse, France, 1986

2-Acaccia, G.; Michelini, R.; Molfino, R. and Piaggio, P. : X-SIFIP : a knowledge-based special purpose simulator for the development of flexible manufacturing cells. IEEE Conf. on Robotics and Automation, San Francisco, April 1986

3-Acaccia, G.; Anselmi, N.; Michelini, R.; Molfino, R. and Piaggio, P. : Expert simulation : a tool for exploiting flexible manufacturing. ASME Conf. in Computer in Engineering. New York, Aug 1987

4-Acaccia, G. ; Michelini, R. ; Molfino, R. and Piaggio, P. : Information data based structures for flexible-manufacturing simulators. Third IFIP Conf. on Advances in Production Management Systems. Winnipeg, Aug 1987

5-Acaccia, G. ; Michelini, R. and Molfino, R. : Knowledge-based simulators in manufacturing engineeering. 2nd Conf. on Applications of Artificial Intelligence. Boston, Aug 1987

6-Acaccia, G. ; Callegari, M. ; Michelini, R. ; Molfino, R. and Piaggio, P. : X-ARS : A consultation program for selecting the industrial robot architectures. Third Int. Conf. on Applied of AI in Engineering. Aug 8-11 1988 p (24)

7-Rieseler, H. and Wahl, F. : Fast symbolic computation of the inverse kinematics of robots. IEEE Conf. on Robotics and Automation, Cincinnati, 1990 pp 462-467

8-Edwall, J. ; Pottinger and Ho. : Trajectory generation and control of a robot arm using spline functions. Proc. of Robot VI, Detroit 1982 pp 421-446

9-Goldenberg, A. and Lawrence, D. : End-effector path generation. Trans of the ASME, J. of Dyn., System, Meas., and Control, 1986 Vol 108 pp 158-162

10-Khatib, O. and Le Maitre, J-F. : Dynamic control of manipulators operating in a complex environment. 3rd CISM-IFToMM, Udine, Italy 1978

11-Khatib, O. : Commande dynamique dans l'espace opérationnel des robots manipulators en présence d'obstacles. Thèse Docteur-Ingénieur, ENSAE, Toulouse, France 1980

12-Khatib, O. : Dynamic control of manipulators in operational space. Sixth CISM-IFToMM Congress on Theory of Machines and Mechanisms, New Delhi, India 1983

13-Khatib, O. : The operational space formulation in robot manipulator control. 15th ISIR, Tokyo, 1985 pp 165-172

14-Meystel, A. and Thomas, M. : Computer aided conceptual design in robotics. IEEE Conf. on Robotics and Automation. 1984 pp 220-229

15-Cotsaftis, M. and Vibet, C. : A derivation of robot control algorithms from the Lagrange formalism. Proc. 2nd Int. Conf. on Robotics and Factories of the Future'87, San Diego. Springer Verlag, 1988 pp 461-466

16-Cotsaftis, M. and Vibet, C. : A new method for the robot controller design. Robotics and Autonomous Systems, 1988, Vol 4, No 1, pp 237-242

17-Cotsaftis, M. and Vibet, C. : Decoupling and control laws generation from Lagrange formalism. J. of Robotic Systems. 1988, Vol 5, No 5 pp 433-441

18-Cotsaftis, M. and Vibet, C. : A direct application of the decoupling method to Euler-Lagrange's formalism. 1989, Submitted to J. of Robotics Research

19-Cotsaftis, M. and Vibet, C. : Modeling of robot system dynamics for CAD based robot programming. NATO ARW, CAD based programming for sensor based robots. 1988, Il Ciocco, Italy. Edited by B.Ravani, ASI, F50, Springer Verlag 493-509

20-Khosla, P. : Choosing sampling rates for robot control. IEEE Conf. on Robotics and Automation, 1987 pp 169-174

21-Vibet, C. : Guide to choosing the sampling rate in computerized controllers. Electr. Lett., 1987, Vol 23 No 7, 1002-1003

22-Paul, R. : Robot Manipulators : Mathematics, Programming and Control. MIT Press, 1981

23-Freund, E. : A nonlinear control concept for computer-controlled manipulators. Proc. of IFAC Symp. on Multivariable Techn. Systems, Fredericton, Canada, Pergamon Press, N.Y., 1977 pp 395-403

24-Freund, E. : Fast nonlinear control with arbitrary pole-placement for industrial robots and manipulators. Robotic Research, 1981, Vol 1,No 1, pp 65-78

25-Freund, E. : The structure of decoupled nonlinear systems. Int. J. of Control, 1982, Vol 21, No 3 pp 441-450

26-Freund, E. : On the design of multirobots systems. IEEE Conf on Robotics and Automation, 1984, pp 477-490

27-Tarn, T. ; Bejczy, A. ; Isidori, A. and Chen, Y. : Nonlinear feedback in robot arm control. Proc. of the 23nd IEEE Conf. on Decision and Control, Las Vegas, Nevada, 1984

28-Bejczy, A. ; Tarn, T. and Chen, Y. : Robot arm dynamic control by computer. IEEE Conf. on Robotics and Automation, Saint Louis, MO, 1985 pp 960-970

29-Tarn, T. ; Bejczy, A. and Yun, X. : Nonlinear control algorithms for multiple robot arms. IEEE Tans. on Aerospace and Electronics Systems, 1988 Vol 24, No 5 pp 571-583

30-Tarn, T. ; Bejczy, A. and Yun, X. : Robot arm force control through system linearization by nonlinear feedback. IEEE Conf. on Robotics and Autom., 1988, pp 1618-1625

31-Bejczy, A. and Tarn, T. : Dynamic control of robot arms in task space using nonlinear feedback. Automatisierungstechnik at, 36. Jahrgang, Helf 10/1988 pp 374-388

32-Vibet, C. : Design of robot controllers. Advanced Robotics, 1987, Vol 2, No 1 pp 9-20

33-Forrest-Barlach, F. and Babcock, S. : Inverse dynamics position control of a compliant manipulator. 1986 IEEE Conf. on Robotics and Automation pp 196-205

34-Khosla, P.; Kanade, T. : Real-time implementation and evaluation of model-based controls on CMU DD Arm II. IEEE Conf. on Robotics and Automation, 1986 pp 1546-1555

35-Cotsaftis, M. and Vibet, C. : Synthesis of the dynamical equations of mechanisms from their related control laws. Computer Methods in Applied Mech. and Engin. 1989, Vol 74 pp 29-40

36-Cotsaftis, M. and Vibet, C. : Simultaneous derivation of control laws and of dynamical model for controlled mechanisms. Real-time Integration Methods For Mechanical System Simulation. 1989 NATO ARW, Snowbird, Utah

37-Vibet, C. : Initial conditions and differential equations. Simulation, 1987 Vol 48, No 5 pp 210-212

38-Zheng, Y. : Collision effects on two coordinating robots in assembly and the effect minimization. IEEE Trans. on Syst. Man and Cybern., 1987, SMC 17, No 1 pp 108-116

NEURAL NET SYSTEM FOR ADAPTIVE ROBOT CONTROL

Kevin Warwick

Department of Cybernetics, University of Reading, Whiteknights,
Reading, RG6 2AY, United Kingdom

Abstract: Simple relationships can be employed in the control of robot manipulator links. In this paper the use of back propagation nets in configuring such controllers is considered. The ability to learn, firstly via training and secondly by experience, is made use of to provide an adaptive structure. An application example is considered in order to show how such controllers can be implemented.

Keywords: Neural Nets; Robot Control; Adaptive Control; Learning Systems; Intelligent Control.

Introduction

The adaptive control of the axes of industrial robots requires relatively high sampling rates and rapid adaptation properties [3, 6, 7]. As well as the possibility of unknown payloads, the nonlinear dynamics of the axes vary continuously with respect to time as the manipulator changes position, and these variations severely compromise possible operating speeds, accuracy and reproducibility when conventional PID controllers are employed. Hence, adaptive controllers are required for improved manipulator performance, but these must remain low on computational requirements.

Considerable improvements in end-effector path accuracy have been witnessed [6], by means of the application of microcomputer-based controllers. However the inherent parallel nature of neural nets leads to the possibility of much greater improvements in speed, than those already obtained. Further, the possibility arises of employing much more complex controllers, than simply PID, operating within the same computational time.

Back-propagation neural nets are ideally suited for on-line, real-time control purposes [2], not only because of their rapid computational features, but also through their ability to learn from the collected sampled sets of input and output signal data. Essentially, the net map is able to provide a feedback control arrangement which adapts (learns) to changes in the characteristics of the robot joints, such that the desired control is applied at any instant in time. The role of the net is therefore specifically to produce a real-time output vector which includes new control signals to be applied as servo power inputs to each robot-joint actuator. However, weightings within the net are adapted in accordance with the manipulator operating conditions.

Although a number of faster applications for adaptive controllers have been suggested in the field of robot control [3,7], practical implementation of such controllers has been considered to require high

NATO ASI Series, Vol. F 71
Expert Systems and Robotics
Edited by T. Jordanides and B. Torby

computational power and thus high-cost equipment. However, for a fairly cheap, adaptive controller for use on plant necessitating high sampling rates, the use of high-computational efficiency and low-complexity algorithms must be considered. In this paper, a description is given of the practical implementation of a low-complexity adaptive algorithm based on a neural-net learning system.

As well as dispelling the rumours that adaptive algorithms are difficult to implement and of limited use on systems with high-computational requirements, this work also indicates the practical usefulness of a neural-net system within an overall adaptive control structure.

In the following sections, we look firstly at the type of neural nets considered, these being Back-propagation nets, and secondly it is seen how such a net can fit into the framework of an adaptive-control algorithm. As a follow on from this, the use of the overall adaptive controller, with neural net included, is shown by means of a Robot Control application.

Back-propagation Net

Basically a neural network, or at least a neural-network model is made up of a large number of small and simple elements, these being linked by connections, each of which has a weight assigned to it. Each element is a node which is considered to sum its inputs, which are weighted connections from other nodes, and to carry out a nonlinear transformation, the simplest of which being thresholding on the sum. Neural-network dynamics are concerned with changes of its form, which generally means changes of nonlinear transformation, or more likely, changes of connection weightings.

The simplicity of individual elements is a major plus point for neural nets, and is a point of difference when compared to processor networks or automation. A number of different approaches to modelling neural nets exist although in this work only one particular type, the back-propagation net, is considered.

As a general principle for neural nets, a large net can be partitioned into sections each of which is acted on selectively. These sections can be regarded as forming layers of nodes such that an overall net consists of a layer of inputs at the bottom, one or a number of intermediate (hidden) layers and a layer of outputs at the top. Neural networks structured in this way are employed when applying a back-propagation training algorithm [10]. In the back-propagation net, it is assumed that there are no connections between the nodes which form a particular layer. The net considered here only makes use of a single hidden layer, mainly in an attempt to keep the overall system relatively simple.

A back-propagation net can be used as the basis for a robot-manipulator joint controller, as shown in Figure 1, in which the output vector contains such information as weighting errors, expected joint angle, and weighting-adaptation variables, as well as a vector of joint input signals - one per each robot manipulator joint.

The back-propagation net itself [10,12] is used to approximate the vector function $y = f(x)$, in which the objective (control) of the net is known, although the net weightings employed are unknown and potentially time varying. Contrary to many published articles in this area, which are concerned solely with steady-state solutions, the approach taken here directly deals well both with dynamic-controller requirements and dynamic-learning procedures. Weightings on the net are taught in the first instance from a collected past-joint, input-output data set, whereby the net output vector contains an estimated joint variable. This estimated value is compared with the true value and the sum of the differences (errors) squared, is used to target the weightings on their best values which minimize these squared errors.

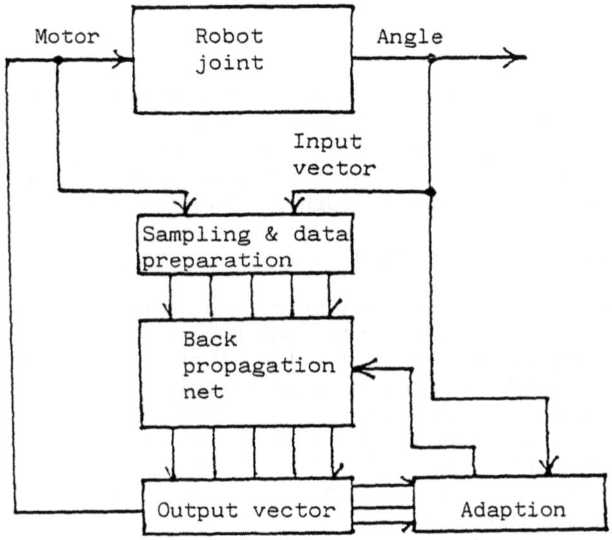

Fig. 1: Neural net for robot joint control.

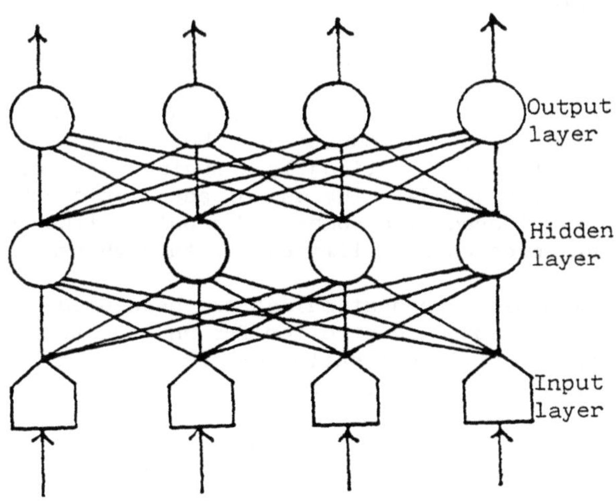

Fig.2: Feedforward net (back propagation part)

The inputs (and outputs) are scaled into (and out of) the range 0 to 1. The net structure contains three layers [9], as shown in Figure 2, an input, hidden and output layer. The input layer takes in a set of present and regressive joint input and output data. The hidden layer and output layer contain neurons in which weightings w_{ij} are applied to produce a sum:

$$S_j = \sum_{i=1}^{N} w_{ij} x_i \qquad (1)$$

The output of each neuron is then either the direct summation or, for example, a nonlinear sigmoid function [2,8].

Updating of the net weighting coefficients w_{ij}, for both the hidden and output layers, is carried out firstly through a linear-least squares teaching period and secondly through a real-time methodology [11] based on

$$w_{ij}(k) = w_{ij}(k - 1) + K \varepsilon(k) \qquad (2)$$

where ε is the error between estimated and actual joint variables, K is a vector of scalar gain coefficients, and k-1, k, etc., signify those variables at particular consecutive-time instants. This updating procedure is thus of the stochastic approximation type.

In concluding this section, it is probably worth pointing out that although a particular type of neural-net has been selected here, namely a back-propagation net, this may well not be the best or most versatile. It was merely selected because of its direct relationship as part of a required learning algorithm, as shown in Figure 1. As stated earlier in this section, many different neural-net arrangements can be formulated [1] and this indicates a fruitful area for further reseach.

In the section which follows, the overall adaptive-control algorithm, employed for robot manipulator control is discussed in more detail.

Application Example

The general adaptive and robustness properties of the neural-net controller are shown by means of an illustrative example, which brings out the controller's high sampling rate and fast adaption properties. this is namely the axis control of an industrial robot, in which the nonlinear dynamics, because of their 'continually changing' nature, severely limit accuracy and operation speeds. The use of adaptive controllers for such robots has been considered by a number of people although their actual implementation has been rather limited due to high cost or implementation problems.

In this work a back-propagation net has been used as the basis for an adaptive, real-time feedback controller for an industrial robot manipulator. Experimental application of the net to a six-axis robot shows excellent results, as depicted in Figure 3. The analog net is operated in a discrete-time fashion as pulses of data are provided as an input vector to the net. Included in the net's output vector are not only the control input signal(s) to the robot joint(s), but also weighting updating parameters and output error terms.

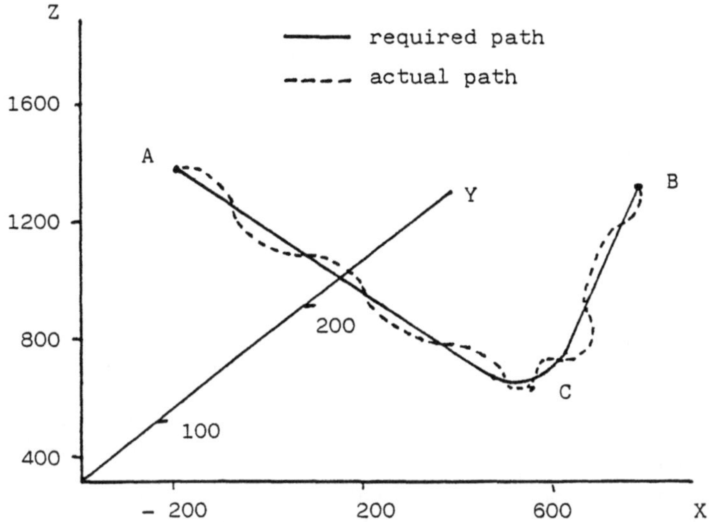

(a) Conventional PID control

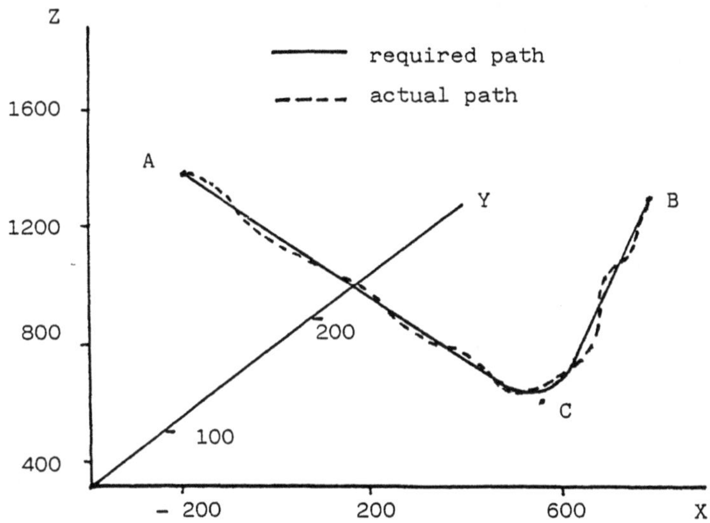

(b) Neural net control

Fig. 3: Six-axis,robot-tip tracking experiment. Path: A to B via C.

The results shown here were gained from the application of six controllers (one per axis), to a comprehensive, continuous-time simulation of the KUKA IR161/15 manipulator [5]. Figure 3 shows the tip position as it is moved from point A to point B, via an intermediate station C. It can be seen that tracking of this relatively simple desired trajectory is much worse when internal fixed parameter PID joint controllers are employed as opposed to the neural-net-based adaptive controllers. The results clearly show (a) advantages in tracking accuracy can be gained when a neural-net-based adaptive controller is employed, and (b) such a controller can actually be used on a real-time application problem.

It should be emphasised that the tracking results obtained, made use of a very simple desired trajectory. As soon as the trajectory becomes more complex, along with high-tip-speed requirements, the performance of conventional controllers becomes much poorer and the improvements gained through an adaptive control scheme are more abundant.

Adaptive Robot Control

The adaptive-control algorithm employed is basically of a self-tuning control form [4], such that it requires an updated model of the plant under control, at all times. In the design presented here, the model is provided by means of a back propagation network. This model is then employed as a starting point from which parameters for a discrete-time adaptive-PID algorithm are obtained.

PID was selected as a controller type due to its fairly simple structure, its popularity in industrial environments and well proven reliability. The pole-assignment method is used to tune the PID parameters thus providing a means by which the controlled response can be tailored. Further, in an adaptive environment, this particular approach has been shown to be more robust and less heavy on control action when compared to many alternative schemes. Although incurring slight penalties with regard to on-line computational complexity, such a scheme makes full use of the system model provided by the neural-net structure, which is a good facility for general monitoring and problem solving - both characteristics to be expected in a practical controller.

As shown in Fig. 1, the net weights are adapted on-line in real-time, indeed the procedure for this was described in the previous section. All that remains is for the neural-net model to be used, in terms of the available measured and estimated joint variables, in order to provide PID controller parameters. This can in fact be done in one of a number of ways, where either the controller parameters are modified every sample period or merely occasionally, when deemed necessary. A description is given here which suggests the former of these, however it should be understood that this is merely one option.

By taking weight values from the net, along with measured and estimated joint variables, the PID controller parameters can be readily calculated in a similar fashion to many self-tuning algorithms [4,6]. The actual control input to be employed is then evaluated merely through the difference equation which results.

Implementation of the overall control algorithm is carried out by taking the following steps every sample period:

1. Collect new data from the joint input and output signals, i.e. collect new joint variable information for each joint.
2. Update the neural-net system models.
3. Update the PID-Controller parameters, using the new models.
4. Calculate the new PID control signals.
5. Transmit the new control signals to their respective joints.
6. Wait for the sample period end (by polling a clock input).
7. Go to step 1.

Obviously, many of the above steps are further sub-divided and the assumption here is inherently made that both the net model and the PID-controller parameters are updated every sample period.

It should be pointed out that to test the controller operation, process diagnostic and simulation modules have been developed [5]. These provide data logging and other facilities such that the performance of various functional blocks, including the neural net and its elements, can be assessed during run-time.

A master shell provides a reasonably friendly user front end which allows for various parameter values, such as initial net weightings and desired closed-loop denominator to be easily set up. This shell also allows for advisable run-time modifications, if these are required, but more importantly its data-processing ability ensures that run-time results can be obtained directly.

Conclusions

A high-speed, practical adaptive controller for use on robot-manipulator control has been described, this being based on a central neural-net feature which learns the relevant, robot joint characteristics. The net is then used as a starting point for on-line, adaptive-controller design. At present both linear and nonlinear, net-updating procedures can be employed, the simplest being a straightforward linear arrangement, of the type described in equation (2), however, various nonlinear arrangements are possible, including sigmoid functions with a dead-zone, although a trade-off must be made between weighting convergence/updating awakeness, and noise rejection.

The tip-tracking application example gives an indication of the operability of such a controller and highlights its fast, adaption properties. These are important points in industrial-manipulator control where the problem is characterized by dynamics which are both nonlinear and time varying. Experiments have also shown the overall control operation to be reliable, in the long-term, thereby offering improved-manipulator performance without sacrificing practical usefulness properties.

References

1. Benes, J.: On neural networks, Kybernetika, 26, No. 3, 232-247 (1990).
2. Bhat, N. and MacAvoy, T.J.: Use of neural nets for dynamic modelling and control of chemical process systems. Proc. 1989, American Control Conference, Pittsburgh, 1342-1347 (1989).
3. Farsi, M., Zackariah, K.J. and Warwick, K.: Adaptive control algorithms for intelligent robot manipulators. Ch. 4 in 'Robot Control', K. Warwick and A. Pugh (eds.), Peter Peregrinus Ltd. 1988.
4. Harris C.J. and Billings, S.A. (eds.): Self-tuning and adaptive control: theory and applications. Rev. 2nd ed., Peter Peregrinus Ltd. 1985.
5. Karam, K.Z.: Fast adaptive control algorithms and their application to industrial robots. PhD Thesis, University of Newcastle upon Tyne, 1988.
6. Karam, K.Z. and Warwick, K.: A high speed adaptive controller for M68000-based PCs. Journal A. 30, 19-24 (1989).
7. Koivo, A.J. and Guo, T.: Adaptive linear controller for robotic manipulators. IEEE Trans. on Automatic Control, AC-28, 162-170 (1983).
8. Lippman, R.: An introduction to computing with neural nets. IEEE ASSP Magazine, 4-22 (1987).

9. Pao, Y-H.: Adaptive pattern recognition and neural networks. Addison-Wesley 1989.
10. Rumelhart, D. and McClelland, J.: Parallel distributed processing: explorations in the microstructure of cognition. 1, Ch. 8, MIT Press 1987.
11. Warwick, K., Karam, K.Z. and Tham, M.T.: Simplified parameter adaptive control. Journal of Optimal Control Applications and Methods, 8, 37-48 (1987).
12. Webos, P.: Beyond regression: new tools for prediction and analysis in behavioural sciences. PhD Thesis, Harvard University (1974).

On-line Scheduling and Planning of Robots and Presses

A. Crespo, F. Barber and P. Albertos

Grupo de Automática e Informática Industrial (DISCA). Universidad Politécnica de Valencia
e-mail: alfons@aii.upv.es, Apdo.22012, E-46071 Valencia, Spain

Abstract: The work reported in this paper deals with the application of Expert Systems (ES) approach to the scheduling and planning of a set of presses and robot feeders in an automobile factory. This work has been usually done by experts in the field with a sound knowledge in all the related topics: Processes, requirements, priorities, constraints, options and decision timing. One of the most important constraints is the response time. So, time constraints have been considered in the design and implementation of the ES solution. In the first section of this paper, a description of the process is presented, pointing out the features of the application, the structure of the ES, and the temporal constraints. In the next section, an overview of the model used to represent the temporal knowledge will show the possibilities of this approach. Finally, the main ideas of this application are discussed, and preliminary results are shown.

Keywords: Planning, Temporal reasoning, Real-time Expert System, Object- oriented Programming.

1 PROCESS DESCRIPTION

The process we are considering was taken from the press section of an automobile factory. It consists of various press lines where pieces are handled by robots. As previously mentioned, time constraints appear at two levels: in the synchronicity of presses and robots and in the planning of orders to be executed. The first case has been treated as an average waiting time, related to the pieces to be pressed, and the second one is considered by the ES to be developed.

The features of the application are the following:

- The presses have similar performances, and the same is true for the robots.
- A number of presses work in sequence.
- Between two press lines, a set of robots is handling the worked pieces. Each robot unloads two presses and feeds the next two.
- The stock of pieces is limited.

NATO ASI Series, Vol. F 71
Expert Systems and Robotics
Edited by T. Jordanides and B. Torby

- To operate each press, a given time is required. The same happens to the robot operation, but the used time is a function of the handled piece, and an average waiting time can be considered to take into account the sharing of the robots between the two lines.
- Each order requires a specific set of tools in the presses.
- Batch production imposes a minimum number of items per order.
- Pieces are priorized.
- There are time restrictions in pieces and worked pieces.
- Any change in the activity of a machine requires some maintenance and preparation time.
- Last-minute disturbances require on-line changes in the weekly planning:
    1. internal, such as breakdown of tools, production failures or high priority demands.
    2. external, such as errors in stocks, delivery delays or external demands/constraints.

1.1 Planning criteria

The goals of the planning are related to quantitative and heuristic criteria. On one hand, typical optimization indices, such as maximum production or minimum storage costs, are considered. On the other hand, time constraints in product delivery, priority in pieces working, and special customer requirements must be handled.

In order to define the knowledge base (KB), an object oriented approach is implemented. The main defined classes are Lines, Presses, Robots, Tools, Pieces, Orders, Agenda, Plannings, and Stocks.

Some temporal attributes can be assigned to these object classes, relating the dynamic evolution of pieces in the press lines or the availability of pieces and worked pieces in the storages. In a similar way, production rules taking into account time constraints and temporal information are defined. The difference between temporal and temporary or variable information, is that the first is related to the time requirements an event or action must satisfy, and will be considered fixed in the running of an application. For that purpose, special tools for real time ES have been developed.

2 TEMPORAL REPRESENTATION AND REASONING

Several approaches have been proposed for temporal representation and reasoning. Temporal logic models, based on modal-logic, have an explicit time representation and the value of an expression depends on the instant it was calculated. Temporal models proposed in artificial intelligence areas, based on time-points or intervals, perform a conceptualization of the temporal expressions and their properties, defining an automatic reasoning method about them. However, they seem somewhat inappropriate for the specific problem of temporal representation and reasoning in knowledge-based systems methodology. The main drawbacks are related to expressiveness and structuring of temporal concepts, specification of temporal dependence (constraints) over problem data and knowledge relations, and the limited causal relationships. In a more specific way, the several frameworks for automatic planning [2,3,7,8] have similar drawbacks as temporal models, and domain expert's knowledge is not explicitly considered.

Several ES prototypes proposed to handle the planning problem [4,5,6], do not form a general solution of the problem and the temporal capacity of the ES has an 'ad hoc' treatment in each case, too restrictive for the wide range of temporal domains. The dynamics of the problem is usually represented by means of: i) 'state model' (pairs of < data, value >, updated successively), where the successive states represent the temporal ordering, and ii) symbolic (trends, change rates, errors, etc.) pre-processing of temporal data, about which the ES reasons.

The application described in this paper uses temporal representation based on the following features: [1]

- 'time points' as explicit representation primitive. In this model, the temporal precedence relationship has transitivity, linearity, finite, and discrete properties over its elements.
- Object oriented approach for dynamic world modeling. The problem data are represented by means of a set of triples <object, attribute, value>, where 'value' can be instantiated on an element of the respective set of attribute value-class. From a temporal point of view, the only interesting attributes are those whose values can vary over the considered time interval. Thus the representation of temporal data is assumed according to a sequence of values of the temporal attributes. These attributes can be classified according to the temporal completion:
  - Event-type: The attribute takes an instantaneous value. An event-token is each one of the occurrences of an event-type taking the value in a time-point.
  - Property-type: The attribute takes a value within a certain interval. A property-token is each one of the occurrences of a property-type taking the value during a given time interval.

  Thus, temporal facts of the problem (occurrences or actions) are represented by means of a set of events or properties-token, as 'temporal instances' of events or properties-type. All these temporal concepts are defined in the appointed reference as Abstract Data-Types (ADT).
- Causal actions and knowledge inferences are expressed as temporal production rules, whose premises and conclusions are related to the representation of the problem dynamics in an expressive manner. Similarly, it is possible to establish temporal constraints about its application in relation to both absolute time or dynamic behavior of the problem (events or properties-tokens represented). With regard to the related application, and with these production rules, it is possible to handle our knowledge over the problem (temporal constraints about actions or tasks) in order to generate a plan.
- Inference time is considered on a temporal-inference method, which allows temporal dependence (parameterization and temporal constraints) in the application of production rules. A temporal-inference control based on temporal precedence of premises completion, is defined.

This temporal model has been integrated in an available rule and frame-based expert system development tool, in such a way that a rule-based Temporal Reasoning Expert System environment is obtained. By this tool, a prototype of the proposed expert system has been developed.

3 KNOWLEDGE BASE

3.1 Facts Base Structure

The KB has been structured in classes and subclasses using inheritance in order to have a general framework of the problem. Specific components in the application will be instances of a class. The system follows a hierarchical scheme.

The structure of each class is described in the rest of this section.

Class: LINE

Defines the common attributes and behavior of all the lines in the system.

Attributes:

- NAME: Identifier of a line
- PRESSES: List of sequential presses involved in a line
- ROBOTS: List of robots used in a line
- DURATION: Working time for the actual piece
- LEVEL_OCCUPANCY: of the orders assigned (not planned yet)
- LEVEL_OVERLAPPING: of the orders assigned (not planned yet)

Temporal attributes:

- ORDER: Order in progress
- STATUS: State of a line (WORKING, READY, BLOCKED, UNDEFINED)
- COST: Evolution of the cost depending of day, hour.

Class: PRESS

Defines the structure of all presses in the system.

Attributes:

- NAME: Identifier of a press
- TOOLS: List of tools that can be assembled to the press
- COST: Evaluation of the operational cost of the press
- DURATION: Required time for one piece.

Temporal attributes:

- STATUS: State of a press (WORKING, READY, DAMAGED, WAITING)
- INSTALLED-TOOL: actual installed tool
- PIECE-IN-WORKING: actual piece under production
- LINE: where the press is assigned.

Class: ROBOT

Defines the common behavior of all robots in the system

Attributes:

- NAME: Identifier of a robot
- SHARED-BY: List of lines that share a robot
- JOIN: List of presses joined by a robot
- COST: Evaluation of the use of a robot
- DURATION: Time consumption of an operation.

Temporal attributes:

- STATUS: Similar to the press object
- LINE: Temporal assignment of this robot.

Class: TOOL

Defines the common behavior of all tools. Tools are installed in a press in order to produce a given piece.

Attributes:

- NAME: Identifier of a tool
- ASSEMBLED-IN: List of presses where this tool can be assembled
- DURATION: Estimated time for the assembly operation
- COST: Estimated cost of the use of this tool
- CONDITION: level of use of this tool, to evaluate the need of maintenance or replacement.

Temporal attributes:

- STATUS: State of this tool
- PRESS: Temporal assignment of this tool.

Class: PIECE

Defines the common behavior of all pieces. Pieces can be considered as an entry to the system, or produced by the system.

Attributes:

- NAME: identifier of a piece
- TOOLS: List of tools to be used to produce this piece
- DURATION: List of estimated time for the working on this piece with the required tools
- MINIMUM-BATCH: Minimum number of pieces to be produced by an order
- STOCK-COST: Daily cost of this piece in sotck
- TOTAL-COST: The average production cost of a worked piece.

Class: STOCK

Specify the common behavior of all stock components.

Attributes:

- NAME: Stock identifier
- PIECE: Piece to be stocked
- COST: Cost of this stock, per time unit and per piece.

Temporal attributes:

- NUMBER: Number of pieces in stock.

Class: ORDER

Set of pieces to be produced under some constraints and requirements.

Attributes:

- NAME: identifier of an order
- PIECE: to be produced
- NUMBER: Ordered number of pieces
- NEEDS: Real needs of the piece

- TIME: constraint about the limit time the pieces should be available
- PRIORITY: level of importance of the piece.

Class: AGENDA

Contains the active lines of a plan, with information related to the allowable time intervals for batch production.

Attributes:

- NAME: identifier of an agenda
- LINE: Line involved
- ORDER: order(s) related to this agenda
- PIECE: in execution
- NUMBER: of pieces to be produced in this agenda
- START: Starting time for this agenda
- FINISH: Expected finishing time

Class: PLAN

Define the objects containing the general information to build a plan taking into account an optimization criteria and incidences in the development of the plan.

Attributes:

- NAME: Identifier of a plan
- ORDERS: List of orders that must be planned
- STOCKS: Number of pieces available in the stocks
- AGENDA: list of agendas defining the work to do
- CRITERIA: Optimization criteria to be applied. The optimization criteria can be some of the following values: PRIORITY, STOCK-COST, SECURITY-INTERVAL, and MINIMUM-TIME.

By these classes, the framework of the application where the particular problem is instantiated, is defined.

It is possible to specify several incidences that can occur during a planning. These incidences can be: PRESS-OUT-OF-LINE, ROBOT-OUT-OF-LINE, DAMAGED-TOOL, DAMAGED-PIECES, PIECES-LOST. They should update the state of the related object in order to run a new planning and the modification of the agenda according to the actual status.

3.2 Knowledge Relations

The plan generation is done according to the criteria defined by the user (several criteria can be choosen in priorized order) and the status of the system (resources status as presses or robots damaged, etc.) changed by incidences report.

From the specified orders, the system verifies the stock of each piece and elaborates the real needs to be planned. With this information, the orders are processed in order to known the estimated duration of the production taking into account the delivery time and the duration of the operations in the working of each piece. All this information is stored in the piece, tool, robot, and press objects required for the process.

The system takes orders (real needs, with a duration) and schedules it. This process is done taking in account the selected criteria and the available resources of the system.

The system topology could allow the production of a piece involving several lines of presses, even using the same press and tool in different lines if they are compatible with the piece at a level. The compatibility should be in tool and time. So, several lines (more than a real set of presses) sharing resources could be scheduled. If we have a topology of m lines and n levels of presses, a complex system of manufacturing could be defined. In a real environment there is not a high compatibility of the presses; there are lines for a subset of pieces.

In order to know if two pieces are compatible (tool and time) a set of rules can be defined. The tool compatibility is easy to determine if there is matching between tools in the same level and the time compatibility is calculated using the temporal information associated (event-tokens and property-tokens) of the presses, robots and pieces, by performing a temporal resoning with it. The rules to determine the time compatibility are constrained by the configuration information of the system (tools to build a piece, tools associated to presses, presses associated to lines, etc.).

With the temporal reasoning in mind, the rules to plan the orders can be organized in subclasses of rules as:

- CONTROL: Fire the different set of rules
- LINE.ASSIGNEMENT: Define the set of lines and their periods related to the production lines.
- INCIDENCE: Use the possible incidences to modify the production plan
- STOCKS: Control of the stocks.
- ORDERS: Choose the orders in accordance to the criteria selected.
- PLANNING: Plan the orders in the selected lines.
- AGENDA: Define the agenda for a plan
- OPTIMIZATION: Analyze the resulting plan in order to optimize it.
- NO.SCHEDULABLES: Solve the problem created when an order can not be planned in the actual selected lines.

The main rules defined in the system are the following (written in natural language)

Rules related to the LINE.ASSIGNEMENT

- An order is initially assigned to all lines defined by the tools and presses involved in the production of the piece
- An order is initially assigned to each line with the just-in-time criteria
- A line is set to READY if all orders are assigned and the level of occupancy of the line has a value between 60% and 80% and the level of overlapping is less than the 25%.
- No more than the maximum number of production lines can be set to READY at the same time.
- All the lines set to READY must contain all the orders.

Rules related to the PLANNING

- An order is planned in the best possible line from the criteria selected if the line is READY in the total or partial duration of the order.

- An order is scheduled in a line if there is time compatibility in meeting its time requirements
- An order has a conflict if there are not UNDEFINED lines and time incompatibility is detected.
- An order has a duration greater, less, or equal to other order scheduled.
- An order can be scheduled inside of a scheduled order if the minimum batch number is accomplished in both ends and the selected criteria makes best this possibility,
- An order is scheduled inside of another as near as possible to its delivery time.
- An order is scheduled before another if it is best from the selected criteria and is not possible to put inside.
- An order could be scheduled in a gap time if the gap time is greater or equal to the duration time.
- An order could be split if the timing gap is greater than the duration of a batch order and maximum number of pieces are planned.
- An order is finished if all pieces are planned.
- An order is split if the result of the selected criteria tell us that it is better to do it than to advance the planned order before to it.
- The size of a split order is as large as possible (duration of the order versus available time) with a remainder not less than the minimum batch size.
- An order can not be planned if there are no lines, or there is no available time (total time) in a line, and a new plan should be developed from this point using a new criteria for it.
- An order can not be scheduled if no more criteria is available.

4 CONCLUSIONS

In this paper, a problem of planning of machines and robots has been stated and a solution based on a real-time ES has been developed. The problem is a general one in industrial environments with large similar products. The temporal behavior of the problem is handled with temporal attributes related to instantaneous events and actions taking place over a time interval. The prototype has been implemented in KEE software, enlarged by Common Lisp subroutines to deal with the temporal attributes and reasoning.

The practical application has been simplified considering independant press lines and robots, being now under validation, but the approach is suitable to handle optimal use of resources arranged in matrix form. Temporal constraints are only considered at the planning level, deleting the real time synchronism problems.

5 BIBLIOGRAPHY

1. Barber, F.: Time-dependent experts-systems: A model for representation and reasoning. PhD. Thesis. Dpt. Sistemas Informáticos y Computación. Univ. Politécnica of Valencia. 1990.

2. Fikes, E.E. and Nilsson, N.J.: Strips: A new approach to the application of theorem proving the problem solving. Artificial Intelligence 26. 1971.

3. Hubert, A. and Becker, S.: Production planning using a temporal planning component Procc. of ECAI-88 Munchen. pp.182-185.

4. Isermann, R. (Ed.): Procc. of the 10th world congress on automatic control. IFAC-7. vo.6. 1987.

5. Laffey, T.L., Coxx, P. A., Schmitdt, J.L., Kao, S.M., and Read, J.Y.: Real-time knowledge-based systems. AI Magazine. Spring 1988.

6. Perkins, W.A. and Austin, A.: Adding temporal reasoning to expert systems building environments. IEEE-Expert Frebuary 1990.

7. Sacerdoti, E.D.: A structure for plans and behavior. Elsevier. 1977. N.Y.

8. Tsang, E.: Elements in temporal reasoning in planning. Procc. of ECAI-88. Munchen. pp.571-573.

KNOWLEDGE-BASED PROCESS PLANNING, SCHEDULING AND ERROR RECOVERY IN ROBOTIZED ASSEMBLY

A. DELCHAMBRE, P. GASPART,
D. COUPEZ, P.R. GLIBERT

CRIF/WTCM - INDUSTRIAL AUTOMATION DEPARTMENT
50 AV. F. ROOSEVELT - BOX 106
B-1050 BRUSSELS - BELGIUM

Abstract : The goal of current researches in robotized assembly is to increase both productivity and flexibility. In this way the trend is now to use two or more robots working together on the same assembly, sharing a part of their working area and a number of tools.

This paper describes a practical approach to the control system of a two-robot assembly cell. This system consists of four main functions : process planning, scheduling, supervising and diagnosing, and error recovery.

The planning phase generates the assembly plans from geometrical and topological descriptions of the product. From a chosen assembly plan and from a rough description of the production means, the scheduling module answers the question : "Who has to do what, when and how?" Finally, when an error has been detected, the error recovery module puts the cell in such a state that the assembly can restart.

In this paper, the specific knowledge and the common sense expertise are specified for each module.

Keywords : *assembly planning - assembly scheduling - error recovery - two-robot assembly cell - off-line programming - specific knowledge - common sense expertise.*

NATO ASI Series, Vol. F 71
Expert Systems and Robotics
Edited by T. Jordanides and B. Torby

1. INTRODUCTION

A lot of research is being done about robotized assembly cells. The main topics of such researches are the robust and efficient control and programming of these complex production tools. Recently, interest has grown for multiple actors assembly cells in which several robots cooperate in order to achieve a specified goal. The introduction of a second robot in a cell can be justified for an assembly product and generally needs actions with different, and sometimes conflicting skills (accuracy, payload, capacity). For some very special assembly operations, the cooperation of two robots may be absolutely required.

The complexity of the problem further increases when the two robots share resources such as grippers, feeders, sensors and have a large part of their working area in common.

The control structure should be flexible enough to allow a quick change of production. On the other hand, it must be reliable and robust to cope with errors that may occur during the assembly.

This paper deals with the control of a two-robot assembly cell with tightly coupled working area that assembles a large family of signalling relays.

2. GENERAL CONTEXT OF ROBOTIZED ASSEMBLY

In figure 1, we present the complete control system of our multi-robot assembly cell.

As we can see, it includes two main parts :
- the off-line system,
- the on-line system.

The <u>off-line system</u> includes two important modules :
- the assembly planner,
- the scheduler.

The <u>assembly planner</u> generates the operational sequences from geometrical and topological information on the final product. General information on the assembly cell is introduced to choose among the assembly plans which group sets of operational sequences.

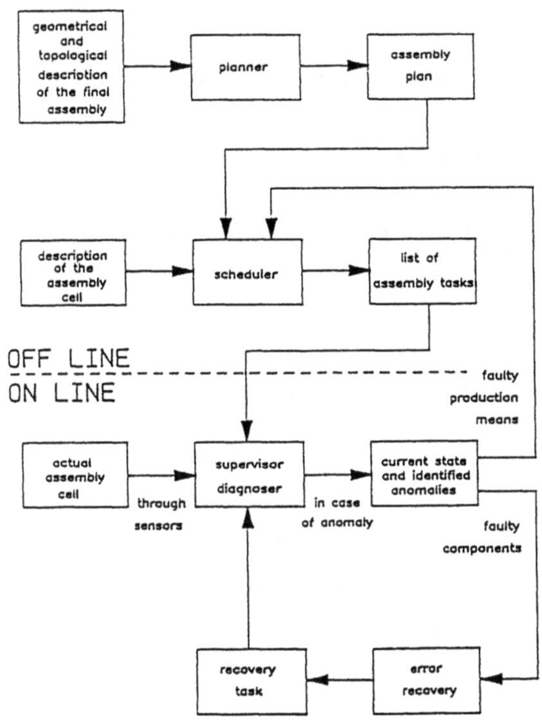

Figure 1 - Complete control system of CRIF/WTCM multi-robot assembly cell

From the chosen assembly plan and from a detailed description of the production means, the scheduler answers the question : "Who has to do what, when and how? "

In other words, it synchronizes and allocates the operations (described in the assembly plan) to the available production means in the cell (robots, fixtures, grippers, specific tools, ...).

The collision-free trajectories generation is performed at this stage. The output of this module consists of a list of assembly tasks allocated to the different robots of the cell with different synchronization points.

The on-line system is composed of two modules :
- the assembly supervision and diagnosing,
- the assembly error recovery.

The supervision and diagnosing module follows up the execution of the different operations. It detects and identifies the anomalies.

When an error is detected, the error recovery module tries to replace the cell in such a state that the assembly can restart.

This paper describes the two first modules and the last one : the assembly planner, the scheduler and the error recovery.
The next section presents the case study we will use to illustrate these different packages.

3. CASE STUDY

The assembly planner has to generate the assembly plans of an electrical relay (21 components). This case has been chosen because it is rather generic. There are not only different directions of assembly, but also several types of liaisons : insertion, screwing, clipping, ...

Figure 2a shows the electrical signalling relay and figure 2b all its components.

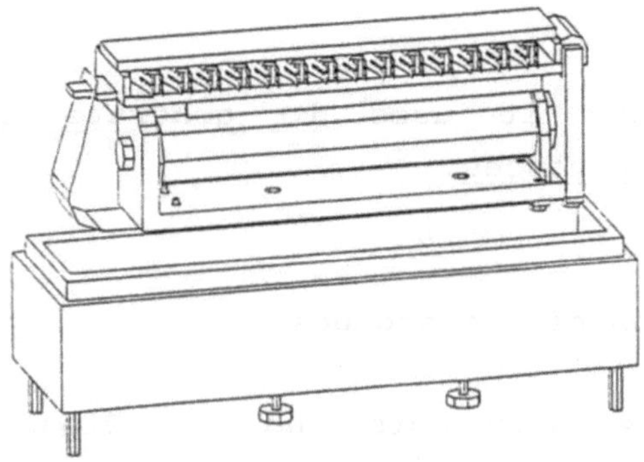

Figure 2a -
Electrical signalling relay

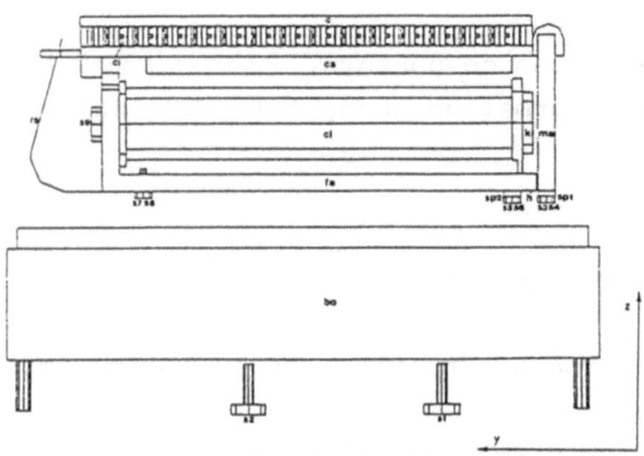

Figure 2b -
Components name

k : kernel
cl : coil
bo : box
rs : return spring
ci : contacts
c : cover
cs : contacts support

fa : fixed armature
ma : mobile armature
sp1 : stiff plate 1
sp2 : stiff plate 2
si : screw i
h : hinge

4. ASSEMBLY PLANNER

This module is composed of two parts :

- the modelization of the product which represents the specific knowledge the planner needs to generate the assembly plans,

- the methodology it uses to generate and evaluate the operational sequences.

4.1. Modelization of the product

It concerns the components and the final assembly. This module includes three kinds of information :

- geometric information on the components and on the final product,
- properties of the components,
- relational information among the parts in the final assembly.

4.1.1. Geometric modelization

This modelization has to give the following information :

- the shape and dimensions of each component,
- the locations of the components in the final product.

At present, we modelize each part by some judiciously disposed parallelepipeds whose edges are parallel to the coordinates axes.

Each parallelipiped is represented by the coordinates of its two extreme vertices as we can see on figure 3.

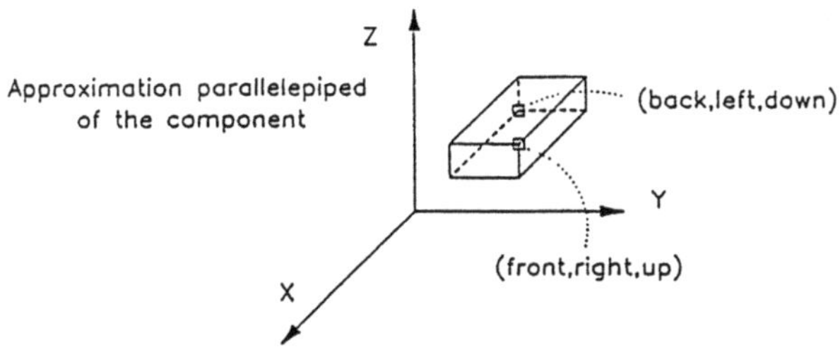

Figure 3 - Approximated subpart of a component

Figure 4a shows the real coil of the signalling relay and figure 4b its modelization.

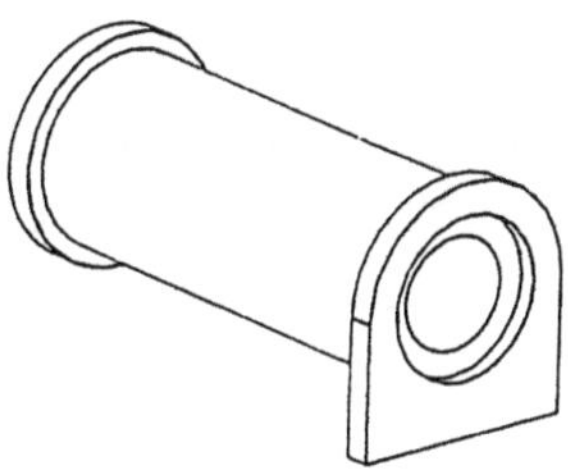

Figure 4a
Real coil

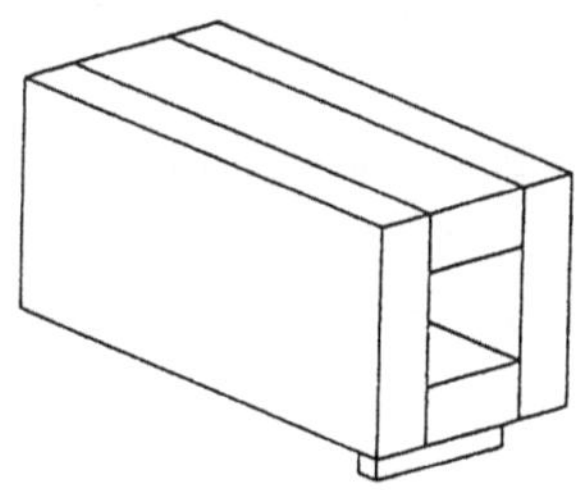

Figure 4b
Approximated coil

This modelization will also be used by the error recovery module as we shall see it later.

4.1.2. Properties of the components

They characterize the specific role of some components in the final product and some physical features which are used during the generation of the operational sequences.

4.1.2.1. Standard categories

Some components are used to fasten other parts or to allow a relative movement among other components.

We have identified the following categories :

- screws,
- clips,
- bearings,
- springs,
- o-rings.

This classification will be used during the generation of the relational information of the operational sequences.

4.1.2.2. Physical properties

At present, we quantify only the manoeuvrability of each component by the handibility factor. This information is needed by the assembly planner and the error recovery modules.

4.1.3. Relational information among the parts in the final assembly

This information is deduced from the geometric modelization and the standard categories of parts. It includes two parts :

- the geometric constraints among the components,
- the product graph.

4.1.3.1. Geometric constraints among the components

From the geometric modelization presented in section 4.1.1, a geometric reasoning module sets lists of parts which obstruct the removal of each component in the final assembly in each of the six assembly directions (parallel to the reference frame).

For example, in the use of the kernel of the relay, we obtain the following facts :

```
obstacles (k,-x,[cl,bo]).
obstacles (k,+x,[cl,bo]).
obstacles (k,-y,[ma,bo]).
obstacles (k,+y,[s9,fa,rs,bo,cl]).
obstacles (k,-z,[s1,s2,s7,s8,s5,s6,
                 sp2,fa,rs,bo,cl]).
obstacles (k,+z,[cs,c,ci,cl]).
```

4.1.3.2. Product graph

The product graph represents the final assembly. Its nodes are the components and its links correspond to the physical liaisons between the parts. The links are labelled and separated in different sub-groups which characterize the type of the liaison and its number of degrees of freedom.

The different categories are the following ones :

- putting on,
- insertion without blocking,
- insertion with blocking,
- clipping,
- screwing.

The graph and its attributes are automatically built from the standard categories and from the lists generated in the previous section.

Figure 5 presents the product graph of our case study.

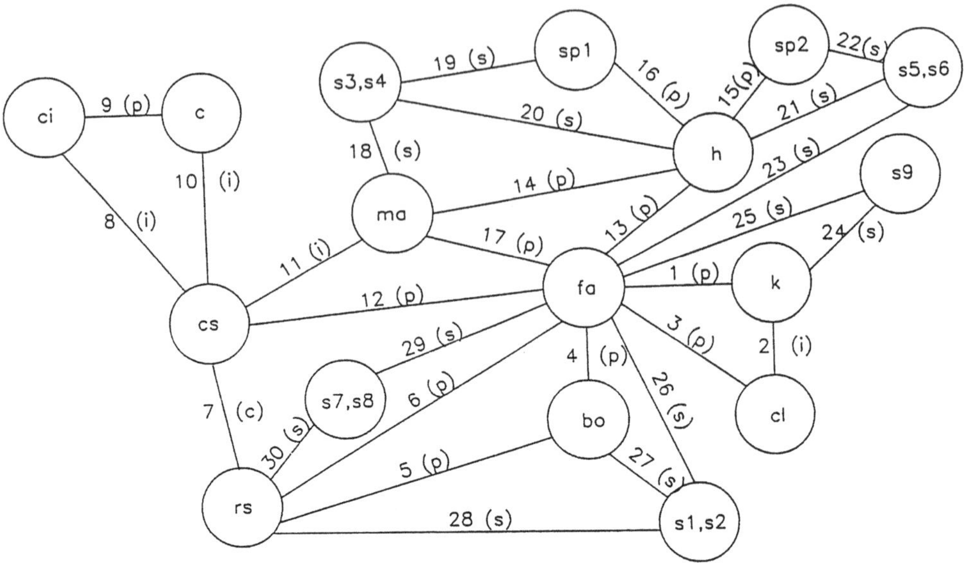

Figure 5 - Product graph

4.2. Methodology of assembly plan generation and evaluation

This module runs in two steps :

- generation of assembly constraints,
- generation of assembly plans.

4.2.1. Generation of assembly constraints

The constraints correspond to precedence orders among the components during the assembly phase.

These constraints are divided in two categories :

- the hard constraints,
- the soft constraints.

The generated operational sequences cannot, in any case, violate the hard constraints. Otherwise they will not have any physical sense.

These constraints consist of two groups :

- the geometric constraints : they avoid the collision among the components during the elaboration of the product.
- the mechanical constraints : they are caused by external fasteners. Before the assembly of the parts fixed by these fasteners, we have to check if the fasteners may be inserted.

The soft constraints can be violated by the generated sequences. They are used to optimize them. We have taken into account :

- stacking constraints,
- technological constraints.

The stacking constraints are caused by external fasteners. If these fasteners fix several stacked components, it is often convenient to impose the stacking order among these parts.

The technological constraints are introduced by the user if he has an a priori knowledge about several production means and if he wants to impose some parts of the total sequence.

All the constraints, except the last ones are automatically generated by our software package. In the case of the relay, 28 constraints are generated.

4.2.2. Generation of assembly plans

With the help of the preceding constraints, our program generates a set of optimal sequences which are grouped in several assembly plans. These plans consist in precedence graphs where the nodes represent the assembly operations and the links, the precedence constraints among these operations.

We try to generate the most parallel assembly plans because we plan to assemble the product in a two-robot assembly cell. It is then quite important not only to extract the independent subassemblies but also the parts which can be assembled on the same base part simultaneously, in the same direction.

Let us explain in some words the method used by this package.

When we have started one branch of the precedence graph, we try always to continue this branch before grouping some parallel branches. One branch can be split in different sub-branches when several operations can be executed simultaneously in the same direction.

When it is impossible to add any components in each parallel branch, we try to group different subassemblies in order to continue the operations.

Figure 6 shows the most parallel plan describing the assembly of the relay.

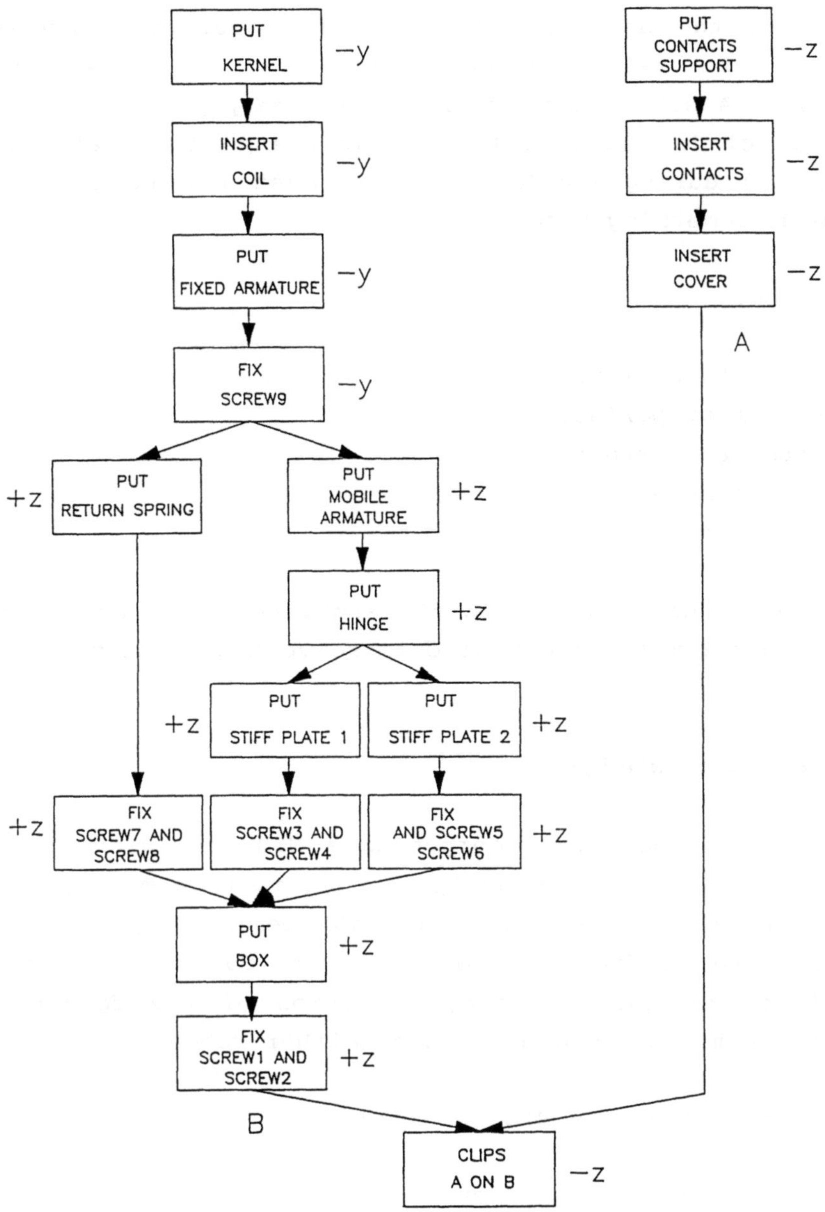

Figure 6 - Assembly plan of the relay

5. SCHEDULER

The input of this module consists of an assembly plan generated by the assembly planner. The operations composing this plan are called NIOs (short for Non Interruptible Operation). A NIO is a node of the assembly plan and consists of a group of low level actions that change the state of the assembly and during which the robot must not be stopped in order to do something else.

For instance,

- *move robot to pallet,*
- *pick part from pallet,*
- *move robot to fixture,*
- *put part in assembly,*
- *move from fixture.*

We will now present the specific knowledge this module needs and the general methodology used to solve the problem.

5.1. Specific knowledge

To execute the operations composing a NIO (screwing, handling of parts, ...) tools were designed that must be mounted on the robot. Furthermore, all NIOs may not be done by each robot. Some NIOs are "dedicated" to one robot due to the physical constraints (payload required, situation of the feeder, ...) while others may be accomplished by either robot.

To describe logically a NIO, we need :

- which robot is able to execute the NIO,
- which tooling is required,
- how much time it takes on each feasible robot.

We need also to describe geometrically a NIO : the trajectories of each robot which executes a NIO are already known (by the use of a simulation system).

Each NIO is divided in two types of movements :

- <u>fine positioning movement</u>. These movements correspond exactly to the pick and place parts of the NIO. They require high precision and are generally rather slow. Furthermore, the robot sweeps a small part of its working space during such phases.

- <u>gross transfer movement</u>. During the transfer of a part from the feeder to the site where it will be added on the assembly, the speed of the robot is very high compared to the fine movement. The volume swept by the robot is on the other hand much lower.

5.2. General methodology

Between two consecutive NIOs, there may exist a tool exchange phase; the robot is improductive, tool exchange must be prevented as much as possible.

The role of the scheduler is then to assign NIOs to the robot so that the precedence constraints are fulfilled while minimizing as much as possible the total time of execution.

Starting from the plan and NIOs description, it must output one linear list of NIOs for each robot. It is also the role of the scheduler to generate the tool exchange phases and insert them in the list of NIOs.

If we look in deep details into the schedule generation for a two-robot assembly cell, we can distinguish two aspects of the hardness of the task :

- the temporal difficulty,
- the geometrical difficulty.

The first aspect is the well known scheduling hardness. The second aspect deals with collision avoidance between two robots. We think facing the two aspects simultaneously is too complex.

Our approach is then in two steps :

- first, try to find a schedule that is optimal or quasi-optimal from the temporal point of view,
- second, patch the schedule found, introducing some waiting times to prevent collision between the two robots.

5.2.1. Temporal schedule

To solve this problem, we use a Branch and Bound (BAB) method and we use the degree of parallelism of the assembly plan to guide the search.

The more a plan is parallel, the more the number of choices at each step in the BAB algorithm, the bigger the chance to find an operation to do, the less we must let a robot idle, and the less the plan lasts.

If at each step of the BAB algorithm, we compute the degree of parallelism of the rest of the plan, we have a clue for choosing between the alternatives, because the more the plan is parallel, the less it will take to finish it.

5.2.2. Geometrical constraints

To solve the geometrical aspect of the problem, we use yet again a heuristic. As we have seen before, each NIO is decomposed into two kinds of movements.

We consider each kind of movements as a whole for the collision test. The volume used by the robot during the execution of a NIO is described as a succession of volumes (fine and gross), we may determine if a collision occurs between the two robots and in which phase (gross or fine).

Then, comes the stretching phase. In this phase, if two trajectories cross each other at the same time, we try to introduce idle times between fine motion and gross motion in order to avoid any collision.
During the introduction of idle times, we must of course take care not to violate the precedence constraints, i.e. the resulting schedule must still verify the assembly plan.

6. ERROR RECOVERY

This module has to solve the following problem : how to put, automatically and economically, the cell in a state where the production can restart, after the detection of an error.

6.1. Scope

We suppose that the error has been already identified. The error type on which our work is mainly focused is the faulty component : this can only be detected indirectly by the failure of a related fastening operation (for example, by the sensing of an abnormal effort during an insertion).

The specific knowledge on the product we need in this module is the same as for the assembly planning. This module needs also a common sense knowledge applicable for the error recovery in a wide class of assemblies.

6.2. The common some expertise

The error recovery has two main goals :

- to generate a recovery task in order to remove all the faulty components, minimizing the removal of other parts;
- to identify the point where the assembly task can restart after recovery.

To achieve these requirements, three chunks of generic expertise were implemented.

6.2.1. Handibility control

The corresponding heuristic is the next one : "When it is necessary to remove a small part (for example, a screw) fastened to a more handible component, it is better to grasp the latter for removal."

6.2.2. Graspibility control

The heuristic is :"When the robot cannot grasp a component, it has to fasten it to a graspable component."

6.2.3. Obstacles control

The heuristic is :"When the robot cannot remove a component because the way is obstructed by another component, it has to remove the obstacle first."

7. CONCLUSIONS

This paper presents a survey on how we have introduced artificial intelligence and expert systems concepts in some of the main control problems in a robotized assembly cell :

- assembly planning,
- scheduling,
- error recovery.

The assembly planning and the error recovery modules are now fully implemented, we have tried it on several assembled products (15 to 25 parts) and the results are greatly satisfactory.

The *Branch and Bound method* used in the scheduler has been implemented and tested on different examples and work well. We are now developing the patcher and collision detection modules.

The error recovery system has been implemented and tested. We have now to specify and develop the diagnosing functionality identifying liably the faulty tooling devices, fixtures and/or components after an indirect anomaly detection.

REFERENCES

[1] COUPEZ D., DELCHAMBRE A., GASPART P.: "The supervision and management of a two robot flexible assembly cell." Proceedings of the 1989 IEEE International Conference on Robotics and Automation, p. 540-550, Scottsdale, Arizona, May 1989.

[2] COUPEZ D., DELCHAMBRE A., GASPART P.: "The scheduling of a multiple robots assembly cell." Proceedings of the 5th CIM Europe Conference, pp. 185-195, Athens, Greece, May 1989.

[3] COUPEZ D., DELCHAMBRE A.: "A branch and bound method for the scheduling of a two-robot assembly cell." Proceedings IFAC-CIRP-IFIP-IFORS International Workshop on Decisional Structures in Automated Manufacturing, pp. 95-98, Genova, Italy, September 1989.

[4] DELCHAMBRE A., COUPEZ D.: "Knowledge-based error recovery in robotized assembly". Proceedings of the 9th International Conference on Assembly Automation (ICAA), pp. 137-154, London, UK, March 1988.

[5] DELCHAMBRE A. , COUPEZ D., GASPART P.: "Knowledge-based process planning in robotized assembly". Proceedings of the SPIE Applications of Artificial Intelligence VII, pp. 768-776, Orlando, Florida, March 1989.

[6] DELCHAMBRE A.: "Data structures for computer-aided assembly : a survey". Preprints of the 6th IFAC-IFIP-IFORS-IMACS Symposium on Information Control Problems in Manufacturing Technology (INCOM'89), Madrid, Spain, September 1989.

[7] DELCHAMBRE A., WAFFLARD A., GASPART P.: "Knowledge-based 'assembly by disassembly' planning".Internal report CRIF/WTCM RO 75 - April 1989.

[8] DELCHAMBRE A., WAFFLARD A., GASPART P.: "An automatic assembly planning approach". Proceedings of the 10th International Conference on Assembly Automation (ICAA), pp 319-326, Kanasawa, Japan, October 1989.

[9] DELCHAMBRE A.: "A pragmatic approach to computer-aided assembly planning". Proceedings of the 1990 IEEE International Conference on Robotics and Automation, pp 1600-1605 Cincinnati, Ohio, May 1990.

[10] GASPART P., DELCHAMBRE A., COUPEZ D., BROUILLARD Ph. : "Rule based procedure for diagnosis and error recovery". Proceedings of the International Workshop on Industrial Applications of Machine Intelligence and Vision (MIV-89), pp 88-93, Tokyo, Japan, April 1989.

COMPLETE GARMENT ASSEMBLY USING ROBOTS

I.Gibson, P.M.Taylor, A.J.Wilkinson, G.S.Palmer, M.B.Gunner.

Robotics Research Unit,
Dept. Electronic Engineering,
University of Hull
HULL, UK.

ABSTRACT

This paper describes the evolution of the problem of robotic based fabric handling as seen by the Robotics Research Unit at Hull University. The third phase of the associated project, following success in ply separation and robotic sewing, is to produce a prototype robotic assembly line for a complete garment. The line is designed for assembling mens Y–fronts and ladies briefs. The project is subdivided into two dimensional component based assembly, three dimensional folding and closing operations, and a supervisory computer system to aid communication and reconfiguration.

The manual operations for constructing these garments are mentioned in this paper, outlining the handling difficulties and the required system goals in terms of cost and performance. The various robotic handling solutions are described in detail, together with an assessment of their performance. Conclusions include the implications of the techniques used in this assembly line on solving the problems associated with products of a similar nature.
KEYWORDS: Automated garment manufacture / Fabric handling / Low cost robotics / Robotic garment assembly / Production supervisor / Hierarchical communications / Flexible assembly.

1. INTRODUCTION

Work is presently being carried out by the Robotics Research Unit at Hull University on automated garment assembly. The associated projects have evolved from earlier studies on the separation of single fabric plies from cut stacks [5], through detailed studies of fabric properties with an aim to automation [9], to individual handling and sewing operations using robots. The successes resulting from this work have led to the present day investigation, involving multiple robot assembly of a complete garment.

This paper describes the construction of two styles of underwear garments. The basic robotic elements that comprise the assembly system are given and discussed together with aspects of their integration. The major elements of the entire system are described in greater detail in related papers [10,7,11].

Present day garment manufacturers are required by variable market trends to be highly flexible. They have to cope with radical fashion changes and must be able to

NATO ASI Series, Vol. F 71
Expert Systems and Robotics
Edited by T. Jordanides and B. Torby

respond quickly to variations in consumer demand. This problem, coupled with size variations within product ranges and variations in material properties, requires a great amount of flexibility in production techniques. Such problems have been considered for the manual system for some time [12]. It follows that any automated garment assembly system that is designed for the future must be equally flexible.

In addition to the flexibility requirement, the Hull University garment assembly system is designed with attention to speed, cost, and maintenance constraints. The speed of the system has to be comparable with the manual process. Also, the initial outlay cost for such equipment as the system may comprise, must pay back within a short number of years. Finally, the system must be easily maintained with a minimum of manual intervention when running and when reconfiguring the system.

2. DESIGN CONSIDERATIONS

The robot assembly line is designed to handle component parts and subassemblies of one style each of mens' Y-front and ladies' briefs. Figure 1 shows the basic elements of the construction of these two garments. The main difference between the styles lies in the final stages of the flat (two dimensional) assembly. In the ladies' brief, the front piece is added onto the other components and subsequently overlock sewn. The two arrow shaped gusset pieces of the mens' Y-front are sewn using binding tape and then added to the front piece for completion of the Y using oversewn tape. There are a number of small variations between these two product types at other stages of assembly, in particular binding tape is generally used in mens' garments whereas elastic tends to be used in the ladies' designs. The operations required to perform these functions are generally the same although sewing machines and attachments may differ.

The fabric used in these garments is usually a knitted, cotton based material. This fabric is supplied in two forms, single and double knit. Both are extremely flexible in nature, the single knit being slightly more rigid but with a tendency to curl at the edge. Handling properties tend to change from material to material. Examples of properties that are considered during garment construction are static cling, linting effects (intertwining), flexibility, and deformability. Manual workers tend to compensate for such factors unconsciously, however quantification of these properties is necessary when considering automated solutions.

The design of each garment in the initial stages incorporates a rear concealed gusset structure. This means that, once the first sewing operation is complete, it is necessary to fold the two gusset pieces around the seam to conceal the stitching. The second seam of the ladies' brief is sewn externally. This varies from the other two most common designs and lies second in order of complexity above flat sewn seams and below double concealed gussets.

Both these designs are volume produced articles and the construction tasks lend themselves to the assembly line principle. This implies that one or more robots will be used for each assembly operation rather than for multiple operations. The need therefore is to consider a large number of low cost manipulators of adequate performance rather than a smaller number of sophisticated, but more expensive robots. The choice of the UMI RTX robot (approximately one fifth the cost of a Unimation Puma) illustrates this approach. Adequate performance means the capability to achieve a desired production rate in the order of 1 garment every 30 seconds to within quality control standards. A typical accuracy requirement is +/-5mm on all dimensions. This is quite simple to achieve for individual operations but the flexible nature of the garment means that small inaccuracies are likely to be accentuated in later stages.

The assembly line is designed with emphasis on reconfiguration. It is necessary to consider the problems involved when a manipulator fails during operation or requires routine maintenance. To avoid all the subsequent operations halting, it is necessary to control the environment to cope with this fact. This is best achieved in an automated

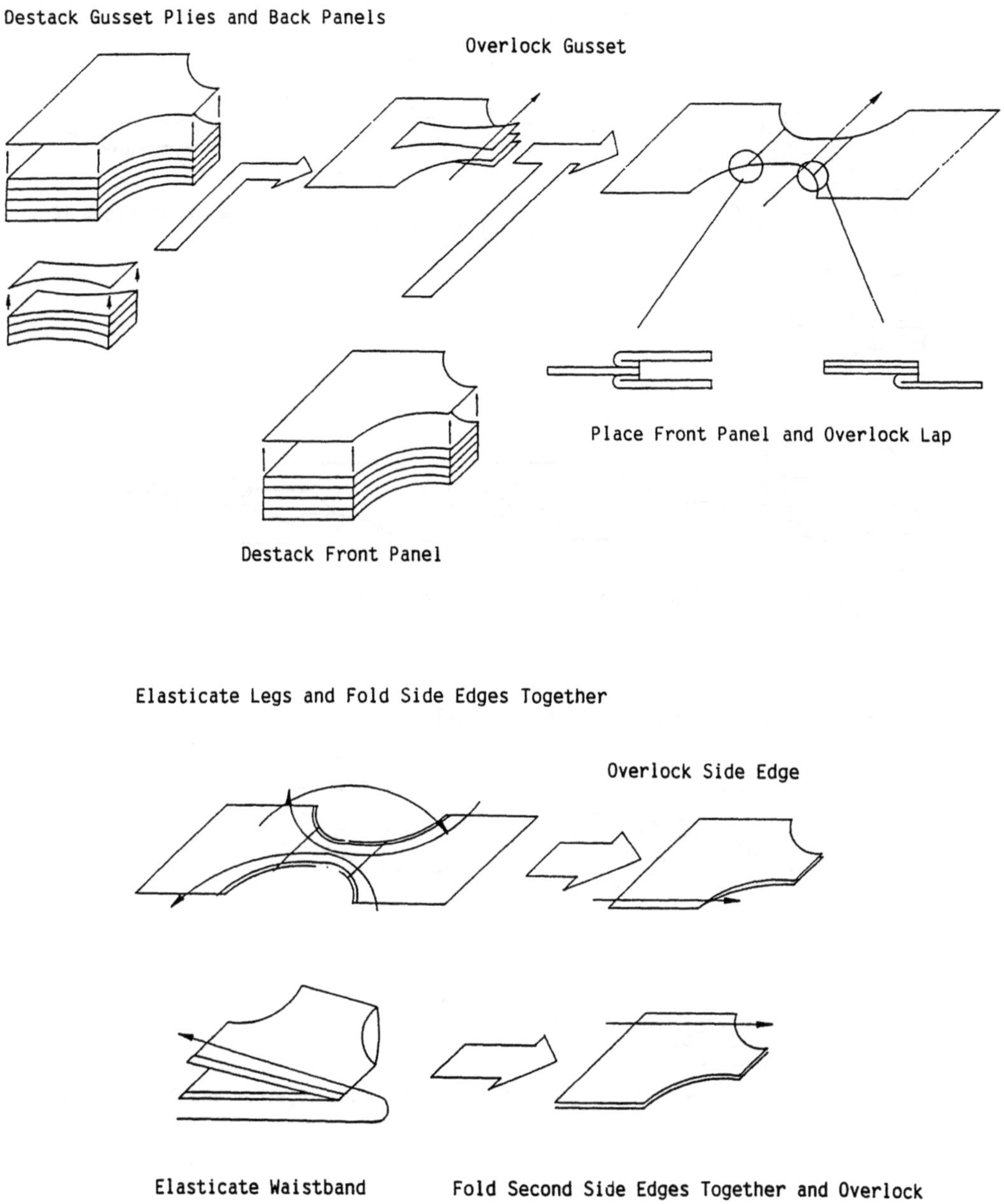

Figure 1(a)
Construction Of Ladies Briefs

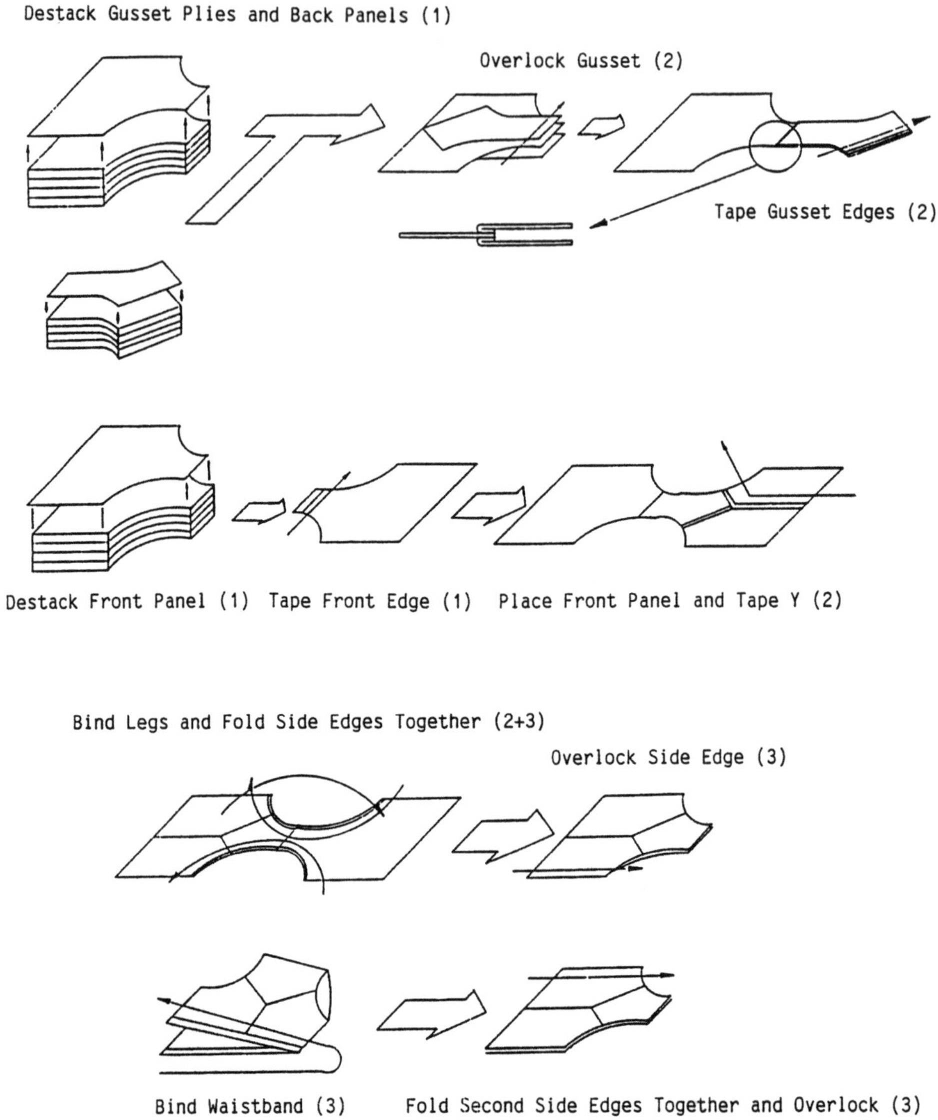

Figure 1(b)
Construction Of Mens Y-Front Briefs

sense the same way as the manual process, by either allowing re-routing of the garment through the system or allowing another manual operator to interact with the rest of the system for short periods of time.

Similarly, these garments are made on a batch basis. The system is designed to cope with variations in size, fabric type, and style. This implies that different machines may be required for different styles and therefore either the route will change or there must be a capability for machine/robot interchange. Either way, reconfiguration for basic changes in size or style must occur within a few minutes.

It will be seen in this paper that most of the operations required will involve an automated solution that uses straight line sewing. Linear feeders can be used to move the garment subassemblies through the sewing machines. Robots can be used to perform pick and place operations, loading the fabric pieces into the linear feeders and transferring the subassemblies between sewing cells. This keeps the required robot gripper designs on a simple basis. Sensor requirements are generally restricted to registering component presence and alignment of fabric pieces and subassemblies.

3. HANDLING CONCEPTS

Figure 2 shows a flow chart of the robotic based operations required to assemble the mens' Y-front and ladies' briefs. Figure 3 shows how this translates into a robotic assembly line.

The fabric component pieces are initially taken from cut stacks in the same way as the manual process. Single fabric plies must be separated from these stacks. An air jet finger device is used to blow compressed air on to the top of the stack. The resulting vibration of the fabric makes it possible to separate off one ply whilst the remainder of the stack is left intact. This is more desirable than the use of pin based mechanisms [8] because of the non destructive effect on fabric. Single plies can be transported around the system on conveyor belts. The air jet destacking unit places fabric pieces directly on to a conveyor belt [10]. Studies have been made on techniques for both reliable placing and removal from belts. Sophisticated methods like conveyor tracking using camera techniques or large flat plate removal devices (like electro-adhesive grippers) are overshadowed by a much simpler and more elegant technique of removal using scoops. This technique uses a thin membrane lying between the moving conveyor and a static, polished surface. The fabric piece slides easily onto the polished surface, larger pieces being impelled by a roller that is in contact with the belt (see figure 4).

Even with a reliable stack placement and destacking technique, single plies are still going to lose their position and orientation on the conveyor belts. The pieces must therefore be oriented prior to sewing. A linear vibrating feeder with flat plate surface is used to impel the straight edge of a fabric piece towards a fixed barrier. The vibrating plate is tilted so that the fabric moves under gravity parallel to the barrier once contact is made. A simple optical sensor is used to halt this process once the fabric reaches the desired position (see figure 5). This technique gives the required placement accuracy of +/-1mm. Once again this is more elegant than other more complex solutions that have been used in the past, involving optical array sensing elements.

Once oriented, components must then be moved into a position to be sewn. This is achieved by using an edge clamping, robot mounted gripper. This is capable of picking up pieces from a flat surface and transporting them horizontally. Placement is performed by dropping the piece from a few millimetres above the target.

Linear conveyor feeders with velocity and positional feedback are used for the first sewing function. Jigs attached to these feeders are loaded at one end by a transfer robot. This sewing operation involves 3 component parts, two gusset pieces with a back piece sandwiched between. These pieces are separated by narrow barriers to facilitate the inversion of the seam after sewing is completed. After sewing, the subassembly is

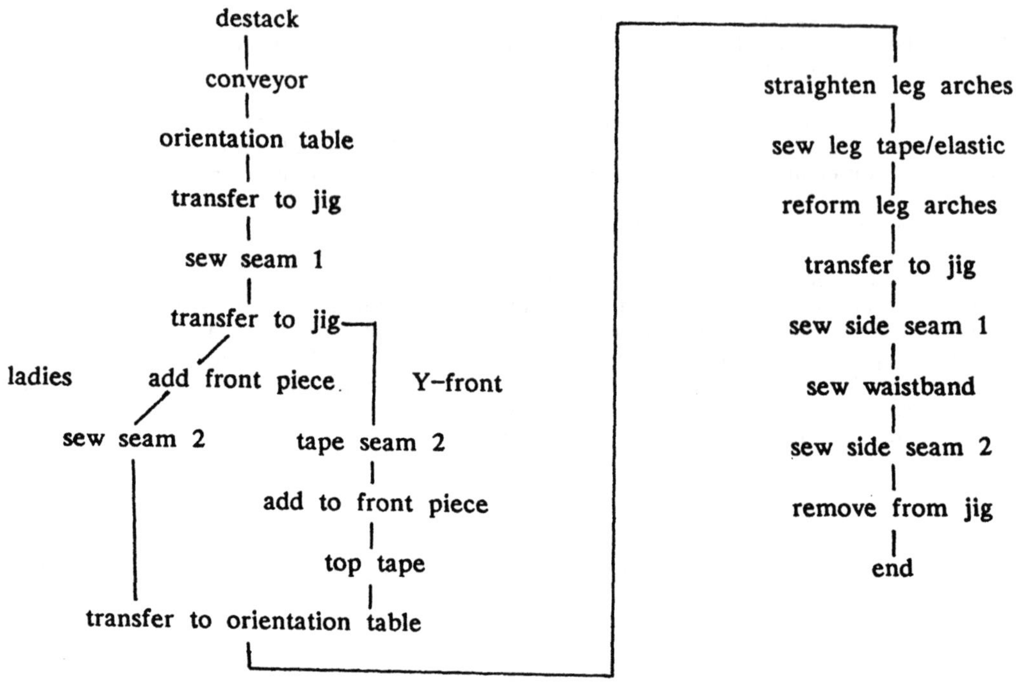

Figure 2. Functional diagram of garment assembly line.

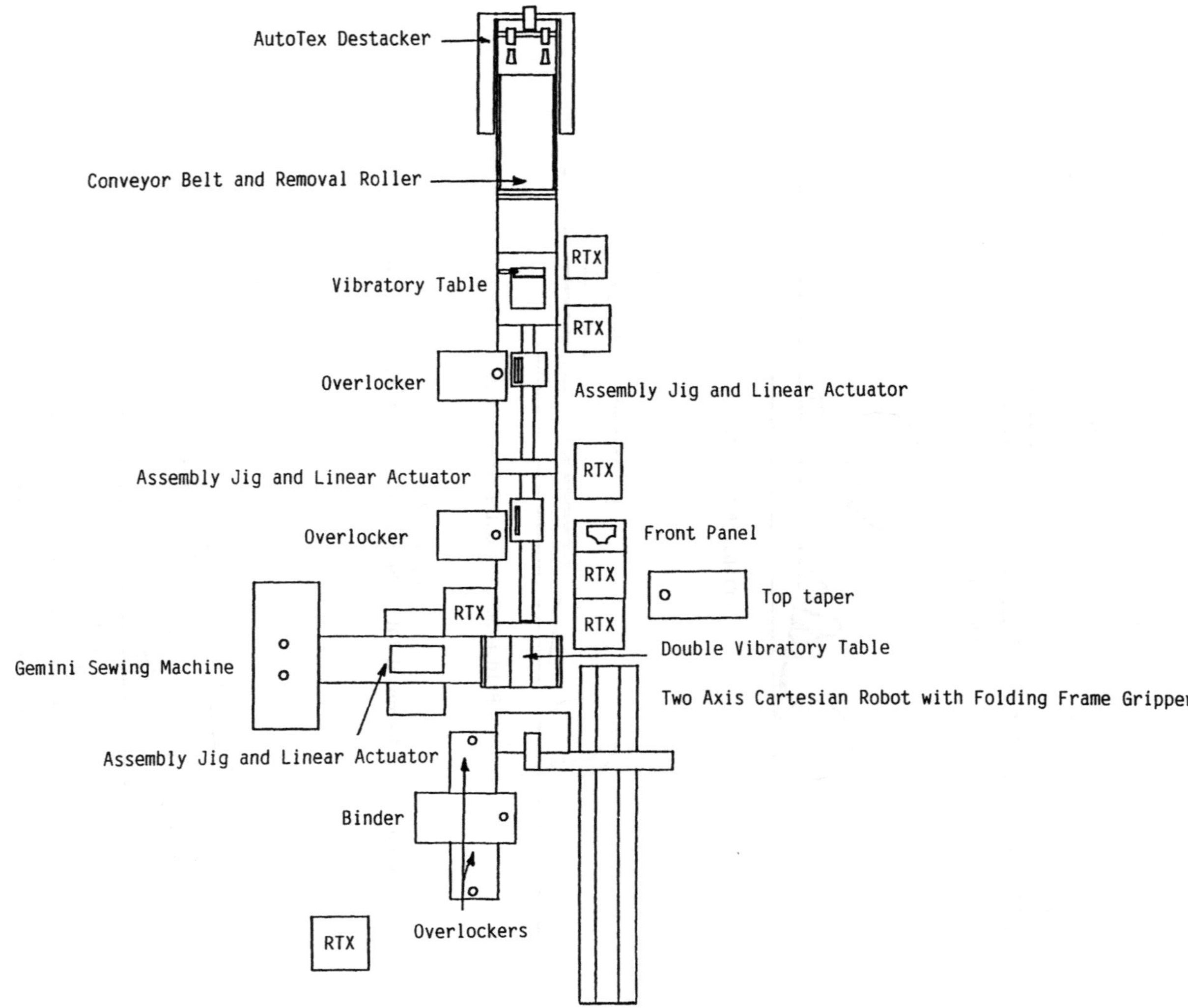

Figure 3 - Schematic Diagram of Assembly System

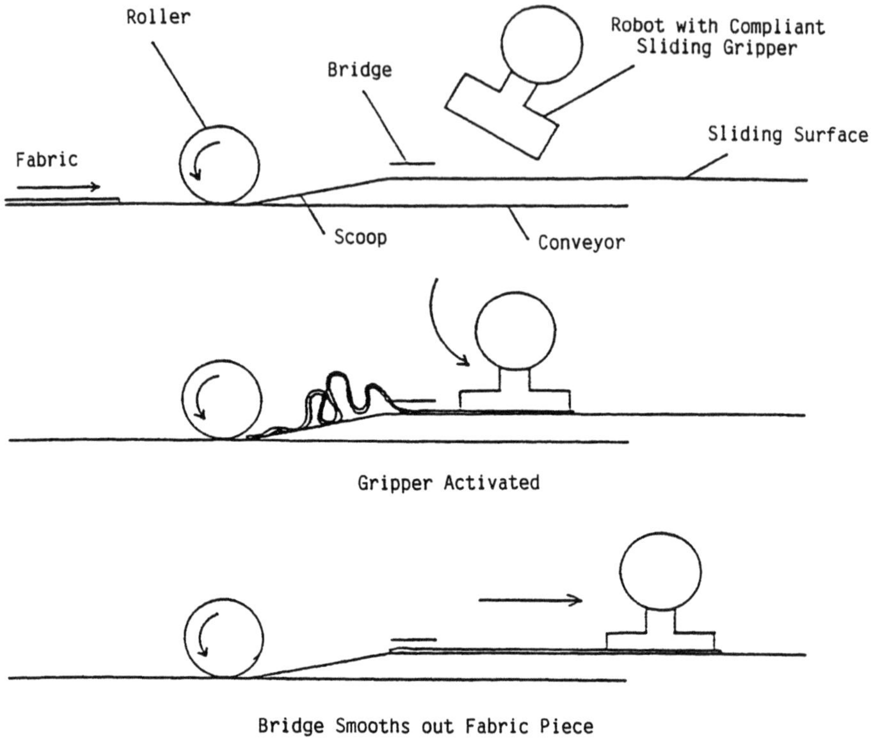

Figure 4.
Device For Removal Of Fabric Panels From Moving Conveyor

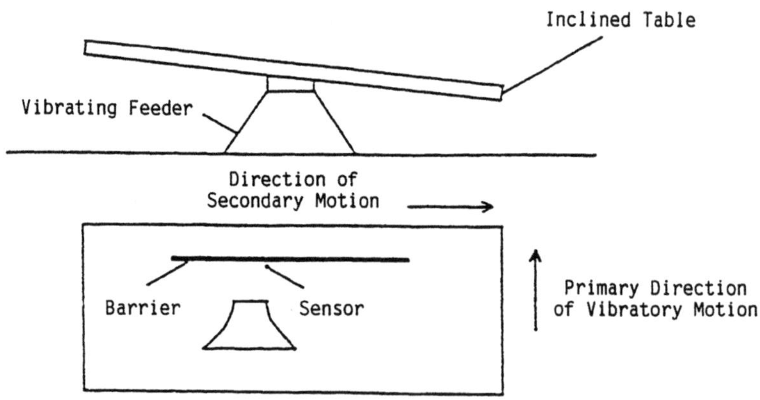

Figure 5.
Vibrating Orientation Table

transferred to a second jig sewing station by another robot equipped with a gripper that affords localised contact on the rear fabric panel. By pulling on this panel, gusset seam inversion is achieved as the gusset pieces unwrap around the barriers.

The positioning of this subassembly varies between the two styles, the mens' being laid at an angle for application of binding. Before sewing the ladies' brief, the sewing jig moves in to a further robot transfer area where a front piece is added.

All the sewing machines are controlled by the same computer that controls the associated linear feeder. This allows for feedback of sewing speed to the feeding jig, providing smooth travel of the fabric pieces through the sewing machines to maximize the quality of the sewing produced.

A robot with a gripper similar to the subassembly transfer stage is used to pull the subassembly out of the second sewing jig. For the ladies' brief a barrier mechanism is once again used to unfurl the front piece, allowing it to be laid flat. This is the end of the two dimensional phase for the ladies' design. The mens' subassembly must be added to a front piece and then top tape sewn to complete the Y shape. This is the only curved sewing operation of the full garment and as such needs a robot with planar motion for this task. Once again there is close control of the sewing speed and robot feed rate, producing a taped subassembly with a minimum of sewing 'pucker' resulting from differences between the tensions in the tape and the fabric.

The finishing stages are split up into two separate groups, one using a leg binding or elastication jig and the second using a side seaming/waistbanding jig. The leg binding is performed using a flexible curve straightener which allows the concave curve of the leg to be sewn using a linear feeder (see figure 6). The curve straightening device is used after sewing to reinstate the form of the garment prior to the following sewing operation. Both sides can be straightened at the same time to allow for parallel sewing machines (like the Rimoldi 'Gemini' double headed sewing machine) to apply the binding or elastic. Vibrating tables are used before and after the leg binding operation to confirm the position of what is now quite a large and unwieldy planar fabric assembly.

The final three sewing operations are performed using the same linear feeding mechanism. The side seams and waistband are sewn whilst the garment is supported by a specially designed reconfigurable former which imposes a shape onto the fabric since it must be folded and unfolded at different stages throughout the operations as shown in figure 7 [7]. Once these three operations are complete then the garment is finished.

4. SYSTEM RECONFIGURATION

A man machine interface is provided for production managers to organise the assembly from a central position. A database/expert system comprises part of the production unit supervisor which is available for confirming resources and providing reconfiguration information for the shop floor.

Linking all the robotic assembly tasks together is a hierarchical download and communications structure [11]. This structure has the main components of system configuration unit, production unit supervisor, and cluster supervisors. The programs for the robot cells are passed through the cluster supervisors which are used to link the autonomous tasks together. Information can also filter back through the structure concerning error conditions, monitor information, statistics, etc.

The cluster supervisors coordinate action between robot cells. The primary functions of this are to avoid robot clashes and to optimise parallel activity. For instance, the orientation and manipulation of a gusset piece can take place whilst the previous subassembly is being sewn. Upon return of the sewing jig, the cluster supervisor can instruct the loading robot to enter the jig area. All this can be determined from monitoring of robot and object position. Flags are set when a robot is in a shared

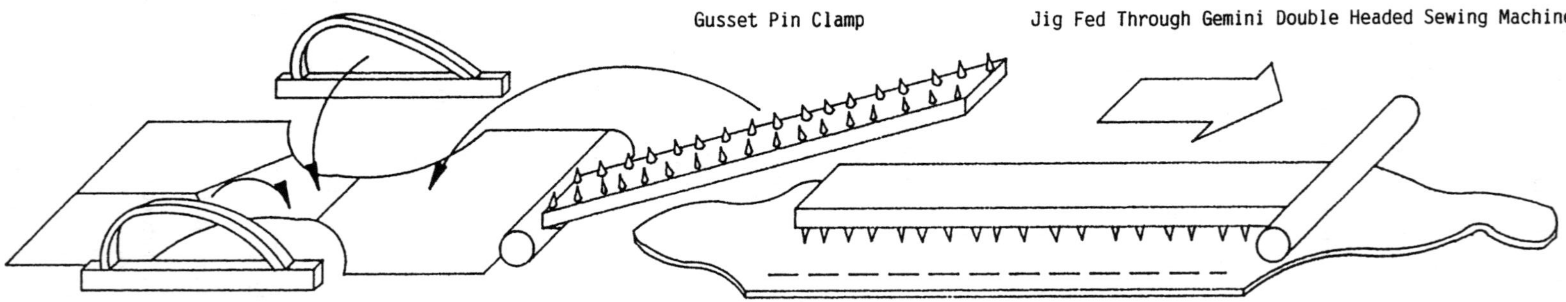

Figure 6.
Curve Straightening Mechanism

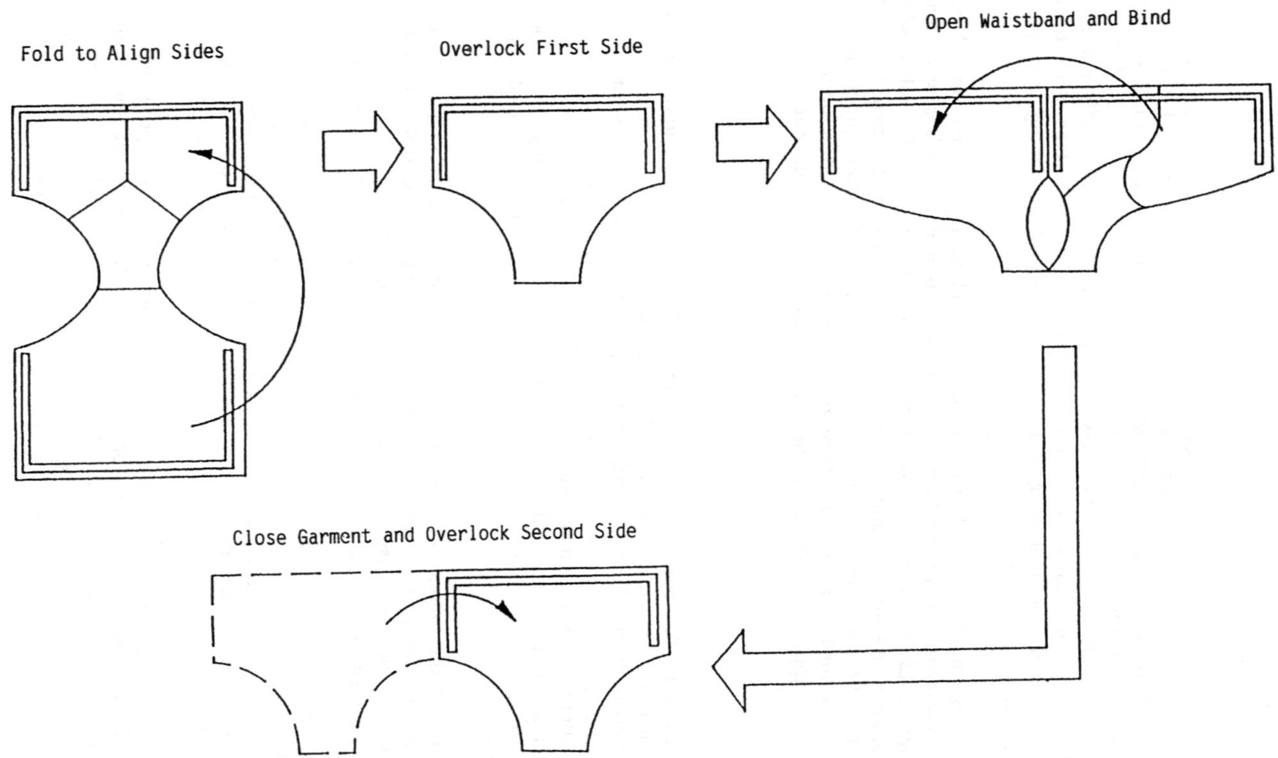

Figure 7.

Folding Frame Gripper Operation

workspace to restrict other robots from entering. Simlarly, objects deposited by robots set the object flag, alerting subsequent robotic stages of their presence, requiring action.

The philosophy behind this approach lies in the use of versatile robotic devices for the loading and unloading of jigs that are linked to standard sewing machines. Assembly tasks use the same robots for handling different machines and/or sewing attachments. This may require different data files to be transferred to the robots. Similarly, although the system is designed to be fully automated, downtime on one cell might affect the entire assembly line. This structure allows for manual intervention at any stage with operators simulating robotic tasks and communicating with the cluster supervisor, using a simple push button interface, to coordinate actions.

Since the sewing machines and robots are not physically linked to allow for interchangeability, coordinate frame adjustment must be considered. Each sewing machine is fitted with a simple LED bar code. This identifies the machine to the robot which is equipped with a simple, medium resolution camera [6]. The bar code also has 3 LEDs that are provided for the purposes of triangulation. A simple calibration procedure is used to adjust the robot coordinate frame, and therefore subsequent recorded positions, to make any robotic procedures appear the same for the new machine as the old one.

5. RESULTS

The system has thus far been partially implemented. The figures quoted below were produced using double knitted cotton material. Ply separation has been a reliable technique for some time, showing high efficiency and reliability. Removal from conveyors is equally reliable and transport to the vibrating orientation table results in a realistic positional error of approximately 10mm in two dimensions and $+/-5^{\circ}$ rotational error from normal. The vibration table actuation time for this greatest error is 15s, resulting in a positional accuracy after transfer to the jig of better than +/−1mm. This lies well within the required specification.

Because of the elastic nature of the fabric, sewing speed needs only to match the feed rate to within 10% without defective sewing seams. With full control over the sewing machine characteristics this is easily obtainable. Actuation of the sewing machine is controlled by the jig feed controller and so can be engaged at the appropriate position.

The inversion of the first sewing seam results in a good alignment of the gusset pieces provided the seam is correctly sewn, using a machine with a well balanced stitching mechanism in the first place. At present no sensors are used to check this alignment but the unpredictable nature of both fabric and sewing machines suggests that they will be required in the future.

Top taping can be performed for the family of curves that comprise the different sizes of mens' Y-front briefs. The angle of rotation and the length of the leading and trailing straights can all be programmed. This operation is mainly dependent on the placement of the fabric pieces prior to the sewing operation but tests have shown that a realistic figure of +/−3mm deviation from the mean curve is achievable [2].

The curve straightening technique used in the leg taping operation has been successfully demonstrated using a robot to load and unload the linear sewing machine feeder. Initial experiments suggest that pucker as a direct result of curve straightening is not evident.

The system is divided into 3 clusters of robot cells. The two dimensional phase is subdivided into 2 equal clusters with one sewing operation each. The three dimensional phase comprises the third cluster. Tests have shown that the robot/object driven philosophy is sound.

All the operations for the assembly line have been simulated using the HOCUS time event software simulation package [4]. A current cycle time of 110s is possible using the demonstration equipment. Eliminating bottlenecks and simulating faster robot motion using an improved nonlinear control system for the RTX robot [3] can bring this cycle time down to less than 60s. This would require some of the slowest operations to be duplicated in parallel to achieve a figure comparable with the manual approach.

The estimated retail cost of this system is around £200,000, including sewing machinery costs for one garment type, a streamlined toplogical layout, and robots operating under nonlinear control. A realistic figure for payback from an individual garment is £0.20 per garment. This means that an estimated 1 million garments must be made to pay back the system costs. If this was done over 3 years then approximately 1000 garments must be made in a day. For a two shift system this works out at approximately 60 garments per hour.

6. CONCLUSIONS

This robotic based, fabric handling system is designed to handle an extremely difficult material. A number of concepts have been mentioned with the ambitious aim of achieving a fully automated garment assembly line. There is no doubt that the work instigated by this project will continue through others. The following conclusions can be made:

Robot feeders are used in preference to robot guided sewing operations. The decorative top taping operation is the only exception, where two dimensional manipulation under the sewing head is necessary. This makes it possible to maintain a modular approach, utilising the simplicity of each individual cell to achieve a complex objective. In this way, low cost, low accuracy robots, in conjunction with simple sensing techniques can be used, resulting in an improvement in throughput and cost effectiveness.

Straight line sewing using curve straighteners is preferable to curve following. Even though this may add operations in some instances, sewing time is reduced and the garment shape is actually easier to recover. All sewing operations are continuous. The manual technique is intermittent, performing adjustments to the alignment of the garment at stages. This is very difficult to achieve using robots but is in fact unnecessary when performing straight line sewing operations.

Single plies are manipulated on the flat, even when transferred through the space. The difficulties in handling fabric using a draping technique [1] show that positioning by dropping the piece from a small height, whilst in a prone state, is preferable.

Simple, low cost sensing is used throughout. This cuts down processing time to a minimum and improves reliability. Complex sensing is only necessary in complex environments. The modular robotic concepts used here do not require such complexity.

The operations described can and will be implemented in systems for other garments and similar products (blankets, bags, etc.). The use of linear feeders, robot loaders, and curve straighteners are particularly generic. Increasing the scale to cope with larger garments (e.g. sweatshirts), should not cause a problem.

Where possible standard sewing machines have been designed into the system. This means that instalation of robotic systems of this nature does not require the added cost of replacing all the peripheral equipment as well, although some adaptation for automation may be necessary.

REFERENCES

[1] Brown P.R., Buchanan D.R., Clapp T.G., "*Large deflection bending of woven fabric for automated material handling*", J. Textile Inst., vol. 81, no. 1, 1990.

[2] Gibson I., Taylor P.M., Wilkinson A.J., "*Robotics applied to the top taping of mens Y-front briefs*", Proc. Int. Clothing Conf. on Textile Objective Measurement and Automation in Garment Manufacture, July 1990.

[3] Gilbert J.M., "*Non linear control of an industrial robot*", Ph.D. thesis, Univ. of Hull, UK, 1989.

[4] Hu X., Gibson I., "*HOCUS simulation applied to the Corah garment assembly system*", University of Hull Internal Report 88/90.

[5] Kemp D.R., Taylor P.M., Taylor G.E., "*An adaptive sensory gripper for fabric handling*", Proc. 4th IASTED Symposium on Robotics and Automation, Amsterdam Holland, 1984.

[6] Palmer G.S., "*An RTX assembly cell calibration system prototype*", Univ. of Hull Internal Report 76/89.

[7] Palmer G.S., Wilkinson A.J., Taylor P.M., "*The design of an integrated robotic garment assembly cell*", Proc. 6th Int. Conf. on Production Research, Glasgow, UK, 1990.

[8] Sarhadi M., "*Practical performance of an automated garment sub-assembly research workstation*", submitted for IEEE Int. Conf. on Systems Engineering, Pittsburgh, 1990.

[9] Taylor P.M., Koudis S.G., "*Automated handling of fabrics*", Science Progress, Oxford, vol. 71, pp 351-363, 1987.

[10] Taylor P.M., Wilkinson A.J., Taylor G.E., Gunner M.B., Palmer G.S., "*Automated fabric handling problems and techniques*", Proc. IEEE Int. Conf. on Systems Engineering, Pittsburgh, USA, 1990.

[11] Taylor P.M., Wilkinson A.J., Gunner M.B., Sawyer A., Gibson I., "*Integration of a flexible workcell for garment assembly*", Proc. 28th Int. MATADOR Conf. on CIM, FMS, and Robotics, pp 131-138, 1990.

[12] The Reece Corporation, "*TSS, 'Just in Time' technology*", PO Box 9168, Waltham, Mass. USA, 1989.

A Model-Based Expert System for the Diagnosis of Faults in a Robot System for Cleaning Castings

Thomas Hübner, Klaus Hörmann

University of Karlsruhe, Institute for Realtime Computer Systems and Robotics, Research Group Prof. Dr.-Ing. U. Rembold, Postfach 6980, D–7500 Karlsruhe 1

Abstract: This contribution presents an Expert System which has been developed within a joint project of the University of Karlsruhe and the Nuclear Research Center Karlsruhe.The long term goal of this cooperation is to develop a generic tool for the diagnosis of faults of technical systems. In order to make such a tool generic enough to be easily adaptable to new applications, a deep-reasoning approach has been developed. A prototype of the system has been implemented using the Knowledge Craft (TM) environment. The prototype has been tested with industrial robots of UNIMATION, Inc. and is currently being applied to the much more complex COMETOS system for the cleaning of large castings consisting of several robots, end–effectors, sensors and different controllers. Both applications are described.

Keywords: deep reasoning, diagnosis, causal nets, structure model, pathological model

Introduction

Many industrial applications require an extremly high reliability of the equipment used. This is the case, for instance, with telemanipulators in nuclear power plants or with industrial robots in production lines. In such areas, faults can cause emergencies or at least high cost. Hence, it is desirable to detect faults as early as possible and to plan recovery procedures. For this purpose, an Expert System has been developed. Based on input by human operators, it is able to support the operator in finding the reason for the fault. This is a non–trivial problem. One fault can produce several symptoms, and one symptom can be caused by several faults. Faults of one component can cause malfunctions of other components, so it becomes difficult to distinguish between original faults and their effects.

The system developed uses the model–based paradigm. This requires a so–called *deep knowledge* or model of the target system. Advantages of this approach are that these models can be easily adapted to new applications. When building a new system from standard components, the corresponding diagnostic expert system can be constructed from a diagnostic library of these standard components. This is in contrast to the traditional heuristic expert systems, which require expert knowledge based on long-term experiences with the operation of the system.

The system has been implemented using the Knowledge Craft (TM Carnegie Group, Inc.) environment. It has been tested with two different applications which are also described in this paper (see also [1], [2], [3]).

Principles of Model-Based Diagnosis

Model–based diagnosis is based on *deep knowledge* of the structure and the functions of a system. One can distinguish mainly between *pathological* models and *functional* models. Pathological models describe malfunctions of the system. They contain an explicit representation of the causal dependencies between diagnoses and symptoms. They describe how different malfunctions are related to each other. Pathological models are mainly used in medicine, where functional models are not known or would be too complex.

NATO ASI Series, Vol. F 71
Expert Systems and Robotics
Edited by T. Jordanides and B. Torby

In contrast to this, a functional model describes the correct behaviour of the system. It comprises a description of structure and behaviour of the system. The structural description contains the system components and their relations to each other, their characteristics, system parameters and so on. Behaviour means the way in which the system functions and how the system parameters are related to each other. In technical systems the prerequisites for functional models are much better than in medicine, because the models can be built during the construction of the system.

The basic idea of diagnosis with functional models is to observe the behaviour of the real system, to simulate its behaviour using the model and to detect deviations between the two types of behaviour. Assuming a correct model and correct observations, one tries to localize the malfunctions by adjusting the model to the real system. This means that one tries to model the malfunctioning of the system. If this succeeds, one possible explanation for the wrong behaviour has been found. This can be done by hypothesizing faults of particular system components in the model. If the simulated behaviour then equals the observed behaviour, the candidate components in the model can be the defective components in reality.

Knowledge representation and inference mechanism

The basic components of our system are the model which contains the diagnostic knowledge and the inference mechanism which comprises the diagnostic algorithm. Both are based on the theory of Reiter [5]. Reiter uses *conflict sets* containing assumptions about the system which can not be true all at the same time. A typical assumption is, for instance, that a particular system component is defective or not. In contrast to Reiter a different modelling technique has been used. This technique allows a priori computation of the conflict sets which makes the system much more efficient. In addition, the system has been enhanced by features for incremental diagnosis based on [4]. This allows a step-wise refinement of the diagnosis through successive input of observations. The overall diagnostic algorithm is as follows.

```
while still too many candidates do begin
    accept next observation
    CONFLICTS := all new minimal conflicts
    for each Q∈CONFLICTS and K∈CANDIDATES do
        if K∩Q = {} then begin
            delete K from CANDIDATES
            extend K with each assumption from Q and add them to CANDIDATES
            delete all non-minimal candidates from CANDIDATES
        end {if}
    report CANDIDATES
    propose observation
end {while}
```

When a new observation is accepted, the old conflicts are updated. Then the old candidates are investigated to see whether they explain the new observation (i.e., the new conflicts) or not. A candidate is a particular hypothesis for how the system differs from the model. More precisely, a candidate is a set of components which are suspected of being defective and would, in this case, explain the system behaviour. A candidate explains a conflict if their intersection is non-empty. If a candidate does not explain a new conflict, it is deleted. This candidate is extended by one assumption from q (which renders a non-empty intersection) and is thus able to explain the conflict. Afterwards non-minimal candidates are deleted. The reason for this is that minimal candidates are of particular interest. The assumption, for instance, that all components of a system are defective would explain every system behaviour but, of course , would have no practical value.

The model representation uses a graph structure with nodes representing entities of the system and arcs representing the diagnostic knowledge. An arc from A to B is interpreted as logical implication : if A is true, then B must also be true. Vice versa, if B is false, then A must also be false. This structure is used to formulate the causal dependencies in the form of complex networks. Naturally, circular structures are not allowed in the model. In spite of this, we will see later on an example of how circular structures can be treated. These causal structures are only the mechanisms offered by the system. It is up to the user to define the interpretations of these structures with respect to reality. In our applications we have used two different interpretations: A causal structure model and a causal fault model.

The **causal structure model** is used to model the propagation of faults within the component structure of the system. If there is an arc from A to B, for instance, then A→B is true. In this case, the truth values of the components a and b are interpreted in the opposite way: A or B being true denotes a malfunctioning of the components a or b. If A or B are false, the components a or b. must show correct behaviour. A→B means that faulty behaviour of a implies a faulty behaviour of b. Vice versa, a correct behaviour of b implies a correct behaviour of a.

Another interpretation is the **causal fault model** (i.e., a kind of pathological model). The model entities are representations of faults. Faults can be symptoms (i.e., faults which are caused by other faults) or diagnoses (i.e., faults which cause other faults). D→S now has the following interpretation. If diagnosis d is true, then symptom s must occur. If symptom s does not occur, then d cannot be the cause. Due to our modelling technique the generation of the conflicts is very efficient. Consider the causal structure model depicted in figure 1.

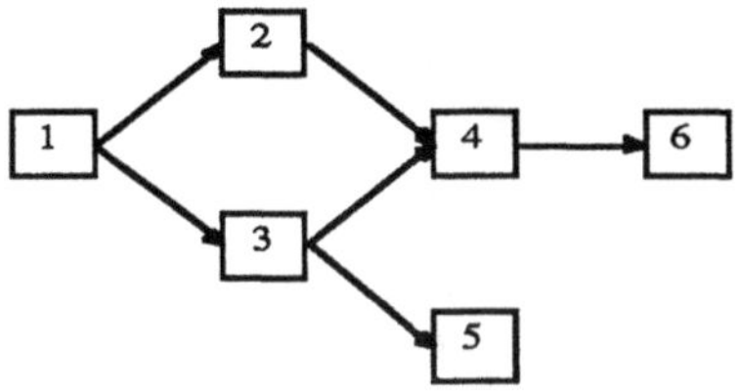

Figure 1: Example of the model

Components in the model can be marked by the user as correct or incorrect. If a component is marked incorrect, then this does not necessarily imply that this component must be defective. The malfunctioning can also be caused by a defect of a causal predecessor component. For instance, if component three is incorrect, then components four, five and six must also behave incorrectly. In this case we will denote four, five and six as deduced incorrect. On the other hand, all causal predecessors of a correct component are deduced correct. For instance, if five behaves correctly, then one and three must be deduced correct.

Let us assume that component four is incorrect. This must be due to a defect of four or of one of its predecessors. Hence the conflict for the observation "behaviour of four incorrect" is simply the set consisting of four and its predecessors.

If we observe that a component behaves in a correct way, this does not result in a new conflict, but an already existing conflict might not be minimal any more. If we observe in our example that component two behaves correctly, then the conflict {1, 2, 3, 4} is not minimal because {1, 3, 4} is also a conflict now. But this is still not minimal because of the deduced correct component one. Thus the new minimal conflict is {3, 4}. We obtain all new minmal conflicts by deleting all correct and deduced correct components from the conflicts computed so far.

Implemented Applications

Up to now the system has been tested for two different industrial applications. The pilot application was the fault diagnosis of a puma 200 type robot. The second application was a complex system (named COMETOS) for the cleaning of large castings.The system consists of three stations, one for measuring and two for cleaning the castings. For each of the two applications, two different diagnostic models have been built: a causal structure model and a causal fault model. We will illustrate this in several examples.

A Causal Structure Model for a Puma 200 Type Robot

Figure 2 shows the cross structure of the robot system.

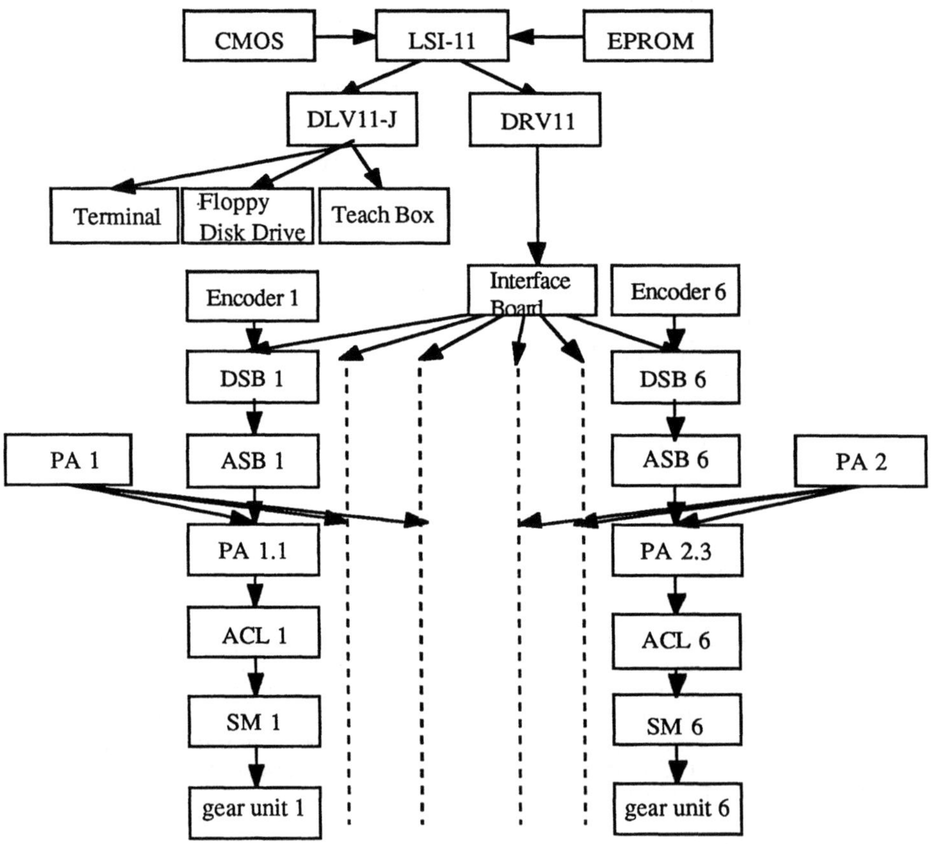

Figure 2: Causal structure model for a puma Robot

Suppose that joint two moves incorrectly. We will feed this observation into the system by marking gear unit two as incorrect. The conflict for this observation is the set {gear unit 2, SM2, ACL2, PA1.2, PA1, ASB2, DSB2, Encoder2, Interface–Board, DRV11, LSI11, CMOS, EPROM}. To focus on the cause for this incorrect behaviour, we will have to gather additional information concerning other components in the model. If, for instance, joint three is also incorrect we can concentrate on the common predecessors of joints two and three. Assuming a single fault only, this will render the following conflict: {PA1, Interface-Board, DRV11, LSI11, CMOS, EPROM}. Suppose now that motor sm6 works properly. Since this implies that its predecessors also work correctly, we can remove these components from our conflict. This results in the single fault diagnosis {pa1}.

A Causal Structure Model for the COMETOS–System

The structure model for the part of the system which has been selected for diagnosis contains about 470 entities and even many more arcs. Normally, it is not possible to describe circular structures using our representation form. The following example will show how this can be done inspite of this. Figure 3 showes the DNC–Interface between the controller and the superior personal computer.

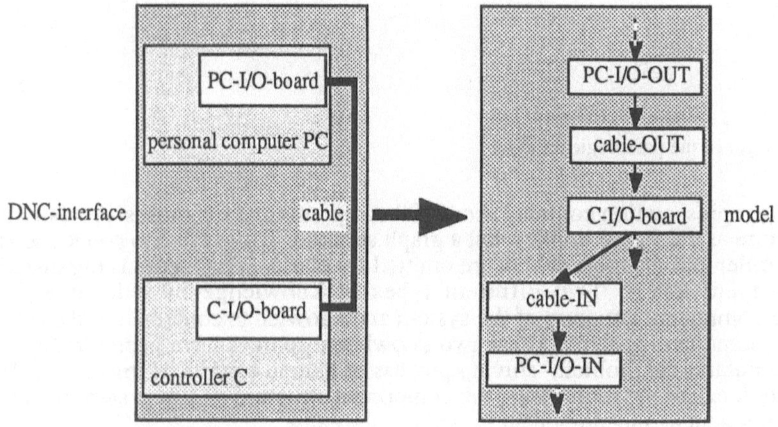

Figure 3: Modelling of a bi-directional data transfer

If an I/O–board in the personal computer is defective it is usually completely replaced by the user. A model representing the two I/O boards and the link between them seems sufficient. But since we have bi-directional data transfer (programs are sent from the PC to the controller and error messages are sent in the opposite direction) using only one model entity for the I/O–board would render a circular structure. Hence we split the I/O–board into two parts: an input part (I/O–in) and an output part (I/O–out). Also, the cable is modeled by two entities: cable–in and cable–out. The right hand side of figure 3 shows the resulting loop-less structure.

Causal Fault Model of the COMETOS–System

Model entities now represent faults of the system. Faults can be symptoms (faults which are caused by other faults) or diagnoses (faults which cause other faults). Figure 4 shows the top level structure of this fault model.

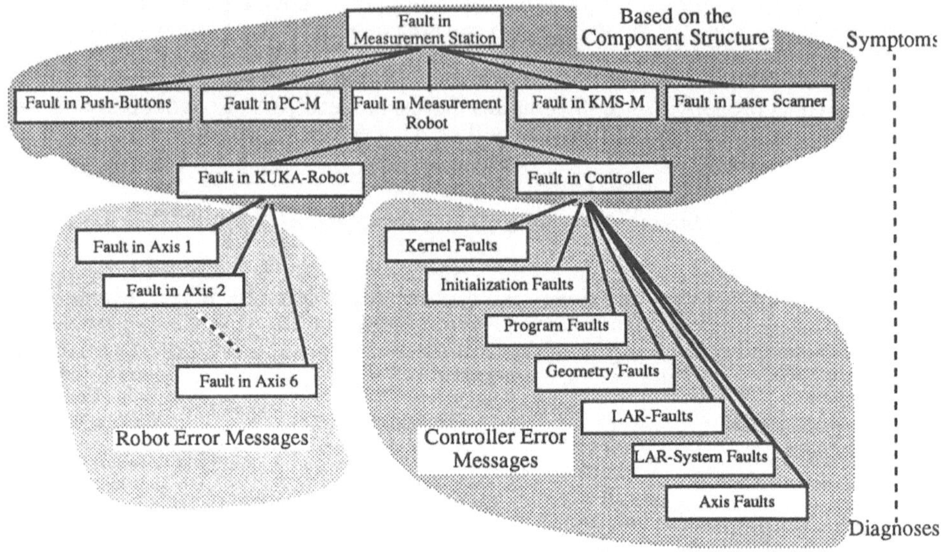

Figure 4 : Top-level layer of the pathological model

The complete representation contains about 300 symptoms and diagnoses. Actually it is not a tree structure as the figure implies, but a graph structure. This is due to connections between the controller and the robot which are omitted for clarity. The model is organized in several subsequent levels. Two different types of knowledge have been used: knowledge of the component structure of the system and knowledge contained in the self-diagnosis of the component systems. These two knowledge sources have been combined in a way which simulates the problem- solving process of human experts. Humans usually first try to roughly localize the fault inside the component structure and then step-by-step descend from component to subcomponent to find out the cause.

Discussion

Compared with the classical heuristical approach ("shallow apprach") it is worth highlighting the following items. Heuristical systems can only solve cases which have already occurred in the past. They are incomplete with respect to all possible cases, and can also contain contradictory knowledge. Theoretically, assuming a perfect model, this limitation is not true for the model-based approach, because every difference between the behaviour of the real system and the model can be detected. Heuristical systems require that experiences have been gathered during operation of the system. Often it is a troublesome process to acquire this knowledge from the human operators. In contrast to this, structure knowledge for the diagnosis can be gathered by the designers of the system. Thus a diagnostic Expert System can be put into operation together with a new machine. Causal models are easier to maintain and to update than the often somewhat confused structures of heuristical knowledge bases.

On the other hand, model-based diagnosis techniques are still in the early stage. In the future development of our system, we will try to combine the advantages of both deep and shallow reasoning. We want to investigate how these can be integrated, and how switching between the corresponding different inference techniques can be realized. With respect to the knowledge-representation formalism we will start to build up a library of descriptions of standard components of our application domain. Another important topic will be the generation of observation proposals which has not yet been implemented.

Acknowledgements

This research has been carried out at the University of Karlsruhe at the Institute for Real-Time Computer Systems and Robotics in the research group of Prof. Rembold. It has been supported by a research contract with Nuclear Research Center Karlsruhe.

Literature

1. Hammes, D. , Hörmann, K.: "Expertensystem zur Diagnose technischer Anlagen und Werkzeuge zu ihrer Entwicklung". Bericht zum Stand der Forschung, Universität Karlsruhe, 1988.
2. Hammes, D., Hörmann, K.: "Entwicklung eines Expertensystems zur Diagnose von Fehlfunktionen an Handhabungssystemen". Abschlußbericht für das Projekt zwischen der Kernforschungszentrum Karlsruhe GmbH und der Universität Karlsruhe, 1989.
3. Hübner, T., Hörmann, K., Cord, T.: "Entwicklung eines Expertensystems zur Diagnose von Fehlfunktionen eines Handhabungssystems zum Putzen großer Gußstücke". Abschlußbericht, Universität Karlsruhe, 1989.
4. de Kleer, J., Williams, B. C.: "Diagnosing Multiple Faults". Artificial Intelligence 32 (1987) 97-130.
5. Reiter, R.: "A Theory of Diagnosis from First Principles". Artificial Intelligence 32 (1987) 57-95.

Genghis: An Intelligent Autonomous Mobile Robot

MARK M. SIMPSON
McDonnell Douglas Corp.
3855 Lakewood Blvd., Mailcode 35-85
Long Beach, CA 90846

TIM JORDANIDES
EE Department
California State University
Long Beach, CA 90840

Abstract-This paper describes the control system of Genghis, a small robot that can track people over rough ground. Genghis behaviors are made of simpler behaviors, each of which is composed of combinatorial logic. Some of these behaviors are connected to sensors and some are connected to motors. All behaviors have a reset, some have memory. All behaviors can run at the same time, while higher behaviors can overide lower behaviors.

I. INTRODUCTION

Robots that move around by themselves and that can do complicated things are appearing more often in the news. Some of these robots can track people [6]. Some of these robots can move heavy objects and avoid obstacles in a factory [9]. This paper shows how the control system of one such robot works.

A. Objective

The objective of this paper is to show how the control system of Genghis works. Genghis is one of MIT's autonomous mobile robots.

B. Approach

The approach to describe Genghis's control system is divided into 4 steps. In Step 1, Genghis's overall behaviors are described. In Step 2, the behaviors's parts are shown. In Step 3, the parts' design requirements and rules are described. In Step 4, each behaviors parts' design is described.

NATO ASI Series, Vol. F 71
Expert Systems and Robotics
Edited by T. Jordanides and B. Torby

C. Scope

This paper is limited to showing how Genghis's control system works. Other parts of the robot are described but in less detail.

II. BEHAVIORS

Genghis was designed to have the following behaviors [5], [6]. It will stay still until someone walks by. If someone walks by, then Genghis will walk towards them. Genghis can walk over rough ground, is battery powered and has a mass of less than 1 kg.

III. BEHAVIORS' PARTS

Genghis's behaviors were divided into simple behaviors using nature as a guide.

A. Nature as a Guide

Nature was used as a guide in dividing Genghis's behaviors into simple behaviors [2], [6], [7]. Evolution shows that higher animals have more complex behaviors. Therefore the lowest animals that behave like Genghis were used as model. In this case bugs were used. This makes a lot of sense because bugs are light and their control system is fairly simple compared to higher animals.

B. Complex Behaviors Are Made of Simple Behaviors

Genghis behaviors were divided into 8 behaviors: standup, simple walk, force balancing, leg lifting, wiskers, pitch stabilization, prowling, and steered prowling [5].

IV. BEHAVIORS PARTS' DESIGN REQUIREMENTS AND RULES

Genghis was designed using design requirements and design rules.

A. Design Requirements

Genghis was designed using 8 design requirements [1], [5]. Many tasks had to be performed. Genghis had to do the most important task first. The robot was required to have different types of sensors, and many of each type of sensor were used. Genghis had to be able to work if some sensors failed or if its environment changed. It had to be able to hand computer errors. For additional behaviors, Genghis had to have additional computers added.

B. Design Rules

Genghis was designed using 7 design rules. Complex behaviors were made using simple controls. The design was kept simple. Interfaces were kept simple. The control system was kept open loop stable. The control system was able to relate to the three dimensional environment around it. Genghis uses vision. Genghis's control system was designed to be self calibrating.

V. BEHAVIORS PARTS' DESIGN

Genghis behaviors parts' design phyical description and control architecture are described in this section. Each behavior workings are also described.

A. Physical Description

A picture of Genghis is shown in Fig. 1 [6]. It is about 35-cm long and has a 25-cm leg span, has a 1-kg mass and has 6 legs. There are 12 motors, 2 for each leg. Genghis has 12 force sensors, 2 wiskers, 2 inclinometers, and 6 passive infrared sensors. It has 4 8-bit microprocessors. The microprocessors are conected by a bus. The microprocessors together have 1 Kbyte of RAM and 10 Kbyte of EPROM. Genghis is powered by 3 silver-zinc batteries [5].

B. Architecture

Subsumption architecture is the control archictecture that was used to design Genghis's control system [6]. Subsumption architecture [1], [4], is a set behaviors. These behaviors [3], [8] run in parallel. The behaviors are divided into increasingly more complex behaviors.

Each behavior is made of a group of augmented finite state machines [5]. An augmented finite state machine is shown in Fig. 2 [5]. This machine is a set of combinatorial logic fed by registers and it may also

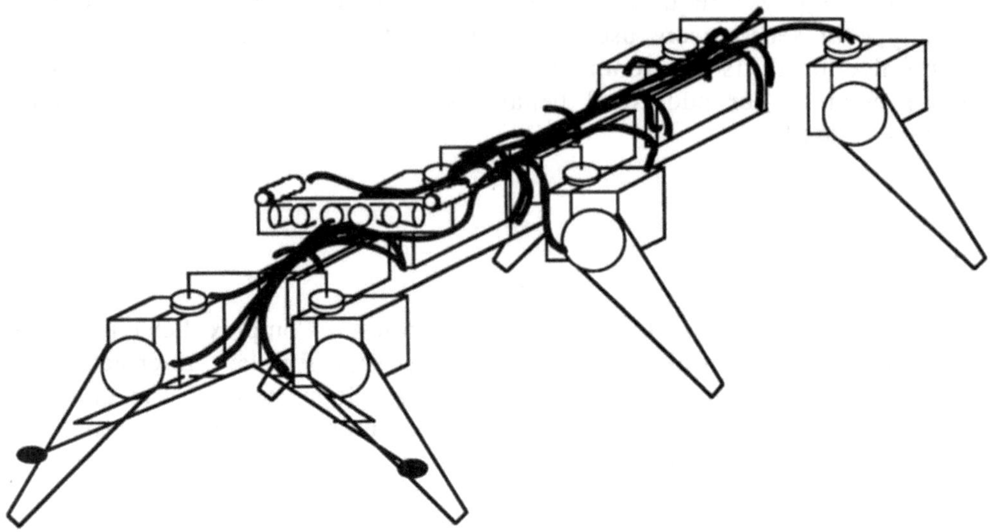

Fig. 1 Genghis : an intelligent autonomous mobile robot

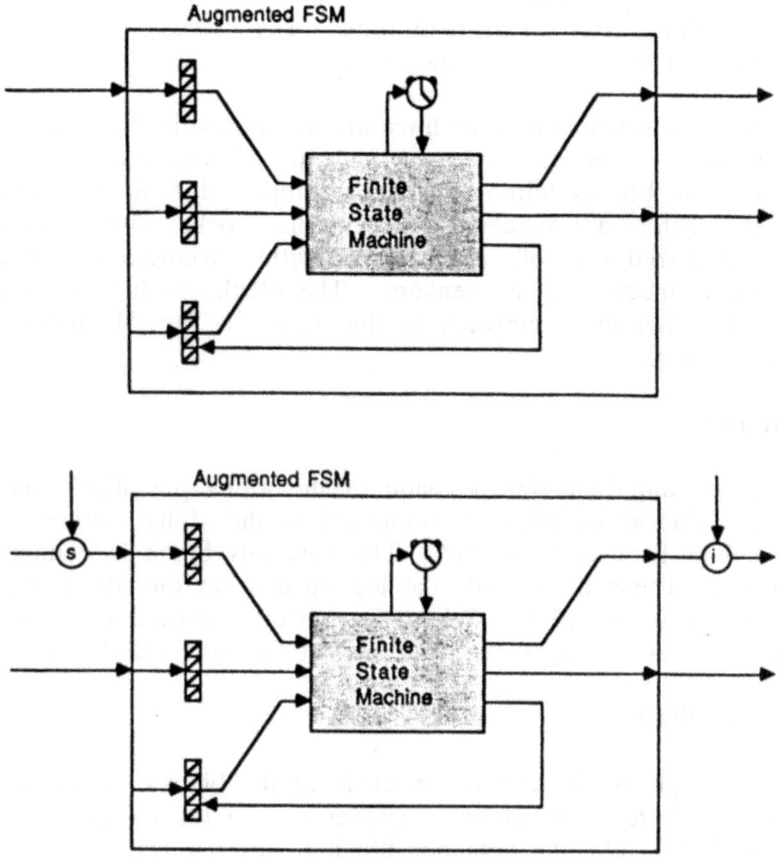

Fig 2. An augmented finite state machine

have some memory. There is an internal timer that will automatically reset the machine after an elasped time. The "s" suppresses the input to the machine. The "i" inhibits the output.

Genghis's subsumption architecture is shown in Fig. 3.
Fig. 3 shows the control system is made of 57 augmented finite state machines. The blocks without bands are repeated 6 times. The blocks with striped bands are repeated twice. The blocks with solid bands provide centralized control. The blocks with a triangle in the upper left corners are connected to the sensors. The blocks with a triangle in the lower right corner are connected to the motors. The "d" turns on the output of a block.

C. Standup

The standup behavior is made of the alpha pos block and the beta pos block. The alpha block is connected to the alpha motors. These motors move the legs back and forward. The beta pos block is connected to the beta motors. These motors lift the leg up and set the leg down. These blocks have a set position. When the power is turned on these blocks move the motors to these set positions so Genghis stands up.

D. Simple Walk

The simple walk behavior is made of the blocks shown in Fig. 4 [5]. The walk block tells the order to trigger the up leg trigger blocks. The up leg trigger block tells the beta pos block to lift the leg. The beta pos block tells the alpha advance block to tell the alpha pos block to move the leg forward. The alpha pos block tells the alpha balance block to move the other legs back a little. This moves the robot forward. The beta pos block tells the leg down block to put the leg down and the up leg trigger to tell the beta pos block to lift the next leg.

E. Force Balancing

The force balancing behavior is made of the beta force blocks and the beta balance blocks. These blocks adjust the beta positon of the beta motors. This keeps the robot level as it walks over rough ground.

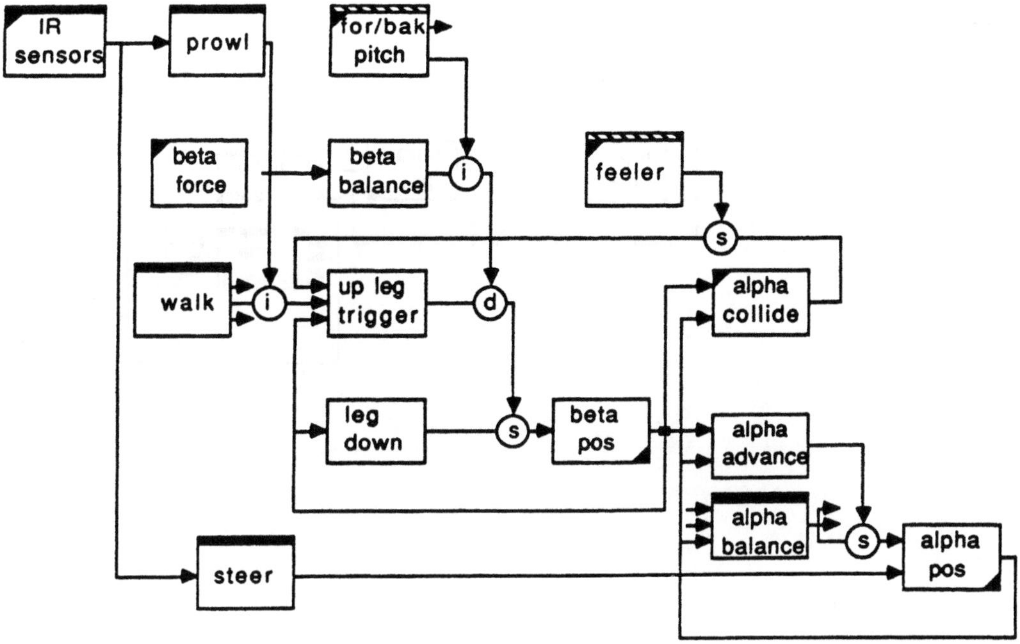

Fig. 3. Genghis's subsumption architecture

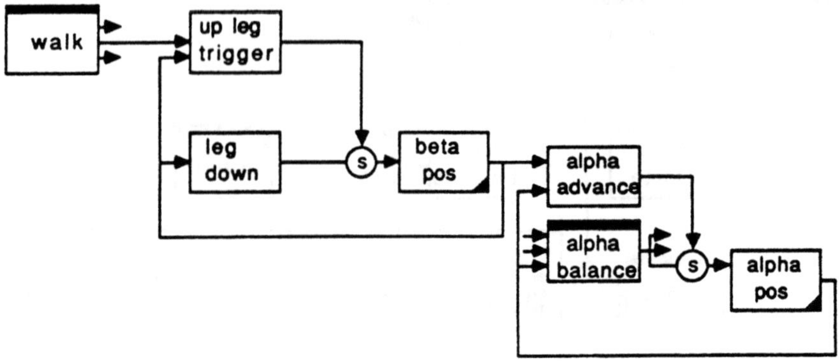

Fig. 4. Simple walk subsumption architecture

F. Leg Lifting

The leg lifting behavior is made of the alpha collide block. This block tells Genghis to lift its leg higher if the leg encounters too much force. This allows Genghis to run into obstacles and then start to climb them.

G. Wiskers

The wiskers behavior is made of the feeler block. This block suppresses the leg lifting to have Genghis lift its leg to start to climb an obstacle. Without this, Genghis would run into the obstacle before trying to climb it.

H. Pitch Stabilization

The pitch stabilization behavior is made of the for/bak pitch block. This block inhibits the force balancing behavior when walking up and down hills. Without this Genghis legs would try to keep itself level. This behavior keeps the robot from sagging.

I. Prowling

The prowling behavior is made of the IR sensors block and the prowl block. The prowl block will inhibit the walk block unless the IR sensors block see something warm.

J. Steered Prowling

The steered prowling behavior is made of the steer block. This block tells each alpha pos block how much to move each leg. This is based on which infrared sensor is detecting the highest levels. This block is what tells Genghis to follow people.

VI. CONCLUSIONS

The control system of Genghis was described. Genghis is a small robot that walks toward people over rough ground. The robot's behaviors and behaviors' parts were described. Genghis's design requirements and rules were described. Each behavior was made of a group of finite state machines. These machines had registers, memory and an automatic reset. Each of these behaviors were described. The control system made the robot appear to behave intelligently.

ACKNOWLEDGMENT

This paper was written as a final project for an EE472 class on the expert system design of intelligent robots.

REFERENCES

[1] Brooks, R. A., "A robust layered control system for a mobile robot," *IEEE Journal of Robotics and Automation, vol. RA-2, No. 1*, Mar. 1986, pp. 14-23.
[2] R. A . Brooks, "Achieving artificial intelligence through building robots," *MIT AI Memo 899*, May 1986.
[3] Brooks, R. A.; Connell, J. H.; and Ning, P., "HEBERT: a second generation mobile robot," *MIT AI Memo 1016*, January 1988.
[4] Brooks, R. A., "Engineering approach to building intelligent beings," *SPIE Vol. 1002 Intelligent Robots and Computer Vision: Seventh in a Series (1988)*, pp. 618-625.
[5] Brooks, R. A., "A robot that walks; emergent behaviors from a carefully evolved network," *MIT AI Memo 1091*, February 1989.
[6] Brooks, R. A., and Flynn, A. M., "Rover on a chip," *Aerospace America*, Oct. 1989, pp. 22-26.
[7] Connell, J., "A colony architecture for an artificial creature," *MIT AI-TR 1151*, August 1989.
[8] Flynn, A. M., "Squirt: the protypical mobile robot for autonomous graduate students," *MIT AI Memo 1120*, July 1989.
[9] Fraser, J., "John Holland and his robots, a love story," *EDN*, 15 Mar. 1990, pp. 233-236.

BENEFITS OF EXPERT ROBOTS: INTELLIGENCE vs SKILL

K. Khodabandehloo

Department of Mechanical Engineering
University of Bristol

Faculty of Engineering,
Queens Building
University Walk
Bristol BS8 1TR

Tel: (0272) 303240
Fax: (0272) 251154

The increasing costs of recruitment and shortages of skilled labour have been the main driving forces behind the recent advances in Automation Technology. Automated systems have brought significant economic benefit through consistency, repeatability, control and improved quality in production within the manufacturing, aerospace and food industries. During the next decade the scale of this benefit to Europe will be even greater as research and technological developments mature into advanced robotic and automated systems for factories currently employing the declining population of skilled labour.

This paper presents the opportunities for cutting and handling automation within the food and composites industries over the next decade.

INTRODUCTION

The potential for the use of robotics and automation in the food sector may be presented in terms of population of labour involved in production, currently estimated to be 2,500,000 people in the EEC, and the value of products consumed each year, currently £240,000,000,000 ECU's in Europe and £30,000,000,000 in the U.K. alone. The world production of meat (Table 1) is perhaps another indicator. Cost reduction and improved quality in a sector that has to cope with short product life cycles and the increase in demand, with fluctuations on an almost daily basis, provides significant opportunities for the automation industry to supply systems to food manufacturers. the main difficulty for the automation industry to overcome has been the development of systems that have the necessary level of 'skill' built in for tasks such as butchery, packaging, handling of composite prepregs or indeed the processing of flexible materials. This paper presents the automation requirements in relation to these tasks, outlining some of the features in the state-of-the-art systems in research laboratories. It is expected that these systems will be developed and used by the food and aerospace industries during the next decade.

Table 1. World meat production (Source: *Poultry International*, July 1988)

	Million tons	
Year	*1985*	*1988*
Bovine meat	47.9	48.3
Sheep and goatmeat	8.3	8.6
Pig meat	59.3	61.5
Poultry meat	31.5	36.3
Other meat	3.7	3.8
Total	150.7	158.5

NATO ASI Series, Vol. F 71
Expert Systems and Robotics
Edited by T. Jordanides and B. Torby

SKILLED ROBOT SYSTEMS

Present day industrial robots are commonly found operating with the aid of sensory devices in order to adapt to a changing work environment including variations in the components being handled. In some cases, artificial intelligence has been used in the form of an expert system to control robots.

Advances in sensory systems such as vision have enabled industrial robots to perform more flexible and complex tasks. The full potential of the industrial robot is still not being exploited fully because of the inability of current system architecture to deal with the uncertain and undefined situations. The existing programming architecture of industrial-robotic systems and computer-vision systems is based on the traditional languages where situations of uncertainty cannot be easily programmed or understood. Operator intervention normally takes place to provide the solution, which involves further programming or indeed conditioning of the environment to a simple and predictable form. An appropriate way ahead may be to provide the system with the knowledge and decision-making ability of the operator by the use of an expert system.

Expert systems are gradually being introduced into many applications including medical diagnosis, system control and others. Languages of expert systems such as LISP and PROLOG are considered to be of a different structure to that of the traditional languages. The basic difference being that the database may be changed whilst the program or the inference remains the same.

At present it is not possible to control a robot directly using expert-system languages, but an expert system may be linked to a robot and vision system to aid the automation of a skilled task [10]. The final necessary information is the knowledge of the expert. Such a system would then be able to operate autonomously by making decisions based on the sensory data. The robot's work envelope is known in mathematical terms, which can be incorporated into the knowledge and used to plan any new joint trajectories that may be required for a solution to an uncertain situation. Should more than one solution be available, then 'supports' can be assigned to each choice and support logic [2] used to determine the most preferable solution.

The system architecture for an intelligent robot system is proposed in Fig.1. The most advanced system produced to date at Bristol University is the Robot Snooker Play [10] which has the configuration shown in Fig.2. A number of other systems have been investigated for skilled tasks such as butchery, poultry packaging and composite handling which are presented here.

FOOD ROBOTICS: Butchery

The food industry has many complex manufacturing and processing automation problems to which existing robot technology cannot easily be applied. Food-processing operations usually involve handling of complex shaped, non-rigid sticky or slippery products. In tasks such as packaging, the manipulation of the product is often required. The automation of such tasks would be beyond the capabilities of current robot technology. Intelligent robots guided with vision and other sensors are therefore envisaged in order to overcome the unpredictable nature of the food environment.

Conditions in the food industry also present other problems uncommon to those in engineering manufacture. Hygiene, contamination and 'cleanability' are key issues when robotic devices are used with food products.

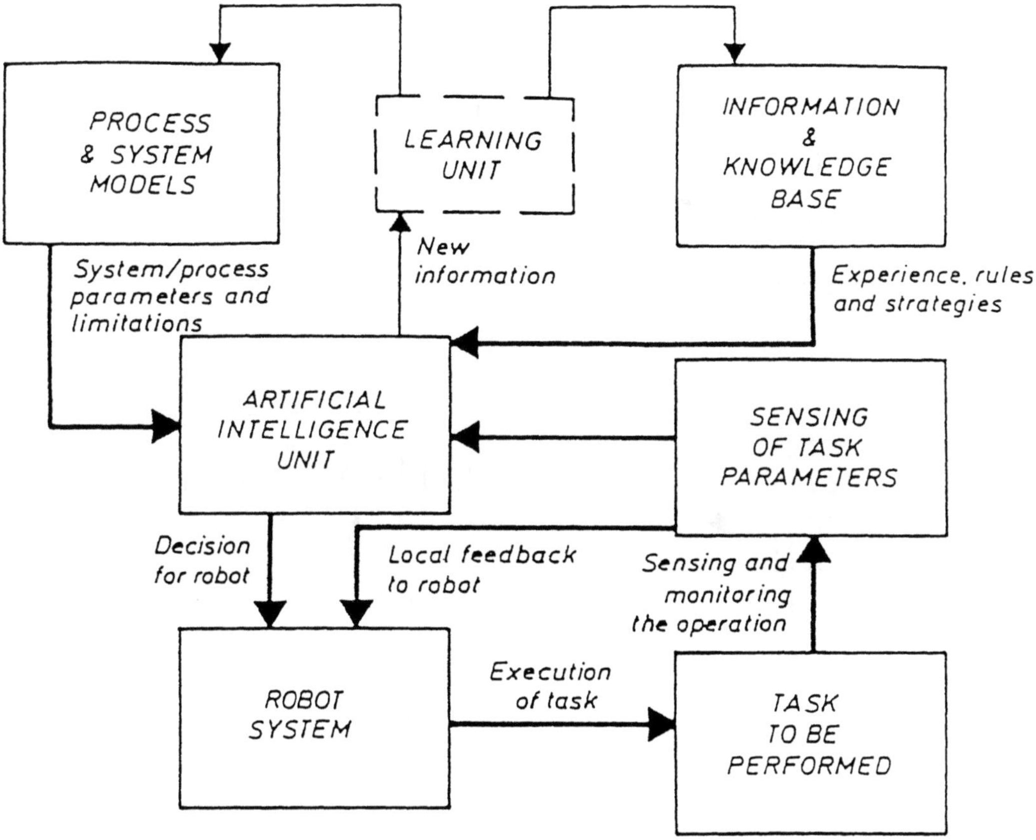

FIG.1 ARCHITECTURE OF AN INTELLIGENT ROBOT SYSTEM

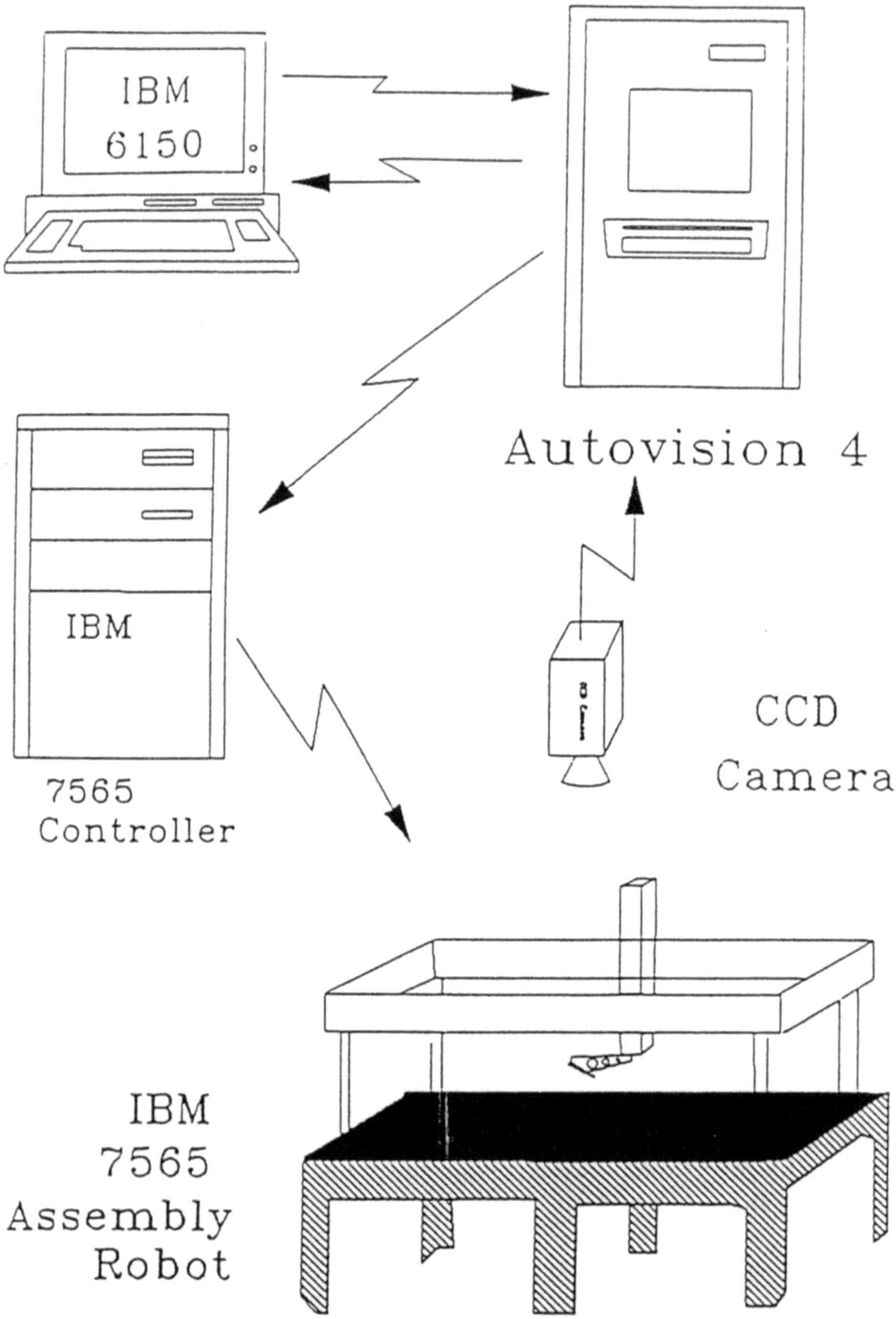

FIG.2 CONFIGURATION OF THE INTELLIGENT ROBOT SYSTEM

Collaborative research between the University of Bristol and the Institute of Food Research-Bristol (IFRB) has shown that the application of robotics technology in the food industry can be made possible with the use of artificial vision and intelligence.

Technical Requirements

Many of the operations in meat production and processing involve handling and positioning of irregular, non-identical pieces of meat of food items. These range in size and weight from whole beef-carcass sides 3m high and weighing up to 200kg, to chicken portions of a few hundred grams, or sliced products a few millimetres thick. As yet, robotics and automation systems have not advanced sufficiently for tasks involving such a range of food products to be automated. Furthermore, the majority of robotics systems currently in use within the manufacturing industry do not meet the exacting requirements of food production systems.

It is necessary to investigate and identify 'universal' characteristics of new handling devices, such as gripping devices and mechanical hands for food products. Such handling devices, once defined and developed, will broaden the range of applications of existing robots and will advance this technology by extending the design of such devices to meet the needs of food-production systems. Sensors ranging from vision systems to tactile or touch sensors will need to be defined and used to increase the capability of handling devices. These capabilities must allow handling of slippery, pliable, sticky, soft or wet products without any noticeable damage to food surfaces. Other requirements include being resistant to microbiological contamination, non-tainting and self-cleaning (or easy to clean). A robotic system used in handling and processing meat or food products must be self-cleaning, self-maintaining, intelligent and guided by advanced sensors, such a computer vision, and capable of being fully integrated into production systems of today and the future. Extensive literature surveys [4,7] or work in the USA, Europe and Australasia have shown that little research is being done to produce systems suitable for use in the meat/food industry. Previous work has been confined to application of existing teach and play-back industrial robots rather than generating new technologies utilising intelligent, sensor-guided systems.

ROBOTIC MEAT CUTTING

Industrial meat cutting is a difficult and unpleasant occupation. It is highly labour-intensive, requires a skilled labour force, and is expensive. The basic carcass structure within animal types is similar, although individual bone positions will vary according to the degree of fat cover, weight and carcass dimensions. Amongst the many types of sensors, computer vision is one of the few that can provide the surface details for an automatic meat-cutting operation. By using geometric models of typical carcasses, held in a data bank, several 2-D images may be used in conjunction with data-mapping techniques for robot guidance. Cutting strategies as part of an expert system will define the cutting path of the robot, whilst sensors in cutting devices will allow 'real time' course changes to avoid unforeseen obstructions, such as broken or displaced bones. To met the production speeds in the food industry, a vision system will be required to operate at high-speeds and in varying conditions.

Under an Agricultural and Food Research Council Grant for a Linked project, the cutting of forequarter beef is being investigated [11]. Initially the work involves the cutting of meat from the forequarter into unspecified shapes or sizes while leaving little meat behind on the bones.

After a thorough examination of the methods used by butchers in cutting meat from a forequarter beef, it was noted that handling of the meat in the same way as the butcher will be beyond the capabilities of robots for many years to come. Many of the operations involve grasping and manipulation of large, irregular and floppy or non-rigid pieces of meat. Definition and implementation of a robotic system that performs the task of meat cutting in the same way as a butcher is considered to be a long term objective. The initial step, however, has been to choose a research direction that leads to the realisation of sensory-guided robots to perform the task of meat cutting following technique more appropriate to the robot [5].

Fig. 3 shows a schematic of the various elements of the system to be implemented. The main subsystems required are :

a) A powered-cutting device attached to the robot arm. This is a reciprocating powered knife specially designed for the task.

b) A robot with sufficient reach and degrees of freedom needed to manipulate the cutter whilst delivering the necessary drive power the cuts is needed.

c) A force sensor providing feedback to the robot controller. This is needed to guide the knife parallel to bones whilst touching and separating meat from it without cutting into the bone.

d) A vision system that uses input from a number of cameras to define the carcass features required for cutting.

e) A system-control computer for deciding the start point, the end point and the rough path of each cut. This is the decision processor which takes the form of an expert system.

e) A database of previously measured carcasses for which the cutting data in (e) is available.

It is intended that the proposed robot system will cut the forequarter beef by moving its reciprocating powered knife through meat, with the cutting blades following a particular bone profile, separating meat from bone at the interface. The determination of the path for each of the required cuts is to be done by the use of the vision system and a database of cutting information for carcasses measured and cut previously (see Fig.3).

Assuming that the features to be measured by the vision system will result in the definition of the co-ordinates of a number of reference points on the carcass describing its geometry, then R_{jk} can be defined as a matrix of (x_j, y_j, z_j) representing the x, y and z co-ordinates of each of the chosen reference points j on carcass k. R_{jk} contains physical measurements of the carcass k in the vision frame of reference. k being the carcass index taking the values 1 to N(T), where N(T) is the total number of carcasses at time T for which R_j has been measured and kept in the database. This database is also to contain the start point, the end point and the rough path for each of the cuts required for the meat cutting operation on carcass k. For a given cut i on carcass k, the start point of each cut is denoted by a_{ik}, the end point by $\acute{a}_{ik}$, and the rough path by D_{ik}. Point vectors a_{ik} and $\acute{a}_{ik}$ are in the robot frame and D_{ik} is a path function joining a_{ik} to $\acute{a}_{ik}$. It is important to note that the cutter is expected to follow D_{ik} as the path for each cut i. However, because the bone positions vary, it is expected that with the use of force feedback the robot will guide the cutter along a different path closer to the bone. This new path is denoted by d_{ik} for k = c where c is the carcass to be cut. Since d_{ik} is closer to the bone being followed in cut i, a higher yield may be achieved highlighting the importance of force feedback. In Fig.4 an appropriate de-boning scheme for use with a robot system is shown. The scheme is also defined in such a way that eliminates handling of cut meat. The database information is essentially defined using this cutting scheme and the work on the expert system is intended to use this data to define the cutting information for the robot. It is envisaged that the operation of the robot system will take the following form (see Fig.3):

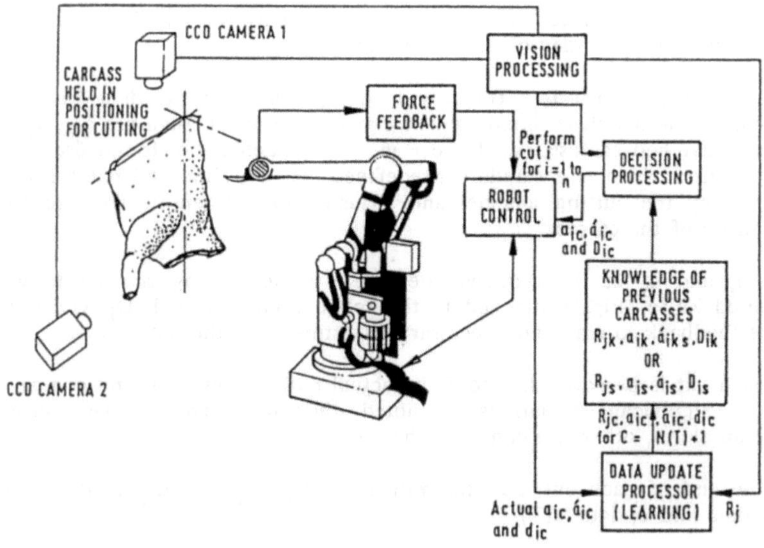

FIG.3 ROBOTIC MEAT CUTTING SYSTEM GENERAL CONFIGURATION AND NOTATION

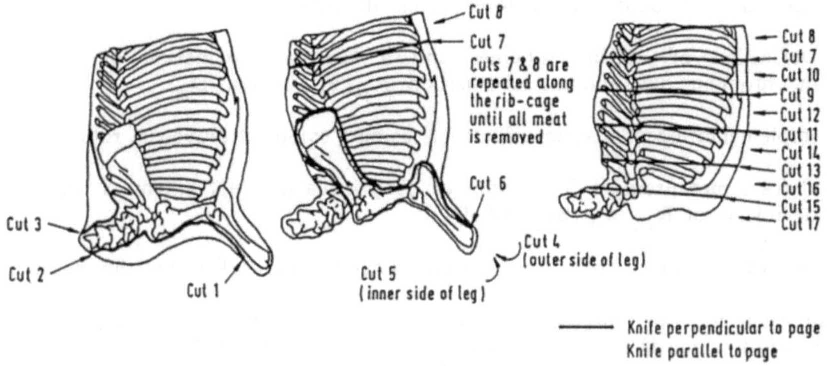

FIG.4 STANDARD CUTTING SCHEMES FOR i=1 to n=17

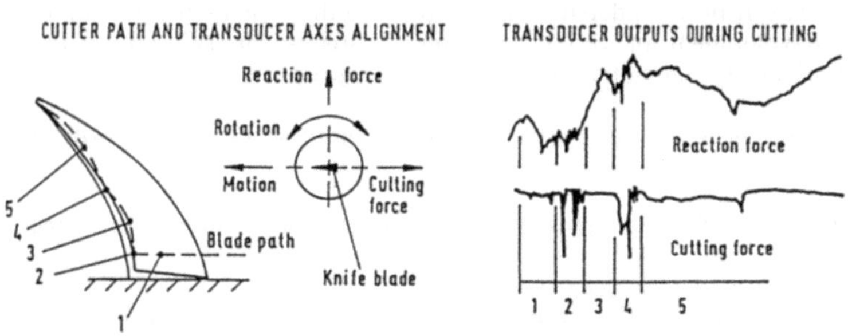

FIG.5 RESULTS FROM CUTTING OF A BEEF FOREQUARTER RIB SECTION

1) Take and process images of the forequarter held by the handling unit and define R_{jc} where c is the index of the carcass to be cut.

2) Compare R_{jc} against R_{jk} for k = 1 to N(T) held in the data base to find a match. If the match is found against a given carcass m, then use a_{jc}, $á_{jc}$ and D_{ic} for the matched carcass, i.e., for c = m. If a match is not found use a_{is}, $á_{is}$ and D_{is}, where s denotes a standard predefined set of points and paths for each cut defined by the cutting schemes and adjusted to compensate for size and shape variations of the carcass.

3) Use a_i and $á_i$ in the sequence defined by the cutting scheme to guide the powered knife (Fig.3) attached to the robot along the path D_i whilst using the force feedback to make any necessary adjustments to the path.

4) As each cut is performed, record the actual path followed d_i and compare this to D_i. If significant deviations are noted, then take error-recovery measures or raise an alarm, otherwise complete the cut.

5) At the end of each cut, add the values of R_{jc}, a_{ic} and d_{ic} to the database, as new data and update N(T).

Research so far has led to the following:

a) Definition of an algorithm for controlling the path of a powered knife moved by a robot along an unknown bone profile, using force control to adjust the trajectory of the cut. Fig.5 illustrates the schematics of the results from a trial cut using the robot to cut a forequarter rib section.

b) Definition of cutting schemes and a first approximation of the standard measurement data and cut information, R_{js}, a_{is}, $á_{is}$ and D_{is}.

c) Implementation of a system for trials using a standard industrial robot.

d) Carcass measurement for a number of forequarter beef carcasses leading to the definition of R_{jk}, a_{jk}, $á_{jk}$ initially.

e) Implementation of software for a vision system for performing the measurements.

It is important to note that the architecture of this system (Fig.3) compares with that shown in Fig.1 which is a general system architecture for an intelligent robot.

Current research is intended to continue to investigate forequarter beef cutting. Extensions of this work are envisaged to include the cutting of primals and meal portions for beef as well as lamb, pork and other such carcasses using robotics. This, however, involves further research into the handling as well as cutting of a variety of meat portions. To date little is understood about the handling characteristics of meat products to help the development of robotic systems. This is the subject of another Bristol University project in Automatic handling of non-rigid products.

In the forequarter beef cutting project, however, more specific work will be pursued in the following areas:

i) Investigation of force sensing to characterise the influence of carcass and meat variations.

ii) The use of vision to define R_{jc} for a given carcass and definition of vision-robot co-ordinate transformation rules.

iii) Study of a larger selection of beef carcasses to determine the variations in R_j, a_i, $á_i$ and D_i.

iv) Investigations of methods using fuzzy relational inferences to define a_{ic}, $á_{ic}$ and D_{ic} from R_{jc} and the cutting data provided in the database.

v) Experimental implementation of a full system to demonstrate forequarter beef cutting and to further investigate the process of updating of the database, or knowledge base learning.

Other past and current food automation related research at Bristol University include packaging of poultry, slicing and processing fish and more generally the handling of non-rigid products. The systems produced so far all involve the use of computer vision and force sensing. Other than the usual labour savings and improved work conditions for skilled workers, significant savings could also be made due to the consistency and accuracy of vision and robotic systems. Future research should aim to improve the techniques and to explore practical methods to cope with the food environment and the production speeds imposed. Consideration of other applications in whole-meal assembly, cooking, packaging and processing of food products is also becoming essential in view of shortages of staff and difficulties in maintaining and improving quality as well as hygiene standards. Despite recent short-term initiatives by U.K. research funding bodies to support selected research programmes, there is considerable need for a more substantial and co-ordinated research strategy in this field, backed up by adequate resources, possibly at a European or an international level.

AUTOMATED PACKAGING OF POULTRY

The world market for poultry is growing by about 6% a year [9] and to meet this demand processing plants, planning to expand cannot do so due to labour shortages. The world wide production of poultry for 1988 was estimated to be 36 million tonnes and with foreseen labour difficulties in this market, automation may be the most appropriate way ahead. the unpleasant, repetitious and hostile working environment of this industry further supports the need for automation.

Off-the-shelf solutions already exist to most of the automation problems that the chicken sector of the poultry industry faces, but the technology involved tends to be specific to a given process such as cutting. Turkey has seasonal market variation and is a lower volume product compared to chicken. Turkey processing is usually done manually. Different birds within a turkey flock may vary a great deal in size to further complicate the problems of automation.

The strengths of robotic technology lie in specific areas and the application of these strengths will yield benefits. Visits to several main poultry plants in the UK has revealed that over 30% of the work force are involved in handling the product. Along the length of the process line, after evisceration, people perform relatively mundane tasks which involve limited decision making. Handling involves a wide range of products including whole birds, portions, (including boneless portions) and quarters. These products may be moved between lines, between conveyors, form bin to conveyor, bin to overhead line, loaded onto machinery, packaged and so on. The packaging of portions tends to be highly labour intensive in most factories. The packaging may mean simply putting thighs or quarters into a bag until it has the right weight and then sealing the bag. Alternatively, it may be the more complicated task of arranging portions neatly in a tray, which is then sealed and priced.

The study undertaken in this project has aimed to define the main requirements as a first step by examining the poultry processing. Fig.6 shows the layout of a typical

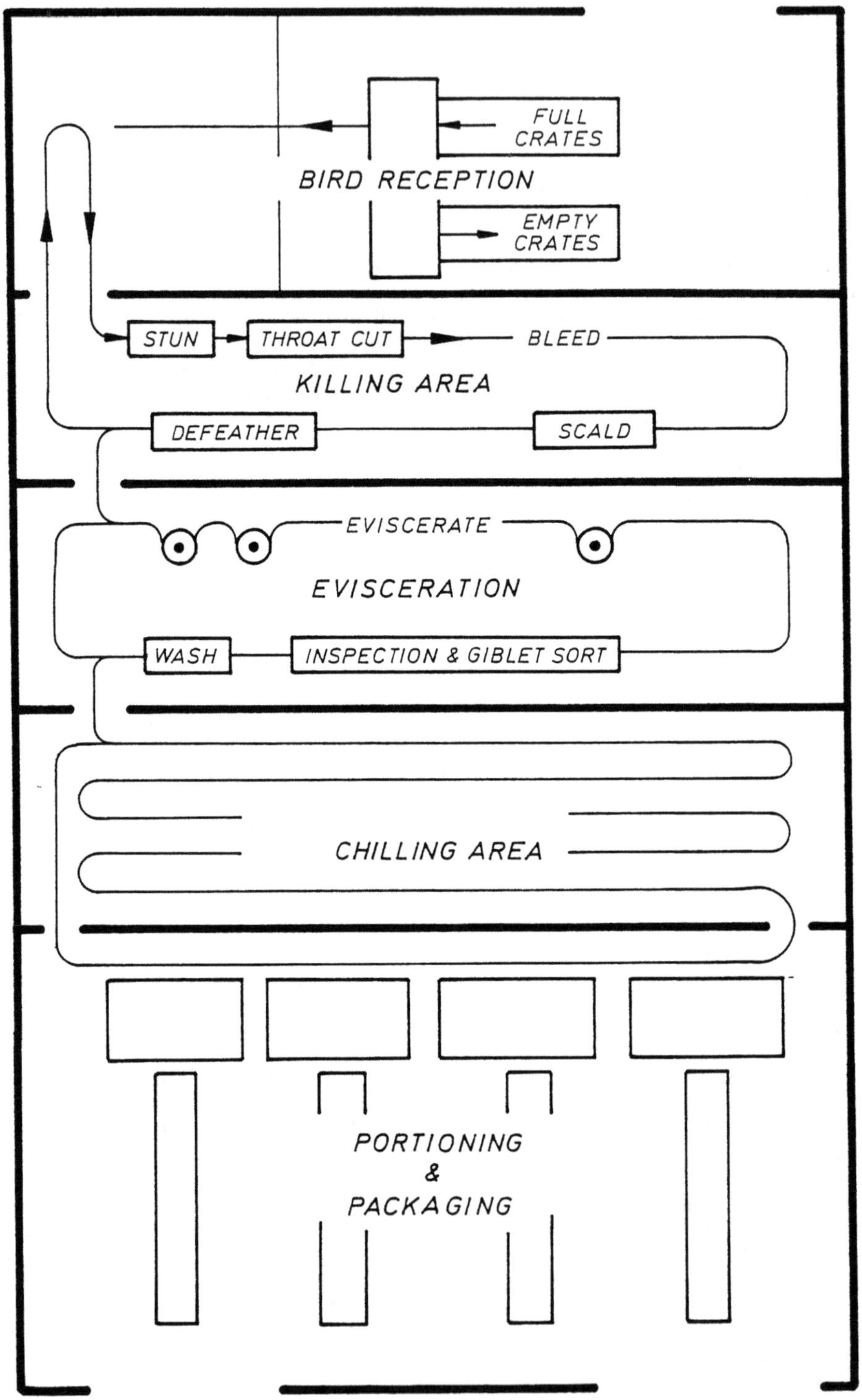

FIG.6 LAYOUT OF A TYPICAL POULTRY PROCESSING PLANT

poultry processing plant. The most labour-intensive area of the plant is in packaging, but this by no means represents the only potential for robotisation.

The packaging process involves the manipulation of a poultry portion off or on the bone from an unknown position and orientating, into a known location within a packet. The task may also require some degree of transformation by straightening or turning the portion as well as translation from one location to another. This process also involves considerable judgement and manipulation skill, not only in handling and packaging, but also in inspecting the quality of each portion.

Much of the work reported in this area is concerned with the construction of dexterous hands. Studies show that no gripper that can be directly applied to handling of poultry portions has been developed. The dextrous hands are unnecessarily complex and the contour-adapting vacuum gripper would cause skin damage. The pneumatic-rubber muscle developed at Bristol University, however, could be adapted for use in this application, giving a degree of compliance as well as requiring a safe and hygienic power supply for use in a hose down, food environment (Fig.7). The speed of response of such a gripper may be a limiting factor. It is therefore necessary for any gripper used for this application to be simple and capable of hygienic use with food. The main features of robotic grippers or hands for poultry handling include:-

Handle non-rigid products without damage
Hygienic and easily cleaned
Handle objects up to 1.0 Kg
Pick up moving objects
Handle different portions
Operate at high speed
Light weight
Inexpensive so that it can be replaced regularly
Rugged and reliable
Deliver portion with final orientation
Easily attached or detached
Capable of accommodating a variety of touch sensors

A pneumatic finger system, similar to that drawn in Fig. 8, has been investigated for this application. The finger is built from a single 25mm diameter and 1500 mm length of red, rubber laboratory tubing, restrained on its inner radius by a coil spring. Under increased air pressure the finger stretches further along its unrestrained side and hence bends.

It is expected that pneumatic fingers of this kind would provide a large degree of compliance. That is, they would give only a small change of force for a given change in angle/end displacement. Experiments have been conducted with a force transducer that has been set up to allow the finger a given maximum deflection angle. The pressure is then raised in steps of 0.5 bar and the force exerted by the finger at that pressure recorded. The force transducer is then moved to the next position to allow a further 2.5 degrees of deflection, and the process repeated.

The results (Force versus Angle with lines of contact pressure) were plotted (See Fig.9). It can be seen from the graph that the contact pressure lines have a gradient corresponding to a change in force of 1N to a change in 5 degrees. For the low forces generated, this cannot be regarded as a large degree of compliance. The advantage of this system is that can be easily cleaned and will be appropriate for use with food.

A vision system is required to act as the eyes of the robot system and it must be able to:

i) take and examine a picture of the scene
ii) identify and inspect the objects in the picture
iii) determine the position and orientation of the objects

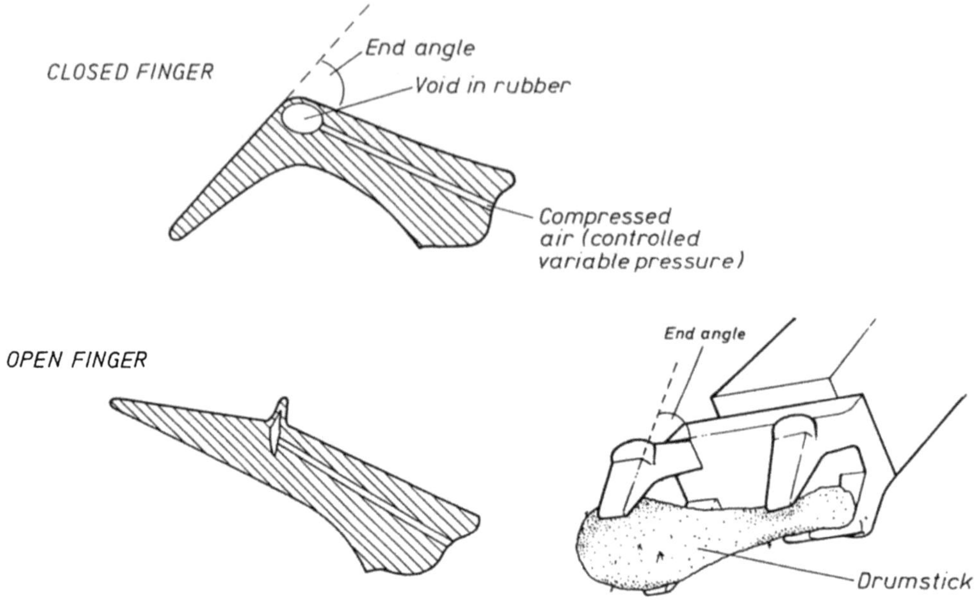

FIG.7 THE PREMATURE RUBBER MUSCLE FINGER AND GRIPPER DEVELOPED AT BRISTOL UNIVERSITY

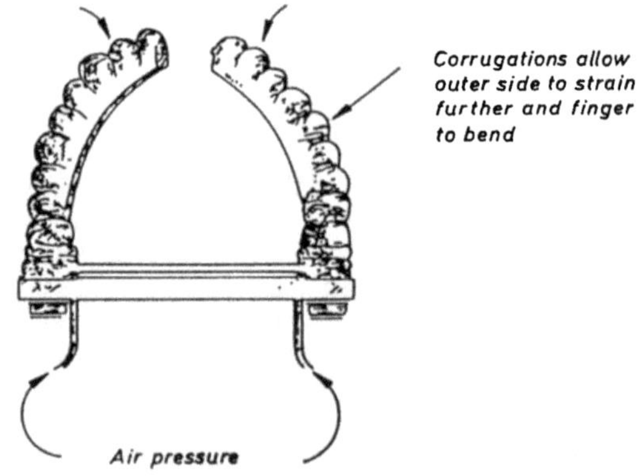

FIG.8 PNEUMATIC FINGERS

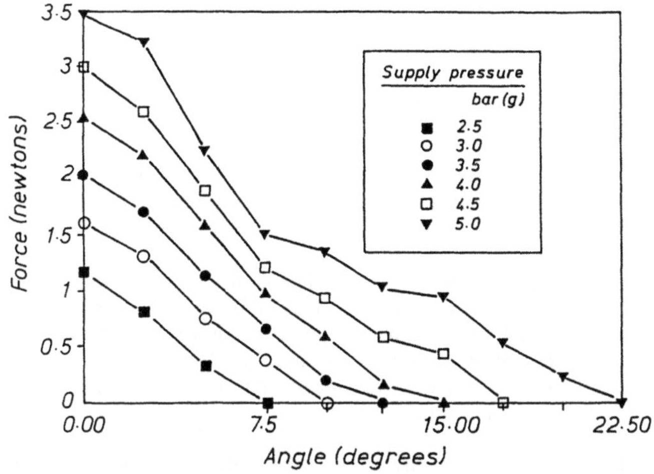

FIG.9 DISPLACEMENT -V- FORCE

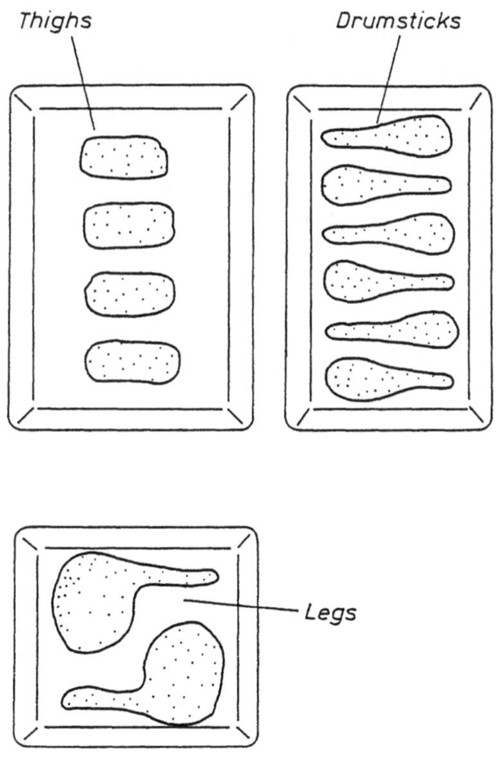

FIG.10 PACKAGING POSITIONS

iv) use this information to guide the robot

The vision system must be able to recognise different objects. In this application, it must be able to tell the difference between drumsticks, legs, thighs, breasts, leg-quarters, wing-quarters and wings.

For the investigations, a vision system providing general software routines has been used. This is the AV3 Vision System from Automatix using the RAIL vision programming language.

Consider how a human recognises objects. Through the experience of perceiving an object on a number of occasions, a picture can be built up of the object in the mind. A mental model of the expected shape and size of object is built. Many thousands of these models, one of each of the many thousands of objects that can be recognised, are stored in the brain. When seeing an object, the closest match between the object and the one amongst the mental models is sought.

These mental models are very sophisticated. First, they are three-dimensional models. An object can be looked at from many different angles and still be recognised. Secondly, they are independent of the distance between the eye and the object. If an object is moved closer or further away, it is still recognised. Furthermore, this is not simply a case of estimating the distance of the object using stereoscopic vision and scaling our mental model accordingly. One-eyed people, or people with one eye shut, do not suddenly stop recognising objects at different distances away. This suggests that shape, rather than physical size, plays a dominant role in the mental models.

The final, and most sophisticated, feature of mental models is the ability to forecast what a model would look like in a particular position and orientation even though the object has never been seen from that angle. This is a highly intelligent thought process which is done tremendously fast and usually subconsciously.

For many applications, the human level of performance is not necessary. For example, the number of models that a vision system would need to deal with in a particular application is small; probably less than ten and certainly not thousands.

A limitation of vision systems is that the majority work in two dimensions and in grey scale. Stereoscopic and colour vision systems are still under development and will probably be significantly more complex than existing systems. There are, however, still many applications, including the case of this project, for which a two-dimensional, grey-scale vision system is considered adequate.

Given the restrictions of a two-dimensional vision system, can the human recognition process be emulated or is there a more suitable method ? Three recognition methods, all using the model-matching technique, are commonly used:

i) boundary model
ii) template model
iii) features model

using method (i), a model of the boundary of a particular type of object is produced. The template is an average of many different samples of the same type of object. It represents a typical image of the object being modelled. It contains information about contour, texture and holes as well as the boundary shape. During the recognition process, the orientation of the unknown object in the pixel buffer is determined, then a suitably positioned and orientated copy of the model is put into a second pixel buffer. the two pixel buffers are subtracted. If the remainder is not significant, then the object is recognised.

Using method (iii), a statistical model of the features of a particular type of object is built up by teaching a number of samples. The model assumes that each feature (e.g. area, length, width) is normally distributed. The model consists of the means and variances of the feature values. During the recognition process, the deviation between the actual and expected feature values is computed for all features in the model (a chi-squared significance test is used). If the deviations are not significant then the object is recognised.

The software consists of three main programs: the first allows new feature models to be taught, the second sets up the vision system prior to normal operation, and the third is the routine that packages poultry portions.

System Integration of Poultry Packaging

In poultry packaging many schemes are used in the manual processes. The main requirement for robotisation is to produce a system that can place poultry portions in trays according to a predefined scheme similar to that shown in Fig. 10. After a detailed evaluation of the market, it has been revealed that the most immediate application for a robotic system of this kind is the fixed-weight-fixed-price (FWFP) packaging of fresh portions of poultry and more specifically chicken. FWFP packaging is a process by which chicken portions of a particular type are packaged in a specific number to meet a predetermined, but fixed weight, say three chicken breasts at a time, into trays which will be wrapped and priced the same. FWFP thus involves the weighting of each portion and selection of a group of portions, say three, that gives the required weight for pricing in this way. It should be noted that the packet weight must not be below the weigh at which it is priced. The main reason for choosing robotics for FWFP is that sorting of portions by weight using sort-weighing machines still relies on the judgement of the human packer to determine the weight of each pack. For people to correctly assemble a given tray with the fixed and predefined number of portions, producing the exact weight, without going under the specified fixed weight of the packet at which it will be priced, can be extremely difficult and time consuming. The advantage of a robotic system is that each portion can be weighed by an appropriate machine and the weight information can be transmitted electronically to the computer controlling the robotic cell (Fig. 11), which optimises the process of FWFP packaging.

It is anticipated that systems of the type described and others for portioning and deboning poultry will be introduced over the next decade.

AUTOMATED HANDLING OF COMPOSITE PREPREGS

In the manufacture of composite components, prepregs are stacked one layer at a time to make up the required structure. Dedicated tape-laying machines are produced on industrial scales within the aerospace industry for simple shape, but large prepregs. The handling and lay-up operation of discrete, and complex-shape prepregs on flat, single-curvature or double-curvature surfaces has not been achieved mainly due to the limitation in automatic handling and robotics technologies. The main handling problems arise from three characteristics of the prepreg composite material; it is tacky to the touch, it is non-rigid and it has a backing sheet which protects the prepreg while it is handled but which must be removed after each composite profile has been layed up. A number of possible mechanisms for the backing removal have been investigated including the use of adhesives, vacuum devices, piercing the backing with a sharp instrument, and leaving tags of material on the edges of the composite profiles. A new design is proposed for handling the profiles by gripping the tags.

The mechanical properties of cured and uncured composites are a result of the combined mechanical properties of their constituents. In this section, the mechanical properties of

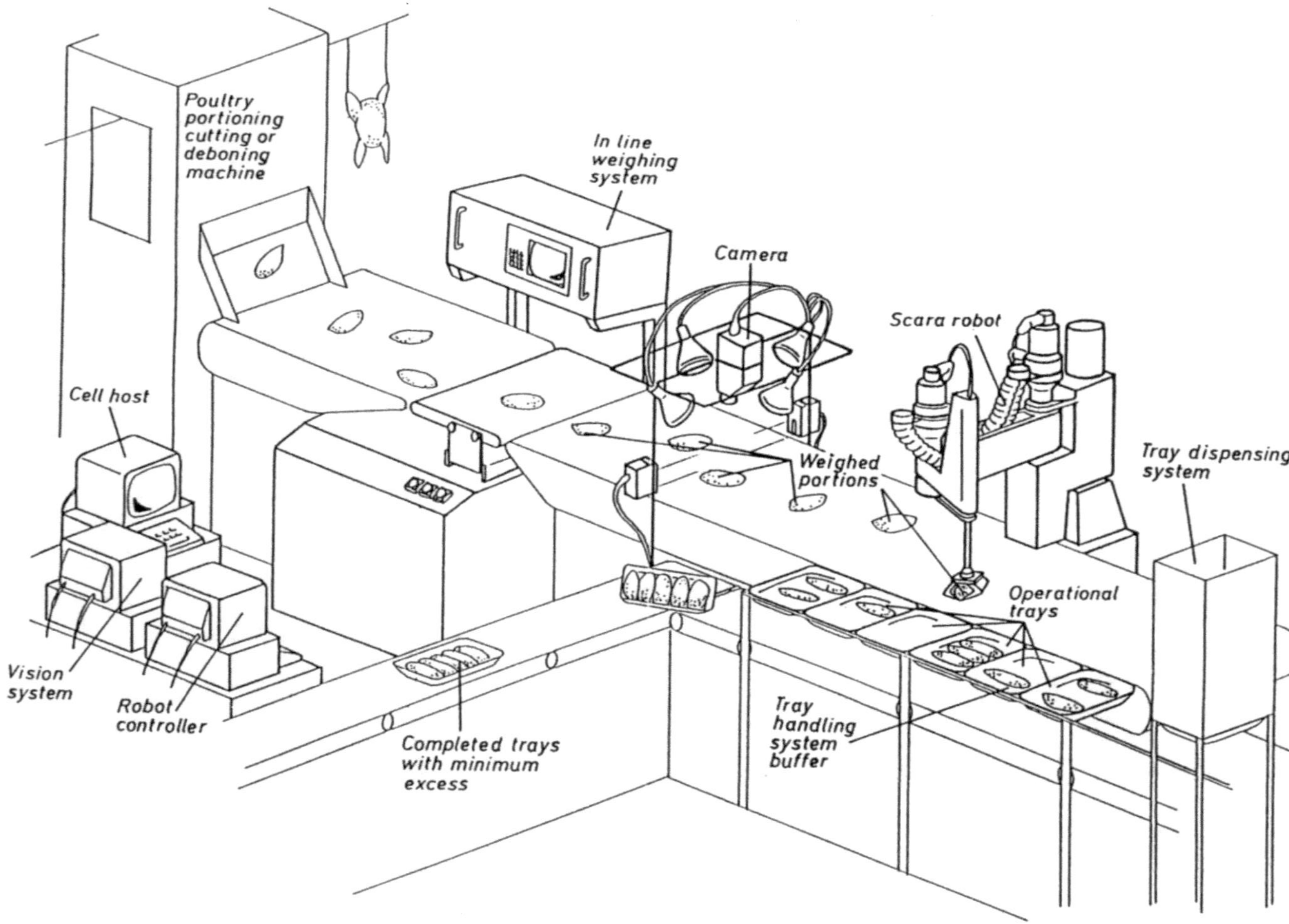

FIG.11 CONCEPTUAL SYSTEM OF A COMMERCIAL FIXED WEIGHT, FIXED PRICE ROBOTISED CELL

the uncured composites are discussed in relation to their handling during the manufacturing processes.

The glass and carbon prepregs are woven and thus behave anisotropically. The stiffness in the directions of the weave, 0 & 90 degrees, is many times greater than at 45 degrees to the weave direction. The resin causes tack between the prepreg fibres. Tack has a negligible effect on the prepreg stiffness in the weave directions but at 45 degrees to the weave it contributes significantly to the overall material rigidity. The strain properties of the prepreg are non-linear due to the non-linear behaviour of both the woven material and the prepreg tack.

The ability of uncured composites to take up complicated shapes allows them to be manufactured using low-pressure moulding techniques. More specifically it is because woven materials can sustain extensive double curvature that gives them this mouldability. The mechanics of double-curvature (detailed by Amirbayat et al [1]) means that multiaxial out-of-plane bending is associated with considerable in-plane strain which can be accommodated by woven fabrics.

The rigidity of the plies, in terms of their resistance to deflections as a result of their own weight, affects the way they are handled. As a first-order approximation, the deflection of a strip of a composite profile under gravity can be modelled using simple-beam theory. (This assumes that the deformation is elastic.) Modelled in this way, the deflection of the free edge of a composite strip is proportional to the cube of the strip length.

The profiles which are to be picked from the cutting table range in size from a maximum dimension of 1.65 m to 0.1 m. The stiffness of the plies depends upon their size and shape. As a general rule, however, the smaller profiles could be adequately gripped along only one edge, side or even one point. The larger profiles would need to be rigidly constrained by the grip and more than one edge or a series of points would need to be gripped to ensure accurate positioning and placement of the profile in the nest.

Tack

Composite material tack is defined as the bonding between two layers of uncured composite prepreg which come into contact with one another. Prepreg-to-prepreg tack or prepreg-to-backing tack are overcome respectively when a layer of prepreg or a layer of backing is pulled from a layer of prepreg material.

The strength of the tack is related to the wetting of the bond contact area by the resin. The following factors affect this wetting:-

1) Resin - the chemical composition of the resin controls the resin viscosity and so affects the tack.

2) Resin content - the greater the resin content of the prepreg, the greater the wetting and hence the greater the tack.

3) Resin distribution - the distribution of the resin through the prepreg affects the wetting at the surface of the prepreg which causes the tack.

4) Temperature - the resin viscosity is related to its temperature and the viscosity affects the resin flow and hence wetting.

5) Contact area (force applied while bond is formed) - as this force is increased so is the microscopic contact area of the wetting by the resin.

Of these only the contact force can be varied by the manufacturing cell. The prepreg-to-prepreg tack is therefore dependent upon the magnitude and duration of application of the brushing force, applied downwards on to the backing during the lay-up operation.

The prepreg-to-backing tack is less consistent across the bonded area than the prepreg-to-prepreg tack of a similar magnitude. The macroscopic contact area between prepreg layers is higher than that between backing and prepreg layers. On a microscopic level, however, the woven texture of the backing causes a lower contact area than the smoother polymer surfaces. These trends, working in opposite directions, account in part for the backing and prepreg tack properties. This may be explained in part by the structure of the backing sheets. In the case of polyethylene backing, the material has a low macroscopic contact area because of its embossed shape. This increases the tendency for only point contact to occur at a number of dispersed positions across the bonded area.

The Peeling Process and Model

A mathematical model for peeling and experimental evaluation of the process has been investigated to aid the development of mechanisms to remove the backing from composite profiles.

The term peel-back is defined in the Mcraw-Hill Dictionary of science and technical terms [6] as follows:-

"The separation of two bonded materials one or both of which are flexible, by stripping or pulling the flexible material from the mating surface at a 90 degree or 180 degree angle to the place in which it is adhered."

This definition is extended to peeling at peel angles other than 90 or 180 degrees. The convention adopted in this report for the definition of the peel angle (θ), peel force P, adhering layer and substrate are shown in Figure 12.

The resistance to peeling of a structure is controlled by the following factors:-

- energy to break the bond between adhering layer and substrate.
- energy absorbed by bending the peeled layer through the peel angle and by the straining of this layer by the peel force.
- energy absorbed in the deflections of the substrate due to the peeling force.
- size and shape of the structure
- the peel angle (the effect of this is described after the peeling model has been outlined).

The relationship between the peeling force, the peel angle, the tensile stiffness of the adhering layer (K) and the work expended in detachment per unit of bonded angle is developed from theoretical considerations by Gent and Kaang [3]. Figure 13 shows the peeling-model diagram. The work done by the peeling force (P) as it moves through a distance (dp) is evaluated in terms of the distance of peeling in the place of the substrate (ds), the peel angle (θ), and the strain of the peeled layer (e). This is equated to the sum of the energy expended in the detachment per unit area (Ga) and the energy expended straining the peeled layer (U).

$$P\,(1 + e - \cos\theta) = U + Ga \qquad (1)$$

Assuming linear straining of the peeled layer with a tensile stiffness of K, this simplifies to become:-

$$Ga = /\ P\,(1 - \cos\theta) + e2K/2 \qquad (2)$$

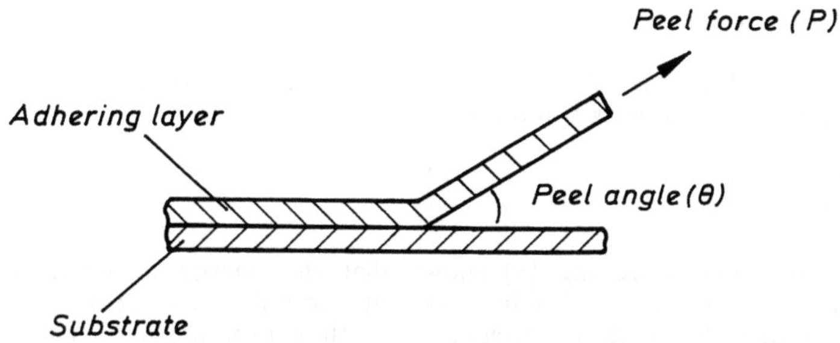

FIG.12 GENERALISED PEELING DIAGRAM

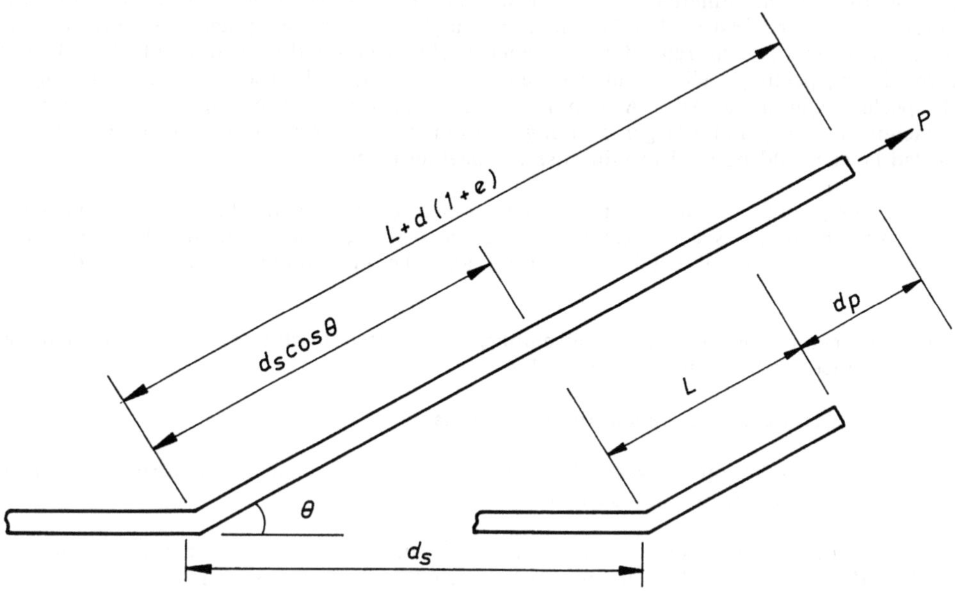

FIG.13 DEVELOPMENT OF A MATHEMATICAL MODEL FOR PEELING

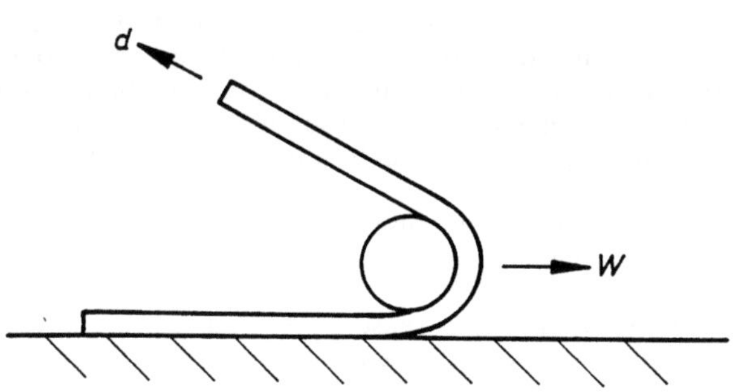

FIG.14 PEELING EXPERIMENT USING A ROLLER

$$Ga = Ke (1 - \cos \theta + e/2) \qquad (3)$$

for relatively inextensible adhered layers and where the peel angle is not small ($\theta > 45$ degrees), this simplifies as follows:-

$$(1 - \cos\theta) >> e \qquad (4)$$
$$Ga = P (1 - \cos\theta) \qquad (5)$$

Practical work by Gent & Kaang [3] shows that the energy of detachment is not independent of the peel angle. Their work with adhesive tape peeling from a rigid substrate showed that the work in detachment with a peel angle of 180 degrees was approximately three times as great as that for small peel angles.

The adhering layers were also peeled off (as shown in Figure 14) with a weighted roller to keep the peeling layer in contact with the roller surface. This arrangement controlled the bending of the adhered layer as it was peeled and the energy of detachment was observed to be reduced to that for small peel angles. It was suggested therefore that the detachment energy was related to the energy absorbed by the bending of the adhered layer during peeling. With small peel angles, the energy absorbed by the stretching of the peeled layer increases. A compromise between a small peel angle to minimise the detachment energy and a larger peel angle to reduce the energy absorbed in straining the peeled layer would be used to minimise the peeling force.

Unlike this peeling of adhesive tapes, the problem of peeling the backing material from composites prepregs is complicated because the prepreg is not a rigid substrate. The woven prepreg material itself is not rigid and the interlaminate bonding in the nest assembly is weak.

The resistance of the prepreg sub-assemblies in the nests to strains in the planes parallel to the surface of the nests is produced by:-

- stiffness of the woven prepreg material
- the adhesion between the prepreg and the nest surface resisting movement relative to the nest surface.

The resistance to movement parallel to the nest surface is greater than the resistance to peeling the plies off. The difference between these two types of ply movement is in the mechanics of the breaking of the adhesive bonds between the plies. When one layer is peeled from another, the peeling force acts on the edge of adhered layer that is being peeled. In the case of parallel ply movement, however, the prepreg transmits the force so that it is resisted by a larger bonded area.

The prepreg must not be disrupted when the backing is removed. This means that the prepreg profiles must not be lifted away nor must they be stretched out of shape or moved in the nest. In addition to these requirements, the bonding between the prepreg layers in the nest must not be weakened by the backing removal; such weakening could cause delamination when subsequent layers of backing are removed.

i) All backing must be removed (because of the weakness that unremoved backing would cause in the completed structure).

ii) Detection of unremoved backing - sensors are required to check that the backing has been successfully removed and to ensure that, unless this is the case, the next ply is not added until it has been.

iii) No delamination of previous composite layer. - the bonds between the prepreg and backing may be strong enough to cause previously placed plies to be delaminated before the backing is removed.

iv) Non-interference with the composite properties - the prepreg must not be weakened by the backing removal. The plies must not be stretched out of shape by the backing-removal forces, nor must there be any penetration of the prepreg by the backing-removal mechanism or any other interference with the weave of the fibres.

An Integrated Cutting, Handling and Peeling Systems using Tags

This section presents a method of altering the profiling of the composite so that tags of backing are left attached to the edges of the composite-profile backing. These tags would overhang the edge of the composite profile as shown in Figure 15 and would be used to grip the profile backing while:-

- the profiles are grasped from the cutting table
- the profiles are placed in the sub-assembly nests
- the peeling of the backing is initiated
- the backing is removed.

Tags of backing could be left on the profile backing if the height cutting with the use of an ultrasonic knife is controlled so that it cut through the prepreg but not the backing. The machine which profiles the woven materials for the EH101 prepregs uses ultrasonic sensors to detect the top of the material and control the depth of cut to within a tolerance of 125 μm. The machine manufacturers now have the capability to manufacture ultrasonic cutting tables which can control the depth of the cut to within a tolerance of $\pm$ 25 μm (Pilkington et al [8] state that the cutting accuracy is 100 μm; contact with machine manufacturers however, revealed that this is a conservative figure). This increased accuracy is possible because a topographical map of the surface of the table is recorded and stored. The depth of the cut is then controlled relative to the surface of the cutting table.

The thickness of the backing is 100 μm. It would therefore be possible to cut through the prepreg but not the backing, allowing tags of backing to be left on the composite profiles. Figure 12 shows a typical profile as it would be cut with, and without, the tags. The solid line around the profile indicates a cut through both the backing and the prepreg and the dotted line shows where the ultrasonic knife only cuts through the prepreg.

The composite profiles must be picked up from the cutting table where they lie backing side down. The backing prevents the prepreg from adhering to the cutting table when a vacuum is applied to "clamp" the composite to the cutting table while it is profiled.

It is proposed that once the cutting is complete, the profiles will be gripped by the tags and lifted away from the cutting table. The number of tags per profile would depend upon the size and shape of the profile; a study of this is needed, but is beyond the scope of this report. At least two tags would be needed to grip the profiles and a minimum of three tags would be required to support and keep the larger profiles flat while they are placed in the nests.

The tags would be gripped on the prepreg and backing sides. There is no problem of contaminating or damaging the tag composite material by gripping the prepreg side of the tags because the tags do not form part of the structure and are removed with the backing. The tag gripping sequence is illustrated in Figure 13 and consists of the following steps:-

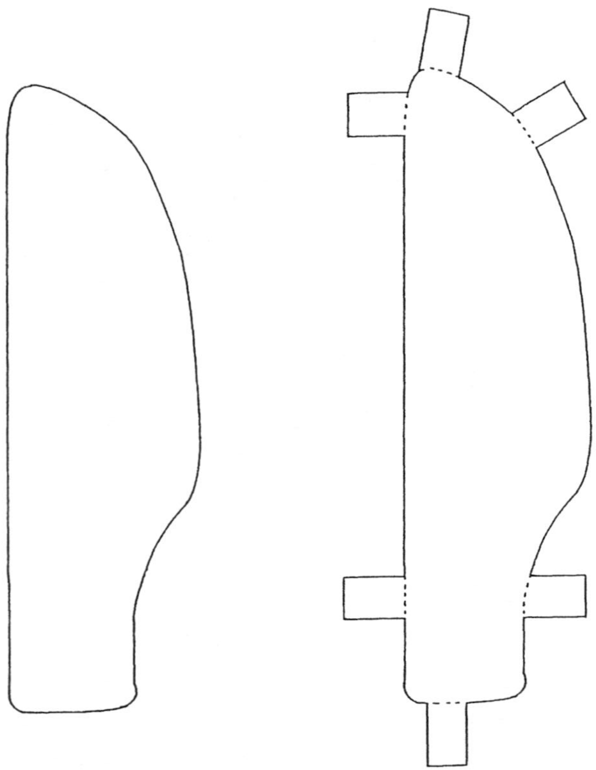

FIG.15 TYPICAL PROFILE WITH AND WITHOUT TAGS

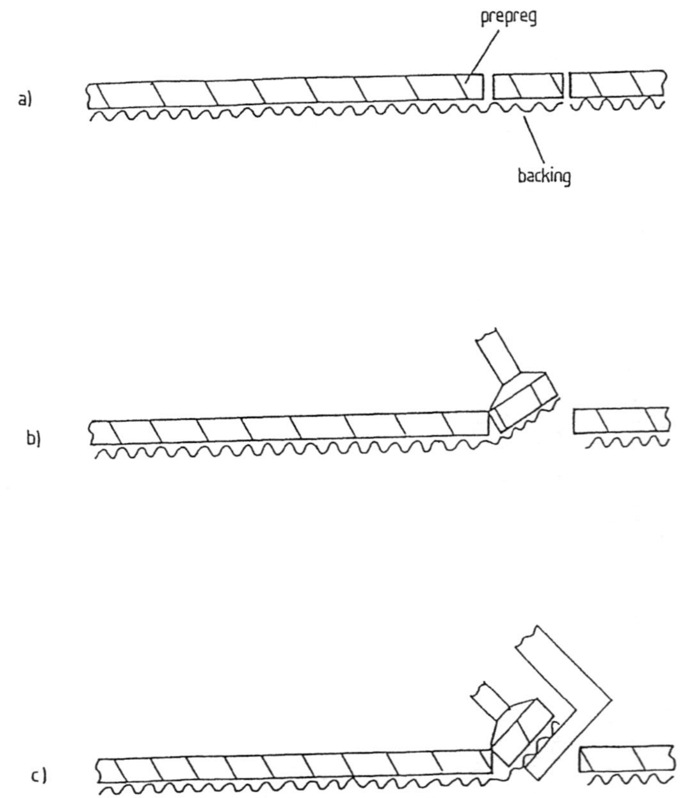

FIG.16 GRIPPING TAGS

(i) The vacuum "clamping" the composite to the surface of the cutting table is switched off when the ultrasonic cutting is finished.

(ii) The free edges of the tags are lifted away from the table using vacuum cups or a needle gripper which penetrates the prepreg or the prepreg tack to grip the tags.

(iii) A second gripper finger is then brought into contact with the backing side of the tags, clamping the tags.

It is intended that each tag is gripped with the use of a suitable multi-arm manipulator using vision guidance. Once all the tags have been gripped in this way, the composite profile is lifted away from the cutting table. The profile is then gripped by the tags and positioned ready for brushing down onto the sub-assembly nests. Fig.17 illustrates this concept and work continues to examine the fundamental problems and the solutions needed to achieve this.

CONCLUSION

This paper has examined the requirements for cutting and handling automation, and the trend of developments over the next decade, within the Food and Composite Industries, based on current research in the field. The concept of 'skill' as compated to 'intelligence' is introduced and the specific tasks of butchery and packaging of food products, and lay up of composite prepregs have been presented as examples of skilled tasks for Robotics Systems. It is necessary to state that there is a fundamental sytem architecture that constitues the basis for 'advanced' robots for skilled tasks. Although the examples presented tend to be specific, the wider implications for the automation of other taks is evident. Amongst these are skilled tasks ranging from simple ones such as cleaning, to complex operations such as surgery. The advancements over the next 10 years will not doubt be signifcantly greater that those in the past decade as our level of understanding increases in the fields of automation and the utilisation of computing techniques and its integration with electro-mechanical systems enables the more complex tasks to be caried out by robotic devices/systems. Among those that will benefit most during the coming decade will be the following industries:-

-Food Processing

-Aerospace

-Nuclear Power

-Oil and Offshore

-Mining

-Medical and healthcare.

There is no doubt that the alignment to research to meet indsutrial objectives and the cultural change needed within these industries, combined with closer co-operation between financial, industrial, government and academic institutions will be among the important drivers enabling progress and international competitiveness for Europe.

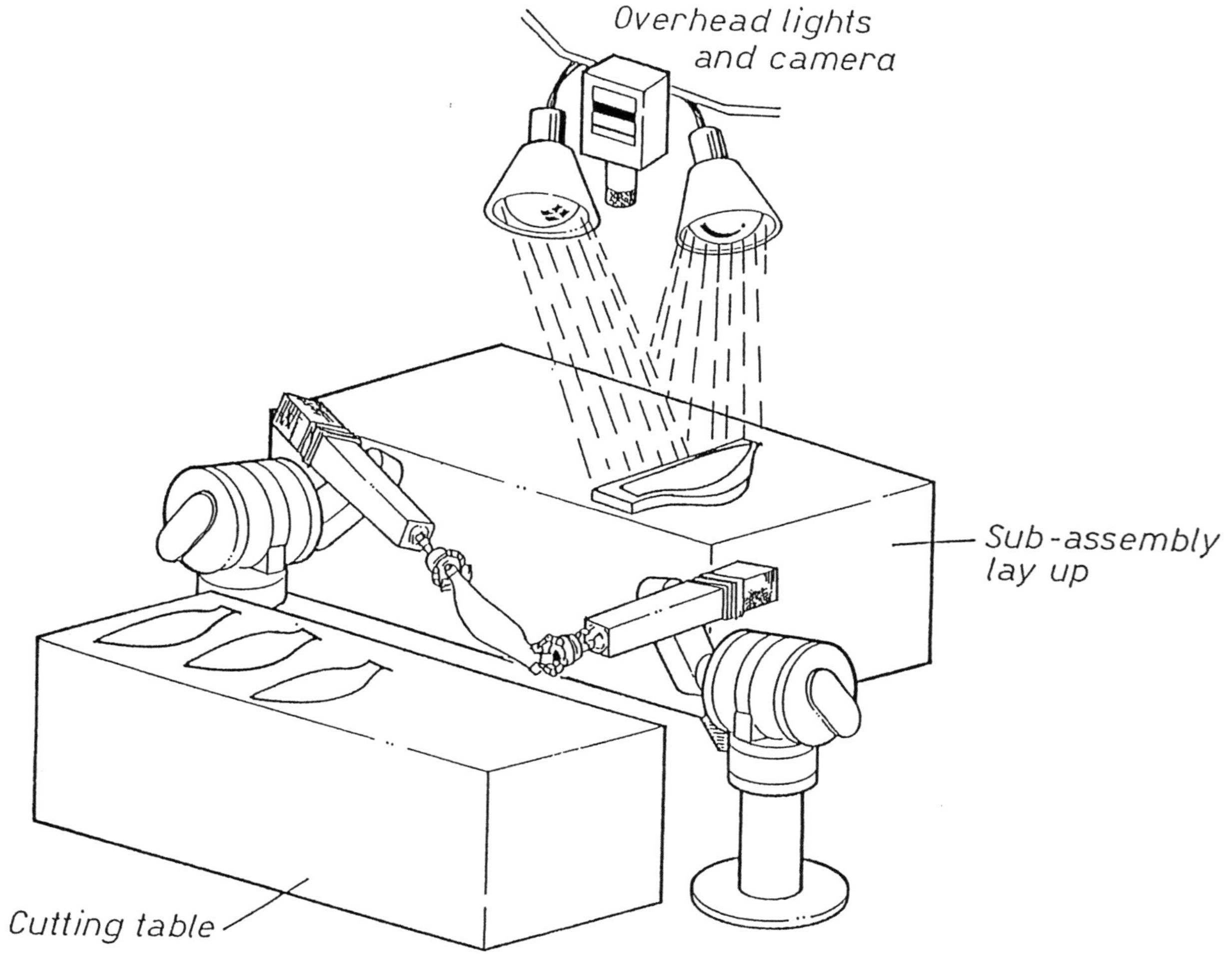

FIG.17 CONCEPTUAL DESIGN FOR ROBOTIC-COMPOSITE-PROFILE HANDLING SYSTEM

ACKNOWLEDGEMENTS

The author wishes to thank the following organisations and people:

-UK Agricultural and Food Research Council (AFRC), The European Commission (BRITE), and the Science and Engineering Research Council (SERC) for funding the related research project at Bristol

-IBM (UK), SKF Automation, Sun Valley, Westland Helicopters, and Automatix UK for their support.

-Dr. Peter Brett, Mr. Colin Bailey, Mr Neil Maddock, Mr. Graham Purnell, Mr. Mark Swain, Mr. Steve James and Mr. Paul Newman for their support and contribution to Food Robotics Research at Bristol University.

-The AFRC Institute of Food Research, Bristol Laboratories for provision of material for testing and access to food processing facilities.

-Last but not least, Mrs. Geraldine Maher and Mrs. Jean Foster for typing this publication.

REFERENCES

1. Amirbayat,J., Hearle,J.W.S. & Leach,C.M., "Complex buckling of textile fabrics".

2. Baldwin,J.F., "Support Logic Programming" (ITRC 80), Information Technology Research Centre, Univeristy of Bristol, England, 1986

3. Gent,A.N. & King,S.Y., Effect of peel angle upon peel force".

4. Khodabandehloo,K., Bailey,C. & James,S., "Intelligent Vision Guided Robots for the Food Industry", 19th International Symposium and Exposition on Robots, Australia, 1988.

5. Khodabandehloo,K., "Getting down to the Bare Bones", The Industrial Robot, International Journal, 1989.

6. McGraw-Hill, Dictionary of science and technical terms.

7. Merchant,J.A., "The Use of Robotics in Agriculture and Food Industries", National Institute of Agricultural Engineering, Silsoe, Bedford, U.K. 1985

8. Pilkington,D.J., Bryce,Dr.G. & Nimmo,W., "Improvements in automated ply cutting and handling", The National Society for the Advancement of Material and Process Engineering Symposium, March 1988.

9. Poultry International, July 1988

10. Rennell,I.J. and Khodabandehloo,K. "Development of Skilled Robots: a new approach in Robotics" - 20th International Symposium on Industrial Robots, Japan 1989

11. Robotic Meat Cutting, Linked Research Group Funding, AFRC reference LRG 114 (1987/91).

EVALUATION OF APPLICATIVE ASSEMBLY DESIGN, SUPPORTED BY EXPERT SYSTEMS

Alberto Rovetta[1], Elio Zamponi[1], Vitor F. Romano[1,2], Remo Sala[3]

[1]Department of Mechanics, Research Group of Mechanics of Robots, Politecnico di Milano, Piazza Leonardo da Vinci 32, 20133 Milano, tel.39-2-23994720 telefax 39-2-2366554, Italy.

[2]Department of Mechanical Engineering, COPPE - UFRJ, Brazil

[3]Consultant

Abstract: This paper deals with the application of an expert system for the design of an assem bly process with robots. The system to be assembled is an electric motor, composed of seven elements. The design process is composed of: geometrical analysis, compatibility of components with robot tools, design of new tools, preparation of logical structures of assembly process, realization of the expert system able to determine the optimal assembly sequence, evaluation of technical characters for different sequences, the final choice, connection with the robot, and actuation of the robot. A second part deals with the expert systems'economic evaluation, in regard to productivity and economic results, until now limited to the realized cases.

Keywords : expert systems / assembly / robotics / logics / economical evaluation

Assembly Characters

This paper deals with the robotized assembly of an electric motor with CAD, expert systems, design for assembly, and high level programming (Figure 1); an expert system is presented as a high level tool for supporting the designer in robotics applications, with particular reference to assembly.

Assembly is one of the most skillful activities in the human operativity. Shape analysis, kinematic choice, dynamic evaluation, logical selection of optimal sequences are actuated by a human operator. In the electric, mechanical, aeronautic and space industry, the largest percentage of a production system lies in the assembly process.

Research in assembly has changed from the determination of the best human utilization to the calculation of the most productive way of assembly. Mechanization in

NATO ASI Series, Vol. F 71
Expert Systems and Robotics
Edited by T. Jordanides and B. Torby

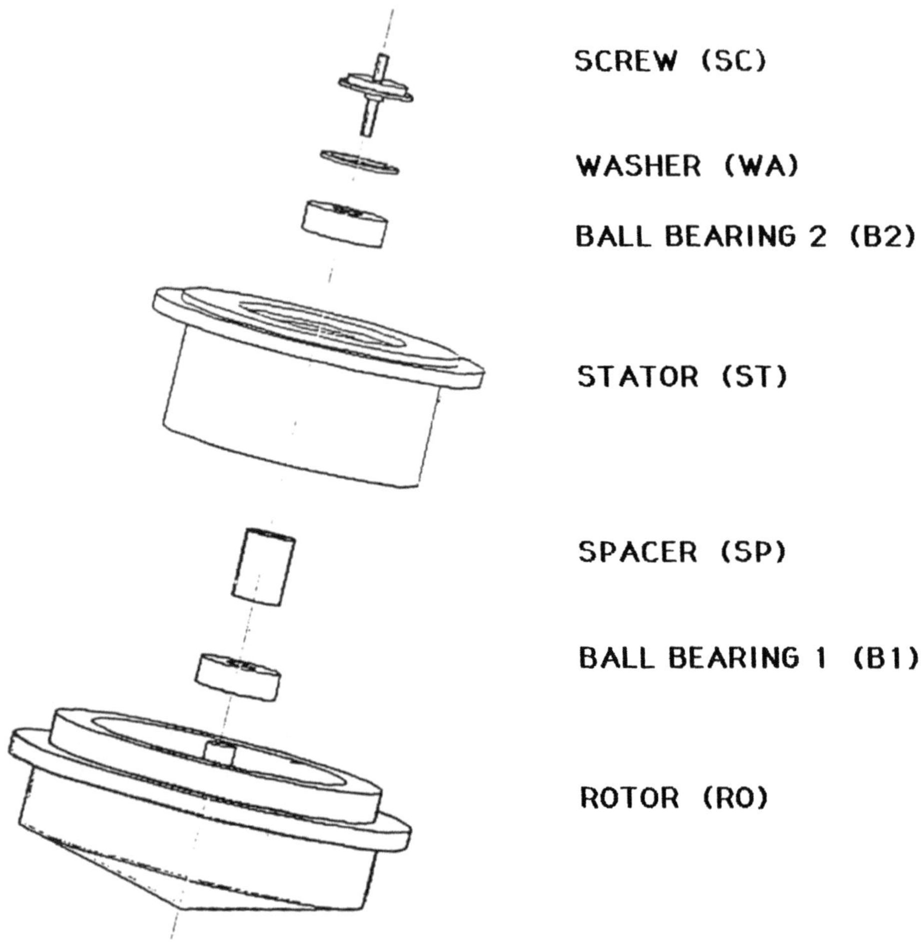

Figure 1: Electric motor to be assembled: 7 components with different sizes, weights and shapes.

many industries gives to the machines the capability of repeating connecting operations, and assembly is performed at high velocity, in a strongly rigid way; the **mechanized system** requires only some attention from the worker, and guarantees a great quantity of similar products.

Automated assembly introduces the integration of mechanics, electronics, controls, sensors, with some adaptability to different working conditions. However, considering real **automated assembly,** its use in industrialized countries is low, compared to the percentage of manual assembly and mechanized assembly that exists (Figure 2); these facts must influence also the spirit of the research, which has to make the connection with the applicative reality, specially when generated from economic necessities.

At the same time, for difficult operations in unstructured environments, like in nuclear plants, underwater works, space operations, etc; automated assembly with self-programming systems may be required. Even if usually, teleoperations are considered the more reliable alternatives.

The paper presents the automated assembly of an electric motor, with the use of CAD for the geometric description, with logical choice obtained from an expert system program written in Prolog language, and with the final operation performed by a Cartesian single-arm robot, where the extremity tool has been redesigned according to the characteristics of the workpieces to be assembled.

Manufacturing Costs

One of the main problems of the productive world is the reduction of manufacturing costs in order to allow the product the best possible market penetration compatible with the capacity for absorption by the market. Given the remarkable progress made by constructive technologies, we now see that there are few possibilities for savings associated with the construction of the single component, while many others are instead to be found in putting components together in the formation of a finished product which is immediately usable as it emerges from the production process. These may be either complete products, ones which are capable of carrying out a certain function entirely, or comprehensive or sub-comprehensive ones, products which are meant for assembly into complete machines and devices and of which they represent functionally complete parts. In substance, the greatest potential for savings can be identified in the activity of assembly which also happens to be the part of the productive process which has been least studied in the past the problems being very diverse and the solutions therefore being more or less empirical.

A good deal of research work has been devoted to the economic justification of assembly-work automation. In general, these are analyses of the technical-economic

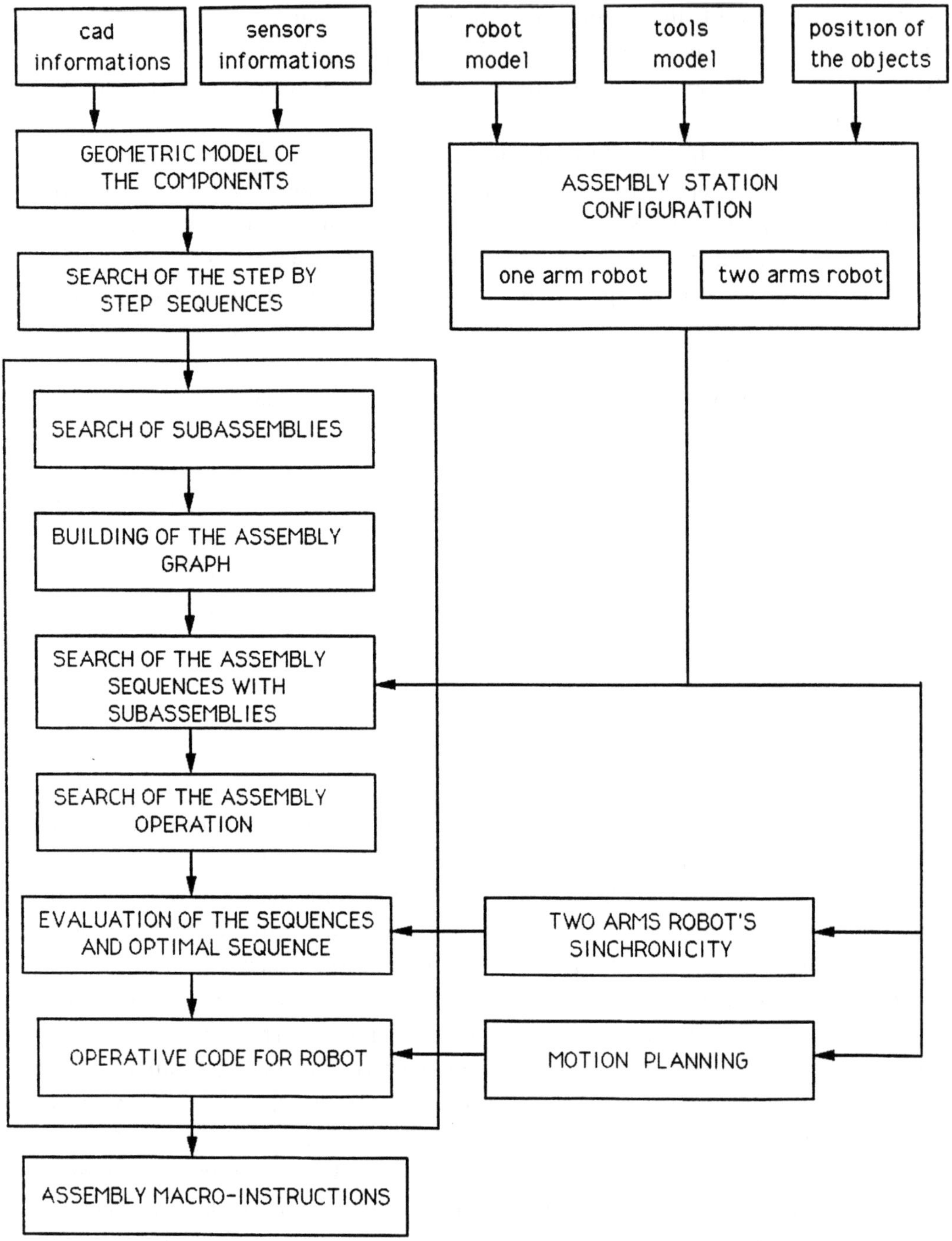

Figure 2: Logic assembly scheme.

kind which provide justifications for the adoption of a certain type of assembly by taking the usual alternatives into consideration, such as manual assembly, automatic assembly with specialized equipment (rigid automatic) and automatic assembly with programmable non-specific equipment (flexible automation).

Emphasis is very often placed on these cost analysis, just as if they were ends in themselves. In reality, they are to be inserted into a framework of business concerns which is wider, more articulated and more likely to engage the interest and involvement of a greater number of services. A single example should be sufficiently illustrative for all of these questions: the opportunity of simultaneously carrying out analyses of possible changes and of product variations to be introduced on the same assembly line. In reality, the choice of assembly procedures has a close and profound bearing on business objectives of the tactical and strategical kind as a whole. Throughout the planning stages of a product, maximum attention must therefore be given to possibilities for automated assembly. Taking the question another step further, we may consider the example of a conceived product already in the phase of planning for assembly with a certain type of technology, while determination of this choice is also guided by the outcome of economic analysis. Quick methods do exist for the approximative evaluation of the investment required by various assembly systems and for the analysis of the sensitivity of results to variations in the contributions of various components.

Clearly emerging from this question is the critical nature of planning and re-planning in the facilitation of the efficiency of the assembly process through careful and repeated application of Design for Assembly (DFA) rules. In practical terms, the product is to be considered in relation to all the components which contribute to the construction process: materials, technology, configuration, selection of assembly equip ment. To these we may also add possible product variants and customization. What we should in fact be thinking about is an approach which integrates the assembly process into the manufacturing process. An approach, in other words, which contemplates the analysis of the component in relation to all of the other technological and economic processes necessary for its realization. Very little research has been done along this line, and what little has been done has barely begun to be coordinated.

Parallelism Concepts for Assembly

In the realization of a product through the process of assembling its components, the choice of the sequence in which the individual parts must be assembled into the product is of fundamental importance.

In reference to the sequence in fact chosen, there are a number of different possibilities which cover the range between completely sequential or serialized solutions and those which are completely parallelized. In the sequential solution, the various operations necessary for assembly are executed in succession, while in the parallel solution these operations are carried out simultaneously.

The possibility also exists for the parallel execution of a certain number of operations with respect to the execution of others. This can be carried out in such a way as to obtain different options for the realization of products through the utilization of various margins of liberty , that is, the possibility of creating alternative solutions which come together with the parallelization of certain operations.

In this perspective, the generation of sequential operations can be seen as the exploration of the procedural itineraries which facilitate the passage from components to finished product by means of the realization, whenever possible or convenient, of certain operations in concomitance with others. In order to do this, it is necessary to understand the characteristics of the parts and the relationships which link them together in the finished product.

Among the various sequences which may generally befollowed for the realization of the assembly process, it will thus be necessary to select the one capable of optimizing the potential of a generic objective function, not necessarily expressed in terms of time or costs alone. Attempts at serious research along this line are naturally directed toward the automation of assembly sequence generation and toward the singling out of ones offering greatest promise with regard to the product to be manufactured. The approaches to the problem are all oriented toward rationalization, understood as the choice of the "best" sequence for the placing of the parts, with respect to a certain spatial reference in order to realize the product.

A good number of enquiries have also been conducted into the design and planning of parts involved in the assembly process in order to attempt the elimination, or at least the reduction , of a certain quantity of the resources, and therefore the costs, necessary for the manufacture of a given product.

The methodologies proposed by published research results and information on the generation of assembly sequences, starting with product analysis, offer analytical tools for planning and replanning the assembly process. Such tools can be of assistance to the planner, but certainly they are no substitute for such procedural planning. In order to reach the final objective it is in fact necessary that the planner intervene, in a decision-making capacity, throughout the application of the procedure.

General Considerations Concerning Robot-Automated Assembly

The planning phases of a robot-automated assembly station have been outlined in Figure 3.

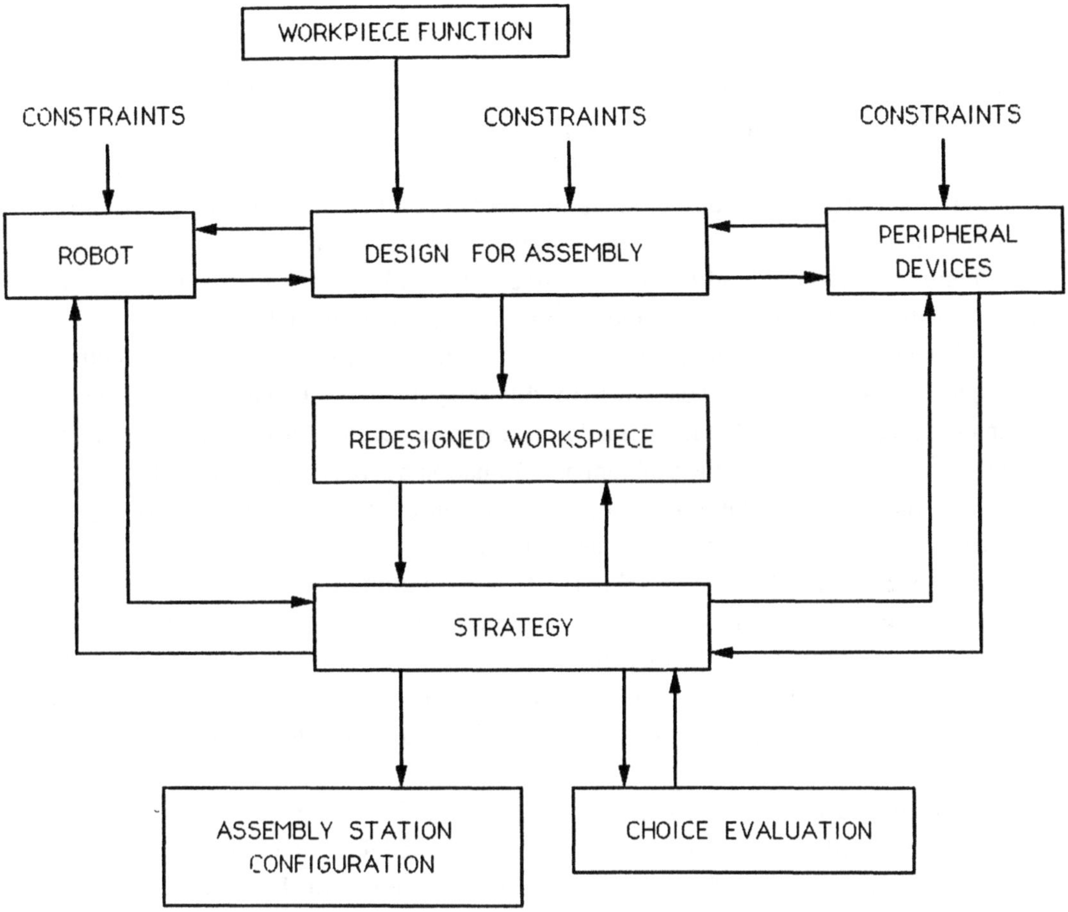

Figure 3: Configurations for assemblies.

The data necessary in order to make a start are:
- **the functionality of the part** or, in other words, all of the characteristics as a whole which ensure that it is likely to be able to carry out the task for which it has been designed. Such characteristics are, as such, to be considered to be indispensable.
- **the constraints**, of both a technical and an economic nature, can have a bearing on: **the part** (available means and production experience); **the robot** and the **peripheral sphere** (functionality necessary for putting the whole together).

Figure 4 shows the working area for assembly and Figure 5 presents CAD for assembly, with simulation of robots, tools, conveyors and components.

Given this basic input, the rules of design oriented toward assembly functions (Design for Assembly), show us how to modify the parts in a way which permits the attainment of versions which a more easily put together. On the basis of these final versions, it is possible to tackle the problem of automation by resorting to the strategy block in which may be found the experience necessary in order to succeed in defining the specific features and functions for the assembly block.

The process which generates the final configuration of the station will in reality follow a pattern which is not rigorously sequential, as the above description suggests, but rather retro-activated (following the itineraries indicated by arrows inFigure 3). In fact , in certain cases the effects of decision made downstream produce their impact even upstream with the imposition of actual modifications to factors previously deemed to have been established.

Evidence supporting this statement can be found by simply considering that the modifications made to the characteristics of parts by the DFA can modify the necessary peripheral sphere (in general, a part is rendered superfluous) or that a difference in assembly strategy (serial or parallel use of robots) can completely upset the implicit order of a station.

The strategy module will have effect by proposing a sequence of solutions in which certain parameters will vary from time to time, eventually to the point of discovering the best possible solution. In order to understand whether the modifications made are positive ones (decrease of the cycle's time requirements and cost parity), a specially-formulated evaluation form will be used which, if the description of the system and of the object to be assembled are accurate, should clearly reveal the attainable minimum cycle time.

Description of Input and Output Parameters

Input factors are used to provide the algorithm with an exhaustive description of the system , the parts , the assembly procedures and the pre-existent program. The last in particular provides the **step-by-step sequences of production of the different parts and the sub-assemblies.** The part-by-part sequences are all of the possible sequences

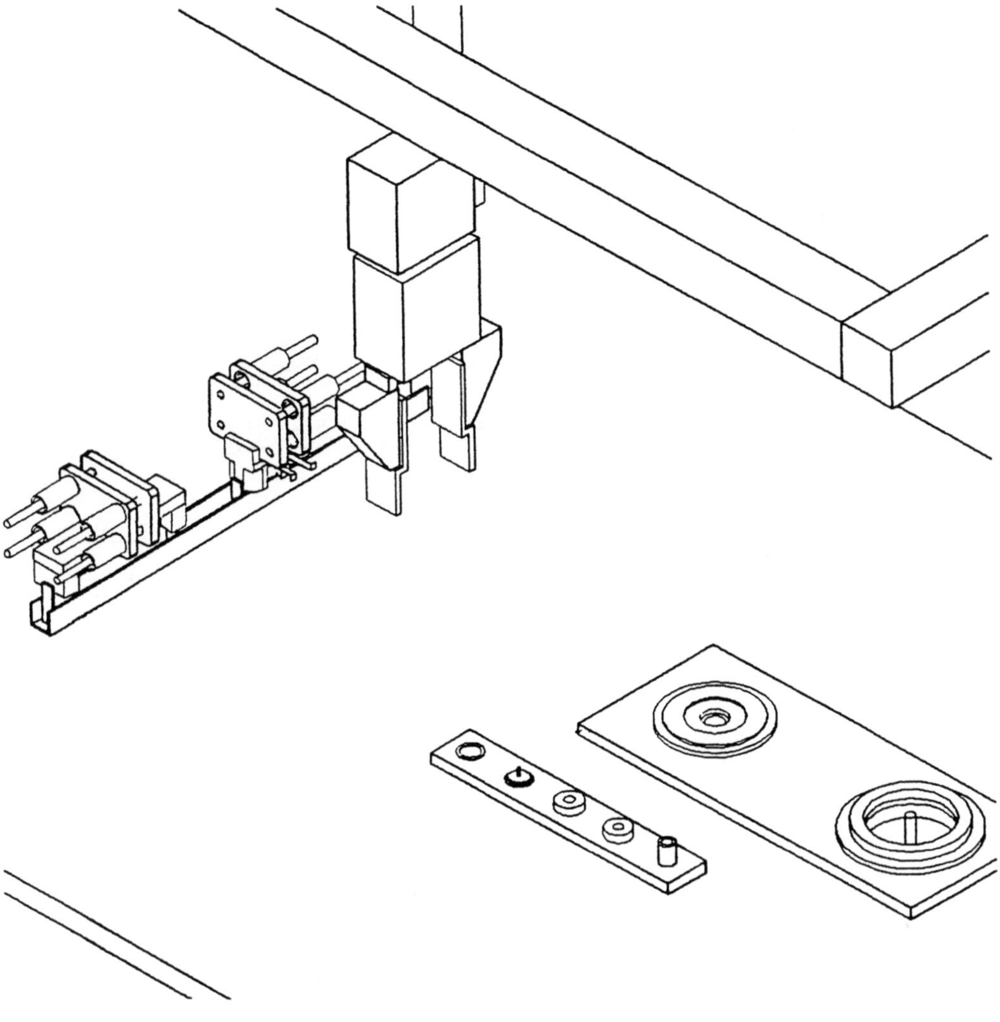

Figure 4: Working area for assembly.

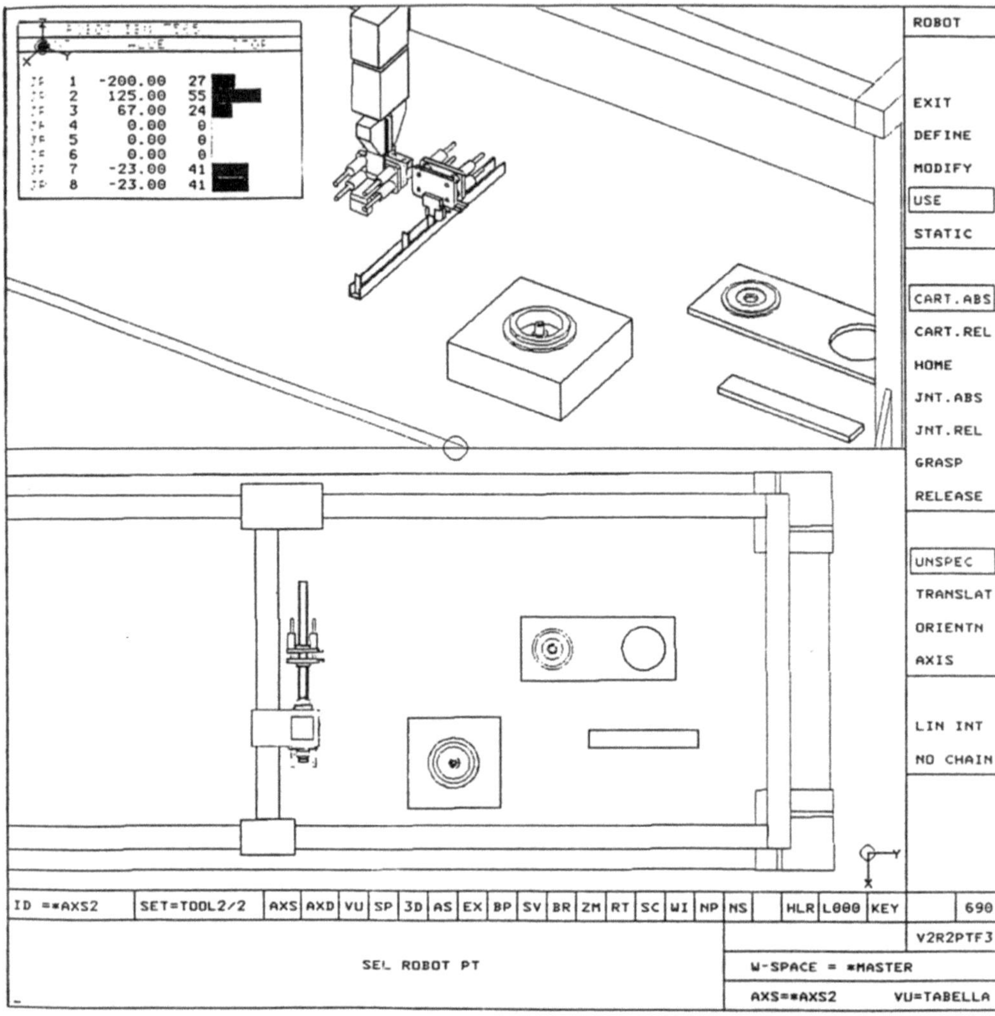

Figure 5: CAD for assembly - simulation of robots, tools, conveyors, and components.

for assembly which allow the finished piece to be achieved through assembly of a single piece at a time. The sub-assemblies are sets of two or more parts of the whole which can be assembled in a single moment and which therefore must be manipulated by the robot. With no further specifications, this means that they can in any case be roto-conveyed through the assembly space without danger of being dismantled. If they are subject to constraints , it must be underlined that they can be manipulated only if certain constraints are respected. An example of a constrained subsystem is a screw with a washer which can be moved only vertically by gripping it at the base.

Other data-bases instead concern: tools, the position and the capacities of which must be noted (where they are found and which parts can be moved); the components which are described by name , by their positions of fastening and assembly and by weight; **the robot** wich is identified by its dimensional features (which describe the work field) and kinematic characteristics (speed and acceleration); **the operations**, in which the execution times for the individual assembly operations are laid out.

Optimal Sequential Assembly Research

In order to represent the procedures of this research more clearly (outlined in the block diagram of Figure 6), Let us make reference to a practical example considering the assembly of a small electronic motor like the one used in the disk-run mass memories of computers.

As shown in Figure 3, it is composed of seven components: the **rotor** (RO), firmly attached to the shaft onto which the two **bearings** are inserted (B1, B2), (separated by a **spacer** (SP)) which support the **stator** (ST). A **screw** (SC) and an elastic **washer** (WA) complete the motor. It must be emphasized that the stator, for construction reasons, is to be inserted onto the rotor shaft before the second bearing.

The fi rst step is to determine all of the possible sub-assemblies, starting with the basic ones. A basic sub-set is to be understood as the list of components which make up such a group (ie. RO, B1). The program elaborates these sub-sets in order to sort out all of the logical permutations: all of the possible procedures with which it is possible to assemble the sub-group (ie. RO-B1 or B1-RO).

We subsequently move on to constructing the assembly graph, initially for the step-by-step sequences alone, and then for those with the subassemblies as well (Figures 7 and 8). This graph visualizes the sequence of assembly which indicates that reaching a junction means having assembled a component with which it is associated, after having executed all of the necessary operations visualized in their entirely throughout the transfer arc between the junctions.

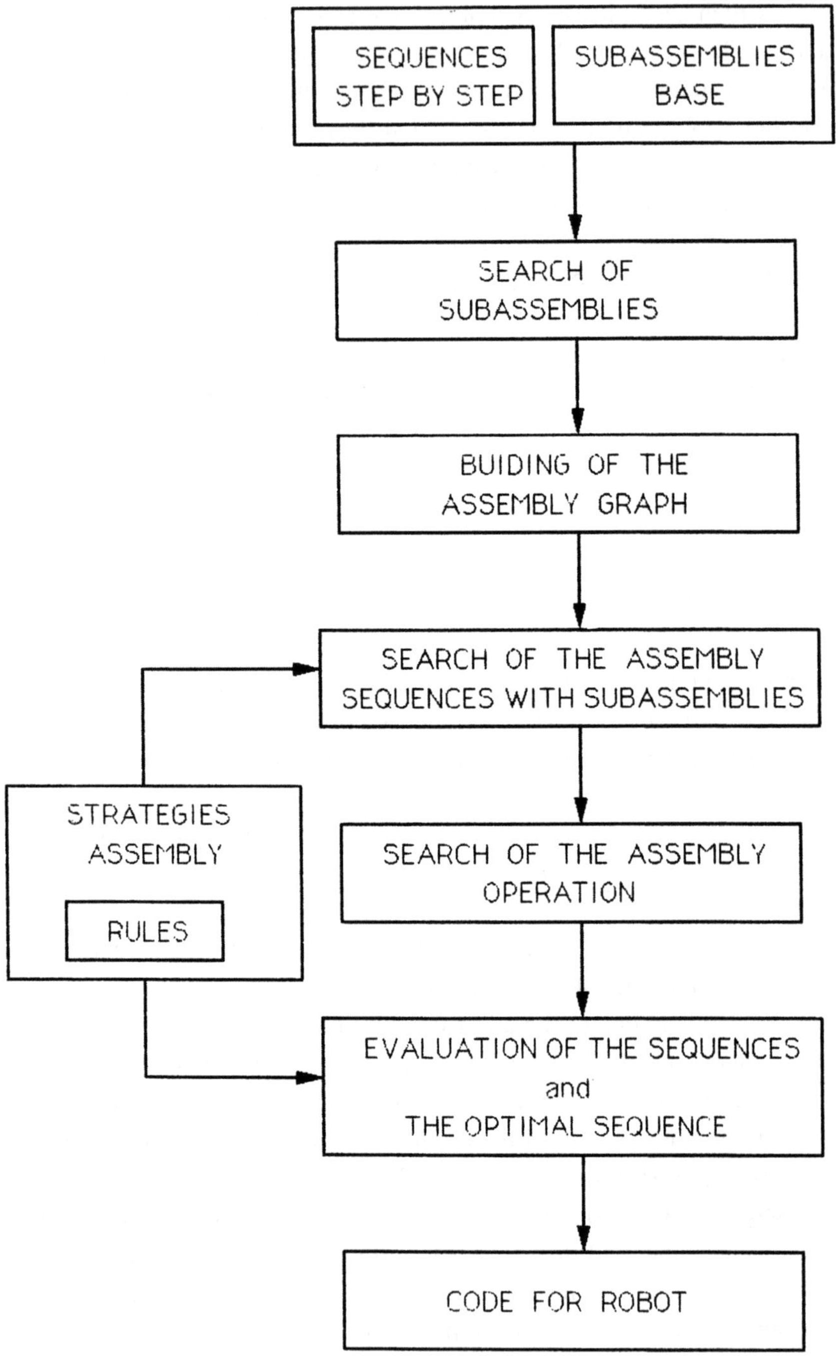

Figure 6: Search of the optimal sequence.

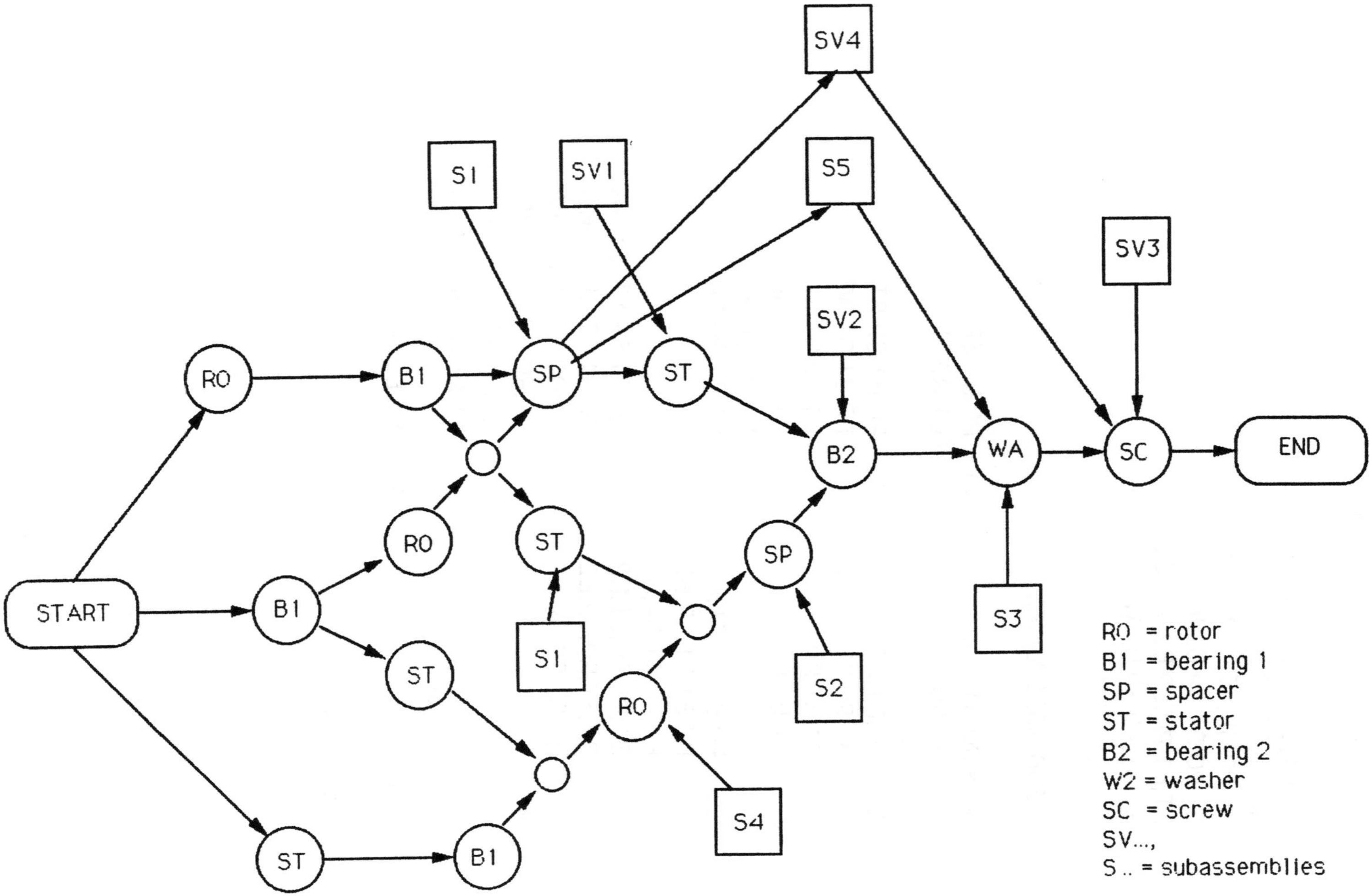

Figure 7: Possible sequences for assembly: scheme for the expert system.

- SEQUENCE STEP BY STEP

RO,B1,SP,ST,B2,WA,SC
RO,B1,ST,SP,B2,WA,SC
B1,RO,SP,ST,B2,WA,SC
B1,RO,ST,SP,B2,WA,SC
ST,B1,RO,SP,B2,WA,SC
B1,ST,RO,SP,B2,WA,SC

- SUB ASSEMBLIES

S1 = RO,B1
B1,RO

S2 = RO,B1,ST
B1,RO,ST
B1,ST,RO
ST,B1,RO

S3 = RO,B1,SP,ST,B2
RO,B1,ST,SP,B2
B1,RO,ST,SP,B2
B1,RO,SP,ST,B2
B1,ST,RO,SP,B2
ST,B1,RO,SP,B2

S4 = B1,ST
ST,B1

S5 = ST,B2

SV1 = RO,B1,SP
B1,RO,SP

SV2 = RO,B1,SP,ST
RO,B1,ST,SC
B1,RO,ST,SP
B1,ST,RO,SP
B1,RO,SP,ST
ST,B1,RO,SP

SV3 = RO,B1,SP,ST,B2,WA
RO,B1,ST,SP,B2,WA
B1,RO,ST,SP,B2,WA
B1,RO,SP,ST,B2,WA
B1,ST,RO,SP,B2,WA
ST,B1,RO,SP,B2,WA

SV4 = ST,B2,WA

Figure 8: Sequences step-by-step with subassemblies.

We then move on to the complete graph in which the sequences using the sub-sets appear, the junctions of which are represented by squares in order to emphasize the difference with respect to those of the single components, represented by circles.

If the sub-set is pre-assembled, it will be treated as if it were a component and therefore its positions must be noted, along with the procedures for contact and assembly, operation times involved, etc. Otherwise, it is assembled in the station itself. Assembly will then give rise to a problem not unlike that of the assembly of the whole and will therefore be undertaken with the same tools. However, once the best sequence has been revealed (by the criteria of less expended time), the subset will at that point be substituted by it and all others will be ignored. Using the complete graph as a basis, an array of rules , comprehensively known as "assembly strategies", will permit evaluation of the sequences so that the best one may be chosen. The analysis of the sequences can be undertaken in an exhaustive manner or according to a limited number of them which are considered to be particularly promising. The guidelines which allow the possible execution of this "pruning" process can be eliminated or modified by the user. For example, decisions may be made as to: the maximum complexity of the admissible sub-sets; whether these are to be considered as preassembled or that the whole must not exceed a certain weight.

Other guidelines must be sought from the cut complete graph, the graph of operations. Here the assembly of each whole part (piece or sub-set) is broken down into elementary operations of which the execution times are already known. Let's take a look, for example, at the path through the graph which, after "START", leads us to B1 after the junction RO. Breaking it down into single operations, we arrive at: - manipulation-tool deposit RO; - manipulation tool socket B1; - transfer of B1 into the field of operations, right up to the assembly zone; - assembly of B1.

A simple summation of operations will be sufficient in order to follow along a generic arc of the graph. By generalizing the procedure, it is possible to discover the time necessary for executing any itinerary which describes the complete assembly of the whole process.

Once the execution times of all patterns to have survived the "pruning" process have been calculated, we can choose, as a point of comparison, the best one. The sequence of elementary operations which describes the pattern, finally permit the establishment and notation of a code for a programme which enables the robot to carry out the assembly operations.

The operative code for the execution of an assembly sequence, when it is automatically generated, can be divided into two conceptually-distinct parts: a library of routines residing with the robot's controller, and a programme which organizes them in order to come up with the correct assembly sequence.

The libraries are sub-programmes, developed in the language of the robot's programming, which realize the elementary operations that are necessary to the assembly process (contact, insertion, and so on...) and these have been labelled as the "parametric assembly macro-instructions" . They are known as "macro-instructions" because the elementary operations which they realize are a sequence of primitive operations of the lowest level, such as shifting and sensor reading.

They are "parametric" because they are not devoted to carrying out a single elementary operation, but rather a class of operations. For example, routine contact-assembly will include parameters which identify the position at which contact must be made, the dimensions of the object to be gripped, the type of socket or contact point, and the sensors to be used for monitoring the operation. The choice of parameters is of fundamental importance since it decides the extent of the class of operations which can be carried out by a single macro-instruction. If few parameters are used, it will be necessary to write many macro-instructions, so that the possibities for their configuration will be limited, while use of many parameters will permit the writing of sub-programmes which are much more generalized. Our experience acquired in the preparation of macro-instructions has suggested that they should be linked with well-defined classes of operations, even if these should be limited. For example, the development of two different macro-instructions has been decided for the external plug or contact point and the interior one of the parts, even though it is possible, a priori, to use a single instruction in which a parameter indicates whether the contact is to be made by closing the pliers (external contact) or opening them (internal contact).

It must be emphasized that macro-instructions are very closely linked with both the tool utilized and the type of robot. For example, again with reference to the contact outlet, the macro-instruction which utilizes parallel grippers will be different from the one using a fulcrum device. By the same token, writing macro-instructions for a robot with four degrees of freedom will be different from writing them for a robot with six.

The automated drafting of the operative code has now become possible so that the elementary operations of assembly have a corresponding macro-instruction. In practical terms, once the desired sequence for the assembly of the final whole has been established, that is, the order in which the individual parts are to be assembled, such a sequence must first of all be developed in the correct sequence of elementary operations that can be put into correspondence with the macro-instructions. After that, the characteristic parameters of each single operation may be identified. This process of splitting up is carried out according to certain fundamental rules which are opportunely integrated, in specific cases, with information provided by the user. For example, when part "A" appears in the assembly sequence, it is supposed that this part must be taken up in a certain position in order to be assembled with another. Therefore, in greater detail, we must verify whether the tool mounted onto the arm at that moment is the suitable one. In the case of a negative answer, shifts and macro-instruction transfers will be undertaken in order to make the necessary change: shifting the robot to the point of supply for the tool mounted onto the arm, release of that tool, shifting toward the supply point for the tool desired and, finally, insertion. We then move on with a shift toward the contact point, selecting and taking up the part, then carrying it to the assembly area and, finally, assembling it. This is the standard cycle established for the assembly of a part and other operations. The cycle's reorientation, for example, can be inserted into the context of such a cycle.

The assembly program will therefore be made up of a sequence of reference points for the macro-instructions, with all of the correct values and parameters and possibly interspersed with lower-level instructions such as those which involve movement.

This program is saved in ASCII file so that it can be directly transmitted to the robot's controller by serial means, for example. The problem proves to be particularly simple. The robot uses a Personal Computer as a working tool, and all the software described above has been loaded on the computer.

Conclusions

The highly robotized assembly process, as developed in the research for an electric motor completely automated assembly, has demonstrated some basic assumptions.

The most significant ones are:

- the **design for assembly** is a necessary phase. The designer may change some elementary characters of the components of the workpiece according to the necessities of the velocity, reliability, simplicity of the grasping and joining process.

Design for assembly is an intellective process. The "new designer" must think like a project man, like a production technician, like an inventor to detect the simplest way to the best solution.The skill of industrial designer, with technological experience, is becoming an integrated capability, and his job evaluation is changing form, from a designing action to an integrated knowledge project.

- **CAD use**, which may describe and visualize the actual and possible positions of workpieces, tools, robot arms, etc (Figures 1, 4 and 5).

The project leader with the use of some CAD methodologies, may save the time of the experimental tests on the systems, because the simulation may indicate all geometric and functional aspects of the assembly sequence. The new designer must be able to use CAD, with adaptation of already existing packages, for the final design of the assembly system. The skill of the technician becomes an improvement of his creativity.

- **computer confidence,** which ensures the use of computers with different languages, structures, in networks, for the support in computation, in the process analysis, in the job evaluation, connected with computer science, engineering and only the integration of design for assembly, of computer aided design, of computer confidence permits expert system developments in robotized assembly at high level of efficiency.

Today industries with large productions or industries with very sophisticated, high quality products can afford to introduce design for assembly, CAD, expert systems, automation also with robots in the traditional process of assembly.

Acknowledgments

This project has been developed in cooperation with IBM Italia, Vimercate, in a research grant.

References

1. Andreasen M.M., Ahm T.: **The Relation Between Product Design, Production, Layout and Flexibility**. Proceedings of the 7th ICAA, Zurich, 1986.
2. Arpino F., Groppetti R.: **ASSYST - A Consultation System for the Integration of Product and Assembly System Design**. Proceedings of the 9th ICAA 1988, Seminar Notes Product Design for Assembly, London, 1988.
3. Behuniac J.A.: **Design for Assembly Automation - the Competitive edge**. Proceedings of the 5th ICAA 1984, Paris, 1984.
4. Boothroyd G., Poli C.: Murch L.E.: **Handbook of Feeding and Orientating Tecniques for Small Parts**. University of Massachusetts, Amherst, 1978.
5. Boothroyd G., Dewhurst P.: **Design For Assembly Handbook**. University of Massachusetts, Amherst, 1983.
6. Boothroyd, G., Dewhurst, P.: **Product Design for Assembly**. Wakefield, 1987
7. Buzacott, J.A.: **Modelling Flexible Manufacturing Systems, operations research models in FMS**. Springer Verlag, 1989.
8. Czajkiewicz Z.J.: **Justification of the Robots Applications, Flexible Manufacturing Systems: Methods and Studies**. North- Holland, Elsevier Science Publishers B.V. 1986.
9. Eversheim W. et al.: **Montagegerechtes Konstruiren mit CAD, IndustrieAnzeiger**. Vol.108 n.20, 1986.
10. Funk, J. L.: **Design for assembly of electrical products**, Manufacturing Review, n.1, march 1989.
11. Jakiela M.J., Papalambros P.Y., Ulsoy A.G.: **Programming Optimal Suggestions in the Design Concept Phase - Application to the Boothroyd Assemblies Charts**, Journal of Mechanism, Transmissions and Automation in Design, Vol.107, June 1985.

12. National Science Foundation: **Research and Development in Industry** 1987, Washington, 1989.
13. Ossipov, V., Rovetta, A., Zamponi, E.: **Design Work with the Support of an Expert System**, Int. Journal of Advanced Manufacturing Technology, n.3, 1988.
14. Rovetta, A.: **Logical Structures for Assembly with Robots**, Proceedings of IEEE Int. Conf. on Automation and Robotics 1986. San Francisco, April 1986.
15. Rovetta, A., Bucci, M., Serio, O., Hansen, V.: **Sensors Redundancy in Production Systems**, Proceedings of Workshop Nato on Robotics 1988, Salo', june 1988.
16. Rovetta, A., Sala, R., Sala, C., Bugo, D., Zecchini, F.: **Assembly with Single and Double Arm Robots by Structured Logic**, Proceedings of VIIth Int. Symp. on Theory and Practice of Robots and Manipulators, Udine, september 1988.
17. Takahashi K., Senba K.: **Design for automatic assembly**, Proceedings of the 7th ICAA, Zurich, 1986.
18. van Brussel H., De Winter D.: **Computer Aided Design of Flexible assembly systems**, CIRP Annals, Vol.34/1,1984.
19. Withney D.E., Nevins J.L., Graves S.C.: **Applying robots in Industrial assembly (Report)**, MIT Industrial Liaison Program, Cambridge 1984.

GROUP FINDINGS: VISION AND IMAGE ANALYSIS AS APPLIED TO INTELLIGENT MODULES

Ian Gibson
Electrical Engineering Department
Hull University
Hull, HU6 7RX
United Kingdom
Tel: (0482) 466-375 FAX: (0482 466666)

INTRODUCTION

There was a common interest in vision systems among the participants with varying levels of expertise ranging from the total inexperienced observer who wanted to learn, to those who had a number of years experience in applying machine vision technology. A full list of participants was not possible, but a partial list was made, comprising the following personnel:

W. Kropatsch (Chairman)	Y. Istefanopoulos
I. Gibson	M. Carlsohn
V. Cagliotti	J. Codina
P. Besslich	J. Frau
J. Raczkowsky	V. Georgopoulos
F. Hamano	Thu-Lan Mai

The overall impression of the meetings was of useful discussion with those attending both contributing and learning from others. A number of topics were discussed in detail, a brief description of the most notable of these follows:

1. A survey of available sensors was made with an attempt to classify them with respect to purpose. Selection criteria were discussed (speed/cost/space/purpose, etc.) along with ways of extending the effect of sensors by including a further dimension (e.g., 2-1/2D vision, 2D-3D inference, shape from shading, etc.). The main conclusion was that it was not possible to clearly differentiate vision systems from other non-contact and contact senses in terms of individual selection criteria.

2. Integration of sensors and actuators was discussed with an aim to producing working systems for industry. Bus structures were discussed, along with an expressed need for some hardware and software standards, along with a required vocabulary for implementing intelligent procedures. This was not anticipated as being realizable within the next five years. An open architecture for all sensor-based systems was considered a necessary requirement.

NATO ASI Series, Vol. F 71
Expert Systems and Robotics
Edited by T. Jordanides and B. Torby

3. Neural networks were discussed with an attempt to ascertain their usefulness in a vision based environment. The conclusion was that they were potentially useful but not directly connected to cameras or frame stores. A level of abstraction was needed to preprocess the data into a less data rich, but more symbolic form.

4. Some examples of existing methodology were presented. Particular problems discussed included moving vehicle tracking systems, and an expert system facility for constructing visual inspection systems.

5. A general conclusion on the development of visual systems as a whole was that there is no alternative to real data. No amount of simulation can account for all instances of operation. It was recommended that vision systems be subjected to a real environment at the soonest possible opportunity.

6. The concept of "intelligence" was discussed in detail with a conclusion that a vision system should have no more intelligence than a robot system. One controlling the other is a great temptation to system designers, but the level of intelligence required should be the domain of a third system which provides the knowledge/expertise in a hierarchical fashion and has control of the data interface between sensor and actuator system.

7. It is clear that vision systems have greater power than their present use. The problem lies in the interpretation of the symbolic data rather than the processing of raw data. This applies even to the less sophisticated, low cost vision systems. Distribution of the computing is essential, along with the incorporation of knowledge based on expert systems. There is still a grey area that remains to be discussed concerning the point at which the vision data becomes abstracted enough to be incorporated into the intelligent control modules.

Computer-Aided Process Planning in Robotic-Based CIM Systems

Lee E. Weiss
Carnegie Mellon University

This paper summarizes the discussions of a working group on computer-aided-process-planning (CAPP) in robotic-based CIM systems. The group consisted of scientists and engineers with diverse backgrounds and expertise in robotics and manufacturing. The discussion focused on defining terminology and outlining the problem scope in order to bring such a diverse group to a common level of understanding. The group participants included:

Lee Weiss	Carnegie Mellon University, USA
Eugenio Oliveira	University of Porto, Portugal
Spyros Tzafestas	NTU Athens, Greece
Fritz Prinz	Carnegie Mellon University, USA
Ali Riza Kaylan	Bogazici University, Turkey
Geleyn Meijer	University of Amsterdam, Netherlands
Asaf Akmehmet	Turkish Glass Works, Turkey
Yulun Wang	University of Calif., Santa Barbara
Urbano Nunes	University of Coimbra, Portugal
Alain Delchambre	U. L. Brussels Ecole Polytechnique, Belgium
Pedro Albertos	University Politecnica de Valencia, Spain
Alberto Rovetta	Politecnico di Milano, Italy

Manufacturing involves several activities including marketing, design, production, and product distribution. For a firm to be competitive, the manufacturing organizations which conduct these activities cannot operate independently. They must act concurrently and synergistically in order to *rapidly respond* to changing market demands[1]. Computer-Integrated-Manufacturing (CIM) (Figure I) is one means to integrate the manufacturing functions in order to achieve a rapid response capability. CIM refers to the use of computers to electronically store and transmit product information and to make the information available to all organizational functions in a *timely* fashion (Figure I). A current trend in manufacturing research involves the development of more *intelligent* CIM systems which also help to interpret and analyze the information, and to help make planning decisions based on this information. The group discussions focused on such planning capabilities in the context of robotic systems used within a CIM environment.

Figure I

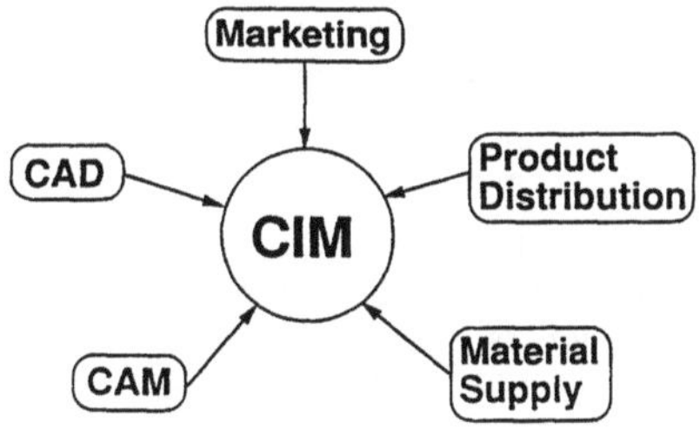

[1]Current world markets are shifting from economies driven by scale (i.e., mass-production of a limited product range) to economies of scope in which the consumer demands the introduction of diverse new product lines in relatively short times

Within the context of computer-aided-manufacturing (CAM)[2], robotic systems may be used to help build parts (i.e., net-shape manufacturing), to assemble them, and test the end-product. Robots were originally introduced into manufacturing plants because of their *promised* capabilities to be both flexible (i.e., the ability for a single cell to build a whole family of products) and to be robust (i.e., the ability to operate in partially structured environments, such as with minimal fixturing, poor tolerances, etc.). Unfortunately, early robotic systems failed to meet these challenges for several reasons:

1. They were not fast enough. Therefore, operational cycle times were too slow relative to dedicated or fixed automation.
2. They were note accurate enough. Therefore manufacturing operations were less reliable and prone to higher probabilities of failed operations.
3. The theoretical understanding of sensor-based robots (e.g. vision and force-based control required for robust operation) was limited and the hardware requirements for implementations were prohibitive.
4. There are several process planning issues involved in the development of a practical robotic system (Figure II). Process development for these often *complex* operations proceeded manually without the aid of computer-aided decision and support facilities. The time required for programmers/engineers to plan systems which would be both *efficient* and *reliable* was often prohibitively long and planning required numerous iterations to synthesize strategies which achieved the desired results.

   A fundamental problem which made the planning problem even more difficult was that products were not designed specifically for robotic manufacturing capabilities. With this lack constraint the search space of possible planning solutions were extremely large. In addition, it often became extremely difficult to find any solution which was satisfactory and ultimately numerous and time-consuming redesign iterations were required to achieve manufacturable designs.

It must also be noted that about fifteen years ago, when the industrial robot revolution started in earnest, there were not many people around who really had any practical experience in this field. Therefore, most manufacturing organizations did not have the manpower to successfully implement and utilize the capabilities of the available equipment. During the past decade there has been concentrated efforts to train people in the robotic arts and sciences and substantial progress has been made toward solving the technical problems, including:

1. Robot speeds are improving with the introduction of low-friction direct-drive arms, robust control algorithms, and faster computing hardware for implementing the algorithms.
2. Robot accuracies are improving with the introduction of direct-drive arms, new and improved calibration techniques, and direct end-effector sensing capabilities.
3. Sensor-based robots are currently being integrated into industrial production environments as a result of extensive research in visual and force sensors, the interpretation of sensory information, and the integration of sensory feedback with robot motion control. Practical implementations have become feasible with the introduction of the high-speed computing hardware required for sensor-based control.
4. Computer-Aided-Process-Planning (CAPP) is one mean which facilitates robotic planning, as well as for integrating the design stage with CAM. CAPP refers to:
   a. The use of computer simulations to help evaluate alternative plans,

[2]CAM systems may include material handling equipment, fixed or dedicated automation, as well as robotic and sensory systems.

b. The use of computer-aided decision systems to synthesize alternative plans, and

c. The use of CAD systems to critique evolving designs for their "manufacturability"

The group discussions on CAPP and are summarized below.

The computer-aided determination of robot control laws, at the lowest level of robotic planning, is a mature field. However, this capability had not found use in industrial environments for several reasons. First, many of the typical industrial applications (e.g., robot spraying or welding) did not require the level of dynamic performance obtainable with sophisticated control laws (e.g., adaptive and computed torque). Second, most industrial arms were not capable of such performance due to excessive frictions and motor limitations. And finally, the computation hardware required for this control was not available. The new generation of robots, including those with low-friction direct-drive joints (i.e., no gear reductions), with rare-earth motors, and in particular force controlled robots, require these sophisticated controllers. High-speed computing hardware is now available to implement the complex and typically computationally expensive controllers synthesized by computer-based systems.

CAPP applied to the higher levels of planning have been limited primarily to computer simulation studies to help evaluate alternative plans. Systems which automatically generate plans do not exist in *common* industrial practice, although a limited number of proprietary planning systems are currently being tested in industry. One problem is that the search spaces of possible solutions tend to be extremely large due to combinatorial explosion. This makes exhaustive search techniques impractical in many cases. Methodologies are required which can extract pertinent geometric information from design models to provide "clues" which help to constrain the search. Even if such form feature recognition capability exists then a second problem arises. For a given application the best plan, or even a satisfactory plan, is a function of subtle differences in both part geometries and desired performance. Therefore it is difficult to formulate plans based upon "generic" strategies. For example, it is difficult to synthesize a single "peg-in-the-hole" mating strategy which works well for all peg/hole geometries.

One methodology which will help to address these problems is to integrate design-for-manufacturability (DFM) information in the CAD system. DFM refers to creating product designs which can be efficiently and reliably manufactured with the available production equipment, and for which it is understand how to plan the required operations. The DFM concept has been in existence for a quite awhile and during the past decade design rules have evolved for robotic-based production equipment. In practice, manufacturers attempt to implement DFM by providing written rules for product designers. However, this manual approach is often difficult to enforce for both technical and organizational reasons. An alternative is to automate DFM, and in CIM one means toward accomplishing this goal is to develop CAD systems which can automatically critique evolving designs for manufacturability and provide immediate feedback to the designer. Automating DFM facilitates CAPP by constraining search spaces to product geometries/functions for which the system knows how to operate on to produce efficient and reliable manufacturing plans. There are two fundamental issues which must be addressed for incorporating DFM into the CAD system and then for generating process plans from the design models:

1. *Geometric form feature recognition* - requires appropriate geometric models from which

geometric form features can be automatically ex.racted[3].

2. *Form/Function reasoning capability* - whereby the the manufacturing methods required to create the particular form(s) feature(s) are automatically generated.

Recent advances have been made toward making these goals a practical reality. Geometric modeling environments are being developed which facilitate geometric reasoning operations[4]. The computational complexity associated with generalized feature extraction remains a challenging problem, however data reduction techniques appear promising[5]. Form-function reasoning is an extremely difficult problem and is hard to automate using purely inductive approaches. Practical approaches operate on constrained search spaces and apply "expert systems" to generate plans. In addition, the geometric models in the CAD system should be structured such that they may be used directly in process planning operations. Each stage of the robotic planning sequences must be able to extract the particular geometric information required to synthesize the plan. The non-manifold environments seems particularly well suited for such operations.

Several additional issues which must be addressed to create successful CAPP systems were enumerated. Unfortunately the group did not have enough time to discuss each in detail. These topics included:

- Automated generation of plans for error and exception handling
- Appropriate process models of equipment and sensors, and
- Adaptive and self-tuning systems which update world models during operation and alter or modify strategies and strategy parameters

In summary, CAPP is emerging as one methodology which will help fulfill the promise of flexible and robust robotic manufacturing capability. CAPP will help manufacturers to more quickly design efficient and reliable robotic systems for their CIM environments. There several research opportunities in this area for engineers and scientists to pursue to provide fundamental and significant contributions to robotics and manufacturing sciences.

---

[3]Feature-based design approaches, which provide prelabeled primitives to design with, are useful. With this approach feature extraction may not appear to be required. However, the arbitrary manipulation of feature primitives creates new and perhaps unexpected features. Therefore a robust system should be capable of feature recognition

[4]For example, non-manifold modeling systems which represent non-homogeneous entities (i.e., of different dimensions) within a single unifying environment (i.e., point sets, wire frames, surfaces, and solids) and permit Boolean operations between entities (i.e., non-regular operations) have been demonstrated as one approach which facilities the required reasoning and planning operations.

[5]For example 3D skeleton models, which are produced by thinning transformations, are easier to reason about.

Current Issues and Future Developments of Intelligent Robots

K. Khodabandehloo

Aim of the meeting: To discuss the above topic and to reach conclusions about the status of research, design, development and use of intelligent robots of the future.

Participants in the group

K. Khodabandehloo (Chairman)
E. Oliveira
B. Triggs
U. Nunes
H. Nguyen-Phu
S. Murphy
R. Blomme

Discussion

It was considered important for the group to discuss and reach a definition of an intelligent robot. Present day robots used in industry cannot be considered intelligent. A robot was recognised to have some of the following elements :-

- Manipulator arm
- Controller
- Sensors
- Task Planner
- Mobility subsystem
- End-effector(s)
- Electronic communication links
- Human interface
- Decision processing
- Programming language

Intelligence and Skill

A major issue raised remains to be the definition of "intelligence" in the context of robots. The group considered this issue and the features of a robotic system that make it intelligent included :

- learning - the ability to derive new knowledge from previous experience/knowledge and new evidence/information.

- interpretation - the ability to recognise and analyse patterns and concepts.

- sensory perception

- decision making capability

- ability to react to unexpected situations.

It was recognised that no system outside research laboratories could be said to have these features although the term "intelligent" is often used with reference to a number of assembly systems. The term "skill" better describes the characteristic of robots that have a specific function such as assembly. Given that a skilled system may be developed by programming using a high level language, it would be appropriate to refer to a skilled robot as possessing intelligence only if it has the capability to improve its skill without programming i.e. by self-programming or learning. In very broad terms, the following definitions were reached:

Intelligent Robots: require no programming in order to continue to perform a task of varying complexity.

Skilled Robots:can perform a programmed task that requires expertise. Where this expertise associated with the task is self-generated (within the robot system) the robot may be considered to be intelligent.

It was noted that a person having expertise in a particular job such as butchery is often considered to be skilled, i.e. the person possesses a quality in his or her job that relies on expertise. This person is also able to perform a number of other tasks, some of which may need to be learned over a long period of time. A robotic system performing butchery can be considered to be skilled and once it demonstrates the ability to increase its quality of skill by adding to its expertise without human support (or programming), it is behaving in an intelligent manner. To this end the definitions above were consistent.

Flexibility and Adaptability

In the manufacturing and other applications of robotic systems, the issues of flexibility and adaptability are closely linked with skill and intelligence. Flexibility refers primarily to the task procedures and its variation. Adaptability however also deals with product variability i.e. geometry, function, rigidity, etc. and the environment within which the task(s) have to be performed.

Environment

In any robot application the environment and its structure has an impact on the level of dexterity and intelligence required. Even a changing environment can be structured to avoid unpredictable situations. In general if a model can be used to describe the work environment of a robot (including the changes influenced by events both within and outside the robot environment), then the robot environment is considered to be structured. In circumstances where a model has not been established the environment should be assumed unstructured.

Autonomy

Autonomy was discussed with the view to establishing a robust definition. It was thought that the term is often used incorrectly an the degree of autonomy possessed by different systems is a measure of our understanding of this concept and thus variations in terminology can lead to problems. A measure of the level of autonomy is necessary in order to demonstrate an improvement when such is achieved. Given that a system has the capability and all relevant inputs and tools to achieve a specific function, but fails to do so, we could at least state that this system is not autonomous. Although the definition of autonomy was not achieved in absolute terms, the definition below may overcome the difficulty to establish if a system is autonomous or not :-

"A robotics system is not autonomous if it has all the necessary inputs/tools and decision capability to carry out a particular task, but fails to do so."

Current status and future issues

Many industrial tasks, welding, painting, assembly etc, have been automated using robotics. The systems currently employed have consumed considerable engineering skill and time in their development. To date no systems that could be said to be intelligent, skilled or autonomous are in use in industry. However, research is continuing to improve our understanding of such systems. In particular, the following issues are being considered by researchers and users of technology :-

- Robot task planning and understanding
- Technology integration
  - hardware
  - software
  - communications
  - man/machine interfaces
  - sensors: . force/touch
    . vision
  - end-effectors

- Safety and reliability
- Learning
- Modelling
  . task
  . product
  . world

- Ability to handle complex and non-rigid materials.

Conclusions

It was concluded by the group that the age of robotics is only beginning. The scientific, engineering and industrial community has a great deal to research and develop. The issues raised point the direction to the future. It has been recognised that no appropriate hardware or software base exists for research and development. An appropriate architecture for the intelligent robot can be defined based on a human model, but further development is needed to establish the hardware/software against this architecture. This will provide the base for research to continue in solving applications and to help the understanding of intelligent robots at a fundamental level. A model-based architecture was considered appropriate as a starting point with subsystems that co-operate and communicate with some inherent knowledge of each other's behaviour and function. Such subsystems may include: the robotic arms with its controller, sensor modules, end-effector modules etc. Software modules for such a system include packages for :-

- sensors
- manipulators
- world representation
- learning
- safety and behaviour
- process/object model(s)
- end-effectors
- task model
- etc.

Finally, extensive support is needed for researchers around the world and there are many uses that robots can be put to. A separate group has reported on the issue of applications. International co-operation is considered to be essential as the resources and the engineering base available in any one country cannot easily help us reach a "Universal" understanding of intelligent robotics. An established programme of international exchange should be encouraged and NATO ASI provides a good platform for this.

K. Khodabandehloo
Director -Advanced Manufacturing and Automation Research Centre
University of Bristol

<u>ASI UNUSUAL ROBOT APPLICATIONS</u>

K. Khodabandehloo
University of Bristol

On the evening of Tuesday, 24th July an informal meeting to discuss Unusual, Creative and Fun Applications of Robots was held. The following were present:

Koorosh Khodabandehloo	Bristol University
Matthias Carlsohn	DST Deutsche System-Technik
Ger Honderd	Delft University of Technology
Mark Baker	Surrey University
Bill Triggs	Oxford University
Steve Murphy	RPI
Ian Gibson	Hull University
Walter Kropatsch and wife	Technical University of Vienna
Thomas Hubner	University of Karlsruhe
Ronny Blomme	State University of Ghent
Fumio Hamano	California State University
Alan Delchambre	CRIF/WTCM, Belgium

The group discussed the above topic in order to produce a list of applications for future robots. The areas mentioned include:

Ironing
Food serving
Lawn mower robot
Cleaning the swimming pool
Plantation care and treatment
Massages
Preparing food (not cooking)
Golfer (that can look like you!)
Car cleaner (insider and outside)
Decorator (painting,paperhanging)
Robot nurse
Road construction
Window cleaner
Driverless vehicle
Intelligent car
Hairdressing
Magician /entertainer
Comparison and decision support system
Driving
Sailing trainer for people
Traffic policemen
Musical instrument player
Cooking
Mending clothes
Gambler
Robot answering machine
Loyal robot!
Chairperson
Cook that prepares a meal from ingredients available
Animal trainer
Artistic painter
Robot (human) duplicate
Super secretary
Administrator
Goal keeper in football
Surgeon
Road sweeper
Fruit picking and grading
Porter (robot luggage ?)
Builder
Boxer
Gardener
Shoe cleaner
Bathroom cleaner
Fireman
Plumber
Butcher
Tennis player
Garment production(inc.ladies' underwear)
American football player
Shopping
Dog walker + cleaning path
Robot to help avoid sunburn
[Good looking robot]
Teacher
Clearing the house
Shaver
Car and machine repairman
Receptionist
Robot to give you a bath
PhD writer
Friendly robot
Shop keeper
Medical specialist
Robot to fold clothes
Laboratory technician
Postman
To tidy up rooms

Note keeper
Electrician
Doctor
Politician
Military robot
Airline crew
Intelligent rubbish collector
(leaves the useful things behind)

Building repairer
Mining
Bodyguard
Perfume designer
Woodcarver
Pet robot
Pollution level controller

K. Khodabandehloo
Lecturer in Manufacturing
Head of Robotics and Manufacturing Systems Research
Director of Advanced Manufacturing and Automation Research Centre

Industrial Needs and Applications

K. Khodabandehloo, University of Bristol

Aim: The aim of this group was to discuss the needs of industry with regards to the Undergraduate teaching of Expert Systems and Robotics. This complemented the discussions of the group chaired by Professor Antonios Grammatikos, University of Patras, which considered the Engineering Curriculum within the University sector.

Present: K. Khodabandehloo - Discussion Co-ordinator
K. Warwick
B. Dimitradios
A. Chassiakos
C. Chiladakis
B. Triggs
R. Blomme
M. Christodoulou
C. Vibet
G. Honderd

Industries

The type of industries that would benefit from the teaching of Expert Systems and Robotics were listed with focus only on the Undergraduate education. The range of industries mentioned included:

. Food	. Mining	. Chemical
. Metallurgical	. Nuclear	. Shipbuilding
. Information Technology	. Construction	. Electronics
. Financial/Banking	. Military	. Process
. Automation Industry	. Postal	. Tourism
. Space	. Textile/Clothing	. Transport
. Power Generation	. Agriculture	. Underwater
. Medical	. Rubber	

What is expected of a fresh graduate in industry ?

In order to become clear about the needs of industry it was suggested that the group considers the industrial expectations from a graduate. the following is a short-list of the main points of discussion.

- Industry requires graduates with a broad engineering background and particularly those who are competent at problem solving.

- A graduate is not expected to solve complex industrial problems soon after graduation. However, his/her education should provide a good base for experience to be gained quickly so that they can contribute within a short time period.

- An undergraduate programme should have a broad base with Expert Systems and Robotics included in the curriculum, rather like the Manufacturing Systems Engineering Degree at Bristol University (UK). It was agreed by all those present that industry does not wish for graduates with expertise only in Expert Systems or Robotics. Specialisation in the subject could be better accommodated at the postgraduate level.

A graduate should have covered as part of the curriculum the necessary theory, methodologies, application awareness, leading to the fundamental understanding of Engineering as a subject which should include Robotics and Expert Systems.

What should be taught ?

Among the many subjects mentioned, the teaching of Robotics and Expert Systems was only part of the wider requirements. This further emphasised the need for a broad undergraduate curriculum. The course titles mentioned include:-

. Robot System Architecture
. Application of different configuration robots
. End-effectors/tools
. Control and dynamics of robots
. Sensors (vision)
. Actuators
. Hardware for robots (medical/electronic)
. Software - including programming
. Systems Design and Simulation
. Performance Evaluation
. Instrumentation
. Static and Dynamics
. Knowledge extraction
. Computer Architecture
. Neutral networks
. Management (value of technology)
. Communication
. Computer Integrated Manufacturing
. Operational Research
. Safety, Reliability & Maintenance
. Operations Management
. Economics/Finance
. Presentation Skills
. Man/Machine interface
. Human systems
. Application of expert systems

How should courses be taught ?

Motivating undergraduates was considered to be a major issue. This was to a large extent related to the way subjects are taught. It was considered essential that both theory and practice should be covered and interactive teaching with the use of examples and case studies should be encouraged even more. The use of simulation software and educational hardware for teaching robotics was raised. It was highlighted that software packages are often too expensive and educational robots do not provide the right control algorithm for teaching purposes. The advantage of having such facilities for teaching was directly linked with their use to motivate students. It was also indicated that early in the course students should be exposed to technology and its use (through visits to companies, use of videos, laboratory demonstrations etc.) to keep them interested and motivated until such time as they reach the final years of their degree to cover the relevant theory and practice.

Conclusions

The main conclusions of the discussion group were

. Undergraduate degree courses must be broad, but expert systems and robotics should be included in the final year(s).

. Industry requires graduates that have a good fundamental understanding of engineering both in theory and practice. Emphasis is on problem solving ability rather than specialisation in a particular discipline such as expert systems.

. Specialisation in Expert Systems and Robotics should be achieved at postgraduate level.

. Emphasis of an undergraduate degree should be on fundamentals covering both theoretical and practical aspects of engineering subjects.

. Motivation of students was considered to be key and teaching methods should expose students to interesting concepts and ideas. Case studies, examples, hands-on practical work and exposure to the technology through the use of videos and visits early in the undergraduate degree programme form an essential part of the curriculum.

The subject of resourcing was raised and it was considered that teaching of Engineering as a whole is expensive to support at University level and Government funding has always been inadequate. It was time for politicians to recognise the needs of Engineering Education and for it to be supported adequately. Engineering forms the only base for wealth creation and subsequently improvement in our standard of living.

K. Khodabandehloo
Lecturer in Manufacturing
Head of Robotics and Manufacturing Systems Research
Director of Advanced Manufacturing and Automation Research Centre

Integrating Diverse Knowledge: New Ideas in Knowledge Processing

ASI Discussion Group on Advanced Knowledge Processing
Bill Triggs[1] (Chairman), H. L. Akin[2], A. Akmehmet[2], G. Honderd[3], K.Khodabandehloo[4], W. Kropatsch[5], and E. Oliveira[6]

[1]Oxford University Computing Laboratory, 11 Keble Rd, Oxford OX1 3QD, England
[2]Computer Engineering Dept, Bogazici University, Istanbul, Turkey
[3]Faculty of Electrical Engineering, Delft University of Technology, Delft, Holland
[4]Dept of Mechanical Engineering, University of Bristol, England
[5]Institut für Technische Informatik, Technische Universität Wien, Vienna, Austria
[6]University of Porto, Portugal

ABSTRACT

There was considerable interest at the ASI in Knowledge Processing Systems (KPS) for intelligent robotics and integrated manufacturing, so an informal discussion group was formed to consider some of the issues raised by current attempts to produce such systems. The group was small but most of its members were actively involved in KPS applications and the lively discussion was continued after the meeting by correspondence.

In order to achieve its goals, an intelligent robot or manufacturing system must *combine large quantities of diverse knowledge* (ie *structured* factual and procedural information) into a useable whole. The key problem in these domains is thus the **integration of diverse knowledge**. The group considered several interconnected topics in relation to this:

- System architecture and control strategies.
- Knowledge representation and accessing.
- Knowledge application strategies.

1 System Architecture & Control Strategies

There was no consensus within the group on the KPS architecture most suitable for the applications considered. The choice was felt to be domain-dependent and it may be that architectures suitable for such complex domains will combine the best features of several systems. Architectures advocated included:

- A central overseer and communications supervisor distributing tasks to worker processes.
- A 'cooperative network' or community of cooperating processes without central supervision.
- A hierarchy of domain experts, with generalists at the top and specialists at the bottom.
- A blackboard system, with competitive processes bidding to be the next to add information to a common database.

- Architectures modelled on human organisations such as manufacturing companies, or on social insects such as ants or bees.

However, we did agree that *some* sort of organisation was essential, even if it was implicit as in the cooperative network model. It was stressed that the key problems of knowledge representation and application appear to be domain-specific and are not really tackled by general architectures such as blackboard systems, which merely use the solutions when they are found.

We felt that flexibility and extensibility were essential properties of such broadly-based systems and should be designed into the architecture and the knowledge representations and applications from the outset. This would be aided by the modularisation of the architecture and knowledge representations although the complex, highly interconnected nature of knowledge in these domains might make this difficult.

Ultimately, support for distribution and parallelism was felt to be necessary for speed and reliability, but again the interconnectedness of the domain knowledge seems likely to make this difficult.

2 Knowledge Representation & Accessing

Again it was stressed that the *integration* of knowledge is the key problem: it is not sufficient simply to append information to the knowledge base, the common structures of and relationships between data elements must be identified and represented so that they can be exploited.

Some of the most important types of relationship were listed:

- Relationships derived from prior knowledge, experience or theoretical models.
- Relationships representing application or functionality.
- Current context, situation or event.
- Causal and logical dependencies.
- Relationships between levels of a hierarchy, such as elaboration or summarisation.
- Relationships representing instance or generalisation.
- Relationships connecting observations to domain models.

We agreed that successful knowledge representations were likely to be based on *models* of objects or processes in the problem domain, that multilevel or hierarchical models would frequently be required, and that modular (eg object oriented) knowledge representations were desirable, provided the all-important relationships between knowledge elements could be maintained.

It was suggested that ease of human understanding of the data representation was particularly significant for advanced knowledge processing systems, since these must make intrinsically complex information comprehensible to humans at the design, application and diagnostic levels. Automatic translation of knowledge to lower level representations might therefore be required.

The large amount of information to be handled suggested the use of a 'knowledge database system' for efficient knowledge access. Classically this would also isolate knowledge applications from the chosen knowledge representation: many of us felt that such a separation would not be feasible in the complex domains being considered, although some argued that it might be possible with a very flexible purpose-designed relational database, perhaps incorporating domain-specific features such as spatial reasoning.

The concept of a 'cooperative database' acting as 'librarian to a community or network of cooperative processes' was suggested. Such a database would clearly have to know something about the applications it served as well as the knowledge it stored.

The ability to merge several knowledge bases into a single integrated system with a wider domain of expertise would be extremely useful, particularly if this process was simple enough to allow the use of standard off-the-shelf 'plug-in knowledge modules'. Although current knowledge processing systems support limited forms of incremental expansion (eg for system development), it is very difficult to merge such systems because:

- The knowledge representations are usually incompatible.
- Duplicated knowledge must be recognised and reconciled.
- When knowledge from several sources is combined important new relationships often arise which must be identified and represented.

It was suggested that it might be possible to develop 'knowledge representation experts' for the (semi-) automatic generation of knowledge representations (probably within restricted problem domains). These would expedite the creation, extension, merging and reconfiguration of knowledge bases by performing some sort of 'knowledge compilation' ('learning') process to link new knowledge into the existing knowledge base.

3 Knowledge Application Strategies

We agreed that it was essential for a KPS in these domains to support a variety of inference strategies, for example:

- data driven/forward chaining
- goal driven/backward chaining
- event driven/opportunistic
- plan driven/top-down or hierarchical,

and that effective schemes for combining these strategies would be extremely valuable.

Efficient problem solving probably requires some sort of 'understanding' of the implications or potentialities of solution states: current systems tend to do large amounts of unnecessary work because they are very poor at this and take only a 'short-sighted' (highly local) view of the solution state. It was not clear how such an overview could be provided: the whole area of the representation and application of *implicit* knowledge seems to be very poorly understood. However we felt that some progress could be made by exploiting the relationships between knowledge elements and hierarchical problem decompositions. At present any method for producing solution overviews seems likely to be specific to the problem domain under consideration.

Several types of inference system were discussed: which was preferred clearly depended on the application and no available system was felt to be suitable for the broadly-based domains considered. It therefore seems that an effective KPS should accommodate a variety of types of inference (both certain and uncertain) and hence a variety of types of uncertainty. Specific criticisms of the various systems were:

- Rule based systems are inflexible and poor at handling numerical data such as sensor or geometrical information.
- Propagation through fuzzy systems often leads to inappropriate degrees of fuzzyness and support.
- Formal probabilistic methods (Bayesian, Dempster-Shafer) are computationally too complex for general purpose use.

- All of the numerically-based theories contain inconveniently many parameters.
- None of the above systems are very effective for reasoning about time constraints.

There was general agreement that for the softer types of reasoning required by diversely based knowledge processing, more qualitative (ie symbolic or non-numerical) measures of support would be appropriate, but it was not clear how these could be provided.

It was suggested that it might be useful to think of knowledge applications as 'operators performing knowledge transformations on a solution-state-space' (for example logical deductions would be simple examples of such operators), and that the mathematical analogy could be carried further by allowing 'functorial operators' which transformed both knowledge elements and the relationships between them. For example, finding that an object is a member of some highly structured class might entail a considerable reorganisation and extension of knowledge about the object, and it seems desirable to treat this reorganisation as a single knowledge transformation.

Summary

Several key points emerge from the above discussion:

- Current KPS architectures and knowledge representation and application strategies do not seem to be flexible enough to support the rich variety of knowledge required for intelligent robotics or integrated manufacturing.
- In particular, current techniques for representing and applying the *relationships* between knowledge elements and *implicit* knowledge are very primitive.
- The diversity of knowledge required in these domains suggests a diversity of knowledge representation and application strategies: there was considerable scepticism that a general purpose 'knowledge processing expert' would be possible or useful, and in applications the emphasis is clearly on *ad hoc* (domain-specific) techniques.
- Much remains to be done on the integration of disparate knowledge sources and applications into single systems.
- Although the interconnectedness of knowledge makes decomposition difficult, the group members consistently advocated an *analytic* approach: holistic methods such as neural nets were felt to be untrustworthy except in strictly circumscribed domains.

In summary, it seems that a considerable amount of work on knowledge integration will be required if we are to represent the richness and diversity of our knowledge of the real world in intelligent robotic or manufacturing systems.

ACKNOWLEDGEMENTS

The members of the discussion group would like to thank NATO and the ASI organisers for the opportunity to attend the ASI meeting, and the chairman would like to thank the other group members for their active participation (they could have been out dancing!).

Intelligent Control

Fumio Hamano

Department of Electrical Engineering, California State University, Long Beach, 1250 Bellflower Boulevard, Long Beach, California 90840-8303, USA

Abstract: This article summarizes the presentations and discussions from the group meeting on Intelligent Control which was held as a part of the NATO Advanced Study Institute on Expert Systems and Robotics, Corfu, Greece, July 1990. The meeting consisted of the short presentations given by Professors G. N. Saridis, S. Tzafestas and G. Honderd respectively on the definition and limitations of Intelligent Control, need for clear definition of intelligent control and need of Intelligent Control from the industry point of view, followed by the discussions by all the participants.

Keywords: Intelligent Control, Intelligent Machine, Entropy

1. Introduction

In this meeting, short presentations were first given by Professors G. N. Saridis, S. Tzafestas and G. Honderd respectively on the definition and limitations of Intelligent Control, need for clear definition of intelligent control and need of Intelligent Control from the industry point of view. Then followed the discussions by all the participants. The summaries of the presentations are given in Sections 2 - 4 in the chronological order of presentation followed by the conclusions from the discussions.

2. Definition and Limitations of Intelligent Control

The presentation by Prof. Saridis is summarized as follows: The term "Intelligent Control" was first introduced in 1971 by K. S. Fu as a method beyond Adaptive and Learning Controls, which is an intersection of Artificial Intelligence and Automatic Control. The notion was later extended by Saridis and others as a discipline which combines Artificial Intelligence, Operations Research and Automatic Control so that it provides intelligence, scheduling and execution capabilities, i.e., so that the system behaves like an organization. Intelligent Machines are the machines that perform anthropomorphic tasks, autonomously or interactively with a human operator in structured or unstructured, familiar or unfamiliar environments, and Intelligent Control defines an architecture and methodology to make intelligent machines to achieve their goals autonomously. As is commonly found in organizational structures, the Intelligent Control System should satisfy the Principle of Increasing Precision with Decreasing Intelligence, which in particular leads to a hierarchical structure consisting of three levels of intelligence, i.e., organization, coordination and execution levels. The system should operate according to a criterion or performance measure. Saridis and Valavanis [2] propose to use "entropy" [1] as the performance measure since it has common mathematical properties for the different levels of the hierarchy, it has a proper physical interpretation for each level of the hierarchy, and the entropies of different levels may be added for the total cost of a complete action. Thus the design of Intelligent Control System can be regarded as a mathematical (probabilistic) problem of finding a right sequence of decisions and controls of a system structured according to Principle of Increasing Precision and Decreasing Intelligence such that it minimizes its total entropy.

3. Need of Clear Definition

Prof. Tzafestas discussed the need for clear definition of Intelligent Control. The following is the summary of his presentation: Intelligent Control is more than a combination of the above three fields. He feels that more development is needed in the future and that we must make it clear what we wish to achieve by the development of Intelligent Control. He further indicated that in order for the Intelligent Control System to be useful it must result from close interactions between the users and designers of the system and that it is an integration of what are traditionally considered off-line and on-line procedures. As such it must have the capabilities of cognition, perception, planning and control. He also thinks that it is necessary to standardize the models so that the methods can be studied and used effectively for a large number of applications.

4. Industry Viewpoints

The presentation by Prof. Honderd was on the industry view points which is summarized below. In 1960 the System Theoretic approaches based on the notion of the state to the control problems became popular together with Optimization techniques and Optimal Control. As the cost of computers and computation rapidly became less expensive, the control methods such as Adaptive and Robust Controls and use of nonlinear feedback, etc. that require much real time computation have become practical. Knowledge-based systems and fuzzy set approaches have also become of interest for industrial applications. In the past, as Prof. Honderd pointed out, too much emphasis was placed on the modeling aspects of the systems such as identification, realization, model reduction, etc.. However, the industry needs more understanding and concrete methodology to deal with safety and reliability, and the standardization is also important in the industrial environments. With regard to reliability, in particular, statistical and probabilistic approaches are not enough in many applications since the results do not guarantee a success, i.e., the probability of failure is not zero. Alternative approaches such as robust control based on deterministic formulation are better suited in the above circumstances. It is also important to study nonlinear systems without relying on linearization since linear analysis fails when "essential nonlinearities" are present in the systems. There are a variety of applications of control engineering in the industry, e.g., avionics, ship navigation, decision support systems, market predictions, etc.. Therefore, control engineers must expand their horizon of applications and methodologies.

5. Conclusions from Discussions

The discussion was conducted on the materials presented as well as the related issues. Based on the debate the following conclusions may be drawn. Intelligent Machines will eventually take over the market where needed. The question was raised in the meeting as to whether Intelligent Machines should have creativity. At this moment, seeking human-like creativity in machines is unnecessary in order to gain benefits from the machines, and it is technologically premature to build such machines. We should rather seek to design machines which have a learning (preferably, fast learning) capability of the environment to be able to cope with unexpected situations. To establish the methodologies for Intelligent Control it is important to integrate different disciplines, particularly System Theory and Computer Science/AI. System Theorists may contribute to discovering mathematically tractable structures and formulations, and Computer Scientists may contribute to expanding the structures as they did in the past. Other fields such as Operations Research, Artificial Neural Networks and Cognitive Science, etc. may also be integrated to determine proper structures of the systems. We recognize that one notable formulation is given by the first speaker and significant activities have been going on in that direction. However, at present, the area is still immature for standardization, and further work is necessary. Reliability is another important issue, and it should have as much focus as stability for traditional control systems. Significant attention should also be paid to social effects due to the development and use of Intelligent Control Systems. Finally, it should be noted that there are applications such as space applications, where Intelligent Control is absolutely necessary since a human operator is not likely to be present on the spot or the machine needs to be remotely operated (or "tele-operated") with a significant communication time delay. Factories and transportation systems of the future may be in similar environments.

References

1. Jaynes, E.T.: Information theory and statistical mechanics. Physical Review, vol. 106, No. 4, pp. 620-630, 1957
2. Saridis, G. N. and Kimon P. Valavanis.: Analytical design of intelligent machines. Automatica, vol. 24, No. 2, pp. 123-133, 1988

LECTURERS

Prof. Gianni Casale (Conference Co-Director)
6 Rue de la Sucrerie
B-7750 Warcoing, Belgium

Prof. Fumio Hamano
Electrical Engineering Department
California State Univ., Long Beach
Long Beach, CA 90840 U.S.A.

Prof. Y. Istefanopoulos
Department of Electrical Engineering
Bogazici University
80815 Bebek, Istanbul, Turkey

Prof. Koorosh Khodabandehloo
Department of Mechanical Engineering
University of Bristol
Faculty of Engineering
Queens Building, University Walk
Bristol BS8 1TR U.K.

Prof. Hao Nguyen-Phu
Cran, Eia-Esstin
Nancy I University
France

Prof. F. Prinz
Director,
Engineering Design Research Center
Carnegie Mellon University
Hamburg Hall
Pittsburgh, PA 15213 U.S.A.

Prof. A. Rovetta
Politecnico di Milano
Piazza Leonardo da Vinci n. 32
20133 Milano, Italy

Prof. G. N. Saridis
Director,
Center for Intelligent Robotic Systems
for Space Exploration
Rensselaer Polytechnic Institute
Center for Industrial Innovation, Suite 8015
110 Eighth Street
Troy, New York 12180-3590 U.S.A.

Prof. Spyros Tzafestas (Conference Co-Director)
Elefteriou Venizelou 91
Holargos, Athens
Greece 15561

NATO COORDINATOR

Prof.ir. G. Honderd
Technological University Delft
P. O. Box 5031
2600 GA Delft
The Netherlands
Phone: 015/78 62 65

FAX NO. +31-15 62 67 38

CONFERENCE DIRECTOR

Prof. Tim Jordanides
Department of Electrical Engrg.
Calif. State Univ., Long Beach
1250 Bellflower Boulevard
Long Beach, CA 90840-8303 U.S.A.
Phone: 213/985-5150 or 5102
714/840-7326 / Home

FAX NO. 213/985-7561

PARTICIPANTS

Akin, Dr. H. Levent
Bogazici University
Computer Engineering Dept.
80815 Bebek - Istanbul
Turkey

Akkurt, Mustafa
Istanbul Technical Univ.
Istanbul, Turkey

Akmehmet, Asaf
Fenerbahce, Tunaman Sok. No:9/8
Kadikoy, Istanbul 81030, Turkey

Albertos, Prof. P.
Grupo de Automatica e Informatica Industrial (DISCA)
Universidad Politec. de Valencia
Apdo 22012, E-46071
Valencia, Spain

Andreakos, P.
Electrical Engineer
Algosystems, S.A.
4, Sardeon Street
Athens, 171 21 Greece

Armada, Dr. Manuel
Instituto de Automatica Ind. (CSIC)
Desv. Km.22.800 N-3 La Poveda
28500 Arganda del Rey
Madrid, Spain

Baker, Mark
Dept. of Materials Science
University of Surrey
Guildford, Surrey, England

Barata, Manuel Martins
Quinta da Torre - FCT UNL
2825 Monte de Caparica, Portugal

Besslich, Prof. Dr. Ph. W.
FB 1 Electrical Engineering
University of Bremen
P.O. Box 330 440
D-2800 Bremen 33, West Germany

Blomme, Ronny M.
State University of Ghent
Electronics Lab
St.-Pietersnieuwstraat 41
B-9000 Ghent, Belgium

Caglioti, Vincenzo
Politecnico di Milano
Departimento di Elettronica
Piazza Leonardo da Vinci, 32
20133 Milano, Italy

Carlsohn, Dr. Matthias
D-2800 Bremen 44
Am Heiddamm 36, Germany

Chassiakos, A.
Dept. of Engrg. and Technology
California State Univ., Long Beach
Long Beach, CA 90840 U.S.A.

Chiladakis, Constantine
Alcor S.A.
Vioprofal S.A.
P. O. Box 80285
GR - 185 10 Piraeus, Greece

Christodoulou, Prof. M. A.
Technical University of Crete
53 Moreos Street
16561 Glyfada
Athens, Greece

Climaco, Prof. Joao C. N.
Universidade de Coimbra
Faculdade de Ciencias E Tecnologia
Dept. de Engrg. Electrotecnica
3000 Coimbra, Portugal

Codina, Joan
C/PAU Gargallo, No. 5
08028 - Barcelona, Spain

Delchambre, A
CRIF/WTCM-Ave Franklin Roosevelt 50
Box 106 - P.4 - 1050 Brussels
Belgium

Dimitriadis, Dr. Basile D.
Technical Director
HITEC S.A.
18 Poseidonos Avenue
Kallithea
Athens 17674, Greece

Ericsson, Prof. Ulf
The Royal Institute of Technology
S-100 44, Stockholm, Sweden

Frau, Juan
C/PAU Gargallo, No. 5
18028 - Barcelona, Spain

Georgopoulos, Voula C.
3 Burlington Street, No. 3
Woburn, MA 01801 U.S.A.

Gibson, Dr. Ian
Robotics - Research Unit
Dept. Electronic Engineering
Hull University
Cottingham Road
Hull, HU6 7RX U.K.

Glentis, George
Inform. and Telecom. Institute
Ag. Paraskevi - Athens
Greece GR-153 10

Grammaticos, Antonios
Division of Systems and Control
Dept. of Electrical Engineering
University of Patras
GR-26001 Rion Patras, Greece

Hatzivasiliou, Fivos V.
Airforce Research Center, NTUA
13 Solonos Street
Kilkis - 61100, Greece

Hubner, Thomas
Universitat Karlsruhe
Institut fur Prozebrechen-
technik und Robotik
Postfach 6980
D-7500 Karlsruhe 1

Jordanides, Dennis
Electrical Engineering Dept.
Univ. of Calif. Santa Barbara
Santa Barbara, CA 93106 USA

Kanakis, Theodoros A.
Democritus Univ. of Thrace
67 100 Xanthi, Greece

Kaylan, Dr. Ali Riza
Computer Center
Bogazici University
80815 Bebek, Istanbul, Turkey

Kaynak, Dr. Okyay
Dept. Elec. and Electronic Engrg.
Bogazici University
80815 Bebek, Istanbul, Turkey

Kouparoussos, George
Elec. Engrg and Computer Science
NTUA, Athens, Greece

Koutras, Prof. Alexander
Electrical Engineering Dept.
California State University
Pomona, California U.S.A.

Kropatsch, Dr. Walter G.
Institute for Tech. Informatics
Technical University of Vienna
Treitlstr. 3
A-1040 Vienna, Austria

Mai, Ms. Thu-Lan
Framentec
Tour Fiat, Cedex 16
92084 Paris La Defense, France

Meijer, Dr. G. R.
P/A Faculty Math & Computer Science
Dept. of Computer Systems
Kruislaan Gog
1098 S.J. Amsterdam, Netherlands

Mertzios, Prof. B.
Electrical Engineering Dept.
University of Thrace
67 100 Xanthi, Greece

Murphy, Steve
Center for Intelligent Robotic
Systems for Space Exploration
RPI
Troy, New York 12180 U.S.A.

Nunes, Urbano
Universidade de Coimbra
Faculdade de Ciencias E Technologia
Depart. de Engr. Electrotecnica
3000 Coimbra, Portugal

Oliveira, Eugenio
Faculdade de Engenharia
DEEC
Rua dos Bragas
4099 Porto Cortex, Portugal

Palassopoulos, G. V.
Military Academy of Greece
41 Blessa Street
Papagos, Athens 15669, Greece

Papanikolaou, Konstantinos
HITEC S.A., 18 Poseidonos Ave.
17674 Kallithea, Athens, Greece

Pires, Fernando Moura
UNL, Quinta da Torre
2825 Monte de Caparica
Portugal

Polydoros, Prof. Andreas
University of Southern California
PHE 414, EE Dept. - Systems
Los Angeles, California U.S.A.

Raczkowsky, Jorg
Universitat Karlsruhe
Fakultat Fur Informatik
7500 Karlsruhe, 12.4.1990
Kaiserstr. 12 /Geb. 40.28
Postfach 69 80 Germany

Somalvico, Prof. Marco
Dipartimento di Elettronica
Politecnico di Milano
Piazza Leonardo da Vinci, 32
20133 Milano, Italy

Torby, Prof. Bruce
Dept. of Mechanical Engrg.
Calif. State Univ., Long Beach
Long Beach, CA 90840 U.S.A.

Triggs, Bill J.
Computing Laboratory
8 Keble Road
Oxford OX1 3QD U.K.

Tzafestas, Elpida S.
Computer Science Division
Electrical Engineering Dept.
NTUA
Athens, Greece

Tzafestas, Mrs. Niki
HTO
E. Venizelou Str.
Holargos 15561
Athens, Greece

Vatistas, A.
Electrical Engineer
Algosystems, S.A.
4, Sardeon Street
Athens, 171 21, Greece

Vibet, Prof. Claude
36 Rue Jean Rostand
91300 Massy, France

Voulgaris, Nicholas
Dept. of Electrical Engineering
Calif. State Univ., Long Beach
Long Beach, CA 90840 U.S.A.

Wang, Prof. Yulun
Center for Robotic Systems in
Microelectronics
Univ. of Calif. Santa Barbara
Santa Barbara, CA 93106 U.S.A.

Warwick, Prof. Kevin
Dept. of Cybernetics
3 Early Gate, Whiteknights
Reading RG6 2AL U.K.

Weiss, Dr. Lee E.
Carnegie Mellon University
Pittsburgh, PA 15213 U.S.A.

Zagorianos, Anastassios D.
Hellenic Airforce Academy
NTUA
9, Larnakos Str.
Athens 11253, Greece

NATO ASI Series F

Including Special Programmes on Sensory Systems for Robotic Control (ROB) and on Advanced Educational Technology (AET)

Vol. 1: Issues in Acoustic Signal – Image Processing and Recognition. Edited by C. H. Chen. VIII, 333 pages. 1983.

Vol. 2: Image Sequence Processing and Dynamic Scene Analysis. Edited by T. S. Huang. IX, 749 pages. 1983.

Vol. 3: Electronic Systems Effectiveness and Life Cycle Costing. Edited by J. K. Skwirzynski. XVII, 732 pages. 1983.

Vol. 4: Pictorial Data Analysis. Edited by R. M. Haralick. VIII, 468 pages. 1983.

Vol. 5: International Calibration Study of Traffic Conflict Techniques. Edited by E. Asmussen. VII, 229 pages. 1984.

Vol. 6: Information Technology and the Computer Network. Edited by K. G. Beauchamp. VIII, 271 pages. 1984.

Vol. 7: High-Speed Computation. Edited by J. S. Kowalik. IX, 441 pages. 1984.

Vol. 8: Program Transformation and Programming Environments. Report on a Workshop directed by F. L. Bauer and H. Remus. Edited by P. Pepper. XIV, 378 pages. 1984.

Vol. 9: Computer Aided Analysis and Optimization of Mechanical System Dynamics. Edited by E. J. Haug. XXII, 700 pages. 1984.

Vol. 10: Simulation and Model-Based Methodologies: An Integrative View. Edited by T. I. Ören, B. P. Zeigler, M. S. Elzas. XIII, 651 pages. 1984.

Vol. 11: Robotics and Artificial Intelligence. Edited by M. Brady, L. A. Gerhardt, H. F. Davidson. XVII, 693 pages. 1984.

Vol. 12: Combinatorial Algorithms on Words. Edited by A. Apostolico, Z. Galil. VIII, 361 pages. 1985.

Vol. 13: Logics and Models of Concurrent Systems. Edited by K. R. Apt. VIII, 498 pages. 1985.

Vol. 14: Control Flow and Data Flow: Concepts of Distributed Programming. Edited by M. Broy. VIII, 525 pages. 1985.

Vol. 15: Computational Mathematical Programming. Edited by K. Schittkowski. VIII, 451 pages. 1985.

Vol. 16: New Systems and Architectures for Automatic Speech Recognition and Synthesis. Edited by R. De Mori, C.Y. Suen. XIII, 630 pages. 1985.

Vol. 17: Fundamental Algorithms for Computer Graphics. Edited by R.A. Earnshaw. XVI, 1042 pages. 1985.

Vol. 18: Computer Architectures for Spatially Distributed Data. Edited by H. Freeman and G.G. Pieroni. VIII, 391 pages. 1985.

Vol. 19: Pictorial Information Systems in Medicine. Edited by K. H. Höhne. XII, 525 pages. 1986.

Vol. 20: Disordered Systems and Biological Organization. Edited by E. Bienenstock, F. Fogelman Soulié, G. Weisbuch. XXI, 405 pages. 1986.

Vol. 21: Intelligent Decision Support in Process Environments. Edited by E. Hollnagel, G. Mancini, D.D. Woods. XV, 524 pages. 1986.

NATO ASI Series F

Vol. 22: Software System Design Methods. The Challenge of Advanced Computing Technology. Edited by J.K. Skwirzynski. XIII, 747 pages. 1986.

Vol. 23: Designing Computer-Based Learning Materials. Edited by H. Weinstock and A. Bork. IX, 285 pages. 1986.

Vol. 24: Database Machines. Modern Trends and Applications. Edited by A.K. Sood and A.H. Qureshi. VIII, 570 pages. 1986.

Vol. 25: Pyramidal Systems for Computer Vision. Edited by V. Cantoni and S. Levialdi. VIII, 392 pages. 1986. *(ROB)*

Vol. 26: Modelling and Analysis in Arms Control. Edited by R. Avenhaus, R.K. Huber and J.D. Kettelle. VIII, 488 pages. 1986.

Vol. 27: Computer Aided Optimal Design: Structural and Mechanical Systems. Edited by C.A. Mota Soares. XIII, 1029 pages. 1987.

Vol. 28: Distributed Operating Systems. Theory und Practice. Edited by Y. Paker, J.-P. Banatre and M. Bozyiğit. X, 379 pages. 1987.

Vol. 29: Languages for Sensor-Based Control in Robotics. Edited by U. Rembold and K. Hörmann. IX, 625 pages. 1987. *(ROB)*

Vol. 30: Pattern Recognition Theory and Applications. Edited by P.A. Devijver and J. Kittler. XI, 543 pages. 1987.

Vol. 31: Decision Support Systems: Theory and Application. Edited by C.W. Holsapple and A.B. Whinston. X, 500 pages. 1987.

Vol. 32: Information Systems: Failure Analysis. Edited by J.A. Wise and A. Debons. XV, 338 pages. 1987.

Vol. 33: Machine Intelligence and Knowledge Engineering for Robotic Applications. Edited by A.K.C. Wong and A. Pugh. XIV, 486 pages. 1987. *(ROB)*

Vol. 34: Modelling, Robustness and Sensitivity Reduction in Control Systems. Edited by R.F. Curtain. IX, 492 pages. 1987.

Vol. 35: Expert Judgment and Expert Systems. Edited by J.L. Mumpower, L.D. Phillips, O. Renn and V.R.R. Uppuluri. VIII, 361 pages. 1987.

Vol. 36: Logic of Programming and Calculi of Discrete Design. Edited by M. Broy. VII, 415 pages. 1987.

Vol. 37: Dynamics of Infinite Dimensional Systems. Edited by S.-N. Chow and J.K. Hale. IX, 514 pages. 1987.

Vol. 38: Flow Control of Congested Networks. Edited by A.R. Odoni, L. Bianco and G. Szegö. XII, 355 pages. 1987.

Vol. 39: Mathematics and Computer Science in Medical Imaging. Edited by M.A. Viergever and A. Todd-Pokropek. VIII, 546 pages. 1988.

Vol. 40: Theoretical Foundations of Computer Graphics and CAD. Edited by R.A. Earnshaw. XX, 1246 pages. 1988.

Vol. 41: Neural Computers. Edited by R. Eckmiller and Ch. v. d. Malsburg. XIII, 566 pages. 1988.

NATO ASI Series F

Vol. 42: Real-Time Object Measurement and Classification. Edited by A. K. Jain. VIII, 407 pages. 1988. *(ROB)*

Vol. 43: Sensors and Sensory Systems for Advanced Robots. Edited by P. Dario. XI, 597 pages. 1988. *(ROB)*

Vol. 44: Signal Processing and Pattern Recognition in Nondestructive Evaluation of Materials. Edited by C. H. Chen. VIII, 344 pages. 1988. *(ROB)*

Vol. 45: Syntactic and Structural Pattern Recognition. Edited by G. Ferraté, T. Pavlidis, A. Sanfeliu and H. Bunke. XVI, 467 pages. 1988. *(ROB)*

Vol. 46: Recent Advances in Speech Understanding and Dialog Systems. Edited by H. Niemann, M. Lang and G. Sagerer. X, 521 pages. 1988.

Vol. 47: Advanced Computing Concepts and Techniques in Control Engineering. Edited by M. J. Denham and A. J. Laub. XI, 518 pages. 1988.

Vol. 48: Mathematical Models for Decision Support. Edited by G. Mitra. IX, 762 pages. 1988.

Vol. 49: Computer Integrated Manufacturing. Edited by I. B. Turksen. VIII, 568 pages. 1988.

Vol. 50: CAD Based Programming for Sensory Robots. Edited by B. Ravani. IX, 565 pages. 1988. *(ROB)*

Vol. 51: Algorithms and Model Formulations in Mathematical Programming. Edited by S.W. Wallace. IX, 190 pages. 1989.

Vol. 52: Sensor Devices and Systems for Robotics. Edited by A. Casals. IX, 362 pages. 1989. *(ROB)*

Vol. 53: Advanced Information Technologies for Industrial Material Flow Systems. Edited by S. Y. Nof and C. L. Moodie. IX, 710 pages. 1989.

Vol. 54: A Reappraisal of the Efficiency of Financial Markets. Edited by R. M. C. Guimarães, B. G. Kingsman and S. J. Taylor. X, 804 pages. 1989.

Vol. 55: Constructive Methods in Computing Science. Edited by M. Broy. VII, 478 pages. 1989.

Vol. 56: Multiple Criteria Decision Making and Risk Analysis Using Microcomputers. Edited by B. Karpak and S. Zionts. VII, 399 pages. 1989.

Vol. 57: Kinematics and Dynamic Issues in Sensor Based Control. Edited by G. E. Taylor. XI, 456 pages. 1990. *(ROB)*

Vol. 58: Highly Redundant Sensing in Robotic Systems. Edited by J. T. Tou and J. G. Balchen. X, 322 pages. 1990. *(ROB)*

Vol. 59: Superconducting Electronics. Edited by H. Weinstock and M. Nisenoff. X, 441 pages. 1989.

Vol. 60: 3D Imaging in Medicine. Algorithms, Systems, Applications. Edited by K.H. Höhne, H. Fuchs and S. M. Pizer. IX, 460 pages. 1990.

Vol. 61: Knowledge, Data and Computer-Assisted Decisions. Edited by M. Schader and W. Gaul. VIII, 421 pages. 1990.

Vol. 62: Supercomputing. Edited by J. S. Kowalik. X, 425 pages. 1990.

Vol. 63: Traditional and Non-Traditional Robotic Sensors. Edited by T. C. Henderson. VIII, 468 pages. 1990. *(ROB)*

Vol. 64: Sensory Robotics for the Handling of Limp Materials. Edited by P. M. Taylor. IX, 343 pages. 1990. *(ROB)*

Vol. 65: Mapping and Spatial Modelling for Navigation. Edited by L. F. Pau. VIII, 357 pages. 1990. *(ROB)*

NATO ASI Series F

Vol. 66: Sensor-Based Robots: Algorithms and Architectures. Edited by C.S.G. Lee. X, 285 pages. 1991. *(ROB)*

Vol. 67: Designing Hypermedia for Learning. Edited by D.H. Jonassen and H. Mandl. XXV, 457 pages. 1990. *(AET)*

Vol. 68: Neurocomputing. Algorithms, Architectures and Applications. Edited by F. Fogelman Soulié and J. Hérault. XI, 455 pages. 1990.

Vol. 69: Real-Time Integration Methods for Mechanical System Simulation. Edited by E.J. Haug and R.C. Deyo. VIII, 352 pages. 1991.

Vol. 70: Numerical Linear Algebra, Digital Signal Processing and Parallel Algorithms. Edited by G.H. Golub and P. Van Dooren. XIII, 729 pages. 1991.

Vol. 71: Expert Systems and Robotics. Edited by T. Jordanides and B. Torby. XII, 744 pages. 1991.

MIX
Papier aus verantwortungsvollen Quellen
Paper from responsible sources
FSC® C105338

If you have any concerns about our products,
you can contact us on
ProductSafety@springernature.com

In case Publisher is established outside the EU,
the EU authorized representative is:
Springer Nature Customer Service Center GmbH
Europaplatz 3, 69115 Heidelberg, Germany

Printed by Libri Plureos GmbH
in Hamburg, Germany